Alfred Wegener

IN MEMORIAM
ALFRED WEGENER

Alfred Wegener

Science, Exploration, and the Theory of Continental Drift

MOTT T. GREENE

JOHNS HOPKINS UNIVERSITY PRESS Baltimore

Printed in the United States of America on acid-free paper

Johns Hopkins Paperback edition, 2018
2 4 6 8 9 7 5 3 1

Johns Hopkins University Press
2715 North Charles Street
Baltimore, Maryland 21218-4363
www.press.jhu.edu

The Library of Congress has cataloged the hardcover edition of this book as follows:

Greene, Mott T., 1945–
Alfred Wegener : science, exploration, and the theory of continental drift.
pages cm
Includes bibliographical references and index.
ISBN 978-1-4214-1712-7 (hardcover : alk. paper) — ISBN 978-1-4214-1713-4 (electronic) — ISBN 1-4214-1712-X (hardcover : alk. paper) — ISBN 1-4214-1713-8 (electronic) 1. Wegener, Alfred, 1880–1930. 2. Geophysicists—Germany—Biography. 3. Continental drift. I. Title.
QE511.5.G74 2015
551.1'36092—dc23
[B] 2014039517

A catalog record for this book is available from the British Library.

ISBN-13: 978-1-4214-2709-6
ISBN-10: 1-4214-2709-5

Special discounts are available for bulk purchases of this book. For more information, please contact Special Sales at 410-516-6936 or specialsales@press.jhu.edu.

Johns Hopkins University Press uses environmentally friendly book materials, including recycled text paper that is composed of at least 30 percent post-consumer waste, whenever possible.

To my wife
Jo Leffingwell

Contents

Preface

This is a book about the life and scientific work of Alfred Wegener, whose reputation today rests with his theory of continental displacements, better known as "continental drift." Wegener proposed this theory in 1912 and developed it extensively for nearly twenty years. His book on the subject, *The Origin of Continents and Oceans*, went through four editions and was the focus of an international controversy in his lifetime and for some years after his death. It was translated into English, French, Italian, Spanish, Swedish, Russian, and Japanese.

Wegener's basic idea was that many (otherwise) intractable problems and puzzles of the earth's history could be solved if one supposed that the continents moved laterally, rather than supposing that they remained fixed in place. Wegener worked systematically over many years to show in great detail how such continental movements were plausible and how they worked, using evidence and results from a large number of sciences: geology, geodesy, geophysics, paleontology, climatology, and paleogeography.

Wegener's idea—that the continents move—is at the heart of the theory that guides the earth sciences today: plate tectonics. This theory is in many respects quite different from Wegener's proposal, in the same way that modern evolutionary theory is very different from Darwin's original ideas about biological evolution. Yet plate tectonics is a descendent of Alfred Wegener's theory of continental drift, in quite the same way that modern evolutionary theory is a descendent of Darwin's theory of natural selection.

Given that Wegener is the progenitor of a major theory governing a modern science, it comes as something of a surprise to discover that no scholarly biography treating his life and scientific work has ever been attempted. His wife wrote and published an appreciative memoir in 1960, and since then there has been a good and useful popular biography by Ulrich Wutzke, the world's leading authority on the provenance and location of Wegener's papers. Wutzke's book, however, contains almost no discussion of Wegener's scientific work and concentrates instead on Wegener's exciting career in polar exploration. Its major narrative line follows closely the book by Wegener's wife. Both of these books are out of print.[1]

Of the many things that make Wegener an interesting story, perhaps the most intriguing is that, although he was the author of a "geological theory" (continental drift), he was not a geologist. He was trained as an astronomer and pursued a career in atmospheric physics. When he proposed the theory of continental displacements (1912), he was thirty-one years old and an instructor of physics and astronomy at the University of Marburg, in southern Germany. He was not "unknown." In 1906 he had set a world record (with his brother Kurt) for time aloft in a free balloon: fifty-two hours. Between 1906 and 1908 he had taken part in a highly publicized and extremely dangerous expedition to explore the coast of northeast Greenland. He was also known—to the much smaller circle of meteorologists and atmospheric physicists in Germany—as the author of a textbook, *Thermodynamics of the Atmosphere* (1911), and of a number of interesting scientific papers on atmospheric layering.[2]

As important as Wegener's work on continental drift has turned out to be, it was largely a sideline to his principal career in atmospheric physics, geophysics, and paleoclimatology, and thus I have been at great pains to put Wegener's work on continental displacements in the larger context of his life and his other scientific work, and to put that life and work into the still larger context of the character of the earth and atmospheric sciences in his lifetime. This is a "continental drift book" only to the extent that Wegener was interested in that topic and later became famous for it. My treatment of his other scientific work is no less detailed, though I certainly have devoted more attention to the *reception* of his ideas on continental displacement, as they were much more controversial than his other work.

Readers interested in one aspect or another of Wegener's career will see that he often stopped pursuing a given line of investigation (sometimes for years on end), only to pick it up later. I have tried to provide guideposts to his rapidly shifting interests by characterizing different phases of his career as careers in different sciences, which is reflected in the titles of the chapters. Thus, the table of contents and the index should be sufficient guides for those interested in some aspect of Wegener's life but perhaps not all of it. My own feeling is that the parts do not make as much sense on their own as the ensemble of all his activities taken together. This is not an unusual standpoint for a biographer to take, but I do urge my readers to try to experience Wegener's life as he lived it, with all the interruptions, blind alleys, changes of mind, and renewed efforts this entailed.

In the most general sense, scientific biography exists to explain how the inner experience of certain individuals becomes the shared "outer experience of the world" that is science. The sciences are not about our inner, private experiences of the world, but only our shared, outer experiences of the world; only in this way can we compare the evidence and decide together what is true and what is not. This is a drastic restriction and a severe rule, but it is what allows the culture of evidence to exist and prevail against mere opinion.

In writing scientific biographies, we generally reconstruct the inner experience of scientists not from their published papers and books but from notebooks, letters, reminiscences of those who knew them, and other such material. Some scientists, such as Newton, Darwin, and Einstein, left mountains of such material behind, letters numbering in the tens of thousands. Others, like Michael Faraday, left extensive journals of their thoughts and speculations, parallel to their scientific notebooks. The more such material a scientist leaves behind, the better our chance of forming an accurate picture of how his or her ideas took shape and evolved. Of course, the sequence of scientific books and papers tells this story, but the story it tells is of the results, and not of the search. Since the seventeenth century, it has been the rule that scientists should report the results and not the history of their investigations: published science is someone's version of a correct answer, with all the false starts, mistakes, and frustrations left out. Were it not so, the history of science as a scholarly undertaking would be nearly irrelevant, as the "story behind the story" would already be there.

What biography also accomplishes that a study of disembodied scientific "movements" or of published papers and books does not is specific knowledge of how certain cultural movements and scientific developments come together in a given time and place. It does this by re-creating the conjunction of these entities, motives, ideas, and events in the life and mind of a single subject. In other words, if we wish to do more than conjecture how events might go together or how they might have gone together—how some philosophy or activity or life experience might have had a part in some scientific development—we have few alternatives to finding them integrated in the mind of a single significant individual and then documenting that integration. One tries to re-create a biographical subject within a historical context and then have the further development of that context explained to some extent by the creative activity of the subject.[3]

This approach, through journals and private papers, makes Alfred Wegener a difficult subject for a scientific biography because only a few hundred of the many thousands of letters he wrote in his lifetime have survived. Deeply introverted and focused on his work, he was not interested in recording his inner life, and he kept no notebooks containing his speculations or diaries recording his activities. He restricted his journal writing entirely to his scientific expeditions, beginning on the first day of such expeditions and ending on the last. He only very occasionally kept copies of the letters he sent and received, and most of these were lost or destroyed at the end of the Second World War. He had few close friendships, was not active (with a few exceptions) in scientific societies, and did not seek to find influence or advance his ideas through professional contacts and politics, spending most of his time at home in his study reading and writing, or in the field collecting observations.

Wegener's story is also difficult for another reason: no other earth scientist has worked successfully in as many fields as did Wegener. He produced important publications in lunar and planetary astronomy, meteoritics, atmospheric thermodynamics, the theory of precipitation, atmospheric acoustics, optics, turbulence, layering, the physics of clouds, the theory of tornadoes, climatology, paleoclimatology, geology, geophysics, geodesy, and glaciology.

The great breadth of Wegener's scientific work created formidable difficulties for me. Most of it took place within sciences (listed above) for which there are no standard histories in which to insert his story. As the anthropologist Claude Lévi-Strauss pointed out some years ago, biography is "weak history" and gets its strength and meaning only when attached to a larger narrative for which it, in turn, provides fine-grained detail.[4] Much of the labor of constructing the story of Wegener's life has been the parallel construction (from the original scientific books and papers) of detailed "master narratives" of scientific ideas in atmospheric physics, geophysics, climatology, and paleoclimatology (and to a lesser extent meteoritics and glaciology) between 1880 and 1930, so that I might be able to assess where the mainstream lay relative to Wegener's work, as well as where he was influential and where he was not.

In the early stages of writing this book I asked my friend, L. Pearce Williams (the biographer of Michael Faraday), what the main line of attack in writing such a book should be. He told me, "Read everything he wrote. Read everything he read. Read as much as you can of what the people he read, read." This counsel of perfection has cost me twenty years of hard work, but I think I have at least approached that daunting goal. I also asked Richard Westfall (the biographer of Isaac Newton) what advice he had about writing a biography. He said, "Try to have your subject on every page, and whenever possible, when explaining his science, let him do it in his own words." I have tried to follow this advice as well.

In the course of researching and writing this biography I traveled to every place Alfred Wegener lived and worked. This took me to Berlin, rural Brandenburg, Marburg, Hamburg, and Heidelberg in Germany and to Innsbruck and Graz in Austria. It took me as well onto the Greenland ice cap. I visited the archives in Copenhagen, Munich, Marburg, Graz, and Bremerhaven, where the majority of his surviving letters and papers are to be found. Readers will find additional details concerning these documents in the bibliographical essay at the end of this book.

Although scientific biography is a nonfiction genre, in which one makes only evidence-based claims, it is also in its own way the writing of a historical novel according to special rules—an observation credited to Umberto Eco. This notion, at first rather odd, improves on acquaintance. The special rules for biography are not numerous, but they are severe: the events reported must really have happened, one must report them in the order they transpired, one may not leave out any significant detail even if it changes the story in a way the author of the biography may not like, and all events reported must be based on surviving written evidence.

I have invented nothing and have made no claims not supported by documentary evidence. I am the sworn enemy of phrases such as "Wegener must have known," or "Wegener was certainly familiar with," phrases to which biographers often resort in the absence of evidence. If the structure of my biography observes the conventions of the *Bildungsroman* (the novel of individual self-development), as indeed does *every biography of a scientist I have ever read*, the content of this biography of Wegener is rigorously empirical. I take no liberties with the historical record.

Nevertheless, no matter how empirically well founded, a scientific biography is a "novel with special rules," and it has literary conventions as well. It should have a plot and should show the development of the protagonist from birth to death. Moreover, because generally we only see biographies of highly successful scientists, the genre convention of the "eureka moment," in which the fundamental important discovery appears in the mind of the scientist in question, is always in the forefront of the mind of the reader, and a biography of a great scientist that does not have such a eureka moment would generally be considered to be lacking.

I am aware of these conventions and have written about them elsewhere in some detail, concerning in particular their dependence on the folkloric conventions of the "hero's quest."[5] I mention them here only so that the reader may know that I am aware of them and how they exert a pull on the biographer's activity. I hasten to assure the reader that Wegener's life has enough danger and drama to sustain such an approach, and the moments of important discovery are clearly discernible in his work. When he thought he had made an important discovery, had done something entirely new, or possibly had revolutionized a field of study, he generally said so in print at that time.

That being said, the reader should also know that I am firmly of the opinion that most of us, Wegener included, are not in any real sense the authors of our own lives. We plan, think, and act, often with apparent freedom, but most of the time our lives "happen to us," and we only retrospectively turn this happenstance into a coherent narrative of fulfilled intentions. This book therefore is a story both of the life and scientific work that Alfred Wegener planned and intended and of the life and scientific work that actually "happened to him." These are, as I think you will soon see, not always the same thing.

Note to the Paperback Edition

I have taken the opportunity of this paperback edition to correct many errors in German spelling. I wish to thank Dr. Thomas Ruedas of the Institute of Planetology, University of Münster, for his painstaking effort in compiling these and sending them to me. Any remaining errors are, of course, my own.

Acknowledgments

I am very grateful to all those who helped me pursue and complete this project. I owe a debt of gratitude to Ms. Kirsten Caning, of the Arktisk Institut (now at the Dansk Polarcenter), for orienting me to Wegener's career in Greenland and for her generosity with documents and advice. She urged me not to be content to chronicle Wegener's career but to understand him psychologically as well—important advice.

I could not have completed this book in its present form without the benefit of the tireless researches of Ulrich Wutzke, also a Wegener biographer, who has uncovered all the surviving documents concerning Wegener and his career. I am grateful to him for meeting me in Marburg and showing me Wegener's various residences in that city.

I thank the staff of the Deutsches Museum in Munich and the Alfred Wegener Institute in Bremerhaven for their assistance in exploring the Wegener archives in these two institutions. I wish also to thank Irina Rockel of the Heimatmuseum, Neuruppin, for her assistance, especially with photographic and documentary resources concerning Wegener's childhood. Walter Hoflechner and Siegfried Bauer of the Karl-Franzens University, Graz, were very generous with time and resources, as well as anecdotes concerning Wegener's time in Graz. I received similar help from Hermann Günzel at the University of Marburg, for which I wish to thank him as well.

I thank the State Department of the United States and the governments of Greenland and Denmark for arranging permission for me to conduct research on the Greenland ice cap. While I was in Greenland, Gösta Lindgren, managing director of the Black Angel Mine (Marmorilik), made resources available to me without which I could not have succeeded. I thank Raymond Bruun of Boliden S/A and Hans Jürgen Hansen of Greenex for their help, which included helicopter travel to and from Marmorilik, far beyond the endpoint of civil aviation. I thank Capt. Fleming, the Dark Angel Mine environmental officer, for pulling me off the beach at Kamarajuk in the midst of a fierce downslope windstorm. Bjørn Thomassen, of the Geological Survey of Greenland and Denmark, helped me to choose a route up the Kamarajuk Glacier and gave me good advice concerning the history of Wegener's expedition as well. I received similar good advice from Jan Lorentz and from Finn Pedersen, the director of the Uummannaq Museum.

I am forever indebted to Henry Frankel, philosopher, historian, and friend, whose conversation, advice, and encouragement often kept me going when I thought I would never finish this book. His four-volume history of the continental drift controversy is as close to the "last word" on the subject as anyone ever gets. I see our work as complementary, and I am especially grateful to him for his detailed comments on my manuscript, which improved it greatly, especially in the later chapters.

My home institution, the University of Puget Sound, in Tacoma, Washington, supported my research for many years. The National Science Foundation and the National Endowment for the Humanities also supported my work in 1983, 1989, and 1999. I thank them all, as well as my colleagues and friends in the Department of Earth and Space Sciences at the University of Washington. Particular thanks to Prof. Jody Bourgeois at the University of Washington, a good historian of the earth sciences and a great geologist, for her friendship and advice. I would like to thank Kelly Vomacka for helping me to proof and prepare the manuscript in its final stages. I very much appreciate the guidance, support, wisdom, and, above all, patience of Robert J. Brugger, my editor at Johns Hopkins University Press. Thanks are also due to Jeremy Horsefield for his excellent copy editing, to Kim Johnson for her expert management of the production of the book, to Glen Burris for the beautiful design, and not least to Alexa Selph, who prepared the index.

My daughter Annie Greene also deserves my thanks for her good humor concerning a project that spans most of her life.

My deepest thanks go to my wife, Jo Leffingwell, to whom this book is dedicated, for her unconditional support and her belief that I would someday finish this book. I could not have done it without her. Thanks, Jo.

Alfred Wegener

I

The Boy

BERLIN AND BRANDENBURG, 1880–1899

> The Berlin of to-day has about it no suggestion of a former period. The site it stands on has traditions and a history, but the city itself has no traditions and no history. It is a new city; the newest I have ever seen. . . . The main mass of the city looks as if it had been built last week. The rest of it has a just perceptibly graver tone, and looks as if it might be six or even eight months old.
>
> MARK TWAIN, "The German Chicago" (1897)

> It is no little thing to get to Rheinsberg from Berlin. The railroad runs past it at a distance of six miles, and only the adroit combining of stagecoach and hired cart leads one ultimately to the long-sought goal. This may explain why a place whose natural beauties are not to be scorned, and which is of great historical importance, remains almost unvisited.
>
> THEODOR FONTANE, *Graffschaft Rüppin* (1862)

Toward the end of October the weather in Brandenburg (North Prussia) turns windy and cool. The late-summer lull that can bring both early morning fog and hot, still afternoons gives way—sometimes quite suddenly—to fresh breezes from the west. This heralds the seasonal inland march of the Atlantic Westerlies across the Prussian plain, and this flat landscape of forest, lake, and farmstead offers the wind little resistance. In Berlin, 200 kilometers (124 miles) inland from the Baltic Sea, the autumn climate is not much different from that of a coastal town like Rostock, though the imperial capital city is less cloudy. By early November the daytime highs are only around 7°C (45°F) and the nighttime lows hover near freezing. The sky is seldom completely clear, and the rain, though less frequent than it was in the late-summer months, is cold and driven by a wind with a sharp edge.

When this autumn wind sweeps into Berlin, it travels from west to east down the second-broadest avenue in Europe, tugging at the miles of carefully ordered shade trees that give the avenue its name: Unter den Linden. It strips their leaves and swirls them past the neoclassical facades of ministries and palaces, and past the pediments and porticoes of the Prussian State Library, the Royal Opera, the Friedrich-Wilhelms University, and the Arsenal, finally releasing them into the old Lustgarten, the great parade ground fronting the Imperial Palace on the Museum Insel (museum island).

The Museum Insel in the River Spree is the heart of Old Berlin and the site of the medieval towns of Berlin and Cölln, founded in the thirteenth century. By the fifteenth century, they had merged (without growing significantly), and the construction of a castle had elevated the town to the "Seat of the Electors of Brandenburg"—the Hohenzollern—a family then just beginning its long climb up the ladder of imperial fortune. Only during the reign of Friedrich I (r. 1688–1713) did the island's aspect change dramatically. Friedrich strove to realize his dream of an "Athens on the Spree" by spending extravagantly on public buildings. From his time on, vigorous architectural campaigns by Friedrich after Friedrich and Wilhelm after Wilhelm steadily transformed the center of Berlin from an undistinguished North European trading town into an Italianate Renaissance and neoclassical metropolis. Among these architectural adventures, that of the Imperial Palace was the greatest and most protracted. An enormous residence with (eventually) almost 1,200 rooms, it was continuously under construction from the seventeenth until the twentieth century.[1] Around it, the remains of medieval Berlin gave way in the nineteenth century to great temples of classical and modern secular culture—the Old Museum (1823–1830), the New Museum (1843–1859), the National Gallery (1867–1876), and the Kaiser Friedrich Museum (1897–1904). This industrious acquisition of an artistic and architectural heritage was accompanied on the Museum Island by massive construction in the service of somewhat more typical princely and royal preoccupations: the monumental, high-Renaissance Berlin Cathedral (1894–1905)[2] and the baroque Royal Stables (1896–1901).

It was in this cold, windy, gritty, imperial construction zone that Alfred Lothar Wegener was born on Monday, 1 November 1880. Alfred was the fifth and youngest child of Richard Wegener (1843–1917) and Anna Schwarz (1847–1919). His birthplace was a converted Austrian embassy at 57 Friedrichsgracht, a scant few blocks from the Imperial Palace, and facing the Spree Canal on the southeastern side of the island. This ample and gracious structure was home to the *Schindlersches Waisenhaus*, a privately endowed orphanage for sons of clergy, teachers, civil servants, landowners, and merchants; the Wegeners had taken over its direction and management in 1873, a few years after their marriage. With its airy, high-ceilinged rooms—the great ballroom in the rear of the building had converted handily to a gym full of gymnastic equipment—the spacious interior of the building was more than adequate to house the Wegener family, the thirty or so orphans in their charge, the two young men (*Adjunkten*) assisting Richard Wegener in the teaching and daily supervision of the orphans, and the resident domestics under the direction of Anna Wegener. Its faintly neoclassical exterior, with four Corinthian pilasters set into the facade, makes common cause with its Palladian neighbor to the north, and together they stand in serene contrast to the surrounding redbrick apartment buildings, which manage only to look sooty and compressed.[3]

The Schindler Orphanage was all but indistinguishable from a small, upper-class boarding school. Perhaps befitting its location in "Athens on the Spree," it was much more an Athenian than a Spartan institution, and thus quite unlike the parsimonious, for-profit boarding schools that George Orwell denounced in his memoir "Such, Such Were the Joys," with their cold rooms, bad food, bullying, and humiliation carefully graded to the class position of the parents. At the Schindler Orphanage there were ample food and heat: the orphans were, after all, upper-class sons of professional and well-to-do landed families, and the mission of the institution was to see that these boys should not lose their hereditary educational and social advantages by a mischance of fate. The orphanage was meant to be a nurturing milieu for boys without parents, a place for the building of their character. The English word "character" is, as often and rightly noted, a weak translation for the German *Bildung*. The latter was a cultural ideal on the order of Athenian *paideia*: a carefully harmonized scheme of personal development seeking to couple good breeding with moral probity, physical prowess, classical education, and high culture.

If the orphanage was a physically comfortable and culturally elite institution governed by lofty philosophical ideals, it was also, and on the emotional side inescapably, a civic institution for little boys without parents. It was a distinctly curious fate for Alfred Wegener and his brothers and sisters to have been born and raised within its walls. Alfred's two eldest siblings, his sister Tony (b. 1873) and brother Willi (b. 1874), appeared in the world soon after their parents arrived to direct the orphanage. They were followed within a few years and in rapid succession by Kurt in 1878, Käte in 1879, and Alfred himself in 1880.

The implications of these unusual domestic arrangements for Alfred Wegener's life and for the lives of the other Wegener children, who shared their parents' attentions day and night with thirty or so supplementary siblings, are manifold and intriguing. When Willi, Kurt, and Alfred were in their primary school years, their lives were to be strictly segregated from those of the orphans. They did not attend class with them, study with them, or eat with them, and the Wegener family apartment was in a wing of the building opposite to that containing the dormitory and schoolrooms. Kurt and Alfred, however, sought the orphans out as playmates. Whenever they were free from adult supervision, they spent time with the orphans in the playfield behind the orphanage and exercised with them in the gym.[4] We are told that Alfred, while still of preschool age, took as his model one of the older orphans, an expert gymnast, and trailed him about the gym doggedly trying to follow his lead in the exercises.[5] One says "after school," but in an orphanage, as in any boarding school devoted to molding the whole child, there was no "after school," only a brief cessation of formal instruction; set routines continued throughout the twenty-four-hour day. It is worthy of note that gymnastic exercise, weightlifting, and sports were not only promoted by the government but also controlled, required, and regulated as meticulously as the rest of the curriculum and treated with (almost) the seriousness of Latin and Greek.

The regulation of the life of schoolchildren and belief in the power and benefit of ordered routine probably reached some sort of historical maximum in Berlin during Alfred Wegener's boyhood. Driven by a serious attempt to scientize and rationalize civil existence, it also aimed to orchestrate, integrate, and harmonize the interior life of these future subjects and citizens to match the external order as much as possible. Prussia was a Kantian state, and this was in many ways a Kantian orphanage (even with a strong fondness for Schiller on the part of Richard Wegener). The day, the week, the year were partitioned by subject and level, with rising and retiring times as faithful as the motions of the planets; sport and play, music and art, languages and mathematics, religion and history, geography and literature each had their apportioned hours and graded tasks.

The orphans, after completing their primary schooling within the walls of the Schindler Orphanage, went on to the Gymnasium zum Grauen Kloster, one of Berlin's oldest (1574) and most prestigious secondary schools.[6] That is, those capable of meeting its standards did so. Those who had fared somewhat less well in primary school went on to a *Realgymnasium*, a six-year course with less emphasis on classics, and those with no discernible academic ability were apprenticed out to craftsmen and left the orphanage altogether.[7] This partitioning by academic performance gave an initial presumption to class privilege but then demanded talent and performance in return; it reveals the perilous and difficult character of the Prussian meritocracy: it was hard to rise in it, and all too easy to fall. The hard-working and capable boys who went on to the

Graues Kloster followed the harrowing and minutely supervised nine-year classical curriculum that assured entrée into the university and eventually the upper ranks of the civil service. The Graues Kloster proudly counted Prince Otto von Bismarck among its graduates (class of 1832), and in 1880 Bismarck was still chancellor and at the height of his power and prestige.

The intellectual and cultural atmosphere of the Wegener family circle was extremely literary and dominated by Richard Wegener's interests in classical and modern drama, poetry, and philosophical theology. He inclined positively toward natural history and physical science, though he insisted that the sciences keep to their proper sphere. He was passionately devoted to languages, with a special fondness for Greek and Hebrew. The details of these interests might not occupy us so much if this were a British or an American or a French family of the same period and social class. But in Germany such preoccupations meant more because the educated bourgeoisie that prized them was, practically speaking, politically disenfranchised. The abundant energy of this vast army of urban professionals, which in other industrial nations was absorbed by party politics, social reform, and sectarian religious disputation, in Germany flowed profusely into the enjoyment and elaboration of high culture and civilization—in a word, into *Kultur.*

The civilizing mission willed by Richard Wegener and men like him had almost nothing to do with the preoccupations of the aristocratic *Junker* landowners who held the real power in the German state. On the contrary, the *Kulturträger*—the bearers of the culture—saw themselves as members of a cooperative enterprise with world-historical implications. The details of this alliance between intense self-cultivation on the one hand and selfless solidarity with all mankind on the other were elaborated in a broad spectrum of speculative metaphysical schemes that functioned almost as political parties of the spirit; these had readerships and followings to an extent unimaginable today, in the English-speaking world at least. Since we know something about the philosophical allegiances of the Wegeners, it is worth peering into the systems of thought prevailing in their home at the time of Alfred's birth.

Alfred Wegener's Family Background

We are born not into our own lives but into the lives of others. We are not even ourselves in any meaningful sense for quite some time. What sort of beings we become even in our childhoods is a combination of what we brought with us into the world—our temperaments, talents, and quirks—and the emotional, physical, and mental structures that were there (or not there) when we arrived on the scene. To understand who Alfred Wegener became in later life, we must look into the character of those into whose lives he was born, above all to that of his parents: the lessons they taught, the things they believed, what they wished for their children, and what they wished from them. It may well be that "the boy is the father of the man." Still, "the father of the boy" is, much more prosaically, the boy's father. One of the most personal and pragmatic rationales for historical study is to understand how generation-long offsets in attitude and allegiance govern the world. Even in a culture of novelty such as we inhabit, our social and cultural ideals are profoundly shaped by what our parents believed when we were children. This commonplace gains depth if we consider that much of what our parents believed and taught us at that time about a "good life," a "good person," and a "good world" was inculcated in them during *their* childhoods, by our grandparents—

meaning that we are walking around with views of the world shaped by events (sometimes quite transient, local events) fifty or sixty years before we were born.

Richard Wegener, who rigorously supervised the upbringing of his children, was a stereotypic embodiment of the cultural and social aspirations of his region, class, and time. Born in 1843, he was ninth of the eleven children of Friedrich Wilhelm Wegener, a hard-working and eventually prosperous owner of a military uniform factory in Wittstock, in the northwest corner of Brandenburg, about 90 kilometers (56 miles) from Berlin. It was only in Richard Wegener's generation that this family acquired wealth and ambition sufficient to overcome the parochial inertia that had held it in rural Brandenburg for 300 years.[8] Richard's brother Otto took one path into the Prussian future, that of rural landed wealth and political power, by becoming the manager of a great estate in West Prussia. Richard took the other: urban education and civil service. Richard realized his father's frustrated ambition to study theology and become an evangelical clergyman. After his seminary study and ordination in 1868, he spent a year as an assistant pastor to a parish in Kolmar, in Posen—the Prussian province centered on the historic Polish city of Poznán. Carefully saving his annual salary and his Christmas bonus, he returned to Wittstock and asked twenty-one-year-old Anna Schwarz to marry him. Anna, Alfred's mother-to-be, was herself an orphan, born in the tiny hamlet of Zechlinerhütte and raised by relatives in nearby Wittstock; she and Richard had met as students.[9]

Anna consented, and for the next five years Richard supported himself and his wife on the salary of an assistant pastor, though his plans and interests were already pulling him far beyond the traditional Wegener orbit both geographically and intellectually. He was drawn to the emerging metropolis of Berlin and, within its cosmos, to three great cultural preoccupations of nineteenth-century Germany: classical philology, philosophy, and poetry. Richard studied Greek, Latin, and Hebrew and earned a PhD from the Friedrich-Wilhelms University in Berlin in 1873, with a thesis entitled *Begriff und Beweis der Existenz Gottes bei Spinoza* (The concept of God and proofs of his existence according to Spinoza).[10]

Richard and Anna in that same year took over the orphanage, and Richard began his parallel career teaching Greek and Latin at the Gymnasium zum Grauen Kloster, the illustrious secondary school that the older orphans attended along with the children of Berlin's cultural elite. Richard fed his other interests and commitments (and supplemented the astonishingly small salary paid men of his rank and education in Prussia at that time) by teaching German literature at a nearby *Mädchenschule* (girl's school) and holding a chaplaincy at the criminal court in the nearby neighborhood of Moabit. At thirty, Richard had come very far very fast. At twenty-five he had been an assistant pastor in a small parish church in rural Poland; five years later, he was *Herr Doktor und Hofgerichtsprediger* Richard Wegener, *Oberlehrer zum Grauen Kloster und Direktor des Schindlerschen Waisenhauses* (Herr Doctor and Royal Court Chaplain Richard Wegener, Senior Teacher in the Gray Cloister and Director of the Schindler Orphanage).

Before we turn from this Homeric credentialing to the substance of Richard Wegener's intellectual and aesthetic preoccupations, of which we know a little, we might pause to summarize the suitability of the Wegeners for their chosen task, since it so strongly affected Alfred Wegener's life and his view of the world. Consider their childhood experiences: Richard knew what it was to grow up in a crowd (eleven); Anna understood in full depth the loneliness of the orphan and the needs of such children for extra care and consolation. Neither of the Wegeners made, nor would they accept, a sharp distinction between intellectual and moral education, in full accord with the convictions of their time and class. In religion they appear to have been devout rather than pious. The Wegener children were baptized in the Werder'schen Church, a few minutes away across the canal. The superintendent of the church, Pastor Steinbach, had been a fellow student of Richard's in his time at the University of Jena. In a brief memoir of Alfred's childhood, written by his brother Kurt sometime after 1950, Steinbach is approvingly described as a man "engagingly free of theological dogmatism." Kurt also noted that Steinbach and Richard Wegener had been in the same fraternity (*Bursenschaft*) at a time when these organizations were informal and voluntary associations, rather than competitive vehicles for social advancement. Steinbach and Richard Wegener remained close friends and visited one another frequently.[11]

In secular matters the Wegeners were devoted in a comprehensive and exacting way to the development of the "whole child"—according to the high standards of Berlin's cosmopolitan version of *Bildung*. This could, of course, have suffocated every child within range under an avalanche of high-minded propriety, had it not been titrated with affectionate temperament and a sense of humor; the Wegeners were affectionate with each other, and they both liked and were comfortable with children—an attitude not universally distributed among teachers and orphanage directors. Yet it would be wrong to leave the impression that the Wegeners were guided in their child-rearing practices solely by some kind of commonsensical, good-natured holism made up as they went along. Berlin in 1880 was a place and an age in which metaphysical theorizing about the nature and destiny of humanity was not only an ordinary preoccupation but also something expected to have concrete significance for everyday life, at least in the households of intellectuals. Richard Wegener was a scholar, but he was also an intellectual, believing, as intellectuals do, that to those who have a sufficient grasp of the context of events and their meaning, the most ordinary acts by the most ordinary individuals exhibit a higher significance. He sought to transmit this conviction of higher significance within everyday activities to both his offspring and the orphans he supervised.

The metaphysical visions that animated Richard Wegener's civilizing mission were within the great mainstream of German idealism. His embrace of idealist philosophy was a part of his journey from Brandenburg to Berlin, from parish to metropolis. By the time Richard Wegener arrived in Berlin to begin his life as a university student, the pendulum of taste had swung away from the ponderous rationalism and cosmic optimism of G. W. F. Hegel toward Arthur Schopenhauer's bracing, frank, realistic pessimism.

Richard bought an edition of Schopenhauer's works and admired his brilliance and honesty, but he could never reconcile himself to Schopenhauer's arrogance and impatience with the common run of humanity. There is an admonition to Schopenhauer in one of Richard's later books: "*Grosses Genie, warum schiltst du immer ergrimmt auf die Kleinen? Würden die Kleinen zu gross, wären die Grossen zu klein*" (Great genius, why did you always rail so furiously against the little ones? If the little were greater, the greater would be diminished).[12] He found a way to hold on to Schopenhauer's honesty, clarity, and humor, while continuing to look for a hopeful and socially conscious program of action, in the philosophy of Eduard von Hartmann (1842–1906), his own exact contemporary and a fellow Berliner.

Richard Wegener found von Hartmann unsystematic and paradoxical but believed that his writing on modern philosophy provided entrée to some very good ideas.

Von Hartmann wanted to do the impossible—to combine optimism and pessimism, rationalism and unconscious will, in short, to combine Hegel and Schopenhauer. The ultimate effect of reading this quixotic effort was to send Richard Wegener back to Kant and Schiller. He was increasingly convinced that all the tremendous activity in metaphysics in the nineteenth century had failed to resolve most of the issues raised by these thinkers at the end of the eighteenth—a conviction shared by a great many of his educated contemporaries. But he carried away from von Hartmann this eudaimonic ethics of constructive work—that happiness resides in and is constantly renewed by work.

Richard Wegener seems to have been most attracted to the thought of Friedrich Schiller—who lived on the borderline of poetry and systematic philosophy, constantly crossing back and forth. In the work of the philosopher-poet-dramatist he found a better means of transforming the biblical narrative of man's fall and redemption—the storybook religion of Sunday school and seminary homiletics—into a great cosmic drama of the individual's alienated separation from the All and his return to it.

The optimism and promise of Schiller's worldview were finally victorious in the Wegener household. Here was the task of a thinker and a philosopher, as Richard Wegener understood it: to search for regularity buried beneath disparate appearance, to believe in and discover a single, real universal order. But there was a greater work awaiting men, beyond this philosophic reconciliation. There was the heroic task of the *poet*. It was not enough to gather what came before one's own time; one faced the creative mission of going beyond one's own finite experience. Aided by hypotheses and inventions, one should struggle to grasp the world as a whole and, simultaneously, struggle to liberate oneself from existing opinion and forge a new view of the world.

In Schiller's version of things, the task of life is to bring reason and sense into harmony through the *Spieltrieb* (spirit of play)—the aesthetic impulse, which Schiller accords full status as a third and coequal aspect of human nature. Human freedom lies not in reason but in aesthetic and creative play; we find our way in the world not by system alone but with the aid of creative insight.

Perhaps Richard's inclination toward Schiller was also an expression of commitment to self-cultivation and to the admiration of the classical ideals that he and Schiller so clearly shared, ideals he worked hard to pass on to his children. Whatever its fount, it supported his lifelong aspirations as a literary and drama critic and as a poet. In 1882 he published an ambitious volume of essays on German literature (his second book), *Aufsätze zur Litteratur.*[13] In 1907, near the end of his career as an author and critic, he published a prize-winning book of drama criticism: *Die Bühneneinrichtung des Shakepeareschen Theaters nach den zeitgenössischen Dramen* (The production of Shakespearean theater in its contemporary setting).[14]

His poetic aspirations, modest but real, culminated in the publication in 1895 of his *Poetischer Fruchtgarten* (A poetical kitchen-garden), a collection of lyric poems in various meters and some epigrams (including the one on Schopenhauer quoted above), both serious and humorous. This material was selected out of a much greater number of poetical efforts made while Alfred Wegener and his brothers and sisters were growing up, and it gives a better sense than even Richard Wegener's academic works of the atmosphere of the household. There is a poem *Theodicee*, and one *Hebräische*; there is a poem *Der Magnet*, and reflections on various bible verses. These are the sorts of poems one might read to family or friends. They are earnest but unpretentious and often self-deprecating. There is a poem *Butter und Ei* followed immediately by one *Wissenschaft und Religion*. The latter two poems are both philosophical, and the former is also amusing. It concerns the envy that a lump of butter feels for egg yolks that, by jumping into a hot skillet, have given themselves a firmer character than the butter possesses. The butter, wishing to have the advantage of the egg, hurls itself into the pan—and disappears:[15]

Dotter ward	The Yolk would
Bald sehr hart.	soon be very hard
Butter sogleich	Butter immediately
Da sie so weich,	being so soft
Wünschend sie sei	wishing it could be
Hart wie das Ei.	as hard as the Egg
Ging in die Pfann'	jumped in the pan
Und zerann.	and disappeared
Mit vollen Lungen	At the top of their lungs
Lehren die Jungen	we hear from the young
Im Unterricht	instruction
Uns das Gedicht:	poetically expressed
Butter und Ei	Butter and eggs are, they find,
Sind zweierlei.	substances of a different kind

The poem on science and religion takes the form of a dialogue between a diamond and pearl in a jewel casket. The pearl marvels at the diamond—once a lump of coal, but made clear by heat and pressure, then ground, polished, and made brilliant to break light into its rainbow components. The pearl is modest, but its luminous beauty is organic and unaltered through time; it grows but does not change. The diamond is science and education; the pearl, faith. The poem ends with a warning:

Science and Religion have each their Crown
Only so long as each remains, free from Envy, in its Sphere content,
And finely honed character must guard against the Reefs of haughty Pride
For even the purest Diamond is kilned from Carbon.
Belief, alone untouched by the Stone,
Without grinding, and like a Pearl,
Rules its Realm.[16]

The fundamental faith that reigned in Alfred Wegener's childhood home was that of all idealist philosophy, once the fine distinctions, then so important and now so forgettable, are swept aside. There was room in the house of *Kultur* for science, religion, and philosophy, and for poetic creation, and plenty of work for willing hands. Individuals would find meaning in existence only by participating in enterprises and ideas much greater than themselves, and they must beware "the reefs of haughty pride." This emphasis on work and creativity, on participation in large structures of thought and action—these were the ideas and concerns that governed life at 57 Friedrichsgracht during Alfred Wegener's childhood in the Schindler Orphanage.

Die Hütte

While enjoying a favored position in the cultural heart of the empire and the pervasive influence of its cultural ideals, the Schindler Orphanage was also a world within walls. It had residences, storerooms, classrooms, kitchens, and laundries. It had a very large cobbled inner courtyard, with a small garden and some Linden trees (just able to grow in the dim light filtering down) surrounded by sitting benches, and a rear portal to a small playfield with natural turf giving access to the adjoining seventeenth- and eighteenth-century buildings on both sides of the block.[17] The Nikolai Haus (1674) behind the site of the Schindler Orphanage gives the flavor of the place with its paneled wood hallways and generous but not cavernous entryway. For the first few years of Alfred's life, except for family outings and trips to church, he rarely left these precincts.

By the time Alfred was born in 1880, Tony (seven) and Willi (six) were already in school, and Alfred spent the day with Kurt and his sister Käte, supervised by Anna and the resident domestics. They had the run of the family quarters, the courtyard, and the large institutional kitchen, but not the school or the dormitory. It was an attractive world, but it was bordered, constrained, and paved. The canal outside the front door gave some relief from the universal presence of stone walls, and the maritime traffic was pleasant to watch from the upstairs windows, as coal-fired steam tugs pulled barges through it all day and into the night. Nearby there was a pleasant little drawbridge across the canal and a sluice gate providing a handy waterfall. A few blocks to the east was the Imperial Palace, and in the opposite direction, at the end of the long block of the Friedrichsgracht, was the River Spree itself, just visible if you stuck your head out the window.

In 1884, when Alfred Wegener was not quite four years old, his sister Käte died quite suddenly after a brief illness.[18] Brother Kurt had turned six in April of that year and was then in school all day, so that Käte and Alfred were the last of the children remaining in the family quarters during school hours. Käte's death deprived Alfred of his closest playmate. There is no written record or even secondhand recollection of the impact this made on him, but it was strong enough that thirty-four years later, when his own second daughter was born in the cold spring of 1918, he named her Sophie Käte.

Alfred's mother spoke about this pivotal event later with the reflective, matter-of-fact stoicism of a woman who risked her life with each childbed and faced immanently the loss of her children from a dozen or more childhood diseases, all today made either improbable or extinct by vaccination. She said that it was likely the close spacing of her last three children (1878, 1879, 1880) that had rendered them less vigorous than the older Tony and Willi, born earlier in the marriage. Moreover, the *Großstadtluft* (metropolitan air) had left the younger ones *"blaß und müde"* (pale and listless); these factors together had made Käte vulnerable and contributed to her death. After Käte's death, Anna turned her grief and loss into increased solicitude for the two youngest remaining, Kurt and Alfred; she coddled and pursued them, worrying about their health and chances for survival.[19]

The killer was Berlin itself. Industrial growth in the last few decades of the nineteenth century brought great chemical works, paper mills, machine shops, and textile and carpet factories to the city and its burgeoning suburbs. Much of this development occurred just to the north and east of the Museum Insel, and these factories were largely fueled by lignite, a poor-quality coal that produced huge quantities of fly ash containing about 30 percent sulfuric acid. In these decades it made a sad jest of the *Berliner Luft*, Berlin's historical claim to fresh and refreshing air. The new industries swelled the population by more than 200,000 between 1871 and 1880, and increasing by another 450,000 over the next decade.[20]

With Käte's sudden death, this concern assumed the proportions of a crisis in the Wegener household, and Anna felt keenly a sense of personal responsibility—while she had cared successfully for the orphaned children of others, her own child had died. The crisis precipitated by Käte's death had been building for some time, and it can be seen in photographs taken earlier that year (1884)—individual photos of the family and a group photo with the *Adjunkten* and the *Zöglinge*, the "pupils," as the orphans were collectively known.

In a group portrait of the residents of the orphanage taken that year, Anna and Richard show the toll taken by years of unrelieved responsibility for the welfare of so many others. Both have aged rapidly and markedly when compared with the photograph taken ten years earlier. Though the prettiness of her earlier pictures is still there just beneath the surface, Anna looks tired; at thirty-seven her hair has gone almost completely gray. Perhaps her memory that the three youngest were "pale and listless" reflects her own sense of exhaustion at that time. Richard's portrait shows him healthy and erect, but he too has gone gray, and a worried, myopic severity has crept into his expression, replacing some of the hopeful earnestness and foppish playfulness of earlier photos. Tony and Willi, in the group photo, are large and strong looking, while Kurt and Käte look intense and pale. Alfred, sitting in his father's lap, has an expression evident in almost every photo ever taken of him: a sardonic, almost mocking, sharply examining stare directly into the camera—eyes that look out but cannot be seen into. If we compare them with his mother's eyes (he has her features more than his father's), hers are searching and open, whereas his are questioning, as they conceal what lies within.

The Wegeners had by now been in Berlin for sixteen years and had directed the orphanage for eleven of those years, and they needed a place to retreat and refresh. But it was more than refreshment: they were homesick. Like many couples before and since who followed ambition from a small town to a large city, they found themselves remembering the advantages of their own rural childhoods, advantages that had weighed little at the time of their departure when balanced against the stifling parochialism of small village life. Now they were successful Berliners, in their forties and with a family, and they contemplated with alarm their own remaining children: well-behaved and well-to-do, certainly, but too often in poor health, and living always in a paved-over world, with scarcely a plant or tree in sight of their home, knowing their animals only from zoos and picture books, with the exception of the cart horses in the streets, squirrels in the park, and puddle ducks in the canal just outside the door. They had a grand residence and access to the manicured parks and immense cultural resources of a great capital city. But when all was said and done, they lived in an institution, and they had no home.

The Wegener family needed a *Heimat*, a true home, and they needed it badly. Emotion and logic led Richard and Anna without hesitation back to rural Brandenburg—the obvious choice. It was a place they knew and understood, and where they would be known and understood. It took them back to the hamlet of Zechlinerhütte, where Anna had been born. It took them finally to a plain but spacious house with extensive grounds, fronted by Linden trees and facing, through many large windows, a lake. Built of oak logs and chinked with masonry, it had been the manager's house of a crystal-glass foundry—an undertaking attracted there in the early eighteenth century by the plentiful fuelwood from the surrounding forests, but no now longer able to compete with the industrial-scale economies of burgeoning Berlin. The enterprise eventually failed altogether, leaving the town to eke out a marginal existence concocted of subsistence farming, woodcutting, fishing, and catering to the wants of vacationing urbanites and their seasonal homes.[21]

The Wegeners purchased the house, the barn, and some adjacent fields for 20,000 marks (about five times Richard's annual salary). The money was provided by Richard's brother Paul, who had taken over the family's uniform factory in Wittstock and was well-to-do, generous, and pleased to have his brother closer to home again. Richard, by the account of one of his own children, "understood nothing of finance" and on his own salary could never have saved enough even for a down payment.[22]

Once the house and grounds were theirs, the Wegeners set about reconstructing their world with great thoroughness. They scoured the surrounding towns to find the mahogany furniture of Anna's parents, scattered and sold years before, and they bought it back. They planted fruit trees and laid out an extensive flower garden. Richard had a meadow cut to make a playfield behind the house, up to the edge of the woods. He also began to collect (and to have the boys plant) rare species of trees unknown in the district, to make a sort of botanical garden. This effort interested the local foresters, as the standing timber around Zechlinerhütte was the monotonous result of the monoculture of a few species of commercially valuable evergreens.

Across the road was a sandy lakeshore with a large area to swim in, clear of mud and waterweeds. There was a "bath-cabin" for changing clothes, and near the swimming beach was a dock with a rowboat and, within a few years, a sailboat as well. The lake offered scope for discovery and adventure as part of a chain of small, deep, and narrow lakes known together as the Mecklenburgischen Seen, all connected by navigable stream channels and canals to even larger lakes surrounding: the Zootzensee to the north, the Grosser Zechliner See to the west, and the Rheinsberger See to the south. All of these sat in forests of fir, birch, maple, and the ever-present, fragrant Linden. Here, 80 kilometers (50 miles) closer to the Ostsee (Baltic Sea) than Berlin, the air was fresh and clean and the sense of quiet isolation very pleasant.

This place, *die Hütte*, as Alfred and the other children called it, was the family home ever after.[23] It was their vacation and summer residence until about 1910, and afterward the year-round retirement home of the parents. It had a life-transforming impact on everyone in the family, almost immediately. Each arrival there from Berlin produced another series of transitions, opening their spirits outward. When the Wegeners set out for "*die Hütte*," they traveled out of Berlin by train through industrial suburbs with their smoke-belching stacks and furnaces, out into the surrounding farmlands as far as Gransee, 60 kilometers (37 miles) north of Berlin. Alfred heard the "clickety-clack" of the train wheels announcing, "*Es geht nach die Hütte, es geht nach die Hütte!*" (It goes to the hut! It goes to the hut!). From Gransee the parents proceeded through country lanes with the baggage wagon, while the children (unsupervised, unaccompanied, and undirected!) hiked the final 20 kilometers (12 miles) from the Gransee Station to Zechlinerhütte through the Menzer Forest, passing only scattered farms and lakes and the minuscule hamlet of Menz on the way.[24]

The children loved the succession of stages in the journey and what they represented—to leave the bustling train station in Berlin with a mountain of luggage and provisions, to disembark two hours later at the village already "at the end of the line," and from there just to *walk away* out of the town, and keep walking until the road diminished into a sandy cart track with a grassy median and disappeared into the "depths" of the Menzer Forest. This great wooded tract, completely cut over in the eighteenth century to feed the glassworks, had sprung back with the especially dense character of second-growth evergreen forest. It was almost oppressively silent for long stretches, with dry twigs underfoot and deep twilight to both sides of the road—the gently swelling and rolling terrain concealing what lay just over the next rise to each side of the road, but inviting one to leave the track, enter the forest, and see. It was with a sense of relief that the children passed these stretches into meadows and sunlight, and it was with a sense of adventure that they entered the next stretch of wood. Small wonder that the area was known as the Ruppiner Schweiz: it was a sort of miniature Swiss vacationland—cool, green, and refreshing. Theodor Fontane sang its praises in his *Wanderungen durch die Mark Brandenburg*, giving it an air of mystery, especially Menzer Forest and the Stechlin See, the dark and "bottomless" lake it conceals.[25]

In the summer of 1888, Alfred, then seven years old, was able for the first time to walk the whole distance from the station at Gransee to Zechlinerhütte. He had already learned to swim and to dive with his father's instruction; Richard was an enthusiastic and powerful swimmer. Alfred had passed the test of leaping from a boat and swimming to shore fully dressed—the rite of passage that allowed him to take the boats out alone.

He had grown a good deal and had good physical stamina. He and Kurt, who had just turned ten in April, shared most of their days together. Tony, at fifteen, already spent much of her time with her mother or alone. She was completely uninterested in what Kurt and Alfred were doing and preferred to walk by herself, closer to home.[26] Willi, at fourteen, was sometimes with Alfred and Kurt, but there is a great divide between ten and fourteen, and Willi, bookish and with his father's talent for languages, was feeling the gravitational tug of an adult world and was more often with Richard or off on his own.

After a morning cup of cocoa (Richard and Anna treated coffee as if it were a poison), Alfred and Kurt toiled grumpily at their obligatory morning hour of study, even in the heat of the summer, while Willi and father Richard were beginning to read Greek tragedy together. When the younger boys' endless hour finally dragged itself to an end, they would leave the confines of the house instantly—most often heading to the boats. They loved to row about and made great voyages of discovery in the lakes, drawing maps and taking soundings.[27]

Alfred and Kurt also did a prodigious amount of hiking and exploring in the woods. They worked at this systematically and within a few years had been everywhere within a 10-kilometer (6-mile) radius of the house. Anna Wegener began to pack lunches so they could go further. They would hike all day, pushing themselves, returning in the evening completely worn out.[28] On occasion they also went squirrel hunting. The squirrels were "the enemy," as they were constantly stealing the walnuts and hazelnuts the boys considered theirs. When they were able to kill one, which happened with some frequency, Anna was obliging and willing to cook the squirrel up for the hunters.[29]

When not with Kurt, Alfred was off by himself. It mystified his parents that a boy who chafed at an hour's homework could lie absolutely still in the underbrush by the lakeshore for more than an hour, his ear cocked, listening to a badger moving underground in its den. He showed the same patience when fishing, or watching and collecting insects. He pleaded endlessly and finally successfully for a terrarium to house and display his captured frogs and worms. His parents found this persistence both remarkable and gratifying. In providing this home they had given scope for all the children, but Alfred and Kurt in particular developed unexpected but strong, and ever stronger, sides of their natures—as wanderers and observers in the natural world. But this is too portentous; they were children, and it was summer, and they had to make the most of their freedom.

A summer outing with the Wegener family and the orphans to an island in the Rheinsberger See, near Zechlinerhütte, in 1889. Alfred is on the extreme left of the photo. From the Wegener family album. Photo courtesy of the Heimatmuseum, Neuruppin.

When September came, they were back in Berlin and back in school before the weather had even turned cool. Back in the suits and ties, buttoned and strapped, at their desks in the stuffy and musty schoolrooms, smelling the chalk dust from the board and the linseed oil in the woodwork, they were already dreaming of the next vacation. This would come at the end of the autumn term, when there was a break at Christmas, and there would be another break at the end of the spring term. During these *Kurzferien*, the Wegeners took the "pupils" with them to Zechlinerhütte, and these holiday excursions were more epic and memorable—such as the time a hired lake steamer took everyone to an island in the Rheinsberger See, where they built a fire to make the previously forbidden coffee and then ran about and played.

These all-too-brief vacation respites from the grind of examinations, drills, and lessons were themselves full of planned activities; there were long group hikes through the woods and athletic contests on the playfield. After supper in the evening, there was singing on the front verandah, and then all the boys were sent upstairs to sleep on straw that had been spread for bedding in the great loft of *die Hütte*.[30] The mice (there were plenty, winter and summer) used to wake them, and they would try to silence them by throwing their boots, which worked well and was very satisfying; there was no such easy remedy for the mosquitoes. The boys were nonetheless agreed that it was much better to be wakened by mice and bitten by mosquitoes in Zechlinerhütte than to sleep soundly in Berlin.

Cöllnisches Gymnasium

Like many young men (and women) who later go on to great artistic and scientific achievements, Alfred Wegener disliked school. Most children have a healthy and biologically based hatred of early rising, tight clothing, hard benches, forced immobility, and repetitive drills. But schooling can be a special agony for physically active children with a desire for an education beyond the covers of books. For students whose talents and imagination lie beyond, school is an obstacle course and a race between talent and boredom, a struggle to survive formal instruction with intuitive skills and gifts intact.

In 1890, at the age of ten, Alfred entered the Cöllnisches Gymnasium, conveniently located (apparently its only redeeming feature) a ten-minute walk from home across a bridge over the Spree. It was a square, five-story brick building with high airy windows, a breezeway, and a large courtyard. It took its name from the medieval town on Cölln, the part of Berlin in which it was located, but it had little of the special tradition and distinction of the Graues Kloster, where Richard Wegener taught. Though the "pupils" in the orphanage, with whom the Wegener boys had all attended primary school, went on to the Graues Kloster for their secondary education, Alfred, Kurt, and Willi went on to the Cöllnisches Gymnasium, a choice determined equally by proximity to home and by a regulation of the Prussian ministry of education (strictly observed) which forbade the attendance of children at schools where their parents were teachers.

The Cöllnisches Gymnasium's curriculum was, like all truly classical *Gymnasien* in Prussia, centered on languages and literature, with a pivotal place given to Greek and Latin. Among the modern languages, in addition to German language and literature, there was instruction in French and English, and students also were taught history, religion, geography, and mathematics. This range of subjects notwithstanding, German schoolboys of this era devoted an overwhelming proportion of their study time to Greek and Latin. During the 1880s, this stress on classical antiquity encountered opposition, as it had already at several points throughout the nineteenth century. Educational reformers in Prussia had been frustrated repeatedly in attempts to shorten, alter, or modernize the curriculum. The controversy broke out again in Berlin just as Alfred entered the *Gymnasium*.

When Crown Prince Wilhelm took the throne in June 1888 to become Kaiser Wilhelm II, the situation changed rapidly. Wilhelm, twenty-nine years old, was impulsive, enthusiastic, intelligent, and sympathetic to everything that Bismarck was not: parliaments, labor unions, social welfare legislation, and modern scientific education—an education suitable for an industrial state that also wished to be a great empire. It was clear from the beginning that he intended to rule as well as reign. In March 1890 he demanded and got Bismarck's resignation. The next few years were a whirlwind of social reform: restriction of work hours, mandatory Sunday holidays for all workers, industrial courts for wage disputes, factory inspections, reduction of universal military service from three years to two, and many other measures that Bismarck had successfully suppressed for many years.

The new kaiser was especially interested in the question of educational reform. By 1892 he had successfully ordered a reduction in the number of hours devoted to Latin and an overall decrease in the number of hours of study per day. These modest reforms set off a fierce and protracted struggle between modernizing reformers and partisans of traditional classical learning, a struggle that lasted the rest of Alfred Wegener's secondary school years.[31]

These changes had some direct impact on Alfred and Kurt, but they had considerable indirect impact as well. That same year, 1892, catastrophe struck the Wegener household. Willi, eighteen years old and his father's favorite, died of an abdominal infection resulting from a ruptured appendix, a condition diagnosed too late for the new scientific-medical operation that might have saved him. Alfred recalled this as a terrible blow to his parents. Willi had identified closely with Richard's aspirations, had excelled in Greek, and had recently, under his father's supervision, begun the study of Hebrew.[32] He was clearly destined for the classical course at the university and may have been considering the ministry as well.

It was natural that the grief-stricken Richard would turn his attention and focus his aspirations on his remaining sons, Kurt (fourteen) and Alfred (twelve), much as Anna had turned to the boys after Käte's death eight years earlier. It was soon apparent, however, that neither Kurt nor Alfred inclined, as Willi had, toward a career in classical studies. Kurt, who was linguistically gifted, stepped obediently into the role of eldest son and was willing to take on additional classical studies with his father. Alfred was adamant that he wished for a career in science, and refused.[33] This was the first clear expression of his vocation, or his first clear memory of the same—unless one wants to treat the collections of frogs and beetles and the mapping of the lakes near *die Hütte* as harbingers of a scientific calling, though these seem common-enough occupations. One may look at the contents of someone's childhood without being therefore obliged to accept point-for-point correspondence between childhood activities and his adult life. If one were so obliged, one would then predict a career for Alfred not just in science but in Arctic exploration, based on his fondness for winter sports such as ice-skating. He did enjoy winter sports, but so did Kurt, his constant companion. If we follow this path and logic, we must then somehow account for Kurt's marked preference for fieldwork concentrated in German Samoa, in the midst of the tropical Pacific. It is wiser simply to take Alfred's word that around the age of twelve he began to think seriously of being a scientist.

Willi's death pushed the choice between classical and modern studies out into the open in the Wegener household, and Alfred was clearly on the side of the moderns. But there was no question of Alfred or Kurt attending anything less than a full classical *Gymnasium* course, even though the alternative schools were more freely available in Berlin than anywhere else in the empire. This insistence on classics was not a matter of antiscientific or antitechnical spirit. Richard Wegener was a modern himself, at least by the standards of his profession. As his poetry from this period shows, he understood that there exists a range of interests and talents, and that butter and eggs are different, and that characters may be firmed by different means. He opposed a technically and scientifically based education for boys because he thought it lacked the character-building and morally edifying possibilities of the classics; he was convinced that the sciences had to stay in their own sphere.

True to this demarcation of studies, but acknowledging the seriousness of his sons' interest in science, he made presents to Alfred and Kurt of an electric motor and a handbook for conducting chemistry experiments. He added to these a book of physics and a good-quality astronomical telescope and left them to teach themselves science.[34] This approach was fully in accord with his official and personal commitment to the classical curriculum while provisionally accepting the sciences and other modern studies as avocations to be pursued on their own time, after the homework was completed.[35] More gently, it affirmed his conviction to advance the constructive possibilities of Schiller's *Spieltrieb*, the spirit of play.

It was fortunate that Richard intervened in this decisive and timely way to help his sons, because the scientific curriculum at the Cöllnisches Gymnasium, even though there was nominal instruction in physics and chemistry, was terrible. Kurt said that of the two mathematics teachers at the *Gymnasium*, one was able to teach, but the other seemed to know less than the pupils, and if a student read the text and did the homework, it was soon clear that he *did* know more than the teacher. It seems to have been a pedagogical principle there not to push students who showed no interest, so competition was not severe, and discipline was relaxed in comparison with the Wegener home.[36]

Because the Cöllnisches Gymnasium offered little support for their scientific interests, Kurt and Alfred pursued them at home, a place they had increasingly to themselves. Tony had already grown into a woman. She was nineteen when Willi died, nearly finished with her education, and though completely dedicated to her parents, she was already an adventurous traveler and away from home for increasing periods of time. The boys enjoyed her letters and descriptions of distant places, and the three of them would remain close for the rest of their lives. But as Kurt's and Alfred's determination to become scientists grew, they created a world of their own within the orphanage. They were able to convince their mother to give them the use of an empty storeroom near the laundry and washhouse, where they established a chemical laboratory. The request came from Alfred. He was clearly his mother's favorite child—he looked more like Willi (whom she missed terribly) than did Kurt, and he was quiet and rather serene, whereas Kurt was nervous and loud. Whenever the boys decided that they wanted or needed something from their mother, Alfred was made the petitioner, as his chances for success were invariably better than Kurt's.[37]

Alfred would later downplay the seriousness of their chemical laboratory efforts—he said that they generally succeeded only in turning their weekly allowance into "fumes and booms."[38] This humorous self-deprecation by a mature scientist should not obscure the reality. In Wegener's life as in many before and since, household pyrotechnics with simple chemicals were the first steps toward a deep feeling for empirical scientific practice.

Modern industrial democracies have tried to protect their citizens and their environments by restricting access to hazardous chemicals, but in so doing they have also closed the door to an inexpensive and serious (if anarchic) encounter with a broad range of basic physical and chemical phenomena. High school and university chemistry laboratories have now moved toward milliliter- and microliter-scale chemistry experiments, where the results register through microprobes and are read out electronically on computer terminals.

Students can certainly learn chemistry this way, but they cannot so easily *see it happen*. It is not the same as manipulating a puddle of mercury, or leaving a piece of phosphorus out to smolder into flame at room temperature. It is not the same as experiencing—and helping others to experience—the effects of the synthesis of hydrogen sulfide, and it cannot compare with the pleasures of detonating the proportional combination of sulfur, saltpeter, and charcoal. At a somewhat higher level of endeavor, modern children miss the beautiful colors produced by mixing reagents almost at random, as well as the mystery of combining two colorless liquids at room temperature to produce a violent geyser that renders the beaker holding them too hot to handle. The same holds for the miraculous appearance of a solid in a beaker when a solution is titrated past a set point.

Unobstructed by well-intentioned safeguards and without the aid of theory, without much mathematics, without problem sets, and, most of all, without adult supervision, Kurt and Alfred were able to explore thermochemistry and thermodynamics,

The great pioneering chemists of the late eighteenth and early nineteenth centuries—Priestley and Lavoisier, Dalton and Berzelius—had made their own momentous first steps in elemental chemistry with no more direction, with no better apparatus, and with incomparably poorer reagents.

As it was with their homemade chemistry lab, their cherished electric engine gave them similar advantages in mechanical technology. Electricity and electrical apparatus moved with astonishing speed in the 1870s and 1880s out of the laboratory and into factories and households—just as industrial and consumer electronics would do a century later.

Most of the fundamental transformations of energy are codified in the great physical theories of the mid-nineteenth century: mechanical work to electricity, electricity to mechanical work, electricity to magnetism, magnetism to electricity, electricity to heat, and heat to light may all easily be accomplished on a tabletop with a small electric motor and some lengths of wire. Add a small chemical storage battery, and the transformation of chemical activity to electricity and vice versa may be included in the list. Before they had studied theory or measurement, Kurt and Alfred enjoyed the pure, qualitative experience of phenomena—seen, heard, and felt. With electricity, of course, the last of these modes of perception makes the strongest impression, though all the senses are continually exercised in such free-form scientific performance.

As their fascination with these novel materials and ideas enlarged, so did their pleasure in the physical experience of experimental play. The contrast with their schoolwork was increasingly stark and invidious, and their reluctance to go to school grew apace. Their father was unrelenting on this point. It was made clear to Alfred and Kurt over and again that their performance at school had to be superlative, not merely adequate. The hours of classical study that had been subtracted at school by the kaiser's decree of 1892 were added back by the father at home. Duty came first, and the first duty was to excel academically. There was scarcely a day in their teenage years when they were not required to study something under duress, winter and summer, term time and vacation: *Nulla dies sine linea.*[39]

Both Alfred and Kurt were also increasingly committed to a program of physical activity. They were fortunate to have the gym in their home, and most days on their return from school they went directly there for an extensive workout. They learned the standard gymnastic apparatus; they could go into a handstand on the parallel bars, they could vault the horse, and they learned the reverse grip that allowed them to do the "giant's turn" on the horizontal bar outdoors in the yard—lacking coaching and proper spotting, however, they stopped short of the flying dismount in this exercise, much to their disappointment.[40]

Only very late in Alfred's secondary school career did his scientific interests find any substantive resonance with his schooling. It 1897, when he was in Secunda, the second grade from the top at the *Gymnasium*, the minister in charge of Prussia's universities, Friedrich Althoff, let it be known that he intended to alter secondary school curricula to link mathematics instruction (always present whatever the educational party in power) to real instruction in physics, allowing physics to become a secondary school subject in its own right.

It appears that Alfred's physics teacher, who was interested in astronomy and had a good-quality refracting telescope, recognized Alfred's talent and interest and invited him to take up the study by joining him in making observations. For the next year and a half, until his graduation, Alfred pursued astronomy whenever time and weather permitted, walking back to the *Gymnasium* in the evenings and observing the heavens with his teacher, from the roof of the school.[41] He began to read the abundant popular literature on this topic. One of his favorite authors, whom he read for many years thereafter with continued pleasure, was Max Wilhelm Meyer (1853–1910), who wrote *The Cosmos—a Popular Astronomy* (1898); Meyer had also edited an edition of another of Wegener's favorites, Friedrich Diesterweg (1790–1866), whose *Popular Astronomy and Mathematical Geography* was a great favorite among young readers.[42] These were books that contained both astronomy and cosmology, with speculations concerning the birth and death of the Sun, the origins of life, the possibility of life on other planets, and fantasies of travel through space. They were instructive but were meant to be inspiring and thought provoking as well, and they portrayed science as a noble undertaking, fully the equal of any other study.

In the summer of 1897, the summer before Alfred passed into Prima at the *Gymnasium*, the family returned, as they had every year for the past twelve, to their *Heimat* in Zechlinerhütte. The boys were almost (but not quite) men and were ready to range farther away from home. They had hiked the country everywhere around; now they turned increasingly to the water and the network of lakes and canals. Richard Wegener had, a few years earlier, taken a fancy to the Norwegian dinghies he had seen used as ship's boats on the freighters that came in and out of the port of Warnemünde, about 120 kilometers (75 miles) northwest of Rheinsberg. These were dories—about 4 meters (13 feet) in length and a little less than 2 meters (7 feet) beam. One person or two could row them, and although they had no weighted keel or centerboard, they could sail very well with a simple gaff or sprit rig, and even better with a jib. Their narrow stems (transoms) made them ideal for rivers because they could be used in rapids both bow and stern first; when loaded, they became more rather than less stable (which is why they were used by fishermen). Best of all, given the ever-slim family resources, they were very cheap to buy. Richard went to Warnemünde, bought one, and had it shipped back by rail and off-loaded at a spot where the rail line passed close to the water, a nice solution since the boat, made of pine and oak, probably weighed 175 kilograms (386 pounds).

It was in this boat that Kurt and Alfred were taught to sail, and during that summer (the last before Kurt went off to university), they made a number of extended voyages through the canals to lake destinations 40–50 kilometers (25–31 miles) away, sailing by day and sleeping in the boat at night. Kurt remembered one particular voyage with Alfred to the Müritz See, the largest of the lakes, some 20 kilometers (12 miles) long. The lock before the Müritz See had a 10-meter (33-foot) fall, and Kurt, not waiting for the lock tender, went to open the upper sluice gate to fill it, while Alfred remained in the boat. He opened the gate too far and the water came rushing in with a roar, creating a huge waterfall and emptying the reservoir, which also functioned as a mill pond, thus disabling the mill for the remainder of the day and infuriating the miller, who was also the lock tender. In the next few days they sailed about having a grand time but managed to get so badly sunburned that their faces swelled up until they were unrecognizable. They could barely see, but at least the miller failed to recognize them on the homeward trip. Back at *die Hütte* their faces blistered and peeled, and they were in such pain that they could hardly sleep. Anna was horrified when she saw them, but, as Kurt recalled, "she was used to this sort of thing from us."[43]

They also talked about their plans. Kurt, who had graduated *Primus Omnium* (first in his class), had startled his parents in bypassing the universities to enroll in the newly created Technische Hochschule (polytechnic institute) in Charlottenburg, at the other end of Berlin. Education Minister Althoff's sweeping (and immediately effective) reforms of 1897 had raised these institutes to the level of the universities, allowing the technical schools to grant a doctoral degree to students who agreed to extend the three-year technical course into four years.

Kurt was committed to experimentation as a means of discovery—he was by his own account more a doer than a thinker—and was also mechanically inclined. He was interested in pursuing a program in applied mathematics including engineering mechanics, geodesy (higher-level surveying), and the statistical theory of errors. Althoff's reform had made it possible for him to follow his inclinations without sacrificing a professional career among the *Doktoranden*. The immediate social consequences were nevertheless real: it would be a school initially filled by students from secondary *Realschulen* (the semiclassical schools) drawn from the lower ranks of the middle classes, men who, in spite of their starched collars and carefully knotted ties, were likely to make their living, in part at least, with their hands. From the standpoint of Kurt's parents it was a perplexing move, the first of many.

Alfred was also ready to make a decision. Though he had another year of *Gymnasium* to go, he was leaning toward entering the University of Berlin to study astronomy. His decision to attend the philosophical faculty of his father's alma mater and seek a doctorate in a university subject was greeted at home with a palpable sense of relief and gratitude. If Alfred's reluctance to become a teacher or classical scholar was now firmly established, the decision to study astronomy was certainly respectable, since it amounted to a decision to seek a career as a university professor; Kurt's future was much less certain. Though the Wegeners soon came to see that the new technical universities would be permanent features of the educational scene in the new Prussia, they could not help but feel that Kurt had taken a step toward abandoning his place in the *Bildungsbürgertum*, the educated mandarinate.

The great relief of Richard and Anna in Alfred's choice had much to do with his apparent allegiance to the privileges of his estate. However sorry Richard might be not to have a son follow him exactly, what a bitter irony it would have been to have both his sons abandon the traditional family status and inherited professional standing he had toiled to preserve for the scores of orphans who had passed under his tutelage in the previous twenty-five years! If Alfred worked hard, his future would be secure. His parents would support him through the university years, as they would Kurt. They planned as well to equip the boys for their compulsory military service. The boys would, of course, be officer candidates: their automatic eligibility for this desirable option was another benefit of their educational and civil service status. The Wegeners also planned to lend some financial support to the boys beyond graduation from the university. Even with their doctoral degrees, it would still take them many years to climb the ladder to fully salaried and pensioned (the magic word!) employment in the educational system.

In the winter of 1899 Alfred passed his *Abitur*, the final and comprehensive examination that guaranteed automatic admission to the university system. As Kurt before him, he graduated *Primus Omnium*, to his parents' great pride and pleasure. He was near his full height (about 5 feet 10 inches) and had as yet a compact build that would only later fill out. He had not finished growing; this was evident from the way his ears stuck out from his rather large head—an exaggeratedly youthful appearance that would persist through his university years. A quizzical smile was often on his lips, as if enjoying a private joke, but perhaps it was only covering the uncertainty of late adolescence in that strange limbo where one is no longer a boy but not yet a man. He was quiet and almost stolid, warming to enthusiasm for his favorite pastimes, but for the rest rather self-contained. This was in marked contrast to Kurt—tall, thin, often with a slightly melancholy or even puzzled expression, and looking at age twenty like a man in his thirties.

Alfred was physically strong and agile from many years of sport and exercise, but he was mentally strong as well. He had mastered what the Germans call *Sitzfleisch*, that special form of perseverance that allows one to stay at exacting tasks for long periods of time, without succumbing to boredom or restlessness in a way that makes work impossible. In an age before typewriters and calculators (let alone word processors or desktop computers) it was an extremely valuable acquisition, because every intellectual task took much longer then than the same task does now.

Most of Alfred's later achievements in science would have been impossible without his exceptional ability to focus and keep himself on task. He had demonstrated this power in school and out of school, and usually in combination with *Geduld* (patience or forbearance). The two are different, but the combination of the two is much more powerful than either one alone. In particular, the ability to concentrate is much impaired by impatience, even if one can stay "in the seat." Alfred seems to have had an uncommon and precocious strength in this area, and he practiced perseverance and patience not only as disciplines but also as virtues, which indeed is the way his parents had taught them to him.

The coming transition to the university from the *Gymnasium* was momentous and exciting for Alfred largely because it meant the freedom to choose his courses and to study henceforth nothing but science and mathematics. It was not to be socially liberating, at least not immediately, for he would still be at home and under parental influence and authority. In any case, the university was not much farther from the orphanage in one direction than the Cöllnisches Gymnasium was in the other, so there was not even much new to see outside. But inside, in what he later liked to call his *Allerheiligstes*—his inner sanctum or spiritual center—things would be different.

2

The Student

BERLIN-HEIDELBERG-INNSBRUCK-BERLIN, 1899–1901

In earlier days every new impression, every new fact entered into my growing individuality as an integral factor, and my individuality grew in proportion to the amount of the facts it took in. Through every new experience I gained a new means of expression; every point of view strengthened my consciousness of self, and therefore it was not senseless if I lived in hope, as it were, of snatching from without what spurred me on from within, though it had not yet revealed itself to me.

HERMANN KEYSERLING, *Travel Diary of a Philosopher* (1918)

Friedrich-Wilhelms University, Berlin

With the opening of the academic year on the first Monday after St. Michael's Day (*Michaelis*), 2 October 1899, Berlin University celebrated its eighty-ninth birthday; Alfred Wegener was a still a few weeks short of his own nineteenth. He had obtained his identity card and the *Anmeldungsbuch* (Registration Book) that would follow him through his academic career. Here his professors would confirm his registration in their courses, as they would note at the end of the semester whether he had attended them.

Alfred walked the kilometer or so from the orphanage past the palace to the University Physics Institute, a spacious three-story facility that occupied a full city block along the Spree—ten to twenty times the size of his old *Gymnasium*. The institute was only slightly older than he was, having opened in 1878. Designed by Hermann von Helmholtz (1821–1894) himself, it had a lecture hall that seated 200 students, a large teaching laboratory, a physics library, and a number of special laboratories for optics, acoustics, mechanics, and electrical work. Until 1870 the library, apparatus, and laboratories of the Berlin Physics Institute had been limited enough to be housed within the residence (a Baroque palace to be sure) of Helmholtz's predecessor, Professor Gustav Magnus (1802–1870). It was very much a statement about the new scale of Berlin physics when Helmholtz specified that the new institute should be big enough to contain an apartment to serve as his residence—which it did.[1]

The Royal Friedrich-Wilhelms University of Berlin was, at the time Alfred enrolled in it, one of the largest universities in the world. It had a student body of almost 7,000 and a faculty of 450 professors, 227 of them in the Faculty of Philosophy—what we would now call the School of Arts and Sciences; the remainder were in law, medicine, and theology. The true size of the teaching faculty was larger, since the German system usually specified only one salaried full professor, called the *Ordinarius*, and one associate professor, the *Extraordinarius*, for each subject. The rest of the faculty was composed mostly of *Dozenten*, a composite group of what would in an American university be the assistant professors and the instructors, though their positions were more precarious and their incomes much lower. The university's place on Unter den Linden, between the State Library and the Royal Arsenal across the way

The Physics Institute at the University of Berlin, where Alfred Wegener began his university career in October 1899. From a contemporary postcard in the author's collection.

from the Palace of Wilhelm I, testified to the importance of the learned professions in the Prussian scheme of things. The late-century transformation of Germany into an empire and a wealthy industrial state, ruled by an education-minded kaiser, flooded the university system with new resources and brought students in great numbers, especially into the sciences.

Alfred's first-year academic program at Berlin must bring a smile of recognition to anyone who has pursued—or thought about pursuing—a vocation in natural sciences: analytic geometry and calculus, physics, and chemistry.[2] This is "year one" of most scientific careers then and now, all over the world. To these fundamental preparatory studies Alfred added a course in "practical astronomy" and a series of inspirational and nontechnical lectures in astrophysics, aimed at a popular audience. This is the program he would pursue until the following April—the end of the "winter semester"—a period of study roughly the length of a North American academic year. The topics of the basic courses were generic, and so was the content. That was the point: in classical physics and astronomy generic content and methods really existed as shared foundations on which one could build. Indeed, in spite of the many revolutions in physical theory since that time, the basic subject matter and contents of these entry-level studies have remained remarkably constant.

If the content of Wegener's first year was generic and ordinary, not so the instruction. Berlin was in 1899 the summit of not just the Prussian but the Imperial German university system; it was the top of the profession, the pinnacle of advancement and promotion. The professors were giants of international reputation, men who ruled their departments like feudal baronies and held their professorial chairs like thrones. Their appointments were for life, and when they died or retired, there were no applicants for their jobs—the ministry of education formed a committee to rank the three current leaders in a given field, and based on this ranking a "call" went out to a specific person, *named* as the successor. The resulting concentration of talent in science and mathematics was awe inspiring. More extraordinary, however, and in contrast to the image of the lofty and indifferent German professor, Wegener's roster of teachers (in his first year and thereafter) was dominated by men devoted to teaching. Among these were a few individuals equally famous for their easy familiarity, their lack of ceremony, and their concern for their students' welfare and progress. The culture of physics, astronomy, and mathematics at Berlin, like that of the Wegener household, was of one of cooperative work, of mission, of participation, and of community. There was, of course, a hierarchy—where is a there a community without one? The professors were also severally quirky, imperious, abrupt, demanding, and tedious: it was, after all, a university. These reservations notwithstanding, the institutional culture demanded careful attention to the instruction of students, whose intellectual welfare was a real consideration.

The course that introduced Alfred to university mathematics was analytic geometry and the theory of limits, taught in a large lecture hall to an audience of 200 or so by Hermann Amandus Schwarz (1843–1921). Schwarz, famous for his work on conformal mapping and Riemann surfaces and in the calculus of variations, was well known in Berlin for his lectures and his relationship to his students. When Wegener first heard him lecture, Schwarz had been in Berlin for eight years and had thrown himself completely into his teaching. He shared Karl Weierstrass's (1815–1897) conviction that the theory of functions had to be established on simple algebraic truths and that the foundations of the theory should be available to students without a knowledge of complex techniques of differentiation.[3]

Schwarz took his time in lecture, not reading from notes or rocketing through the designated material, as did some of his colleagues. His explanations were detailed and discursive. To illustrate his points, he drew beautiful diagrams in colored chalk, completely unhurried. He stuck to the fundamentals of analytical geometry, not moving rapidly to those topics that the advanced mathematics students were waiting to hear about. The slow pace reinforced the importance of the basic algebraic and trigonometric concepts, of logarithms and exponents, sequences and series, rehearsing the linkage between the world of algebra and that of geometry, and returning again and again to the basic concept of the limiting value—the discovery of the point at which some mathematical function or physical quality reaches a maximum or a minimum.[4] Schwarz tried to get all of his students to join the Mathematical Society, urging them to come to the meetings and get to know and meet with their *Dozenten*—a means to learn mathematics and to develop a sense of community in a large and impersonal university.[5]

In sharp contrast to Schwarz's lectures were those of Johannes Knoblauch (1855–1915), who taught Wegener introductory calculus. He was a clear lecturer but moved very rapidly and developed the differential calculus with a very abstract and arithmetical approach that was too sophisticated for most of his young students. Knoblauch was at pains to give them exercises and problem sets in order to apply the principles, but the problems were generally too hard.[6] Alfred had worked diligently at his school mathematics in the *Gymnasium*, knowing that he would be going on to the university for astronomy and physics, but he found Knoblauch's course hard sledding. He had no special gifts in mathematics—something he knew early on and freely acknowledged to family, friends, colleagues, and later his students. Neither had he any taste for the sort of mathematical abstraction that gives some physicists such intense aesthetic and intellectual pleasure; in later years he exhibited a strong antipathy toward it, and one may suspect that this antipathy was born here.[7]

More compelling and exciting were the lectures in experimental physics given by Emil Warburg (1846–1931). Warburg lectured to audiences of 300 or more, crammed into the largest hall in the institute. His lectures were formal and showy demonstrations in which the experiments were performed by the professor and the instructors before the class, rather than a laboratory-based study in which the students performed the experiments themselves; there was no opportunity for discussion. But Warburg put tremendous energy into this teaching, and some of the demonstrations were wonderful and are still performed today, all designed to be as striking as possible and to link the phenomena under examination to a unified mechanical picture of the world through the laws of motion. This bringing of physics together under the concepts of mechanics had been the goal of Berlin physicists in the previous generation under Helmholtz and Kirchhoff, and it had been affirmed by their successors, including Hertz and Boltzmann.[8]

The lectures by Julius Scheiner (1858–1913) on "Popular Astrophysics" had the merit of being about the subject that Wegener had come to the university to study. Scheiner had come to Berlin as "Extraordinary Professor of Astronomy" in 1894, though Berlin was not a center for astrophysics, but rather one for classical observational astronomy. Scheiner was a somewhat distant figure who spent a good deal of his time in the observatory at Potsdam, where he was part of an international project to create an "astrographic chart" of the heavens, and where he worked in developing practical techniques for celestial photography and spectrography.

Given Wegener's intense excitement about moving on to the university, there was something very unsatisfactory about all of this. The classes were huge, and the course work consisted in listening to lectures, taking notes, looking at demonstrations, and reading the professors' textbooks. There were no real laboratory exercises, and the problem sets were limited and unsatisfactory, or too hard. The professors were clearly well-meaning and abstractly concerned for students, but they were not available for consultation outside of the *Sprechstunde*, the one office hour per week that each professor was required to keep. Even for that minimal contact, one had to stand in line to get one's name on the sign-up sheet when it was posted, and one would be lucky to speak to the professor for a few minutes in the course of an entire semester. Wegener's studies were interesting and necessary, but it was not scientific work; it was *schoolwork*.

There was one wonderful exception: the course taught by Adolf Marcuse (1860–1930). Marcuse had come to the university only two years before, as *Privatdozent* in astronomy.[9] His course for the 1899–1900 year—small, intense, and pragmatic—was entitled "Practical Astronomy," and it was exactly that. Finally, there was something to *do* rather than to *listen to*. The course had three segments, and each of them had a profound effect on Wegener's thinking and his plans for a career. The very first part was "Theory and Use of Astronomical Instruments, Especially for Geographical Position Finding."

Marcuse, at thirty-nine, was the youngest of Wegener's teachers in that first year, and he was an expert on the use of astronomical observations to make precise determinations of latitude and longitude. In connection with this work, he was a leading observer throughout the 1890s in a program to measure the amount and direction of the very slight oscillations of Earth's axis of rotation, now known as the Chandler wobble (after Seth Carlo Chandler, 1846–1913). These motions had first been detected in Berlin in 1884–1885, though they had been predicted theoretically by Leonhard Euler in 1765. In the years 1889, 1890, and 1891 Marcuse worked at the Royal Observatory (Berlin-Potsdam) with a universal transit instrument—a telescope mounted in such a way that it could rotate in any direction and swivel so that it could look straight up, and equipped with vernier scales and micrometers capable of extremely delicate adjustments and very fine scaled measurements of angles. Marcuse was looking for repeated, minute variations in the apparent positions of pairs of stars. These variations could be interpreted as indications of slight variations of the latitude of the observatory and were what one would expect if Earth's pole of rotation was slightly and continuously displaced. The phenomenon was very interesting in its own right, but it also provided a necessary correction when preparing maps of star positions, and the Royal Observatory was at that time preparing such a map. Finally, it was a technical tour de force involving both instrument design and observational precision.

In 1891, Marcuse had traveled to the Hawaiian Islands and made the same observations. Hawaii lies between about longitude 155° and 160° west and is nearly halfway around the world from the Berlin Royal Observatory (Potsdam), at longitude 13° east. If Earth's pole were wobbling in an irregularly circular path over a period of months, the same latitude variations should be seen in Hawaii that were seen in Berlin, but they should be in the opposite direction. Marcuse showed that this was the case: the measurements confirmed an oscillation of Earth's axis of 0.3 arcseconds over a period of fourteen months. With 60 nautical miles (111 kilometers) for each degree of angular measure, this was a displacement of the pole of just under 10 meters (31 feet).[10]

Astronomers greeted these measurements with great interest, and the amplitudes of the oscillations were exactly confirmed by those measured by Seth Chandler at Harvard Observatory. Chandler had also made such measurements in 1884–1885, but he published them only in 1891. He also made an interesting and successful attempt to confirm the motion historically by going back to look at observatory records of the past 150 years indicating similar shifts of star positions, including the observations of James Bradley (1693–1762) at Greenwich Observatory, an astronomer renowned both for his accuracy of observation and for his strenuous efforts to remove every source of error. These historical researches uncovered a twelve-month variation superimposed on the fourteen-month period and set off a controversy even more absorbing than the measurements themselves. Euler had predicted an oscillation of 305 days, and Chandler's period was considerably longer, at 434: what could be causing the variation? Since it was a "free motion," it should ultimately "damp out" and cease, unless re-excited, but 150 years of records showed no damping, and there was no candidate mechanism for the excitation to keep the motion active. The phenomenon was of sufficient interest that in 1899, the same year Alfred entered the university in Berlin, an agreement was reached to establish the International Latitude Service, with four observatories to be constructed in the Northern Hemisphere 90° of longitude apart from one another, all four at the same latitude, to measure continuously the shifting of the pole.

Wegener had landed not just in an astronomy course in which he could do astronomy, but in one that implied that doing astronomy sometimes involved expeditions to distant places, and signifying a kind of work where a relatively short series of very precise measurements could be decisive in a long-standing controversy—and in this case could open new questions not only about the sky but about the earth as well. Marcuse took Alfred and the other students on field trips and taught them to level and orient the transits, telescopes, alt-azimuths, and other instruments. He taught them how to calculate instrument errors and how to correct observations for temperature (expansion and contraction of the instrument itself) and for the relative humidity, since the amount of water vapor in the air changed the way the light was refracted, causing a measurable (and correctable) angular displacement. He regaled them with stories of expedition science, both from his work in Hawaii and from his more recent trip to German Samoa.[11]

In the second wing of the course Marcuse took the students through a general survey of the fundamental ideas and achievements of modern astronomy, and here again was a pleasant novelty—the lectures were illustrated with lantern slides. Marcuse was a prolific and expert photographer, with a missionary zeal on the subject of photographic illustration. He taught every course using slides and believed that all subjects benefited from profuse illustration.[12] When he published a popular travel book about Hawaii, written while on the latitude expedition, he illustrated it abundantly with his own photographs, even including a color frontispiece.[13] Marcuse's use of photography in the classroom on this scale was as great and powerful an innovation in 1899 as classroom projection of computerized images directly from the Internet became 100 years later. The use of photographic illustration in science teaching is so uniform, and the slide format so ubiquitous (the most commonly repeated phrase at any gathering of scientists was, for some decades, "next slide please"), that it is interesting to remember that this approach to teaching was something that had to be pioneered and defended.

Wegener was entranced by the photographs, and he immediately and completely embraced both the technology itself and Marcuse's passionate advocacy of its use. There was certainly a scientific case to be made for the intellectual impact of visual images unfiltered by profuse verbal description. In a textual presentation of some aspect of the world, the author reveals the object of study in a series of steps and is in control of the unveiling. Even if the verbal text is accompanied by drawings, these inescapably reflect both deliberate and unconscious selection of detail. A photographer can do this too, of course, by selecting the illumination, angle, film speed, exposure, and so on, but the editing of a photograph is a matter of filters and thresholds. In a drawing, only what is put in goes in; in a photograph, everything goes in that is not screened out.

Marcuse made it clear to Alfred and the other students that photography was not solely an illustrative and pedagogic device but also a potent means of scientific discovery. In the third wing of the course, the first-year astronomy students accompanied Marcuse to the Royal Observatory, where they watched him and the other staff astronomers demonstrate the photographic methods used to document their observations. The students were put to work with practical exercises of observation, photography (including the preparation of photographic plates and darkroom work), and measurements of the shifts in the plates thus produced. These were not part of the institute's regular documentation; they were practice exercises—but they were nevertheless real observations and real calculations based on real measurements.

Wegener liked this approach to the world—he liked the small groups, the tactile and practical character of the work, and the emphasis on standardizing methods of doing things, as opposed to the standardized manipulation of abstract principles. He was exhilarated by the direct connection of observation and discovery and found that the emphasis on repetition gave him a certain advantage because of his habitual tenacity. By this route he had discovered much from the telescope on the *Gymnasium* roof which was new to him; now he was moving into a situation where really new discoveries were actually possible and even likely.

The summer semester of the year 1900 was coming soon, and with it a chance to alter his academic program. It was typical at that time for Berlin students to leave for the summer term, from May to August, especially during their first years. Consequently, enrollment at Berlin dropped by about 2,000 in the summer. Students headed generally for smaller and rural universities, like Göttingen, the choice of Alfred's student friend Walther Lietzmann (1880–1959), who wanted to pursue higher mathematics but also wanted to have a chance to hike and do some field botany and geology.[14] For his own first semester away Alfred settled on the university in Heidelberg, and he looked forward to it with great anticipation—not least because he was nineteen years old and had never been more than a few miles or a few days away from his parents.

The freedom to move about in this way was built into the German university system, operating on principles quite unlike those of Britain and North America at the time. In Germany, admission to any university at all was admission to all the universities in the system. Moving to a new university required less trouble and expense than most American students face in acquiring a certified copy of their academic transcript. This immensely rational approach to higher education allowed students to move on to whatever university or universities offered the concentration of disciplines most useful and congenial to them, no matter where they had begun their study. It allowed them not just to major in a subject, but to study that subject with different teachers in different locations, and to see how a subject matter is transformed by the particular approach to the subject a professor chooses, as much as by differences in the contents and methods themselves.

Ruprecht-Karls University, Heidelberg

Taking a summer semester in Heidelberg offered Alfred Wegener a number of advantages. First and foremost, it was far to the west and south of Berlin and therefore away from his home and his parents. In fact, other than Munich and Passau, there were no German universities farther away. Heidelberg lay among hills of forest and vineyard on the south bank of the River Neckar (a rapidly flowing tributary of the Rhine) and about 100 kilometers (62 miles) south of Frankfurt am Main. It was an old and picturesque city, most of it in 1900 still running parallel to the river on the Hauptstraße and clustered to the south of one of the larger town squares (the Ludwigsplatz, with its giant equestrian statue of Wilhelm I). It had been extensively rebuilt in the middle 1880s with the addition of a great new library in the Renaissance style to house a collection of more than 500,000 volumes.

Heidelberg was a good choice for another reason: it had acquired considerable fame as a scientific and medical university in the middle of the nineteenth century. It was here in Heidelberg that Robert Bunsen (1811–1899) and Gustav Kirchhoff (1824–1887) made the fundamental advances in spectroscopy which allowed the analysis of the composition of stars by study of their absorption spectra. The university also maintained a new and modern astronomical observatory on the Königstuhl, 335 meters (1,099 feet) above the town. It was much smaller than Berlin, having in 1899 about 150 faculty and 1700 students even at full enrollment in the winter; there were many fewer in the summer. It was therefore easy to find lodgings, and Alfred took a room in a lodging house at No. 5 Heinrich Ingrimmstrasse, very near the university.

Heidelberg in 1900 had its share of famous physicists, astronomers, and mathematicians, most of them with some sort of a tie to Berlin. There was Leo Königsberger (1837–1921), famous for his work in both mathematical analysis and analytical mechanics. In 1899 he had just finished writing a major textbook on differential equations, and his ties with physics gave his lectures on differential and integral calculus a quite different feel from those of Knoblauch at Berlin: it was very much a course of "calculus with applications," when Wegener took it in the summer of 1900.

In addition to registering for the calculus course offered by Königsberger, Wegener signed up for the course on experimental physics given by Georg Hermann Quincke (1834–1924), a classical, almost premathematical physicist in the tradition of Michael Faraday and the first physicist in Germany to insist that beginning students do practical laboratory experiments in order to learn physics.[15] Here finally was a chance for Alfred, almost the first since his and Kurt's freelance work in the orphanage washroom, to *do* physics rather than read about it or see it done by others. Quincke built the laboratory exercises around the simplest of materials—pieces of cork, wax, coins. He had done this out of necessity in the early years, not having money for laboratory materials, but he continued it with some pride later in his career.[16] Like Faraday, whom he greatly admired, the old physicist delighted in exhibiting fundamentals of physics with everyday materials, in the spirit of Faraday's *Chemical History of a Candle*; this sort of deliberate unpretentiousness appealed to Wegener.

The best academic reason for a trip to Heidelberg in 1900 was certainly to study astronomy. Max Wolf (1863–1932), a student of Königsberger, was supervising the completion of a new astronomical observatory on the Königstuhl, for which he had assembled the financing from a combination of royal and private patronage. Both Wolf and his coworker Wilhelm Valentiner (1845–1931) were pioneers in astronomical photography, and like their colleague in Berlin, Marcuse, they used the technique both as an investigative instrument and as a teaching tool, illustrating their lectures with lantern slides. Valentiner had published a photographic atlas of the solar system in 1884, with many of the plates devoted to the Moon. When Alfred signed up for his lectures on general astronomy in the summer of 1900, Valentiner was in the midst of editing a five-volume encyclopedia of astronomy.[17] Both Valentiner and Wolf had made extensive studies of near-earth objects such as comets and asteroids, though much of their later fame came from studies of double-star systems and nebulae.

In the summer of 1900, however, Wolf turned out to be teaching not astronomy but meteorology; Alfred signed up for the lectures anyway. It was a mild disappointment perhaps, but he knew something of this subject secondhand from Kurt. Moreover, astronomy and meteorology, now completely separate and found in very different parts of universities, were then parts of the same subject. The etymology of the word "meteorology" groups together all the things "in the air," including precipitation, tornadoes and waterspouts, and storms, as well as those astronomical objects entering the atmosphere from space for which the word "meteors" is now generally reserved. Added to the historical reasons for grouping them together and having the ensemble taught by astronomers were practical reasons: one had to consider atmospheric conditions in every astronomical observation, and especially carefully in positional astronomy, where one was trying to determine *where* something was as much as *what* it was. These atmospheric influences meant that would-be astronomers had to learn how to measure and record temperature and humidity and to correct for their effects. Moreover, whatever one could learn about weather prediction and prognostication was also useful to an astronomer: a falling barometer, high cirrus clouds at noon, and a wind backing from the south to the northwest were strong indications in Heidelberg that an evening planned for astronomical observations might well be spent instead socializing with friends or working in the darkroom.

This was the second time in Wegener's brief university career that astronomy was presented to him as a subject in which one had to learn first about the earth in order to learn something about the sky. He had begun astronomy in secondary school studying planets and stars. He had read popular works on cosmology, great sweeping stories about the universe as a whole and its life and death. These had provided him scientific nurture where school gave none, as well as a means to transcend the extremely constricting boundaries of his life in Berlin while physics and chemistry, doing what they could, were still trapped in the orphanage washroom. This transcendental power had much to do with his decision to pursue the study of astronomy at the university level. But his first two semesters of university, 1899–1900 and 1900, had shown him an astronomy almost entirely planetary in character, and as much about the earth below as the sky above. Marcuse used the sky mostly to study Earth; Wolf and Valentiner were as yet occupied with Earth's near neighbors and occasional visitors. All his instructors stressed, in practice if not in theory, that Earth was part of a continuum and a family of similar objects with a common history—the solar system. Moreover, their astronomy was extraordinarily and pragmatically local: not the sky and the universe, infinite and eternal, but this portion of the sky, tonight on this date, here in this place, under these atmospheric conditions, in this moment, recorded on this photographic plate—a universe not seen in the mind, as with the eye of God, but seen through a telescope with the eyes of one local observer.

Being thus brought to earth would have had a greater impact on Wegener had he been a serious young man paying close attention to what his instructors were saying, but there is some evidence that he was not. It appears, rather, that Wegener made only two discoveries that summer in Heidelberg: one was sabre fencing, and the other was beer. German universities in 1900 still had the same clientele that all European universities had had for the past 1,000 years: boys turning into men. At Heidelberg as elsewhere in Germany, they gathered in bands distinguished by the colors they wore: *Farbenbunden* (color bands). These organizations, semimilitary in character, were an important part of university life—an exclusively male preserve. Close counterparts of North American fraternities, they were then as now fundamentally contentless vehicles for male bonding expressed in initiation rituals, sports, organized nonlethal aggression and threat displays, singing, and the consumption of beer.[18]

The universities in Germany, as in the United States, France, Britain, Austria, and the other great industrial states, were also the principal training grounds for the officer corps of the armies and navies, as much as for the learned professions. Given the scale of military preparedness and the capacity for mobilization necessary to win a war at the end of the nineteenth century, European powers could no longer depend exclusively on an aristocratic and hereditary officer caste. The latter might make up the corps of career, active-duty officers for the standing army, but there was also a need for a corps of reserve and replacement officers. The empire therefore encouraged the adoption of military rituals and mores by doctors, lawyers, magistrates, pastors, professors, and the upper strata of the commercial bourgeoisie—bankers, merchants, and the indispensable legion of civil servants who kept the empire moving forward in orderly fashion. These youthful deposits of fraternity, patriotism, and mutual trust were banked against the inevitable day when these men's deaths would become more valuable to the state than their lives.

So Alfred Wegener joined a *Farbenbund*—which color no one any longer remembers. He went to the fencing hall with his color-brothers, put on the heavily padded tunic and gauntlets and the protective headgear, and learned to cut and slap with the heavy and dull-bladed cavalry sabres used in fencing competition. There was no flexible foil, no thrusting épée, just a broad-bladed sabre, meant for slashing and cutting, making a thunderous clash and clang as the pairs of combatants up and down the hall parried and cut to the encouraging shouts of the assembled brethren. In the evenings, after supper and study or bypassing study altogether (it was summer, after all, and examinations were years away), they repaired to the beer halls for singing and, of course, beer and tobacco. They drank their beer in half-liter and full-liter quantities, served off rolling carts running the aisles between the long tables and benches, along with potato salad, sausages, onions and radishes, and bread. They smoked black shag tobacco in long pipes with ceramic bowls, or smoked cigars, more often cheroots than the fat stogies that were thought to be a smoke for old men. The room grew hot, and they sweated their collars and grew red in the face. The band in the stand in the middle of the hall, on a raised platform like a prize ring with no ropes, played thunderous choruses of polka and martial music and marching songs, and the *Farbenträger* bellowed the lyrics and pounded the rhythm on the table with their fists or their beer steins.

On one such night, between Tuesday and Wednesday, 3 July 1900, at about three in the morning, Constable Eiermann of the Heidelberg Police was summoned to the Marktplatz to answer a noise complaint. He intercepted the perpetrator on the Hauptstraße and issued the following summons: "The aforesaid Alfred Wegener, Heinr. Ingrimmstrasse No. 5 is accused of disorderly conduct and disturbing the peace, in that he, wrapped in a white sheet, was proceeding down the Hauptstraße toward to Marktplatz and provoking thereby, through his too-loud shouting, an unseemly disturbance. On the basis of paragraph 360 of the Municipal Code, a fine of 5 Marks will be assessed against him. He must also pay costs. Failure to appear and pay the fine carries a jail sentence of two days."[19] Alfred appeared at the police station on Friday and paid his five marks and court costs of 20 pfennigs, and he was issued a copy of his summons and a receipt. He kept the document for the rest of his life as a sort of declaration of independence. The discipline of the parental home was exceedingly strict, and he had finally been able to shrug it off, even if only temporarily.[20]

Berlin, 1900–1901

Alfred traveled at the end of the semester from Heidelberg to *die Hütte*, for some vacation time with Kurt, Tony, and his parents. The latter were certainly not informed of the matter of the bed sheet and the predawn meeting with Constable Eiermann, though they no doubt got news of the fencing hall and such other matters as were suitable for parental consumption. It was time to hike and talk with Kurt and plan for the second year at Berlin. Kurt was progressing well in meteorology at the Technische Hochschule in Charlottenburg; perhaps because of Kurt's account of these experiences, as well as Alfred's own introduction to the subject from Max Wolf in Heidelberg, Alfred thought about adding meteorology to his program. What he heard from Kurt was rather different from what he had studied at Heidelberg: it was much more comprehensive and three-dimensional, and much more like doing physics than astronomy. But there was still the preparation in physics and the mathematics needed to do it, so these had to come first.

The schedule Alfred planned for the winter semester of 1900–1901 was more rigorous than that of his first year. The mathematics course was differential equations with Lazarus Fuchs (1833–1902). Fuchs's approach to the subject was an excellent means of continuing the work on calculus Alfred had done in the previous semester. Differential equations are the essential language for much of mathematical physics. In solving a physical problem—like the rate of change of temperature of some object as it is heated in the open air—the heart of the solution is the writing of the differential equation that expresses the relationship of the various physical quantities that enter into the problem. Of course, one needs measurements of physical quantities to put into the equation as terms, if there is to be a specific rather than a general or purely mathematical solution to the problem. Indeed, the terms of such differential equations—temperature, pressure, density, cross-sectional area—presuppose physical processes and measurements. The study of the construction of differential equations is the gateway to doing physics, since most problems in physics are answered with an equation—a concise expression of fundamental relations of quantities in mathematical terms.

Just as challenging as Fuchs's course in differential equations was that of Max Planck (1858–1947) in general (theoretical) mechanics. This subject, since the onset of the quantum-mechanical worldview, is now called classical mechanics or analytical mechanics. It is a unified approach to physics through concepts of motion and is still offered in modern university physics curricula. Then as now, a course in differential equations is usually a prerequisite or corequisite, and the organization and treatment of the subject are historical. Students began with the development of a physics of force and Newton's laws of motion. This treatment was then extended to the ideas of work, energy, and the "conservation laws"—particularly the conservation of momentum and the conservation of energy. Such a course usually ended with a physics of energy based on the formulations of Joseph Louis Lagrange (1736–1813) and William Rowan Hamilton (1805–1865). Along the way students were introduced to the mathematical treatment of classical problems: central-force motions and the orbits of planetary bodies, oscillations and harmonic motions (pendulums, springs, floating objects bobbing up and down), and the motion of rigid bodies, where the objects treated are no longer considered as point masses with a location only, but have a shape and an orientation (such as a center of mass). It was fortunate for Alfred that he had moved as far along in mathematics as he had: his friend Lietzmann had attempted this course without differential equations and had not understood much.[21]

The autumn of 1900 also brought the first of Alfred's many courses with the astronomer Julius Bauschinger (1860–1934), at forty already one of the world's leading figures in the very specialized field of determining the orbital paths of heavenly bodies. Bauschinger had come to Berlin from Munich as professor of astronomy in 1896, after the death of Friedrich Tietjen (1834–1895). Bauschinger was interested in orbits and only in orbits, and he fitted in well at Berlin, where, with the exception of Prof. Scheiner, everyone worked on the measurement and calculation of locations, orbits, and distances; the calculation of planetary tables (ephemerides); and the refinement of methods for correcting observational errors.

Berlin astronomy in 1900 was classical celestial mechanics with a vengeance: very exact, very technical, very powerful, and very narrow. The astronomers at Berlin, led by Wilhelm Förster (1832–1921), who had started his teaching career there in 1858 and would continue to lecture until 1920, were strongly oriented to the history of astronomy, and therefore some aspect of this topic was offered almost every semester. The faculty was small and the viewpoint severely limited. Between 1899 and 1904, bracketing Wegener's time as a student in the department, there were really only three active professors, Förster, Bauschinger, and Scheiner, and two docents, Adolf Marcuse and Hans Battermann (1860–1922). Battermann never taught more than one course per semester, and it was almost always a technical course on observing the Sun. Scheiner taught astrophysics and observation of stars, and Marcuse taught only introductory astronomy and technical courses on position finding. This left Förster and Bauschinger to do all the rest of the teaching, and thus the courses rotated through their sequence very slowly.

Wegener's first course with Bauschinger was "Celestial Mechanics: Older Theories." He had missed "Introduction to Celestial Mechanics" (summer 1900), and there was almost nothing else he was equipped to take. He did sign up for another year of geographical position finding and celestial navigation with Marcuse, and this looked to be valuable. It consisted of both demonstrations and exercises, and Marcuse had designed the exercises specifically to simulate the kinds of problems one would be likely to run into on a scientific expedition. After selecting this course, Wegener ran out of reasonable choices in astronomy. Förster was teaching a popular course on the history of astronomy and a course on the mathematical reduction of astronomical measurements; the first was too elementary for Wegener, and the second too advanced. Bauschinger was also teaching a course in chronology and one on the construction and use of planetary tables, but both were too specialized for someone with—so far—only a summer-semester introduction to general astronomy.

Bauschinger's "Older Theories" course began with Johannes Kepler (1571–1630), who made the first determination of the laws of planetary motion, and Isaac Newton (1642–1727), who had spent much effort on determination of cometary orbits. It then passed through the elaboration of celestial mechanics by Pierre Simon de Laplace (1749–1827), Joseph Louis Lagrange (1736–1813), and Wilhelm Olbers (1758–1840) and finished with the methods of Karl Friedrich Gauss (1777–1855). In all, it covered the period from 1600 to about 1850. This approach allowed Bauschinger to develop the course as a history of orbital calculation and error reduction. He was in the midst of writing what would become the standard text on orbital determinations, *Die Bahnbestimmung der Himmelskörper* (1906), and was also spending much time documenting the history of the field in Germany and producing translations of hard-to-find earlier works, especially those of Johann Heinrich Lambert (1728–1777) and Johann Franz Enke (1791–1865).[22]

Bauschinger was a prodigiously active observer and generator of planetary tables and ephemerides and of mathematical tables for the use of astronomical calculators. His most famous work, coauthored with Jean Peters (1869–1941) and regularly reprinted, was *Logarithmic-Trigonometrical Tables to Eight Decimal Places, Containing the Logarithms of All Numbers from 1 to 200000 and the Logarithms of the Trigonometrical Functions for Every Sexagesimal Second of the Quadrant* (1911).[23] The number of decimal places in the tables was a signal of the lengths to which it was necessary to go when questing for accuracy in orbital determinations; the late date of its last reprinting (1970) indicates also that with Bauschinger's work the quest had nearly reached its practical limit.

The fundamental problem in this branch of celestial mechanics was to find the orbit (and thus the position relative to Earth at any given time) of an object such as a comet or an asteroid that was moving around the Sun, and to find it with a minimum number of observations—usually three. Restricting observations to the absolute minimum also minimized the immense amount of trigonometric calculation required to solve the equations involved in the problem. One took the celestial latitude and longitude (actually the right ascension and declination) of a celestial object on three successive occasions and, using these coordinates and the time of observation, generated a total of nine equations that had to be solved for nine unknowns.

There were three principal approaches to the problem, all treated in Bauschinger's course: Wegener was obliged to learn the theory and the methods and to perform the calculations in a variety of ways. Each approach had strengths and weaknesses, depending on the character of the orbit being studied, especially its eccentricity and the number and character of perturbations. Multiple methods were required of students, as one may observe in a sample problem of the period: "Take three observations of an asteroid not separated from one another by more than 15 days, or three of a comet not separated from one another by more than 6 days, and compute the elements of the orbit by both the method of Laplace, and also that of Gauss."[24]

As the academic year progressed, it seemed to fall into two quite distinct parts. The mathematics, physics, and celestial mechanics courses were mutually reinforcing and had considerable overlap, and they also shared the sense of mature, finished endeavors, nearly complete before Alfred had arrived in them. Their historical orientation, reaching back hundreds of years, seemed to include in its roster of predecessors nearly every great name in physical science. The other part of the year's program, however, felt much less finished. If there was nothing new in the contents of Marcuse's navigation and position-finding course, it had at least the sense of the possibility of novelty and adventure in its deliberate assumption that the skills being learned were for the use of explorers on expeditions. Alfred was gaining an instrumental technique that might help him find something really new—somewhere. On this same side of the line, the unfinished side, also came his final course enrollment for that year: Bezold's course in general meteorology.

Wilhelm von Bezold (1837–1907) was near the end of his career but still very active. He was the first professor of meteorology in Germany, the head of the Prussian Meteorological Institute, president of the German Physical Society, and past president of the German Meteorological Society. He had worked with Helmholtz and Planck and was a pioneer, along with Heinrich Hertz (1857–1894), in the study of the dynamics and thermodynamics of the atmosphere. He was an international leader in the establishment of meteorological station networks and of the exchange by telegraph of information about pressure, temperature, and wind observations. He was a leader in creating and interpreting daily weather charts as a basis for forecasting. He had a prominent place in every meteorological success of the past twenty-five years. Yet the story he told of his science was not at all like the triumphal narratives of the physics and astronomy classes. It was the story of something that was just coming into being. For this newborn science Bezold was a crusader, a recruiter, and a prophet. For all the tremendous attention paid to the atmosphere in the second half of the nineteenth century, he argued, all the major questions remained unanswered. What drives storms—what is their energetic, their thermodynamic foundation? How do centers of high and low atmospheric pressure interact? How do clouds form, why are there different kinds of clouds and cloud shapes, and why does it rain and snow and hail? Are there rhythms and cycles longer than the seasonal year? Where do tornadoes and waterspouts come from? The questions went on and on.

Meteorology as presented by Bezold was about not weather forecasting but the struggle to create a physics of the atmosphere. One of the strongest claims to public (and government) attention for this new field was the role of manned flight in attaining the necessary data for atmospheric physics. Bezold was president of the Verein der Berliner Luftschiffahrt (Berlin Aeronautical Society) and was an enthusiastic promoter of manned ballooning for scientific purposes.[25] Working together with Richard Aßmann

(1845–1918) and Arthur Berson (1859–1942), Bezold had requested a grant from the kaiser of 25,000 marks in 1892 to support manned flights from Berlin; the young kaiser, enthused by the project, gave him 50,000 marks instead. Most of these flights were eventually made by Berson between 1892 and 1898 using Aßmann's instruments.[26]

With few exceptions, meteorology in the previous fifty years had tried to study the three-dimensional structure of the atmosphere using only two-dimensional methods of observation. There was, by 1900, a globe-girdling network of meteorological stations in the Northern Hemisphere, but the information it gathered was information about what was happening at the surface of Earth or at best a few meters above it. It had been possible to expand this network vertically by building and manning meteorological mountain stations, as advocated by the Austrian meteorologist and climatologist Julius Hann (1839–1921), who had an observatory on the peak of Sonnblick (3,106 meters [10,190 feet]) in Southern Austria, but few observers had as much success as Hann in adapting the results obtained to the understanding of storm systems. Manned balloon flights would and did allow the investigation of the three-dimensional structure of the atmosphere up to very great heights in the free air—away from the topographical and thermal effects of mountains. Bezold was certain that this information would allow the theoretical unification of meteorology as a physics of the ocean of air—all that was required was for young men of vision and courage to step forward and carry the program out.

Wegener was yet in no position, and also under no pressure, to make any decisive move toward meteorology. He belonged at Berlin to a varied group of students who made their university home in the Akademisch-Astronomischen Verein (Academic Astronomical Society). The A^{2}V, as it was known, was one of the "black" scientific associations, in contrast to the "color-bearing" fraternities of the sort Wegener had joined in Heidelberg. It was affiliated with the Mathematics Society, though it was much smaller. The members were so close that, as Wegener's friend Walther Lietzmann later remembered, it seemed more like a real fraternity than a scientific club. The members of A^{2}V were well-enough known for their liberal hospitality and gregariousness that members of the Mathematics Society often used to attend their evening lectures and social hours. Lietzmann recalled, tongue in cheek perhaps, that Wegener already exhibited an inclination toward polar matters—he was always willing to oblige the group with a lusty rendition of "*Von dem Eskimann und von der Eskifrau.*"

Alfred was finding his way, acquiring friends and peers, but there is little doubt that this Berlin life was making him restless and that he was once again feeling constrained. At Christmas, he and Kurt hatched a plan: they would take the 1901 summer semester together at the University of Innsbruck, in Austria. They would register for field geology and botany and go exploring. When the term ended in April, they were packed and ready to go immediately. That summer there would be no hiking at *die Hütte*; they were headed for the Alps.

Innsbruck

Innsbruck is the capital of the province of Tirol, in southern Austria close to the Italian border. It sits in the middle of the wide plain of the River Inn at an altitude of about 600 meters (1,969 feet) and is completely surrounded by high mountains that seem to come right up to the edge of the town. In 1900 it had a population of about 28,000 and a university student body of just over 1,000. The mountains that surround Innsbruck are part of the central chain of the Eastern Alps, a 400-kilometer (249-mile) series of

Innsbruck, Austria, where Alfred and Kurt Wegener studied botany and field geology in the summer semester of 1901, and where they learned Alpine mountaineering and climbed extensively. From a contemporary postcard in the author's collection.

peaks from the Swiss border to the outskirts of Vienna. These mountains, largely of crystalline rock and covered with snow, ice, and some large glaciers, include more than fifty peaks above 3,048 meters (10,000 feet) and are the cradle of European alpine mountaineering.

Kurt and Alfred arrived in Innsbruck and registered for the summer term. Their program had a strong sense of purpose and place: it was not just in Innsbruck and the Alps, but about them, and it was a liberal cross between a university term and a summer scientific field camp. They registered for "General Botany" (lecture) and "Exercises in Identification of Flowering Plants, with Special Attention to Medicinal Plants" (lab), both taught by Emil Heinricher (1856–1934), an authority on wild iris and primroses growing at altitude. His lecture and laboratory were preludes to his course "Botanical Excursions," in which students learned field identification and collecting of common and rare alpine wildflowers.[27] Alfred and Kurt also registered for "Geological Tour of the Tirolean Alps" with Josef Blaas (1851–1930), who had spent his entire life hiking, exploring, and mapping the Alps and had just sent to press his seven-volume *Geological Guide to the Tirolean and Vorarlberg Alps* (1902).[28] The microscale of geology was handled by Alois Cathrein (1853–1936), a specialist in crystal symmetry who taught an introduction to mineralogy, followed by a mineralogical field course: "Mineralogical and Petrographic Excursions."

Alpine botany is a surprise even for those who have some experience in plant identification, as Kurt and Alfred both had. As a very rough rule of thumb, every 200 meters (656 feet) of elevation is equal to a degree of latitude when considering the environment of the flowering plants. That is, in the continental interiors, a plant found near sea level at a given latitude will, for each degree of latitude one moves south, appear 200 meters higher in the landscape (and thus in the same approximate temperature conditions). Innsbruck was located 600 kilometers (373 miles) south of Berlin, so the plants within Innsbruck and its immediate environs on the flood plain of the Inn still contained wild species familiar to Alfred from summers at *die Hütte*. But a hike of 300–400 meters (984–1,312 feet) in elevation brought plants with no counterparts in North Germany at all, and a further 600–700 meters (1,969–2,297 feet) brought a whole new flora, whose nearest relatives at sea level would be found in Bergen or Stockholm. At the tree line, 1,000 meters (3,281 feet) higher up, one attained a tundra environment similar to that at sea level at the Arctic Circle, found in lowland Europe only at the North Cape of Norway and Sweden.

In the course of a day hike one might pass through a number of vegetation zones, often very sharply demarcated. In early spring to midsummer one actually walked through time as well, for spring moves up the hillsides slowly and steadily, with summer in the valley, late spring in the high pastures, and early spring above the passes, with the dwarf and miniature trees and plants rushing to flower in their short season of growth, and where a chill hangs in the air, even on the sunniest days, with the slightest flush of wind.

Field geology in alpine terrain offered a similar range of contrasts with the lowlands of the North. To begin with, most of Prussia had no other geology than the glacial geology and geomorphology of unconsolidated tills, sands, and other glacial outwash, all of it recent. The only sizable rocks available for view were the large, erratic boulders carried south from Scandinavia with the great ice sheets. In Innsbruck, except for the alluvium of the valley floor, the alpine pastures, and the moraines of the living glaciers, this dirt was out of the way and the rocks available for inspection.

Alfred and Kurt's principal field guide, Josef Blaas, was an expert on the Brenner Pass region immediately to the south of Innsbruck, a place combining ease of access and amenities, with exposures of fresh rock created by blasting to widen the road on this major trade route to Italy. Because Blaas and his colleague Alois Cathrein worked in crystalline (metamorphic) rocks as opposed to sediments, the field trips did not stop when the strata were overtopped. Even into the 1920s many classical geological field trips involving the traverse of a mountain chain would stop instruction for the day when the crystalline rocks near the crest were reached. The students would hike up over the summit ridge and descend, to begin the "geology" again the next day when the sedimentary rocks resumed.[29]

Swiss, Austrian, and French geologists in the preceding twenty-five years (1875–1900) had made sensational discoveries in the Western Alps of massive folds and overthrusts (*nappes* in French, *Decken* in German) on scales of tens and hundreds of kilometers. These were structures so huge and so physically improbable (how could rocks maintain coherence thrust over such distances?) that most geologists disbelieved the published field reports and had to be convinced, as it were, on pilgrimage—literally taken to the top of the mountain to undergo conversion. Whereas all the excitement in European geology even a decade before had been in glacial geology, by the turn of the century the hottest topic was Alpine tectonics—the folding and thrusting of these huge mountain ranges. For introductory geology students to be introduced to dynamical and global theorizing was not as common 100 years ago as it is today, but in the Alps it was inescapable—there was almost nothing else to talk about. The Eastern Alps were puzzling, complex, contradictory, and still to be unraveled. Even if Alfred and Kurt could not grasp all the implications, there was a sense here of something new happening on a large scale, of major new interpretations, of great discoveries.

In between the field trips, Alfred and Kurt went hiking, scrambling, and climbing on their own. They arrived in Innsbruck severely underequipped and had to buy a complete mountain kit.[30] Out of their quite modest allowances they procured the necessaries. First, they bought climbing boots with soles 15 millimeters (0.6 inches) thick (for rigidity) nailed with "triple-hobs." Hobnailed boots provide traction on rock and are ancestral to the Vibram sole found today on almost all hiking and climbing boots. Because hobnails cannot find good purchase on hard snow and ice, Alfred and Kurt also each acquired crampons—sets of spikes that can be strapped underneath the boots. To this equipment they added ice axes and climbing rope. The ice axes they used are what today would be called "Alpenstocks," staves of hardwood about the size and length of a broom handle, shod at the bottom with an iron spike and capped by a cast metal head with a pick at one end and an adze at the other. It was a multipurpose tool: it provided stability when hiking with a heavy pack, it served as a probe for crevasses, its adze blade could be used to cut steps, and the pick could be dug into the ice to arrest a fall.

They started climbing during the Whitsun (*Pfingsten*) vacation, which in 1901 came in the last week of May, so on their first "Alpine tour" they encountered glaciers still deeply covered with snow. The planned sequence in which they explored the region indicates they had either good advice or immense good luck.[31]

Climbing in the Austrian Alps in those early days before mountaineering became a popular sport was a wild and isolated pursuit, even if this sort of Tirolean trekking was not technically difficult by modern standards. Alfred and Kurt were on a series of expeditions and explorations, since they often followed trails and climbing routes that were scantily marked or not marked at all. They had their footgear and ropes and ice axes, but no climbing aids—no pitons, anchor bolts, ice screws, chocks, or wedges. They had access to few of the fixed cables and ladders available on these routes today, nor had they hard hats to protect them from falling rock and ice, nor slings and harnesses or any other of the wonderful devices modern alpinists enjoy.[32] In return, they got the solitude, the sense of being alone, the wonderful (surely false but wonderful) feeling that perhaps no one had ever stepped *here* before, and the joy of self-reliance, of their own pace, of unscheduled progress, and of simple living.

By the end of the summer of 1901, Kurt and Alfred were climbing continuously and joyously in the Zillertal region to the east of the Brenner Pass—the best snow and ice climbing in Austria. The geology and meteorology they pursued were of a practical sort—safe and unsafe rock and ice, and the endless variety of mountain weather. At altitude, conditions change with an exciting and sometimes alarming suddenness. A walk along a ridge in bright sunlight can give way in a few moments to a shrouding and bewildering fog, caused by only a fractional change in temperature and humidity. Summer thunderstorms develop rapidly, and one can feel the updrafts and, as the rumble comes closer, the static electricity, making the hair stand up all over one's body, with the rock all around glowing with a crackling blue aura. Sunbathing and a peaceful calm can give way in minutes to a drizzle or a snow flurry in a sharp wind, plunging the windchill below freezing. There is, on the other hand, the regular and delightful experience of passing through cloud layers into the sunshine and looking down on the clouds below—often several thousand feet below. Understanding, accepting, and overcoming these challenges and then enjoying the pure sense of being alive constitute the essence of mountaineering.

While Kurt and Alfred had never been very far or very long apart either physically or in spirit, the summer brought them closer than ever. They were both scientists now in fact as well as aspiration, both drawn to physical activity and exertion. They loved this sort of life—outdoors, trekking, climbing. For Alfred especially, the thought of an immediate return to the university was not pleasant; he had been in school almost year-round in every year of his life from the time he was five.

At some point in the summer, early or late, Alfred made the decision to complete his year of compulsory military service, starting in September—that is, immediately on his return to Berlin. Certainly he would never be in better shape for the rigors of basic training. He would be in Berlin again, of course, but living in officer-cadet billets with his regiment as he trained, and not at home. Certainly there would be no trouble obtaining uniforms—not with Uncle Paul's uniform factory in Wittstock. Moreover, Europe was at peace and had been at peace now for thirty years. For Alfred military service was only another discipline to master, another credential, another set of skills and regulations: in short, another inevitability of Prussian life. It would certainly be less dangerous than what they had spent the past few months doing—though, for that same reason of course, much less fun, but at least it would break the routine of schooling.

Grenadier Guards

In September 1901 Alfred Wegener reported for duty to the headquarters of the Queen Elisabeth Grenadier Guards, in Westend—at that time a pleasant suburb of Berlin near the end of Unter den Linden, and close by the Technische Hochschule in Charlottenburg, where Kurt was almost finished with his degree. Given a choice, he would doubtless have been in Bavaria, training with an Alpine regiment, but such latitude of choice was not in the German scheme of things. One registered for service where one lived, because in future wars victory would depend on rapid mass mobilization, which in turn depended on having the ready reserves assemble close to their residences. The young Herr Wegener lived in Berlin; therefore, he would belong to a Berlin regiment. Alfred was enrolled in Company No. 4 of Queen Elisabeth Grenadier Guards Regiment No. 3, with the expectation that he would, upon completion of his training, become a reserve lieutenant of infantry in the Xth Guard Reserve Corps.

A generation before, to be a "Guards Officer" would have carried considerable social cachet and have been a good acquirement for an ambitious young man desiring good connections and rapid promotion in the civil service, but as the army reorganized in the face of universal conscription in an expanded and unified Germany, "The Guards" came to include all the military units of the German Second Army, regular and reserve, stationed in and around Berlin—all generically designated as the "Guard Corps" without reference to any particular regimental traditions.

It so happened, however, that Alfred's particular regiment was still mostly an elite unit, as was its "brother" regiment, the Kaiser Alexander Grenadier Guards, stationed at Potsdam. Both regiments had members of the royal family as honorary officers, even though the regiments' missions no longer entailed protecting the persons of the royal family (the original rationale for designation as "guards"). Since they were the infantry units closest to the capital city, they were regularly called upon to march in an endless succession of parades celebrating national holidays—the number of which had been continually increased in recent years in order to cement the sense of a unified and patriotic nation.

Alfred, like Kurt before him, was exercising his right, as a member of the educated upper middle class, to avoid conscription into the army for a period of two years,

Kurt (*left*) and Alfred Wegener in their Grenadier Guards dress uniforms; photograph taken at the Wegener family home in Berlin in 1902. Author's photograph from the original in the Wegener-Gedenkstätte, Zechlinerhütte. Photo courtesy of the Heimatmuseum, Neuruppin.

by volunteering instead for one year. Such men were referred to approvingly as *Einjährig-Freiwilligen* (one-year volunteers). They began their training under the command of noncommissioned drill instructors and learned the immemorial lessons of barracks life, with everything at double time. They went through physical training and learned "spit and polish," marching, field exercises and infantry tactics, marching, weapons training, military etiquette, marching, military topography and map reading, logistics and mobilization drill, and marching.

Officer cadets of infantry units were (and are today) trained first as if they were enlisted or conscript infantry soldiers, and only thereafter trained as officers, which meant learning all the menial and minor duties of enlisted men, but above all how to march as members of the rank and file of the national army. The kind of marching Alfred learned in the barracks yard (close-order drill) was no longer of any practical significance in actual combat, but endurance marching along highways, carrying full field pack and weapons over long distances, was a crucial element in German military strategy and tactics.

The German forces, like those of France, Austria-Hungary, and Russia, had swollen to such immense size that a vast amount of military planning went purely into the logistical nightmare of getting the bulk of the army to the front line or frontier in time to make any difference in the outcome of a war. At regular intervals in the training, therefore, especially in the summer months, the cadets were turned out on short notice for forced marches, along highways in company with their cavalry and field artillery units, but also cross-country. Some of these marches lasted twelve hours or longer, and the troops moved as rapidly as the officers could manage it, and often far into the night. The cadets also joined in large-scale, coordinated maneuvers designed to test the readiness of Guard Corps units for full mobilization, practicing rapid movement from dispersed encampments toward a single location, as if massing for a real assault.

The German army, like everything else in Germany in 1900, came equipped not just with traditions, regulations, and procedures but with a philosophy. The philosophy of the hour was laid out in a book by Baron Colmar von der Goltz (1843–1916) and entitled *Volk in Waffen* (The nation in arms). First published in 1883, it was a tremendous popular success and very influential at home and abroad, with translations into French, English, Spanish, Italian, Russian, Japanese, and Turkish. By 1901, when it became assigned reading for Alfred during his military service, it was already in its fifth edition. It is a fascinating book. Its sober preparations for mass mobilization and rapid deployment by rail, for heavy artillery bombardment followed by human-wave assaults, and for bunker and trench warfare, as well as its expectation of huge casualties and the need for a ready reserve as large as the standing army, all belie popular accounts of the First World War as some new form of warfare that caught unsuspecting nations by surprise. In Germany they had been planning it for forty years, and other European nations had read of these plans and followed suit on their own. Period photographs, even from wars that few outside Germany have ever heard of, such as the German-Danish War of 1864, show that the "unparalleled ferocity" of the First World War, with its moonscapes of total destruction, was already visible a half century earlier.[33] These landscapes were the battlefields of the future in von der Goltz's book. Even an army passing through a district on its way to the front must lay waste to the countryside: "the crowded roads, the deterioration of the roads in bad weather, the confusion and friction between columns on the march, the picture of desolation which is spread when hundreds of thousands, like a swarm of locusts, pass over a district. The atmosphere is full of dust, smoke, and smells of burning. Those in the lead may find it bearable. If, however, the passage of troops extends over two, three, or four consecutive days, the hindmost must march through a mere wilderness."[34]

The whole point of von der Goltz's book was that true military preparedness consists in expectation more of failure than success: one must expect almost everything to go wrong from the start. "The commander-in-chief most undoubtedly will, in spite of the changing fortunes of war, always keep the main object in view, but the means by which he hopes to attain it can never be sketched out with any certainty long beforehand. . . . No plan of operations can with any safety include more than the first collision with the enemy's main force."[35] Given the scope and scale of planned hostilities, no other outcome was at all likely. Therefore, officers must be trained to expect adverse developments, to maintain in adversity disciplined self-control and a fierce resolve to win, and then to transmit this fitness and courage to the troops under their command. Moreover, because of the expectation of heavy casualties, "it will thus be unavoidable that many companies of the line will be in the hands of reserve officers immediately after the first few battles. . . . In the course of a great war . . . eventual success depends on the capacity of this class, for only good leaders produce good soldiers."[36]

The core of Alfred Wegener's officer training, then, was first to teach him to live and work and fight as an enlisted man and to understand how war looks and feels to the men who do most of the dirty work and the dying. The next stage was not to instill in him initiative and flexibility in making battlefield decisions, but rather to teach him how to maintain discipline and order among his men in great adversity. The fundamental idea and aim to achieve this was *Kameradschaft*—a sense of mutual trust and fellowship. The infantry company should be like a family, with common aims and interests. It was with this expectation and for this reason that officers were rarely transferred: once Alfred was commissioned as a reserve officer, he would continue to command many of the same men over the period of his active service in the reserves—the next fifteen years (until age thirty-six).[37] When in the field, he was told, he should often visit the men in their quarters; he should be in charge of their clothing, food, and drink; see to their equipment; inquire after their health; and work to bolster their morale.

The explicit philosophical basis of this approach to military discipline was none other than Charles Darwin—and not the Darwin of *The Origin of Species*, with its struggle for survival, its "nature red in tooth and claw," but the Darwin of *The Descent of Man*, with its attempt to explain the value of altruism, fellow feeling, and mutual dependence for the evolutionary success of the human species. General von der Goltz was quite explicit about this fellow feeling as the basis of the discipline that can win wars:

> The best explanation of discipline and its marvelous power is found in the saying of Darwin, contained in his *Descent of Man*: "The superiority which disciplined soldiers show over undisciplined masses is primarily the consequence of the confidence which each has in his comrades." This absolute confidence is, beyond all doubt, the prime means by which discipline works, and it most appropriately explains what we really understand under this trite word. . . . Every man in the ranks knows from experience that his officer does not, under any circumstances, leave the company to which he belongs. . . . Thence springs that confidence of which Darwin speaks, and in which the great judge of human nature finds the superiority of disciplined armies. . . . This inward force, exercised by a feeling of relationship, will endure, when order produced by law fails.[38]

This may well have been Wegener's introduction to Darwin, at least in a formal setting. This may seem odd, for Darwinism spread rapidly in Germany after the translation of *The Origin of Species* in 1860, and evolutionary thought, especially progressive evolution, was a mainstay of German philosophical thought throughout the nineteenth century. However, Darwinism was not taught in Prussian secondary schools in the 1890s for a variety of reasons. Foremost among these was the impossibility of treating the theory at all without a frank discussion of sex, and sex was not discussed in Prussian schools.

In the same year that Wegener began to learn the details of Darwinian theory in a formal setting (1901), the Association of German Scientists and Physicians voted for the first time to press for a renewal of biology instruction in the schools, but even here there was no agreement that this instruction should include evolution. Indeed, the few secondary biology texts that existed (none of which were used in Prussia) excluded not only Darwinism but reproductive physiology altogether.[39]

A few passing references to Darwin might not have had so powerful an effect on Alfred in some other context, but here they came mixed with a rhetoric of mission and sacrifice, of pride and strength. Moreover, for once, Alfred was presented with a unity of philosophy and task in which he was expected to be a leader, not a follower, and a father rather than a son—to his troops at least. Here were adult and concrete questions about life and death, not abstract philosophical problem sets, and he was stirred, sometime in the autumn of 1901, to a curiosity about life "in general" in a way he had not been before. It was a combination of factors: the sense of personal strength and confidence he had gained in Innsbruck, the military training itself and its weighty expectation for him to be "a man" and a leader of men in war, and just the process of maturation itself in this third year past high school; Alfred was now beginning to move away from home and family mentally and emotionally as well as physically. There was nothing abrupt in this nor any strong impulse to rebellion. He was a loving and dutiful son, though he was resolute in pursuing his inclinations.

Not burdened by school work for the first time in his life, he could pursue a course of elective reading, and he began to read more widely in popular philosophy and evolutionary theory. Needing some structure and guidance, he took advantage of his proximity to the university to enroll in a year-long course of popular lectures in philosophy. This enrollment was both permitted and encouraged by his military superiors, who had no objection whatever to having the university's lecture halls salted with uniformed officer cadets. In the winter term of 1901–1902 these lectures would be given by Wilhelm Dilthey (1833–1911) under the title "History of Philosophy and Its Relationship to Culture," and in the summer of 1902, by Friedrich Paulsen (1846–1908) as "History of Modern Philosophy with a Consideration of Contemporary Culture."

Alfred was doing something new for him, but something that most people, at least those offered any choices in life at all, come to at this point in their lives. He was finding out who he was and what his place was in the world—not "place" in the sense of calling, for he knew he would be a scientist, and probably a professor, and a reserve officer. Rather, it was time for him to find out what sort of person he would be and what principles would guide him. The world he had been born into and the worldview practiced, preached, and enforced within it were still strongly present within him, but he was becoming conscious of the need to sort things out, deciding what to keep and what to discard, as well as the need to *affirm* something. None of this was very clearly articulated, of course, but an inarticulate striving is, initially at least, a necessity if one is to pursue questions of existence and meaning. If everything seems very clear and simple, the impulse to investigation is rarely strong enough to cause much of an awakening. So if this first search was still vague and rudimentary, it was a search for meaning nonetheless, and Alfred brought to it both his native enthusiasm for discovery and his very considerable tenacity.

The Astronomer

BERLIN, 1901–1904

> We are not called upon to solve the meaning of life but to find out the deed demanded of us and to work and so, by action, to master the riddle.
>
> J. W. Hauer

The Generation of 1880

One notices in photographs—Berlin street scenes, for instance—taken just after the turn of the century a change in the appearance of men, especially the younger men. Their coats are shorter and worn casually open; they are beardless, and some even completely clean-shaven. Top hats are gone, and bowlers are giving way to soft fedoras. Black is still the world's favorite color, but one sees waistcoats in contrasting colors, and neckties are wider and softer under rolled collars, and not knotted like nooses under sharply starched wings. Seventy years earlier the romantics had shaved off their beards and loosened their ties; now they were back and youth was once again in fashion.

From the beginning of the twentieth century to the outbreak of the First World War, European intellectuals (both literary and scientific) spoke a language exalting life, youth, vital freedom, will, energy. The philosophical vehicle for these slogans and tendencies was called "life-philosophy" (*Lebensphilosophie*) by advocates and opponents alike. *Lebensphilosophie* has endured bad press since the middle of the twentieth century, having been blamed for a range of cultural developments from the terrible to the merely disconcerting. It has been accused of inciting and strengthening war, colonialism, fascism, nationalism, racism, and xenophobia. It has been held equally responsible for irrationalism, spiritualism, the new-age movement, environmentalism, anti-intellectualism, and a general collapse of manners and mores.

This "philosophy of life" has vanished from the scene so completely that it is hard, especially in the English-speaking world, to get a sense of its overwhelming dominance from about 1880 to 1930. The sociologist George Simmel said at the time that the idea of "life" had come to dominate philosophy in the twentieth century in the way that concepts such as God, nature, being, or ego had in other periods.[1] Philosophy of life clearly contained a number of contradictory tendencies that have since come apart—how else could it be blamed for so many different phenomena? It has been accurately (if abstractly) described as a vitalistic variant of pragmatism.[2] It emphasized dynamism, activity, and striving and aimed to increase the qualitative intensity of experience. It placed humanity within nature, not above it, and saw humans as at once thinking subjects and natural objects. It emphasized a world full of possibility waiting to be actualized, so much so that there was no need of a "beyond." It was naturalistic and this-worldly, but it insistently opposed the reduction of everything to material mechanism.

In subsequent decades its ethical message became part of existentialism, and its emphasis on "vitality" passed into biology. The pragmatic elements of the philosophy survived as a theory of scientific practice based in experience and giving special attention to contexts and relations. In 1900, however, these various components still seemed integrally connected. Exponents of "life" held to the coherence of science and arts and to classical means of investigation, newly accompanied by exaltation of a "vital freedom" that was somehow to be bound into the process of making art and doing science.

A number of philosophers, artists, and writers of stature devoted themselves to the elucidation of this prewar yearning for "life," and a surprisingly large fraction of these turn out to be Alfred Wegener's exact contemporaries. Among them is the Austrian novelist Robert Musil (1880–1942), born five days after Wegener. Musil's great epic of the last years of the Habsburg Empire, *Der Mann ohne Eigenschaften* (The man without qualities), portrays the agitated sensibility of a generation coming to maturity at the turn of the new century:

> Out of the oil-smooth spirit of the last two decades of the nineteenth century, suddenly, there arose a kindling fever. Nobody knew exactly what was on the way; nobody was able to say whether it was to be a new art, a New Man, a new morality or perhaps a re-shuffling of society . . . these were different and had very different battle cries, but they all breathed the same breath of life.
>
> Something at that time passed through the thicket of beliefs, as when many trees bend before one wind—a sectarian or reformist spirit, a blissful better self arising and setting forth, a little renascence and reformation such as only the best epochs know; and entering into the world in those days, even in coming around the very first corner, one felt the breath of the spirit on one's cheeks.[3]

This sense of ferment provided abundant material for other writers, all of them concerned with young men's search for meaning. Hermann Hesse's (1877–1962) spiritual seekers in *Demian* (1920) and *Siddhartha* (1922), Roger Martin du Gard's (1881–1958) fiery young journalist in *Jean Barois* (1913), and even Thomas Mann's (1875–1955) convalescent Hans Castorp in *Der Zauberberg* (1928), however different from one another, were all young men seeking life, health, and spiritual renewal in periods where crude and commercial materialism was on the rise and where outmoded creeds and ideologies offered no nourishment. All three of these writers, and many others who took up the theme, were a part of the "generation of 1880," as was Wegener himself.

Also of this generation were the "philosophers of decline," Albert Schweitzer (1875–1965), Oswald Spengler (1880–1936), and José Ortega y Gasset (1883–1955). The latter was, of course, not a German, but he was educated in Germany, and his premonitions of decline, as well as of the need for the best and the brightest to make a strenuous and vital effort on behalf of the survival of high culture, were based on his experiences in Germany as a student of philosophy in the first decade of the twentieth century.

One important effect of this life-philosophy movement was to shift the center of thought, in Wegener's generation, outside the universities and the official organs of culture where they had comfortably resided for fifty years. Among those that answered the call to "life," almost none had ordinary careers in the stolid Wilhelmine sense of "ordinary." Albert Schweitzer was a virtuoso organist and musicologist who left Europe forever in 1913 to become a medical missionary in Africa. Ortega y Gasset was a newspaper editor and cultural commentator in Madrid, where he gave philosophy lectures to huge audiences in theaters, and whose books were read not by scholars but by a broad, nonacademic public. Spengler was a *Gymnasium Oberlehrer* (senior teacher) who had failed his PhD exams in philosophy the first time through and then, having obtained his degree, quit his position within a few years to become a writer. Hermann Hesse was a high school dropout and bookstore clerk who bypassed university altogether and went directly to writing. Robert Musil studied successively to be a military officer and civil engineer and then finally took a PhD in philosophy from Berlin—though he passed up a university position in Munich to become a freelance author. None of them followed their fathers; none of them were "pensioned," and none of them stayed home.

Life and Evolution: Philosophy at Berlin in 1901–1902

If "life" ruled outside the German universities, it was Kant who still ruled within, with his intellectualism and concern for categories and first principles. The only exception was Berlin. In Berlin, in Wegener's student years, Wilhelm Dilthey (1833–1911) held sway, and his subject was "life." For Dilthey the whole of "The History of Philosophy and Its Relationship to Culture" could be summed up in that one word: "life." Life was all that existed: all the acts, dreams, and ideas of all the men and women who ever lived. All laws and all art, all philosophies and all science were not pale reflections of some greater reality beyond; they were "all that exists"—they were "life." Dilthey, like his French contemporary and counterpart Henri Bergson (1859–1941), directed his students to the reality and intensity of their own experiences as a proper starting point—not to some categorical structure that lay beyond them, nor to some historical origin long before they were born.[4]

In the well-attended lectures that Alfred Wegener heard at the University of Berlin in the fall of 1901, Dilthey maintained the mildly shocking heresy that the conventional and traditional way of life of Wilhelmine Germany was not necessarily a good way of life; it was simply the way things were. If one's parents and teachers and leaders tell one that something is "good," that is the result of their view (*Anschauung*) of the world and the result of their education and their experiences. But *your* task (Dilthey told his listeners) is not only to inherit but to fashion for yourselves a *Weltanschauung*, a "worldview."

Here was something that Alfred could grasp, employ, develop. He learned that we should and we must expect change because the world is nothing but change against a changing background. Fundamentally, Dilthey's message was not radically opposed to the philosophical lessons Alfred had learned at home. All those schoolroom recitations of his father's poem about butter and eggs—that what will do for one character will not suffice to firm up another—are more solemnly proclaimed here. Neither was the "openness to the future" of life-philosophy so different from Richard Wegener's (and Schiller's) view that we must bring the poetical spirit of play and creative novelty into things, to make the world our own, to make it better.

If Dilthey's message was to some extent compatible with the philosophy Alfred had learned at home, it also had crucial differences that Alfred would come to embrace. Here no longer did the "diamond" of analytical science bow before the organic and perfect "pearl" of faith. Science, it turned out, was no less an organic development than religion—it was another *Anschauung*—another view of the whole. The character of such *Anschauungen* was to differ from one another and to struggle in Darwinian competition. These worldviews were the product of not just our thinking but the totality of our experience of life.[5] Freedom was not simply the ability to think privately about truth and meaning free from coercion, as in Kant. Freedom, to be real, had to find expression in action: we are not what we think, but what we do.

This life-philosophy, in spite of its later reputation for irrationalism, exerted a tremendous influence over Alfred Wegener's generation precisely because of its relationship to science and its connections to recent advances in the sciences. The vital dynamism of evolutionary biology and the principles of conservation and transformation of energy indicated a world of processes and movement in time, not one of fixity, stasis, or eternity. Many members of the "generation of 1880" besides Wegener were moved by this scientific message. It was no accident that Spengler wrote his thesis on Heraclitus, who had maintained that the only permanent and eternal thing is change. Nor was it happenstance that Robert Musil wrote his doctoral dissertation on the scientific ideas of Ernst Mach, who held that concepts of science are no more than economical and convenient fictions that help us to think about and fix the immensely complex and various flux of phenomena that are the real.[6] Each and all of them—Alfred Wegener included—grasped science as activity, a kind of life, something that was meant to grow, to evolve, to change, as they were to evolve and change with it.

If Dilthey rephrased for Wegener the question "What should I believe?" as "What should I do?" he gave no answer.[7] More promising in Wegener's search for an answer was the work of Friedrich Paulsen (1846–1908). In the summer of 1902 Alfred found himself with considerable free time. The demands of his military training were much lighter than in the previous six months, and except for several weeks of muddy maneuvers and the obligatory marching and parading, he was at leisure. He therefore enrolled for the summer semester, at Berlin, for a single course in philosophy with Paulsen.

Paulsen was much more empirically minded than Dilthey, and there was none of Dilthey's relativism in his presentation. Paulsen had been (for decades) a fighter for educational reform in schools and universities, especially for the return of sciences to the secondary curriculum, and for a return of science to the center stage of philosophy. His course "History of Modern Philosophy with a Consideration of Contemporary Culture" was outlined in his plainspoken, modest, and, consequently, hugely popular books such as *Einleitung in die Philosophie* (*Introduction to Philosophy*, 1895).[8] The book, like Paulsen's lectures, aimed at a broad public outside the university. Paulsen addressed contemporary social and ethical problems, and in so doing he set a style of presentation that dominated the teaching of philosophy in Germany for the next forty years.[9] William James wrote an effusive preface to the book for the American edition of 1895, and he was so excited by the book's promise as an undergraduate text that he used it while it was still in proof sheets. He remarked, "In a long experience as a teacher, it is one of the very few text-books about which I have heard no grumbling."[10] He also defended it against its critics: "Professor Paulsen makes philosophy and life continuous again; so the pedants of both camps among us will unite in condemnation of his work. Life lies open, and the philosophy which their intellects desiderate must wear the form of a closed system. We need ever to be reminded afresh that no philosophy can be more than a hypothesis."[11]

Paulsen's candor was electrifying. Philosophy, he announced to his hearers, had lost ground in Germany throughout the nineteenth century, and it had not been doing very well in the eighteenth. The English had figured out more than 200 years earlier that the job of philosophy was to help out the natural sciences where it could, as Locke had tried to help Newton.[12] Every German attempt to do the same had been thwarted by theologians and speculative metaphysicians, who succeeded in convincing the majority of listeners that philosophy was one sort of activity and natural science entirely another.[13] The latter part of the nineteenth century had taken revenge on philosophy by progressively ignoring its swollen claims, in favor of the sciences. Yet, warned Paulsen, at the end of the century the lush and rampant growth of the sciences that had pushed speculative philosophy to the wall now threatened the scientific worldview with disintegration. Every science, proud and specialized, took the full attention of each of its practitioners, as each scientific field asserted its independence from every other branch of science.[14]

A generation later José Ortega y Gasset would make an industry of decrying the "barbarism of specialization," but he learned of the danger from Paulsen's introductory college text. The danger was real, and by 1900 Paulsen had been warning of it for twenty years. Paulsen offered a concrete plan and a procedure to remedy this defect. The plan was that philosophy would henceforth be a structural means of bringing the various parts of science together: "Modern science is its starting point and precondition, while the universal reign of law in natural occurrences is its fundamental idea. Whatever is not in accord with this thought lies outside the sphere of modern philosophy."[15] Philosophy will not be a "special science" but the name for all the sciences taken together.

Paulsen's history of philosophy aimed to show that this had always been the case from the Greeks to the present: "Philosophy," he wrote, "cannot be separated from the sciences; it is simply the *sum-total of all scientific knowledge*. This never-completed system, which the ages are building, is philosophy. Each particular science investigates a definite portion or cross-section of reality."[16] Note that there are no "worldviews" here, but a series of cuts through a single reality: the world is one, not many.

Paulsen's plan was to continue the construction of the never-completed system, and this dictated a common procedure, at once pragmatic and personal. In the common and joint effort to build a picture of all of reality, Paulsen argued, we must be problem solvers: we must find our problems, analyze them, and construct the range of possible solutions. Next, we must look to the historical development of thought on each problem. Then, from among the competing adequate solutions, we should choose the most inclusive, the one that brings the most phenomena together.[17]

However abstract this formula sounds, there is no mistaking its profound effect on Wegener, and its signature is visible in every piece of scientific work he ever did. Certainly the elements of Paulsen's "procedure" can be found in any description of scientific method, but there are special emphases here: the classical picture of science as an evolving collaboration and not as individual achievement, the insistence on the enduring historical importance of previous work, and the exhortation to synthesize, to bring together, to unify, to generalize, to ensure the survival of philosophy.

Wegener accepted this view of science, of philosophy, and of the world. It was to some degree the scientization of the philosophy of his childhood lessons and home life, but its call to action, its avoidance of intricate dialectic and fine categorical preconditions, its disinterest in theology, its historical sweep, and its universal scope all suited his mind and his temperament. In addition to this view of science, Wegener obtained two additional things from Paulsen. The first was a more extensive introduction to the Darwinian theory of evolution, a theory that Paulsen accepted as absolutely descriptive of the world and as governing all life and all aspects of human life, including politics and morals.[18] The second thing that Paulsen provided for Wegener was an inspirational rhetoric of spirit and soul that allowed him to use these concepts without ever straying far from the detailed study of physical phenomena.

Pressing the theory of evolution to its full extent, as Paulsen did, had an important psychological consequence as well: what was true of science was true of our own lives. We try to plan and shape our lives, but they will be shaped and determined by the world in spite of our plans, and we must understand this as a natural outcome, welcome it, and accept it: "Plan and design do not play a very important part in the history of the mind. The same law of development prevails in the mental world that prevails in the organic world. Organic creations are produced in nature and in history not by forethought, but by the spontaneous unfolding of germinal beginnings. Thing are not made, they grow; that is the fundamental law of reality. Even the works of the human mind are on the whole the results of unintentional growth. A planned outcome is a rare variant form of such growth."[19]

On the inspirational side, Paulsen's rhetoric of soul and spirit "ran cover" for Darwin's theory in the same way that many other popularizers of Darwin in Germany had employed it, blurring or ignoring previous demarcations between nature and God, between science and religion.

Such expressions at the turn of the century did not even stop short of seeing science and scientific laws as "holy," insofar as they expressed the universal harmony of nature.[20] Recognizing the power and persuasiveness of this sort of philosophical vision can help us make sense of statements (otherwise puzzling) by Wegener and his contemporaries, including the oft-quoted Albert Einstein (1879–1955), one year Wegener's senior. When Einstein called Euclid's *Elements* the "holy geometry book" or spoke of physics as finding out "the secrets of the Old One," or said that "God is subtle, but He is not malicious," he was using the common language of this movement in a way that all his contemporaries understood. When he said that "science is a free creation of the human spirit," a phrase emblazoned across many an Einstein poster, he was quoting Friedrich Paulsen.[21]

"Literature of the Sort That Has Engaged Me Intellectually"

In secondary school Wegener had read books of popular astronomy, and in his university years he still found them intellectually engaging and inspiring. Added to his old favorites, Max Meyer and Friedrich Diesterweg, were the popular works of one of Wegener's teachers, the Berlin astronomer Wilhelm Förster, who wrote extensively for popular magazines and gave public lectures at Urania, the astronomical observatory he had helped to found. Förster periodically collected these lectures and articles and published them in book form, a practice as old (and effective) as popular science itself; he enjoyed a very wide circulation.[22]

In his popular lectures as well as his university teaching, Förster was famous for his ability to interweave expert scientific knowledge with aesthetic and philosophical remarks and insights.[23] A lecture might begin with an examination of our knowledge of some part of astronomy and end up with a meditation on freedom of the will, or some other philosophical point. Förster's efforts, like most popular science in the nineteenth and twentieth centuries, also aimed to encourage the educated public to support science, both intellectually and financially. Every article or lecture, no matter how varied in the astronomical novelties and curiosities it contained, always included *obbligato* reflections that we learn about ourselves by studying nature because we are a part of nature, that science is a grand adventure, that we are living in an astonishing time, and so on. The lectures had a moral tone and an idealism that sound a bit earnest to the modern ear, though they were perfectly ordinary for the time; readers of the many popular writings of Carl Sagan (1934–1996) would immediately recognize and be at home with the mix of astronomical fact and inspirational exhortation.

Sometime around 1901–1902, Alfred began to read in another area of science: evolutionary biology. In a brief journal entry made a few years later while in the field, he recorded the following lament: "How I miss literature, especially the sort that has engaged me most intellectually, and that I believe to be perfectly suited to such expeditions, especially since one has the opportunity of an exchange of ideas afterward about what one has read. I mean principally Darwin, Haeckel, Chamberlain, Bölsche, Meyer (his popular astronomy), Diesterweg, Förster (several of his). I don't think novels are similarly suitable—I find no release here in reading a novel."[24]

This brief remark provides one of the few clues Wegener bothered to leave about himself and his interests outside his technical scientific work and the *only* clues we have about Wegener's early interests, other than school transcripts and the reminiscences of his brother Kurt and two or three fellow students. We can deal presently with the larger figures named in his list of authors, Darwin and Haeckel, and we have already looked into the popular astronomy works. Let us turn here to the less well known. Houston Stewart Chamberlain (1855–1927) was Richard Wagner's son-in-law and is remembered today chiefly for the virulence of his anti-Semitism. In 1899 he published his best-known work, *Die Grundlagen des 19. Jahrhunderts* (The foundations of the nineteenth century). It was broadly influenced by Wagner's ideas on the sources of human creativity and became an instant best seller in Germany: it sold out eight printings in ten years, upward of 60,000 two-volume sets.[25]

The basic plot of Chamberlain's *Grundlagen* is a Darwinian struggle emerging between two concepts of life: human knowledge and enterprise. On the one side was the dead hand of the past, of book learning, and the "repetition by rote of antiquated wisdom in dead, misunderstood languages." On the other was scientific observation, discovery, poetic creation, and "all truth and all originality."[26]

What about this work engaged Wegener? Certainly from the standpoint of a young science student it would have been the exhortation to perform "great deeds in science and life." Chamberlain was himself a failed scientist (botany), but unlike many failed scientists, he admired science and never turned against it. Among the "great deeds of thought" he proposed was the scientific comprehension of the world as a whole, an activity reserved for a great individual scientist who would also be an artist.

Chamberlain characterized this scientific-poetic activity as the special genius of the Teutonic peoples, setting them apart not only from the Jews but from the ancient Greeks: Teutons were supposed to have a passionate impulse to pure intellectual discovery closely related to the artistic and religious impulse. As theorists they had no great claim to importance, but as discoverers they had no rivals.[27] The secret of this gift for discovery lay in their nature, though this had long been masked, buried, and thwarted by classical education.[28]

In addition to a genius for discovery, Teutons were supposed to have a special gift for organizing and combining the results of discovery and thus for science: "All scientific systematizing and theorizing is a fitting and adapting; of course it is as accurate as possible, but never quite free from error, and, above all, is always a humanly tinted rendering, translating, interpreting." In contrast (once again), the Greeks were forestalled from much useful mathematics by a demand for geometrical perfection: "But for this introduction of approximate values our whole astronomy, geodesy, physics, mechanics . . . would be impossible."[29]

This concoction of national genius, rough-and-ready pragmatism, and poetical striving, taken together as legitimate and nearly essential elements of scientific work, got an even stronger push in the works of one of Wegener's other inspirational guides, Wilhelm Bölsche (1861–1939), a pioneer of German naturalism who began a hugely successful career in the 1890s, popularizing natural science with a series of pocket editions aimed at workers and students, in books that always maintained the connection between poetry and science. Bölsche's books were manifestos not only for realism and science but even more for the holiness of science. Like Chamberlain, he located the source of science in the soul, the same source as that of religion and of art, and he portrayed modern science as a dynamic outgrowth of religious evolution, in the spirit of Novalis.[30] Science, specifically evolutionary naturalism, would be a "third testament," attached to the first two, but moving beyond and correcting them.[31]

In Bölsche's *Naturwissenschaftlichen Grundlagen der Poesie* (Natural scientific foundations of poetry, 1894) there is a chapter entitled "Darwin in der Poesie," in which the struggle of poets and thinkers is declared a Darwinian contest between the young and strong and the old and weak. Moreover, to be a poet like Darwin is not merely to get the names of things right, or to shout slogans, but to make a serious study of how the world works. Poetry, like science, has its basis in the observation of the real and not in some fantasy world. One's work—and poetry and writing are treated as real work—should develop organically, like the rest of the world of which it forms a part.[32]

Less didactic and more "poetical" in its execution was Bölsche's *Die Abstammung des Menschen* (The evolution of man, 1904). Like many of his other works, this is a "wandering through time." He urges the reader to imagine himself (rifle in hand!), a million years ago, walking across the "immense prairies" of southern Europe (looking much like the interior of modern Africa), tramping for weeks across a "green ocean of grass," then turning north and walking through "impenetrable, primeval forest" where today great civilizations rise.[33] Then the reader is urged to set his footsteps back through time, past man's anthropoid ancestors, searching for man's deeper ancestors—where is man hiding, in what form is he concealed? The search takes us through changing climates and landscapes without moving from where we stand, past Neanderthals and Pithecanthropus, past the tarsiers, by comparative embryogeny down to fish, past amphioxus to the worms and to Hydra: "Might it be possible we could follow man down to this stage?"[34] Apparently so, and on down through the protozoa to the origins of life. All of this is presented as a great poetic drama, with all intellectual invention and human understandings being emergent characteristics of natural evolutionary relationships.

Chamberlain and Bölsche qualify as antiromantic realists not because of their views of science but because of their views of art. The aestheticism of this movement—exaltation of vitality, holy science, science as art, science as poetry—was paired with its Faustian ambition to totality. Here is Bölsche, at the end of *The Evolution of Man*:

> Poetry did not die when it became known that it is not the sun which actually rises in the east, but the earth which revolves toward it. Genuine religious feeling is truly something very human. . . . A cold fact from the history of human evolution cannot dampen this spirit. . . . But we should not be worthy of this triumph if we did not have the strength to dominate the spirits of the past with the calmness of the master who can look at them serenely and say: "You are of the past and the struggles of the past belong to you; but *I* am, and above me are *my* stars."[35]

This sort of aestheticism was impressed on Wegener again and again in his reading, not least because it was an important part of the philosophical structure erected by Ernst Haeckel (1834–1919), a figure inseparable from Darwin in the German imagination and in German cultural history. Haeckel, a marine zoologist, "converted" to Darwinism in 1859–1860, as soon as he read *On the Origin of Species*. He was a wide traveler, was physically robust and energetic, and illustrated the vividly written descriptions of his travels with his own watercolors. He was also an enthusiast of photography, especially microphotography, an endeavor that he proclaimed would open a new world and awaken a new aesthetic sense.[36] He was part of the group of German theorists who took Darwinism out of biology and made it a dynamic principle of all cultural and social change.

In 1899 Haeckel summed up his beliefs for his wide readership (Wegener included) in a book called *Die Welträthsel* (The riddle of the universe), a mix of biology, human cultural evolution, cosmology, psychology, and theology, intended as a worldview for a new century. It was unequivocal in its rejection of supernaturalism and revealed religion, but it was a credo nonetheless.[37] Herbert Schnädelbach called it "the Bible of the Free-Thinkers" and noted that Haeckel's group of monistic, vitalist, evolutionary metaphysicians would later be classified by the Nazis as *gottgläubig* (believers in God) in spite of their nontheism, because of the spiritualistic flavor of their metaphysics.[38]

Based more in observation than in theory (and in this sense like Chamberlain, who was also influenced deeply by Goethe), Haeckel's was a philosophy for natural historians, for doers, not for armchair types—rejecting the reading room for the "room of nature." While Haeckel's views caused some very violent disputes at the time of their publication, they now form, as Rollo Handy first remarked, the basis of the worldview of many educated people, especially in their emphasis on seeing man as a part of the ecological web of nature.[39]

We have begun here only with a list of names from a brief diary entry that Wegener made in 1907, but these authors and their works all point the same way. Taken together, they place Wegener, in his early twenties, within an optimistic "human-potential" movement in the sciences. Inside this movement, human life in both its biological and historical dimensions might be tragic—for life was brief and fraught with great difficulties—but it was something to embrace with energy and resolve, facing one's responsibility to fulfill the creative destiny of the world by exploiting one's own potential to the utmost. The movement's biologism and flirtation with eugenics and race theory were a part of its embrace of human life as a Darwinian predicament.

German Darwinism was determined not to accept alienation from life, and it would not postpone comprehension of the whole of the world: *at every moment* a synthesis of all of science was deemed possible and necessary. This required technical skill and knowledge, but even more, it needed courage and persistence, leavened with some good fortune. At this higher (if not quite transcendental) level all the really great achievements of science were to be found, and as Wegener returned to university life in the fall of 1902, this possibility was placed directly before him, and in a way that he had never before seen or imagined.

Planck, Thermodynamics, and Theory

Alfred was released from full-time military service in September 1902, on the completion of his "free-will" year. He still had to complete a series of six-week field courses before he could be commissioned as a lieutenant in the reserves, but he was confident that these could be fitted into the longer school vacations without slowing his scientific work. Taking home leave in August, he departed Berlin immediately for a holiday with the family at *die Hütte*; in October he returned to the university eager and rested, ready to resume his training as an astronomer.

If the parade ground drill was finished, he found that in his scientific work the toil of "basic training" continued unabated. In this third year of university study Wegener was plunged into concentrated and highly technical studies in positional astronomy and data reduction, in the form of Julius Bauschinger's "Seminar on Scientific Calculation," where he learned the details of orbital calculations, perturbations of orbits, and the calculation of the timing and path of solar eclipses.

Wegener was also finally able to take a course (the first of many) with Wilhelm Förster, on techniques for calculating meteor trajectories and cloud altitudes. This combination of objects today found in different sciences (meteors in astronomy and clouds in meteorology) preserves, as we have noted before, the sense of meteorology as the comprehensive study of all the phenomena that occur within the envelope of the atmosphere. Simultaneously with these technical courses, Alfred began his rotation as a student assistant in the Berlin observatory and center for popular astronomy, Urania (founded by Förster); this combination of observing and "public outreach" work at the observatory was expected of all the doctoral candidates.

The observations Wegener participated in during his work at the Berlin observatory had a strong component of atmospheric physics and geophysics. Förster was deeply interested in the physics of the upper atmosphere—what we would now call "solar-terrestrial physics"—especially the study of Earth's magnetic field by observations of the aurora borealis, a subject of much research in Germany in the middle of the nineteenth century. He had established a regular program of observations in the 1870s and had joined in the planning for the first International Polar Year (1882–1883) in cooperation with Georg von Neumayer (1826–1909), the founder of the German Marine Observatory at Hamburg. Förster was also interested in establishing the height of Earth's atmosphere by whatever methods were available, and this led to a series of observations of the altitude of the aurora borealis.

These studies of the upper atmosphere received a tremendous boost after the catastrophic explosion of Krakatoa in 1883 injected millions of tons of dust and gas into the stratosphere, leading Otto Jesse (1838–1901) in 1885 to the discovery of "noctilucent clouds." These ghostly objects, visible at night because of their great altitude above Earth, could be photographed from astronomical observatories, and in the decade after their discovery many observers used these clouds to discover the "full" height of Earth's atmosphere—which Jesse was able to establish by 1896 as near 100 kilometers (62 miles).[40] In this context we can better understand Förster's emphasis on "cloud altitudes" in conjunction with meteor paths—these were both upper atmospheric phenomena, to be studied in conjunction with the aurora.

Förster's program of atmospheric research at Berlin was an important stimulus to the field of *aeronomy*—the portion of geophysics devoted to the study of the atmosphere (the name is no longer in general use as a separate field). During Wegener's time there, its observational arm, "aerology," the investigation of the atmosphere by sending or carrying instruments aloft in balloons, was just becoming established in Berlin. It proceeded in conjunction with Förster's work, though under the direction of others, and served to make Berlin a major center for atmospheric research through most of the 1890s.

Along with his seminars in astronomy in 1902, Wegener signed up for Max Planck's course of lectures on thermodynamics and thermochemistry. He had expected to study thermodynamics, and it was unthinkable that one could gain a degree in physics without it, but he had certainly not counted on this series of lectures becoming an experience that would profoundly and irrevocably alter his view of science and shape his scientific career, perhaps more than any formal study he ever undertook. In October 1902, when Wegener wandered into his physics lecture hall and opened a notebook, Max Planck was forty-five years old.[41] A few years before, in 1897, he had gathered together his work on thermodynamics for the first time in the form of a textbook, *Vorlesungen über Thermodynamik,* which later went through many editions and was still the standard textbook thirty years after its initial publication. It was translated several times, including versions in Russian, Japanese, Spanish, and English.[42] Perhaps as importantly, it also set a style of presentation and ordained a range of topics adopted by most subsequently influential thermodynamics texts, including, for instance, that published in 1937 by the nuclear physicist Enrico Fermi (1901–1954).[43]

If one looks at Wegener's work in the decade that followed, one can see that he adopted into his scientific practice both Planck's treatment of thermodynamics and the philosophical commitments that came with it (as a sort of scientific credo) in Planck's introduction to the course of lectures. In the foreword to his lectures, Planck announced his decision to present thermodynamics without any reference to molecular or atomic motions—that is, to the kinetic theory of gases or the mechanical theory of heat. This was directly contrary to the position taken by Emil Warburg in Wegener's first university physics course—Warburg had created a plausible mechanical story for every physical phenomenon he described. Planck, however, declined to employ even Herman von Helmholtz's conservative version of the mechanical theory, in which heat was assumed to be a consequence of motions but the motions were not specified.

Planck argued that since there were still formidable obstacles—both mathematical and theoretical—to the full integration of thermodynamics with the mechanical view of nature, there was no sense in pretending that it had been achieved. If Helmholtz's approach had the satisfying effect of suggesting a linkage between them, it did not allow "a foundation of sufficient breadth on which to build a detailed theory."[44] That is, there was nothing to be gained *in physics* by adopting such a position: all you got was confirmation of some general laws that you had already deduced in other ways directly from experience.

Planck instead proceeded to develop a general framework for physics out of thermodynamics by the formal deduction of a variety of physical and chemical laws beginning from a few empirical facts, foremost among them the two fundamental principles of thermodynamics: the principle of the conservation of energy, and the principle of increasing entropy.[45] These principles, often called "laws," are not explanations, but rather descriptions. When we talk about energy in thermodynamics and thermochemistry, we are actually only applying an accounting system to nature. The science has an empirical feel because we are always patching and doctoring the equations of state which describe the systems we analyze, in order to make them work. Classical thermodynamics is a scheme, a set of rules to show the changes in energy in a system.

That being said, thermodynamics is still the most powerful and general theory in physics. It covers everything from the very small to the very large, it applies over any time interval, it covers both living and dead things, and it applies equally to classical physics, quantum mechanics, and relativity—precisely because, as Planck saw so clearly, it is not dependent on any microscopic model of reality for its accuracy.[46] Rather, it serves as a check on the accuracy of microphysical models, which, when scaled up to macroscopic size, must give results consistent with thermodynamics.[47]

Thermodynamics led one to meditate on the measurable bulk properties of things—their mass, temperature, pressure, and volume—and the transformations of these variables through time. In Planck's treatment, one studied moving systems (machines or gases or chemicals in solution) as things passing through a variety of states of equilibrium, on their way to a state of maximum stability.[48]

It is perhaps notable for Alfred Wegener's impressions of the subject that Planck introduced thermodynamics via a variety of real atmospheric processes and mixtures of gases. Planck was also extremely careful about definitions, constantly reminding the students that to define the state of a substance (and therefore its energy), one needed to know not only its mass and chemical nature but also its pressure and temperature.[49]

Planck explained to his students that this streamlined—and what we would now call "phenomenological"—approach to thermodynamics corresponded best to the actual state of the science. It was appropriate to what was known; it did not overreach or deceive. The caution and intrinsic modesty of this physical theory, with its professed indifference to causal mechanisms (always treated as subsidiary hypotheses), would stay with Wegener for the rest of his life and characterize all his work. It would also have important implications for the later consideration of his work in the English-speaking world. In British and American physics there was in the late nineteenth century and first part of the twentieth century a fascination with and a commitment to mechanisms and mechanical analogies: some force or some process had to be named as the cause, or the theory was incomplete; not so in Germany.

Albert Einstein's insistence on physics as, first of all, acts of measurement by human observers was a part of the cautious phenomenological strategy he learned in his university years, no more uniquely his own than his conviction that science is something freely created by human minds. So it was for Wegener and every other physics student at Berlin. To do physics in this fashion, one had principally to overcome, as Erwin Schrödinger later remarked, the custom, inherited over thousands of years, of thinking *causally*.

Wegener took away from Planck's teaching a strong commitment to brevity in the service of clarity. He adopted a caution, bordering on aloofness, in offering mechanical models and causal explanations, especially of the sort that only confirmed what one knew from experience without adding anything to the facts. He became convinced that the best avenue of approach in presenting a theory to an audience was to begin with some facts of experience and then deduce some physical and chemical laws governing the system from these facts—pushing the formal derivation of consequences from the facts as far as possible, testing by applications, and looking for (and expecting) exceptions. Most importantly, he heeded Planck's injunction never to consider any *form* of a theory as final, and to think of "good theory" simply as that mode of treating phenomena that corresponded to the actual state of a science at that moment—and never to one's aspirations for it.

Cosmic Physics

It is clear from the selection of courses in his third year of graduate study that even as Wegener returned to the technical training in astronomy, his focus had begun to shift from the observational astronomy that had brought him to the university to topics in physics, geodesy, and meteorology. There is nothing very remarkable about this. Students generally know at an early age that they want to be scientists (this was certainly true for him), but their initial choice of a science is controlled by what they see in secondary school. Most biochemists and geneticists turn out to have had, in high school, some idea of medical school, and they only found their actual vocation while taking "premed" courses in college. Very few geologists know they are going to be geologists or even scientists until they get to the university. For subspecialties the process repeats itself, and a typical scientific education goes through several apparent changes of field before the student finds the proper mix of talent, opportunity, and inclination.

The Berlin curriculum in astronomy and physics allowed one to stitch together a program in what was then called "cosmic physics." There was no such degree, nor any university professorships in the subject: "cosmic physics" was not a field so much as a point of view. When we talk of cosmic physics today, we mean astrophysics or space physics—the study of processes and substances with no counterparts on Earth because of the extremities of temperature, pressure, and radiation required to manifest them. The cosmic physics of 1902 was something rather different and, perhaps surprisingly, something more ambitious. It was an attempt to bring the study of the heavens together with the study of Earth, including its oceans, its atmosphere, and the intense and hostile regions of severe temperature and pressure below its surface. To this it added geology, geography, and biology, the latter existing in a fragile envelope extending a few hundred meters above the surface and a few meters below it—the biosphere. It was a physicist's counterpart to the *Kosmos* of Alexander von Humboldt (1769–1859), with its grand vision of the unity of nature and of mankind's place in it.

This cosmic physics, for which Wegener felt a growing affinity, was fueled by a number of intellectual tendencies and discoveries since the middle of the nineteenth century: it seemed to combine, in its vitality, the fundamental ideas of both thermodynamics and evolution. Among the discoveries that advanced this viewpoint was the spectroscopic study of light coming to Earth from planets and stars, and the determination that the stars and planets are made of the same elemental substances as those found on Earth. Cosmic physics was equally propelled by the study of organic chemistry, biochemistry, and physiology, which pressed the conclusion that we (and all life) are made of the same elements that we see present in the most distant stars and nebulae. Svante Arrhenius's (1859–1927) textbook *Lehrbuch der Kosmischen Physik* (1903) was only the most popular and widely read of many thick treatises attempting a unified presentation of the physics of nature.[50]

Wegener was now being pulled, albeit gently, in conflicting directions. He had still to complete the course work required for his astronomy degree, his emerging interests notwithstanding. He had not completely relinquished the ambition to work as an astronomer, nor was he yet committed to a scientific vocation in the terrestrial side of cosmic physics, though he found it highly stimulating. Therefore, as a test of vocation, in spring and summer of 1903 he pushed his work in astronomy beyond the solar system and, under Bauschinger's direction, extended his studies to celestial mechanics, double-star systems, and calculation of stellar ephemerides—tables of star positions. With Förster, he studied the history of Greek astronomy, theory of chronometry, and selected topics in the theory of errors, continuing his work at the Urania observatory.

Förster's historical emphasis was congenial to Wegener, who enjoyed contextual work that was classical in motive. Physics education in Germany trained students to see themselves and their work in historical context, and as part of a large enterprise, though not often to the extent that characterized the curriculum of Berlin astronomy. On Förster's part it was a way to make scientific study integral to the ideals of classical education in Greek and Latin, and it was in this way a holdover from Förster's struggle to establish astronomy and science generally in Germany as part of *Bildung*. His first course of lectures in Berlin, forty-five years earlier in 1858, had been on Copernicus, Kepler, and Tycho.[51]

Förster's history of science attested the close connection between advances in astronomy and the debut of novel instruments and methods of calculation. Historical study, when it contained detailed excursions into the technical details, was not a triumphalist exercise in celebrating progress but a chronological record of the evolution of thinking about the universe and about the intricate relationship of perceptions and concepts.

Wegener made sure in the summer of 1903 that there was room in his schedule for his new interests, though he was again assigned to the Urania observatory. He enrolled in Wilhelm von Bezold's course in theoretical meteorology, that year concerned with the thermodynamics of the atmosphere. Bezold was in 1903–1904 preparing for retirement and was summing up his work from the previous twenty years and preparing it for publication, including his extensive work on atmospheric thermodynamics. The contrast between Planck's approach to thermodynamics and Bezold's was striking. Planck's presentation of the science cut to the bone, seeking elegance and simplicity after the mode of theoretical physics. Bezold, on the other hand, had a narrative loquacity and anecdotal approach characteristic of his role as an experimenter and observer, and his writings preserved the signs of the strenuous activity that had produced the results. He had completed much of this research in the later 1870s and early 1880s and had published it (long since) in the proceedings of the Berlin Academy of Sciences between 1880 and 1892.[52]

Bezold's thermodynamics of the atmosphere was a pathbreaking effort, but he knew better than anyone that it begged to be redone and modernized in many respects, not least of which was the pressing need to get more information about the "free air"—the atmosphere kilometers above Earth—away from mountains and their self-created weather. A serious application of physical theory to weather data was also needed, to integrate the rapidly changing picture of the vertical structure of the atmosphere and of the energetics of storm systems, in terms relevant to research that was appearing in print, for the first time, even as Wegener took this course.

In his final—and most intense—year of study in Berlin (fall of 1903 through October of 1904), Wegener's work was almost entirely given over to detailed studies of measurement: history, theory, and technique. In the fall and winter he continued his work with Bauschinger—a history of celestial mechanics, and a seminar on the design and use of planetary tables. With Förster, he took the second half of the two survey courses he had begun in 1903—following the history of Greek astronomy with a history of Arabic and Medieval (European) astronomy, and following the course on chronometry with one on the theory of stereometry (i.e., volumetric as opposed to planimetric measurements).

Wegener also took a course in fall 1903 entitled "The Method of Least Squares," given by Friedrich Helmert (1843–1917), professor of advanced geodesy at Berlin and head of the Prussian Geodetic Institute. In tandem with this course Wegener studied "Topographical Surveying" with Helmert's student Hermann Eggert (1874–1944). In 1903–1904 Helmert was at work on means of correcting measurements of gravity and on studies of the movement of Earth's rotation axis (given by Helmert as the explanation of measured latitude changes at German gravity stations).

Not ony did Helmert's course function as an introduction to probabilistic techniques of estimation and error reduction, but in its concrete examples it furnished an introduction to a range of geophysical problems: the measurement of variations in gravity, the reality of latitude displacements, and displacements of Earth's axis—or "pole wander."

This problem of the displacement of Earth's axis, pursued by Friedrich Helmert, kept turning up in Alfred's studies: it had been there from the first practicum with Adolf Marcuse in 1899–1900. The motions proposed for the pole were not large on an annual basis (several meters per year), but no one knew whether there was a secular (i.e., long-term) accumulation of these annual displacements, and since the International Latitude Service had only been up and running for about a decade, the question was unresolved.

At some point in the 1903–1904 academic year, rather late in the game, Alfred made his final decision to switch from astronomy to cosmic physics. In an anecdote reported in 1931 by his longtime colleague Hans Benndorf (1870–1953), Wegener supposedly told student friends, "In astronomy, everything has fundamentally already been dealt with, and now only exceptional mathematical gifts and sophisticated equipment in astronomical observatories can lead to new discoveries. Besides, astronomy offers no opportunity for physical exertion."[53]

In the fall and winter of that year he found the physical activity he longed for in Bezold's "Meteorological Practicum" accompanied by a course of lectures entitled "Wind and Weather." There was no doubt about it: meteorology was an outdoor science, and most aspects of it required physical stamina: hiking to recording stations to retrieve records, and certainly everything connected with sending up and retrieving balloons and meteorological kites at the station in Templehof. In the spring of 1904, Alfred joined Bezold's Meteorological Colloquium, where Bezold and the advanced students met to read and discuss current topics and literature at the research front of atmospheric physics.

It was in Bezold's practicum that Wegener met and befriended Walter Wundt (1883–1967), at that time finishing his degree at Berlin in physics, but with a thesis in meteorology. Wundt later recalled that Wegener told him in 1903–1904 of his ambition to join in polar science and exploration, specifically to go to Greenland and traverse the ice cap following a route to the north of that taken by Fridtjof Nansen (1861–1930) in 1888.[54] This squares well with fellow student Walther Lietzmann's memory of Wegener expressing his "polar ambitions" at this time to the members of A^2V, the astronomical fraternity at Berlin. It would not be surprising for Alfred to have had such ambitions. This was the great age of polar exploration, and neither pole had yet been reached in 1904. Moreover, Bezold had been part of the planning committee for the German South Polar Expedition (1901–1903) led by Erich von Drygalski (1865–1949), who had also led a major expedition to West Greenland. Drygalski was, in 1904, lecturing about his polar exploits to scientific groups all over Germany, and it is probable (though by no means certain) that Wegener had heard him speak.[55]

Now that he was no longer considering a career in astronomy, he decided against some otherwise interesting courses: "The Three-Body Problem," "The Temperature of the Sun," and "Solar Physics." Since he was in Bezold's meteorology working group, he had no reason to attend the Astrophysical Colloquium. He retained an active interest in the history of science and elected to complete the full historical survey of astronomy with "History of Modern Astronomy since Newton." A practical course on measurement of celestial angles by theodolite (with Förster) looked useful, and he signed up for that as well. With Bauschinger, he took up a subject that strongly complemented both his work with Bezold and that with Helmert, "Potential Theory with Applications to the Figure and Rotation of Heavenly Bodies," and its practical counterpart, "Introduction to (the Art of) Calculation." Since the global circulation of Earth's atmosphere and the direction of prevailing winds in different latitudes are largely controlled by Earth's figure and rotation, this final course in astronomy could be employed in atmospheric and cosmic physics; nearly everything in Wegener's last semester had a direct application to his future work outside astronomy.

The steady march of classwork, reading, and calculation in his student life was matched by a profoundly stable and quiet home life. As a matter of finances, inclination, or both, Alfred was still residing with his parents at the orphanage on the Museum Insel, close to the university. As the student years came to an end, however, other changes were in the offing as well. After thirty years of directing the orphanage and, in Richard Wegener's case, thirty years of teaching high school Greek, Latin, and literature, the Wegeners were retiring and moving out of the orphanage. Richard had turned sixty the previous year and was now eligible for his pension; with their own youngest child ready to finish school, it was time to leave.

The Wegener family (Alfred too) moved at the end of the Easter term in 1904 to a house at the far western edge of Berlin, in the neighborhood called—then and now—"Halensee," after one of the small lakes dotting this gracious, green, and quiet neighborhood. Alfred had been here before for his military training—the Queen Elisabeth Grenadier Guards Regiment was stationed at Westend, about 2 kilometers (1.2 miles) to the north. A walk or short streetcar ride of the same distance to the south and west brought one into the Grunewald—Berlin's huge and wonderful forest tract, complete with walking trails, riding paths, and a large lake, the Grunewaldsee, on which Alfred and Kurt had skated almost every winter of their lives. The Wegener family took a house at 20 Georg Wilhelmstraße, just off the great boulevard of the Kurfurstendamm.

At the end of the 1904 summer semester, the family made the long-awaited annual pilgrimage to *die Hütte*. Alfred knew that he was saying good-bye to this period of his life, and he was savoring it to the end. He would return here often in later years, but this was the last summer with his parents, as well as the last long vacation, and he knew it. In mid-August he (and Kurt, of course) took a three-day sailing tour as far as Neustrelitz, 30 kilometers (19 miles) by lake and canal to the north, and wrote up their adventures as a poetic "Edda" for the rest of the family. It was a gentle parody of a great sea voyage, including a survey of the "natural history" of the region and a roster of the expedition's equipment and provisions (wine, bread, cheese, etc.).[56]

A week after their return from the first boat trip, they took off once more, this time for a week, armed with a camera and some navigational equipment. They sailed west, as far as they could—about 60 kilometers (37 miles) to Plau on the Plauer See. This voyage produced a longer (seventy pages) and more serious literary effort, illustrated with Alfred's own photographs and entitled *Sieben Tage im Boot* (Seven days in a boat). It has the flavor of Theodor Fontane's popular and reflective travel writing about the same scenes and sights, *Wanderungen durch die Mark Brandenburg*. It is a first attempt at an exploration diary by a young man longing for an expedition—mixing breezy anecdotes with sober disquisitions on when one should and should not unstep the mast, or how one should take bearings from a lakeshore to plan a dead-reckoning course. It is as if he were cutting a template for himself, or producing an explorer's "apprentice piece" to show his command of the skills, making something that is not a toy, but not the real thing either.[57]

The Alfonsine Tables

There was one matter still to be seen to before he could (as he ardently wished!) bring his student years to an end: the doctoral dissertation—an apprentice piece of another kind. Having decided to abandon astronomy, he knew that it made little sense (for him or the astronomy department) to pursue dissertation research on a problem in observational astronomy, since he would have to be given valuable observing time at either Berlin or Potsdam in order to make his measurements. The usual dissertation for Berlin astronomy students at this time was an orbital determination for an asteroid or comet; these were the sorts of problems on which Bauschinger was expert. The second most common was some study of the motion of the Moon and planets, or observations of the Sun; Förster supervised these.

Wegener's (assigned) topic, supervised jointly by Bauschinger and Förster, was quite different and indeed unique in the annals of the department: Wegener was directed to undertake a historical and critical study of a set of astronomical tables—the Alfonsine Tables—commissioned by Alfonso X ("The Wise") of Castile in the thirteenth century. In the recent past (reckoned from 1900) there had been at Berlin two historical-philosophical theses: in 1903 Abraham Hoffmann had submitted a dissertation on theories of the universe leading up to the work of Descartes, and in 1904 Max Jacobi had produced a thesis on the cosmological theories of Nicholas of Cusa. Neither of these, however, had any substantial technical content. Wegener's problem required a good deal of historical knowledge, planetary theory, and calculating technique and provided a humorously gratifying, if unexpected, scientific employment for all those years of Latin.[58] The latter skill perhaps sealed his fate—as the son of a classical scholar who had drilled him in Latin almost every day for fourteen years, Alfred was perhaps the only student in the history of the department to have his languages still largely intact years after *Gymnasium*. Though there is no record of collaboration with his father, one can hardly imagine that father and son would not have discussed this work, or that Richard would not wish to have a hand in solving the difficult textual questions that loomed.

As for the work itself, the original (thirteenth-century) purpose of Alfonso's planetary tables was both astrological and astronomical; they allowed one to find the position of the Sun, Moon, and planets at any hour and minute of any day of any year. Astrologically, the positions were used to cast horoscopes; astronomically, they were used for navigation and time reckoning. The creation of a modernized edition of these tables in the twentieth century was, however, not an antiquarian exercise, though it might appear so today: it actually had applications to a problem in astronomy in which both Bauschinger and Förster were keenly interested.

Here is the problem that Wegener's work was intended to solve. About thirty years before, in 1870, the great American astronomer Simon Newcomb (1835–1909) had discovered that the set of tables he was using to predict the position of the Moon showed increasing deviations from the Moon's actual position. This was disturbing to Newcomb because the tables in question had been prepared by the German astronomer Peter Hansen (1795–1874), probably the greatest master of celestial mechanics since

Laplace, and one of Newcomb's heroes. Hansen had worked by matching his theory of the Moon's motion to observations stretching back to 1750, and he had extrapolated earlier positions by calculations based on theory. To figure out what had gone wrong with Hansen's predicted positions for the Moon, Newcomb traveled to Europe to study even older tables of the Moon's motion. He found that the farther he worked back before 1750, the greater the discrepancy became. He conceived the notion, based on his absolute confidence in Hansen, that the reason for the discrepancy had to be a variation in Earth's rate of rotation. Newcomb worked steadily on this problem until the end of his life, assembling astronomical records back into preclassical antiquity to try to get the longest possible series of lunar positions, so as to determine the pattern of rotational variation.[59]

The Alfonsine Tables figure into such investigations because they were the tables of reference throughout Europe from about 1330 until 1551, the year in which Erasmus Reinhold's Prutenic Tables appeared—the latter being the vehicle whereby Copernicus's theory first appeared in print. The Alfonsine Tables quickly fell out of use once the more accurate tables became available, having by 1900 been out of general use for almost 350 years. With the growing interest, around 1900, in what we would now call "long time series" of observations of celestial motions, there was now a good reason to refer to and correct these tables to establish the Moon's position in the middle of the thirteenth century. Moreover, by looking at fifteenth- and sixteenth-century editions of the tables, one could examine the corrections introduced by later astronomers to overcome emerging divergences between predicted and observed positions of the Moon. This might also provide valuable hints about the Moon's actual (vs. predicted) position and therefore information about Earth's rotation.

The difficulty, as Wegener himself explained in his introduction, was that these Alfonsine Tables were now rare books available only in great university libraries. They were written in a difficult form of Medieval Latin and couched in terms of the Ptolemaic (Earth-centered) solar system, with its immense geometrical complexities and unfamiliar terminology. Moreover, the numerical values in the planetary tables employed the sexagesimal system (base 60) rather than the decimal system both for angular measurement and for date and time reckoning. The modern system uses the sexagesimal system for degrees, minutes, and seconds (and minutes and seconds of time) but then proceeds decimally. In brief, the Alfonsine Tables were generally unavailable, and where they were available, the difficulties of calculating or even reading them rendered them nearly useless.[60]

Wegener began his work in September 1904, comparing the (six) existing principal Latin editions to eliminate printer's errors. He then translated the text from Latin to German. Then (in October and November) he converted the tables from sexagesimal to decimal values, a task involving about 9,000 calculations. That was the heart of the task, but there was a good deal more. He had to provide extensive notes to give astronomical calculators the means to use them. The calculators in question were not machines, of course, but observatory staff whose job it was to perform the actual calculations.

Alfred provided a correction for the difference between Toledo Spain, the 0° of longitude in the tables, and Greenwich, England, the 0° of longitude for modern astronomy. He uncovered a systematic sixteen-minute offset in the tables resulting from a discrepancy between the Alfonsine way of calculating the "mean time" of a transit and the modern method of doing so. Finally, he devised a formula to eliminate a correction in the tables meant to account for the precession of the equinoxes which employed a (nonexistent) celestial motion called a "trepidation," which the modern user must discount.

Wegener also had to show the calculators how to calculate using the tables, and he gave worked-out examples of such calculations. It required several explanations: the tables for the Sun are used differently than the tables for the Moon, and there are separate sets of tables for each of the planets. Latitudes are calculated in a way quite different from the calculation of longitude and distances, and these require separate tables and figures showing the geometry of the relationships (Wegener drew fourteen figures for his text). A sample calculation for the use of the Mars Tables and a copy of Wegener's first page of figures give the flavor of the work.[61]

He also drew up a concise glossary of Latin technical terms for which there were no German counterparts, since the concepts in question had vanished from astronomy before German had become a scientific language.

This was indeed the work of an apprentice: straightforward and, though useful, much less valuable for itself than for what it revealed about its maker—self-discipline, attention to detail, a spirit of repetition, stoicism in laborious calculation, and an interest in rationalizing, modernizing, and easing access to data and theory, especially where long historical time series of measurements were involved. It was yet another practicum for Alfred in understanding the deep penetration of theory into scientific observation and, more importantly, by understanding the theory, how one learned to extract and discard it—an extremely useful scientific acquisition. Wegener's *The Alfonsine Tables* was not scientifically important and led to no program of research for its author, though its technical proficiency was admirable, its explanations models of clarity, and the tables themselves still useful today for the purposes they were meant to serve. It was a parting gift to his mentors, Bauschinger and Förster, and also a gift to his parents and the classical world they inhabited, as reflected in its dedication: *Meinen Eltern* (To my parents).[62]

From another perspective, though, it was a serious underemployment of an exceptionally able and well-trained scientific intelligence. Wegener's research produced a great deal more material than could be fitted into the scope of an edition of the tables, and he also found a good deal of error and sloppiness in the work already done on the tables. He wrote in his introduction,

> It is worth while emphasizing that, on the other hand, the 1863–67 edition of the "Libros del saber de astronomia del Rey Alfonso X. de Castilla etc." edited by the Madrid Academy of Sciences were completely useless for the present work. In the fourth volume of this work the Castilian original text of the [Alfonsine] Tables is included, but all that appears in the very place where the tables themselves are supposed to be is (erroneously identified as "numerical fragments of the Alfonsine Tables") a special kind of Ephemeris for the tabulation of dates, later known as a "perpetual almanac," which has nothing whatever to do with the original Alfonsine Tables. For the most part the contents of the Spanish publication are dedicated not to the planetary tables, but to a newly discovered edition of the work of Alfonso X on astronomical instruments.[63]

Wegener's impatience with a careless error, combined with a too-limited project, seemed momentarily ready to project him into a new edition of the works of Alfonso of Castile, but he soon recalled that he was trying to get out of astronomy, not into the

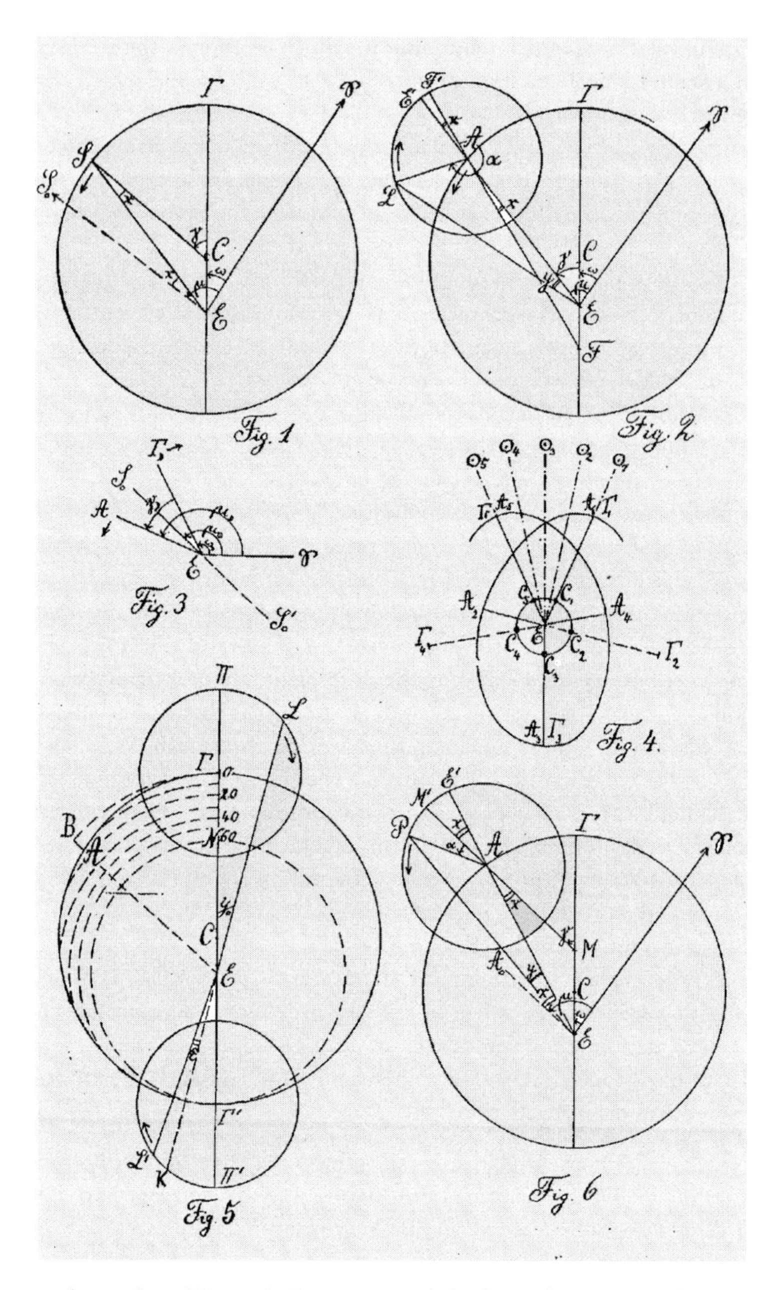

Explanatory figures from Wegener's dissertation to help the reader visualize the relationship of Earth to the Sun and other planets. The Earth-centered Ptolemaic system, on which the Alfonsine Tables are based, requires a number of extra geometrical devices to explain the observations while preserving the demand for perfectly circular orbits. In this illustration Fig. 1 is the orbit of the Sun; Fig. 2 is the orbit of the Moon, with the Moon's epicycle centered on A; Figs. 3–5 are aspects of the changing distance of the Sun, Moon, and Earth from one another; and Fig. 6 is a composite picture of the geometric relationships among Earth, Jupiter, Mars, and Saturn. From Alfred Wegener, *Die Alfonsinischen Tafeln für den Gebrauch eines modernen Rechners: Inaugural-Dissertation zur Erlangung der Doktorwürde genehmigt von der philosophischen Facultät der Friedrich-Wilhelms-Universität zu Berlin* (Berlin: E. Eberling, 1905).

Beispiel: Gesucht die wahre Länge des Mars für 1477 Sept. $20^d\ 6^h\ 1^m\ 36^s$ M. Z. Toledo, d. i. $1477.0 + 263^d\ 6^h\ 1.6^m$ oder 1477.72.

Wir entnehmen aus der Tafel der mittleren Bewegungen IV und V:

	$\mu_♂$	$\mu_\odot$	
1470.0	71.11	288.896	
7^a	260.04	0.298	
200^d	104.81	197.129	
60^d	31.44	59.139	
3^d	1.57	2.957	
6^h	0.13	0.246	Tafel I: $\pi = 19.546$
1.6^m		0.001	Tafel III: $\omega_0 = 115.204$
	$\mu_♂ = 109.10$	$\mu_\odot = 188.666$	$\omega = 134.750$
		$\mu_♂ = 109.10$	$\mu_♂ = 109.10$
		$\alpha_♂ = \mu_\odot - \mu_♂ = 79.57$	$\gamma_♂ = \mu_♂ - \omega = 334.35$

Damit sind μ, α, γ bekannt, und wir gehen nun an die Berechnung der Ungleichheiten.

Mit γ gehen wir in die Tafel X ein und entnehmen die aequatio centri:

$x = +4.54$, so dass wird $\gamma + x = 338.89$

$\alpha - x = 75.03$

Mit $\gamma + x$ entnehmen wir ferner:

min. prop. $= 0.93\ l.$

Mit $\alpha - x$ entnehmen wir:

aequatio argumenti $y_0 = +28.53$, sowie divers. diam. $l. = 1.92$

pars proport. 1.79 $\times 0.93 = 1.79$

$y = +\ 26.74$

dazu $x = +\ 4.54$

$\mu = 109.10$

$l_♂ = 140^0.38$

Wegener's simplified explanation of how to use his modernized tables. The text reads, "Example: Find the true longitude of Mars for September 20, 1477 at 6:01:36 [a.m.] M[ean] T[ime] Toledo, decimal 1477.0+263 days, 6 hours, 1.6 minutes or 1477.72. We draw from the table the Mean Motions IV and V . . . [calculations follow yielding the true Mars astronomical longitude of 140.38°]." From Wegener, *Die Alfonsinischen Tafeln* (1905).

history of astronomy. After all, it could not have been lost on him that no one had remarked on the error in the forty years since its publication, so this was clearly no "hot corner" in astronomy or its history. The experience with the bungled Spanish edition resulted instead in a publication (nearly as long as the dissertation itself) in a leading journal of the history of mathematics, cataloging and analyzing Alfonso's astronomical works.[64] Wegener was taking the opportunity, limited as it was in subject matter, to develop an approach to research in general, finding what for him would be a comfortable depth of understanding of a text or a topic. This depth, here and afterward, turned out to be quite great, as we shall see.

It is also fair to say that Wegener could not resist the temptation to show off a little. He did not like to work without publishing, then or later, and almost every piece of work he ever did found its way into print somewhere and somehow. In a burst of festive file clearing in 1905 as he prepared to leave the university, Wegener collected his preliminary background studies on the relationship of observation to theory in cosmology from the pre-Socratics down to the time of Laplace and published them in the Berlin scientific monthly *Mathematisch-Naturwissenschaftliche Blätter*.[65] He concluded that a survey of the history of cosmology shows a constant repetition of a few

Alfred Wegener at age twenty-four in the summer of 1905, in a portrait made to accompany his dissertation on the Alfonsine Tables. From Wegener, *Die Alfonsinischen Tafeln* (1905).

ideas—a disk, a vortex, nested spheres, or hemispherical heavens. The "right answer" appears again and again, but only as speculative philosophy. What separated modern (here Descartes and Newton) cosmology from the ancients was not new ideas, but the harvesting of fruitful conceptions—especially those of Anaxagoras—and then treating them mathematically.[66] Newton would probably have agreed with and approved of this judgment: he held originality of conception to be of little weight and the mathematical explication to be the real scientific accomplishment. Wegener was defending his own work on the Alfonsine Tables and stating a credo.

The Last Classical Physicist?

Wegener's university studies came to an end just as a number of profound changes swept through physics, astronomy, and cosmology. Einstein's work on special relativity, the photoelectric effect, and Brownian motion—all in 1905—changed the meaning of space and time, divided light into discrete packets, and made the statistical interpretation (as opposed to determinate interpretations) of atomic behavior inevitable. Between 1905 and 1906 a number of stunning developments in quantum theory, physical chemistry, and microscopy drove even the most outspoken opponents of the atomic theory into accepting the reality of atoms and therefore of the granular character of the universe—putting back into basic physics what Planck had worked so hard to remove a few years before.

In astronomy, the work of J. C. Kapteyn (1851–1922) in 1904–1905 and that of Ejnar Hertzsprung (1873–1967) in 1905–1907 opened the door to the study of the evolutionary history of stars and of the dynamics of the universe and completely revitalized observational astronomy, albeit, as Alfred recognized, only for those with access to giant telescopes and those with great mathematical talents.

One is tempted to tell a story about a "fateful timing" in which Wegener was born at the very end of classical physics and, like a Lohengrin missing the swan boat, was left in the backstage of physics to smoke and think while the action went on without him, under the dramatic stage lighting of history. This would have a spark of tragic poetry, but it is too falsely theatric to contemplate for more than a moment. For one thing, there were other equally powerful and fundamental developments in these same years which created a great deal of new theoretical and practical turmoil in the regions of science Wegener was moving into, among them radioactive transmutation, a new theory of the formation of the planets, and other developments we shall discuss. But even had there not been these changes, it would have come out the same for Alfred. He began his adult scientific career in a world of observation, experiment, and data collection, in a world of macroscopic, thermodynamic physics, and this world remained virtually untouched in its character and practice by these fundamental changes in the view of the microscopic world—just as Planck had predicted. Timing had nothing to do with it. Had he been born a little later, had he taken those courses in stellar evolution, it would have made no difference; he still would have wanted to be outdoors. We need not divide up physicists into "theorists and experimenters" to explain his decision to abandon astronomy because we have recourse to another equally valid partition of humanity: those who do not need to feel the wind in their faces, and those who do.

4

The Aerologist

LINDENBERG, 1905–1906

The balloon of experience is in fact of course tied to the earth, and under that necessity we swing, thanks to a rope of remarkable length, in the more or less commodious car of the imagination, but it is by the rope we know where we are, and from the moment that cable is cut we are at large and unrelated.

HENRY JAMES (1909)

The chief danger attending ballooning lies in the descent.

Encyclopedia Britannica (1910)

On 1 January 1905, Dr. Alfred Wegener joined the scientific staff of the Royal Prussian Aeronautical Observatory, a complex of buildings rising rapidly on a 28-hectare (~70-acre) tract in the middle of some wheat and barley fields 60 kilometers (37 miles) southeast of Berlin. The observatory was a little to the north of the main rail line, though served well enough by a station and siding a kilometer away in Lindenberg.

Of course, Alfred wasn't really Dr. Wegener yet, and would not be until the publication of his dissertation in March, but then the observatory wasn't really an observatory yet either. It had been under construction since June 1904, and while the pace of the work had been aided by an exceptionally dry summer, the scientific station was still far from complete. The headquarters building, the staff residence, and the director's house were finished, but the machine shops, the hangars, a powerhouse for the diesel generators and the massive array of electrical batteries, the central steam-generating plant, the dry ice machine, the large electrolytic apparatus for obtaining hydrogen and oxygen, and many other utilities and support buildings were in various stages of incompletion; the full scientific program could not begin until April.[1]

Alfred's appointment, even if only as *technischer Hilfsarbeiter* (technical assistant), was his portal to a new career in a new science, and it was also the first real step toward an independent adult life. Both were welcome, and the latter perhaps overdue for a twenty-four-year old man who had finished his doctoral degree and the bulk of his military service without ever residing more than a few kilometers from his parents—if one excepts the Heidelberg semester and a summer in Innsbruck. The tether to home and family was still secure, but it was time to pay it out a little.

Alfred was clearly pleased with the position, made possible by the accelerated shift from astronomical to meteorological subjects in the final year of his doctoral work at Berlin and by Bezold's recommendation. He had no regrets about taking his degree in astronomy and recognized that it was actually a form of security, since prospects for academic employment in meteorology were the poorest of all the fields of physical science. Indeed, many academic scientists and most physicists doubted that meteorology had enough theoretical content and practical rigor to qualify as a physical science at all. In consequence, there were only two real university professorships of meteorology in Germany and Austria, created for distinguished individuals who had beaten the odds that "a physicist who goes into meteorology is lost"—the often-repeated opinion of the physicist Friedrich Kohlrausch.[2] There was Wegener's own professor at Berlin, Wilhelm von Bezold, holding the first and only professorial chair of meteorology in the whole German Empire—and this, in part, as a courtesy appointment befitting his status as head of the Prussian Meteorological Institute. There was also the chair in meteorology and geophysics created for Julius Hann (1839–1921) at Graz in Austria, which persisted after Hann moved in 1900 to Vienna for the chair in cosmic physics.

With the exception of Bezold and Hann, all the other great meteorologists in German-speaking academia worked in nonacademic positions in state-supported institutes: Wladimir Köppen (1846–1940) at the German Marine Observatory in Hamburg, Hugo Hergesell (1859–1938) at Straßburg, and Richard Aßmann (1845–1918), Wegener's own superior officer at the Lindenberg Observatory. Under the direction of these institute leaders and section chiefs were a large number of workers variously styled "collaborators," "assistants," and "helpers." These were modest employments (in terms of compensation and job security) but often filled by scientists of international distinction, such as the great balloonist Arthur Berson (1859–1942), Aßmann's collaborator for twenty years at Berlin and then Lindenberg. The list also includes the Austrian Max Margules (1856–1920), the great theorist of atmospheric energy processes, who never got farther than a staff position as an assistant at the Vienna Zentralanstalt für Meteorologie and took early retirement in 1906 on a minuscule pension, despairing of better employment.

In England, Russia, and Scandinavia the situation was much the same. A meteorologist as distinguished as Nils Ekholm (1848–1923), a collaborator of Svante Arrhenius and a great polar traveler, climatologist, and student of weather forecasting, worked as an assistant at Uppsala and Stockholm from 1876 until 1898 and then supported himself as a mathematician for a life insurance company from 1898 until 1913, when, at the age of 65, he finally obtained the professorship that went along with the directorship of the Meteorological Institute in Stockholm.[3]

Still farther from the ranks of university and government posts there were independent scientists like the French pioneer of upper atmospheric research, Léon Teisserenc de Bort (1855–1913), and his American counterpart and friend A. Lawrence Rotch (1861–1913). They combined personal means and private sponsors (including, in Teisserenc de Bort's case, Prince Albert of Monaco) to erect independent institutes where they pursued their own programs of research.

Wegener's job as a technical assistant was, therefore, not necessarily the sort of self-limiting position we now call a "postdoctoral" appointment, nor was it realistically a stepping stone on the path to the security and status of the sort of employment that the Germans called *pensioniert* (pensioned). Yet, notwithstanding its clear limitations and the almost nonexistent chances for advancement, it was at the time a sort of minor miracle: a real, immediately available scientific employment in meteorology, in a completely new area of scientific study—trying to describe and understand the vertical structure of the atmosphere.

There was more good fortune at hand for Alfred: the only other technical assistant hired as of the first of the year was his brother Kurt, who had just taken his degree in meteorology at the Physical-Technical Institute in Charlottenburg. Here was a chance to continue to work together toward plans and goals that, for the moment at least, seemed to be converging. They had hiked, climbed, and studied together in the Alps, done their military service back to back, and passed every university holiday either touring or relaxing at *die Hütte* with Tony and their parents.

Now they were to share the beginning of their scientific careers, as well as new, modern, clean, spacious, and even luxurious accommodations on the third floor of the headquarters building—with indoor plumbing, hot and cold running water, and a living room and bedroom of their own. On the second floor were the scientific workrooms, and on the ground floor, in addition to the administrative center, were a dining room and a *Rauchzimmer* (smoking room), where they could, respectively, be served their meals by the cooks and waitstaff and indulge an already extravagant fondness for tobacco. They could, of an evening, enjoy a quiet beer here with colleagues, discussing their work, with their own steins kept on a shelf like any neighborhood tavern.

Alfred and Kurt were happily oblivious to the disarray and general confusion caused by construction at Lindenberg. They had spent most of their lives in Berlin surrounded by massive and noisy construction projects, and it was nothing new to have everything torn up and nothing finished. Moreover, the station rising in the middle of the sparse landscape had an almost expeditionary character and excitement to it.[4]

Lindenberg was to be the main German center for aerology—the investigation of the three-dimensional structure of the atmosphere by sending up (to altitudes of several kilometers) meteorological recording instruments via "captive" balloons and kites, tethered to the ground by steel cables. Aerologists also sent aloft to much greater altitudes "free balloons" (designed to parachute back to earth with their instrument packages). Finally, the scientists went aloft themselves in manned balloons capable of carrying several investigators and a large and varied array of sensing and recording devices. It was specifically to pursue these aerial investigations that the Wegener brothers had been recruited. As the technical assistants, they were to work directly with the "observer," Arthur Berson, and with the director of the station, Aßmann, in conducting flights of these experimental aircraft and the even more experimental instruments they carried. Supported by a staff that was soon to number almost fifty technicians, machinists, engineers, clerks, and cooks, they were to be an atmospheric science counterpart to the scientific staff at German Marine Observatory in Hamburg, which had been founded decades before as the principal station and clearinghouse for German oceanographic and marine meteorological research.

The unprecedented sums of money available for this new scientific station were a result of direct royal patronage. Kaiser Wilhelm II aspired to be a patron of science and technology on the pattern of his friend Prince Albert of Monaco—long a patron of oceanography, meteorology, and marine biology. The kaiser had an interest in aeronautics and meteorology, intensified by his general staff's conviction that long-distance and controlled flight was imminent and would have important military consequences.

Lindenberg Observatory was intended to be the aeronautical version not just of an oceanographic station but also of a research vessel, carrying a crew to study the "ocean of air." Aerology was borrowing terminology and methods from the preexisting scientific and technical field of oceanography—just as aeronautics was busily borrowing the vocabulary of merchant shipping. The first flying machines—the zeppelins—were "air-ships," and they landed and took off from "air-ports." They had captains and a crew, as well as pilots and navigators on the "flight deck" or in the "cockpit" wearing nautical-style uniforms and gold braid. They measured their speed in knots (nautical miles per hour), and they had a fore and an aft, port and starboard sides, and soon also passenger cabins and cargo holds. The choice of terms was obvious and deliberate, but it was also apposite to a degree that makes it a pleasure to reflect on even a century later, when we are long accustomed to it.

Of course, the metaphor of Lindenberg as a research vessel had its limitations, for ocean-going ships study the ocean as they move about, while this "vessel" was anchored permanently in Lindenberg; but what matter if the ship move over the water, or the water under the ship? It is actually more convenient to have an observation platform in one place, and not always to have to subtract or add the station's direction and velocity to calculations of currents and flows, as one must do on a moving ship. In this way Lindenberg's situation was also like that of an astronomical observatory, with the sky passing in review each night. There was plenty of weather passing over Lindenberg and consequently a great many interesting atmospheric phenomena arriving overhead to investigate.

The Work of the Observatory

The heart of the observatory, and of Wegener's work there, was the *Windenhaus*, a large, octagonal metal and glass gazebo standing on the highest point of the station grounds (127 meters [417 feet]) surmounted by a cupola containing three 1,500-watt beacons visible at night for 20–30 kilometers (12–19 miles).[5]

The *Windenhaus* itself reflected Richard Aßmann's perfectionism and love of devices. It was mounted on a rotating platform, supported on four rollers seated on a circular iron rail. Inside the building was the *Winde*, the large motorized winch with its gearboxes and electrical motor capable of paying out 20,000 meters (65,617 feet) of wire. Inside the *Windenhaus* was a workstation holding the station logbook, with a chronometer and an anemograph that recorded the wind speed and direction from the anemometer atop the cupola.

The workstation also featured another of Aßmann's myriad instrumental monotypes: an aspiration meteorograph on a design he had conjured twenty-five years earlier in Berlin. Air was pulled into the instrument by an electric fan through a pipe that penetrated the wall on the windward side of the building. As the air passed through the pipe at 7 liters (1.8 gallons) per second, compact instruments (of Aßmann's design or modification) measured the barometric pressure, air temperature, and relative humidity and recorded these on a scrolling sheet of graph paper which could hold a week's observations.

The aspiration meteorograph served, before and after each flight of a balloon or kite from the *Windenhaus*, as a reference standard for the instruments and recording devices sent aloft. Before and after each flight, the scientist standing watch in the *Windenhaus* placed the kite or balloon meteorograph on a meter-long horizontal strut, fastened to the outside wall of the *Windenhaus*. This strut could be moved to whichever was the windward side, by virtue of regularly spaced mounting brackets on the exterior walls. It was also fitted with an electrical lamp so that readings could be taken at night. Once mounted, the instrument destined to go aloft was allowed to run for a set time to measure and record the temperature, humidity, and atmospheric pressure, while ventilated by an electrically driven fan. This procedure served to calibrate the instruments relative to a reference ground station and to provide a correction factor to add or subtract from the recorded aerial data.

The scientific staff, Wegener and his colleagues, calibrated the tiny anemometers in the instrument packages sent aloft using a "Scirocco-Ventilator" in the laboratory—a

Windenhaus (winch house) at Lindenberg, showing a tethered balloon ready for launch. Note the recording instruments trailing the balloon on the cable tether (here just above the roof line). From Richard Aßmann, *Das Königlich Preußische Aeronautische Observatorium Lindenberg* (Braunschweig: Friedrich Vieweg & Sohn, 1915).

variable-speed wind generator. They also regularly checked barometers and barographs in a vacuum chamber and recorded each instrument's peculiarities and standard errors. There were corrections for the corrections: Aßmann's aspiration meteorograph could not help heating up in still air and strong sunlight, so it had to be corrected yet again, using a separate aspiration psychrometer, before, during, and after each flight. Moreover, all these readings had to be compared to the observed values in the Sun-shielded ground station that was the official meteorological station record of the Lindenberg Observatory.[6]

As scientific work, it was painstaking and even finicky, but it appealed to Wegener's mechanical inclinations, his pleasure in collecting data, and his love of calculation, correction, and data reduction. It is no accident that all the great advances in aerology at this time were made by scientists with a love for instruments and instrument technology. Aßmann, Léon Teisserenc de Bort, and Hugo Hergesell all employed instruments they had designed and built, and each had a variety of instruments named after him. Much of the work of this new science was the struggle for every marginal increment of accuracy and durability in meteorological instrumentation, and each of the competing designers hoped that his instrument would become a professional standard.

While at Lindenberg, Wegener's principal employment was to help send these carefully calibrated instruments aloft seven days a week (holidays included) at set hours. The ambitious schedule included a flight from 7:00 to 10:00 a.m., one from 2:00 to 6:00 p.m., and one from 9:00 p.m. until midnight. At set times during the year "International Flight Weeks" meant the addition of a flight from 2:00 to 5:00 a.m. Once a month (usually the first Sunday) an instrument package went aloft for twenty-four hours of continuous observations.[7] It actually took many years to reach these ambitious goals. In the sixteen months Wegener spent at Lindenberg, there were about 400 kite ascents and about 140 captive balloon ascents, with balloons going up every third day (on average)—with the frequency determined by the wind speed, or lack of it, since kites could not be flown in still air.[8]

Meteorological kite flying was only a decade or so old when Wegener took it up at Lindenberg. An Australian scientist named Lawrence Hargrave (1850–1915), who designed and built model flying machines, had invented, around 1885, the kite design we now call a "box-kite," characterized by a high angle of flight and great stability in the wind. Part of the design was tinkering, but part was a result of advances in the theory of the optimal loading of sails for the great clipper ships that dominated Pacific commerce. Until the advent of wire for the standing rigging that held the masts stable, the rope rigging of such ships caught so much wind that it was extremely difficult to calculate the optimal size, position, and trim of sail. With the advent of thin-wire rigging, catching almost no wind, sail design and ship rigging jumped ahead, and by the later 1880s the theory of sail was beginning to be available in the technical literature. One fallout from this was kite design, since a kite is nothing but a sail or series of sails rigged on a wooden frame and tethered by a line to the ground.

Scientific kite and balloon flying was a field that for several decades attracted both professional and amateur interest—the young Ludwig Wittgenstein (1889–1951), on his first trip to England in 1908, built and flew meteorological kites for the Kite Flying Upper Atmosphere Station at Glossop, near Manchester.[9]

The Hargrave kite made aerology possible as a systematic enterprise, because it dramatically lowered the cost of sending instruments into the air. When reeled out on a cable and carrying the lightweight and immensely durable meteorograph designed by the American Charles F. Marvin (1858–1943), a Hargrave kite allowed precise meteorological observations to be taken and recorded at altitudes of several kilometers, day after day.

It was, of course, possible to do this with free balloons, as Teisserenc de Bort had done for years at his own station at Trappes, in France. These were delicate, expensive things made of silk, kerosened paper, or goldbeater's skin, a fabric made from strips of cured ox-gut glued together. All of these balloons were capable of carrying meteorographic packages of instruments up to 14 kilometers (9 miles) above the ground. When they reached a certain altitude or time aloft, a mechanical hook tore a strip from the balloon, transforming it into a parachute to return the instruments to earth. The instrument packs were marked with Teisserenc de Bort's name and address, and he paid a substantial reward for their return.

The great advantage of kite flying, even though kites never reached such impressive altitudes as the free balloons, was the virtual certainty of the return of the instruments. Moreover, using the sort of winch that Wegener had access to at Lindenberg, it was possible to control very accurately the rate of ascent and descent of the instruments. There was often a significant lag time (upward of forty-five seconds) before a recording instrument could respond to a change in temperature or humidity. Therefore, a slow, controlled ascent, with the ability to pause the instruments briefly at any altitude, gave a much more accurately detailed record of atmospheric conditions.

On days when there was not enough wind to take the kites aloft (winds of less than 6–7 meters per second, or about 15 miles per hour), Wegener and the other technical assistants and helpers sent aloft, using the same cable and winch arrangement, large rubber balloons (*Gummiballon*) of Aßmann's design, twice as tall as a man and filled with 20–30 cubic meters (706–1059 cubic feet) of hydrogen generated with an apparatus at the station. These balloons flew well only when the wind speed was less than 3 or 4 meters per second (about 7 miles per hour), so there were days when no flying could be done at all—too much wind for a balloon, not enough wind for a kite. Moreover, as anyone knows who has been in an airplane or watched clouds drift by on a sultry summer day, winds shift velocity and direction rapidly with altitude, and it is possible to have still air at the ground and a respectable breeze only a few hundred meters up—so the flight plans for the balloons were often frustrated in this way as well.

There were other alternatives: sounding balloons could be released (without instruments) and followed visually to very great altitudes, in order to study shifts in wind velocity and direction. Meteorologists, on partly cloudy days, had for many years studied winds aloft with theodolites using the direction and velocity of clouds as the data, and using triangulation from observers far apart to find the altitudes. But sounding balloons allowed observations of motions in clear air as well. Aßmann had worked out a system of underinflating the balloons to keep them from rising too rapidly. Using the theory of the rate of expansion of hydrogen gas with altitude and the corresponding resistance of the rubber balloon fabric (a theory developed by Hugo Hergesell in 1903), Aßmann was able to calculate how fast a balloon was rising at each altitude.[10] A balloon rising at 1 meter per second (2 miles per hour) at an altitude of 6 kilometers (4 miles) would be rising at 8 meters per second (18 miles per hour) at 15 kilometers (9 miles) and 8.5 meters per second (19 miles per hour) at 20 kilometers (12 miles).[11] One could theoretically use these numbers during a flight to calculate the air pressure at various altitudes by plotting the calculated velocity of ascent against the actual, though the main reason for sending up these balloons was to observe the wind speed and direction.

In 1905–1906 the Lindenberg scientific staff attempted thirty-nine sounding balloon ascents.[12] Observers (usually two or more) followed the balloons with theodolites of various designs, and they constantly tested new ones—work in which Wegener took a vigorous and active role. Balloon theodolites are compact telescopes capable of extremely precise angular measurements and are set on mountings that swivel both up and down and from side to side (technically, swivel in both elevation and azimuth). They are fitted with an eyepiece with a right-angle prism so that the observer always looked horizontally into the instrument no matter what its orientation or inclination. As one observer followed the balloon with the theodolite, the other observer used a graphing apparatus to mark the elevation and direction of the balloon at intervals of several seconds. Using two pairs of observers, the angular data could give the actual flight path of the balloon.

Armed with these instruments, the Lindenberg staff studied every aspect of the atmosphere in the first few kilometers above the surface: atmospheric layering, winds, temperature and humidity, vertical atmospheric motions, cosmic radiation, polarization of light, atmospheric electricity, atmospheric particulates, cloud types, and photographic documentation of atmospheric phenomena of all kinds. The emphasis was definitely not on development of new theory; it was on testing theories by observation and data collection, and on the refinement of observation by improved instrument design and employment.

Discovery of the Stratosphere

Richard Aßmann himself was not a theorist, nor did he intend to be; he had made his reputation as an organizer, observer, editor, and instrument designer. He did, however, make observations of tremendous theoretical significance. One of these had been the determination in November 1884 that clouds are made up of microscopic droplets (not bubbles) and that these droplets do not have dust particles as condensation nuclei. He had determined this on a mountain outside Berlin, the Brocken, where for several days he had worked in a cloud layer about 50 meters (164 feet) thick just below the summit, with a microscope of about 200 times magnification, collecting the droplets on glass slides, studying them under oblique illumination, and measuring them with a micrometer. This series of several hundred careful measurements answered unambiguously a question that had been in the meteorological literature for many years and in so doing inaugurated the study of the microphysics of clouds.[13]

Aßmann's other great triumph (and also his greatest near miss) was his role in the discovery of the stratosphere. This event was of immense significance for meteorology and inspiring to Wegener, who just missed taking part in it: Aßmann in Berlin and Teisserenc de Bort in France performed the work, between 1900 and 1903. Much of Wegener's own work in the next decade was an outfall of this discovery, which dominated the research program at Lindenberg during his year there in 1905–1906 and for years thereafter. The work leading up to this remarkable finding, as well as the problems it created, led the Russian historian of meteorology A. Kh. Khrgian to declare, "No other chapter in the history of meteorology contains as much of the element of drama as the chapter we are coming to now. Heroic, very dangerous flights, difficult observations, remarkable and unexpected discoveries . . . and the constant attention of the press and general public, were all part of the period."[14]

The discovery of the stratosphere is one of the most influential discoveries in modern physics, as well as one of the most overlooked. It ranks with the theory of radioactive transmutation as an example of fastidious, delicate experiments, many times repeated, leading a scientist (or scientists) to reject a universally accepted theoretical principle in favor of an empirical result. In the case of radioactivity, the theoretical and laboratory work of Ernest Rutherford (1871–1937) and Frederick Soddy (1877–1956) on the elements uranium and thorium led them to the unbelievable but inescapable conclusion that they were witnessing the active transmutation of one element into another, thereby disproving one of the most fundamental principles of physics: the immutability of elemental species of matter. This led within a few years to the discovery that elements, rather than distinct substances, were simply different ways of organizing the real fundamental stuff of the universe—protons and electrons.[15] Within three years it also led to a means of determining the age of a rock by comparison of the ratios of uranium and lead they contained, understanding the lead to be a decay product of uranium.[16] This revolutionized both geology and evolutionary biology almost instantaneously, showing that Earth was billions, not millions, of years old, and that there was, after all, sufficient time for even the slowest and most gradual sorts of changes to shape the world into the form in which we see it.

The discovery of the stratosphere, occurring the same year as the work of Rutherford and Soddy (1902), was epochal and far-reaching in its own way, leading to profound changes in meteorology, oceanography, and geophysics, all within a decade. Aßmann had a role in the discovery, but it was not in him to take the leap of judgment or faith that would have let him enjoy sole credit for uncovering this aspect of the natural world. That distinction belongs to Teisserenc de Bort, working from his private laboratory near Paris.

Teisserenc de Bort, like Aßmann and Hugo Hergesell, designed and built meteorological recording instruments and sent them aloft on balloons and kites, also of his own design or modification. On 9 September 1899, he had come briefly into the public eye when one of his kite flights—a large kite carrying a meteorograph and ten other "helping-kites" supporting the tethering cable—broke free and, trailing 7 kilometers (4 miles) of cable, cut a narrow but astonishing swath through Paris, where it stopped traffic, disabled a train, and then "cut off all telegraphic communication with Rennes on the day when all France and most of the rest of the world were anxiously awaiting the result of the famous Dreyfus court-martial at that city."[17]

Under the circumstances, it seemed wise to Teisserenc de Bort to suspend kite flying for a while and return to a series of upper-air experiments he had begun in 1898 with free balloons. He used balloons of several designs but preferred working with those made of kerosened paper. These were filled with hydrogen and trailed a (very short) cable holding the meteorograph and a bag of sand ballast, which dribbled out at a steady rate to control the balloon's rate of ascent. Teisserenc de Bort's aim was to get these all up as high as possible—often over 11 kilometers (7 miles), and occasionally as high as 14 kilometers (9 miles). When the balloon reached its maximum altitude, the instrument pack was parachuted back to earth.

When Teisserenc de Bort looked at the temperature records on flights that reached altitudes of 10 kilometers (6 miles) or greater, he noted that the recorded air temperatures failed to decrease with altitude, though theory predicted that they should. The measurement of the lapse rate of temperature with altitude, performed by James Glaisher (1809–1903) in the 1860s and corrected by Aßmann (with much fanfare) some years later, showed that the temperature should decrease by about 6°C (11°F) with each kilometer of ascent. If one followed the assumption of William Thompson (1824–1907) that the atmosphere was in convective equilibrium (and everyone did), one expected the temperature to decrease indefinitely with altitude until it approached the zero of temperature on the Kelvin scale (−273°C [−460°F]). In Teisserenc de Bort's records, however, the temperature showed no such decrease between 10 and 14 kilometers (6 and 9 miles).

Aßmann had measured the effect several times between 1894 and 1897 in manned balloon ascents, but he was reasonably sure that it was due to the warming of the instrument packet by solar radiation, and he was delighted to treat the phenomenon as another problem in instrument design—leading to an improved, ventilated housing for his thermograph. Hergesell agreed that the phenomenon was probably due to instrument warming, and so did Teisserenc de Bort; all of them tried various ways to shield their thermometers from reflecting or absorbing surfaces on the instrument packets that contained them, in order to keep the thermometers from false readings.[18]

Teisserenc de Bort took the investigation further. He solved all the instrumental problems he could think of. No longer could the strange temperature recordings be

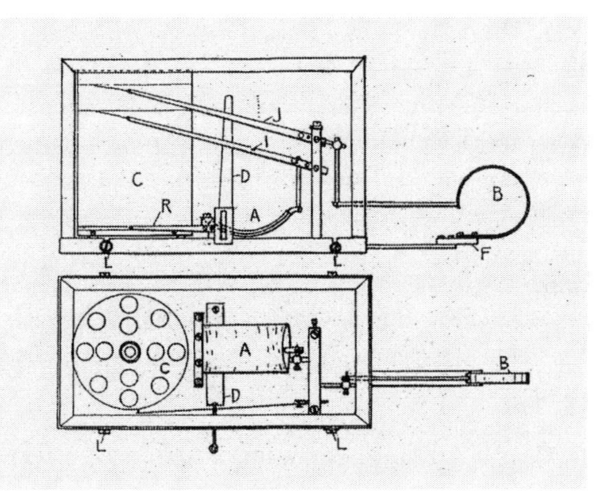

Drawing of Teisserenc de Bort's modified balloon meteorograph. The thermometer (B) was composed of two dissimilar metals that expanded differently with changes in temperature, attached to a stylus (J) that recorded data directly by scratching the smoked paper on the revolving drum. Air pressure was measured by the expansion and contraction of the "Bourdon tube" (A), attached to its own scribing apparatus (I). Teisserenc de Bort had moved the thermometer outside the housing to eliminate the possibility of heating within the box of the apparatus.

blamed on frozen ink in the recording pens—his pens scribed directly onto smoked paper without the use of ink. Neither could the thermometer reading be blamed on warming by reradiation from the thermograph itself, since he had boxed the thermograph with cork and moved his thermometer outside the housing: he had never attributed the effect to direct solar radiation, but to reradiation from the instruments, from Earth itself, or from clouds. The thermometer fluid could not be a problem because there was none—his thermometer was a curved spring of two strips of dissimilar metals that expanded and contracted at different rates, moving the recording pen directly. Aware of the possibility of a radiation effect, he began to fly balloons at night. When these night flights also showed a steady temperature at around 11 kilometers (7 miles), he began a systematic program to see whether there was a seasonal variation—the "radiation effect" should be smaller in the winter than the summer if this were the answer to the subtle but persistent temperature anomalies.[19]

The problem was difficult because the effect was elusive—the zone of steady temperature moved up and down between 8 and 12 kilometers (5 and 7 miles), though it was most often at 11 kilometers (6.8 miles), and it was stronger and weaker over a range of many degrees. It was clear that the predicted lapse rate was not consistent in what he came to call the "isothermal zone." By 1900 he had the record of 146 ascents to report to the Académie des Sciences in Paris, but he still postponed discussion or conclusions concerning his ample measurements of temperatures at altitudes above 10 kilometers (6 miles), unsure as to the character of what he was seeing.[20]

He worked for two more years on the problem, and in 1902, with an accumulation of 236 ascents to buttress his argument, he made the plunge and publicly rejected the prevailing theory of temperature decrease with altitude. He asserted the existence of an isothermal zone of varying thickness, in which the adiabatic lapse rate (of temperature with altitude) diminished to zero, usually at an altitude of 11 kilometers (7 miles), after which the temperature would remain constant for several kilometers. The layer was higher over anticyclones (high-pressure centers) and lower over cyclones (low-pressure centers). In consequence, he suggested, the problem of the general circulation of the atmosphere would have to be reconsidered, since such an isothermal layer meant that the atmosphere as a whole was not in convective equilibrium, but only that part bounded above by this "isothermal zone."[21]

It was an astounding and transforming discovery, though it remained far from the public eye. Aßmann realized immediately what he had missed and rushed to claim a share of it for himself, based on the analysis of six balloon ascents at Berlin in 1901; if he could not get ahead of the discovery, he could at least show that he had the relevant data himself even before Teisserenc de Bort's announcement. He endorsed Teisserenc de Bort's results but modified them by asserting that his own superior rubber-balloon technology, with a controlled rate of ascent and more sensitive recorders (both true claims), could prove that this was not simply an isothermal zone but a true "inversion zone": the temperature not only stopped decreasing in this region but actually increased for several kilometers before decreasing again. Moreover, he thought he could detect (he was right) an upper boundary to the region of temperature shift in a layer at about 15 kilometers (9 miles), meaning that there was both a "lower inversion" and an "upper inversion."[22]

The discovery that the atmosphere was distinctly layered, with a permanent boundary layer near 11 kilometers (7 miles), was momentous. It required an entirely new approach to the study of atmospheric circulation and the energetics of storms, as well as a new (but by no means well-understood) picture of the vertical structure of the atmosphere. In the new picture there was, between the surface of Earth and the altitude where the lapse of temperature diminished to zero, a shell of turbulent air, soon named the "troposphere." The term was Teisserenc de Bort's, coined in 1908, and pleasing to the classically trained ear. It was the "sphere of change," of rising and descending air, of precipitation and clouds; it was the zone of weather. Its upper boundary was a mobile surface, later named the "tropopause" (this was Aßmann's "upper inversion"), and above that boundary surface was the "stratosphere," a zone of stable or rising temperature without clouds, moisture, turbulence, or convective mixing and characterized by a laminar (flat and circumglobal) airflow: the atmosphere's weather zone had thus a ceiling as well as a floor. Storm systems might now be seen as coherent masses of air trapped between the defining boundaries of Earth's surface below and the tropopause above; that the latter was invisible made it no less real than the former—it was a true barrier and a boundary surface.

These events provided for Wegener and others of his generation an unparalleled opportunity. It meant that the atmosphere had to be "remapped" from pole to equator and through its full vertical height. No one knew how the tropopause changed with latitude, or how it differed over land and over the oceans, or even why it was there: was it a matter of temperature and pressure alone, or was it the result of a chemical differentiation of the atmosphere with altitude? A number of major theoretical predictions had been overthrown, and the real structure and dynamics of the atmosphere would now be decided by intensive field research—most of it in distant (polar or equatorial) locations. The bounded layers—no one knew how many—in the upper reaches of the atmosphere might even hold the key to the motion of storms and their sources of energy (the grail of nineteenth-century meteorology). In any case, meteorology now had a goal in which weather prediction formed only a subsidiary part: this was the development of a comprehensive physics of the atmosphere.

Wegener's Scientific Work at Lindenberg

For Wegener, the discovery of the stratosphere was a momentous change—something big enough and far-reaching enough to make it seem that he had come into the science just as it was reaching its "modern form"—not its culmination, the sad fate of the planetary astronomy of his day, but that moment in the life of meteorology when it had moved past some threshold, some barrier, beyond which lay an ample freedom to change the overall structure of the science. Most scientists can point to such an event in their own training and early work, something that galvanized their community, generated new research opportunities, opened new vistas.

So this was the deed demanded of Wegener, his scientific task, and the riddle for him to undo—to understand the layered structure of the atmosphere, and to understand the physics that made it and moved it. This was why he had come to Lindenberg, where, in Aßmann's words, "the tasks of the observatory are primarily scientific, and directed to the investigation of the physical relationships and processes in the upper layers of the atmosphere."[23] Not upper *reaches*, but upper *layers*.

Alfred and Kurt set to work on this task immediately after they took up their positions. At Kurt's request, he was permitted to occupy and use the mountain observatory at the summit of the Brocken, outside Berlin, from 1 January until 8 February 1905 to carry out a program of kite ascents, as well as to examine the appearance and behavior of the upper surface of the cloud layers. Since the cloud layer that blanketed Berlin in January was usually lower than the summit of the Brocken, he should be able to observe motions and patterns in the upper surface of the overcast. While Kurt worked there, Alfred would take part in the daily round of observations at Lindenberg.

The problem that Alfred and Kurt had in mind—hazily at first—was based on an often-repeated opinion of Bezold's: that one had to think of the upper surface of the cloud layer at any given time as a sort of "second surface of the earth" and think of the atmosphere between Earth's surface and the top of the clouds as a region with its own dynamics.[24] The discovery of the boundary layer at 11 kilometers (7 miles) intensified the search for dynamic layers elsewhere in the atmosphere, and cloud layers were the most visible evidence of such layering. The bottoms of clouds are typically flat, as they are the thermal signatures of the altitude(s) at which the atmosphere becomes saturated with moisture and the excess condenses into clouds. The upper surfaces of clouds may be similarly flat but are often billowy—reflecting the presence of vertical (convective) motions in the atmosphere. Sometimes, however, the upper surfaces of extensive cloud layers (and, less frequently, their lower surfaces) show a regular up-and-down billowing, resulting in long rows of peaks and troughs that look like frozen waves. They are indeed waves—and in Wegener's time they were known as *Helmholtzschen Luftwogen*, "Helmholtz air waves."

Their interest in the problem of air billows, or Helmholtz air waves, or, as they are known today in the English-speaking world, "Kelvin-Helmholtz instabilities," grew out of their quotidian observational work for Lindenberg Observatory in the first months there. The Wegener brothers were responsible for working up, out of the records of the individual balloon and kite flights, a monthly summary and map of the temperatures at various altitudes above Lindenberg. These appeared in the meteorological journal *Das Wetter*, edited and controlled by Aßmann. Alfred and Kurt were allowed to publish the data for April 1905 under their own names—their first real scientific publication.[25] In working up the data from individual kite flights, Alfred occasionally noted (as had his colleagues) a pattern of slight temperature oscillations in the record of kites flying on a fixed length of cable, and he wondered about their cause.[26] He asked the other observers and assistants about the phenomenon and collected their comments.

The course of his investigation of these temperature oscillations followed a pattern often repeated later in his career, when a question brought forth by a small but interesting anomaly led to another, and then another, and soon the investigation of some phenomenon was pulling him along willy-nilly into a deep and broad search, though his original idea had been a mere technical note. In his own words, "I originally only had the idea of bringing together and drawing attention to the observations of my aeronautical colleagues (to whom the phenomenon had appeared) on the cause of these small temperature oscillations; in working them up, however, I stumbled on a number of issues that made a wider investigation necessary, and their discussion did not seem to be completely without interest; consequently the investigations have far outgrown my original plan."[27]

As he studied the problem of the temperature swings, the daily work of the observatory continued. He became an expert in the rigging and flying of kites and balloons, with their special lashings, knots, clips, and cable ends, and in the daily rhythm of observing and recording meteorological data—calibrating the thermographs against the station records, keeping the logbook in the *Windenhaus*, and standing watches day and night while the instruments were aloft. It really was very like being at sea, an impression not lessened by the rippling of the surrounding wheat and barley as the prevailing northwesterly wind passed across the plain, or by the creak of the cable against the winch drum, or, as often as not, the drumming of rain on the roof of the *Windenhaus*—in this and many respects like the wheelhouse of a merchant steamer.

On this "expedition," as on any other, the interruptions, snafus, and assorted emergencies provided much of the interest and excitement. Kites snapped their cables and sped off and away with alarming frequency, occasionally cutting telegraph lines, but more commonly making a mess of farm fences. These truants had to be pursued, the cables and equipment salvaged, and the instruments recovered. Once, during a kite flight on which Alfred was standing watch, a lightning strike incinerated the kite aloft and sent 9 kilometers (6 miles) of wire into an incandescent meltdown, creating a beautifully spiraling smoke trail across the sky and causing an impressive brush fire when the incinerated apparatus crashed to the ground. It was a harrowing escape at the tether end: had the cable not parted aloft, he might easily have been electrocuted, becoming not a subject for a biography but a grim footnote to Ben Franklin's injunctions about flying kites in bad weather without a lightning rod. Thereafter, on days with a danger of lightning, the scientist rigging the kites and clipping on the "helper-kites" (as needed) stood inside yet another Aßmann invention—the *Blitzschutzbügel*—a rectangular lightning-rod enclosure attached and grounded to the *Windenhaus*.[28]

Aßmann liked Alfred, not least because he found in him a kindred spirit in the matter of instrument design and tinkering. From the time he arrived at Lindenberg, Alfred gravitated more and more toward work that involved the theodolites used to track the free balloons. He had enjoyed his work in navigation and position finding with Marcuse at Berlin, and his doctoral thesis had led him to study the forerunners of these astronomical instruments. He had become proficient in both plane and geodetic surveying, and in the latter he had employed zenith telescopes (a form of astronomical theodolite) to locate the terminal points of the geodetic baselines, by fixing

Only surviving photograph of Wegener at Lindenberg, here rigging a kite tether from within a *Blitzschutzbügel*, a rectangular lightning rod designed to prevent electrocution in a lightning strike to a kite (a common occurrence). Note the "Hargrave" box kite on the ground behind him: Wegener is preparing to send it aloft. From Aßmann, *Das Königlich Preußische Aeronautische Observatorium Lindenberg.*

the positions of stars directly overhead. These were extremely fine measurements, and the accuracy of an entire triangulation network depended on the accuracy with which the baselines were anchored. Alfred was familiar with these instruments and their design, but he was pleased that the work at Lindenberg offered a different set of design problems—aimed at instruments that could be swiveled easily and read off rapidly; he spent much time in studying ways to improve the combination of speed and accuracy for these specialized tools.

Aßmann put Alfred to work testing new instruments as they came to the station, and one of these was a theodolite for following free balloons, designed by Swiss meteorologist and geophysicist Alfred de Quervain (1879–1927). De Quervain, though only a year older than Wegener, had already been on a polar expedition, participating in Drygalski's "Gauss" Expedition in 1901–1903 and making meteorological observations on the island of Kerguelen in the far South Atlantic.[29] Wegener's review of de Quervain's instrument is a nice demonstration of how tinkering and experimentation can go on while still providing a scientific result, and it gives some of the flavor of daily life at Lindenberg and of Wegener's earnest enthusiasm for his work.

The test took place on a beautiful, clear January morning, with a light surface wind from the south, a temperature of −8°C (18°F), and unlimited visibility—perfect conditions. The morning kite flight was already aloft and registering when the balloon was released at 8:00 a.m. from the *Windenhaus*. Wegener manned the theodolite, while his assistant called out at thirty-second intervals for him to mark the altitude and azimuth. To accomplish this, he had to take his eye from the telescope and read the vernier scales, but he succeeded in keeping track of the balloon—mostly. At an altitude just over 1,000 meters (3,281 feet), the balloon, which had been traveling almost due north, suddenly veered 90° to the east "and disappeared, behind the Windenhaus, from the view of the completely unprepared observer [himself]. The theodolite had to be set up again, but we were lucky to find the balloon and thereafter the observations could be carried on without further disruption."[30] He followed the balloon for eighty-five minutes in its ascent and descent, collecting 134 observations: a few were lost in moving the theodolite, and for most of the last kilometer of descent the parachute on the packet of instruments was invisible—though Wegener guessed its trajectory and caught it again 150 meters (492 feet) above the ground to fix the landing site.[31]

The flight aimed to test the theodolite, but also to compare two thermographs aloft simultaneously on a single balloon—one of these was a modified design of Hergesell's, the other an instrument built by Teisserenc de Bort. Moreover, Aßmann had decided that they might as well experiment with a new method of plotting balloon tracks (de Quervain's), in which the balloon's altitude was plotted along a vertical axis and the time along a horizontal axis, the latter also scaled in kilometers. One could in this way get a picture of the flight path in both its vertical and horizontal dimensions in a single view and, from the plot and the distance between the points, calculate the wind velocity (as well as the direction) at every altitude. The test chart also featured a compass rosette centered on the *Windenhaus*, so that one could read the direction of flight readily. This sort of planning and economy of effort, this packing in of information and results, was typical of Aßmann, who had ample resources but stretched them to the limit at all times—a valuable lesson for an apprentice aerologist to learn.

Wegener got a good result out of this flight and was pleased. After recovering the thermographs (the balloon landed 20 kilometers [12 miles] away), he determined that there was an overnight inversion (pooling of cold air near the ground) of almost 10° at an altitude of 1 kilometer (0.6 miles). Even more exciting was his confirmation of the passage of the balloon into the permanent inversion zone. Just above 7 kilometers (4 miles) the balloon, until then traveling steadily northeast for some time, suddenly began to wander in a circle and slow its velocity, and it continued a 360° turn during its passage through an altitude of 10 kilometers (6 miles). After that, it stabilized and continued in a new direction at a lower speed. Finally, after reaching a maximum altitude of almost 11.5 kilometers (7 miles), an hour into the flight, the balloon deflated and the parachute opened. Wegener was able to track the parachute descent and observe the same 360° turn on the way down, later verifying—on both thermographs—a true inversion of 1°C (33.8°F) on the Hergesell instrument and 1.5°C (34.7°F) on Teisserenc de Bort's, occurring in the last 500 meters (1,640 feet) of the ascent. This was, Wegener wrote, "clearly not explainable as a radiation-effect, rather, it was as it appeared."[32]

Wegener's comments confirming the presence of the inversion were of course directed to Aßmann, who was ever the confirmer more than the discoverer. The turbulent eddies just below the inversion zone, combined with a marked decrease in wind velocity, ought to have interested Aßmann, but they did not: his desire to enhance his supporting role in the discovery of the stratosphere was leading him away from another discovery—the turbulent eddies witnessed by Wegener were evidence of the lower surface of the tropopause at 7 kilometers (4 miles), not unheard of for a deep barometric depression in wintertime in temperate latitudes. The juxtaposition of low- and high-pressure areas (a deep low to the northwest and a high to the southwest) on that day indicates that the jet stream (as yet undiscovered) was probably nearby and the tropo-

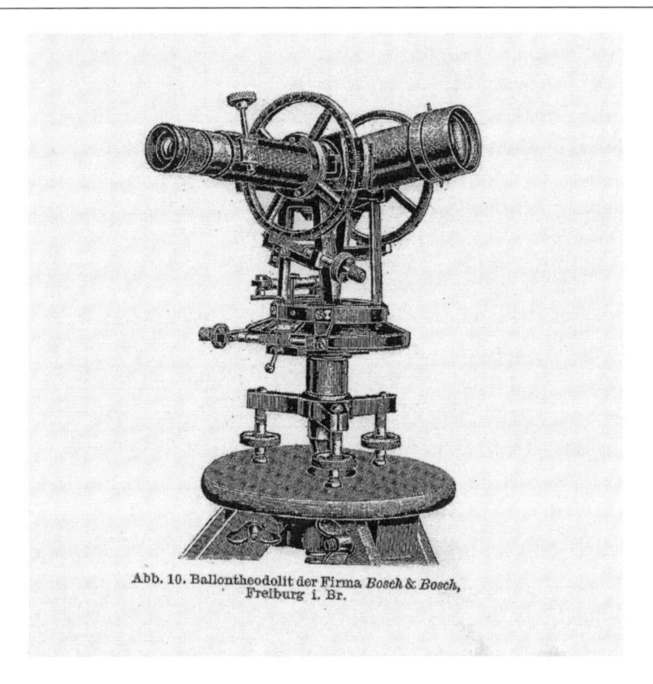

De Quervain balloon theodolite that Wegener tested and reviewed at Lindenberg. From Alfred Wegener, "Über die Flugbahn des am 4. Januar 1906 in Lindenberg aufgestiegenen Registrierballons," *Beiträge zur Physik der freien Atmosphäre* 2 (1906): 30–34.

pause complexly layered. Wegener saw the unusual structure (the eddies) and could relate it to the distribution of pressure across the middle of Europe.[33] What made the eddies especially interesting to Wegener was that he had seen them in clear air, rather than in cloud formations. He was, however, an apprentice, writing to his superior within his superior's scheme of things. A deeper look could be reserved for another time.

Wegener seems to have really liked the design of the de Quervain theodolite.[34] Aßmann was pleased with the work and gave Wegener permission to write it up for publication: this was his first solo scientific paper in a refereed journal, and the official beginning of his scientific career and the possibility of a scientific reputation. The journal was *Beiträge zur Physik der freien Atmosphäre,* founded the year before and edited by Aßmann and Hergesell as a venue to draw attention to the physics of the "free atmosphere," the region beginning a few hundred meters above the surface of Earth. Aßmann was anxious to make a place for atmospheric physics separate from that of descriptive meteorology—a subject more suitable to *Das Wetter* (The weather), the journal Aßmann had edited since 1884.

In October 1905, there came a milestone of sorts: it was Alfred's first October in twenty years without being registered somewhere as a student. Still, with the fall, after a summer of vigorous activity, he had a desire to do some real physics, in addition to his observational work. Anyone close to academic life feels a tug at this time of year: in the university and scientific world it is much more the New Year than the first of January. The stimulus of fall weather was reinforced for Alfred and Kurt in midmonth by the arrival at Lindenberg of Arthur Coym (b. 1875), transferring from the Meteorological Institute in Berlin, as the aerological operations there were finally shut down in favor of Lindenberg.[35]

Their new colleague was a Berlin physicist turned meteorologist (PhD in 1903) and had been helping Bezold (by then seriously ill) finish editing his collected scientific papers for publication; Coym was still editing these and reading proof in Lindenberg during the fall of 1905.[36] Coym had a sweeping view of what needed doing in meteorology, and Kurt and Alfred welcomed his company and conversation and evidently benefited from it as well.[37] The three of them were the only scientists on staff other than Berson and Aßmann, who were much their seniors. With Aßmann as the captain and Berson as executive officer, the three were more or less the lieutenants of Lindenberg, and they lived, worked, and stood watch together.

There was one difference, subtle but fateful, between Coym and the Wegener brothers: Coym arrived with the rank of *ständiger Mitarbeiter* (permanent staff member), whereas they were only *technischer Hilfsarbeiter* (technical assistants); thus, he outranked them. With Coym's arrival, Berson could be promoted to *Observator*, and this compact shifting of chairs foreclosed permanent, salaried, and pensioned employment for either Kurt or Alfred at Lindenberg, as long as Coym wanted to stay: there was room for only one *Mitarbeiter*, and it was to be Coym. The arrival of Coym was perhaps more a blow to Kurt than to Alfred. Alfred had elected a path leading to academic employment early on; Kurt had rejected it in favor of a technical degree likely to lead to a career as a government staff scientist somewhere, whereas Alfred's degree had prepared him for immediate academic employment, were he to seek it.

It was clearly time for both the Wegeners to look to their resumes. They both had to get out as many publications as possible, and to get as much practice in every part of aerology as they could, if they were to have a chance at jobs where they could pursue meteorology. One area of expertise they could develop, by far the most exciting and exhilarating, was flying aloft themselves—in the great observation balloons made available for meteorological research to Aßmann and his staff by the German army. Flying anything at all was a rare and dangerous skill and immensely popular with the public; manned flight was as miraculous and exotic then as is manned spaceflight today: in 1905 balloon and zeppelin pilots had something of the status of astronauts. That may have been part of the allure for Kurt and Alfred; they were, after all, experienced and devoted sailors and athletic outdoorsmen who enjoyed alpine mountaineering, and they were not afraid to take risks. Ballooning was the perfect job for them—a complete fusion of science and sport.

Ballooning

The opportunity to go ballooning had been from the beginning one of the main attractions that brought Alfred and Kurt Wegener to Lindenberg. Bezold had been for many years a great exponent and urged his students to enroll in the Luftschiffahrtsverein (Aeronautical Society). Berlin flattered itself a world center for aviation under the lavish patronage of the kaiser and the army, and Kurt and Alfred had, many times in their childhood, looked up to see the great balloons soaring over the city.

In 1905 heavier-than-air flight was only two years old, and the maximum times aloft for these primitive and rickety craft were still measured in minutes; the altitudes were negligible, often only 10 or 20 meters (33–66 feet) above the ground. Ballooning, on the other hand, offered time aloft measured in hours, altitude measured in kilometers, and the prospect of serious scientific work in flight. Airplanes were loud, were cramped, and required constant use of both hands and both feet to steer and propel; they vibrated madly and were in every way unsuitable as research platforms. Balloons were nearly perfect platforms for studying the vertical structure of the atmosphere: they were silent, they went easily and directly up and down, and they were able to hover as conditions permitted; they were roomy enough to move around in, were made of nonconducting materials, and could be flown "no hands." They were also tremendously exciting to fly in. The only drawback (besides the danger of the flights themselves) was the immense expense involved.

Wegener had his first flying lesson with a master of the craft: Arthur Berson, Aßmann's and Bezold's scientific coworker since 1890 at the Meteorological Institute in Berlin. Berson was now Aßmann's top aide at Lindenberg and world-famous for his exploits as a balloonist, though he was also professionally distinguished as a meteorologist. In 1903 he and Aßmann had shared the Buys-Ballot Medal of the Amsterdam Academy of Sciences—an award given once every ten years to the "individual(s) giving greatest service to meteorology in the preceding decade."[38] He had flown more than 100 times and was a public hero in Germany for having taken a balloon up more than 10,000 meters (32,808 feet). He was certainly the most celebrated scientific aeronaut since James Glaisher, the British meteorologist whose dramatic, life-threatening balloon ascents in the 1860s had established the rate at which a rising parcel of air cools with increasing altitude—the so-called adiabatic lapse rate. Berson was forty-six years old and near the peak of his career as a scientist when he took Wegener aloft for the first time.

Berson and Wegener got an early start on 11 May 1905, lifting off just after 8:30 a.m. from the aerodrome at Reinickendorf (in Berlin) in a hydrogen-filled balloon. Berson was interested (on this particular flight) in measuring the electrical conductivity of the atmosphere and its change with altitude; this was part of an ongoing research program he shared with Aßmann, and Wegener helped with these measurements. Alfred's real job, however, was to be the navigator and determine their geographic position at regular intervals, as well as keeping the journal and handling the ballast.

On this flight Wegener was able to find their position with an error of 10–15 kilometers (6–9 miles).[39] To anyone who knows celestial navigation with a sextant, that sounds dreadfully inept—and it sounds even worse in an age of global positioning systems and satellite navigation, in which a modestly priced handheld instrument can give a reliable and instantaneously corrected position within a few meters, anywhere on Earth. At the time of these first attempts by Wegener, an aeronaut could give only an intuitive guess at his position by observing the direction of flight and airspeed, or by recognizing landmarks below. Moving laterally at several meters a second and changing altitude almost constantly, and knowing one's altitude by reference to atmospheric pressure alone, put much uncertainty into a measurement in which (at sea) one corrected not only for the height of the deck of the ship above the sea but for one's own height above the deck to avoid errors of hundreds of meters. From this vantage it was not such a bad first approximation, and it showed Wegener immediately how a sextant or theodolite had to be modified to make more accurate determinations possible. Wegener and Berson flew for ten hours, got up over 5,500 meters (18,045 feet), and made a perfectly controlled landing in Gleiwitz in East Prussia.

Wegener's second balloon flight, which took place about three months later on 30 August 1905, was a somewhat different story. The plan was to observe a partial eclipse of the Sun and, from the faintly visible stars, provide an opportunity for Wegener to practice astronomical position finding while aloft. The balloon's pilot was Hans Gerdien (1877–1951), an expert on electrical phenomena in the atmosphere, working on the same sort of measurements that had occupied Berson and Wegener in May.[40]

The two aeronauts cast off at 10:30 a.m. and rose rapidly to an altitude of 1,000 meters (3,281 feet), where they encountered a cloud layer that drenched their balloon, causing it to descend precipitously to about 500 meters (1,640 feet) and forcing them to drop some ballast. They then rose again rapidly to between 1,300 and 1,400 meters (4,265 and 4,593 feet). "There we floated," wrote Wegener, "between two cloud layers, and we could again see the sun, so that at noon we were able to measure its altitude and investigate the eclipse as it progressed. The truly remarkable reduction of the sun's brightness was just right for our mission, but it also made a strong emotional impression."[41]

The clouds were so beautiful that it was hard to concentrate on the scientific work. When they finally dumped enough ballast to rise above the overcast layer, the vista was stunning: a sea of brilliant white clouds with dark, almost black plumes convecting rapidly upward, making a fantastic landscape. They could hear thunder from every direction, and eventually a huge thunderhead—black as iron—built up near them: Wegener said that it was like riding at anchor next to a towering island. Now there was a brief respite, giving time to work with the instruments as the balloon continued its ascent. Gerdien opened the valve on the gas cylinder to inflate the balloon further, and a little after 3:00 p.m. they reached an altitude of just over 6 kilometers (4 miles). It was cold—almost −24°C (−11°F), and as soon as the gas in the cylinder was exhausted, the balloon began to sink.

While they were still packing up their instruments, the balloon descended into the cloud layer and, soaked with water once again, began to fall rapidly. Gerdien jettisoned two sacks of ballast, but this did not entirely control the descent. "We were completely out of the clouds and could already see the earth when I lowered the anchor," Wegener reported. "We were barely finished packing up the instruments when we were already dangerously close to the ground." The winds near the surface were brisk—about 15 meters per second (25 miles per hour)—and the balloon was moving very fast across the plain below. They knew they were in for a rough landing, as the means of stopping the balloon consisted entirely in tossing the anchor overboard and waiting for it to catch something and jerk the balloon basket to a halt. At that second the crew was to release the remaining ballast to keep the balloon upright and keep the basket from smashing into the ground—too rapidly. Just before the basket hit the ground, one "pulled up" into the rigging—holding fast to the ropes overhead and lifting one's feet off the floor of the basket (schoolyard gymnastics to the rescue) to keep from absorbing the full impact.

Seconds after tossing the anchor, Wegener felt the basket tip and realized that the anchor must have caught; he let go the remaining ballast. Suddenly, there was a tearing sound and the balloon picked up speed again—the anchor rope had torn completely away! "Gerdien, in this critical moment, did exactly the right thing," remembered Wegener. "He pulled the ripcord and deflated the balloon, and then shouted: 'Pull up!' and then came the impact. All I could think about was holding on, and I saw, in a flash, everything go topsy-turvy, and realized that I was being dragged along the ground—then I felt a sharp pull on my body, and saw that my left foot was tangled, and then my left boot pulled off and—I was free. . . . My stiff collar dug itself into the earth like a plow and my head was covered with dirt."

He got unsteadily to his feet and went looking for his boot, and he was delighted to find both it and Gerdien, who had a worse landing—he had hit his knee very hard and wrenched his left shoulder. Searching the area around, Wegener managed to find his house keys and eventually his hat as well. Limping to a nearby farmstead, he and

Gerdien learned that they had crashed near the village of Novy Miastov, almost 500 kilometers (310 miles) to the east of their home base, and well inside the borders of Russian Poland. They spent the night quartered in the "disgusting and filthy" village school.

It took three days to get out of Poland. They packed up the balloon after their sleepless night in the school building, and it was dark before they set out under escort for the nearest Russian military outpost. The road was terrible, and their cart driver, unable to see anything in the gloom, hit a deep rut and pitched them into a roadside pond. Drenched, muddy, and sore (Gerdien had landed on his bad shoulder), they eventually found their instruments and their day packs, containing their only clean clothes, in the deepest part of the pond. At the Russian fort there were endless formalities ("of which we understood not a word"), and they were finally conducted to a room and allowed to sleep. On the next day they were examined by the post surgeon (at their request), and it appeared that the Russians were finally beginning to understand who they were. The surgeon took them to lunch, and that night the colonel threw a full Russian banquet in their honor with (as Wegener ruefully noted afterward) "the obligatory overabundance of hard liquor." On the following day, transit visas arrived from Warsaw, and they were allowed to return to Germany.[42]

Wegener was allowed to write up both the scientific results and the adventures for the *Illustrated Aeronautical News*, a national magazine devoted to all aspects of aviation and avidly read throughout Germany. It must have been a thrill to have not just his scientific work but his "exploits" featured in a magazine of which he had been, until then, merely an enthusiastic reader. Such exploits and the chance for them were few and far between—as he and Kurt had almost complete responsibility for the daily kite and balloon ascents from the *Windenhaus*, working on alternate days. With Coym's arrival, manned ballooning opportunities became even scarcer.

Changing Course

By the end of the summer, Alfred had come to see what sort of place Lindenberg was, and what sort of scientist Aßmann. Whether or not the latter was afflicted with *Bauwut* (building mania), as Knowles Middleton suggested, his career shows that the best is indeed the enemy of the good.[43] While Teisserenc de Bort had moved ahead to discover the stratosphere with a controlled series of ascents using equipment of fixed design that he knew well, Aßmann was trapped in an endless sequence of instrumental prototypes. This impetus to push instrumentation to the "edge of resolution" is indeed part of science, and in this case it shows that Aßmann understood his legacy (instruments) as well as Teisserenc de Bort understood his (discoveries about the atmosphere). It was, however, clear to Alfred and Kurt and to everyone else at the station, except perhaps Aßmann himself, that Lindenberg was not going to be a place where major new discoveries would be made. It would be a place where they were elaborated and confirmed, and their underlying evidence more precisely measured, but it was not a place where anyone was likely to see something not seen before.

Toward the end of October 1905, and shortly before his twenty-fifth birthday, Alfred read a newspaper feature story about a Danish polar explorer, Ludwig Mylius-Erichsen, who was preparing to mount a two-year expedition (1906–1908) to the far northeast of Greenland, to map the last uncharted section of Greenland's coast and carry out an extensive scientific program. The account given was of a great public meeting in the Copenhagen Concert Hall on 17 October 1905. Mylius-Erichsen, just back from a two-year expedition to the west coast of Greenland, spoke feelingly about the nobility of the quest and the importance of the opportunity—almost the last chance offered to any man or nation to fill in the final blanks on the map of Earth. The meeting in Copenhagen had made a big splash—on the podium, in addition to famous naval officers who had explored Greenland, were representatives from parliament, the university, scientific societies, and the press, as well as even Crown Prince Frederick (soon to be King Frederick VIII).[44]

The alacrity with which Alfred responded to this news is an index of his desire to depart from Lindenberg and also of the depth of his ambitions to take part in the exploration of Greenland—the vision he had shared with fellow students at Berlin. A part of the Mylius-Erichsen expedition plan, as described in the press, included the possibility of a four-man team crossing the ice cap well north of Nansen's route, from the east coast to the west—exactly what Alfred had told friends he wanted to do.[45]

On his birthday (an auspicious beginning), 1 November 1905, Alfred sat down to compose a letter of application. Not knowing how to contact Mylius-Erichsen, he wrote instead to Professor Adam Paulsen (1833–1907), a distinguished Danish meteorologist and director of the Danish Meteorological Institute, who had considerable experience in Greenland, dating back to the International Polar Year of 1882–1883. Alfred, taking great pains with his penmanship, sent off this letter to Paulsen:

> Distinguished Herr Professor
>
> I have read that Herr Erichsen will undertake a Greenland Expedition in the coming summer, and not knowing his address, I take the liberty of asking you to please tell me whether, and under what conditions I might still be able to become a member of the expedition.
>
> Since you do not know me, please allow me to acquaint you with the following particulars about me. I am 25 years old and since 1 January of this year a technical assistant at the Royal Aeronautical Observatory at Lindenberg, near Beeskow. I majored in astronomy in which I already have my doctoral exams out of the way, so that I have mastered enough astronomy to carry out the essential position-finding tasks of an expedition. (I have most recently occupied myself with astronomical position finding in balloon flight). My actual present occupation is that of meteorologist. I have already taken part in a number of scientific balloon ascents, and I have been entrusted at the Aeronautical Observatory with carrying out the daily kite ascents. In geology I'm only an amateur, to be sure, but I have pursued it sufficiently on my own that I believe I may be able to be of some use in this area. . . .
>
> I would also like to emphasize that I have been a keen alpinist, and together with my brother have carried out a large number of glacier ascents- unguided—in the Tirol. This past winter I have made my first attempts at skiing, on the Brocken.
>
> Should the expedition offer the possibility of making kite ascents, I hope to be able to assemble a scientific kit for this purpose, though this opportunity is not a condition of my taking part in the expedition.
>
> I ask you again to forgive me for bothering you about this matter. If it should be possible, through your mediation, that I should be able to take part in Herr Erichsen's expedition, and thereby make a wish long fostered in me, come true, I would be deeply in your debt.
>
> I am with deepest respect, sincerely yours,
>
> Dr. A. Wegener[46]

Paulsen did not respond to Wegener, but on 3 November he forwarded the letter to Mylius-Erichsen with a brief note—no advice or recommendation—merely saying that he had received a letter from a Dr. A. Wegener who hoped to take part in the expedition to Greenland.[47] Wegener could not know how indistinct the expedition plans actually were and how far from picking a scientific team Mylius-Erichsen was at this point. Wegener's phrasing in his letter to Paulsen, "whether I might *still* be able to take part," indicates that he thought that time was short and the expedition well under way; in reality, other than a paper plan, there was no expedition yet for him to join.

Mylius-Erichsen was, at the time of Wegener's letter, in a white heat of fund-raising following his public appeal. Jens Christensen (1856–1930), a leading officer in the Danish government, had promised that if Mylius-Erichsen could raise half the money necessary to fund the expedition, Christensen would see that parliament provided the other half. Mylius-Erichsen had to come up with 130,000 kroner (his share) in a very short space of time.[48] It is hardly surprising that Wegener's first letter got no response.[49]

Alfred waited two weeks, and when no response was forthcoming, he wrote again, this time to Mylius-Erichsen directly, assuming that his letter to Paulsen had gone astray. He sent the same letter he had sent to Paulsen, this time assuring Mylius-Erichsen that he would certainly provide a full scientific kit for atmospheric research, but again emphasizing that he would come on the expedition whether or not there was any need or desire for such research: he wanted to go, and he was willing to suspend his career to do so. This letter apparently brought no response either.[50]

He was not to be deterred. He wrote to Mylius-Erichsen again around 26 November, reminding Mylius-Erichsen that he had written "about two weeks ago" and had received no answer. "I can well imagine," Wegener wrote, "that you are overwhelmed with work, but could you get back to me as soon as possible?" Wegener explained to Mylius-Erichsen that he was in a bind: if he were to go on the expedition, he would have to give up his place at Lindenberg on 15 April 1906, which entailed giving notice by 1 January 1906; without such notice, he ran the risk of not being released to join the expedition, even if chosen. Moreover, he had to straighten out his military obligations, so that they would not hinder his participation in the expedition, and he needed some sort of answer from Mylius-Erichsen to accomplish that. "Under these circumstances," he wrote, "even a provisional and non-binding response would be of great value to me."[51]

Still hearing nothing from Mylius-Erichsen, Alfred could stand it no longer, and he obtained a brief leave to travel to Copenhagen, where he arrived on Sunday, 17 December, and checked into the Hotel Victoria. Disappointment awaited him here as well—in quantity. He called at Mylius-Erichsen's home the next day, only to be told that Mylius-Erichsen was in Norway until the following Friday. Not sure how to proceed, he then called on Paulsen, at the Meteorological Institute. Paulsen's news was bitter: Mylius-Erichsen was billing the expedition as an all-Danish affair, in order to assure a good response to the public appeal for funds, and in the hope of arousing patriotic sentiment and royal patronage. No non-Danes, whatever their suitability and experience, could currently be considered. Even though "Wegener" was a common Danish name, it would soon come out that Wegener was German, and there wasn't really much hope. Wegener returned to the Hotel Victoria and penned a disconsolate note to Mylius-Erichsen: "Prof. Paulsen has informed me that, as a matter of principle, only Danes will take part in your expedition. I must hereafter assume that no prospect remains for me to take part in the expedition. If you should, however, change your mind, I hope to hear from you as soon as possible."[52]

That was that. Alfred returned to Lindenberg and went back to work. When the New Year arrived, he did not give a notice of departure to Aßmann; there was no need. He would be at Lindenberg through the summer, at least. He could console himself that he was not the first man ever to be turned down for an expedition. In this case, he was a victim of politics and circumstance, and not left out by any fault or lack of his own, save that he had been born German and not Danish. He would have to bide his time by the ticking of Lindenberg's chronometers, while he waited for "another ship."

His hopes for a polar adventure dashed, he threw himself back into the scientific question he had begun to pursue in the fall, the cause of the temperature oscillations recorded on the kite thermographs. Throughout January and February 1906, he studied the daily records of the kite flights, looking for the pattern. While doing so, he began to reread and review the literature on the subject, most of it actually quite recent. The hunch he was following was that the temperature oscillations were the result of the kites riding on atmospheric waves of the kind predicted by Helmholtz. There was some solace here in a kind of work in which he excelled, as well as the chance to pursue a very interesting and potentially important problem.

The problem Wegener pursued concerned a phenomenon that we are all familiar with from airplane travel: that of "clear-air" turbulence, when the plane suddenly begins to buck and shudder, and the pilot turns on the seat-belt sign and announces that he will ascend, or descend, to find smoother air. The knowledge that moving the plane up or down a few thousand feet will solve the problem is empirical, but the theory behind it is easy to understand. "Within the atmosphere," Wegener explained, "just as on the surface of a body of water, when a warmer and therefore lighter layer of air moves across a colder, denser layer, waves must form."[53] The idea is that when one layer of a fluid slides over another, the difference in density suppresses mixing, but the difference in velocity encourages it. The mixing is the cause of waves, which draw their energy from the velocity difference between the two layers, and if the wavelength is right, the waves will continue to grow. If the velocity difference is great enough between the two layers, the waves can even break—creating severe "clear-air" turbulence.

Within the atmosphere, the phenomenon is tied to inversion layers—warm layers above cold layers, in defiance of the normal lapse of temperature with altitude; the greater the temperature and velocity differences at the boundary surface (or discontinuity) between the two air layers, the greater the likelihood of mixing and the formation of waves. If conditions are just right, clouds will form at the wave peaks (though not the troughs), and these billow clouds can be seen from the ground in parallel lines at a common altitude—as if a number of cumulus clouds had been stretched enormously in the same direction.

Helmholtz's work on this problem was pathbreaking (in Wegener's estimation) precisely because of the promise of generality: "using the principle of mechanical analogy . . . allowed one to calculate the wave-lengths of all the waves in a family sharing the same wave-form, for *any* air-density and any wind velocity, if one knew all the numerical values for one single case."[54] Actually, Helmholtz had achieved this result, Wegener knew quite well, by disregarding a number of problems, and the form in which he left his solution was very little use to meteorologists because of its lack of generality. Helmholtz had, however, put his student Wilhelm Wien (1864–1928) on the problem, and using analytical methods, Wien had been able to develop Helmholtz's results into a much more useful form. This was where Wegener came in: "The theory [of Wien] is scarcely known in meteorological circles," he wrote, "and because

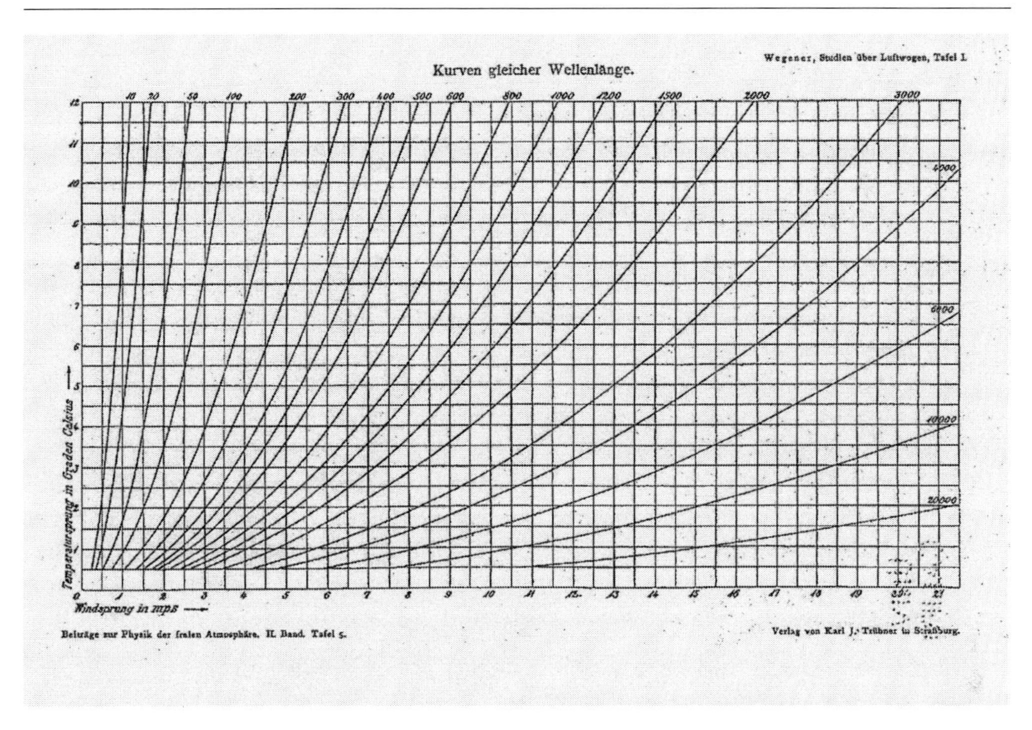

Wegener's attempt to plot the wavelength of "Helmholtz air waves" as a function of the difference in temperature of two successive layers of the atmosphere (*vertical axis*) and the wind speed (*horizontal axis*). From Alfred Wegener, "Studien über Luftwogen," *Beiträge zur Physik der freien Atmosphäre* 2 (1906): 55–72.

of its purely mathematical character, is accessible only to those well-versed in analytical methods." Shorn of its formality, the plain language of this statement was, "Here's a means for solving an interesting and important problem that has been around in the literature for a decade till a meteorologist came along who could understand enough of the mathematical physics involved to use it."

Wegener actually had only some relatively light lifting to do to adapt the equations for meteorological use. Wien's extension of Helmholtz's work used the example of wind waves on the ocean, though his equations contained terms not readily available in meteorological observation. The real theoretical work for Wegener was the graphical representation of Wien's theory in terms of "curves of equal wavelength." To do this, he plotted the results of his (many) calculations with the vertical axis showing the temperature difference between the layers and the horizontal axis showing the wind-speed difference. The idea was to provide, for his analytically challenged meteorological colleagues, access to the predicted wavelength at the boundary surface, for any combination of air temperatures and wind speeds.

With the graph in hand, the next part of the investigation was to compare temperature observations on kite flights with the theory as embedded in the chart, thus creating a role for kite and captive balloon aerology in the study of these air waves. There were a variety of methods already in use for finding and studying the waves, all requiring relatively unusual conditions. If one were on a mountaintop or in a balloon above a cloud layer, one could see them directly from above; if there was sufficient condensation at the wave crests, one could see them from Earth as clouds. For invisible (clear-air) waves, manned balloon flights could provide a record of the waves via recorded temperature oscillations, or, when there was no vertical convective motion ("vertical wind"), one could discover waves in the up-and-down motion of the balloon, registered on a barograph as oscillations in atmospheric pressure. Finally, one could, in clear air, look for oscillations in a record of atmospheric refraction, since the air on one side of the discontinuity was less dense than that on the other. Compared to any of these, kites and captive balloons seemed to offer a less expensive and exacting path for study.

The observational difficulties were formidable, however. The contrast between the grand sweep of theory and the niggling minutiae of observation stands in sharp relief in the accompanying figure, showing Wegener's calculated curves of equal wavelength and the raw data off the meteorograph from the kite flight he superintended on 6 December 1905. From the latter, after numerous corrections, Wegener had tried to extract a wavelength by comparison of near-invisible oscillations of temperature between 0.1° and 0.7° and then to compare it to the predicted value on his chart. The results were not encouraging here or in succeeding months. By the end of February, he had only been able to find three useful records out of nearly fifty flights, each of them punctuated by their own maddening peculiarities and vagaries. Moreover, the agreement between theory and observation was terrible—an error of 25–30 percent. At least it was consistent—the observed wavelengths were all shorter than the values predicted from the table and persisted across an order of magnitude (the error for a 200-meter [656-foot] wave was nearly the same as the error for a 2,000-meter [6,562-foot] wave). The best possible construction he could put on the outcome was that "there seems to be a systematic discrepancy between observation and theory, insofar as the very small number of observations permit a conclusion."[55] It appeared that this required a long series of observations and some intricate correcting—it would last perhaps the rest of the year, and certainly the rest of his time at Lindenberg. He could see the implications if he had any success: some further understanding of global atmospheric circulation, and some marginal advantage in weather forecasting—but that seemed a long way off.

The days began to blend into one another—there was heavy overcast most of the time, cold rain, and a round of increasingly demanding work: not only did Alfred and Kurt do most of the kite and balloon flying, but it was up to them to extract the observations and interpret the meteorograph records—sometimes two or three a day. In the evenings, as time permitted, they had many discussions with Coym, who had a number of practical hints for Alfred about tricks for squeezing more data out of his flight records. Kurt and Alfred also talked privately about what they were going to do after Lindenberg—Kurt was negotiating (quietly) for a job as assistant at the Physikalischer Verein in Frankfurt beginning in the fall, where he would be involved in an intensive program of flying, soon to include airplanes as well as balloons. Alfred was seriously considering making a move at the same time.

On 23 March 1906, Alfred was in the smoking room of the headquarters building at Lindenberg reading the *Tägliche Rundschau*, a major daily newspaper from Berlin, when his eye fell on a telegraphic notice about the Mylius-Erichsen expedition. It announced that the Danish government had indeed decided to match the funds raised by Mylius-Erichsen: Christensen had made good on his promise, and Mylius-Erichsen had collected sufficient sponsors. The article also contained the following astonishing paragraph: "Outside of the Danish members of the expedition it is likely that Dr. phil. A. Wegener of Germany will take part as physicist and meteorologist, and Dr. phil. Baron von Firicks from Russia as geologist: the negotiations with these two scholars, have not been concluded, however."[56]

Alfred dashed upstairs and sat down to compose a letter. He knew better than to try to get an answer from the ever-elusive Mylius-Erichsen directly, so he wrote to Paulsen, describing what he had read "just this minute" in the paper. When he had read his name and that of a Russian along with the news of the award of funds by the Danish parliament, he "could not help thinking that this meant that Herr Erichsen now found himself in the position of putting his undertaking on a broader basis, and eventually of relinquishing the principle that only Danes should be allowed to take part." Alfred indicated to Paulsen that he badly needed an immediate response, adding, "I'm in the middle of negotiating the terms of a new position, negotiations that I'll have to break off [if I am being considered for the expedition], so I'd be very obliged if you could tell me as soon as possible what you know about this."[57] He apologized to Paulsen for bothering him again, begging him to take it as an expression of his deep interest in the expedition.

Paulsen received Wegener's letter the next day, and he wrote immediately to Mylius-Erichsen, telling him that he had yet another letter from Wegener and urging him to make up his mind about the matter immediately and to write to Wegener and tell him whether he was to be a part of the expedition. Mylius-Erichsen very much wanted to make this a Danish expedition, and he had been able to put together an all-Danish cartographic and scientific staff from those who had applied (though no contracts had yet been signed). With the money in the bank, the pressure for an all-Danish expedition had lessened at least enough for him to send up the "trial-balloon" in the newspaper about Wegener and Firicks, and it is clear that he had to.

In March of 1906 the expedition still lacked a geologist and someone to do physics and meteorology—glaring lacunae for an enterprise billing itself as a major scientific expedition. It had been easy to find a hydrographer and a marine biologist—chosen from a large number of eager suppliants—and there were plenty of eager ornithologists. The expedition's successful candidates for physician and botanist had even spent several years in Greenland. Yet they were all young men, many still in school. Even the older candidates mostly lacked degrees or academic affiliations; only a few came with expedition experience. Mylius-Erichsen was an unknown quantity, and the planned expedition, in spite of its towering ambitions, was also a shoestring operation with low salaries. It was therefore difficult for him to hire credentialed scientists, and in the end he still lacked the scientists needed to study most of what they would see: rock, ice, and weather.

Time to prepare for an expedition was running desperately short by any measure. The funds had been voted on by parliament on 22 March and announced on the next day. On the twenty-fourth, the expedition's governing committee officially took up its duties, setting a departure date of 24 June 1906, a scant ninety days away—and the expedition did not yet even have a ship! Mylius-Erichsen decided that he could get along with a geology student and settled on young Hakon Jarner (1882–1964), a Danish student at the Polytechnic Institut—thus no need to bring the Russian baron (Firicks) along. Wegener, however, was the only expedition candidate with a PhD and the only professionally employed scientist who had applied for any of the jobs. He was a German, true, but had a Danish-sounding name, and Prof. Paulsen, who had been unfailingly helpful to Mylius-Erichsen with advice and support, really wanted a meteorologist on the expedition. Wegener had been remarkably persistent and had expressed unfailing interest for five months.

Alfred, once again unable to stand the suspense, took a train the next day (Saturday) to Copenhagen, arriving on Mylius-Erichsen's doorstep within hours of Paulsen's letter. With money in hand and little time to waste, Mylius-Erichsen was now prepared to take Alfred's participation seriously, and when they finally met face to face, Mylius-Erichsen must have liked what he saw, because he made a decision on the spot and formally offered Alfred the job of physicist and meteorologist on the expedition, with the provision that he would also be expected to make geological observations and to take part in the cartographic and position-finding work. They signed a preliminary agreement and discussed salary. Mylius-Erichsen told Alfred that he should prepare his personal and scientific kit immediately: the expedition would leave in less than three months.

Wegener was too happy with the outcome to be very analytical about it on the train ride back to Lindenberg, but it was a wonderful object lesson, seen from outside, in how dogged persistence can aid chance in producing an event. His odds for getting on the expedition, objectively, had always been slim—if any Danish meteorology student with ten fingers and ten toes had shown up, he'd have been out of luck. He had not made success a precondition for effort, and he had remained aggressively in pursuit of his goal from November until March. It had "happened to him," yes—but he had *made* it happen to him, forcing all visible contingencies in his direction, and in some real sense he had the experience of bending the world to his will. Here was that philosophy of "vital freedom" made real. He had obtained the results of such vital striving—and they felt very good indeed.

5

The Polar Meteorologist

GREENLAND, 1906

> There is no such thing as friendship on an arctic trip. Not friendship of the right kind, where worth is equal and individual partners really mean something to each other. The arctic air, which is said to be free from bacteria, certainly contains one contagious germ: personal ambition.
>
> LUDWIG MYLIUS-ERICHSEN, Journal (13 July 1903)

The Danmark Expedition

The Danmark Expedition to East Greenland, for which Alfred Wegener was to be meteorologist and physicist, aimed to be the largest polar-scientific endeavor ever mounted. Its immediate predecessor, Drygalski's "Gauss" Expedition of 1901–1903 to the Antarctic, had employed a scientific staff and ship's crew of twenty-two, though this number included everyone on board, including stokers and deckhands; there were actually only seven scientists.[1] Ludwig Mylius-Erichsen's Danmark party was to be twenty-eight, of whom sixteen would perform real scientific work; all of the latter had agreed to work as deckhands and even as stokers if need be, in order to keep the ratio of scientists to support staff as favorable as possible.

Of the forty or so major polar expeditions mounted between 1890 and 1914, Mylius-Erichsen's exceeded all others in size and scientific ambition. Even so, most readers outside Denmark will never have heard of it, perhaps because the reputation of polar expeditions and their scientific worth are generally in inverse proportion. The only two expeditions that came close (in size) to Mylius-Erichsen's were the Australasian Expedition to Antarctica in 1912, led by Douglas Mawson (1882–1958), which had planned to land twenty-six men, and the celebrated Endurance Expedition in 1914–1915 of Ernest Shackleton (1874–1922), which got to twenty-eight only by counting Perce Blackboro, a deckhand who stowed away. Of these twenty-eight, only four were scientists, and the crushing of the *Endurance* by the Antarctic sea ice effectively destroyed their scientific program almost before it began.[2]

Wegener, with no polar experience, was hired for the Danmark Expedition with a late and brief interview. He felt fortunate to be included in such an undertaking, but he was better prepared than most polar novices. He had mountaineering and glacier-climbing experience, he could sail, and he had meteorological skills the expedition needed. In any case, men with scientific credentials and serious polar experience were quite rare, even in Denmark, where national sovereignty extended to all of Iceland and most of Greenland. Mylius-Erichsen, choosing among novices, picked more carefully than many better-known expedition leaders. Shackleton appointed Leonard Hussey meteorologist for the Endurance Expedition, even though Hussey knew nothing about meteorology, on the grounds that he "looked funny," and it appealed to Shackleton's sense of whimsy that Hussey had applied for the expedition while working as an ethnologist in the Sudan. Shackleton hired Reginald James as physicist for the same expedition because he had good teeth, had a sense of humor, and could sing.[3]

Part of the secret of Alfred's successful candidacy was that while the expedition airily aspired to be "scientific," it had no specific scientific aims beyond the most obvious and essential—geographical discovery and making maps of the newly discovered terrain. Mylius-Erichsen was an adventure traveler and literary intellectual; if science was to be done on his expedition, it would be up to his subordinates to do it. The expedition contracts specified that the Committee of the Danmark Expedition would pay for all the scientific equipment, but declined to specify what that equipment might be. Each scientist had to plan his own program and then build, buy, or borrow the instruments to carry it out. "The Committee" (i.e., Mylius-Erichsen) would have to approve the purchases, but the scientists were responsible for acquiring their equipment and getting it to Copenhagen by the middle of June. Each man had a budget: the more ingenuity he showed in stretching that amount, the more instrumentation he could take and the more science he could do. Alfred had come forward not just with expertise but, as evident in his first letter to Mylius-Erichsen in November 1905, a specific scientific program in mind. Mylius-Erichsen's reluctance to allow a German scientist a role in a Danish undertaking gave way to his desire to claim abundant scientific results for the expedition and, in this way, to secure his own place in the history of *scientific* exploration.

Alfred plunged into action from the moment of selection. He knew he had just ten weeks to make the preparations that would govern his work for the next two years. He sat down and began to write, sending letter after letter of inquiry and appeal. The letters, filled with ambition and naïve delight, convey both urgency and youthful deference. On 28 March 1906, a few days after his appointment to the expedition, he wrote the following letter to Wladimir Köppen (1846–1940), head of the Meteorological Department of the German Marine Observatory at Hamburg:

> Esteemed Professor:
>
> I will participate in the polar expedition of Mylius-Erichsen (to East Greenland) as meteorologist and physicist and I intend while there to carry out, as well, kite and balloon ascents. Given the lack of time (the expedition departs Copenhagen at the end of June) and the scarcity of means, it will only be possible for me to accomplish this program if I can obtain generous assistance from those [scientific] institutes that carry out such work. Although I will receive a number of kites from Herr Geheimrat Aßmann, I presume to ask whether you would be inclined to sell me for a modest sum some of the kites of your design, or to have some new ones built for me. . . .
>
> Hoping you will pardon this importunate request, which comes only from my earnest wish to carry the aerological program of the expedition to a successful conclusion, I am with highest regards
> yours very truly.
>
> Dr. A Wegener[4]

This was the first of dozens of such letters, and in the following weeks support flowed in. Aßmann immediately gave his blessing and promised twenty kites and three of the precious varnished-cotton captive balloons, all at cost. Arthur Berson was willing to enlist the entire aerological fraternity to equip his young subaltern for this pioneering effort in arctic meteorology. In response to a letter sent with Berson's endorsement, Teisserenc de Bort not only agreed to sell Wegener two of his meteorographs at cost but also made him a present of two additional meteorographs, "in the cause of science."[5] Alfred's partner in balloon flight, Hans Gerdien, agreed to loan him instruments for measuring atmospheric electricity, with instructions for their use. If Alfred could operate them and teach others to do so, Gerdien would find someone else to work up the results on the return—the important thing was to pack as much science as possible into the trip.[6]

Alfred was exhilarated to put together his part of the expedition, but the call had come suddenly and unexpectedly, and he still had a burden of work at Lindenberg, peaking just at this time. He (and Kurt) had been selected by Aßmann and Berson to represent Lindenberg and the German Empire in the Gordon Bennett International Balloon Competition, scheduled for 4 April 1906. Teams of aeronauts all over Europe were to compete for time, altitude, and distance. For the Lindenberg team this was not just a stunt and a sporting event, though it was that, to be sure. They were to exploit their night flight to practice navigation by star sightings. Kurt Wegener, with five previous flights, was the more experienced balloonist and the qualified pilot, as well as the elder brother—he would direct the flight. Alfred (two previous flights) would serve as navigator, instrument monitor, and ballast heaver.

There was another task in the offing for Alfred, more daunting even than the planned balloon contest: he had to tell his parents (who knew nothing whatever of his expedition plans) that he was leaving in a few weeks for a two-year journey to an uncharted and completely unknown portion of Greenland. On 29 March, Alfred and Kurt left Lindenberg for Berlin to arrange for the balloon flight. Arriving in Berlin, they traveled first to the borough of Reinickendorf, where the army's Airship Battalion was headquartered, and from which they would lift off the following week.[7]

That evening, Alfred (with Kurt standing by for moral support) broke the news. Richard listened in grim-lipped silence to Alfred's enthusiastic description of the expedition and then exploded. He was completely opposed to the idea. It was bad enough, he said, that they persisted in their life-threatening and daredevil balloon antics—Kurt had even flown across the channel and crashed in England! Now Alfred was planning to run off—for years!—on an expedition, at precisely the time he should be finding a permanent, pensioned employment and starting a family—something that Kurt should also be considering. Richard pleaded with Alfred to abandon the plan, to no avail. It gradually dawned on him, however, that Alfred was not asking for permission, but announcing a decision. Alfred wanted his parents' blessing, but he did not need their consent. This was, he told them, a legitimate scientific undertaking, and he was a trained scientist, planning to do what scientists did.

Richard Wegener's anger cooled as Alfred explained the scientific program he was mapping out. He would be doing the same work he did at Lindenberg and would be paid a salary for it—he would just be doing the work in Greenland instead of Germany. The expedition was a public-private joint venture under the protection of the Danish Crown, and Alfred's supervisors, Aßmann and Berson, had already given their professional consent to his departure, even without the requisite notice. Alfred pointed out to his father that he would not even have to resign his post at Lindenberg, but merely to take a leave of absence from 15 April 1906 (two weeks hence) to 1 January 1909. Richard remained uncomprehending—why would anyone do such a thing? Why would one earn a PhD in astronomy at Berlin and then run off to fly kites at the North Pole? He was opposed, but he was resigned to his willful son's decision, and in the course of the evening, he began to inquire how Alfred would equip himself. The salary for the expedition, Alfred told him, would be paid only on his return, and though the expedition would buy the scientific equipment, he still faced considerable expense preparing himself for the trip. Richard, as a doting (if baffled) father, finally offered to advance Alfred what he needed to outfit himself for the journey. Moreover, he invited Alfred to move back into the house at Halensee while making his preparations—an offer that Alfred gratefully accepted: he was still pleased to be a son of the house.[8]

In the next days and weeks, though his head was full of Greenland plans, Alfred's hands were full of his daily duties at Lindenberg and the preparations for the balloon flight on 4 April. He was pushed to the limit just to manage his multiple roles: government meteorologist, balloon navigator, reserve cadet officer, and now expedition member—all of these had to be played out and resolved officially in a matter of weeks. There was also his own scientific program of research and publication, including the observational and theoretical work on Helmholtz air waves, woefully incomplete but promising—what to do with that project now? Kurt pushed Alfred to write up what he had for publication and promised that, if necessary, he would finish the drafting and editing of the text and see it through the press. Arthur Coym also pressed him to get the work out.[9]

Alfred had too many tasks to see through to the end with too little time, but the crush of work appealed to his sense of struggle. He liked working long hours, and even the labor of calculation; for him, science was a matter of life, not just of work. That is perhaps not the best way to put it. Rather, work was already his life, and he spoke to others of *die Arbeit* (the work) as if it were a cause and a mission, not some job to be got out of the way. He did not much care to drink and socialize and, somewhat like Wagner's Siegfried, had scarcely ever spoken to a woman to whom he was not related. He had successfully integrated his appetite for physical labor into his science, in a way that was about to accelerate remarkably. His physical amusements and hobbies—including the excitement of balloon flight—were part of his work as well. He was overwhelmed by work, even buried in it, but had never been happier.

Aeronautical Interlude

On 3 April, after three more days at Lindenberg performing the dozens of numbingly routine but essential tasks that keep a meteorological station functioning, and after writing many more letters, Alfred traveled with Kurt back to Berlin to prepare for the balloon flight. They carried in their luggage the balloon's meteorological instrument package and Alfred's own balloon theodolite for night navigation, as well as a camera and glass plates. They arrived at Reinickendorf early on the next morning to discover that their balloon, the 1,200-cubic-meter (42,378-cubic-foot) "Brandenburg," a lovely golden-black globe made of cotton varnished with linseed oil, had sprung a leak while filling and would have to be repaired. The other available cotton / linseed oil balloon owned by the observatory (the "Meteor," 850 cubic meters [30,018 cubic feet]) was too small to carry the load and have a chance at the competition, so hurried arrangements were made to borrow one of the army's 1,200-cubic-meter military balloons. This needed to be unrolled and inspected and could not be made ready until the next morning.[10]

The substitute military balloon, a mass-produced, tested, completely unexceptional design, was unpacked overnight and, in the early dawn hours of 5 April, filled rapidly and uneventfully with 1,200 cubic meters of hydrogen. The instruments were lifted on board and secured to the basket rim and to the trailing cable: the meteorograph had to be suspended away from the balloon basket if it was to be read properly. The ballast sacks, thirty-eight in all, were slip-knotted to the inside of the balloon basket, with their pull toggles at the level of the basket rail. The provision sack for the flight, containing 0.9 kilograms (2 pounds) of chocolate, four smoked pork chops, 2 liters (0.5 gallons) of seltzer in rechargeable siphon bottles, and two oranges, was gently stowed in a corner.[11] Setting their watches by the station chronometer and checking the engagement of the catch on the meteorograph recording drum, they jumped aboard, nattily attired in summer suits and their fedoras, and waited for their scheduled liftoff.

At 9:00 a.m. the balloon took off rapidly and flew northwest. It headed directly for Brandenburg and flew right over the Ruppiner See, just a few kilometers from *die Hütte*. From there it proceeded on to Wittstock, Richard Wegener's birthplace and the destination of many childhood visits; Kurt and Alfred were able to photograph it from the air at an altitude of 500 meters (1,640 feet). From there, still heading northwest, the balloon passed over the Plauer See (scene of many boyhood sailing adventures), then on to the Baltic, and then headed straight for Denmark, crossing into Jylland (the larger, northerly peninsular part of Denmark) at about dark. As the wind dropped, it began to drift east, over the Kattegat (between Denmark and Sweden), and to gain altitude, passing 1,000 meters (3,281 feet). With the coming of night and the increase of altitude, the air turned sharply colder, and it was only then that Kurt and Alfred realized that in the midst of their careful preparations they had left their overcoats in Berlin. The air temperature fell below freezing, and soon their shivering made sleep impossible. They moved about to keep warm as much as they could, and Alfred was able to make two position fixes; monitoring the weather instruments also helped to take his mind off the cold.

By morning they were both nearly frozen. In the dry air they had risen to 2,500 meters (8,202 feet) and were drifting slowly back south in slack wind. Not until noon on 6 July did the wind pick up again, and then the balloon began to rise and fall in the convecting air: down to 300 meters (984 feet), up to 1,000 meters, down again. They began to drift away to the west, and at about 8:00 p.m. they passed back over the Danish coast. They had now been in the air for thirty-five hours—and they had been up most of the night previous to the flight, while the balloon was being filled. They were dehydrated, very cold, and very tired. It looked that evening (6 July) as if the balloon was going to head directly west across Jylland to the North Sea, and they knew that once past that coast, it was 500 kilometers (311 miles) of open ocean to England. Consequently, they began to pack the instruments and prepared to set down. Just then, however, the balloon changed course and began to gain altitude and fly south toward Hamburg. It continued on this course through the night, traveling at very low altitudes, sometimes sinking to 100 meters (328 feet); this was hair-raising, as it put them repeatedly within a few seconds of crashing to the ground. At least it was warmer at the lower altitude, and this was fortunate, for they were weakening badly and beginning to cramp in their arms and legs from dehydration and almost forty hours of bracing against the swaying of the basket.

As dawn broke, they passed over Kassel (halfway between Hannover and Frankfurt). They were by then shivering uncontrollably, having been at −16°C (3°F) for three hours or more. They were out of food and water. They decided, exhausted as they were, to make a final push for altitude before ending the flight. Their target altitude was above 5,000 meters (16,404 feet), and they calculated that they could make it in spite of the cold and their rapidly dwindling stamina. However (as Ulrich Wutzke has remarked), this time cold and hunger were stronger than will: they were so weakened that when they tried to drop ballast, neither of them could push a sack over the rim of the balloon basket. They were done, and they drifted to a landing near Aschaffenburg, east of Frankfurt. They sent the following telegram to Aßmann: "Today at one-thirty landed very smoothly at Laufach near Aschaffenburg after 52 hour flight over Aalborg on Jutland [Jylland] around 3000 [meters] minus 16 degrees."[12] They had stayed aloft for fifty-two and a half hours, having broken the world record for time aloft by seventeen hours.[13]

Their flight was a major event in the early history of aeronautics and was treated as such by newspapers around the world. Everyone involved got a share of the rewards—Aßmann was delighted, as was his principal patron, Kaiser Wilhelm. The German army was pleased that their balloon had accomplished the feat while a civilian balloon had proven unreliable. It was a public delight in Berlin that the record broken by these novice German balloonists had previously been held by the accomplished French aeronaut, the Comte de la Vaulx (1870–1930). It gave Kurt Wegener a tremendous boost in landing the job he desired at Frankfurt, and it gave Alfred something priceless at the moment: widespread name recognition while he was writing letters to numerous officials who had never heard of him, asking them for large and expensive favors. It gave him a distinctive achievement as an explorer before he had even gone anywhere. Alfred and Kurt were the first to recognize that their exploit was in part a lucky accident, especially the grand looping flight path that had allowed a sustained north-south flight. Yet their skill and daring, their parsimonious use of ballast, and their willingness to come dangerously close to the ground had also determined their success, as had their hardiness in the face of physical discomfort and privation. Whatever luck they had enjoyed, they had struggled fiercely to make it work for them.

Hurried Preparations

Returning to Lindenberg on 9 April, still very tired but very pleased with himself, Alfred found a great pile of correspondence—from individuals, firms, and institutions, all quoting prices and asking for further specifications on equipment. On 10 April he wrote to Mylius-Erichsen asking for the right to make the purchases of kites and balloons and instruments without asking for permission each time: "We just don't have enough time, I believe, for you to place the order personally in each case."[14] This was far from the last time that Mylius-Erichsen's passion to control everything at all times asserted itself. Passion for control and the micromanagement of affairs are evil twins, especially in leaders (like Mylius-Erichsen) gnawed by an insatiable craving for honor and fame. Such leaders fear that without them something will go wrong, and that they will have to accept responsibility for a failure. Even more, however, they fear that without them something may go right, and that they will have to forego credit for a success.

Time, or the absence of it, was on Wegener's side in prying some small purchasing authority away from Mylius-Erichsen. Alfred had been conferring with Erich von Drygalski (1865–1949), Germany's Antarctic hero and a great Greenland explorer as well, to see what of the Gauss Expedition's equipment (since parted out to German scientific stations) might be reassembled for the Danmark Expedition.

On 15 April, with the application for leave from Lindenberg approved, Alfred packed his clothes and papers and left for Berlin, moving back into his parents' home at Georg Wilhelmstraße 20. The reprieve from daily scientific duties was an intense relief, freeing him for travel, correspondence, and negotiation. Aßmann remained helpful, interceding two weeks later to arrange the purchase of the hydrogen gas canisters from the military. He was very pleased with his young assistant and slashed paths through bureaucratic formalities that Alfred could never have cut himself.[15]

Alfred's plea for assistance aroused the interest and concern of the senior meteorologist at Hamburg, Wladimir Köppen, from whom Alfred had requested the right to purchase some kites, and with whom he had wished to discuss his planned program for Greenland. In May and early June Alfred exchanged several long letters with Köppen, requesting advice and help on getting his kite winch finished, asking him to examine instrument correction curves for error, and asking for samples of Köppen's field-notebook format for observations—the sorts of things he needed to know but had little time to find out. He wanted to come to Hamburg to talk with Köppen, though he feared that the crush of work would prevent it. Yet he was clearly touched by Köppen's generosity with time, advice, and materials, and he realized that Köppen had a genuine interest in him and his work, far above and beyond any pro forma professional courtesy or national pride.[16]

By early June he was almost ready to go. His commission as a lieutenant in the reserves came through in late May, and he now had the essential permission from the military to proceed.[17] Friedrich Bidlingsmaier (1875–1914), the young geophysicist who had accompanied Drygalski in the *Gauss*, helped Wegener to acquire a self-recording declinometer. This iron-free instrument, to be housed in a prefabricated iron-free hut, could record variations in magnetic declination (the horizontal angle between true and magnetic north) through twelve- and twenty-four-hour periods, using photographic recording paper on a clockwork drum.[18] Bidlingsmaier also arranged for Wegener to be able to purchase the ice corer that Drygalski had used in Antarctica.[19]

The scientific program was taking shape. There would be a full meteorological station wherever they finally landed on the northeastern coast of Greenland, with multiple daily measurements of temperature (including twenty-four-hour maxima and minima), humidity (by two methods), barometric pressure, wind speed and direction, precipitation, and cloud cover. There would be a full aerological program with as many kite and balloon flights as Alfred could manage, looking for layering, inversions, and air waves and measuring temperature, humidity, and wind speed and direction aloft. Alfred also planned to direct the magnetic observatory, to study atmospheric electricity, and to record the aurora and other optical phenomena photographically. He planned to photograph ice crystals, clouds, ice, and rock formations; he was very keen about the photography and pleased that the expedition would travel with its own prefabricated darkroom.[20]

The only major stumbling block in the preparations was getting a powered kite winch. The factory that was building the winch was in Eimsbüttel, a district of Hamburg, where, until recently, Wladimir Köppen had lived with his very large family. This had made it natural for Köppen to act as a go-between in the construction. Though the winch was being built in Hamburg, the motor for it was being built in Copenhagen, and there were all sorts of problems putting the designs together.

For this and other reasons, Wegener traveled to Hamburg in June, to meet Köppen and try to work through the difficulties. Köppen, generous and gregarious, ran a perpetual "open house" at his residence in the Violastraße—a beautiful and rustic suburb close to his meteorological kite station, and some kilometers (by streetcar) outside Hamburg.[21] Neither Wegener nor Köppen kept a record of this initial meeting and visit, but it is clear that they took an instant liking to one another and that Alfred enjoyed the generous openness of the Köppen household.[22]

From Hamburg Alfred traveled on to Copenhagen, to superintend the final loading of his scientific equipment on board the expedition's ship. His parents were to meet him shortly to tour the ship and would stay to see him off. He wrote excitedly to Köppen on 21 June—his last piece of correspondence before the scheduled departure three days later—to inform him that the packages of kites had arrived from Hamburg and that the canisters of hydrogen gas for the balloons had been safely stowed on board. The winch had still not arrived, though it had apparently been shipped from Hamburg on the nineteenth. Even if it did not arrive by the twenty-fourth, Alfred wrote, "all will not be lost. There is a ship departing for Iceland three days after we leave, to carry late arriving packages and etc. for us."[23]

Wegener's final letter to Köppen, hurried and smudged, contains both his deep thanks for all of Köppen's efforts on his behalf and an avowal of an obligation incurred: "Now it is up to me, upon the expedition's return, to be able to lay good results before you." He added, "Finally I wish to add my hope that on my return home I will again find you yourself and your honorable family full of health and vigor," closing with, "and I hope you will remember me most kindly to your esteemed family." The Köppen household—bustling, informal, filled with scientists and scientific work—made a powerful and lasting impression on Wegener, astonished to have passed in a matter of hours from a casual visitor to a "friend of the family."[24]

Voyaging

The last few days before the expedition's departure from Copenhagen were frantic—and thus no different from any other expedition. Their ship was a retired Norwegian sealer, the *Magdalene*, rechristened the *Danmark*. She was built of very thick oak and had a greatly enlarged forecastle to provide additional sleeping room and storage; rigged for sail, she also had an ancient and lumbering but still powerful one-cylinder engine. The crew and scientific staff, many already living on board, were working feverishly to stow three years' worth of provisions and equipment below decks, though they were interrupted and even besieged by streams of onlookers, reporters, well-wishers, and dignitaries. Mylius-Erichsen had done everything he could to whip up enthusiasm and excitement about the expedition, writing articles emphasizing its importance and dangers. These newspaper reports brought hordes of curious spectators to Copenhagen's long wharf, the Langlinie, where the *Danmark* was moored.

Mylius-Erichsen was anxious to depart on schedule, even though many necessary instruments had not arrived, the stowing of the tons of material on deck was far from complete, and little was secured below: he had orchestrated the departure as a great public event, and he had the additional spur to depart on time (unbeknownst to the crew) that he had wildly and recklessly overspent his budget and feared that the expedition committee would see the bills and cancel the expedition before it left.[25]

The ship departed on schedule from Copenhagen on 24 June before a cheering crowd. Wegener wondered (with a literary flourish) in his first journal entry of the voyage, "Will they also be so happy, when we return?"[26] His parents, who had come to Copenhagen to see him off, were in the crowd along the quay as the ship set out, but he could not see them. His own mind, so long preoccupied with family, documents, packing, and preparations, could now turn to the voyage itself. He found himself assigned to share a cabin with Lt. Johan Peter Koch (1870–1928) of the Danish army, the expedition's cartographer. Koch was a seasoned polar traveler, having made surveys in East Greenland in 1900 with Georg Amdrup (1866–1947) and having since worked for three years (1902–1904) surveying southern Iceland; he was also an experienced seaman and a qualified ships' master.[27] This cabin pairing of Koch and Wegener was a great kindness on the part of Mylius-Erichsen, as Koch spoke excellent German and would be Wegener's closest scientific colleague on the trip. Language was still a problem for Wegener, and on the evening of this first day he turned again to his journal: "My ignorance of Danish is really annoying. I sit around most of the time without understanding a word, like a deaf-mute."[28]

The decks of the *Danmark* were a chaos of bales, boxes, and lumber, and there were already thirty Greenland sled dogs on board, barking and running about unrestrained; two of these already had litters of pups. The crew worked hard stowing and securing the loose cargo, and they did their best to get some sleep in their four hours of free time between watches. Koch and Wegener stood the second watch together; all hands ("the sailors and the scientists both," noted Wegener) were to stand "watch and watch"—four hours on and four hours off, round the clock.[29]

The *Danmark* sailed north up the Øresund, between Denmark and Sweden, and out into the Kattegat, over which Wegener had flown just three months before. Around noon the next day (25 June) the sea rose and the wind picked up sharply from the west. As the sails filled under their first real pressure since the refitting, the fore-topgallant yard (supporting the middle tier of sails on the foremast) snapped and was left dangling above the deck. It took the inexperienced crew until evening to lower the broken yard to the deck and assess the damage. The next morning, they were barely under way when the ship's jib sheet parted and the sail, flapping about, tore from top to bottom. The *Danmark*, heavily overladen and out of trim, wallowed and rolled horribly, and all but the most experienced sailors were violently seasick and barely able to work. The topgallant yard, as it turned out, was rotten and would have to be replaced. Mylius-Erichsen decided to put in to the commercial port of Frederickhaven at the northern tip of Denmark and make repairs.[30]

This embarrassment turned out to be a blessing in disguise. It was better to have things break and fail where repairs could be made rather than in Greenland, where they could not. The problems with the rigging were annoyances but easily made right. Other expeditions had had much worse luck. The *Gauss* had gone halfway round the world before Drygalski discovered that she was underpowered and unstable, that her winch was unable to raise the anchor, and that someone had neglected to mention to the designers that using pitch as insulation inside and out would be a disaster for a ship that had to sail through the tropics—where it all melted and ran.[31] Roald Amundsen (1872–1928) was 2,000 miles from home on his way to the South Pole before he discovered that his specially commissioned snow boots, designed not to leak at the knee, were sewn so narrowly at the top that they could not be pulled over the men's calves.[32]

The most serious problem for the *Danmark* was the weight of her cargo. Over the next week, as the *Danmark*'s captain, Alf Trolle (1879–1949), superintended repairs to the yardarm and inspected and strengthened the rigging, he also ordered the ship to be drastically lightened. He put ashore twenty-four tons of coal that had been stored in the 'tween decks in sacks, along with six tons of sand ballast. He then had a barge brought alongside and had the ship completely unloaded and reloaded, this time with the cargo secured and the boats tied down.[33]

When the ship left Denmark on 2 July, she was much more seaworthy. By 8 July she had covered half the distance to the Faeroe Islands, and the crew was winning the battle against seasickness. Wegener was finally able to write, "I'm very glad not to have been seasick. Early on I was queasy almost all the time, but I haven't missed a watch or a meal. And this when the ship works so that everything that isn't riveted or nailed down is constantly tossed about. . . . But now," he added proudly, "even the most violent movements of the ship cause not a bit of discomfort."[34]

By 10 July the shipboard routine (and Wegener's stomach) had settled to the point where he could begin to think about his scientific work. The stop in Frederickshaven had made clear how badly they had overestimated their cargo capacity, and yet they had more to take on board in both the Faeroes and Iceland. Many things would have to be left behind—there was simply no place to put them. The original scheme, involving manned stations at different points on the coast, would be abandoned, since the prefabricated hut needed to establish the second station would remain onshore in Iceland. This meant that Peter Freuchen (1886–1957), then a nineteen-year-old medical student who had signed on as a stoker, was out of a job, since the plan had called for him to man the second station. Mylius-Erichsen rechristened Freuchen as Wegener's "meteorological assistant," and, on 10 July, Wegener began to train Freuchen in the use of the instruments, beginning with the simplest—the anemometer. This effort seems to have reinvigorated Wegener: "In Iceland, I'll set up the thermometer hut, and from there on take regular observations. I will also draft an observational program."[35]

The next morning, 11 July, they anchored in Transvaag Fjord on Syderø, the southernmost of the Faeroes, where they took on several kayaks and 100 more Greenland sled dogs, brought from West Greenland by the three remaining members of the expedition, all Greenlanders. The senior of these, Jørgen Brønlund (1877–1907), had traveled with Mylius-Erichsen in West Greenland in 1902–1904. With the arrival of the dogs, the fragile sense of order on board disappeared for the remainder of the voyage. There were no kennels or pens, and the exuberant, quarrelsome, half-wild creatures wandered everywhere, dominating the life of the ship. "The deck," as Peter Freuchen recalled years later, "was soon covered with their discharges."[36] Almost no provision had been made for their food, and none at all for water; they were so ravenous and thirsty that nothing remotely edible could be left unattended for a moment. Nor were they safe themselves—in the first day of heavy sea on the way to Iceland three of them were swept overboard and lost.[37]

From the Faeroes the *Danmark* plodded northwest to Iceland, arriving at the east-coast trading station of Eskifjördhur early on 18 July. The supply ship was there, as well as the answer to Wegener's greatest worry: his kite winch was on board and his aerological program secured. Other delights awaited—there was mail from home, and a chance to wash their clothes. Wegener was overcome with the beauty of the place: "The mountains on the fjord are so beautiful, basalt terraces covered with green turf. In the harbor, just the two ships at anchor, a colorful picture."[38] He wanted very much to photograph it and immediately opened a box of his photographic plates. He was learning three-color photography, using sets of tinted glass plates that produced red, yellow, and cyan images when developed (this is what color films do today). These plates, when projected together on a screen, would give a composite color image. The chaos on board prevented him, however—the darkroom was filled to the ceiling with fur clothing hidden away from the dogs, and there was no place to store or develop the plates. Later in the morning, Lt. Koch took him ashore to buy necessities—including a set of oilskins and an Icelandic wool sweater. After lunch, Koch invited Alfred to accompany him and one of the expedition's artists, the painter Aage Bertelsen (1873–1945), in the harbormaster's launch to see the end of the fjord.[39] It was (and is today) a stunning sight—a deep green valley extending from the end of the fjord westward far into the mountains, with great torrents plunging into the valley from some of the highest peaks on the east coast of Iceland.

Back at the ship it was otherwise—endless, noisy, barely coordinated loading and unloading amidst the dogs. The crew was in deep mourning over Trolle and Mylius-Erichsen's decision to send back to Denmark 8,000 bottles of beer, a gift of the Karlsburg Foundation. There was no room on board for the almost 400 cases of beer, weighing almost five tons. While they planned to carry (and did) into the Arctic prodigious quantities of alcohol, this would now be limited to spirits, some cases of port, and a little champagne for the most festive occasions. The oversupply of champagne (many cases thereof) was to be sent back to Denmark as well, but the crew vigorously protested and prevailed on Mylius-Erichsen to let them break it out and drink it all.[40] The party lasted for three days and nights and gave Wegener his first view of the anarchic vigor of his Danish companions.

The departure from Iceland on 21 July marked the real beginning of the expedition. They sailed northeast for several days and then struck a northerly course, reaching the Arctic Circle on the twenty-fifth. Wegener noted disconsolately in his journal, "This evening we crossed the Arctic Circle completely without ceremony."[41] He had been anticipating the sort of organized horseplay associated with "line-crossing" ceremonies at sea, in which "King Neptune" appears to demand tribute from all first-time line crossers. This usually involves having one's head shaved, or being forced to drink some sort of vile concoction, or having to kiss the "Sea-Hag," or, in the case of the Arctic Circle, having one's nose painted blue. None of this materialized, however; the crew was still trying to recover from the festivities in Iceland.

Day by day the wind was sharply colder. Even in July the sea temperature just north of Iceland is only a degree or two above 0°C (32°F), as the East Iceland Current brings polar water southward; the temperature of the air is only a few degrees warmer. The rapid cooling was all the more noticeable because they were making terrific headway, climbing a degree and sometimes two degrees of latitude per day. By the twenty-eighth they were almost 72° north, and the nights were becoming bitter in the unheated cabins. The crew was issued reindeer pelts as blankets—something of a milestone in itself. Fridtjof Nansen—and there was not a man on board who had not read Nansen—had praised their warmth and utility; they had become a sort of talisman of polar adventure. "I was issued a reindeer pelt yesterday," wrote Wegener, "but Koch cooled my rapture by informing me that I had most likely inherited a hundred lice with it. In his opinion, all Greenlanders have lice, and all furs that come from them. We have therefore brought the corpus delecti under medical supervision by dusting it with insect powder."[42]

On 30 July, at about 75° north, they encountered the outer fringes of the pack ice that guards the coast of Greenland. "Yesterday we saw the first pack ice! These are gorgeous tableaux," he wrote, "these wonderful, lustrous shapes! I was amazed by the beauty of it all."[43] He was captivated by this world from the first moment he saw it, and he wanted to capture it, as well. "I can't get it on film," he fumed; "it's too dark." He was gratified to see that Bertelsen shared his aesthetic excitement and had rushed on deck to paint it, immediately.[44]

Captain Trolle turned the ship due west, and they made remarkable progress toward the coast of Greenland, covering half the estimated width of the ice pack in a single day and night between 31 July and 1 August. Karl Ring (1870–1918), the ice pilot, sailed and powered the ship through open leads of water between the scattered floes, passing an occasional iceberg, but never close to danger. "No expedition has ever had such an easy passage," wrote Wegener excitedly. This was true: the German North Polar Expedition of 1869–1870 had landed in these latitudes but had lost one of its ships, the *Hansa*, to the ice. Nansen's expedition in 1888 had tried to force the ice here and had been swept hundreds of miles to the south in the Greenland Current.[45] On the *Danmark*, for the moment, there was little for Wegener to do but admire the scene: "In the evening and all night a thin, uniform layer of fog lies over the sea everywhere, and the midnight sun floods it with a pale yellow light—a wonderfully moody atmosphere."[46]

On 4 August 1906 the *Danmark* entered the thick pack ice, and there began the dangerous *pas des deux* between ice and ship, as Ring continually forced the ship forward, backed off, and pushed again. Wegener could not stop looking at the ice. "Today," he noted breathlessly, "I spent an hour in the crow's nest. The vista in clear weather over the ice is beautiful beyond all measure. There is nothing that does not radiate color! The blue water, the clean, white ice, the yellowish-red sun, the light green of the floes under the water—enchanting. Out of the ice comes every shade of color. We saw an ice floe turned an intense violet by the sun. We photograph now madly, it is all too fantastic."[47]

Their progress slowed on the fifth and sixth, and the euphoria began to give way to some realization of the danger. Wegener attempted to get some barometric and thermometric recordings and start his meteorological protocols, but he found his recording instruments knocked off their calibration by the constant vibrations of the deck. "Perhaps I can also improve things by removing the bundles of dried codfish sitting on top of the thermometer-hut—then it can ventilate better."[48] Unbeknownst to himself, he had become a comic figure to the crew. The fun centered on his hat—an indescribably ugly garrison cap in an improbable shade of green, with a border of green astrakhan wool. As he came up the gangway three or four times a day to check his instruments, the hat appeared first and always gave a start to Knud Christiansen (1876–1916), one of the seamen. Christiansen assured everyone that the hat was actually "the devil responsible for all the bad weather we've had on the voyage." Achton Friis (1871–1939), one of the expedition's artists and later the author of the most famous book about this Greenland adventure, noted that it was hard to believe that the young, smiling face that emerged onto the deck beneath that hat was in charge of 100 steel canisters filled with hydrogen—"enough explosive power to send fifty ships like ours to the bottom of the sea in a few seconds."[49]

The humor went both ways, however. Wegener had succeeded in setting up a hygrometer (to measure relative humidity) and a recording barometer in his cabin, and he noted indulgently in his journal that "the hygrometer is read off on each watch with large satisfaction by the sailors. In the ship's log a column for humidity is provided, and they are as proud as Spaniards that we can now fill out this column. Also the barograph is eagerly used. I owe partially my position among the sailors to it. It is consulted daily about five times, and my cabin, where it hangs, is thus a sort of inner sanctum."[50]

On the morning of 7 August the smooth sailing and the accompanying carnival atmosphere of photography, shooting seals, and sightseeing came to an abrupt halt as the ship was beset and pressured severely in the ice. "Today," Wegener noted, "just before noon we were trapped between two gigantic floes pressing in on us. It looked as though we would end up wintering here."[51] The ice released them in the late afternoon, and they managed to find a lead of open water, but the incident prompted Mylius-Erichsen to schedule an "abandon-ship" drill for the next day. It was a standing order that sleeping bags, guns and ammunition, extra tools, clothing, and cooking gear be always at the ready in case the ship had to be abandoned. Kept near the three ship's boats were boxes of provisions and water casks ready to load at a moment's notice—though none of this had ever been put to the test.[52]

On the morning of 8 August, in the middle of the forenoon watch, Knud Christiansen went to the wheelhouse and (on Trolle's signal) began to hand-crank the ship's siren to announce the boat drill. The eerie wail gave the maneuver a surprising sense of urgency, and the crew rushed to their stations, tossing their sleeping bags and gear into the boats and loading the boxes of provisions. The boats were lowered, and after rowing around in the ship's wake for a few minutes, the men made for some nearby ice floes, where they pulled up onto the ice and unloaded the boxes and gear. The three groups waved and shouted to each other across the open water and then began to load up to row back to the ship.

The leader of the third boat party, which held Achton Friis and Peter Freuchen, decided to open one of the provision crates before rowing back to the ship—just to see what sort of food they were supposed to subsist on. The box contained 80 pounds (36 kilograms) of mixed pickles—nothing else. Of the six boxes of "emergency rations" on board the boat, three were nothing but mixed pickles—320 pounds (145 kilograms) in all. The men grinned at each other, relieved that they need not survive on mixed pickles and that dinner waited in the ship, swinging at anchor 100 meters (328 feet) distant.

As they rowed back to the ship, Achton Friis noticed that it was uncannily quiet, and he realized that he could not hear the dogs quarreling and barking, nor could he see any dogs aboard. How odd, he thought. He saw the first boat pull up to the ship and two men jump over the rail. The silence was suddenly splintered by a howling and wailing from the ship that went on for several minutes. What on earth was happening? As Friis hauled himself over the rail, everything became clear, as dogs were spilling madly from the doorway to the mess room, driven by yelling men with dog whips. Whoever had gone last over the side in the boat drill had, in the confusion and rush, neglected to latch the door to the mess, and all 120 of the dogs had gone below. Inside, the ship was a madhouse. The dogs had swarmed into the galley and the larder. They had tipped over the cooking pots on the stove, holding 20 pounds (9 kilograms) of meat and gallons of cooking water, and had eaten it all. They had somehow turned the petcocks on the water pipes over the stove and doused the cook fires and then had broken into the larder and eaten 20 pounds of butter, all the bread, and a large variety of staples, including a sack of flour, another of salt, and a crate of sterno fuel. They had chewed and torn all the leather they could lay their teeth to. They had overturned a large tureen of fruit soup and then run through it and the flour. As they rampaged through the ship, wherever they ran—on desks and tables, on the crew's bedding, in the machine shop, in the engine room, in the hold and the bilge, even on the keyboard of the piano in the main salon—they had left their floured, syrup-soaked footprints.[53]

Mylius-Erichsen was beside himself with rage at the carelessness. Wegener did not mention it in his journal, nor Friis in his, but many years later, when most of the principals were long dead, Freuchen told the story:

> This was the occasion of Mylius-Erichsen's first fit of hysterical fury. We were to experience many of them later, but this was the first demonstration. He screamed and shouted, stamped on the deck and nearly wept in his fury. "I am in command here, I am the owner of the ship and everything in it! Whoever left that door open will get a punishment he will not forget! I'll put him in chains!" he shrieked. Afterwards we could laugh at it but at the time the impression was rather grim. With this lack of balance this man was to be our leader for the next three years, the head of an undertaking which was far beyond his powers. He had no experience in organizing such a group of people and he lacked the authority to make the men respect his orders. We all grew very fond of this man who was more of a poet than an explorer, but we first had to get used to the fact that what he said today did not hold true tomorrow.[54]

Mylius-Erichsen was neither mad nor mentally unstable but merely overwhelmed by what he had undertaken. He was only thirty-three, and he had never commanded anything larger than a dogsled. He had traveled to Greenland only once before, in a four-man party that included Jørgen Brønlund, the artist Harald Moltke, and Knud Rasmussen, the greatest Greenland traveler of all; Mylius-Erichsen had been the least experienced of the four. Now he was in charge of a ship and twenty-seven men, many of whom also had greater experience than he. Authority was parted out between himself, Captain Trolle, and Ring (the ice pilot); Lt. Koch, who had already been on four scientific expeditions, overshadowed his reputation as an explorer.

If Mylius-Erichsen was anxious, he was still capable of decisive action. When the ship broke through the pack into an open shore lead on 13 August, he had the ship turned north and fought for every mile of "northing" in the open coastal water. On the fourteenth they steamed past Cape Bismarck just above 76° north, the northernmost point reached by the German Expedition of 1869–1870. "I was asleep, unfortunately, when we passed Cape Bismarck," wrote Wegener that night, "and thus I did not see the memorable point where the German expedition abandoned its efforts [to reach a "farthest north"]. It is so strange that we steam along here effortlessly in open coastal water, where the Germans, with almost inconceivable effort, pulled their sledges over the ice."[55] Though this cape held a small bay that seemed an ideal winter harbor, they steamed north of it for another day and night, finally stopped by impenetrable pack ice on 15 August at latitude 77°30′ north. They had reached farther north in Greenland than any expedition with plans to overwinter, even though the coast at this latitude had been visited by other expeditions and by innumerable sealers and whalers in the past 300 years.

This was as far north as they could sail. They unloaded a quantity of stores onto the land as a depot against the dogsled journeys north the following spring. Mylius-Erichsen detailed Koch to take the motor launch and a whaleboat in tow and set stores as far north as he could—the boats being able to navigate in narrow shore leads where the *Danmark* would have gone aground. Every kilogram not landed by boat would have to be hauled by dogsled with great effort. The *Danmark* immediately turned south and by the next day steamed—with Wegener at the helm—back to the winter harbor at Cape Bismarck, christened that day "Danmarkshavn."

Greenland Autumn

Their winter harbor and home for the next two years was a snug bight facing south-southeast, protected from drift ice and ice pressure by Cape Bismarck (6 kilometers [4 miles] to the east) and from the prevailing northwest winds by low, undulating

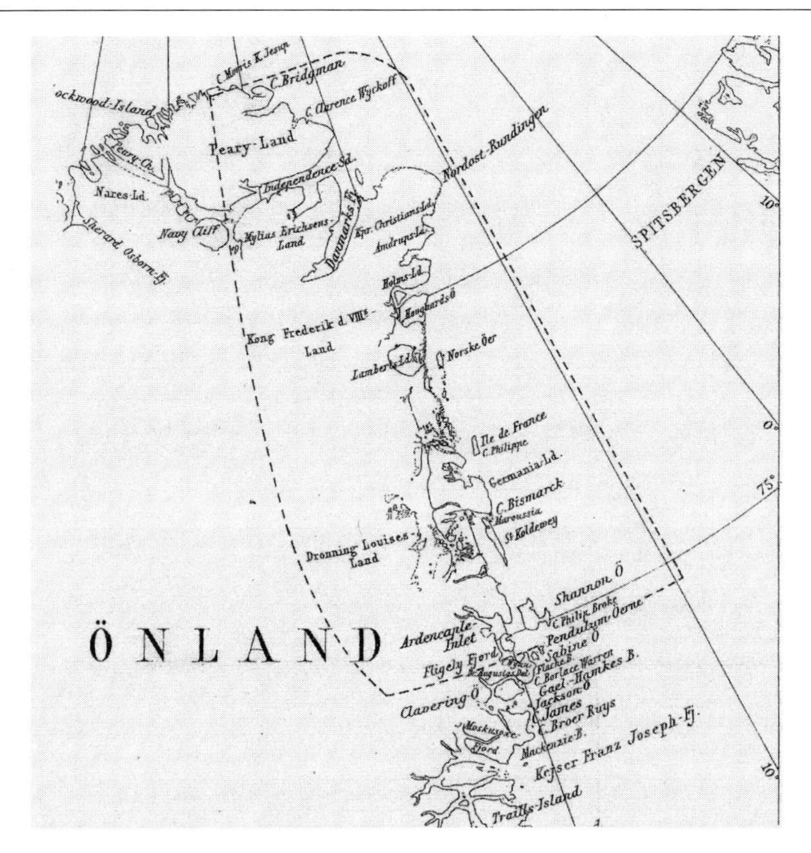

Northeast Greenland. The area investigated by the Danmark Expedition is outlined by a dashed line. Danmarkshavn is at the tip of Kap Bismarck. From G(eorg). Amdrup, "Report on the Danmark Expedition to the North-East Coast of Greenland 1906–1908," *Meddelelser om Grønland* 41, no. 1 (1913).

hills. To the west lay a vast region of lakes and fjords—all unexplored—and beyond that, 60 kilometers (37 miles) distant, the margins of the Inland Ice, last remnant in the Northern Hemisphere of the great ice sheets of the Pleistocene glaciation. The landscape was gray and somber gneissic rock, enlivened to the north and east of the harbor by extensive carpets of Cassiope—that lovely, intensely green shrub with its nodding white and pink flowers. Beyond that was a swampy bog with a scant but welcome growth of sedges and reeds. The site for the shore station was a low, sloping flatland of rock and gravel, flanked by two rushing streams providing abundant freshwater, promptly christened the "West River" and the "East River." In all, it was a lucky spot—an ice-free shore, abundant vegetation uncommon in this latitude, and evidence of game animals; this last feature was crucial as they planned to live as much as possible "off the land." This might be an ideal from the standpoint of the men, but it was vital with regard to the dogs; there was no other way to feed them.[56]

Wegener was satisfied with his new home and anxious to get settled: "The area for scientific work is ample, and seems good to me. The rock is crystalline, undulating and severely and even grotesquely scoured—no problem with the electrical and magnetic measurements. With the width of the flat land and the low slope of the mountains, I can make kite ascents . . . now the scientists can stop being sailors and begin

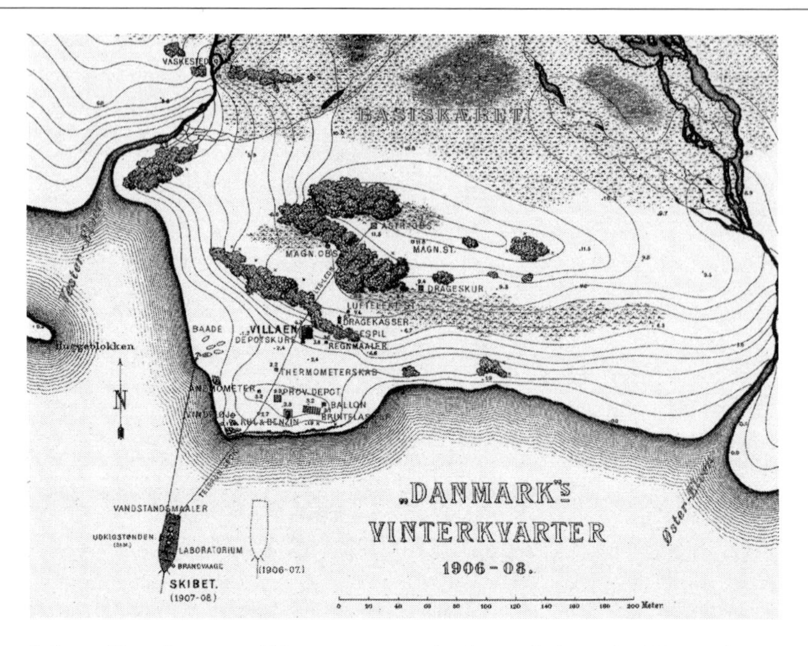

Danmarkshavn. Note that most of the structures on land were dedicated to Wegener's science: the thermometer hut, electrical and magnetic observatories, precipitation gauge, and storage for the kites (*Drage*) and the kite winch. From Amdrup (1913).

their scientific work."[57] This was true, but not, perhaps, as he had imagined. Before going north with the motor launch the day before, Koch had stunned Wegener by putting him in charge of the reconnaissance of the winter harbor and of preparations for the geodetic triangulation and survey of the coast, including the crucial point of anchoring the observation network. Wegener tried to put a brave face on it: "It is probably good to connect the pleasant and the useful. A tour of the surroundings is important for my own work as well."[58]

With the ship anchored and the dogs unloaded, Wegener was glad that work on the station's scientific buildings could begin, but to his surprise (and dismay), this did not happen. Mylius-Erichsen immediately left by boat with all the field scientists, the ship's carpenter, and some seamen to investigate the islands and inlets of Dove Bay to the west, leaving only the ship's real crew—captain, pilot, mates, cook, engineers, and two deckhands—along with Wegener, Friis, and the Greenlander Hendrik Olsen, the latter to see to the dogs. Friis and Wegener were to be free to pursue their "own special investigations," but what that could mean for Wegener wasn't clear: all his equipment remained in crates on shore, and he had as yet no place to hang it or use it.[59]

Wegener seems not to have realized that Koch, with the orders to "make a reconnaissance," had offered him a generous opportunity to go exploring, too. He might have ranged widely around the area, climbed and even named some mountains, taken off in a boat, and in any of these ways have had a share in the general exhilaration of exploring an unknown region in the glory of a High Arctic summer, but he would have none of it. He was obsessed with his aerological work and could think of nothing but that. Friis, on the contrary, immediately got the picture of what was (not) happening at Danmarkshavn and, after a day or two, took off after Mylius-Erichsen.

Wegener stayed doggedly on. After six days of helping with the unloading of the ship, he found his winch, his kites, and his balloons. The last of these were badly stuck together, and he agonized over peeling them apart, fearful he would tear and ruin them. He managed on 25 August to inflate one, but he found this process painful: sand and pebbles had glued themselves to the sticky balloon surface as he unrolled it on the ground, and, on inflation, these were shot at him with some force. He tried to clear an area free of sand and pebbles (to avoid punctures and being hit by rocks again), but every time he got an area completely clear, the dogs would invariably come dashing through and everything would again be covered in rock and sand—the dogs had left the ship and had wasted no time establishing their dominion on land.

He was also discovering what it was to be on his own. He had no assistants and no assistance. He begged some help to wind the wire on the winch and had some dream in mind to start a full aerological and meteorological program on 1 September, but he soon saw the folly of these plans in the face of the general inertia and his own sense of strain. "After Sunday we have only five days [until September]. Shall we succeed in getting everything built in this space of time? It seems impossible that I can be in the house on the first, when no one has made the first move toward building it. No one here has the slightest interest in it but me, and all work proceeds at a dreadfully slow pace."[60]

Everything was different than he had supposed. The coast was low and uninteresting, not Alpine, glaciered, and majestic; in fact, there was not a glacier in sight—in Greenland! He was anxious to begin his observations but needed help to set up the instrumentation, and none was to be had; everyone was away, exploring in the west. Finally, on the morning of 1 September, he managed to test his winch and get a kite in the air, though he only let out 800 meters (2,625 feet) of wire before the wind died. This seems to have improved his mood, as now he had at least one observation: "about 20 minutes after the kite landed the wind came from the opposite side! Obviously the core of a small low pressure center, recognizable only by clouds, barometer and thermometer, and not by precipitation, had passed over us."[61]

Wegener was floundering. He had not made the reconnaissance tours Koch had requested of him, and when Koch returned from the north with the motor launch, he immediately swept up Wegener and headed off east and south to Cape Bismarck and Koldeway Island to build some observation points for the geodetic network. Wegener (finally) wanted to explore a bit, but now it was too late; Koch moved very fast and single-mindedly: "This forced-march touring with Koch is interesting, and one learns how to make the best use of every moment, but it is impossible to combine any second purpose with one of his trips. If I want to photograph these beautiful vistas I will have to come back here alone."[62]

Koch kept him close after their return to Danmarkshavn, and Wegener did his best to keep up with him—much as he had tried to keep up with the "bigger boys" in the exercise hall in the orphanage many years before. He was in awe of Koch: "He is wonderfully suited for this work. He is strong as an ox, with a body of iron, and this incomparable energy, which I so admire. The projects are always well planned. When he is trying to accomplish something he is ruthless in pursing it, and this I admire as well. I know I can learn a lot from him, even living with him is by itself useful since his enormous energy is contagious, and buoys me along in each new task."[63]

There were plenty of tasks, as all the prefabricated instrument huts had to be assembled, their foundations dug, and their guy-wires strung and anchored. It had already begun to snow—the first measurable flurries came on 4 September—and yet Wegener did not have a full meteorological station established: "yesterday was the first snow, but who knows when the instruments will be ready to use."[64]

With Koch's help, he began soon after to make progress in his work, and began to think of Koch as a coworker as well as a mentor. He helped Koch make the observations necessary to establish the cardinal points of the compass and the true north, and then Koch helped him set up the anemometer. He helped Koch set up his astronomical observatory and cement the transit instrument's granite pillar (brought from Denmark) to a secure foundation; in return, Koch assisted him with his second kite flight, on 11 September. Wegener continued to be amazed at the pace of the expedition and how different it was from what he had read of such things and from what he had imagined an expedition to be. "I am learning so much here," he wrote. He continued,

> I am particularly impressed by the extraordinary mobility of this expedition, which is in sharpest contrast with the German South Pole Expedition, where everyone stayed with the ship. Mylius-Erichsen has already been roaming to the west of us in the fjord for almost a month, and Koch is constantly away with the motorboat if this is not needed for other work, and people are continuously going back and forth to Mylius-Erichsen, hiking overland, in short, the expedition has an astonishing radius of action, and we have not even begun to use our main conveyances—skis and sleds. It is important to remember that these small enterprises are extraordinarily facilitated in this region because of the continuous good weather, and by the fjords being completely ice-free all summer. In South Polar Regions it would be much more difficult. I hope, in any case, to learn here enough self-sufficiency that someday I may play on a German expedition the same role that Koch plays here.[65]

This passage is interesting because it makes clear that Wegener already had ambitions to go to Antarctica, but while dreaming about the glorious future in his journal, he also had some time to reflect on the present. In the middle of a (rather typical) diary entry on 14 September, talking about the problems with his mercury barometer, his need for some solvent to clean his air-electrical instrument, and progress on the (woefully insufficient) prefab structure for the magnetic measurements, he suddenly stopped and began to write about himself—the first of a rare series of introspective passages.

> It is remarkable what trivial things one records here in a journal. But it is symptomatic. If I report everything I did during the day it is a sort of justification, it is a victory, that I succeeded in occupying myself productively. For me this expedition is unquestionably very valuable. In the past I have had of course to some extent a certain energy, which I would like to call here, for the sake of contrast, moral energy, but here I am learning practical energy, the energy of occupational work. All of those things which appear insignificant to one, for example washing every day, disposing of any outstanding task, no matter what it is—all these little things, which make up daily life, are the means by which one can learn practical energy.[66]

He was, in other words, overworked, anxious, and slightly depressed, having to push himself even to wash and accomplish small tasks—none of these being uncommon phenomena for expedition novices. He was also a little homesick. On the previous day he had written, "Isn't today Papa's birthday? I really cannot remember for sure. They're probably at home, receiving visitors, maybe Kurt is there too, and they will surely be talking about me."[67] There was no relief from the work, however, and it was as well that he did not have much time for such musings. It snowed often now, and the bay was freezing over. The nights were very cold and the daytime temperatures below freezing as well.

Setting Up and Falling Apart

With the carpenter, Gundahl Knudsen (1876–1948), back from his travels with Mylius-Erichsen, work had finally begun on the large hut that would be the main shore station. This building (5 meters square) and the magnetic and astronomical observatories had been completely prefabricated so that they could be set up and bolted together in a matter of hours. For the smaller buildings this worked as planned, but during the outward voyage all the sections of the larger hut had been tightly lashed over the main hatch of the *Danmark*, and constant cycles of wetting and drying had warped and damaged these panels so badly that Knudsen had to completely disassemble and then painstakingly reassemble them.[68]

By the time the large building was finished in the middle of September, the ground had begun to freeze, and it was exhausting work raising an insulating earth-and-stone wall around it. Scientific research and journal writing ceased for a week in the race with the oncoming winter. "Dead tired," scrawled Wegener on 21 September. "The whole day, strenuous earth-work on the house." This house had no sooner been raised than it had been christened *Villaen* (the Villa) by the crew, and its four residents—the botanist Andreas Lundager (1869–1940), the painter Aage Bertelsen, Wegener, and Koch—were known henceforth as the "aristocrats."[69] Someone found an old rack of reindeer antlers and mounted them on the north gable, completing the illusion of an aristocratic hunting lodge. Wegener and Koch, of course, had to be on shore to be close to their instruments, since winter weather would make multiple trips back and forth from the ship dangerous or even impossible, and summertime trips would have to be a series of rowboat voyages; the stern of the triple-anchored *Danmark* was 60 meters (197 feet) from shore. Bertelsen and Lundager were there simply because there was room for four, though they enjoyed themselves and the relative privacy the Villa afforded.

The "Villa" where Wegener would live for most of the next two years was a Spartan accommodation but quieter, cleaner, and less odiferous than the ship—and it had no rats. It had a single large room 5 meters square (about 270 square feet), with two ample bunk beds flanking the door. At the south end, facing the only window, was a large worktable for four. A small cooking stove stood in the middle of the room, with its chimney passing up through the storage loft above, where sledge materials, tents, and boots were kept. The stove, they soon found, was too small to heat the space and, consequently, had to be constantly fed. There was room to work, though, if one could keep warm.

Wegener installed (in the southeast corner of the room) a barometer and a barograph, a normal clock (i.e., not a chronometer), and his air-electrical recording instrument, with its insulated wire leading via a brass rod to a tellurium-pointed sensor on the roof. There was shelving along the walls for the other instruments in their cases, as well as shelf room for notebooks and the astronomical reference works, including the stellar ephemerides, the mathematical tables, and the nautical almanacs. There was even a working telephone that allowed calls to and from the ship.[70]

The environs began to look "scientific" as well. Along the phone line leading to the ship Wegener set up his thermometer hut, with the instruments in a louvered box

First snowfall at Danmarkshavn, September 1906. The "Villa" where Wegener and Lt. Koch lived from 1906 to 1908 is in the foreground. To the left is the kite winch. Wegener may be seen beyond the Villa at the "thermometer hut," changing the paper on the recording drum. One of the crew is standing guard with a rifle to watch for polar bears. From Achton Friis, *Im Grönlandeis mit Mylius-Erichsen: Die Danmark-Expedition 1906–1908*, trans. Friedrich Stichert, unaltered 2nd (1913) German ed. of 1909 Danish original ed. (Leipzig: Otto Spamer Verlag, 1910).

about 2.5 meters (8 feet) above the ground. These consisted of a thermograph, maximum and minimum thermometers that would register the highest and lowest temperature in each twenty-four-hour span, one of Aßmann's aspiration psychrometers (to measure the humidity), and a hair hygrometer to measure humidity when the wet-bulb thermometer in the Aßmann instrument froze. Closer to the ship were the anemometer, to measure wind velocity, and, on a separate mast, the wind vane. The rain gauge was a few meters from the entrance to the Villa, where someone could reach and empty it daily.

The hydrogen canisters for the balloons were stacked along the shore near the petroleum depot (fire was a constant and terrible danger, and flammables were stored far from the ship and the huts). Wegener's air-electrical and magnetic observatories stood a few tens of meters inland from the Villa, as did a "kite-house" made of provision crates, snugged under the lee of the only rising ground in the area in a mostly vain attempt to protect it from the prevailing northwest wind. This makeshift structure for storing the assembled kites (and sometimes an inflated balloon) was necessitated by the great labor of lashing together kites for each flight and the near impossibility of disassembling them after winching them down, when their lashings were frozen. Surmounting the small rise that protected the kite-house was the astronomical observatory, which featured (miracle of modernity) electric lights powered by dry-cell batteries in the Villa—a creation of Bendix Thostrup (1876–1945), the ship's third mate who, in the view of everyone on board, could make anything.[71] With the meteorological instruments set up—both in the area immediately around the Villa and at the second station slightly inland on an exposed hilltop at an altitude of 132 meters (433 feet)—Wegener felt he could really begin his aerological program.

On the equinox (21/22 September) he was finally able to get a balloon launched: "After all the work, finally, we have come to the desired result. We have made the first balloon ascent!" It was no little thing to get it airborne, though, and his description gives a hint of what it requires to attempt science under these conditions. The balloon was stuck together and frozen, and he had to pull it apart three or four times before it would lie flat. "We worked at this for hours, and the 'end of the song' was that we tore a hole in it. It had to be repaired. The rubber patch would not stick on the mirror-smooth frozen varnish, and so I had to sew it—this was a real delight, outdoors, at −7°C (19°F)."[72]

The repair of the balloon took until noon, when, in Danish Greenland as in Denmark itself, everything stopped for lunch. After lunch, Wegener wanted to hurry back to the balloon work, but a polar bear had wandered across the new ice toward the ship. Everyone on the expedition was wild to shoot a bear, and they all plunged below for their guns and then dashed out on the ice and began blasting away at it. Wegener had none of this sporting fervor and ambition, nor any desire to be killed by the astonishing display of poor marksmanship. "The bear was unbelievably stupid, and walked along a line of hunters stretching almost a hundred meters. I had to wait till someone shot it, and then I could finally go clean the instruments—it is awful, the number of things one has to consider when one is making all the attachments [of the apparatus] the first time."[73] He could not get the inflating tube into the balloon until finally his housemate Lundager happened by with Niels Høeg-Hagen (1877–1907), the other cartographer besides Koch, who took pity on him and helped him inflate the balloon and get it aloft. He was delighted: "even the recording apparatus worked."

The following day he was able to make a second balloon flight and observed something that absorbed his interest for the rest of the expedition—the presence of a sharp temperature inversion. A few hundred meters aloft, above a fog layer, the temperature was 14°C (39°F) warmer than at the ground. He had seen inversions earlier in the month, but usually only 1°–3°—this was a striking example and worthy of further study. Moreover, he was beginning to grasp the weather pattern. Whatever the wind at the surface, above 500 meters (1,640 feet) it almost invariably blew from the west-northwest very constantly.[74]

At the end of the day on the twenty-third he got some long-hoped-for news from Koch: he was to be included in a major sledge trip in November and in the trip to the north the following spring. Mylius-Erichsen had, to this point, left Wegener completely out of the planning for the exploring trip to the north. For this undertaking, already bombastically titled "The Great Sled Trip to Map the Coast of Northeast Greenland," preparations had begun as soon as the station was established—including the building of dogsleds on the Greenland model, the plaiting of dog harnesses, and the training of the Danish novices, supervised by the Greenlanders, in driving sled dogs. This was not only a cartographic endeavor but a political one—to extend Danish sovereignty along a coast also claimed by a recently independent (1905) Norway. Wegener, mentally and physically tied to his weather instruments, had no part in any of these preparations.[75]

Koch, as chief cartographer, had successfully talked Mylius-Erichsen into a major—and risky—dogsled trip in November even as the winter night began. He had convinced Mylius-Erichsen that it would be necessary, if the expedition's mapping of the north in the following spring were to be accurate, to correlate the longitude of Danmarkshavn with the longitude of Germania Haven, some 245 kilometers (147 miles) to the south. This longitude had been repeatedly measured by the Germania Expedition in winter quarters in 1870–1871, and Koch had great confidence in its accuracy. Mylius-Erichsen gave in, partly because he had already committed to a smaller trip himself, depositing messages and locating caches of food to the south as part of their escape plan, should the *Danmark* be crushed in the ice, but he had grave misgivings.[76]

Koch claimed he needed Wegener's skills with the astronomical instrument (a zenith telescope) on the trip, and he had gone to Mylius-Erichsen and insisted that Wegener be included. Wegener and Koch had, it would seem, hatched this plan together,[77] and Wegener was pleased—he had come to Greenland to be an explorer, had he not? Yet with this gratification came a sort of unease about his scientific program, and he confided to his diary, "So I am to take part in the great sled journey, and I also am bound for the northern tip of Greenland! I am very satisfied by this thought, though I doubt that anyone will make meteorological observations in the meantime."[78]

Spurred by the knowledge that he would soon be unable to make observations for some weeks, Wegener pushed forward his meteorological observations and struggled to get some aerological work done. He managed, in the rest of October, to get off eleven kite and balloon launches, but not without his share of disasters. On 4 October he lost one of his precious recording instruments—one of two he had obtained from Hergesell—when the balloon tore in its descent and hurtled to the ground, utterly smashing the instrument on the rocks.[79] He took this disappointment philosophically, noting in his diary that without the sort of assistance he had at Lindenburg, such things had to be expected, and one can see him revising his expectations downward: "If I can just carry forward a [limited] observation program, and gain experience in sled trips and cartography, these two years will be well repaid."[80]

Working anxiously and in haste to achieve even limited observations, he was repeatedly frostbitten, mostly through his own carelessness. On 7 October he volunteered to stand the night watch on board the *Danmark*, which required him to ascend the frozen iron ladder 90 feet to the crow's nest every other hour to take observations. Bertelsen had loaned him a revolver, in case he met a bear walking in the dark to the ship, but in touching the frozen metal parts of his instruments repeatedly with an ungloved hand, Wegener froze his right index finger so badly that the gun was rendered useless as a weapon on the way back to the "Villa." On 10 October he froze his left ear, and on the twelfth the fingers of his left hand, from constantly taking off his gloves to adjust the instruments—nighttime temperatures were already nearly −20°C (−4°F).

On the fifteenth he had his worst day of the expedition thus far: "A Tycho Brahe day." Fumbling with frozen fingers while adjusting his magnetic recording instruments, he dropped and smashed one of Koch's precious pocket chronometers. He was terribly embarrassed. In all the preparations for Greenland, he had forgotten to bring a timepiece of his own. Now that this chronometer was broken, he had to find another and borrow it whenever he wished to make any observations (i.e., many times each day), which was a source of constantly renewed mortification. On the eighteenth he frostburned his face quite badly while winching down a (very successful) kite ascent to 2,400 meters (7,874 feet); the hoarfrost on the kite cable showered him as he winched it onto the drum—by hand.[81]

These repeated accidents frustrated him and left him feeling vexed and inadequate. He was obsessed with having broken Koch's chronometer, and on top of everything else, Mylius-Erichsen paid him a visit and suggested that Wegener should stand night watch regularly, like everyone else. Wegener was appalled at the realization that, in the eyes of the leader, he was not pulling his weight on the expedition. On the nineteenth he fell into a black depression, and on the twenty-second he wrote,

> The last few days have been really bad. Moody! I am brought low by two different things—above all my misfortune with the clock (the balance is broken) and also my clumsiness in not freely volunteering to take a regular part in the night watch, but waiting until Mylius-Erichsen had to ask me if I would. So the last few days I've been really down. I could probably work my way through the worst of it, and I want to work, but I can't keep myself from being intolerable to everyone around me. So I've read a lot the last few days, and now I'm over the hump.[82]

When he did return to work on Friday, 26 October, he found himself able to function but overwhelmed by all he had to do: "This business with the chronometer really took its toll on me, and now all these observations! Today is clear and still. So I should of course make a tethered- balloon ascent, send up a sounding balloon, make some magnetic observations, begin the air-electrical measurements, and take part in the astronomical work. It is all too much! But I believe I can make it through this whole episode, if I can just keep calm."[83]

Fortunately, he was saved from his indecision by Mylius-Erichsen. If Wegener was to take part in the sled trip to the south in mid-November, he had to be able to drive a dogsled, and Mylius-Erichsen wanted to find out if he was up to it. Taking advantage of the waxing Moon, which would be full on the twenty-ninth, Mylius-Erichsen had cooked up a little trip—himself and Wegener and two dogsleds—to Hvalrosødden (Walrus Spit), some 50 kilometers (31 miles) to the east along the coast of Germania Land. The ostensible reason was to check on the cache of walrus meat from the animals Mylius and the others had shot and slaughtered in August and September for dog food—with the hides reserved to make overboots. But it was quite clear to Wegener that this was a test: "Mylius-Erichsen obviously wants to see how I conduct myself on a sled-trip. I haven't had any training, and recently have been nearly immobile. But I'm glad to go. I have to admit it's about time! Other than the two or three days I spent with Koch in the motorboat, I haven't been out of sight of the masts of the ship."[84]

From the moment they left on 27 October, Wegener was stunned by the speed with which the dogs moved. Mylius led the way with his team, followed by Wegener driving Koch's dogs (these were superbly trained, and Koch was stacking the deck in Wegener's favor). They moved along at a trot, the dogs pulling the lightly loaded sleds enthusiastically and with ease. When they hit deep snow, Wegener followed Mylius-Erichsen's example and jumped off the sled and trotted along with the dogs. As the miles unrolled, Wegener thought constantly of the Koldeway Expedition in 1870–1871. It was along this stretch of coast that they had abandoned their attempt to reach the North Pole. Where the German expedition had struggled and sweated, without even skis, and pulling their ungainly and balky sledges through deep snow, Wegener was gliding almost effortlessly along, alternately riding and running at an easy jog—the kind of pace one could easily keep up all day.

That night, at their first bivouac, almost 24 kilometers (15 miles) from Danmarkshavn, Mylius-Erichsen commended Wegener on how well he had coped with the dogs, admitting to the bashfully pleased novice that he was surprised at his success. Wegener, aware of his position, thanked Mylius-Erichsen for his help and professed his earnest desire to learn from him how to be a good dog driver. He wrote in his diary, "This little bit of flattery was not too far from the truth."[85] He had to win Mylius-Erichsen's good opinion if he was to have any chance of participating in the sled journey to the north, or in any forays onto the Inland Ice; all his efforts were directed to making a good impression.

On the evening of the second day, they arrived at the Walrus Spit and disturbed a pack of wolves "in the midst of their evening meal," having torn their way into the remaining cache of walrus meat. Their howling, Wegener learned, was exactly like that of the dogs—"just as if there was another expedition nearby." Their time at the spit was brief, and they loaded what they could carry of the walrus meat on the sledges and set out again the next morning.

Wegener, on the return trip, actually exhibited a modest talent for driving dogs. He learned to harness them to the sledge in the fanwise array used by Greenlanders. He learned how to use commands to make the dogs run in good order (and to use the whip to keep them from crossing and tangling their traces). He learned how to make the team jump up and run, as well as how to stop. Rather than congratulating himself, however, he was filled with admiration for the dogs, astonished to discover how much they liked to work. "It is unbelievably interesting how anxious they are to pull the sled. It is quite moving. Hour after hour they run like this without stopping, as if their only thought was to pull the sled."[86] He also learned, with Mylius's help, to tell one dog from another and to see their individual personalities. All this was to the good, in and of itself. It brought him down from numerical abstraction to the world around him and anchored him in everyday life, a life he had to master if he was to have any significant part in the exploration of Greenland.

What might have passed as a triumphal return was canceled by events at Danmarkshavn. While Mylius and Wegener had been in the interior, safe under the lee of Germania Land's mountains, the first great winter storm had come down on the coast from the north. Freuchen, Wegener's assistant, informed him as soon as he climbed off his sled that, in the hurricane-force winds of the past two days, the entire weather station they had constructed on the "Thermometerberg" at 132-meter (433-foot) elevation a half kilometer away from the "Villa" had been blown away and all the instruments destroyed. There was more. The kite-house Wegener had made of packing crates had collapsed, demolishing three kites. His main thermograph and hygrometer had stopped working. There were other casualties—the leather roof on his magnetic station had blown off, and his instrument was now buried in snow and frozen; the same had happened to Koch's astronomical observatory. He could not even get into the Villa by the doorway, as the entryway had drifted full; he had to enter through the loft window until they could dig the house out. It was three days of constant and exhausting work before he could get back to his diary to note, ruefully, "Yesterday and today we saw—probably for the last time—a trace of the Sun."[87]

He was energized by his trip with Mylius, however, and even the destruction of some of his instruments could not dismay him. He also noticed, with some relief, that he was not the only one suffering depression. He looked at his bunkmates in the Villa with new eyes. Aage Bertelsen, the painter, was so depressed that he couldn't even get out of his bed. Lundager (the botanist) could interact easily, but he didn't rise until noon and was very quiet. Even Koch was having a taste of it, after a big fight with Mylius-Erichsen.[88] After this shouting match, Koch fell into a funk that lasted several days—a long time for him—and he and Alfred discussed the psychology of the winter depressions, how they could be set off by some trivial encounter or event and send one into a tailspin of indefinite duration in which work became agonizingly difficult. Koch's opinion was that it was a direct effect of the darkness and that one could not expect to get through the winter without adapting to the darkness by sleeping more—and that this was acceptable as long as one could get up and do some work each day.[89]

Alfred put most of his energy in the following days into preparing for the trip south. He sewed himself a pair of Kamiks (fur overboots) out of reindeer skin "that would have fit an elephant," and he extended the length of his sleeping bag, having discovered at Hvalrosødden that he couldn't stay warm in it. He worried about his instruments, which had suffered frost damage, and worried about having to suspend all his observations for several more weeks, but mostly he worried about getting frostbitten on the sled trip. He confided to his diary, though not to Koch, that he thought the trip too long, too dangerous, and too late in the year. Yet he still had to prove himself to Mylius if he wanted a part in the northern exploration the following spring. "It is going to be hellish . . . the trip will be seen as having been especially difficult—and rightly so. Our normal temperatures are already −20 to −22 degrees [−4° to −8°F], and by 1 December (we have to count on being gone that long) it will go drop to −25 [−13°F]. There will certainly be days of −30 and −35 [−22° to −30°F]."[90] These predictions turned out to be harrowingly accurate.

Sled Trip to Germania Haven

Wegener's sled trip south to Sabine Island, the site of the Koldeway Expedition of 1870–1871, lasted from 13 November until 4 December, and it gives a good snapshot of the hardships and dangers endured in expeditions of this kind, in return for the most meager sort of scientific results. It was also one of the very few (and perhaps even the first) "winter-night" trips undertaken by European explorers; Mylius-Erichsen always had his eyes open for any bankable "first." Wegener was to come back from it a changed man—harder, more confident, and much more realistic about what he could hope to accomplish.

The trip started on 13 November, having been held up for nearly a week by storms. The party consisted of six men, each driving a sled. Wegener's group incorporated himself, Koch, and Gustav Thostrup, the second mate of the *Danmark*. The other group had Mylius-Erichsen, Karl Ring (the ice pilot), and Jørgen Brønlund, Mylius's longtime Greenland traveling companion. They began well, and with the wind at their backs and the dogs rested, they flew south.

Wegener had the usual troubles of a novice and felt keenly his uselessness the first night as they camped. He still knew almost nothing about setting up the tents, finding the appropriate gear in the pitch dark, assembling the stove, managing and feeding the dogs, or stacking the provisions and gear where the dogs could not get to them. He also found, on the next day, that his imagined "talent for driving dogs" was a still-unrealized possibility; he had trouble restraining his young team. When the party reached a food depot at Cape Pechsel, 80 kilometers (50 miles) south along the coast, an extra crate of provisions was added to his sled to slow his dogs down. Now his inexperience with sled driving created the opposite problem, and his dogs slacked off and fell back, so that Koch and Thostrup constantly had to stop and wait for him. In the midst of his troubles, he found moments of aesthetic pleasure: "It was fascinating to travel at night, completely alone, when one could see, at most, 100 meters. The sleds glided soundlessly across the flat snow and ice desert. Above all I had the impression that I was floating in a boundless void. I had to stare at the 16 hard-working hind legs of my dog team to convince myself that we were actually moving across the snow with our usual speed."[91]

Conditions worsened on the third day, and before long they were in the midst of the "hellish" trip he had feared. Mylius had prudently led them away from the coast onto thicker ice; even in November the action of the tide was capable of breaking up the inshore ice and leaving dangerous open leads of water. The consequence of accepting the safety of older, thicker ice was, however, that they soon ran into "pressure ridges"—walls of ice several meters high and badly broken, like stone rubble. As they approached the inshore island known as "Haystack," Wegener's inexperience took its toll, and he toiled miserably. Koch was patient, noting charitably in his diary that "the ice around Haystack was easy enough to get over, but presented difficulties to those unpracticed in managing dogs and sledges."[92] Even where Wegener could follow the others into gaps and low spots in these ice ramparts, his sled repeatedly lurched forward from the pull of the overeager dogs and then went off balance and slid backward, smacking into his legs and running over his feet.

In the afternoon of the third day, as he struggled to catch up to Mylius and Brønlund, he was only able to overtake them because their sleds had collided and their dogs had become entangled. As they untangled the harnesses, the dogs signaled the presence of a bear, and Mylius and Brønlund grabbed their guns and ran off to shoot it, leaving Wegener with three sleds of excited, wildly baying dogs. Brønlund shot the bear and called for Wegener to come help them. Suddenly Wegener found himself, in the dark, at −20°C, helping to skin and butcher a polar bear. As he worked, he noticed that the wind was picking up and the temperature dropping, and within moments it was snowing hard. With a storm in the offing, they had now to find shelter, and so they drove the sleds toward the shore of Haystack to camp on land. They pushed their sleds uphill, looking for shelter, and finally pitched their tent in the lee of some large rocks. The wind gusts were frequent and powerful, and as they were unloading the sled, a blast of air picked up Wegener's sleeping bag and carried it away. He ran after it and found that it had blown up onto a rock overhang, where it fluttered in the wind like a flag, and he had to wait there until it dropped back to earth.[93] When he got back to the camp and turned to the sled again, he noticed that one of his prized "elephant-kamiks" had also blown away.[94]

The next day, the sixteenth, was worse still. It was overcast, pitch-dark, and windy, with blowing and drifting snow. Brønlund advised staying put, but Koch, anticipating the next storm, wanted to get farther south, and Mylius agreed. They set out along the shore of the mainland on the "ice foot," the belt of ice that forms along a shore between the high- and low-tide marks. Such travel, when available, had the advantage of snow cover, and therefore stability for the sled runners, without the struggle entailed by pressure ridges farther out over the water. At one point, however, an obstacle forced Mylius to leave the ice foot and drive onto the new, glassy, inshore ice, and before he could do anything to prevent it, the wind took his sledge and began to blow it away from the shore across the nearly frictionless ice, dragging the dogs with it. Wegener and Thostrup, following closely, soon found themselves in the same predicament, unable to get traction for their sledge runners, "at the mercy of the wind and with no semblance of control."[95]

If Wegener's group was having a bad time, Koch, Ring, and Brønlund were worse off. No sooner were the sleds under way again than Koch had an accident. "On continuing through the screw-ice [pressure ridge] to reach the land, Koch was so unfortunate as to fall down from the top of a hummock over his own sledge. One of the sledge uprights struck him in the breast just at the spot where he was carrying 2 chronometers in a bag against the bare chest. The glass of one of them was broken with the blow, and the chronometer at once stopped."[96]

If Brønlund had not been there, the groups might never have found each other. It was snowing hard and blowing a gale, and Koch's and Ring's eyelids were so glued together by the snow that they could barely see. Then Brønlund stopped and shouted to them that his dogs had scented something (not a bear) that could only be the tracks of the others. Koch thought he was making an ill-timed joke, but Brønlund left them (and his sledge) and disappeared into the dark, returning in fifteen minutes with the news that he had found sledge tracks on a patch of snow-covered ice that led onto the land. Following a trail only Brønlund could see, they soon arrived at the camping place: "We had just pitched our tent in the lee of a huge boulder, and were making some coffee," wrote Wegener, "when to our joy, up came the other party, who had found us by virtue of Brønlund's keen senses."[97]

Things could have been worse, and they soon were. That night a blizzard set in and blew furiously for two days, trapping them in their tents. At least this allowed some variety in their diet. They had been living throughout the trip on boiled dried codfish, bread, butter, pemmican, and coffee. On the seventeenth, with time to kill, they made a blood pudding with macaroni and had some dried fruit—a seemingly small thing, but under such pressure any slight variety becomes a notable event.[98] Toward evening of the eighteenth they were able to emerge and repair their sleds, the lashings of which had worked loose and were in need of daily attention.

The nineteenth was clear and calm, and they felt the bitter chill in the air as the temperature dipped to –28°C (–20°F). It took two hours of walking about with a lantern to round up the dogs, which had scattered to find shelter from the wind; eventually they found most of them, but not all—two dogs had run away completely, a bad sign.[99] When they finally got going again, they had good luck, driving south across relatively smooth ice for almost 85 kilometers (52 miles), to the southwest corner of Shannon Island, in a single day. Here the groups, by prearrangement, split up. Mylius went east with Brønlund and Ring to search out depots and food caches from previous expeditions on Shannon Island. Koch, Thostrup, and Wegener headed south to Germania Haven.

Wegener was learning rapidly about traveling in these conditions. He had seen the outcome of trying to push on in bad weather: danger, disorientation, overexertion, and little progress. He liked and admired Koch, but he recognized the danger latent in his impatience and his tendency to press on when it was neither wise nor safe to do so. (Wegener could also rest easy about that chronometer incident a month before—now that Koch had smashed one as well.) Mylius-Erichsen, for all his faults as an expedition leader, was an experienced polar traveler. "Mylius-Erichsen's Principle—and I think it a very good one—is to travel only under good conditions, not stay out too long a time, but travel as fast as you possibly can."[100] This last condition was important now, as it stayed bitterly cold and it cost them much to move. Wegener was already losing weight—he could feel it—and only some of this was dehydration; the rest was the stark inability of their rations to match the metabolic cost of their exertions.

Pushing their reluctant dogs hard, by midnight they were within a few miles of their destination, and on their way to learning one more lesson about the danger of "staying out too long." Koch, anxious to reach Germania Haven to make camp, took the group off the difficult ice foot along the east coast of Sabine Island and onto the new, glassy ice on Pendulum Strait. With the wind at their backs they moved very fast—Koch far out in front, Thostrup off to Wegener's right. Wegener noticed that the ice was getting thinner (darker) but holding them well—then suddenly he heard Koch yelling, and Koch appeared out of the blackness and threw himself on Wegener's dogs to stop them, as Wegener pulled back on the sled, stopping it just before a lead of open water. Koch's dogs had seen it just in time and stopped, but his sled had swung around them and was now in the water, floating. Everything on it was soaked—his sleeping bag, the food crates, the toolbox. Worse, without thinking, Koch had tried to grasp the iron runner of the sled to pull it out of the water, and the runner, at −20°C, had blistered the inside of all his fingers on one hand—freezing it and rendering it useless. With Thostrup's help they hauled the sledge out of the water. Of course, everything wetted froze solid in a matter of minutes. They had a terrible time getting ashore—the ice foot was so broken up that the sleds could not pass over it, and they had to unload all three sledges and carry their loads, one box at a time, over the ice jumble at the water's edge, often falling through thin ice into the shallow water. It was hours past midnight when they threw themselves down to sleep.

Wet and exhausted by their struggle to set up a bivouac, they slept until near midday; when they awoke, they used the faint light off to the south to look for a way past the open water. There was none—lead after lead opened away across the full width of the strait. It would be impossible to reach Germania Haven from the east side of the island. It would be necessary instead to go completely around it: back north, then west, then south again, hoping to approach Germania Haven from the west—a trip of at least 60 kilometers (~40 miles). There was no alternative. After a hurried meal of pemmican and coffee, they set out.

They drove back north along the shore of Sabine Island, and things went well until afternoon, when they turned west and south into Clavering Strait, between the island and the shore of Greenland. When, after more hours of struggle, they reached the south shore of the island, a few miles west of where they had stood twenty-four hours before, they found themselves once again on new, black ice—now so thin that each footfall of the men and the dogs caused the microorganisms in the seawater to phosphoresce—"which added an air of pure fantasy to the whole situation."[101]

Koch was afraid that they were going to fall through the ice and perish, so he pulled them inshore as far as they could go, stopping frequently to probe the ice thickness with his large knife. At 11:00 p.m. they finally saw the ruins—like old barrows—of the German observatory buildings looming out of the darkness. They pitched their tent next to the ruin of the German magnetic observatory. As they were bedding down at 2:00 a.m., Koch, exhausted, pulled out his two remaining chronometers to wind them—his nightly ritual—and discovered, to his shock and immense frustration, that they had both stopped. Nine days of intense, dangerous struggle—for nothing. Without chronometers it was impossible to make a geodetic measurement of the longitudes. The entire trip was, from his standpoint, futile, wasted, superfluous.[102] They had arrived at Germania Haven and not fallen through the ice, but the geodetic work and the principal scientific point of the trip had been lost. What to do? Koch was severely hypothermic and frostbitten and could not get warm in his wetted/frozen sleeping bag.

The next morning, a Thursday (as if such a thing mattered there and then, and if it could be called a morning, with day consisting of a faint band of red on the far southern horizon), the party set out to do something, anything, to get themselves out of the tent and put some meaning back into the excursion. Koch's program was, as Wegener noted in his scrawled diary, already "drowned," but there was still work that could be accomplished.[103] This was important to Wegener, who still measured everything by the amount of scientific work performed. They hiked a short distance along the peninsula on the east side of the harbor and searched throughout the forenoon for the cairn built by Captain Koldeway in 1870, purported to contain record of the Germania Expedition. From a scientific standpoint this was not essential, but the leaving and the finding of such communications had enormous emotional importance for members of polar expeditions. It was a way of making contact with history, of creating a human context for their own action in the midst of a dark void. If the *Danmark* group could not, on this try, extend Koldeway's geodetic measurements, they could at least connect with his history and perhaps leave a mark of their own for others to find. In this aim, however, as so often on this trip, luck deserted them. They found the spot, but the cairn was broken open and there were no documents inside (they were never found, though many subsequent expeditions made the attempt to discover them).

That night a storm came down from the north and pinned them in the tent all the next day. As difficult as it was to move about in the cold outdoors, it was harder staying in the tent in their sleeping bags; they were cold, hungry, and dispirited. Wegener did not even bother to make a diary entry.

As the storm blew itself out toward evening on the twenty-fourth, Wegener and his companions had a pleasant surprise as Jørgen Brønlund appeared suddenly at the tent flap. The news was good. Mylius-Erichsen had located the depot at Bass Rock. This was no ordinary depot. It had been prepared in 1901 as a cache of emergency supplies for the Baldwin-Ziegler Expedition and had been checked by a support ship in 1905. Brønlund produced tobacco, of which he had found a huge parcel; he was delighted: like all the members of the Expedition except Peter Freuchen, he was a heavy smoker. More important, from the standpoint of Wegener and the others, was the hoard of chocolate candy Brønlund delivered. They were starved for fat and sugar and had been living on coffee, sausage, and dried fish for weeks. They stuffed themselves with as much as they could eat.

Brønlund and Thostrup left at noon the next day to help Mylius-Erichsen transfer the contents of the depot to a place where they could be ferried back to Danmarkshavn. Now it was just Wegener and Koch, and they could begin to do some science together. Wegener especially wanted to measure the magnetic declination (horizontal variation) and inclination (vertical variation) at Germania Haven. The German expedition in 1870 had made very accurate and extensive measurements of these two quantities, and Wegener wanted to learn the amount of variation in both after thirty-five years. Most of this interest was scientific, but some was also a desire to be part of the (short but very proud) history of German polar science. There was an additional personal reason: Wegener's old professor at Berlin, Förster, had helped plan the 1870 Germania measurements; they were an extension of his interest in the aurora. Wegener, in preparation for this work, had spent a great deal of time in early October at Danmarkshavn working with a self-recording declinometer, making adjustments and corrections. That instrument, he hoped, would run continuously while he was in the south, and he could then compare simultaneous readings of declination at Danmarkshavn and Germania Haven.

The study of Earth's magnetism intersected with a subject Wegener knew a great deal about and cared about as well: the motions of Earth's poles. The geographic North and South Poles of Earth do not coincide (except accidentally) with Earth's magnetic poles, and continuous measurements had shown a long-term and erratic migration of the magnetic poles—much larger than the modest oscillations of the pole of rotation Wegener had observed in Berlin with Marcuse. Declination measurements aimed to grasp Earth's magnetic behavior by mapping its motions and currents, just as in the study of the circulation of the ocean and the circulation of the atmosphere, but polar offsets of all kinds, and especially long-term migrations of the poles, were of interest in their own right.

The instrument that Wegener and Koch employed was a magnetic theodolite. This was, in most respects, a standard astronomical theodolite, and Alfred knew very well how to use these instruments; he had evaluated them and even constructed them. The modification that allowed this instrument to perform its special function was a tiny magnetic needle suspended from a fiber inside a glass tube perpendicular to the theodolite mount. Glued to this fiber was a tiny mirror with a scribed line across its face that reflected a scale marked off in submillimeter divisions of the 360 degrees of the compass. To use the instrument (in this case), one aimed the theodolite telescope at Polaris, the North Star. Then one released the magnetic needle to rotate freely, and as it came to rest, one read off the angular difference between the true north and the magnetic north, visible through the telescopic eyepiece as a reflected compass reading.

The observations were simple in theory but quite difficult under these conditions, as Wegener recorded in a trembling hand in his notebook. It was so cold (–15°C [5°F]) that their hands shook as they tried to adjust the instrument: "It was agony to observe in the cold weather." Because it was an older theodolite without an electromagnetic damper (to lock the needle of the compass immediately once it came to rest, pointing to the magnetic north), every bump or jostle caused the needle to oscillate, and they had to wait for it to come to rest again. Sometimes even the wind started it oscillating again, and there was a good deal of wind. Usually such measurements were taken in a protected area like a hut; here they were out in the open. It was, moreover, so dark that they could barely see the reflection of the mirror scale in the light of the small lamp on the theodolite base.

Koch did what he could to help, but after about two hours he became severely hypothermic and began actually to go into shock, and Wegener had to take him back to the tent. Koch shivered uncontrollably and could not get warm in his frozen sleeping bag, wetted in the accident fours days earlier. There was a little heat from the paraffin stove, but not enough to get him warm. Finally, in the early morning hours, in the shelter of the tent, and with the help of a lot of coffee, Koch began to revive.

Alfred took the time to assess the measurement program in his notebook and to make the calculations from their scant observations (eight in all). Only four of these were at all reliable, and they showed a lot of variation—indicating not compass variation but "more noise than signal." Without chronometers (they had all stopped) they could neither measure the longitude nor be sure of the time, and Koch seemed too ill to be able to help without risking his life.[104]

The next morning Koch was significantly better, and they spent a full day measuring the declination and inclination of the compass. That night Wegener took a walk in the moonlight alone: "It made an extraordinarily strong impression on me. I had, here, for the first time in my life, that feeling of desolate, inconsolable loneliness that has so many times descended upon men in polar regions, and brought their work to a paralyzed halt."[105] This feeling of loneliness was intensified by the surroundings.

Wegener later told Achton Friis that he had thought constantly about the German expedition while at Germania Haven, where the ruins of the old magnetic observatory stood in the dark twilight like the shattered battlements of some old castle. He also told Friis how he had thought "how often, from this place, in the hostile winter night, must their [the Germans'] eyes have longingly swept the southern horizon, where that glowing-red band of light told them a comforting fairy-tale of the mild sunshine of their faraway homeland."[106]

Wegener and Koch were intensely relieved when Thostrup arrived on the evening of the twenty-eighth and broke up their brooding reverie. Thostrup had wonderful news: time to leave! Moreover, he was laden with spoils from the American depot: dry sleeping bags, blankets, socks, and even finnesko—those marvelous reindeer-skin overboots of Lapp design, which could fit over several pairs of socks and were lined with a dry grass that wicked moisture away from the feet. Best of all, Thostrup had brought another hoard of candy, on which they feasted before sleeping dry and warm for the first time in many days and nights.

The trip back, begun the next morning, turned into a race against the cold. The temperature hovered near −30°C (−22F°). They drove relentlessly, covering nearly 60 kilometers (~38 miles) a day, and on one day more than 80 kilometers (~50 miles) in slack wind and clear skies. The dogs were pulling well, and the moonlight behind the clouds created, for imaginations starved of visual stimulus, "a fantastic scene. Koch hit the nail of the head when he said he wouldn't have been surprised to see fire issuing from the muzzles of the dogs [with their breath coming in great clouds]. The other sleds slid by noiselessly in the darkness, like ghosts, scarcely visible, and the pale light like that one sees at home before a breaking tempest."[107] Wegener, exhausted, was repeatedly jarred from this Faustian dream as he fell asleep while driving his dogs and was pitched off his sled, waking with a jolt and having to rise and run after it. They arrived "home" near midnight on 4 December, having covered 315 kilometers (~200 miles) in five days.

Reflections at the Turn of the Year

The rest of December was a curious time for Wegener. His mood swung wildly between episodes of furious energy and bouts of apathetic immobility. He pressed on with his scientific program, but he was becoming uneasily conscious of the fact that his scientific results so far were meager and that he had not made a single real discovery. He could see now what Koch had seen long since: that the structure of the expedition worked against serious science; the venture was only nominally scientific and was in reality entirely directed to Mylius-Erichsen's dreams of great geographic discovery in the following spring. Wegener was learning a great deal, but already, with twenty months still to go before the *Danmark* left Greenland, he began planning his *next* expedition.

He spent much time observing himself and his companions and thinking about the winter-night depression that plagued them all and that was proving so deadly to systematic scientific work. "It is a remarkable phenomenon," he wrote. "It all turns on a single point: deprivation of visual stimuli. One finds such relief if one detects, at midday, the mountains of the surrounding environment, even if only in outline—what an astonishing impulse to action one draws from even a single glimpse!" He found the same phenomenon indoors: "It is extraordinary the lengths to which one goes to find some visual stimulation. With all-consuming interest one examines the photos one has made, and pages restlessly through every available book. How one loves the pitiful petroleum lamp, which hangs over the desk—and how one hates all outdoor work in the blackness.—I think that for any future expedition one should plan, for the winter-night, work that must be accomplished by lamplight." Thinking of his color camera, he imagined what relief they would all obtain from a slide projector: "It would provide the stimulation we so lack."[108]

When not thinking about the psychological constraints on arctic science, he had ample time to reflect on the state of his aerological program. October had been good, though the eleven flights had been costly in effort and frostbite. November, however, had produced but a single ascent, and when he tried a kite ascent on 17 December, the first in more than a month, the results were depressing—he could not get the kite aloft, in spite of tremendous efforts. Looking back over his experiences, he felt lucky to have any winter observations at all: "if you add bad weather to the prevailing darkness—it is almost impossible [to send up kites]. . . . I completely underestimated the difficulties of the winter night. Objectively, these are quite extraordinary, due to the primitive state of my equipment. If I had a decent kite-house, one I could work in, that would help a lot." Not for the first time he bemoaned the lack of help: "If I had here an interested assistant, as I had in Lindenberg, then I could be sure that in his free time he could take care of the little details, the sum of which lead to insurmountable difficulties. Koch is also annoyed at the lack of a real assistant."[109]

Wegener's hypothesis of lack of "visual stimulus" as a cause of *Energielapsus* (loss of vitality) in the winter night got a good test on 15 December with a brilliant display of the aurora borealis. Even Brønlund, who had lived in Greenland all his life, said he had never seen anything like it. Wegener reported that in the next days he was "murderously energetic," and this energy led him, indeed, to one small but real discovery.

On the night of 21 December—the winter solstice—he spent some time observing the Moon, then in its first quarter. He noted that its margins were sharply deformed and that the stars were twinkling (*szintillieren*) with exceptional brilliance, more than he had ever seen. Something similar had happened near the end of September, and he had noted it in his diary.[110] Now he took down off the shelf his copy of Marcuse's *Handbook of Geographic Position Fixing for Geographers and Explorers* (1905) and copied out, into his weather diary, Marcuse's remark that "experience had shown that the higher the latitude, the more the stars twinkle, and more in winter than in summer, and that this must be accounted for in calculating a star's zenith distance."[111]

Wegener had concluded that atmospheric inversions in the first tens of meters above the ground were causing the apparent deformation of the Moon's disk and the exaggerated twinkling of the stars. He had noted such inversions in kite flights from September on; they were among the most interesting findings in his aerological work thus far. "It is small wonder that in winter and in polar regions such inversions occur, which, with their multiple layers and wind speeds, cause variation in the refraction [of the incoming Moon or starlight]." This was pleasant enough, as well as of some import for his and Koch's geodetic work, and even allowed him to mention his own work on Helmholtz waves to himself. What interested Wegener most, however, was the sense that he was right and the experts who had written his textbooks were wrong. "The generally accepted method [for reducing the refractive anomaly], as in Pernter's *Meteorl.[ogische] Optik* and also Markuse [*sic*] by comparison of the air temperature and water temperature for observations made at sea is completely idiotic. One must see . . . that the air and water temperatures have nothing to do with one another, whether measured at the height of the deck or in the crowsnest. The correction-chart on page 58 of Marcuse, reprinted from Koss, with azimuthal wind speed and air-water [temperature] difference is the summit of ridiculousness."[112]

The mood that produced the above remarks, at once euphoric and combative, stayed with Wegener the rest of the month of December. He was transfixed by the aurora and tried feverishly to figure out a way to study it. The night watch in the crow's nest, previously such a burden, now seemed a wonderful opportunity:

> I've seen the northern lights often enough now that I have a vivid impression of them. When one sees this magnificent phenomenon for the first time, even while being mesmerized by its beauty, one has the sense of being hopelessly overmatched. What am I to measure of this never-still optical phenomenon, playing across the whole sky? No sooner do I aim my theodolite at the exact spot, than the Northlight is gone before I have even made an adjustment, only to reappear in a completely different part of the sky. If I expose a photographic plate, it darts away before making any impression on the plate. It defeats all our instruments![113]

Wegener's desire for an auroral picture became an even bigger joke among the crew than Jørgen Brønlund's and Achton Friis's as-yet-frustrated desires to shoot a bear. Peter Freuchen, already harboring the literary ambition that would make him one of the most popular nature writers of the next half century, was in charge of orchestrating the Christmas celebration, and this included a special number of the expedition's "newspaper," the "*Polarpost*." His effort comprised a variety of goofy and entirely spurious communications, such as the following notice, purportedly from Mylius-Erichsen: "Given the currently worrisome shortage of food for the dogs, we will hold a lifeboat drill at 9 am sharp tomorrow."[114] Wegener came in for some kidding, too. Freuchen had aided Wegener in developing the blank negatives that always just missed the aurora, and on the first page of the *Polarpost* Freuchen therefore included a black square entitled "*Kohlenschleppen*" (Coal-Hold) with the note, "From the Polarpost's photographer, Dr. Alfred Wegener, whose specialty is color photography, we have the below-presented extremely well captured photograph of the inside of the coal hold on board the Danmark."[115]

Wegener was not normally thin-skinned, but he seems not to have been amused, nor did he partake willingly in the Christmas festivities. On Christmas Day he wrote, "I survived Christmas Eve with my stomach reasonably intact, but 'one shouldn't praise the day till evening comes.' Tomorrow is another day off and the day after too. Only then will this excessive gluttony cease."[116]

After the midday meal, he returned to the Villa and worktable, but now his thoughts were of home, and he was struggling with conflicting emotions even here, which led to the following passage in his diary:

> Today they [my parents and Toni] will all be home in the usual way with friends, and Kurt will naturally have taken the holiday with them. What will that home look like to me when I return? I think I will send my first telegram with some apprehension. But I also feel that any sentimentality [I experience at my return home] will hardly suffice to hold me back from coming out here again. Out here there is work worthy of a man, work that gives life meaning. Let the weaklings stay home and get all the theories in the world by heart; here outside, to stand face to face with Nature and

look it in the eye, to test the keenness of one's sense in probing Nature's riddles, gives life a meaning I had never even dreamed of.[117]

That afternoon he took a long walk alone. The ship's company had trooped over to *Villaen* to see the Christmas tree that Aage Bertelsen had put up and to drink coffee—but Wegener was in no mood for company and went outside.

Today I conferred for a long time outside with the dead-still air, (it is now blowing again) and enjoyed the silence of the polar night. How cold and silent lie these hard rock hills—once polished by enormous natural forces! Nothing moves. Even the sea lies in icy rigidity glittering under the moonshine that penetrates, with difficulty, through a veil of ice crystals. So it is along the whole endless extent of the East Coast of Greenland. Only in that black speck down there, in the *Danmark*, the sight of which, with its masts leaning over to starboard, is so familiar to me, is there life and movement. Otherwise silence, nothing but silence—dead silence. Only one natural force is at work here, and it works quietly but incessantly—the cold. Its goal is to turn all Nature to stone. Slowly but inexorably the ice-crystals grow, and the running drop freezes. The air itself becomes more and more sluggish. At the moment, it seems as if this work is complete. A living animal in the picture before me would be unthinkable, outrageous. The ice begins to groan and ache. The tide is coming in. The pulse of the sea still penetrates this icy armor. But will the cold succeed in silencing even this life spring? The dogs suddenly make their presence known, destroying the picture. Now I notice that in my light clothing my ears are almost frozen, that poetry must give way once more to reality. But I plan to afford myself this pleasure in the future more often. For this to happen, of course, I must be alone, completely alone."[118]

6

The Arctic Explorer (1)

GREENLAND, 1907–1908

> This is really a gypsy life we lead. We are pinned down by bad weather, and have eaten an unbelievable amount, from which fact one can see that our daily rations, morning and evening, are insufficient, at least for conditions like this. So I stuff myself as full as I can. My foot wraps under my kamiks have, remarkably enough, held together. I can now change them for the dry pair. My finger [badly frostbitten] is now better—most of the skin has sloughed off. Countless minor injuries, all healing well.
>
> Wegener's Journal, 18 April 1907

In late December 1906 Mylius-Erichsen had accepted and reviewed Wegener's scientific plan for 1907, but in January Mylius told him that he would not, after all, play a significant role in the "Great Sled Journey to the North," the excuse being that there were "not enough good dogs." Wegener would instead be allowed, at most, to make some maps of the regions immediately north or south of Danmarkshavn. This was simply another in an endless string of changes of mind and of plan on Mylius-Erichsen's part, and Wegener, resigned, took the disappointment in stride: "It will be interesting anyway, when I get to do some independent cartographic work." What was harder for him to bear was a growing sense of isolation from the rest of the expedition. He still could not comprehend the Danish way of doing things. They had partied incessantly at Christmas, but "the New Year began without ceremony, just punch and cake and 'Happy New Year and thanks for the Old.' There were no speeches. I cannot imagine on a German expedition the leader not seeing and taking the opportunity to say something about the results achieved and the prospects and hopes for the expedition."[1]

Wegener drew up a new year's assessment of his own work and congratulated himself on what he had attained. He was proud of his meteorological station and data collection, as well as the monthly twenty-four-hour measurements in the crow's nest (taken also simultaneously on the deck and at sea level—with the stark temperature contrasts in the first 50 meters [164 feet] above the surface). He was certain they would arouse interest in Europe. The yield of aerological data, from his kites and balloons, was "respectable," and he hoped to crack the problem of midwinter flying to get more records in the coldest months in winter 1907/1908. The atmospheric electricity measurements had been a disappointment, but there was still hope, and he had the solid magnetic observations from Germania Haven. Then there was the harvest of photographs of cloud forms and inversions, as well as landscapes. "Of everything that 1906 brought me, these [magnetic observations and photographs] matter most to me. And is not this personal scale ultimately the only standard?"[2]

Yet, notwithstanding his Christmas resolution to "go it alone," he fretted at his marginal place in the expedition, a separation reinforced by his inability to speak or be understood in Danish. He could see how things could be improved and knew he could be playing a substantial role if only "I had the use of my voice!" Unable to play a steering role in this expedition, he dreamed of expeditions to come. He indulged himself frequently in reveries of what it would be like on a future *German* expedition: "What a joy it would be to work in those circumstances." And what a joy it would be to be listened to, and someday even to be obeyed. He fantasized about the independence of command but admitted to himself, "When I get back I'll still be too young to lead an expedition, but perhaps I could go with Drygalski to the Antarctic, and make sled trips there."[3]

He was, indeed, still too young to lead, as he often had reason to note. Even Freuchen, his nominal assistant, went his own way. Freuchen was in charge of changing the paper on the thermograph at the remote station, the "Thermometerberg" on the 132-meter (433-foot) hill inland from the shore station. Because Freuchen (like Wegener) was stubborn about obtaining measurements, he had badly frostbitten his forefinger. Wegener then ordered him to wait for a break in the weather before changing the paper, so as not to injure his hand further, but Freuchen paid him no heed, even enlisting Mylius-Erichsen to go with him—and of course he again badly froze his finger, which reddened and lost all feeling. Wegener just gave up and rationalized it away: "He'll live through it somehow," he wrote in his diary; "everyone has to learn through his own experiences."[4]

The work was always his solace. In January, he managed to get kites with meteorographs aloft on six different days, and each time he got his apparatus back with an intact record. He was determined to obtain a full calendar year of aerological observations in 1907, and to this end, as a part of the scientific plan he filed with Mylius-Erichsen, he had enlisted the ship's engineers in helping him man his kite winch. Both Ivar Weinschenck (1880–1963), the first engineer, and Andreas Koefoed (1882–1951), the second, would work with him throughout 1907 in maintaining and operating the winch. Once trained in the scientific part of the enterprise, they would be able to send up some balloons or kites while Wegener was away doing independent cartography.

Although the aerological results were encouraging, they were brutally difficult to obtain, as the following suggests:

> Even those familiar with the special difficulties of night ascents accomplished in Europe can scarcely begin to imagine how much force of will it takes to carry out these measurements here—the primitive equipment, the complicated manipulations and tinkering connected with the hauling out and assembly of the kites, the attaching of the measuring apparatus, the temperature and wind measurements and so on, all of it done in pitch darkness and deeply drifted snow at temperatures more than 20° below zero [−4°F] with winds that, even if they are rarely more than 10–15 m.p.s. [23–33 mph] still in Arctic regions have to be classified as storm-winds. The results are invariable: after a short time one is forced, with the lantern blown out, with frozen fingers, toes, or nose, and with eyes glued shut with snow, to stumble back to the house.[5]

Such difficulties were daunting, but the work outdoors was preferable to what had now become the bane of Wegener's existence: sewing. Mylius-Erichsen had set everyone (except those already in the north laying down depots) to work at sewing—sleeping bags, provision bags, clothing. It was one of the expedition's plotlines that they should "live off the land," and they had all agreed to work at nonscientific tasks on board the ship. Yet this was more than Wegener and many of the others had counted on. In early February the morale of the expedition, already low with winter depression, fell lower still, and the discontent with Mylius's leadership, or lack thereof, was as obvious as it was ominous. Wegener, in his diary, put his finger on the basic problem: Mylius-Erichsen was not a scientist. He was interested not in science but in geographic discovery and exploration and the planning and recording of his own exploits. He didn't know enough science—any science—to talk with the scientists on their own terms and thus encourage and advance their scientific work. Because he was not interested in science, he sacrificed it at every turn to advance his own goals. "It is a fundamental misuse of the expedition's resources," wrote Wegener, "and has forced everyone indoors. Koch sews like his life depended on it, and cares not at all about his science. The painters don't paint, they sew."[6] At least, he consoled himself, he was allowed to do his sewing in the "Villa," far from the turmoil and racket of the ship, and beyond earshot of the increasingly vocal arguments taking place there.

The time spent with needle and thread at least gave an opportunity for thoughts of the future. Wegener was making "career plans" and, as is often the case in the young, these shifted easily and often. The magnetic observations, with which he had recently been so pleased, now seemed a bore. "Meteorology, especially meteorological optics and flying kites, is still the field in which I excel, and which I pursue with the greatest willingness and enthusiasm."[7] He was also increasingly drawn to the idea of going to the South Pole and was making expedition plans with Koch. As they sewed their way through the month of February, Wegener grew more and more entranced with the idea and put together an elaborate expedition plan, in which Koch encouraged him. While Koch was an experienced dog driver, his Icelandic cartographic experiences had given him even greater respect for the capacities of the Icelandic Pony. He had tried to bring them along on this trip, though the attempt was forestalled by the lack of room for transport. (Shackleton would use them in Antarctica with success.) It was a great pleasure for Koch and Wegener to plan a compact, highly organized expedition with scientists in charge—a fitting contrast to the sprawling and often rudderless enterprise in which they were currently enmeshed.[8]

Wegener was now convinced that his future lay in polar exploration and travel, and his inspiration was a blend of personal ambition and nationalistic fervor. Lack of polar success had been a sore point in Germany since the 1860s, when August Petermann had funded the ill-fated Germania Expedition, in whose faltering footsteps Wegener had already trod. As late as 1899, at a meeting in Berlin to promote Drygalski's Antarctic plans, the great geographer Ferdinand von Richthofen (1833–1905) lamented that "men of German nationality have been no more than supporting players" in polar exploration, and he bemoaned equally "the paucity of candidates in Germany suitable for the task of leading an expedition, willing to risk everything, absolutely determined, yet of solid scientific worth."[9] Wegener himself wondered, as he sewed, "Is it really impossible for Germans to carry through a successful polar expedition? I believe," Wegener confessed to his journal, "that my determination to take part in this expedition will be decisive for the course of my life. The winter night is now over [the Sun reappeared on 9 February] and in place of my weariness I am convinced, more than ever, that I will 'stand to the colors.'"[10] In his own mind, he was cast in a new role: a German polar expedition leader-in-training, someone who would someday raise the German flag in polar regions.

Wegener's patriotic and nationalistic feelings turned his mind toward the future, but the Danmark Expedition already had a nationalistic purpose of its own. Overtly, it was designed to produce "Glory for Denmark" and so on, but less obviously it was also there to lay sovereign claim for Denmark to this huge stretch of the coast of Greenland and its adjoining waters. It was the latter aim that had helped finance the expedition and gained for it royal patronage. With Norway's independence from Sweden in 1905, Norway was in a position (as a sovereign state) to lay claim to Northeast Greenland, where Norwegian sealers, whalers, and fishermen had toiled for centuries. Moreover, the American explorer Robert E. Peary (1856–1920) had for years been moving relentlessly up the northwestern coast of Greenland to its tip (Cape Morris Jessup, N 83°39′, in "Peary-Land") and had established a verifiable U.S. claim east of it at Cape Clarence Wycoff (N 82°05′). Mylius was aware that Peary had left the United States in summer 1905, to make a bid for the North Pole from Greenland in spring of 1906. These territorial disputes were substantial. These close competitions go far to explain Mylius's willingness to suspend all scientific work in winter 1907, in the service of logistical support for his plans of exploration and "conquest."

In spite of his sewing duties, as the daylight increased, Wegener's ability to pursue scientific work grew with it, and he seized every moment, including the time he might otherwise have devoted to his journal. This document, never ample, holds only about ten cursory entries for the first three months of the year. This literary drought could be expected given his rising spirits: expedition journals and diaries, in general, become prolix when things are going badly, and they dry up when the work is intense and things are going well. In February Wegener managed ten kite ascents, and the Teisserenc de Bort meteorograph, in use since October, continued to perform beautifully. On the last day of the month he fulfilled his quota of sleeping-bag sewing, spent five hours sending up and winching down a kite, and then stood the monthly twenty-four-hour weather observation watch in the crow's nest. This duty would have fallen normally to Peter Freuchen, but his hand was still so badly frostbitten that he was incapable of taking the readings.

In March, in the midst of final preparations for the trip north, Wegener still pushed the aerology program fiercely. He was avidly investigating that part of the atmosphere we now call the "boundary layer" or the "friction layer," where wind speed and direction are strongly influenced by the frictional drag of the surface of Earth and the temperature is influenced by the heat flux (and convective turbulence) close to the ground. This layer typically extends up about 600 meters (1,969 feet), above which height one enters the "free atmosphere." At Danmarkshavn, Wegener was discovering that the boundary layer usually topped out at about 500 meters (1,640 feet), above which the prevailing northwest wind blew steadily. Consequently, he could lower the altitude of the average flight, saving both time and excruciating struggle with the rimed and frozen cable, and still obtain an accurate record of the prevailing wind and, where the air was stable, of the layering. His thermograph was still picking up dramatic temperature inversions, a record enhanced by the need of the men to rest frequently as the kite was winched down, giving the thermograph residence time at successively decreasing altitudes.

Of greatest interest to Wegener was the complexity of the atmospheric layering, which made a mockery of the notion of a convectively mixed and adiabatically simple lower atmosphere. He had, at Lindenberg, played a minor but real part in the documentation of the tropopause and the stratosphere, and he had written a paper on the turbulence at the boundary of an inversion a kilometer or so above the ground. Now he was extending those observations to atmospheric phenomena much closer to Earth's surface. On 21 February he had observed, in the face of the rising Sun, a series of refraction anomalies and associated mirages caused by complex layering in the first tens of meters above the surface: the Sun was distorted and appeared to be made of several parallel bands offset from the disk. Such observations were a way of documenting the layering very close to the ground, where the aerology apparatus was useless and one could only observe, sketch, and photograph—the last being his favorite method.[11]

At the very least his observations were providing ammunition for those (like his teachers) who argued for the absolute necessity of the aerological approach to atmospheric physics. Surface observations reflected the effects of the topography and its temperature and were no guide at all to the "free atmosphere" a kilometer above, or even the character of the boundary layer more than a few meters above the measuring station. One could not proceed on the basis of "theory"—one had to measure. More than this, however, Wegener was carrying out the most thorough aerological investigations ever attempted in high latitude and extreme (surface) cold. Even if these observations brought no theoretical novelty (though they might yet do so), they were a station record that would give the first annual summary of the behavior of the atmosphere—up to a kilometer above the surface—within 15° of the pole. For the time being, his task was not to sort out the phenomena and find the simplicity behind the complex appearances, but rather to document apparent complexity to a degree that it could not be ignored.

With the arrival of March, Wegener found himself suddenly caught up once more in the expedition. His "minor role" was now to be a "major role": Mylius-Erichsen had changed his mind again. Wegener had signed on to this expedition as a physicist and a meteorologist, but he had admitted to having done some geology, and it was in the latter role that he now found himself cast as a player in the ten-man team making the "Sledge Trip to the North." There was a real geologist along on the expedition, Hakon Jarner (1882–1964), and neither the expedition records nor Wegener's diary indicate why Wegener rather than Jarner was chosen for the role of geologist in the northern trek. Jarner had made a number of sledge journeys in the fall while Wegener was hanging about Danmarkshavn. It would appear that Koch, who shared with Mylius-Erichsen the choosing of the sledge teams for the push north, had his way here and had intervened strongly on behalf of his friend and now, in some sense, his protégé.[12]

Wegener found himself wishing at this time that he had brought along on the expedition something to read. He didn't mean novels; he meant evolutionary theory and cosmology, subjects of great scope "especially suitable for expeditions such as this, and especially when you have the opportunity, after reading, to discuss with others what you have read." He wished he had some Darwin to read, some Haeckel, or even Bölsche's popular evolutionism, as well as some astronomy and cosmology books.[13] Wegener was an idealist and a *Kulturträger*—a "bearer of the culture"—and at Danmarkshavn he found himself lacking the literary and philosophical tools he needed to integrate this great quest he was on with these larger cultural and historical themes. He wanted to see his work not just as "exploration" or "physics" but as a part of the great struggle of mankind to understand its place in the world: something he and his generation could say to themselves without a hint of irony; he missed the books that would set his activity in this expansive context. "I don't have any use for novels," he wrote; "I don't find any release in them."[14]

His desire for books was then not just the desire of anyone on long foreign travel to hear or read his own language, but a more specific need. In Germany, every one of his acquaintances was a cosmopolitan intellectual trained in the same way as he, men who had read the same books and thought the same thoughts: it had never occurred to him that he would end up on a scientific expedition without any other intellectuals, because the distinction between scientists and scientific intellectuals had no counterpart in his experience. His Danish companions were competent, strong, brave, hardworking, well trained, humorous, skilled, energetic, and intelligent, but with the exception of Koch, there was no one interested in large scientific themes or ideas, and there was, to be honest, some doubt about Koch's interest in larger themes, too. Wegener wanted his work to be good work, but it had to fit some larger scheme than a map or a chart. Mapping and charting were, however, just what he was now scheduled to do.

The Sled Trip, Spring 1907

Mylius had set departure for 28 March. This trek north was the point of the entire expedition, and excitement and anxiety increased apace with the proximity of the event. There were to be four teams: two exploration teams and two support teams, ten men in all. The exploration teams were to be led by Koch and Mylius-Erichsen. Koch was to drive rapidly to one of the farthest points reached by Robert E. Peary, Cape Clarence Wyckoff (N 82°57.7′, E 23°09′).[15] Koch was to find Peary's cairn, retrieve his notes, and deposit evidence of his own arrival there. Mylius would strike a little farther south and east to the other farthest point east reached by Peary: "Navy Cliff" (N 81°37′, W 34°05′, at the head of "Independence Sound").[16] Wegener was to be a part of one of the support teams, laying down depots, hunting game for dog food, and filling in blanks and details left by the high-speed reconnaissance of Koch along the coast. This push north aimed to cement Danish control of all lands to the east of those reached by Peary and, in so doing, to map the last uncharted stretch of Greenland coast. One question Mylius hoped to answer was whether Peary Land was an island or actually the northern tip of Greenland itself. Peary's 1900 map showed a channel continuing through Independence Sound right across the northern tip of Greenland, so that Peary Land was itself a substantial island in the Arctic Ocean cut off from Greenland.

Accounts of polar travel are, in general, stories of danger and disaster or near disaster, of high expectations dashed, of unanticipated obstacles, and of amazing achievements in spite of everything. In this sort of work every day is, without any exaggeration, a matter of life and death. So it was in the Great Sledge Trip to the North. Where to begin this tale of wonder and woe? Half the drivers were inexperienced, and the men were out of shape after winter quarters and easily injured. The sledges were overloaded, and the dogs were undertrained and so underfed that they ate the sledge harnesses and their own traces during the nights; only the handful of experienced drivers could repair them.

The snow was much deeper than they had anticipated, the pressure ridges in the sea ice the worst they had ever encountered, and the ice foot along the shore sloping and dangerous: it broke the sleds daily. They tied their skis to the sled runners to help the dogs in the deep snow, and the skis broke. They tried the Greenlanders' method of turning over the sleds and urinating on the runners to give them an ice coating, and they froze their fingers while smoothing the freezing liquid. They had expected to cover 60 kilometers (37 miles) a day, and they were fortunate when they made half that distance. They had expected to hunt along the way, but the musk oxen and bears (for dog food) failed to materialize in the numbers they had estimated and counted on. Using Peary's wildly erroneous description of the trend of the coast, Mylius had underestimated the length of the outward march by almost 300 kilometers (186 miles)—an error of ten days' travel at their rate of progress.[17]

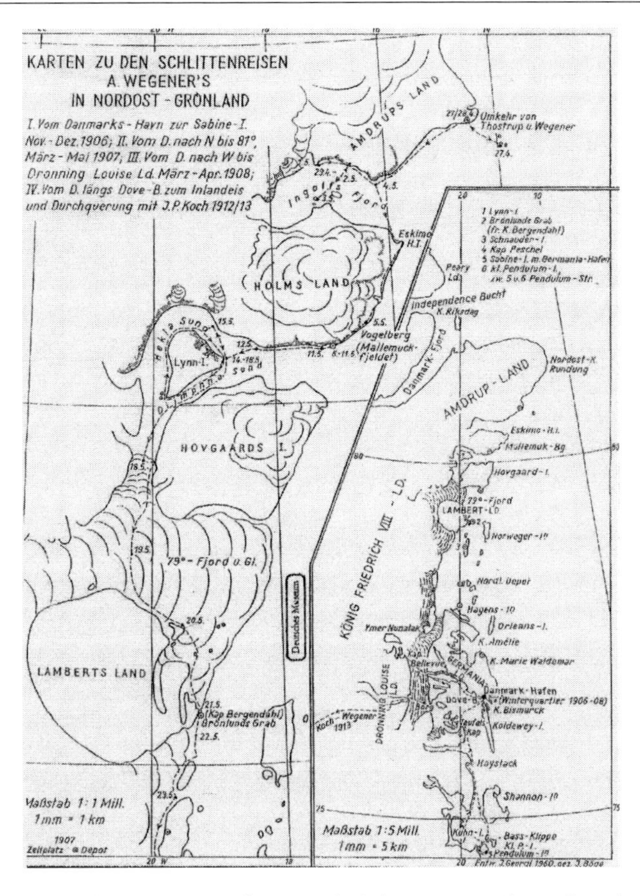

Sketch map drawn by Johannes Georgi of Wegener's sled trips in Northeast Greenland. The legend reads, "I. From Danmarkshavn to Sabine Island Nov.–Dec. 1906; II. From Danmarkshavn N to 81° March–May 1907; III. From Danmarkshavn to Dronning Louise Land March–April 1908." There is also reference to a planned trip by Wegener and Koch at a later date. Wegener Nachlaß. Courtesy of Deutsches Museum, Munich.

When they reached latitude 78° north, they encountered for the first time the Inland Ice flowing as glaciers directly into the ocean—and these huge tongues of ice rose and fell with the tide, constantly splitting and making huge crevasses, while simultaneously shattering the ice foot, making the shorewise traverse harrowing and exhausting. They were trapped in time between two crucial climate events. They had to get north before the Sun began to melt the snow cover, making it impossible to travel with fully loaded dogsleds, and they had to return south before the ice went out along the shore—and there were already places in mid-April where they could see open water only a few hundred meters from the land. The going was made more uncertain by the prevalence of mirages. Wegener noted that the pressure ridges in the sea ice gave the impression that great glacial fronts coming off the Inland Ice were everywhere—but this was mostly atmospheric distortion.[18] Koch constantly had to go ahead and reconnoiter and then come back and guide the rest of the party.

The journey began as badly for Wegener as for the rest. He had a good sled but poor dogs, pulling more weight than they had ever attempted. Anxious and careless the first day while fixing his sled in bad weather, he froze his fingers and blistered them severely: "I was too embarrassed" [to say anything]."[19] By the time his fingers had begun to heal, his foot, which had developed a fissure on the heel the first day out, had rapidly worsened and within a week was infected, swollen, and painful. There was nothing to do but go on, and on 10 April, after fifteen days of wet, cold, exhausting, and dangerous slogging, he limped along with the rest, covering twenty-five miles on foot behind his dogs and arriving at Cape Bergendahl (Lambert's Land, N 78°33′) at midnight.

The next day they paused. They were all badly beaten up by two continuous weeks of forced marching. Mylius had a swollen knee, and Hagen and Ring—among the most stoic and uncomplaining of men—said that their feet were too sore to go on, which Wegener reported in his diary with grim satisfaction, as he had not himself complained in spite of the terrible pain in his foot. The sleds were a mess and, because the starving dogs had gnawed the lashings, were many times retied with hemp, which loosened as it got wet instead of tightening; in consequence, the sledges were falling apart. On top of this, the dogs were exhausted. The party spent the "rest day" building a cairn on the point, both to lighten the sledge loads for the dogs and to ensure a landmark where the returning parties could find food.

If there were any doubt how laconic a writer Wegener was in matters other than his admiration for the dogs, the majesty of the landscape, and the varieties of weather, the events of the next two days would have settled the point. Wegener's journal for 12 April reads as follows: "We found fresh bear tracks. The Greenlanders Brønlund and Tobias took off at once with empty sleds and shot a bear and two cubs. Stopped. Big dog feeding. I saved a good piece of bear meat for my team."

Here is what actually happened: Crossing the sea ice in front of the huge glacier extending out from the fjord at 79° north, Mylius saw bear tracks. He told Brønlund and Tobias to follow them and hunt them down even if it took all day. The other eight sleds went on and camped 18 kilometers (11 miles) further north. Later in the day, the Greenlanders returned with the bears on their sleds. The dogs (seventy or eighty in all) were, as in every camp, free of their traces, and as the men began to cut up the carcasses, the dogs commenced to lunge at them. To keep the men who were doing the butchering from being bitten, six men stood in a ring, fending off the dogs with their whips, while the remaining four men cut up the bears. The dogs became so frantic to get at the meat that several eventually attacked the men like wolves, ignoring the whips and blows and trying to bite them. Suddenly all eighty dogs rushed the carcasses at once, and the men had to flee. The dogs stripped the bears to the bones in fifteen minutes. That is covered in Wegener's diary by the phrase "Big dog-feeding." Later that night, when Wegener heard his dogs barking outside the tent, he grabbed his gun and crawled out of the tent, only to find a polar bear coming straight toward him. He managed to get his gun up just in time and killed it with a single shot. The dogs were so full from their afternoon gorge that the men, roused by the gunshot, could cut up this bear in peace. This entire episode, which might for many outdoorsmen be the anecdote of a lifetime, was covered in Wegener's diary by the sentence "I saved a good piece of bear meat for my team."[20]

On the other hand, the exploits of the others, especially the Greenlanders, often thrilled him. After a good day's travel on hard snow, the fourteenth of April ("travel with these sleds when you can ski beside them is wonderful"), they ran into a difficult pressure ridge, and in climbing it, one of Koch's sled runners not only broke but split on its long axis. Koch had no idea what to do, but Tobias Gabrielsen and Jørgen Brønlund came up with their sleds and, quickly unloading Koch's sled, turned it upside down.

Holding their rifles to the split runner, they shot two holes through it, quickly threaded leather dog harness through the holes, and bound the pieces tight. Then they drove small wooden wedges into the remaining gap, flipped the sled over, whipped up their dogs, and sped off.[21]

With the going somewhat slower during the next week, if no less difficult, Wegener had some time to write. In addition to loving the new land they were seeing, the giant basalt mountains, their first views of sedimentary rocks with fantastic forms and colors, he made a few notes about expedition life and even about himself, mostly reflecting on the poor rations and bad weather, but also proud of his increasing strength and endurance.[22]

On 20 April, as the weather cleared, Koch took Wegener out to do some geology—to show him what needed to be done in the way of collecting—while the others built depots and tried to keep their sled lashings and harnesses away from the still-ravenous dogs, and while Mylius and Brønlund were trying to find a way north across very bad ice. Wegener knew quite enough geology to do what was required—distinguishing a volcanic rock from a sediment, or a sandstone from a shale. The essential thing was a good field description—some hammering of samples and collection of fossils, bagging and tagging. Most of the world was mapped stratigraphically in this straightforward way, and Greenland would be no exception: the finds need not be evaluated immediately. They collected a number of "astonishing" fossils, which was exciting because fossils, and not the fabric or mineral content of the rock, would determine (on expert inspection in Denmark and Germany) to which geological period the rocks belonged.

Pushing north on the twenty-fourth, they passed into land dominated by high sedimentary cliffs backed by even higher alpine peaks. The shore was indented now with alternating headlands and fjords, the cliffs footed by slopes of loose pebbles swept clean of snow by the powerful and constant wind. There were numerous fossils and even sandstone pyramids that reminded Wegener of similar structures in Saxony. With so much geology to do, and a major new fjord to put on the map, Mylius decided that Wegener and Gustav Thostrup (1877–1955), the *Danmark*'s second mate, should turn back in two more days. The expedition was low on dog food (as usual), and decreasing the pack by sixteen dogs could make a big difference in the provisioning. On the twenty-fifth they discovered an even bigger fjord, named by Mylius "Ingolfs Fjord," and pushed on to the next headland, reaching a latitude of 80°42′ north. This was as far as Wegener and Thostrup would go, and on the twenty-seventh Wegener, in a gesture of support for those going on to the north, gave his good sledge to Aage Bertelsen, the painter—one of his bunkmates in the "Villa"—in return for Bertelsen's badly beaten-up sled, and he also offered a good dog to Lt. Høeg Hagen in exchange for his weakest dog, named Ajorkpok, who was quite old and feeble.[23]

Whatever disappointment Wegener may have felt at the order to turn back, it was soon swept away by the exhilaration of his freedom. He liked poking around, not dashing forward in a race, which always seemed to be Koch's and Mylius's goal. On their first day of freedom, he and Thostrup struck out over the clear open sea ice to a series of islands, only about 25 meters (82 feet) high. Wegener scoured the pebbly shore for fossils and was rewarded by some good examples of fossil corals. They were sure that they had reached the farthest east extension of Greenland (they were wrong), and this added to the excitement.

They spent the next week mapping the interior of Ingolf's Fjord. Thostrup was taciturn, completely absorbed in the cartography. It was simple and laborious plane surveying with a theodolite. They zigzagged back and forth across the fjord ice: sighting on a distant point, then sledding to it and sighting back the other way. Wegener helped Thostrup each day with the noon sighting and calculation of latitude and azimuth, but he spent most of his time wandering. Their ability to stay out and explore depended crucially on their ability to find food for the dogs. Thostrup was not a keen hunter, so Wegener offered to hunt by himself, to which Thostrup assented. Wegener was thrilled to be on his own. He had already collected marine fossils along the north shore of the fjord, which turned out (on later examination) to be from the Permo-Carboniferous boundary. Sledding deep into the fjord and onto the land, Wegener discovered a small herd of musk oxen—seen first through his telescope. He was able to shoot two of them but found he had no more ammunition, so he returned to camp and rousted Thostrup; together they killed the remaining musk oxen—animals that, having never seen humans before, showed no fear, not even shying at the sound of the guns.[24] This was good luck, since it meant at least seven more days of exploring—measured in quantities of available dog food.

On 3 May they crossed the ice of Ingolf's Fjord to the south side, where they found some small islands to use as a measuring station. Thostrup, perhaps pleased by Wegener's industriousness, uncomplaining character, and silence (the latter a result of Wegener's lack of Danish but congenial to the taciturn Thostrup), named these "Wegener's Øer." Wegener was in an expansive mood. He was drawing good geological profiles, doing original cartography, and making a good collection of rocks and fossils. He had proven himself as a hunter and was accepted as a team player. He was getting used to living with deprivation and danger and learning the land, and he could note in passing in his journal, "Clouds visible in the west, we think the weather will turn bad, petroleum for only about two more days." He could begin to read the signs of game as well: "When we got here, saw fresh bear-tracks, a mother with two cubs. If we were Greenlanders we would start following their trail, as a mother with cubs can't travel very fast or very far, and the tracks are at most a day old." However, he added wryly, "As civilized Europeans, we would rather lie around and sleep."[25]

Wegener was also photographing constantly, keen to get color photos of the sediments, especially the sandstones; he was soon down to his last plates. He possessed a strongly visual imagination and intuition, and one reason for the sparseness of his diary entries was his preference for keeping a visual diary in the form of photographs. Of the 9,000 photos in the Danmark Expedition Archive, Wegener took a significant portion. He was also excited to be doing geology. He had collected so many samples that he could barely move the box on and off his sledge—"and I keep on finding more and more!"[26]

On 5 May they got past what they thought (mistakenly) would be the most difficult part of their southward journey, a shattered ice foot at the southern cape of Holm Land, under a looming 500-meter (1,600-foot) cliff they had named Mallemukfjeld. As it turned out, colder temperatures, while Wegener and Thostrup had been in the north, had sealed the leads of water, but the ice was still bad, and Wegener's sled came completely apart; Thostrup needed all of his fabled ingenuity to get it back together. They were almost out of food now, and facing an even greater catastrophe—they had run out of coffee, then as now the national beverage of Greenland, and something they dreamed of whenever they weren't actually drinking it. "Yesterday we had coffee and soup mix. Today, an infusion of musk ox meat, chocolate, and used coffee grounds. Our menu is quite uniform: 'soup mix with pemmican, and pemmican with soup mix, in alternation.'"[27]

They spent three weeks getting back to Danmarkshavn. For the most part they got on well, though over time Thostrup's silence and stubbornness began to wear on Wegener. It was a hard trip, harder than they had anticipated. Wegener was indefatigable in his geological work, and Thostrup shared his enthusiasm for specimens. Of their struggles during these weeks, several exploits stood out in the annals of the expedition and are worthy of mention. For example, on a single day, 9 May, they hunted down and killed four bears, stashing the meat in a depot for the returning northern parties.

Exploring deeply into Djimphna Sound and Hekla Sound, which they discovered and mapped in the course of a week, they proceeded south with a dangerous traverse over the crevasse fields of the Inland Ice for four days, trying to shorten their return. This desperate adventure was forced on them by the melting of the snow above the sea ice. Trying to get off the fjord ice in Hekla Sound, they fell continually through treacherous snow crust into running meltwater. Their feet were wetted inside their kamiks, and they could feel them beginning to freeze, with more than ten days of travel still to go. Soon everything was wet—dogs, sleds, harnesses, sleeping bags. Sleep was almost impossible, food scarce, energy receding. They were exhausted. Thostrup, who had smashed one of his fingers under a falling boulder building a cairn for the bear meat, was driving with one arm. Wegener, whose Greenland-style sun goggles had slipped, was snow-blind in one eye, with the shocking pain that a sunburnt eye yields its owner. He was so burned on the surface and even the inside of his nose that it hurt to face into the wind, making it even harder to see where he was going.

Toward the end it was a race to the ship, hastened along by Thostrup's desire to have done with it. Wegener thought they were moving too fast—faster than was good for the dogs. "It is unbelievable," he wrote, "how exhausted the dogs are. I have to confess, that I side more with Mylius-Erichsen's methods. . . . Thostrup and Koch don't work the dogs properly—they lack 'Dog-Sense.'"[28] Wegener was brooding, as he would for some weeks, on the death of Ajorkpok, his oldest dog. After he had arrived at the ship, he had a long conversation about it with Achton Friis:

> I had one great heartbreak on this trip, when old Ajorkpok died on the way back here from Lambert's Land. I had exchanged a better dog for him with Hagen, when we parted. He was thin, deaf and blind, and had lost every tooth in his head. And he just couldn't make the forced march through the sound south of Mallemukfelsen. One day he broke down altogether. We stopped on his behalf and put up the tent, but there was nothing we could do: it was too late. The next day he tried vainly to follow the sled, making use of the sled track. All day long he struggled to put himself back into his old place in the span, but he just couldn't keep up. Finally I hauled him up on the sled. But he didn't like it there, and he always struggled to get down among the other dogs. Finally, on the third day, he died right next to me on the sled. How many miles did that dog slave in the service of men? All you can say is that for a few years he had been a good strong dog. Now he was finished, of no further use and finally freed from his exhausting service only by a wretched death in it. We buried him deep in a crevice in the sea ice at the North depot where he died.[29]

On 27 May Wegener broke through the ice, and his feet were soaked through the holes in his kamiks; he spent the next three days with very cold feet, though with the temperature suddenly above freezing, his feet were merely wet—pure luck. He was so exhausted that the whip fell from his hand three different times, and consequently he

Alfred Wegener (*right*) and Gustav Thostrup on 1 June 1907, the day after their return from the sixty-five-day sled trip north. From Friis, *Im Grönlandeis mit Mylius-Erichsen* (1910).

was wetted through on his back and shoulder from falling off the sled: "In short a most unpleasant day."[30] But his thoughts were elsewhere, still dreaming of the South Pole and how he would travel there: "On a sled trip in South Polar regions, when you are alternating between sea ice and Inland ice this combination of Greenland sleds and skis is absolutely crucial."[31] He was sorry the trip was coming to an end; he was having a fascinating time, mishaps notwithstanding. To the end he was photographing, measuring, sampling, and dreaming: "If only my photographs of the North turn out! The first thing I do when I get back will be to develop the plates. I've got to use the time before Koch gets back, and Hagen and Mylius-Erichsen, all of whom will have photographs."[32]

Wegener and Thostrup arrived back at the ship on the last day of May, having been out for sixty-five days. They posed the next day for their official return photo—exhausted but in one piece, enjoying their first cigars. Thostrup looks slightly stunned, and one can see the rough bandage on his smashed finger. Wegener looks like he would like to smile but hasn't the energy. As a "civilized European," however, he has bothered to change clothes and shave.

Summer and Fall 1907

Wegener had now proven one of the things he had come to Greenland to prove: that he was one of the strong, not one of the weak. He not only had survived a grueling and dangerous trip but had distinguished himself. He had done good science, hunted musk ox and bear, explored the coast and the fjords, and mounted the Inland Ice. He had a part of Greenland (small, but real) named after him. It was not just any German name on the map, but his name. In the first weeks after his return, Thostrup's stories of the trip included quiet praise of Wegener's hardiness, cooperation, courage, and determination. Achton Friis, always working up notes for his "expedition book," questioned Wegener closely about the trip and was delighted and moved by Wegener's solicitude for the old dog, Ajorkpok. Alfred was not so much "Dr. Wegener" anymore as "Wegener," a comrade. Such shifts of position and perception are common on expeditions of all kinds, as the members accumulate successes and stories of success, or the reverse. Wegener had been in the North. He was now one of the explorers, not the "German scientist." His virtual silence on the sledge trip, largely due to his inability to converse in Danish, had been read approvingly by Thostrup as a mirror of his own taciturnity and was approved in turn by the others.

Wegener's exploits had won him the respect of those who had stayed behind, and he enjoyed the sense of inclusion. Yet the problem of getting his science done was still there. In spite of his training of Weinschenck and Koefoed in aerology, as well as their promises to keep the flight program going in his absence, they had made only four flights in April and but a single ascent in May, nearly undoing Wegener's design of flights in every month of the year. There was nothing to be done about it. Koch had not yet returned, and Wegener made use of the darkroom to develop his photographic plates and store them away. He indexed and packed his geology specimens, though there were still some boxes to be recovered from the depot immediately to the north. The Danmarkshavn station party had at least kept the magnetic observatory going, and he had a continuous month of declination readings.[33] He was fairly certain they would amount to nothing, but his psychic equilibrium depended mostly on whether something was being attempted, not whether something was being achieved: the activity was itself an achievement.

Danmarkshavn in June took on a gentler aspect, and the world once again had color. The ice was going out in the outer harbor, and one could see blue water and the lovely light green of the submerged floes. Flowers were appearing everywhere—white heather, pale green saxifrage, the soft red of ice ranunculus, all in rapidly expanding carpets. Scattered across this apron of color, the scientists—in their dark shirts and broad-brimmed straw hats—looked like large children in summer camp uniforms, practicing nature study. Arner Manniche with his bird specimens, Andreas Lundager from the "Villa" with his botany, and Fritz Johansen with his microscope and jars of marine microorganisms all gave the shoreside camp a sense of studious, if incongruously juvenile, occupation.

Over all of this activity, like guardian deities, flew Wegener's kites and balloons. "Wegener's work," wrote Friis, "goes up and down, insofar as it is attached to his kites and balloons. These strange creatures float above us from morning to night, and we have grown so accustomed to them that they seem to belong to the landscape and to be characteristic of the region we occupy. The immense patience required to carry out these investigations under the adverse weather here, in all seasons of the year, can scarcely be imagined, unless you look through Wegener's observation journal, in which he has itemized every mishap."[34]

The kites and balloons were a visible sign of Wegener's role in the expedition and, even more, a symbol of his constancy of purpose. Friis remarked that Wegener seemed unstoppable once he had set his mind to a task. Once and only once had he ever seen Wegener give up on something. The case involved one of the dogs, a dog so averse to work and so independent that he had early in life been dubbed "Misanthrope." Misanthrope, other than during mealtimes, spent the entire summer sunning himself in the lee of the Villa. He liked to sleep sitting up on his haunches, swaying back and forth as he dozed. As Friis tells the story, Wegener, in the course of some errand to the Villa, noticed Misanthrope in his characteristic spot and determined to get a rise out of him. He walked up to him and pulled on his muzzle. The dog never even opened his eyes. Wegener tried to knock one of his forelegs out from under him; Misanthrope calmly returned it to place. Wegener yelled into his ear: nothing. Finally, looking around, he found a feather, inserted it into the dog's nostril, and for a minute or more twirled it delicately back and forth. A tear appeared in the dog's eye closest to the feather, but Misanthrope never opened his eyes. Finally, after another minute, he gave a great yawn, and Wegener, defeated, gave it up as hopeless, dropped the feather, and walked away.[35]

The return to "station science" and the aerology program was a step back into something more familiar than exploring, but no less arduous and, for Wegener, hardly less interesting. In June he flew as many kites and balloons as his stamina, the weather, and the endless string of technical mishaps would allow. He managed to get something aloft about every third day. The work of winching, still exhausting, was made easier by continuous daylight and moderate temperatures—now rarely below freezing. The stark temperature contrasts with the winter (June at Danmarkshavn has a mean daily temperature 25°C [45°F] higher than February) made possible interesting comparisons of the structure of the atmosphere, particularly the way it would cool (or warm) at higher altitudes in summer. He set a goal of doubling the altitude of his kites and balloons, in spite of the additional hours of winching involved. In June he managed a mean altitude of 1,183 meters (3,881 feet) for his ten flights, twice anything achieved in the previous three months by his assistants, and got one balloon, in nearly still air, up to about 2,000 meters (6,562 feet).[36]

As this work went on, the last of the expedition was still coming back from the North. Koch returned to Danmarkshavn at the end of June and announced that Mylius, from whom he had parted a month earlier, should be along in a few days. He'd have to hurry, reported Koch, as the ice was almost out at Mallemukfjeld and they had barely made it themselves. Koch's return was good news for Wegener, who now could converse in German with someone again. Koch enlisted him immediately in geodetic work, as he was anxious about the exactitude of his longitude measurements for Danmarkshavn. He needed assurance that the geodetic part of his work in the North would stand, since his longitudes and descriptions were very different from those of Peary. Koch had made it to Navy Cliff and to Cape Morris Jessup, at the very northern tip of Greenland. In these places he had made a major discovery: the "Peary Channel," supposed by Peary to cut across Greenland completely at the tip, was no channel at all, merely a very deep fjord that Peary had misread from afar. This was important for the expedition's political as well as scientific aims. When Koch had told Mylius of his discovery, Mylius had been unable to let the matter rest, and instead of turning south, as conditions indicated he must, he had insisted on going north himself. Koch understood: it was not that Mylius disbelieved him; it was that Mylius could not stand the idea of a major discovery in which he had no role. He would go and "confirm it."

July brought warmer weather, more flowers, and millions of mosquitoes. In a Greenland summer, as in parts of Alaska, the only way there could be more mosquitoes is if some of them were smaller. Breeding 50 million to the hectare (2.5 acres), they made every outdoor activity a considerable challenge, and the residents of Danmarkshavn found themselves dreaming of the winter as they had just finished dreaming of summer, more or less proving that the one constant of a polar trip is discomfort: the source may vary, but the intensity remains the same.

Trying for ever-greater heights in the aerology program (Wegener wanted to reach 3,000 meters [9,843 feet] that summer) meant that Koefoed and Weinschenck, the engineers, had to share additional hours of winching the kites and balloons up and down. They worked like galley slaves—six and seven hours of winch work in a day. Wegener had found that anything much above 1,500 meters (4,921 feet) was so exhausting that it wasn't worth the effort, but still he lacked data above a kilometer or so, and this worried him considerably.[37] On 10 July he had witnessed and photographed the evolution of three very large waterspouts reaching from the stratocumulus layer above the station to the surface of the sea—and a kite at 3,000 meters could have provided material of real scientific interest, as little was known about the formation of these *Wasserhosen*.[38]

Wegener's winch motor had never arrived, or had been left in Iceland—no one knew which. But one piece of equipment that had made it to Greenland was the expedition's automobile. A fragile-looking buggy, steered with a joystick and with the engine mounted behind the rider (a true and early horseless carriage), the automobile was to have been a minor featured "story" for the expedition—the "first mechanical overland travel in Greenland." The automobile was indeed a marvel, but it had not actually been anywhere or done anything significant until the engineers and Wegener dismounted one of the rear wheels and slipped a drive belt over the rear axle, thereby using it to power the kite winch. The drive belt had been manufactured aboard the ship (and the wooden chassis fashioned) by Harald Hagerup (1877–1947), whose skill as a leather worker had already proven itself in the form of the durable footwear he had cobbled from walrus hides. The automobile winch worked like magic, and with it they managed thirteen ascents in July and, by early August, a record altitude for a kite of 3,110 meters (10,203 feet).[39]

August marked the first anniversary of the expedition's arrival in Greenland, but all the fireworks were provided by the weather. Early in the morning of 1 August, without warning, Danmarkshavn was hit by a violent windstorm from the north, which tore the ship's anchors from the shore and set the *Danmark* adrift. The wind blew the ship out against the ice margin, crushing the two ship's boats tethered to the lee side. Everything not battened down (and that was almost everything) was blown about on the ship and on the shore, including two of Wegener's better kites and the kite-house—they were completely destroyed.

The storm was unlike anything Wegener had ever seen, including the cloud forms—great cloud rolls stretching from east to west, from the sea all the way to the distant Inland Ice, gray-blue and red-brown, laden with dust, and tearing and reforming, only a few hundred meters above their heads, hour after hour. The surface wind would die down to a flat calm and then suddenly gust to 30–35 meters per second (70–80 miles per hour), as the clouds continued to roll and billow overhead. It was their first experience of the *Piteraq*, the sudden and ferocious winds of East Greenland that blow downslope from the Inland Ice. The phenomenon is known in Europe and elsewhere (as in the case of the fjord winds of Norway), but only in Greenland and Antarctica do these winds blow with such force and on such an immense scale.

The weather of the next few days provided exciting confirmation of character of the phenomenon. On 3 August Wegener's meteorograph record from the kite at altitude of 3,110 meters confirmed relatively warm air aloft, with adiabatic cooling of

Wegener launching a captive balloon in August 1907. He is standing just to the left of the balloon cable tethered to the winch, and he is rigging the meteorograph. Note the expedition's automobile, with a drive belt around the axle to power the winch. From Friis, *Im Grönlandeis mit Mylius-Erichsen.*

only half the normal value and a temp of only −8°C (17°F) at the maximum altitude, where normally he would have expected about −25°C (−13°F). That day and the next the sky was covered with the serrated ranks of stratocumulus clouds, row after row of cotton puffs. To the south, above Koldeway Island, were large "Mushroom-cap clouds" (*Pilzhüten*) with flat bottoms—the sort that form in the lee of a mountain or other obstruction that produces a "standing wave" in the air flowing over it. All this was evidence of a sharp inversion and strong layering.

This was the most interesting scientific problem Wegener had yet encountered in Greenland, and it provided an avenue to employ his theoretical training with his practical skills. The phenomenon of the föhn wind, the warm downslope winds common to mountainous areas in all parts of the world, was a preoccupation of his teacher at Berlin, von Bezold. The particularly violent föhn of Greenland had been described by the Danish glaciologist Hinrich Rink (1819–1893) in 1862; Rink, however, had to be content with visual observation and ground-station temperatures. What Wegener had seen in August clearly matched the outlines of Rink's description of the phenomenon, though Rink's explanation—of a warm wind blowing across Greenland from east to west—was ruled out by Wegener's knowledge of the prevailing upper-level winds and by the experience of the föhn at Kap Bismarck, which had blown in the opposite direction from the north and from the west: it was clearly a downslope wind, an "air avalanche." Wegener saw that a series of careful observations, made simultaneously inland and at the coast, could provide the materials for a definitive description and open the way for a complete physical analysis of an outstanding meteorological question.

With Mylius-Erichsen still gone in the North and unlikely to return now until the sea froze over, Capt. Alf Trolle found himself in command of a difficult situation. Without Mylius's strong direction, the expedition was beginning to come apart. The democratic mixing of crew and scientists, a feature of the story Mylius had planned from the start, was the first casualty. Lunchtime aboard ship, which required two sittings, quickly broke into a crew lunch and a scientist lunch. While the latter were in good spirits and discussed their work over their food, the crew's conversation at midday was descending into squabbles and tittle-tattle—or, as Wegener dubbed it, *Klatsch*.[40]

Wegener's suggestion to Trolle that they begin immediately to construct and staff (over the winter) a second weather station inland was therefore both welcome and opportune. Trolle agreed almost immediately. The transport of materials and construction of such a station would require a substantial expedition, and its provisioning and fueling would require continuous activity through the winter. The ship's crew would thus be occupied through the autumn at least. The proposed location was the deep recesses of Mørke Fjord, some 70 kilometers (43 miles) to the west, in an area reconnoitered the previous autumn by Mylius himself. The halfway point was the Hvalrosødden, where they had hunted walrus the previous summer and to which the crew all knew the way.

Trolle put the plan to work immediately. The materials for the hut would have to be cut and packed, food and fuel apportioned, and a coal stove fabricated, all of this to be accomplished by the otherwise restive crew. Meanwhile, Wegener doggedly pursued his electrical and magnetic measurements, and with the aid of the automobile-driven winch, he managed a kite flight almost every third day in August, reaching a total of sixty ascents for the year before the end of the month.[41] Meanwhile, the waning of the light and the coming of winter led to an interesting and delicate calculation concerning the construction of the inland weather station. It would be ideal to have the hut finished before the ground froze, but the transport of materials would be much eased by the ability to travel on the ice, rather than overland. This indicated a window of opportunity in the first half of September.

On 25 August Wegener and Trolle set off on foot along the shore to reconnoiter ice conditions at Storm Bay (10 kilometers [6 miles] distant) and Hvalrosødden (20 kilometers [12 miles] further along). If the ice conditions were favorable, the Mørke Fjord Expedition (as it was grandly overtitled) could avoid much rough terrain and cut across the new ice at both these locations. They returned on 31 August, and Trolle gave orders for the "expedition" to set out the next day.[42]

On 8 September the "expedition" party returned, having got only as far as the hook in the coast known as Snenaes, barely a third of the distance to Mørke Fjord. There a great quarrel had broken out. If the ice conditions were good enough to transport stores to the west, argued one party, they were good enough to support a rescue expedition to find Mylius, Hagen, and Brønlund in the North. This group therefore refused to proceed further, insisted on dumping the stores intended for Mørke Fjord, and forced the entire party to return to Danmarkshavn. Wegener could barely contain his fury and disappointment, turning to his diary for release:

> The Mørkefjord expedition has returned after having carried the materials for the station the least distance possible—as far as Snenäs. . . . That's how it is with these

> people! Now it will be impossible to build the house before the ground freezes. Their anarchistic relations with one another have a destructive effect on every cooperative endeavor. When left alone to do their individual work everything is fine. But as soon as it is a matter of working together, everything falls apart, everyone does what his own head tells him to, no one does anything except what he wants, no one submits his own preferences to the needs of the others.[43]

Perhaps they were really worried about Mylius; maybe they just wanted to do what they wanted to do. In any case, a meeting was held, after which a relief expedition with six sleds was sent north. Over the next forty days they would suffer terribly, starve and freeze, kill or ruin a large number of dogs, and accomplish nothing—being met by just what Koch said they would meet: open water at Mallemukfjeld, the same open water that was keeping Mylius and his companions in the North. Koch had quite pointedly declined to go.

Maybe those that went north were the lucky ones. The depression of the winter night came early that fall to everyone at Danmarkshavn, even before the darkness was complete. It hovered over them on the wings of their own fear of it, sapping their will to work. Wegener felt it come over him:

> We all suffer in the same way from the endless uniformity and the lack of sensory input. In the mornings we are unbelievably sleepy, and have too little desire for work. Even though I am continually occupied by my tasks, I must admit to myself that I accomplish unbelievably little work. What one accomplishes, feels, sees, discusses here in the course of a week, one could do, feel, see, and discuss at home in the course of a day, without needing even to hurry. I think the only thing we work at harder here than at home is sleeping. And yet Koch and I must count ourselves as among the most active here of all the expedition members, and I am very steady in making my observations, indeed am ridiculous enough as a model of bustle, yet I am all the while acutely conscious that I am doing shamefully little.[44]

After working all day at his own measurements, he took to assisting Koch in the astronomical work. None of this was actually essential—it was Koch's own way of fighting the inertial drag of winter night. They did surveying work by day, to check the accuracy of the leveling apparatus, and by night pursued measurements of stars to determine the coefficient of refraction of the night air under different conditions. It was all, as Wegener admitted, a form of therapy, a "patent medicine"—anything to keep active. They rebuilt the kite-house, the one destroyed in the August storm. They cooked up a "24-hour observation expedition" in which Koch, Lindhard, Freuchen, and Wegener stayed up for twenty-four hours on 12–13 September making air-electrical, temperature, and astronomical measurements, drinking so much coffee in the process that the Greenlanders dubbed it an "observation party" (*Observations-Mik*) after their own tradition of a *Kaffee-Mik*.[45]

During all this time, nothing was happening toward the construction of the inland weather station. Dog food was in short supply, and many of the best dogs were off in the north with the relief expedition. In later September it finally snapped cold, and the ice began to form in earnest. Someone had the bright idea that the motorcar, heretofore employed only as a kite winch, might be good for something more. On 25 September, Freuchen, Hakon Jarner, and Ivar Weinschenck took off on the ice hauling an empty sledge, with the idea of transporting the construction materials the rest of the way to Mørke Fjord with this mechanical contraption. It began as something of a lark, but their rate of progress astounded them. They arrived at Snenaes in a few hours—a day's slog by foot or dogsled—loaded the sledge, and hauled everything to Hvalrosødden. The next day they set out and once again covered an amazing distance—halting at the mouth of the Mørke Fjord only because they could see open water ahead inside the narrow fjord.

Greenland's first auto trip was followed immediately by its first auto accident. Weinschenck got stuck in a snowdrift on the way back to Kap Bismarck and had to abandon the car and sledge. The car was buried in snow and was not seen again until the end of November: it had fallen through the ice and sunk to the bottom—a total loss.

Somehow, in spite of this and other mishaps, miscalculations, and a good deal of stormy weather, the station got built over the course of the next month. Wegener took no part in the building or provisioning and had become quite fatalistic about it—either they would get it done or they would not. He stayed at Danmarkshavn and sent up kite after kite, though he was by now down to his last working kite meteorograph.[46] When, by the first of October, Mørke Fjord had not yet frozen over, Gundahl Knudsen (1876–1948), the ship's carpenter, made the decision to build the station in a narrow bight just to the south of where the last piles of lumber for the hut had been dropped—a place christened *Pustervig* (wind gap). From Wegener's perspective, this was an unfortunate spot, as it sat in the lee of the Monumentberg, which blocked the prevailing wind from the northwest. It was, moreover, 15 kilometers (9 miles) seaward from the end of the fjord, where the station was supposed to have been, and thus almost 25 kilometers (16 miles) from the ice margin.

When Wegener arrived with the meteorological instruments in early November, he found his assistant, Peter Freuchen, already snug in his cabin at Pustervig. The place was so well set up and provisioned that Wegener could almost forget about the unfortunate location. Over the next few days he compared the weather in Mørke Fjord, on the other side of Monumentberg, with that at Pustervig, trying to get a feel for the temperature and wind differences. There was no way around it—in order to make the station work and get any significant data, Freuchen (or whoever else staffed the station) would have to climb the Monumentberg every day and take a temperature reading at 400 and 800 meters (1,312 and 2,625 feet). Wegener set up the temperature measurement locations and climbed the mountain several times on 6 and 7 November to mark out the path.

Wegener's ascent of the Monumentberg on 7 November coincided with the arrival of a föhn wind. As he stood on the tabular summit of the mountain, he had one of those experiences that made his Greenland journey seem worth all the effort, anxiety, and frustration. As the northwest wind swept up the mountain on the Mørke Fjord side, a "föhn wall" formed at the edge of the summit plateau, a line of precipitating clouds parallel to the axis of the mountain. As the clouds arched up and over the summit, they thinned and began to dissipate as they descended on the Pustervig side. This was a thermodynamics text come to life. Bezold (and many others) believed that the warmth of the föhn wind was caused by the adiabatic compression of descending dry air—the air having lost its water vapor by condensation and precipitation on the windward side. The heat generated in compressing dry air is familiar to anyone who has ever held the shaft of a bicycle pump. Wegener's temperature measurements on the

Observer's hut at *Pustervig* ("wind gap"), where Wegener's assistant Peter Freuchen spent the winter of 1907/1908. In the background is the Monumentberg that Freuchen had to ascend each day to make observations. Wegener stayed here in the summer of 1908, alone. This is Wegener's photo to document the parhelia of the Sun's light refracted through ice crystals. From Friis, *Im Grönlandeis mit Mylius-Erichsen*.

descent from the mountain showed a stronger-than-expected temperature increase. It was an amazing stroke of luck. Now he could alert Freuchen to the phenomenon and get many more records of the same sort.[47]

Freuchen, who was left entirely alone with the departure of Wegener and the others on the eighth, paid the price, in the course of the following winter, for the station's poor location. The harshness of the conditions, however, and the dangers of the work were more than anyone had counted on. Freuchen kept a team of seven dogs with him, but because these could not come into the small hut, one by one they were killed and eaten by the large white polar wolves that frequented this valley. These wolves followed Freuchen in the darkness when he climbed the Monumentberg to make his daily temperature measurements, a trip that never took less than four hours. Though he would spend most of the next twenty years in Greenland, his fear and hatred of wolves, developed in these few months, never left him. The mountain was slippery with ice, and he risked his life each day simply climbing a prearranged path. At Christmas Wegener sent him a puppy for company, and in January a set of crampons (ice teeth fitted to the bottom of a boot) to make his ascents of the mountain safer, but poor Freuchen had a bad time of it nonetheless. Condensation from his breath inside the hut froze to the walls, producing an ice shell more than 1 foot (0.3 meters) thick on all interior surfaces.[48]

Supplying coal and food for Pustervig became a major struggle for the expedition. Like much of what one learns on any expedition, it was mostly an example of what not to do. There wasn't enough game in the vicinity of Kap Bismarck to make food for the dogs needed for the supply runs ("don't build an expedition around hunting"). The ice conditions in winter 1907–1908 were terrible, with open water in January along the coast going west to Pustervig ("don't plan to have last year's ice conditions"). It was necessary to travel to Pustervig every other week, because of the need to supply Freuchen with coal ("never, ever heat a remote hut with coal!"). The wolves were a nightmare for the supply teams throughout the winter night, attacking the dogs of the coal and provision sleds at every camping stop. The indomitable Greenlander Hendrik Olsen, leaping from his tent at Hvalrosødden to break up a fight among his dogs, found himself one winter night plying the whip to the snout of a huge arctic wolf in the midst of the fray.[49]

Back at Danmarkshavn, things were not much better. If Wegener, by his own assessment, was wintering better in the second year, the same could not be said for many of the Danmark staff and crew. The early November return of the expedition to the north in search of Mylius had produced no news of any kind. The end of this hopeful effort, with the would-be rescuers sick and exhausted, their dogs ruined, their sleds broken and worthless, depressed everyone at Danmarkshavn. Wegener returned from Pustervig on 11 November to find a bleak mood everywhere.

One bright spot was his ability to interest Gustav Thostrup and Aage Bertelsen in a program to make the study of the auroras (*Polarlicht*) more useful. Earlier Wegener had fretted about a way to represent them. He and Thostrup decided on a polar projection using the station zenith as the center and the horizon as the outer (circular) border and plotted the auroras on a star chart (the easiest way to locate their position). Because the color could be used to infer the composition of the ionized gases, Wegener was overjoyed to have Bertelsen paint the aurora on a stretched canvas.

Wegener wrote to Freuchen on 12 November, asking him "if he had the time and the energy" to observe the aurora, noting the direction, elevation above the horizon, and time, and to make a sketch of its shape. He urged him to make a star map of his own and to plot the course and shape of the aurora in "three or four sketches, as it goes through the different phases of its development (e.g. before it gets to the zenith, one in the zenith with a corona, and finally after it passes) this would be very useful material."[50]

Wegener was interested in the aurora as a beautiful display and photographed it many times. He was also interested in what caused it, and his work shows that he was on the winning side in the interpretation—ionization of atmospheric gases by a flux of solar radiation. His observation program—assembled with such difficulty because the expedition members (a) didn't know what they were doing or why and (b) didn't like freezing in the polar darkness doing what they were not required to do and did not understand—had nevertheless a chance to make a real contribution.

When he could get everyone on board, and he did occasionally, he could chart the course of the aurora, get measurements of the variations in atmospheric electricity as it passed overhead, record the declination of the magnetic field using the magnetic theodolite, get Bertelsen to record its colors in paint, and, with Freuchen's observations of the same zenith passage 60 kilometers (37 miles) distant, have a baseline long enough to get a quite reliable estimate of the altitude. Excited by the usefulness of these results, he urged Freuchen again and again to measure the aurora whenever possible. On more than one occasion, Freuchen and Wegener were able to see the curtain from the aurora descend below the elevation of the Monumentberg and to measure its approach to the ground—even though such accounts are still disputed.[51]

Though he had wanted to spend Christmas with Freuchen, he had to beg off. It was the time of year when everyone had to report the results of their work from the preceding year. Wegener had done a tremendous amount of work, and that meant "a frightful amount of writing (to cover the eventuality that I don't make it back from the expedition)."[52] At the request of Achton Friis, he copied out extracts from his journals, especially of the trip to the north, and produced for Friis a twenty-page manuscript (translated into Danish) summarizing the principal meteorological findings of the expedition.[53]

When the New Year came, Wegener, as a gesture of solidarity and achieving a long-sought goal, began to keep his diary partly in Danish. This would also save copying and translating later on. He had been able to converse a little in Danish for some months; though he kept his instrument records in German, his correspondence with Freuchen at Pustervig alternated between Danish and German, and he seems to have acquired a mastery of the language.

His increasing comprehension of what was being said gave him a deeper insight into the dynamics of the expedition. Much of what he had assumed the year before to be spirited conversation turned out, when one understood it, to be bitter argumentation. There was no deep animosity in play, simply boredom and confinement. "It is terrible," wrote Wegener, "when men accustomed to hard work have neither any work nor anything to interest them. The result is an endless succession of quarrels."[54] Were he in charge of an expedition (a frequent and favorite winter reverie), he would pay close attention to keeping his men occupied in the winter. They should have a slide projector along, and the scientists should commit themselves to giving illustrated lectures, or there should be prepared courses of study that could be accomplished while cooped up indoors.[55]

The boredom of winter was wearing on Wegener, too. "My circumstance, my position in the expedition, is working out very unfavorably. I haven't held a dog whip in my hand since Mylius took me with him to the north. And this spring will bring not a hint of novel personal experience."[56] One need not take this too seriously: he had reached his winter nadir the same day in the previous year. His fear of underemployment had the same consequence as the year before—he turned to long reveries of future expeditions. He had already planned the South Pole expedition extensively; now he thought about crossing the Greenland ice cap. It would be a small expedition—himself and another European if they used man-hauled sledges, with Hendrik Olsen along if they chose to use dogs. The more he thought about it, the less he liked it, though. "There isn't much in this preliminary plan to tempt me. The autumn would be fast and stimulating, but then there would be a long overwintering, and the traverse itself would be a mindless grind [*Stumpfsinn*]. Nevertheless, on practical grounds I can't help thinking that it has a lot to recommend it."[57]

Perhaps Wegener discussed his inland-ice plan with Koch, though his diary does not record it. It is an interesting coincidence, in any case, that Koch assigned Wegener to make a month-long sledge trip in March with Aage Bertelsen, Weinschenck, and the expedition's doctor, Jens Lindhard (1870–1947), to investigate the snow- and ice-free land that had been observed on the far side of the great tongue of the Inland Ice, Storestrømmen, which bordered Germania Land to the west. This trip would be man hauled (no dogs), as had been the journey to carry the instruments to Pustervig in the previous November.

Man-hauled sledges (*Zugschlitten*) had advantages and disadvantages, Wegener decided. The greatest advantage was that the entire time you were under way, nature was before your eyes, rather than the rear ends of a team of dogs. Because you didn't have to keep your eyes on the dogs (and your whip ready), you could contemplate what was before you. Sledges also were clearly superior in uneven terrain—dogsleds of Greenland design were heavy and intended for flat sea ice, not inland hills and valleys.

Where there were no dogs, there were no dogfights, no wet and impossibly snarled dog harnesses to untangle. The food boxes and leather goods did not have to be stacked at all times so the dogs couldn't get at them. One didn't spend every third or fourth day hunting for something to shoot and butcher to give to the dogs. The result was that "you have a lot of time at your camp site. Setting up and knocking down the tent takes no time at all, and leaves a lot of time for taking photos, making measurements, sketching, writing in one's journal."[58]

On the other side of the books was the reality that if there are no dogs along to pull the sled, you become the dogs. Four men pulling a single sledge move at a pace no single one of them desires. The weight is still the weight. On this particular trip they set out with a sledge and gear and provisions weighing 350 kilograms (770 pounds), which meant that each man was pulling 87.5 kilograms (192.5 pounds). It was too much, and on every substantial climb (there were many) they had to break the load in half to be able to pull it.[59] Other than the weight, their troubles were the usual troubles so familiar that they endured them without comment: four men trying to sleep in a three-man tent, cold that was just bearable, the reindeer sleeping bags that grew more sodden each day, the monotonous food and not enough of it.

Storestrømmen itself, their first taste of the Inland Ice, was something new. They ascended a glacier on the ninth day of travel and found the surface of the main ice stream to be very rough; it looked like the surface of a sea with heavy swell that had been suddenly frozen. As they moved west, the surface became rougher—all the normal features of a glacier were exaggeratedly huge. They met ice hummocks 10 meters (33 feet) tall—features that normally were only 1 or 2 meters (3–7 feet) high. The official record of the expedition noted laconically that "the ice was full of cracks and fissures covered with snow, so that the men could not see them until they fell in."[60]

Eventually, after five days of hard hauling and frequent falling, they established a base camp on the ice stream, unable to budge their sledge further. Wegener and Weinschenck went west on foot. The Inland Ice ended in a cliff adjacent to the ice-free land, a frozen vertical scarp of about 25 meters (75 feet). After much searching, they found a way down and hiked to and climbed one of the *nunataker*—the steep, rocky hills of what was apparently a huge tract. In the far distance there were many more mountains and a great inland lake filled with icebergs.

Already in March there were alpine plants to collect, fossils were abundant, and the unusual rock itself was eagerly sampled. As in the previous spring, when he had been in the North, these small gestures of natural history seemed so important. The collecting, the photographic documentation, and the cartography were the only things that separated them from some sort of extreme tourism, wildly exhilarating as that might be. Wegener, in particular, seemed to want a scientific rationale to mark each step. He believed deeply in *scientific* exploration. In any large frame of analysis the scientific results of this and most of the life-threatening trips he took that year were entirely negligible—nothing fundamentally new turned up on any of them. It was the aura of science that connected his daily activities to the great energetic themes of his Berlin *Lebensphilosophie* and made them meaningful.

They returned to Danmarkshavn on 3 April, to learn the news that they had expected but dreaded—Mylius-Erichsen was dead, as were Lt. Høeg Hagen and Jørgen Brønlund. Koch had gone north with Tobias Gabrielsen in early March. They found Brønlund's body at the depot in Lambert's Land. The dispute over where Mylius and

Hagen died, why they died, and whether they might have been rescued by a timely effort began immediately and (in Denmark) still goes on today. Brønlund had arrived at the depot in Lambert's Land at the end of November, a month *after* the relief party had come looking for him there. At that time Mylius and Hagen were already dead. But all three men were still alive when the search party set out, and even after its return to Danmarkshavn. Indeed, had the search party, stymied by open water at the Mallemukfjeld, tried to go further north by land behind the headland, by mounting the Inland Ice . . . what then? This sort of speculation haunts every expedition fatality in the Arctic even after all the principals have passed on.

The remorse, recrimination, and second-guessing tore at the crew's coherence and morale. Plans were floated to send an expedition north to find and recover the bodies, but these had to be given up—not enough dog food, and too far to travel by man sledge. Wegener was deeply moved by Brønlund's suffering and by his courage as he waited for certain death, making a last entry in his journal, and leaving it and Hagen's maps where they might easily be recovered. He had used some of the food and fuel at the depot, but his feet were too badly frostbitten to go on, and he couldn't find his way in the dark. This much was known because Koch had translated Brønlund's journal from Greenlandic to Danish on the way back, and Wegener later copied the translated extracts into his own journal.[61]

It occurred to Wegener that this would be a good time to relieve Freuchen at Pustervig. The plan had been to have a companion for Freuchen there at all times. This worked well in the fall and early winter but had failed in the depths of the winter dark: no one wanted to face the wolves, among other things. Freuchen had begun to suffer badly, battling hallucinations and talking to his kitchen implements. Wegener was not exactly at loose ends, but the aerology program had ended with the loss of the last working meteorograph the previous November. He had spent the winter tinkering, on and off, with the remains of two broken meteorographs, trying to assemble a working instrument from them, and he had succeeded.[62] Yet there was no appetite for work at Danmarkshavn, and cooperation was hard to come by. With the death of Mylius-Erichsen the decision had been made by Koch and Trolle to go home as soon as the ice broke—the attempt to cross the Inland Ice would have to wait for another expedition.

Wegener was actually taking his own advice. He was of the opinion that a change of scene at regular intervals, with new work, was an antidote to winter depression, and he would even have encouraged competition between teams of scientists rotating in and out of the station to see which team could do the most scientific work. These suggestions he prudently confided to his diary, however: they would have struck his fellow expedition members as the ravings of a madman.[63]

His first day completely alone at Pustervig was 9 May. "It is a strange feeling, to be so utterly alone under these conditions. Today I heard the cry of a snow grouse while out on a short ski trip, and this was my only encounter with the organic world of our planet—that's not much!"[64]

On the next morning, around 11:00 a.m., the valley fog lifted and he made a snap decision to climb the mountain and make the 400- and 800-meter temperature observations before coming back to complete the daily 2:00 p.m. station log. He put on his Tyrolean hobnailed boots and took off. He made it to the 800-meter station and back in two hours. "The rock-climbing was child's play, and in spite of the fact that I took the time to admire the wonderful vistas, I was back more than an hour sooner

Wegener standing in front of the margin of the Inland Ice of Greenland in late March 1908, encountered while traveling with the group that discovered Dronning Louise Land, a large ice-free area within the inland ice cap. Note that the ice appears as rock, with sedimentary layering, dark coloring, and flow bands. From J. P. Koch and A. Wegener, "Die glaciologischen Beobachtungen der Danmark-Expedition," *Meddelelser om Grønland* 46 (1912): 1–79.

than the others [who have made the climb] though naturally you cannot do what I did wearing kamiks."[65]

He was bursting with manic energy and competitive spirit, not the first Artic traveler to feel that rush of enthusiasm and possibility when twenty-four-hour sunlight replaces twenty-four-hour darkness. In the afternoon, after his observations, he went out again, this time with his crampons and his ice axe, to try a more difficult route to the top. He found a couloir (a steep gully) in the mountain wall, filled with compact snow, the fruit of many avalanches. It was "murderously steep," but the snow conditions were perfect for his crampons and axe, and the climb was "like going up a staircase." He made it to the top in three hours and descended in half that: "coming down I was completely one with the crampons and the axe, and descended with a speed that was at times remarkable.[66]

Two days later, he decided to climb it at night, or rather the deep twilight of the midnight hours, so as to be able to take pictures of the sunrise from the summit. He was taking unbelievably reckless chances climbing alone and untethered under these conditions, but he seemed not to care. He found another couloir to climb, "steeper even than the last one." He ran out of snow halfway up and had to pick his way over a section of rock, but when he arrived at the top, he decided it was "all in all an easy and delightful route." He remembered that the northwest face of the Monumentberg was a sheer and overhanging cliff that he had admired from below when mapping in the fjord. He decided to go and have a look from above. He got as close to the edge as he dared (about 2 meters [7 feet]) and looked down. "I believe myself to be entirely free of vertigo, but it spooked me good and properly when I looked down from this unbelievably sheer cliff. It was by a wide margin the most amazing piece of rock I have seen on this expedition. . . . I rolled a big rock over the edge. It disappeared immediately and I listened attentively for the sound of the impact. It seemed to take an eternity until I heard it land. It gave me goose bumps all over my back."[67] He was in love with the place and climbed it again and again, taking photograph after photograph, framing up vista after vista for his color camera.

He was a man on holiday—from the expedition, from his preoccupations with work, from himself. Kept off his mountain by several days of heavy weather, he read a novel—"the only one here"—though a year before he scorned all novels as providing no release. But something much deeper was at work in him.

> Mostly I sit here in the hut and ruminate, smoking pipe after pipe of this awful expedition tobacco. I wonder often about myself, how can I sit here for so many hours and do nothing? I smoke, I listen to the cheerful hiss of the spirit lamp that is my stove, and my reveries carry me from one end of the world to another, from the South Pole with its unexplored continent, to Zechlinerhütte where the lilacs will be blooming just now, to Berlin, to my parents, to Lindenberg. Whether everything will look the same to me when I get back home? Where is Kurt? How are my parents? And then I'm off again, the Chilean Andes, South Africa, New Zealand. It is going to be very hard for me going back there—to bourgeois society, and that indoor clutter, home. It is going to be horrible, to be presented here and there like a polar bear with a ring in his nose. I will in any case see that I do whatever I can to keep from giving any public lectures, or writing newspaper articles.[68]

He was discovering that he didn't want to go home, and perhaps not even back to Danmarkshavn. "I'm now completely at ease with being alone, and can't find anything unpleasant in it."[69] There was so much to see everywhere he went. He wrote in his journal like a man awaking from a dull dream to the fullness of possible experience. "The summer is coming on with force. The dwarf willows have fat buds and even catkins, and just now I saw in the midday sun two fat flies tumbling over each other on the wall of the hut."[70]

His diary on the return trip from Pustervig to Danmarkshavn reads like a dream journal. Lundager met him at the halfway point with a sled to help transport the scientific equipment. They traveled at night through dense ground fog that created fantastic mirages. There was a thick coating of hoarfrost on the sea ice—cruciform needles more than a centimeter in length. He wanted to stop and photograph them. "On a future expedition, one could study all the forms taken by the snow cover on the sea ice, especially the ones formed by the wind. It appears here in every possible form from great regular wave systems to the most delicate textures, like moiré fabric."[71] Here was his old fascination, the development of disturbances at any sharp and level boundary surface between two substances of different densities. Here it was the density difference not of ice and snow, but of snow and air. Snow was a form of ice light enough to be sculpted by wind. "When I work up my meteorological results, I'll have to pay careful attention to the different forms of precipitation."[72]

Back at the ship, no one had made any measurements (other than simple station meteorology that almost did itself) since his departure. "I have got the air-electrical apparatus going and have hounded Hagerup into making the magnetic measurements."[73]

Wegener started up his kite and balloon flying again with the cobbled-together instrument he had remaining. Even in the mild June weather he had trouble getting anyone to help him, finding himself once again in opposition and an outsider. "Ever since we lost Mylius-Erichsen," he wrote, "there are some members of the expedition who speak of the [scientific] work with the greatest possible contempt. You can't cheer them up and they won't help. I can't say that I'm all that enthusiastic about the work myself . . . but between feeling that, and saying loudly to all and sundry that your work is Humbug, is a big step, and, it seems to me a fatal step for men forced to live together in the narrow confines of an expedition.[74]

With Mylius dead, Wegener's circle of friends was small indeed. Koch had taken Freuchen with him to do more glaciology on the Inland Ice, knowing how keen Freuchen was for the experience. Lundager, his companion in the Villa, was strong enough to help with the winching, but as Freuchen later remarked admiringly, "he was good company, but I have never met a stronger, nor a lazier man."[75] In any case the window for botany on such an expedition was small, and Lundager was taking it. Wegener was thus often left to this work himself; never one to shirk, he pressed on with the aerology doggedly, making eight ascents in June and five in July.[76] He also got one record for free. Hendrik and Tobias, hunting on the drift ice southeast of Cape Bismarck on 1 July 1908, found the meteorograph ("Teisserenc de Bort No, 334") lost the previous November when the cable had snapped; though the instrument was ruined, the record was readable, and this brought his expedition total to 125 flights spread over eighteen months.

The *Danmark* came free of the ice on 2 July, and the mosquitoes arrived on the third. They had fashioned bags of mosquito netting to wear over their heads, and Bertelsen and Wegener had taken one of these and made a butterfly net from it. "Since yesterday evening the two of us have worked intensively together—catching butterflies. We have caught nine different species."[77] With the breakup of the pack ice proceeding apace, it was now a waiting game: they would leave when they could. On the tenth three Norwegian sealing ships hove into view and anchored off Cape Bismarck—their first contact with the outside world in two years. There was mail for the expedition, including a letter for Wegener from his parents. "Crammed full of news, most of it good. Thank God things at home are this good, it makes me ever so much happier about the prospect of going home."[78] On 13 July there was open water; waiting until the last moment, Wegener packed up his electrical and magnetic instruments and built a cairn on the site of his magnetic observatory. While waiting for Trolle to decide whether he had enough open water, he began to work up the scientific results of the expedition and produced a concise summary of them and sent them off on the seventeenth with the Norwegian sealers.[79]

It wasn't until the twenty-first that Capt. Trolle saw enough open water to suit him and sailed out of Danmarkshavn, not to the east, and home, but straight north. Admiral Georg Amdrup, in his terse, relentlessly upbeat, and laudatory official account of the Danmark Expedition, published in 1913 in English for the widest possible circulation, said of this decision, "The energy of the Expedition did not fail, right up to the very last. Trolle desired to supplement his hydrographical series as much as possible . . . and he specially wanted to make these investigations as far north as possible."[80] It doesn't take much imagination, however, to see that they were going north to find Mylius's and Hagen's bodies and recover their journals. According to Brønlund's journal, their bodies lay on the ice at the Nioghalvfjerdsfjord (N 79°) between Lambert Land, where Brønlund had died, and Hovgaards Ø to the north.

Trolle made good progress going due north, covering 70 kilometers (43 miles) in the first two days. By the twenty-fourth they were already at latitude 78° north, and in the early morning fog, turning east to avoid pack ice, they rammed a large floe, the shock of which cracked their boiler and brought them to a halt. Weinschenck worked constantly on the boiler for the next few days, as they alternately looked for an opening in the ice through occasional dense fog and moored to large floes to make major repairs. Trolle made the decision to give up the "hydrographical investigations" and sailed the ship toward open water in the southeast. With the boiler functional but leaking, and the *Danmark* already twice pinched in pack ice, he had the survival of the ship to think about and turned for home.

On the way home Wegener did what he always did—took meteorological measurements; there was no time when he ever seems to have found the weather uninteresting. As they proceeded east, the condition of the boiler worsened, and on 15 August, low on coal and with their boiler cracked and rusted out, the ship was towed into the harbor at Bergen Norway. Freuchen recalled, "The first day on shore we went wild and behaved like savages." While they continued to celebrate, they began to realize, via the steady stream of visitors, reporters, and dignitaries, that they were famous. One of the first to make a visit on board and congratulate them was the great Norwegian explorer Roald Amundsen (1872–1928), who was preparing, he said, an expedition of his own with the *Fram* to reach the North Pole.[81] This, of course, was a ruse—he had already decided to go south, to beat Scott to the South Pole, though he had told no one.

The news of the expedition's return and of the deaths of Mylius, Hagen, and Brønlund had created an immense sensation in Denmark. Trolle received a telegram from the expedition committee: a tug had been dispatched from Copenhagen to bring them home, and they should prepare themselves for a hero's welcome. The poor old *Danmark* was leaking badly and had to be pumped day and night in the four-day voyage to Copenhagen. When they arrived, the tumultuous greeting stunned them—a crowd of many thousands, waving flags and cheering loudly, lined the Langlinie as the ship was towed past. After a formal greeting at the harbor, the expedition members, wearing their musty street clothes, were whisked away to the university for a great banquet with laudatory speeches. King Frederik VIII praised their courage, shook their hands, and decorated them one and all with the Danish Royal Order of Merit.[82] Whether Wegener experienced this decoration as the insertion of the ring in the polar bear's nose, we don't know. He stayed on for a few days, said goodbye to Koch and Freuchen and the others, boarded the train for home with a case full of scientific results, and headed to Berlin. He had much to be proud of and much to do. He was, perhaps only for the moment, and other than Erich von Drygalski, the most famous Arctic explorer in Germany.

7

The Atmospheric Physicist (1)

BERLIN AND MARBURG, 1908–1910

> I don't know whether this is just my own idiosyncrasy, but everywhere I turn I stumble over these typical boundary surfaces . . . this idea seems to me to offer an extremely useful perspective.
>
> WEGENER TO WLADIMIR KÖPPEN, 13 November 1909

Back in Berlin

Much has been written about the struggles of men to be chosen for Arctic expeditions; less well chronicled are their struggles to be released from them after they are over. Expedition contracts, eagerly sought and signed with alacrity, sometimes remain in force for years after the homecoming and may include burdensome and obstructive restrictions and obligations.

In the case of the Danmark Expedition, there were several important restrictions. At Mylius-Erichsen's insistence, all members of the expedition had agreed to refrain, for a set term of months, from giving public lectures without the prior consent of the expedition committee. Mylius, as we have seen, was fatally obsessed with the fear that someone might make a discovery to which he could not lay claim. He had also planned to superintend, after the return, the matter of who should talk about the expedition, as well as when, where, and to whom. His aim had always been to employ the expedition to advance his own credentials as an explorer, and he had planned, as well, to retire the expedition's debts, make his living, and raise funds for a new expedition by giving illustrated lectures featuring his own exploits and those of the others under his command. Mylius-Erichsen's death certainly lessened the severity of the prohibition, but it did not change the contractual obligation.

In addition to the limitations on public lectures, there were also restrictions on writing about the expedition, and for the scientists these included a codicil that they should restrict their use of the Danmark Expedition's scientific results (in print) until after these results had been worked up and published in the expedition's proceedings. The scientific "results" of the expedition, meaning the reports of the uninterpreted but reduced and ordered data—physical, meteorological, geological, botanical, zoological, oceanographic, and archeological—were already committed to a planned series of volumes of *Meddelelser om Grønland*, scheduled to appear between 1908 and 1911 under the title *Danmarks-Ekspeditionen til Grønlands Nordøstkyst 1906–1908*. This was, incidentally, a breakneck pace for this venerable but sometimes leisurely journal, indicative of the importance Denmark accorded this scientific venture.

The reciprocal of the restriction on writing about the expedition was Wegener's obligation to work up all the scientific material for publication before he could be released from his expedition contract. Initially this posed no problem for him, since he had no immediate plans for anything else. He had returned to Berlin and moved back in with his parents at 20 Georg Wilhelmstraße, in Halensee. He wanted to pursue his scientific work.

Catching Up

In order to pursue a scientific career, however, Wegener had to reestablish relations with his scientific colleagues, and this could not wait until his expedition responsibilities were completed. In the two years he had been out of touch, meteorology, aerology, and aeronautics had all moved fast and far, and the professional landscape had altered quite markedly. Among Wegener's closest senior colleagues and sponsors, the ranks were now rather thin. Bezold, who had been in poor health when Wegener left, was dead. Aßmann was very supportive but was tied to his instrumental prototype work in Lindenberg, which was still—unbelievably(!)—under construction. Arthur Berson was leading a major aerological expedition in East Africa, having departed only a month before Wegener's return.[1] Wladimir Köppen was at Großborstel, the aerological station near Hamburg, and working on his climatology more than his meteorology.[2]

The real action in the mapping of the vertical structure of the atmosphere had, in Wegener's absence, moved outside the borders of Germany. Teisserenc de Bort had been continually on the move: while Wegener was away, he had sent up sounding balloons from the *Princesse Alice*—the yacht of his patron, Prince Albert of Monaco—in Spitzbergen (1906) and the tropical Atlantic (1905–1906, 1907). During 1907 and 1908, he had done the same in land-based expeditions to Swedish Lapland. The study of continental interiors was also moving ahead: Russia had mounted aerological expeditions to Central Asia in 1907 and 1908. In maritime Asia, the Dutch had begun an aerological program in their colonies at the Batavia Observatory, under the direction of Willem van Bemmelen. In the summers of 1907 and 1908, a coordinated program of observations at sea had been the focus of the "international aerological weeks." The mapping of the atmosphere, for which Bezold had been an advocate and prophet, was moving forward very fast. There were new, exciting, and ample data emerging on the structure of the atmosphere in both polar and equatorial latitudes up to almost 20 kilometers (12 miles).[3]

While Wegener had to catch up with what the most senior and well-funded scientists were doing, he had also to follow the work of his contemporaries and measure their efforts against his own. Two meteorologists of his own age, in particular, had emerged into prominence while he was gone: the Swiss scientist Alfred de Quervain (1879–1927), and his own countryman Heinrich Ficker (1881–1957). Wegener knew of de Quervain, having published an evaluation of his balloon theodolite in 1905. That instrument, with modifications, had become standard during Wegener's absence. De Quervain had observed and published on a phenomenon Wegener had also seen in 1905 during pilot balloon ascents—a sharp discontinuity in the wind speed and direction at the boundary of the "inversion layer." Moreover, de Quervain had, in 1908, worked out (in tandem with Hugo Hergesell) an important correction factor for the velocity of ascent of a pilot balloon—a question to which Aßmann and Hergesell had previously devoted a great deal of attention.[4] Hergesell had the bright idea to release and follow balloons inside Straßburg Cathedral (with de Quervain observing with his theodolite), and he discovered that in such still air, balloons rose 0.5 meters per second (1 mile per hour) slower than outdoors. This indicated that the actual rate of ascent outdoors was some combination of air-pressure lifting and turbulence pushing the balloon.[5] This was a very substantial correction: almost 50 percent in the first kilometer above the surface—that part of the atmosphere on which Wegener had done most of his own work. It was easy for Wegener to see that had he not gone to Greenland, he, and not de Quervain, would have been working with Hergesell. Moreover, de Quervain was rumored to be planning a meteorological expedition to Greenland, building on credentials established with Drygalski in the Antarctic in 1901–1903.

The other young meteorologist making an impact, Heinrich Ficker, had completed his dissertation at the University of Innsbruck in 1906, and in researching it, he had made more than 100 climbs near Innsbruck (the heart of Austrian "mountain meteorology") to study the föhn. His work, not yet published but widely known, provided a strong empirical confirmation of Julius Hann's theory of föhn winds and led Ficker to a job at Hann's Zentralanstalt für Meteorologie at Vienna.[6] Indeed, Wegener returned from Greenland to find that the question of the föhn, on which he had lavished so much effort and attention, was now generally considered to be nearly solved.[7] If true, this would be a real dent in his accumulation of novel results, reducing him, in an area on which he had worked hard in 1907–1908, to the helpful (but minor) role of confirming another's discovery.

The Hamburg Meeting

Wegener could see that he must get his Greenland results out quickly or, at the speed things were moving, there wouldn't be any *original* results to get out. Here he was in luck. The triennial meeting of the Deutsche Meteorologische Gesellschaft (DMG) was to be held from 28 to 30 September 1908. Moreover, it was a special meeting: the twenty-fifth anniversary of the founding of the DMG. The meeting was to be held in Hamburg, returning to the site of its founding for the first time in a quarter century. While it would have been unlikely for Wegener to miss this meeting wherever it occurred, it was unthinkable for him not to attend a meeting arranged by Wladimir Köppen, the very man who had done so much to make possible his career in Arctic meteorology.

In spite of the Danmark Committee's restriction on public lectures and on the reporting of scientific results before the expedition reports were out, Wegener submitted a proposal to Köppen to present a lecture at the Hamburg meeting on his research in Greenland, to be illustrated with lantern slides. Köppen was, of course, delighted. He immediately gave Wegener a prominent slot on the program and took the draft program for the meeting from his office in Hamburg to the family home in Großborstel. He always, as his daughter Else recalled, wanted the whole family to share his excitement in the development of his "young science." He was especially enthusiastic about Alfred's participation, telling his family that "we need in meteorology these days the kind of minds coming our way from physics; it is now time to comprehend and explain, from the standpoint of physics, the atmospheric processes we have discovered in the course of our kite and balloon ascents."[8]

Köppen was himself vitally interested in developing a physics of the atmosphere. Though the bulk of his own efforts had been in descriptive climatology and in the study of climate cycles and periods of various kinds, he hoped that a global network of meteorological observatories would lead not only to a map of the world climate but to an accurate picture of the three-dimensional structure of the atmosphere. The latter would provide the observational basis for an atmospheric physics, and that physics, in turn, would lead to a theory of global atmospheric circulation and to an understanding of cyclonic storms—the great whirling high- and low-pressure systems that dominate the weather of the middle latitudes. Köppen had written much on the topic in the 1890s and had devoted a good deal of his own observational work at Großborstel to elucidating the problem; he had been willing to advance Wegener the instruments for Greenland precisely to extend kite and balloon observations of atmospheric behavior (well known at latitudes 30°–60° north) to latitudes above 75° north.

Now that Wegener was on the meeting program, he had to decide what to talk about and how to shape it for maximum effect. The only thing really in shape to go forward in a public arena was the aerology, and so he decided to concentrate on a description of the heart of his observation program, the kite and balloon ascents, to supplement that with photos of mirages (many of them spectacular), and to talk about his hopes to explain at least the *polar* föhn.

Wegener's talk, given on 28 September, created quite a stir. It was a wonderful audience for him to face in his first major public lecture: the entire German meteorological community was there, as well as many foreign visitors, including Teisserenc de Bort. They were, of course, accustomed to hearing about voyages and expeditions: they all went on them; but this report of a two-year effort in the high Arctic was an unusual event even for them. Wegener certainly looked the part of the Arctic explorer. His frame was stocky and muscular, his hands large and rough. His skin was deeply tanned from two years' exposure to sunlight and wind—not brown, but almost copper red, like an Inuit—an impression made more startling by contrast with his pale, gray-blue eyes.

When his turn came, he walked to the podium and, without preamble, began to speak in a clear, firm voice:

> At the main station of the Danmark-Expedition, at Cape Bismarck at 76 3/4° north latitude, on the Northeast Coast of Greenland, between summer 1906 and summer 1908, somewhat more than 100 kite ascents were carried out, up to an altitude of 3100 meters, and 25 captive balloon ascents, up to 2300 meters. This is the first time that an aerological program has ever been successfully carried through to completion on a real Arctic expedition. I must note in passing that lack of funds forced me to work with extremely limited and scanty equipment, and you may easily imagine the difficulties experienced in carrying out such experiments, especially during the winter night.[9]

He sketched out his results quickly: the wind was remarkably constant from the northwest in all seasons and föhn-like. Winds from the east were rare and limited to the lowest 500 meters (1,640 feet). Temperature layering was very sharp, with many inversions in the first few tens of meters above the surface—in the spring there were inversions of up to 8°C (46°F) in the first 30 meters (98 feet). He then turned to the question of mirages, noting that this was the first systematic photographic study ever made of mirages, and launched into an excited technical discussion of their causes—a combination of inversions between 100 and 1,000 meters (328 and 3,281 feet), hyperadiabatic cooling in the lowest few meters, and, in winter, the release of heat by freezing water at the surface of the sea. After this discussion, he turned his attention to the establishment of the Pustervig station and his study of the föhn winds under circumstances in which mountain measurements could be made at elevations of 400 and 800 meters (1,312 and 2,625 feet) while simultaneous airborne measurements were under way on the coast, 60 kilometers (37 miles) distant.[10]

As is often the case in science, what looks like a bare summary of observations is actually a focused series of contributions to a set of ongoing controversies. If one were to rephrase Wegener's talk with the questions put back in, it would sound something like this: "My extensive and original polar observations over two years confirm that the polar easterlies, like the low-latitude trade winds investigated (while I was away in Greenland) by Rotch, Teisserenc de Bort, and Hergesell, comprise a shallow frictional layer only a few hundred meters deep, with prevailing winds above the friction layer coming from an entirely different direction. Additionally, I can confirm that at very high latitudes, as at very low latitudes, the atmosphere is complexly layered up to about 1,500 meters (4,921 feet), and I have documented this both instrumentally and via photography—my pictures of mirages. These conclusions confirm our intuition that surface observations are extremely misleading in understanding atmospheric structure and dynamics, and that aerological investigations are required everywhere and at all times. I extended this observation program in the second year to take up the question of the föhn. I made wind and temperature observations both within and above the friction layer, at 400 and at 800 meters, simultaneously at an inland mountain station, and at the coast using kite observations. I should, from these data, be able to make a definite contribution not only to the study of the föhn but to the ongoing controversy in Germany, Austria, and elsewhere about the relative value of observations of temperature and wind at mountain stations and in the free air at the same altitude." Wegener did not say this, but it is what his professional meteorologist listeners heard him say.

When Wegener had completed his brief summary (and implicit conclusions), he asked for the lights to be dimmed, and then he narrated his spectacular slide show. He showed the technique of kite and balloon launching in the cold and the dark, following this with a long sequence of his best mirage photographs. He had, as well, a number of pictures of cloud forms and some beautiful close-up pictures of hoar frost; he concluded with a stunning color photograph of a Sun ring 800 meters in diameter, projected by local atmospheric conditions on the cliff face of the Monumentberg.[11]

It was a great performance, and he was right to be pleased by the effect, but more important to him than the praise and congratulations was the sheer intense pleasure of being once more in a group of people who understood what his work meant and with whom he could discuss it. He was starved for scientific conversation. He spoke at length with Aßmann, who was very interested, as usual, in correction coefficients: Wegener had some ideas about a new way to extrapolate temperatures, taken at 1,000 meters, up to a height of 1,500 meters.[12] He spoke with Hugo Hergesell, who solicited his Danmarkshavn temperature data for a map he was preparing of the circumpolar weather in the Northern Hemisphere in July 1907, and who urged him to speak at the geographers' meeting at Straßburg in November.[13] He also spoke with Reinhard Süring (1866–1950), who was being promoted from Berlin to take over the meteorological station at Potsdam, and who invited Wegener to come see him in Berlin for further conversation in October.[14]

The most important reconnection of Wegener's Hamburg trip was, however, with Köppen and his family. Köppen's daughter Else, then sixteen, got permission to stay away from school to attend Wegener's lecture. Returning home on the afternoon of the lecture, she spoke to her godfather about how wonderful and exciting she had found Wegener (and his lecture). Later that day, to her surprise and delight, her godfather presented her with an invitation to the official banquet—the first time in her life she would attend a formal dinner; she was to be Wegener's guest at table. She was extremely nervous and unsure of herself, and Wegener did what he could to put her at ease. When the speeches began (in Germany, as elsewhere, the prescribed punishment following every official meal), Wegener pulled a stack of expedition photos from his pocket and passed them to her one by one under the table.[15]

Wegener stayed on after the meeting for the field trip to the Kite Station at Groß-borstel, and he accepted Köppen's invitation for an informal dinner with family and friends at his home. Köppen invited his old colleagues Aßmann and Hergesell, of course, but also went out of his way to include a number of the younger meteorologists. Köppen had a serious demeanor at official functions, but he liked to let his hair down at home and to meet with the younger scientists as equals. He listened to their concerns and gave them advice. He fed them and drank with them, and he treated them, after dinner, to a concert of *Meteorologenlieder*—popular song melodies he had adapted with meteorological lyrics—the sort of corny clowning that only academics truly love.

After dinner and singing that Sunday evening, Köppen invited Wegener to sit down for a chat that, as it turned out, lasted far into the night. August Schmauß (1877–1954), who was then an assistant at the Central Meteorological Station in Munich, was there that evening, and he recalled their conversation:

> Anyone who met Köppen carried for the rest of his life the memory of having come upon one of the most interesting personalities in our meteorological world. I had that good fortune, and an important lesson, at the Köppen home after the meeting of the German Meteorological Society in Hamburg in September 1908. It so happened that A. Wegener was also there. I found these two men, who were completely consumed by their scientific work, in the midst of the most animated discussion of problems in aerology, and debating global climatology. It was very important to Wegener to be able to express his ideas in the presence of this insightful critic, while Köppen, though a mature and famous scientist, was visibly inspired to hear and to take in what his younger colleague had to say.[16]

For Wegener, the Hamburg meeting was pleasant, auspicious, and productive. In a single stroke, he had made his mark and had tied his work to the frontline problems of the day. Yet it was clear to him, on his return trip to Berlin, that he still had to solve pressing and immediate problems. Where, for instance, would he live? How would he make a living while working up his *Danmark* results? He could not camp forever with his parents; already he found Berlin crowded, noisy, oppressive, and, in no small measure, absurd—as the bourgeois world often appears to those who have lived for a time free of its constrictions and conventions. Wegener had to make up his mind what to do. Brother Kurt had moved ahead decisively. He had become, in Alfred's absence, one of Germany's best-known aeronauts, with more than fifty balloon flights under his belt. Further, he had been picked to head the Observatory in German Samoa and would leave almost immediately; the Reich Treasury had voted this tropical observatory the princely stipend of 25,000 marks per year for the period 1905–1910. Kurt would direct construction and superintend meteorology, seismology, and magnetic studies.[17]

Everyone else in the family also seemed to have found a footing and a place. His sister Tony, living year-round at *die Hütte*, was making a reasonable living and a solid reputation as a painter. His parents were enjoying their retirement immensely, shuttling between Berlin and Zechlinerhütte, and father Richard had just won a major drama prize for his book on Shakespearean theater.[18] This book had a practical as well as scholarly significance: Alfred's cousin, Paul Wegener (1874–1948), on his way to becoming the greatest actor of his generation, was employing the concepts chronicled in Richard's book for a series of triumphant performances on the Berlin stage as

Richard III, Othello, and Macbeth. All of this was a source of pride and pleasure to Alfred, but it also increased pressure to find a place for himself.

Die Arbeit

The immediate problem, however, ever squarely before Wegener and dominating his mind, was how best to work up the mountain of scientific data he had accumulated in Greenland. He had begun to organize it, but he had no clear sense of how large a task this would be, nor how long it would take. He therefore accepted Reinhard Süring's invitation, proffered in Hamburg, and took a selection of data to him at the Meteorological Institute in early October. Süring studied the data and talked with Wegener about the Danmark Committee's plans for the publication of the expedition's science. Süring estimated that, given the volume of material, if he (Wegener) did no other scientific work at all, took no job, and worked at it full-time, he would require a minimum of two years to get just the meteorological material into publishable form; he would need four years if he worked on other projects and took a government or academic position.[19] This estimate of time did not include the electrical, magnetic, geological, or glaciological material, just the meteorology.

Wegener seems to have been disconcerted and even shocked by Süring's estimate of the time to complete the work. On 12 October he wrote to Professor Johannes Warming (1841–1924), the great Danish botanist and plant ecologist, who was the lead scientist on the committee of the Danmark Expedition. Warming had himself been on long expeditions to Greenland, Brazil, and elsewhere and might be able to help him with his dilemma. Wegener told Warming of Süring's estimates of time and of the number and complexity of the tables and illustrations required for the work. He then gave vent to his anxiety, adding, "While the rapid publication [of this work] is my deepest wish, I would hate to be obliged to go two full years without a job, and without pursuing any other scientific work. The best possible outcome, from my point of view, would be to have the question resolved thus: I will immediately begin devoting all my time to this work, with the stipulation that later on I may be permitted either to pursue other scientific work, or to accept a position."[20]

This suggestion seems to have been met with generosity on Warming's part, and Wegener's negotiations with the expedition committee went smoothly and well. Moreover, by the end of October, he had succeeded in off-loading part of his burden: Georg Lüdeling (1863–1960), a meteorologist at the Institute in Berlin, agreed to work up the electrical measurements, to be completed by spring 1909. Wegener then had his electrical instruments and barometer shipped from Copenhagen to Berlin to be calibrated against the instruments at the Meteorological Institute, and he attended, in the next few weeks, to the dozens of small details surrounding the project, such as formatting instructions for the tabular data and similar matters.

In late October Wegener signed a memorandum of understanding with the Danmark Committee, including a pledge not to publish the results before they should appear in *Meddelelser om Grønland*. Here, however, he made an important stipulation: that he should be allowed to publish, in the scientific literature in Germany, articles on his kite launching and on the climate at Cape Bismarck, so long as he did not publish the data in extenso. "In the case of these brief articles," he wrote to the committee, "I do not consider myself bound to see that they appear *after* the larger work, as their whole point is to stimulate interest for the larger work, and if they were to appear after it they would be essentially worthless."[21]

Wegener's was an interesting balancing act, of two kinds of science and two ages of science in the world. If one looks at the history of exploring expedition reports, one sees that these sometimes appear not just years but even decades after the expedition ends. The material overwhelmingly consists of description of things encountered in some particular place and detailed mathematical-cartographic work establishing exactly where the place is. The descriptive dimension here trumps the explanatory: in extending the map of the world, of its geology, of animal and plant species, and generally in documenting the natural history of the planet, it is deemed more important to get it right than to get it written. Wegener's scientific work belonged to this classical tradition, but only in part. His documentation of the climate at Cape Bismarck was as real and durable as the documentation of the fossils, the birds, the plankton, or the Inuit ruins. But Wegener also belonged to a much faster moving enterprise—modern physics, where publication delay of months or even weeks could mean that someone else claimed a discovery that was, in some sense, one's own. His urgency to pursue these traditions simultaneously lent a feverish intensity to his activity in the next few months.

Such external circumstances explain his situation, pulled in two directions, but not his drive to go in both directions at the same time. Wegener felt his obligation to the expedition—the need to find a place, to make a mark—but he appears to have been more powerfully driven from within by the structure of his own imagination, an imagination readily inflamed and yet also sustained by his intuitive talent for making connections. Wegener's work in late 1908 and early 1909 clearly shows the interplay between his processing of a large volume of raw data and then the extraction of a manifold of interesting phenomena and subsequent relentless pursuit of these in parallel lines of attack, based on intuition, but requiring subsequent confirmatory research.

In mid-October 1908 he had moved out of his parents' home and taken rooms down the street at 15 Georg Wilhelmstraße. There, as in each subsequent residence he would inhabit, he set up a large worktable and rolled out the polar bear rug he had brought back from Greenland—the most prized of his few possessions. He had moved out of his parents' home because he needed the solitude to work and think, because he kept long hours disruptive of domestic routine, but above all because of his need to smoke incessantly as he worked, sometimes as many as ten cigars a day, with the smoke eventually forming a blue stratus layer in the room, hovering just above his head.[22]

The task at hand for Wegener in these months was the interpretation of the kite and balloon aerology of the expedition. This was his best chance for significant results; he had publicly, at the Hamburg meeting in September 1908, hinted at the possibilities latent in his records. Now he had to produce them. The raw data consisted of the station records for the Danmarkshavn meteorological instruments, the monthly "24-hour measurements" (from the shore, deck, and crow's nest of the *Danmark*), and the 125 flight records taken by his kite and balloon meteorographs. The flight records he had interpreted to some extent in Greenland.

He knew what he was looking for: layers and layering in the first 2 kilometers (1.2 miles) of the atmosphere. It was there, of course, in the form of complex suites of inversion layers revealed by the mirage photos, mostly within the "friction layer." Wegener was looking for something else, though: for stable layering higher up, layering as consistent as the "inversion" layer discovered by Teisserenc de Bort (who had just proposed a new name for it: the "tropopause"). The latter discontinuity, the tropopause, separating the "troposphere" below from the "stratosphere" above, everywhere marked by a temperature inversion, had now been detected in all latitudes from the equator to 70° north and 75° south and was clearly a global phenomenon. At the equator it was higher, and in polar latitudes somewhat lower, than in temperate mid-latitudes, but it was a real and physically significant boundary layer, defining the structure of the atmosphere.

Wilhelm von Bezold had insisted years before, in Wegener's first meteorology seminar in Berlin, that there were additional boundary layers in the atmosphere just as real and persistent as the major inversion discovered at 10–12 kilometers (6–7 miles). Bezold believed that these other layers often coincided with the lower or upper surface of clouds, but that they would be discovered at the same altitudes in clear air, if conditions were not right for condensation.

Wegener was looking for these "other" layers in the Greenland data. The signatures of novel conditions and phenomena (these layers included) in these records would be miniscule, certainly depending on meticulous graphical interpretation and careful numerical reduction of the data. Results of significance would be uncovered, just as at Lindenberg three years before, in the form of small but characteristic blips or nicks in the pen tracings, indicating unusual modulations and reversals of conditions at some altitude, amounting only to fractions of a degree of temperature, or fractions of a meter per second of wind speed. Wegener was aware that Richard Aßmann's failure to believe his own instrument readings had cost him credit for the discovery of the stratosphere, and he did not want to fall victim to such a mistake, if his own data were to contain such treasure.

The accompanying figure shows a kite-meteorograph record from one of Wegener's flights in Greenland. The recording instruments, like those used by Teisserenc de Bort, depended on the expansion and contraction of various substances. The temperature was measured with a bimetallic strip thermometer capable of very great accuracy. In this thermometer a strip of brass is bonded to a strip of Invar, an alloy of nickel with an extremely small thermal expansion coefficient.[23] Expansion and contraction with temperature forced the strip to curve and allowed it to drive a scribing apparatus and record temperature changes directly. The meteorograph barometer operated on a similar principle: it was a "Bourdon tube," a thin curved tube of elliptical cross section filled with liquid. As the air pressure increased or decreased, the tube would contract or expand: this movement drove (directly or indirectly) the scribing apparatus. The measurement of the atmospheric pressure, corrected for surface station pressure, could be read off as the altitude directly. Relative humidity was measured by a "hair hygrometer," an apparatus that magnified and recorded the contraction and expansion of a length of human hair. Finally, there was some sort of a device for registering the wind speed, either a rotating or a "pressure-tube" anemometer.[24]

Wegener should thus have had access to four direct streams of data: temperature, air pressure, humidity, and wind speed. The reality of his Greenland data was, instead, rather more meager. In winter (i.e., a third of the time) the hair hygrometer froze and could not record. Almost as bad was the status of the wind data: of his six meteorographs, only one ("Hergesell 104") had a working anemometer, and that instrument lasted only five flights in summer 1907 before it was lost when a cable parted; it functioned principally to help him calibrate his procedure for estimating wind velocity aloft *without* an anemograph.[25] This meant that for 120 of 125 flights, wind direction and speed had to be estimated from the ground by observing the motion of the kite as it swung on its tether (direction) and using cable tension and the angle of flight as proxies for wind speed. Alfred had worked with his brother Kurt at Lindenberg to produce a

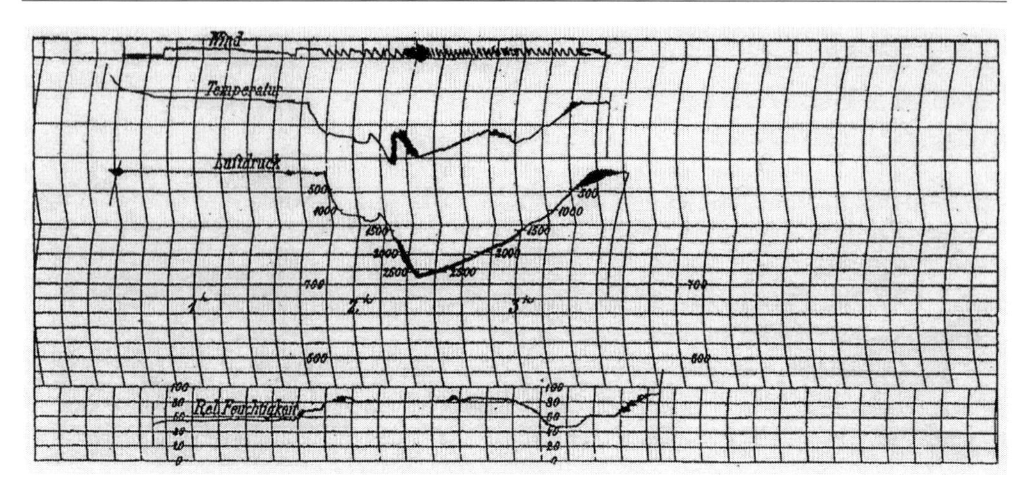

Kite meteorograph from Wegener's work in Greenland. The four records from top to bottom record the wind speed, the temperature, the barometric pressure (a proxy for the altitude), and the relative humidity. The recording represents a flight of just under four hours.

very useful tabular system of estimating wind speed by this method, but it was some distance from such estimates to actual recorded data.[26]

Working up the kite and balloon data for publication took six weeks of strenuous work, from mid-October to the end of November, and more than 400 cigars. Wegener's exceptional capacity for disciplined work served him well here. He had kept good station records, but he also needed to review the actual meteorographs and check his instrument readings against the interpretations he had made of them in Greenland, in the cold and the dark and (sometimes) in the grip of his winter depression. The accompanying figure gives the narrative summary of the first three of his Greenland kite flights, as published in *Meddelelser om Grønland* in 1909.[27] The hardest work, clearly, was the determination of the altitude from the air pressure records. Comparisons for theoretical purposes relied absolutely on accurate measurements of the altitudes; this accuracy is reflected here in the instrument records, with pressure readings to the tenth of a millimeter of mercury.

Wegener's obsessive attention to detail while in Greenland had produced a mountain of data. Each flight record included readings for time, altitude, pressure, temperature, humidity, wind direction, and wind speed. He had tabulated these instrumental data (reduced from graphic to numerical form) for altitudes of 5 meters (16 feet), 200 meters (656 feet), 500 meters (1,640 feet), and then 500-meter intervals thereafter, up to the maximum altitude, which was recorded to the nearest 10 meters (33 feet). Each flight record was accompanied by remarks about sky conditions and any unusual phenomena—mirages, inversions, surprising wind shifts. Taken together, it made an attractive and informative publication of a useful set of data. But what did it amount to?

Wegener reduced to numerical form his principal finding from two years of aerological work in Greenland. His table contains four columns of data. Reading from left to right, these are as follows: the fall of temperature per 100 meters (328 feet) of ascent, the increase (or decrease) in wind speed, the rotation of the wind relative to the direction at the ground, and finally the increase (or decrease) of the relative humidity. The numbers given are mean values for these quantities at six different levels of alti-

II. ERGEBNISSE DER AUFSTIEGE.

Erklärung der Zeichen und Abkürzungen.

* Fallender Schnee.	△ Graupeln.
⇉ Fliessender (treibender) Schnee.	≡ Nebel.
⊠ Schneedecke.	⊕ Sonnenring.
● Regen.	Mondring.
⊥⊥ Reif.	⊖ Sonnenhof.
V Rauhreif.	∈ Mondhof.
∾ Glatteis.	⌒ Regen- und Nebelbogen.
→ Eisnadeln.	Nordlicht.

Die Stärke der Erscheinung wird durch die Exponenten 0—2 gekennzeichnet.

Die Bewölkung (Bew.) wird von 0 (wolkenfrei) bis 10 (ganz bedeckt) angegeben. Der Exponent gibt die Dicke der Wolken an (0 = sehr dünn, 2 = sehr dick).

Wolkenarten: ci = Cirrus, ci-str = Cirro-Stratus, ci-cu = Cirro-Cumulus, a-cu = Alto-Cumulus, a-str = Alto-Stratus, str-cu = Strato-Cumulus, cu = Cumulus, str = Stratus, fr-str = Fracto-Stratus, fr-cu = Fracto-Cumulus, ni = Nimbus, cu-ni = Cumulo-Nimbus, fr-ni = Fracto-Nimbus, P. B. = Polarbanden (langgestreckte parallele Wolkenstreifen, die perspektivisch an 2 Punkten des Horizonts polähnlich zusammenzulaufen scheinen).

Wind: W = West, N = Nord, E = Ost, S = Süd, C = Calme, WzN = West zu Nord.

Nr. 1. 1. September 1906. 1 Drachen (4 m²), 750 m Draht.

Zeit	Seehöhe m	Luftdruck mm	Temperatur C°	Rel. Feuchtigkeit %	Wind m p. s.	Bemerkungen
9^{a}51	5	775.1	+5.7	..	NW 7	Bew. $6—8^{1}$ a-cu, str-cu, letztere sehr langsam aus NNW, zunehmend, Basis 800 m (Gipfel der 1000 m hohen Koldewey-Insel ist verdeckt). Wind unten abflauend, gegen Schluss auf kurze Zeit nach S umschlagend. In der Maximalhöhe starke Windabnahme, so dass der Drachen nicht höher zu bringen ist.
10 06	200	757	+3.9	mit der Höhe zunehmend	NNW 8	
10 30	360	742	+2.0		NzW 6—5	
10 40	200	..	+2.9		NNW 8	
11 13	5	775.4	+4.2	..	S 3	
Nr. 2. 9. September. 2 Drachen (8 m²), 3120 m Draht.						
9^{a}06	5	762.5	−2.3	71	WzN 4	Bew. $2—1^{0}$ str-cu im S, Fahne über der grossen Koldewey-Insel, Höhe ca. 15—1800 m. Bei ca. 800 m starke Windabnahme. Wind unten veränderlich, frischt gegen 10^{a} auf, flaut aber später wieder ab.
9 10	200	744	−3.0	72	WzN 6	
9 16	500	716	−3.3	63	WzN 9	
10 30	950	677	−5.0	54	WNW 4—7	
10 58	500	..	−3.3	56	WNW 8	
11 06	200	..	−1.9	(> 58)	WNW 5—6	
11 18	5	762.0	−0.8	72	WNW 5	
Nr. 3. 10. September. 1 Drachen (4 m²), 1850 m Draht.						
9^{a}36	5	759.7	+0.8	..	W 15	Bew. 1^{0} a-str. Inversion zwischen 350 und 500 von −1.0 auf +3.7 (im Abstieg von +0.2 bei 280 m auf +4.0 bei 500, hier jedoch in 2 Stufen geteilt, die durch Temperaturabnahme getrennt sind). Beim Abstieg sind Luftwogen bei der Inversion registriert. Oberhalb 500 m erhebliche Windabnahme.
9 43	200	742	+0.4	55	W 20	
9 46	500	714	+3.7	(< 42)	WzN 19	
10 00	1000	671	+1.7	33	WzN 16	
10 32	500	..	+4.0	31	WNW 18	
10 46	200	..	+0.5	(> 38)	WNW 18	
11 04	5	760.2	+1.2	59	W 12	

Data from Wegener's first three kite flights in Greenland in September 1906. All the numerical data are reduced from meteorograph records like that above. *Right column*: *Bemerkungen* (Remarks) includes information on cloud types and altitudes, precipitation, presence of inversions, changes in wind direction, and so on. From Alfred Wegener, "Drachen- und Fesselballonaufstiege ausgeführt auf der Danmark-Expedition 1906–1908," *Meddelelser om Grønland* 42, no. 1 (1909): 7.

tude: all the intervals are 500 meters except the lowest, which Wegener broke into two components of 5–200 meters and 200–500 meters. The latter was clearly a way to separate out the extreme anomalies he recorded in the first tens of meters above the ground: the procedure here was the same as that in his descriptive summary of the flights noted above.[28]

What the table purports to show is indicated by the numbers clustered near the 1,500-meter mark. Wegener was convinced that he had discovered in Greenland evidence of a stable layer in the atmosphere at about 1,500 meters—a surface of discontinuity, whether or not clouds were present. It might be a little higher or lower, it might not always be present, but it was, he was certain, a persistent structural phenomenon. The evidence for this conclusion is indeed there, but it is extremely slight, and it bears careful scrutiny.

Of the four sets of figures in the table, the most reliable were the measurements of humidity and wind rotation: the increments and decrements of humidity are

Höheninterval	Temp.-Abn. pro 100 m	Zahl d. Beob.	Windzun.	Zahl d. Beob.	Winddrehg. + rechts	Zahl d. Beob.	Zunahme d. R. Feucht.	Zahl d. Beob.
5—200 m	−0.20°	120	+1.6 m.p.s.	124	+11°	113	−7.0 %	83
200—500	+0.19	99	+0.1	103	+ 6	92	−2.6	72
500—1000	0.36	56	−0.6	59	+ 7	53	+0.5	47
1000—1500	0.32	27	+0.5	30	+ 9	27	−4.1	24
1500—2000	0.37	10	+0.9	12	− 1	9	+0.1	11
2000—2500	0.36	2	+0.8	2	0	2	+2.5	2

*) Auf der Danmark-Expedition 1906—08, unter Leitung des dabei verunglückten Mylius Erichsen. Beobachtungsort: Kap Bismarck, auf 76¾° N.B. an der Ostküste Grönlands.

Beiträge zur Physik der freien Atmosphäre III. 3

Wegener's principal result from two years of work in Greenland. The shift in three meteorological elements (temperature, wind speed, and relative humidity) away from the trend of the values lower than 1,000 meters and those above 1,500 meters indicated to him the possibility of a new layer in the atmosphere. From left to right, the columns are as follows: Altitude Interval, Decrease in Temperature, Wind Speed, Wind Direction, and Relative Humidity. From Alfred Wegener, "Über eine neue fundamentale Schichtgrenze der Erdatmosphäre," *Beiträge zur Physik der freien Atmosphäre* 3 (1910).

large—often several percentage points—and the wind rotation is easily read by the compass direction of the kite or balloon cable relative to its direction at the time of release. The wind speed data are somewhat suspect, because almost all of them are estimated. Wegener was claiming that he could detect, from cable tension, angle of flight, and the general behavior of the kite, variations in wind speed of the order of 0.5 meters per second (about 1 mile per hour).[29] This may be the case, but even with highly developed craft skills, which he possessed, these are very fine discriminations. Then again, in the age of sail that Wegener inhabited, ship's officers and crews were capable of such discriminations of wind speed by similar methods, such as feeling the tension of the rigging and looking at the bellying of the sails; the claim is not outlandish.

The temperature data, though they appear almost fanciful on first inspection, are probably real. Wegener's bimetallic thermometers were capable of this accuracy. While normal ground-station thermographs with bimetallic thermometers are generally accurate only to about 0.5°C (1°F), calibrations of the kind Wegener carried out meticulously before and after each flight would have allowed ten times that accuracy. This would mean that he could discriminate with confidence an interval of 0.05°C.

Here is the result Wegener thought he had obtained: between 1,000 and 1,500 meters, the decrease in temperature with altitude suddenly slowed and even reversed slightly. At the same altitude, the wind speed, which had been dropping, increased again. The wind's rotation, which had been decreasing, increased again, and the relative humidity, which had been increasing, dropped sharply, almost 5 percent. Wegener took these four trends as independent lines of evidence for his proposed surface of discontinuity in the atmosphere.

Before Wegener could make this crucial plunge into his data, however, he had to turn again to preparing the full ensemble of measurements to meet his expedition obligations. With his completed manuscript of the Greenland aerology data in hand, Wegener traveled to Denmark in late November 1908.[30] The manuscript was much too valuable to consign to the post; it was his ticket to freedom from the expedition contract. Wegener interpreted the act of delivering this manuscript as proof he had completed his contractual obligation to work "full-time" on the scientific results of the expedition. This was in spite of the fact that he had only been at it for six weeks, and that everything else he had done in Greenland remained to be worked up. This new attitude grew out of a conversation with his friend and expedition comrade Johan Koch.

In talking to Koch after his correspondence with Prof. Warming in October, he had discovered that the Danmark Committee had let Koch "off the hook" on publication, asking only that he immediately complete the results of the "Great Sledge Journey" which would document the territorial claim to that section of Greenland. All of the other material—the geodetic triangulation, the detailed maps, the maps of the regions south and west of Cape Bismarck, and the repeated determinations of latitude and longitude at Danmarkshavn (in all of which Wegener had a hand)—would be published outside the volumes devoted to the Danmark Expedition. Wegener was annoyed by this imbalance in publication demands on himself as opposed to Koch, perhaps inflicted by his own dutiful naïveté, and resolved not to be bound any more severely than his closest expedition colleague.[31]

A New Atmospheric Layer

When he returned to Berlin, he closed himself off from everyone and everything and set to work. Meteorology has often been a passion of solitary men, but perhaps never more so than in Wegener's case. He might as well have been at Pustervig still, or on the Moon, for all the contact he had with the rest of the world in these months. He calculated and reflected. He wrote. He smoked. From time to time he saw his parents. For the rest, he had no friends; he sought no amusements, and he rarely went out except to consult the library at the University of Berlin, to drink coffee, or to buy tobacco.

In one sense, the search for a context was quite simple. He had made a series of observations that indicated a discontinuity in the atmosphere at an elevation of about 1,500 meters. The simplest possible context for such a set of observations would be the record of other observations that exhibited the same layering. This is the first move in a theoretical exercise: it is the expansion of the data at hand, as well as the demonstration that one's own empirical results are not anomalous or idiosyncratic. To find such confirmation, Wegener turned to the most obvious and nearest materials, beginning with the widely known results of Aßmann and Berson's Berlin balloon observations from the 1890s. From these he moved on to an extensive set of observations made by Süring at Potsdam from 1904 to 1906, and then to Süring's collection of the results from seven different meteorological stations around the world, published in 1907.[32]

The numbers (from Süring) that Wegener used to verify the existence of a fundamental discontinuity at 1,500 meters did not initially look very promising. Süring had been measuring cloud altitudes and trying to find the modal altitude for the different cloud forms. These indeed had particular altitudes: stratus most often at 0.6 kilometers (0.4 miles), fracto-cumulus at 1.6 kilometers (1.0 miles), stratocumulus and/or nimbus at 1.7 kilometers (1.1 miles), cumulus at 2.1 kilometers (1.3 miles), altocumulus and altostratus at 3.0 kilometers (1.9 miles), and so on up to cirrus at 9.0 kilometers (5.6 miles). Süring's averaging of the results of the seven other stations around the world gave varying but comparable values: stratus at 0.5 kilometers (0.3 miles), cumulus at 2.0 kilometers (1.2 miles), but then altostratus and altocumulus at 4.3 kilometers (2.7 miles)—a rather large departure from Süring's own observations. Note that, to this point, the numbers cannot have been much help to Wegener—a characteristic altitude of 1,500 meters is nowhere in sight. However, Süring, to reduce the discrepancy between his own cloud altitudes and those of the other stations, had made a separate calculation: without regard for the type of cloud, he had calculated the altitudes at which clouds of any kind were most frequently found. These altitudes were 1.6, 4.4, 6.8, 8.8, and 10 kilometers (1.0, 2.7, 4.2, 5.5, and 6.2 miles, respectively).

The latter numbers, the frequency maxima for the appearance of certain types of clouds, gave Wegener an altitude sufficiently close to his surface, which, after all, was itself an average of a number of values spread across the interval of several hundred meters (a mean for 1,000–1,500 meters, and a mean for 1,500–2,000 meters). He recognized that 1,500 meters could only be defended as a mean level—that his discontinuity was mobile—and he took this as his first task.

The type case for atmospheric discontinuity was, of course, the "upper inversion," the tropopause. Via its ubiquity and its strength, it had created a benchmark idea of what a structural discontinuity in the atmosphere should look like. Wegener's argumentative starting point was, therefore, that the tropopause was not the *typical* case but the *limiting* case of atmospheric discontinuities.

Underneath this "great laminar boundary," Wegener argued, from the surface of Earth up to the limit in the region near 10 kilometers, "the atmosphere is populated by a whole family of surfaces of discontinuity, which appear as stable layers, and which reveal themselves either as cloud surfaces, or in the absence of condensation, as stepwise jumps in the value of the various meteorological elements."[33] Unlike the tropopause, however, these layer boundaries were incompetent—easily broken through by adiabatically ascending parcels of air. As a result, the atmospheric phenomena in this lower region lost their simplicity, their consistency, and their completeness. They grew weaker and more rare with increasing altitude, and in consequence, "the law governing these layers is, in many instances, often completely concealed."[34]

This framing of the phenomena allowed Wegener to use the statistical method he championed: "It is, in fact, a fruitless undertaking to try to fixedly locate these lower surfaces of discontinuity by investigating any single case. On the contrary, it is obvious that any law-like distribution of these layers will only be revealed in the mean alteration of the meteorological elements with [increasing] altitude."[35] This characterization of the situation, with these layers becoming less visible with altitude, allowed him to argue for the importance of "his" 1,500-meter layer.

He immediately pushed forward, in a companion paper running parallel to the first, to build this idea into the practice of meteorology by establishing it as part of an observational paradigm. The means of this advance was a novel correction factor in extrapolating the adiabatic cooling of the atmosphere, from an altitude where the values had been observed (say, 1,000 meters) to an altitude where the values were unknown (say, 1,500 meters). The so-called *Zustandskurve* (state curve) of the atmosphere was the basis of a standard graphical aid in meteorology and aerology, a nomogram containing two curves representing the fall in temperature (with increasing altitude) of a given volume of air. One of these curves represented the fall of temperature in dry air, and the other in saturated air. Such a diagram, in which these curves were printed on graph paper with intersecting isotherms and isobars, allowed simultaneous calculation of a variety of meteorological elements.

In a paper entitled "Über die Ableitung von Mittelwerten aus Drachenaufstiegen ungleicher Höhe" (Averaging the mean values for kite ascents to unequal altitudes) Wegener argued that in extrapolating from a known value to an unknown value, rather than using the dry adiabatic curve or the wet adiabatic curve, there was a third curve, intermediate between them, which represented not one or another *ideal* state of the atmosphere but the *actual* structure of the atmosphere. Rather than use the theoretical extrapolation for the full atmospheric column based on a uniform change in the meteorological elements with altitude, one used the observed mean difference, over many observations, between the elements at 1,000 meters and the elements at 1,500 meters. He showed that the mean value of the change between 1,000 and 1,500 meters, when applied to the 1,000-meter value, always gave more accurate values for the meteorological elements at 1,500 meters than theoretical extrapolation alone.[36]

This does not sound like much, but it makes an important point: purely theoretical calculations taken by themselves are misleading and need to be empirically modified. Wegener is here arguing that the empirical *Zustandskurve* of the atmosphere contains a knick point at 1,500 meters, which ought to be graphically represented every bit as much as the upper inversion—the discontinuity between the troposphere and the stratosphere. His point was not the importance of any given layer, but rather that the future advance of the science of meteorology lay in turning an idealized, generalized, homogeneous, theoretical atmosphere into a real, specific, layered, empirical atmosphere—via aerological discovery.

Aßmann's enthusiasm for Wegener's correction factor (Wegener had described it to him in 1908 at the Hamburg meeting) was very much the driving force for writing this paper immediately. In late January, Wegener wrote to the committee of the Danmark Expedition asking for permission to publish it, because "in this article a problem is discussed for which I have worked out a new calculation method which Herr Geheimrat Aßmann will recommend for general adoption at the international meeting, on April 1 in Monaco, of directors of aerological institutes."[37] This, of course, was great news for Wegener, because it was a validation not only of the existence of his layer but of Wegener's more general, semiempirical and statistical method of determining discontinuities within the atmospheric structure.

Marburg

With one year of postdoctoral work at Lindenberg and two years of field experience in Greenland, Wegener was working well within the conceptual frames he had inherited from Wilhelm von Bezold and Richard Aßmann. On the theoretical side this meant the extension of atmospheric thermodynamics through the study of boundary layers and discontinuities. On the empirical side it meant kite and balloon aerology and manned balloon ascents using standard instrumentation in standard ways.

Wegener's appointment as a technical assistant at the Aerological Observatory at Lindenberg expired on 1 January 1909, along with the leave of absence granted in April 1906. The Danmark Expedition committee was still paying him a stipend, though this also would expire in a few months, leaving him without employment or funds. He had some savings from his expedition salary, but much of that sum had been eaten up beforehand in equipping himself for the expedition, and the remaining savings would not last long.

There was, he had concluded, no real future for him at Lindenberg as a technical assistant. Berson was apparently solidly in place as the *Observator*, and whatever Aßmann's original plans for staffing the institution, the turnover of technical assistants was very rapid, with most of the helpers staying only a year or two. This mirrors the turnover of support staff at Lindenberg at all but the very highest levels, and by the time Wegener returned from Greenland there was no longer any expectation that he would remain in his position, nor was their much intellectual continuity or promise of community.

The obvious alternative, and in fact the only alternative, was a university position as a *Privatdozent* (instructor). At some point in January 1909, Wegener decided in favor of pursuing this option at the University of Marburg. Marburg was a city with a population of approximately 20,000 located in Prussia, in the state of Hesse-Nassau, about 100 kilometers (60 miles) north of Frankfurt, on the main line of the Prussian State Railway. The university was a Protestant shrine of sorts, the first Protestant university in Germany, and in 1905 had about 1,500 students. Contemporary photographs, as well as the early twentieth-century paintings by the Marburg artist Karl Bantzer (1857–1941), show a completely rural landscape surrounding the *Oberstadt* (the hill containing the old town, the palace, the Lutheran church, and the university), with some of the nearby agricultural open land already returning to forest. In Wegener's time there were mature and extensive forest tracts around Marburg, in spite of the region's relatively dense population.

At the time of Wegener's decision to move to Marburg, in 1909, the professor of physics and director of the university's Physics Institute was Franz Richarz (1860–1920), well known for his work on the interaction of light and electricity. Richarz had made persuasive speculations about atomic charges and had edited textbooks on both theoretical physics and Earth's magnetism. Richarz had also been a close colleague of Paul Drude (1863–1906) and, at about the time of Wegener's application, had become interested in using manned balloons to map variations in Earth's magnetic field, with a fledgling program of this kind under way around Marburg.

This was a good fit for Wegener. Richarz was known to be sympathetic to the claims of meteorology to be recognized as a science. Moreover, many of the doctoral students in physics at Marburg were training to be high school teachers (*Oberleher*), and this meant that the department was also sympathetic to the claims of cosmic physics (at this time a standard high school subject but not one in which professorial positions were available at the university level). This was also a plus for Wegener. Finally, the physics department's close connections with both Berlin and Göttingen also implied a strong sympathy for geographical exploration and expedition science.

The application process for the Marburg position created an entire new stratum of work and correspondence for Wegener, to be completed immediately, on top of the consuming activity of his original scientific work and his continuing work on the Greenland data, including the minute and laborious correction of proofs from the Greenland aerology manuscript submitted to the Danmark Committee the previous November.

Wegener's chief recommenders for the Marburg job were his old professor from Berlin, Wilhelm Förster, and his recent supervisor at Lindenberg, Richard Aßmann. Förster sent a positive but entirely perfunctory recommendation to Richarz on 8 February 1909 endorsing Wegener's *Habilitation* at Marburg, describing his dissertation as "solid" and "valuable," and expressing the likelihood that "his work as an investigator and teacher will be competent."[38] Aßmann, on the other hand, sent a superlative recommendation. "Dr. Wegener," he wrote, "is in every respect an excellent man and it is extraordinarily regrettable for my observatory to have lost him. Solidly trained, and endowed with sharp understanding and a rich scientific imagination, he is industrious, energetic, and persistent, the prototype of the most strenuous, most ambitious sort of young scholar, who, unless I'm much mistaken, will produce very significant work."[39] Aßmann went on to praise Wegener's personal qualities: "Moreover he is on the personal side, a pure type, a high-minded and likable man, the sort of person who would be desirable in any academic setting."[40]

Aßmann's enthusiasm for Wegener was real and reflected in other ways as well. While Wegener was in Greenland, Aßmann, together with Hergesell, had advanced a new journal, *Beiträge zur Physik der freien Atmosphäre*, that would be concerned not just with meteorology but with atmospheric physics. This was a new and promising venue for Wegener, and he submitted to Aßmann the two manuscripts he had been working on: that on the layering of the atmosphere, and the associated work on the averaging of mean values, discussed above.

The first of these two manuscripts was acceptable as written, but the second involved Wegener in his first scientific controversy and his first priority fight. He had contacted Arthur Berson in an attempt to gain data on the altitude of the upper inversion in Africa, data that Berson (recently returned) had not even had a chance to publish yet, and with this request Wegener had informed him of the work on the averaging of values for atmospheric elements with kite flights of unequal altitude. Berson's reply was combative: he denied that there was anything new in Wegener's method, claiming that he had developed it himself and had used it since the 1890s. He also expressed a desire that he, and not Wegener (though the latter had been Aßmann's wish), should present a proposal for international adoption at the meeting in Monaco in April.[41]

Wegener therefore inserted into the manuscript an agreed-upon statement that Berson had already employed the method, and he readily surrendered participation in the conference at Monaco, though hoping that Aßmann would make the presentation himself. Wegener insisted, however, that one had to accept the very great difference between using an empirical correction factor such as Berson's (which was not new) and, on the other hand, providing an exact formulation of the scientific reasoning for the general adoption of such a correction factor, based on a comparison of several large data sets, some of which did not belong to Berson. It was the latter contribution that Wegener insisted was his own. He included this claim, with slightly different wording, in the preface to the paper itself, following the acknowledgment of Berson's priority.[42]

Wegener does not seem to have been substantially diverted from his work plan by the scuffle; things were moving along quickly in Marburg: he had sent along a great packet of reprints of his papers and the required summaries of his credentials and professional accomplishments, along with his plans to teach. He proposed to give lectures in meteorology, astronomy, and cosmic physics. He would offer his Greenland aerology as his *Habilitationsschrift*, that piece of writing beyond the dissertation that establishes the candidacy of a scholar to teach at the university level. He suggested as candidate topics for his *Probevorlesung* (a demonstration lecture to determine whether someone can actually teach) a lecture on new methods of investigation in meteorology, a lecture on new methods of position finding from a balloon (something to please Richarz), or perhaps a presentation of his historical work on the history of astronomy dating from his graduate school days. Finally, he offered a slide show of his investigations in Greenland.[43]

The appointment in Marburg was a virtual certainty. The dean of the faculty, the celebrated geologist Emanuel Kayser (1845–1927), had moved the appointment before the Marburg faculty at the earliest possible date and had scheduled the probationary lecture for 8 March. With this knowledge in hand, Wegener began to make his plans to move. He was soon informed that the Marburg faculty had elected to hear "Results and Aims of the New Methods of Aerological Investigation in Meteorology," which

Wegener then prepared on short notice. It contained mostly material he had written, and written again, though here he had a good chance to put his work in context, a context he could create himself.[44]

Defining His Place in Science: The "New Bezold"

Wegener's "demonstration lecture" on 8 March shows how strong the historical instinct was in him. He had, after all, written a dissertation in the history of astronomy, albeit one of value to working astronomers. The research tradition of Berlin astronomy was embedded in an explicit historical context, and its pedagogy was also historical in structure. Berlin's students were to understand not just the techniques of research but also the evolution of their field up to the time of their own work. This historical approach was characteristic of German astronomy and physics in general, but nowhere more strongly than in Berlin.

At the time of Wegener's visit to Hamburg in November 1908, Wladimir Köppen had delivered an address on the evolution of meteorological science; this was published in January 1909 in *Meteorologische Zeitschrift*. Wegener took Köppen's address as his starting point for his Marburg lecture. In the history of meteorology, Wegener declared, there is a rhythm of progress in which periods of introduction of new techniques alternate with the theoretical integration of the results of research. In meteorology there had been four such periods. The first was from 1650 to 1750, during which most of the meteorological instruments had been invented, and through which meteorology could be said to have originated. The second powerful impulse had been the development of climatological maps in the first half of the nineteenth century, which led to the emergence of climatology. The third new perspective emerged in the third quarter of the nineteenth century, with the foundation of synoptic meteorology and the development of the synoptic weather maps and their associated aids. Finally, the most recent revolution, barely twenty years old, consisted in the introduction of methods and aids that had given the science of meteorology a powerful impulse toward its true aim: its transformation into a physics of the atmosphere. These new methods, said Wegener, have been christened "aerology," though this name, however well established, scarcely characterized the nature of the new branch of scientific research. This new period was still in the phase of discovery, of accumulation of facts: "the theoretical exploration of the insights thus gained has only succeeded in the most limited fashion, and the principal work in this area is reserved for the future."[45]

Wegener's timeline and capsule history gave a convergent series. With successive periods of 100, 50, 33, and 20 years (thus 1, 1/2, 1/3, and 1/5), Wegener suggested (numerically) that the completion of the next phase should occur within perhaps a decade. This also converged on the career of the lecturer, Wegener himself, whose work was poised between discovery of new facts and the first development of theory out of them.

Having located the history of meteorology in converging segments of time, Wegener went forward to locate the new work in space, as a complement to its temporal evolution. His associated figure, a "complete profile of the atmosphere," is an extremely important datum in understanding Wegener's concept of his own future work and career, laid out here before his audience.

Wegener began his account of the atmosphere at the outer limits, the extent of which could only be inferred by its ability to support the aurora borealis. These auroral phenomena appeared at altitudes of 400–500 kilometers (249–311 miles), 200 kilometers (124 miles), and 60–70 kilometers (37–43 miles). In addition to the aurora, meteor trails also gave evidence of atmospheric air at very great altitudes, principally between 150 and 100 kilometers (93 and 62 miles). From the phenomenon of twilight could be observed the existence of air sufficiently dense to reflect light at about 70 kilometers (43 miles) and the phenomenon of "noctilucent clouds."

Just as Wegener had made the lecture converge in time, so he now made it converge in space, moving much closer to Earth's surface, to the significant region known as the "Weather Zone." He reminded his audience that water condensed out in the atmosphere in only an extremely limited zone, below about 10 kilometers. Measurement of the atmospheric pressure showed that about half the mass of the atmosphere lay in the lowest 5 kilometers (3 miles) (and the next 25% at altitudes between 5 and 10 kilometers). Wegener described this region of the atmosphere as the *Schauplatz*, the "stage" of meteorology, and thus produced a classical unity of place, time, and action converging on the hero himself. All those years of high school Greek and Latin, of Greek tragedy, of classical philosophy, were not to go to waste, nor were they wasted in being presented to a classically trained audience.

Wegener had now brought the lecture out of the past and down from the sky, to Earth and to his subject—that the aim of aerology was to advance meteorology toward atmospheric physics. To do this, it must not only discover but also integrate empirical results into a theory. With Bezold deceased and Aßmann and Berson still at work, the open role in this drama was clear: who would now play the part once held by Bezold? This was precisely the role he proposed to fill, and the next section of his lecture was a tacit assertion that he was the "new Bezold."

Wegener asserted that the most striking discovery of recent years was the structural discontinuities in the atmosphere marked by temperature inversions. In this way he linked his own modest work on a 1,500-meter inversion to the discovery of the upper inversion (the tropopause) by Aßmann and Teisserenc de Bort. Wegener emphasized the stability and global character of the latter discontinuity and admitted the more uneven and labile character of the layers below it, but he still insisted that the principal theoretical aim of meteorology was the understanding of these discontinuities and the laws governing them: "If this great boundary layer shows such a stable configuration, may we not also, in the apparent chaos of the surfaces within the Weather Zone, be able to discern a law governing them?"[46]

This, he continued, was a problem too difficult, until recently, for aerology, and its solution and understanding had required aid from another quarter: the study of clouds. The discontinuity surfaces are identical to the surfaces of the clouds, and the variety of these discontinuities and their behavior are given by cloud classifications. This had been Bezold's contention, and here it led Wegener directly to his own work once again. In this address, all roads of theoretical and practical advance led unashamedly to the work of Alfred Wegener.

So, from Wegener's standpoint, the principal task of aerology is to determine, with great exactitude, the mean change of the meteorological elements with altitude, not just in Europe but also from the equator to the poles. We must learn, he said, the mean thickness of the layers and their variation with altitude and all associated phenomena. We may then, he argued, approach the next question, which is how these "typical" layer boundaries persist in special cases, cyclones and anticyclones, and the laws governing the ways in which these layers are breached. "Behind all these investigations wait, however, the deeper question to be answered: what is the cause of such layering? What power places, for example, the alto-cumulus level, despite frequent displacements and even destruction, again and again at 4000 m? The same for all the others!"[47]

All fundamental problems and solutions of meteorology lie in understanding the laws and causes of discontinuities in the atmosphere. This includes the upper inversion at 11 kilometers (7 miles) and the probability that another *Sprung* (jump) occurred at 70–80 kilometers (43–50 miles). This understanding of the vertical structure of the atmosphere would provide a unified point of view leading to the understanding of global atmospheric circulation.

Wegener concluded his address with three observations. First, to do justice to what was known already from aerological research would require not an article or two but a book. Second, in the study of cyclonic storms, aerology had, within the few years of its existence, demolished every theory ever proposed to explain them. A clear physical understanding of these phenomena, Wegener argued, was nowhere in sight. Third, while the aims of meteorology can be furthered by far-flung stations and a loose global network of expeditions, the understanding of cyclonic storms could only come from a dense net of aerological observing stations (Europe the likely candidate) through which daily observations, like those at Lindenberg, would replace "random tests."[48]

We have no record of how this lecture went over with the audience, but at a distance of a century or more it is a remarkably clear and accurate picture of the state of atmospheric physics at that time. Since most of the history of meteorology has told the story of the theoretical understanding of cyclonic storms, the first decade of the twentieth century seems (as in Wegener's characterization and prediction of major problems) exactly the sort of "prelude to discovery" of the character of cyclones that it turned out to be in standard (and correct) histories.[49]

A history of meteorology in the twentieth century constructs the story for us: the nineteenth century theorized the energy of storm systems as thermal convection throughout the full atmosphere. The discovery of the stratosphere by Aßmann and Teisserenc de Bort constrained the weather—as in Wegener's account—to a limited zone from the surface to 11–15 kilometers (7–9 miles) above it. From knowledge of this limitation, Max Margules, at the Central Meteorological Station in Vienna, was able to write a series of papers in atmospheric physics showing that the energy of storms was not convective but *advective*, not vertical updrafts and downdrafts but horizontal motion—as air masses moved from areas of high pressure to areas of lower pressure, trapped between Earth's surface and the "upper inversion" (the tropopause).

The motion of these air masses, sliding over one another and past one another on surfaces of sharp discontinuity, led to the postulation of "frontal weather" by Vilhelm Bjerknes (1862–1951) at Bergen. Bjerknes, who was later Alfred Wegener's friend and colleague, using the results of a dense network of European and North Atlantic aerological stations (which Wegener had insisted was necessary), was able to produce a series of synoptic maps showing the possibility of weather prediction out several days by analysis of the evolution of storm systems. The latter were to be mapped as waves moving along the "frontal boundaries" between coherent, adjacent air masses. It was this hydrodynamic theory that led forward to modern meteorology as we daily experience it on television, in print media, and on the Internet. Indeed, this work by Bjerknes led directly to an attempt by Lewis Fry Richardson in 1917 to explore the prediction of the weather from "meteorological elements" alone, by calculation of the motion of air masses and by solution of a series of fundamental equations governing the flow of air around the world.[50]

From the standpoint of weather prediction and dynamical meteorology, Wegener was not on the "main line." But he *was* on the main line of advance in atmospheric physics, whose future he predicted with remarkable perspicacity. It is important to keep in mind that Wegener was never, except temporarily and accidentally, a forecast meteorologist, but rather an atmospheric physicist who saw his work as a contribution to a general physics of Earth and indeed the whole cosmos.

With his demonstration lecture completed, Wegener could now move forward with the more mundane details attending his new life. He arranged lodging for himself from 1 April, rooms in a pleasant Georgian brick house at Wilhelm Rosestrasse 9. From this residential neighborhood on the north side of the *Oberstadt*, it was a vigorous ten-minute walk, up a winding, cobbled alleyway flanked by gardens and cottages, to the Physics Institute at Renthof 6. Walking uphill, one had an agreeable view of the *Schloß* (palace) and the buildings of the old town. Walking downhill, one had the engaging prospect of the rolling hills and farms just beyond Wegener's residential neighborhood.

Exploring the environs, he discovered that the prevailing winds were from the northwest, often driving clouds up and over the palace. Beyond the university and over the hill was the market square, with a pleasant breeze filtering through the narrow streets flanked by their fifteenth- and sixteenth-century stucco and timbered houses. The *Rathaus* (city hall), with its large clock, complete with a bronze rooster flapping its wings as it crowed the hours, had the sort of droll, "not-taking-itself-too-seriously" attitude that was extremely congenial to Wegener. It seemed a calm and quiet place. Away from the train station a mile outside of town, the city was nearly silent; in the *Oberstadt*, above the city, there was almost no vehicular traffic.

With these matters concluded, Wegener took leave of his new department chair, Richarz, and returned to Berlin. There was much to prepare and to wrap up, including a planned visit to Copenhagen for eight days, to oversee the printing of his Greenland work, to visit with his friend Koch, and to have a face-to-face meeting, long postponed, with Prof. Warming, the head of the Danmark Expedition Committee. There were bundles of proofs to correct, including articles by his brother Kurt, then in Samoa, which Alfred had promised to handle in order to save the turnaround time.

On 22 March 1909, Wegener wrote to Aßmann, sending along corrected proofs and promising that if everything went well, the rest of these would be dispatched before he left for Copenhagen on Friday, 26 March. But things did not go well. On the twenty-third he began to feel tired, and by evening he was feverish. On the twenty-fifth he sent a brief note to Aßmann telling him that he had fallen ill, would not be going to Copenhagen, and for the rest of his time in Berlin would be living with his parents.[51] He was, in fact, seriously ill with influenza, not a mild flu but an epidemic version that swept through Europe in the winter and spring of 1909. It was the most severe influenza to appear since 1889, and it struck everywhere: the pope, the king and queen of England, and a variety of lesser notables canceled trips and took to their beds. The fatality rate was high in Berlin and even higher in England, where there were 9,000 *recorded* dead that year from the flu.

Wegener had been pushing ahead strenuously to tie everything up before his move, and he was vulnerable on several counts. He had been sedentary for many months after years of physical activity, he was smoking heavily, and, not least, he was immunosuppressed, as are all returnees from the antiseptic air of the Arctic. His condition worsened rapidly, and by the twenty-ninth he was completely bedridden and so weak that he could not lift his pen. He dictated to his mother a letter to the committee in Copenhagen postponing his voyage there, pleading for the rapid publication of his Greenland work, required for his employment in Marburg, and asking that his stipend for the remaining Greenland work be continued beyond 1 April. His relocation to Marburg had left him financially embarrassed, a situation that would persist for a long time.[52]

By the following week he had recovered enough to finish the page proofs of his Greenland manuscript (working from his bed) and to send them on to Copenhagen, but he was still extremely weak. He wrote to Professor Richarz asking whether his formal inaugural lecture scheduled for 20 April might be postponed a week, as his doctor had urged him. This was good advice: it was the pattern with this particular influenza (like that of 1899 and even more so in the Spanish Influenza of 1918) that victims would feel better, return to work, and then relapse and die suddenly—a pattern noted by a number of Berlin physicians in 1909.

Professor Richarz obligingly postponed the lecture until 7 May, which was fortunate for Wegener, because recovery was slow.[53] By the end of April he was able to make the long-postponed trip to Copenhagen to help see his complex manuscript to the press.[54] The pace was depressingly slow, and even this trip did not solve all the difficulties in formatting the tables and graphs.[55] Spending the first few days of May packing his belongings for shipment to Marburg, he traveled there on 6 May and the next day delivered his lecture. It was the easiest of subjects for him, his history of astronomy from the pre-Socratics to the present, and, like his demonstration lecture in March, it was shaped to appeal to a broad audience.

Here it was, then—the transition from scientific laborer to university instructor, the path so strenuously advocated by his father. Wegener's appointment was official on 8 May, but there were no duties in the offing until October, when the *Michaelis* semester (winter semester, from October to April) would begin. He unrolled his polar bear rug in his study at Rosestrasse 9 and, with the coughing from his flu nearly abated and smoking cigar after cigar, set again to work.

His illness and convalescence in March and April had put a crimp in his ability to write and smoke, but not in his ability to think, and since the lecture on 8 March, he had been increasingly drawn to the problem he had proposed in that lecture as the prime question in atmospheric physics: the causes of the major fundamental layers of the atmosphere. Their existence was not in question, but there was no physical explanation: what caused these inversions and discontinuities? Was the "upper inversion" (tropopause) really a unique layer in its prominence and solidity, or were there others? Was the "upper inversion," in fact, really a permanent feature of atmospheric structure?

Thoughts on the Tropopause

This last question, the permanence and stability of the upper inversion (tropopause), had occurred to him while working in January on his paper "Layering of the Atmosphere." Then, in March, he had noted in his Marburg lecture that the altitude of the upper inversion shifted with latitude: it was high in the equatorial regions and lower near the poles. This suggested that there was a link between the altitude of the tropopause and the prevailing average temperature at the surface. If this were so, he asked himself, would there not be a marked seasonal effect? Should it not be higher in summer and lower in winter, especially in temperate latitudes?

The immediate spur to continue this work came to him in the June issue of *Meteorologische Zeitschrift*, containing two papers summarizing the work of August Schmauß (1877–1954) and Teisserenc de Bort on the "upper inversion." These papers were electrifying for Wegener: "the first really excellent overview of the striking law-like behavior of this phenomenon." Even so, Wegener continued, "I find missing here, as well as in all other similar work I have seen on this theme, an assessment of the simple fact that the temperature on the underside of the 'upper inversion' or, following Teisserenc de Bort, on the boundary surface between the troposphere and stratosphere, is lower in direct proportion to the altitude at which this surface lies."[56]

Wegener was looking here at data that (for him) were not only law conformable but begging to be expressed in law-like form. The relationship between temperature and altitude for the location of the tropopause had led Schmauß to postulate three mean "central temperatures" for the upper inversion, at their different altitudes (low, middle, high), leading to three empirically generated state curves for the atmosphere expressing the decline in temperature with altitude. "However," Wegener plunged on, "the law I have just articulated [the greater the altitude the lower the boundary temperature of the tropopause] finds expression in the three mean 'initial temperatures' [given by Schmauß], but to my knowledge the investigation of this initial temperature as a direct function of the altitude, has never been carried out by Schmauß or anyone else who has studied the problem until now, and yet it seems to me that this statistical methodology is quite suitable as a means to produce law-like behavior."[57]

Wegener's attitude expressed here is as interesting as it is characteristic of him. It is not so much his impatience (though that is there) as an astonishment that such well-respected researchers could produce such exciting and trustworthy work and then fail to make the "obvious" connections, connections (to him) as plain as day, leading directly to explanatory hypotheses and even to scientific laws.

The paper that Wegener wrote in June 1909 on the upper inversion is almost a template for what would become his standard method of attack in all his theoretical work. It goes something like this: "I have just read something. It immediately occurred to me that such and such a relation followed from the ideas presented—but I cannot find any hint of this relation stated here." Then a quick search of the relevant literature shows him that no one has produced such a relation. Then follows a speculative foray using whatever fragmentary data he can get his hands on. He develops a hypothetical relation and tests it using the data he has found. His hypothesis does not come from these supplementary data, but as an intuition based on the initial reading of the inspirational article or articles by others. He then works through the available data and finds, in spite of the sparseness and the fragmentary and provisional character of such studies as exist, that the relationship hypothesized has a clear signature in the data. A working scientist would today describe the sort of relationship Wegener thinks he has found as "robust," a major structural fact of the world, insensitive to detailed departures from the central tendencies of the data. Wegener then proposes his relation be accepted provisionally and urges others to produce supporting studies, giving a clear formulation of what sort of measurements would provide confirming or disconfirming instances.

He would think, write, and speak in this specific manner for the rest of his career. This method of work and of attacking problems had its roots in his training in astronomy

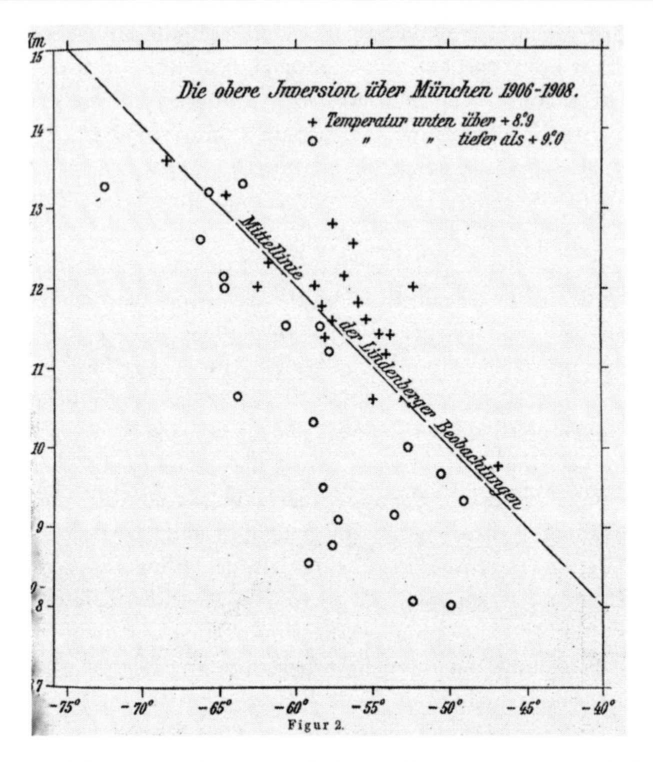

Wegener's analysis of the seasonal alteration in the height of the tropopause. While the data here are from Munich, the dashed line separating the circles and plus signs (different temperature regimes) is from the Lindenberg data. The point is that the line that separated the Lindenberg data also neatly divides the Munich data. See discussion in the text. From Alfred Wegener, "Über eine eigentümliche Gesetzmässigkeit der oberen Inversion," *Beiträge zur Physik der freien Atmosphäre* 3 (1910): 206.

in Berlin, where he had been taught to calculate cometary orbits from no more than three observations; it also expressed his own speculative and sometimes reckless cast of mind. It was, at this time in his career, certainly also an artifact of his playing "catch-up" with the major advances in aerology while he had been gone. Finally, it was also an expression of ambition (always with him) to discover something really important. In atmospheric physics as in Arctic travel, his chosen role and his great love were exploration, reconnaissance, and the mapping of new territory.

If we now return, in light of this imaginative strategy, to the contents of the paper he wrote in June 1909 on the upper inversion, the following elements fall into place both logically and rhetorically. The material "just read" is the three-level temperature scheme in the paper by Schmauß. The "missing conclusion" is the determination of the precise relationship of temperature and altitude with the location of the lower surface of the inversion layer (the tropopause). The "fragmentary data" at hand are reports of pilot balloon ascents from Berlin and Lindenberg.

Wegener's approach also consisted of three parts: First he graphed the data, plotting altitude versus temperature in a scatter plot. Wegener then analyzed his plot by drawing a line at 45° as the mean line of the data points, passing through −50°C (−58°F), at 10,000 meters (32,808 feet) (or very nearly) dividing the data points into two groups. He used the standard value for the lapse of temperature with altitude (0.5°C per 100 meters) established by Hann in 1903. The data points then fell into a band 3,600 meters (11,811 feet) wide, across a span of temperatures of about 18°C (64°F). The altitude of the tropopause could (thus) vary by about 3.5 kilometers (2.2 miles) and (thus) exhibit a temperature variation of 18°C.

Here it becomes interesting. Wegener contingently, and not quite arbitrarily, divided the data in another way. Beginning again with the Lindenberg results, he identified winter ascents with a small circle and summer ascents with a small cross. He knew, from reading Schmauß and Teisserenc de Bort, that the height of the tropopause was measurably different in winter and summer. He defined winter as an Earth surface temperature on average below +9.0°C (48.2°F), and summer as an Earth surface temperature above +8.9°C (48.0°F).

He then generated the temperature at the altitude of the tropopause as follows: $T = -50° - (h - 10{,}000) \times 0.005 \pm 9°C$. That is, the temperature at the tropopause will be −50°C, minus the variation in the altitude from 10,000 meters, times a coefficient for the lapse rate, ±9°C—the latter representing the width of the strip of tropopause temperatures clustered around the mean line at 10,000 meters.

With the data set thus divided, one saw the frequency maximum for tropopause altitudes in the summer above the mean line and its maximum altitude in winter below the mean line. These maxima were there in the Lindenberg data, but not enough to support the argument by themselves. To strengthen the argument, Wegener took the Munich data and, rather than plotting a mean line of the data for Munich, used the mean line of the data for Lindenberg (10,000 meters, −50°C) superimposed on the Munich data. Here the result was striking: almost all the winter points are below the mean and the summer points above it. Once again, the zone of variation in altitude is approximately 3,600 meters wide, and there are clear frequency maxima for the summer and winter altitudes as well. Finally, once again using the Lindenberg mean across a chart of scanty data for three other stations at latitudes 38°, 45°, and 51° north, he established that the data points also lie within a 3,600-meter-wide stripe.

Wegener concluded that in spite of the fragmentary data, there is a clear signature: records from six aerological stations show that "it seems accordingly, that this broad stripe is produced by an annual parallel displacement (*Parallelverschiebung*) of the mean line [of the altitude of the tropopause]."[58] Wegener was buoyed in his conclusion by the knowledge that the data sets available for his inspection were accidental, and he had no choice among them as they were the only data that existed: "the results of this purely accidental compilation in which there was no choice, seems to me to be interesting enough to publish in spite of its intrinsic uncertainty."[59]

Then to complete the "template," he noted that confirmation of his relationship would await further studies at the pole and at the equator concentrating on both the spread of temperature and altitude and trying to determine whether his formula would apply to the whole globe and not only to the middle latitudes. He concluded, "Unfortunately I have not been able to examine the values which Herr Berson has recently obtained in equatorial Africa. It would have been very interesting to see whether they would have fit (or not) the schematic pattern set out here."[60]

The presentation, then, is that of an open-minded but percipient observer who has used up all the available data in establishing a hypothetical relationship that he urges be treated as hypothetical until further studies can confirm it. This sort of formulation ends every speculative paper he ever wrote, and there are a great number of them. The remark about Berson's refusal to show him the data reveals that Wegener was still smarting from their dispute. Quite coincidentally, an explanation of Berson's refusal was unexpectedly forthcoming just as he had sent his paper off to Aßmann for publication, on 18 July 1909.

An Unexpected Offer

On 20 July Wegener received a long and astonishing reply from Aßmann. Aßmann had found the paper, on an initial reading, extremely interesting: "so far as I can see in a quick overview of [your paper] no one has ever before carried out an analogous study concerning the upper inversion."[61] Wegener's astonishment stemmed not from this comment but from what followed: Aßmann was actually writing to offer Wegener Berson's job at Lindenberg!

"Since October or November," wrote Aßmann, "Berson has been in the grip of a rapidly worsening neurasthenia, and I am in great need of finding his successor, for which I must, by all means have a theoretician, because the current theoretician, Coym, does not have the capacity to reach a sufficient level of care and exactitude."[62] Arthur Coym, who had joined the staff at Lindenberg during Wegener's short tenure there, was the person who Wegener had come to believe stood in the way of his obtaining permanent employment at a higher level than a mere technical assistant. Now it appeared that his work had been found unsatisfactory, and with Berson's illness, the door was once again open at Lindenberg for him.

"Neurasthenia," Berson's complaint, is a diagnosis (no longer generally in use) for a variety of symptoms that in the nineteenth century were thought to be a disorder of the nervous system. But when a man like Berson returns from equatorial Africa and begins to manifest fever, lassitude, aches and pains, and inability to concentrate, it is easy to imagine that he had contracted one of the chronic diseases of the tropics, most likely malaria. In any case, Berson, who would yet live another thirty years, was incapacitated and imminently taking a disability leave or retirement; the job was now open and waiting.

Aßmann made a persuasive case, running through the other likely candidates and rejecting them one by one: this one wanted too much money, that one is a social democrat, another is not enough of a theoretician. "I know it will not be easy for me to persuade you, my dear Dr. Wegener, but I believe that you are the right man for me and the right man for the job."[63]

If Wegener were to return to Lindenberg, Aßmann would raise the salary from 2,700 to 3,200 marks. "Of course as my first assistant, you would no longer have to work the [kite] winch, and, aside from some unavoidable administrative duties, you would be free to pursue your purely scientific work without interruption." Moreover, Aßmann continued, "you can see that this is an opportunity that will not likely occur again . . . obviously other [meteorological] stations—first at Danzig—will one day be established, and the post of Director-Superintendent would open up—and Danzig has a Technical University!"[64] Aßmann was as much as offering him the directorship of the new station at Danzig when it opened, if he would take the job of *Observator* at Lindenberg now.

It was a stunning and completely unexpected offer that, if accepted, would throw Wegener into an entirely different life and career track. Most of the other young meteorologists (including Kurt) were in technical universities or at observing stations, not at regular university posts. The job was clearly a way up into official rank in government service, rather than university life.

The day after receiving this letter, Wegener left for Göttingen to give a lecture at the Aeronautical Club there, his head spinning. The morning after the lecture (23 July) he wrote to his parents and asked for their advice. He told them he had asked Aßmann for time to think and would ask Köppen, his brother Kurt, Richarz, and others about their assessment of future prospects and positions before deciding.[65]

Yet he had no sooner returned to Marburg than it seemed clear to him that this was not the path he wished to follow. His parents' reply has not survived, but we know how strongly Richard Wegener had championed university positions over government ones and "real" universities over the technical institutes. On 27 July Wegener wrote to Aßmann thanking him for the opportunity but urging instead that Aßmann pursue Friedrich Bidlingsmaier (1875–1914), a geophysicist and specialist in geomagnetism who had aided Wegener in 1906 to prepare himself for Greenland.[66] Bidlingsmaier was a docent in Berlin and had shown some interest in coming to Lindenberg, but he was reluctant to give up his Berlin appointment. Moreover, he had been called up for a term of naval service. Wegener pointed out to Aßmann that he himself had been called up and would be on maneuvers with his regiment from August until late September.[67]

Aßmann persisted, however, and again appealed to Wegener to take the job. Wegener wrote to him on 18 August from the military barracks at Charlottenburg, profusely thanking him, and asking him to specify the conditions for his entry into the employ of the observatory at Lindenberg once more. Wegener could see, he thought, that a year or so hence when the Greenland results were published, his position would be quite different, and that the conditions of employment would have to be extremely favorable to overcome his reservations.[68] He was in a quandary, not knowing which way to go, and asked for yet more time to consider. At this juncture both men appear to have decided to let the matter rest for a time.

At the end of September, following his military course at Charlottenburg, Wegener visited his parents and Tony briefly at *die Hütte* before returning to Marburg. Much of his attention was taken up with ballooning. His experience as a balloonist was one of the main reasons that Richarz had wanted him at Marburg. Richarz had access to a balloon free of metal in its rigging which could fly cross-country trailing a magnetometer and produce profiles of Earth's magnetic field. Marburg was becoming in 1909 (and continues today to be) a center for civil aviation, both scientific and recreational, and had just established its first airfield and aerodrome.

Wegener, who had joined the International Aeronautical Federation and been certified as a pilot in ballooning, used these flights in September 1909 not just to assist Richarz but to adjust and improve a balloon theodolite of his own design for finding positions while in motion, something he had been working on mentally ever since his record-breaking flight with his brother Kurt in 1906.

While enjoying these aerial excursions, he was still deeply immersed in work on the Danmark material. He had finally finished the aerology, the only part he cared about very deeply, and his colleague Lüdeling was proceeding well with the electrical measurements. He still had the magnetic work to account for and finally was on the verge of talking Walter Brückmann (1878–1960), a geophysicist acquaintance from Berlin, into taking them on in return for publication credit. He wrote to the Danmark Committee that the negotiations were proceeding well. This, of course, was a means of "offloading" this work and of speeding up the process of getting out from under the expedition work.[69] The expedition work was both psychologically and intellectually draining for him. He desperately needed the stipend being paid him to finish the work, but, of course, if he finished the work, the stipend would end. It was, nevertheless, not just advantageous but really necessary for him to get free of all of this as soon as possible.

The theoretical development of his ideas on the physics of the atmosphere, forced on him by his desire to catch up with the work that had been done while he was in Greenland and by the necessity to give lectures at Marburg, was inspiring and encouraging to him. He was getting back into a familiar routine, getting back to the world of books talking to other books, where he had lived most of his life. But even as he did so, his very divided feelings about being back in an entirely bookish and academic setting were becoming evident.

He would have to give lectures in the winter semester, now that he had committed himself to an academic career, and he chose to offer a single course entitled "Astronomical-Geographical Position Finding for Explorers."[70] That this course should have come first, rather than astronomy or meteorology (in which he had actually habilitated), may seem odd, but there was a reason for it.

Antarctica?

However strong his impulse to have an ordinary academic career, sit at a desk, please his parents, and make a success as a university scientist, he was anxious to get back into the field. Even as he published his papers (or sent them off to Aßmann—he had written so many that spring that some would not be published for another full year) and announced lectures for the fall, he was in the process of applying to be part of a proposed 1911 German expedition to Antarctica to be led by Wilhelm Filchner (1871–1957).

Filchner was a military cartographer who had led an expedition to Tibet and one to Russia, and he was extraordinarily well placed, as an aide to the German General Staff, to assemble both military and civilian patronage in Berlin and elsewhere. He very much wanted to lead an expedition to Antarctica, even though he had no polar experience whatever. There is an interesting historical sidelight here: a contest between the supporters of Drygalski and those who sided with Filchner, with the sense among the partisans of the latter that Drygalski's Gauss Expedition had produced very little in terms of national honor and "names on the map," for all the scientific results collected. Most of the staff for this new expedition, including the scientists, had been long since chosen: Filchner had made most of these decisions in the spring of 1908 while Wegener was still in Greenland. There was as yet no firm date for departure of the expedition to Antarctica, and in the meantime there was a planned pre-expedition to Spitzbergen, in the winter of 1910, so that the members of the expedition might acquire some polar experience. Of course, Filchner might have acquired polar experience by hiring some of the members of the Drygalski expedition, but the controversy, to which he was a party, clearly prevented this.

Filchner's expedition was certainly well financed. He had purchased a ship, the *Bjorn*, and rechristened it the *Deutschland*. The expedition had the endorsement of Shackleton, who had just returned from his first expedition to Antarctica, and of Roald Amundsen. Wegener's chances of getting on the expedition were very slim. Like most expedition leaders, Filchner was reluctant to take anyone along with more experience than himself—meaning, in this case, anyone with any polar experience at all. Moreover, as someone who had explored in Greenland, Wegener was viewed as one of the "Drygalski people." Still, Wegener reasoned, he had made it onto the Danmark Expedition at the last moment; maybe he could finesse this again.

Wegener, of course, was dreaming of Antarctica while still in Greenland in 1906–1908. Now, with the very public dispute between Robert Peary (1856–1920) and Frederick Cook (1865–1940), both of whom claimed to have reached the North Pole in 1909, the excitement about Antarctica was nearing its peak. Nation after nation was throwing money at Antarctic explorers, not least for the possibility of claiming these lands as part of their empire.[71] These include the Shackleton expedition from Great Britain in 1908–1909 and proposed expeditions by Japan, Great Britain (Scott), Australia (Mawson), Amundsen's (as yet well concealed) Norwegian team, and Filchner himself. Germany, always a latecomer in the imperial sweepstakes, was late getting started here as well, and this was all the more reason for Wegener to believe that he could attach himself at the last minute to this expedition.

Running Away to Sea

Life in Marburg, pleasant enough and filled with his relentless pursuit of his own scientific work, was nevertheless beginning to seem confining before he had even begun to teach. Ballooning was exciting and never without danger, but Wegener was an explorer, not a thrill seeker, and his balloon trips were not really "expedition travel." He had returned from military maneuvers in August and September in top physical condition and wanted very much to be out of doors even as his work called him constantly to his desk.

It was in this restless frame of mind in mid-October, with the looming of the *Michaelis* semester only weeks away, that he decided to postpone by a few months his entry into teaching and decided instead to "go to sea." The International Commission for Scientific Aerology was about to launch a coordinated aerological experiment at sea. Hergesell and Teisserenc de Bort had been studying the global atmospheric circulation and had come to absolutely contrary positions on the character of the trade winds, especially at higher altitudes. These were two leading figures in the study of the structure of the atmosphere, and one of them had to be wrong; this experiment would determine which one.

The question was absorbing and of immediate interest to Wegener. While he had been in Greenland, Hergesell and Teisserenc de Bort had made observations in both the tropical Atlantic and the Mediterranean. These had established (as Wegener had remarked in Hamburg in the fall of 1908) that the trade winds were a shallow frictional layer above the boundary layer (300–500 meters) and below 1,500 meters, and within this altitude range, these two senior investigators had found a complicated structure of temperature and humidity with very strong inversions. This was, in Wegener's eyes, his "1,500-meter layer."

The problem that the "Experiment at Sea" was intended to solve was the layering of the atmosphere (and the wind direction) up to an altitude of 4 kilometers (2.5 miles). Hergesell championed the idea of "counter-trade winds" blowing in the opposite direction from the trades at greater altitude, and Teisserenc de Bort disagreed. What excited Wegener about this debate was the possibility, of which he was already himself convinced, that the issue had been misconstrued by these famous principals: the issue was not the wind *direction* and its constancy in some direction or another, but the *constancy of the layers within which these winds blow.* For Wegener, as opposed to his senior colleagues, this was a problem to be solved by thermodynamics, not dynamics—it was about the permanent structure of the atmosphere in the vertical direction, not its motion horizontally.[72] This debate was one, therefore, in which he very much wanted to participate and in which he felt he could play some sort of a role.

In the experiment, in addition to dedicated research vessels (mostly the yachts of wealthy patrons), the idea was to put a number of aerologists on commercial freighters as supernumeraries, from which vantage they could make pilot balloon observations of the direction and velocity of the wind. Wegener had applied to Hergesell to be part of this and been accepted into the experiment, scheduled to run from mid-November to mid-January. Subsistence allowances and equipment were to be provided for the observer.

Of course, nothing happened in Wegener's life without a list of official permissions. On 1 November 1909 Wegener wrote to the Danmark Committee for permission to set aside his contract and the work on the Danmark Expedition results for two months and push his remaining stipend forward for two months. He had, the week before, written to the committee to confirm that Brückmann would take over the magnetic observations, and with this gesture of good faith in place, the required permission in this quarter was immediately forthcoming.[73]

Wegener had also to obtain permission from his university to postpone the beginning of his winter semester lectures until the end of January.[74] He also had to ask the dean (the geologist Emmanuel Kayser) to intercede on his behalf with his regimental commander, to excuse him from yet another extended session of military training and maneuvers scheduled to run from February until mid-April 1910.

Wegener asked Kayser to tell the military commander that "he could not be spared" for the maneuvers, and Kayser, who appears to have liked Wegener a good deal, was glad to oblige.[75] Then, also in early November, Wegener wrote to Köppen to tell him of his plans to take part in the experiment. Wegener had not been much in correspondence with Köppen since the previous year, though he had used Köppen's ideas about the advance of meteorology in his inaugural lecture a few months previously. Wegener was to travel aboard a Lloyd steamer, the *Tübingen*, from Bremerhaven to Montevideo and then to Buenos Aires, by way of Madeira. The core of the experiment was to take place in the first half of December, when a broad range of observations would be made by many observers sending aloft pilot balloons from aboard ships to analyze the airstream of the trade winds at various altitudes.

Still in the midst of his preparations for sailing on the *Tübingen*, Wegener received a letter from Köppen including a recent paper and some information on recent advances in aeronautics. Most interesting to Wegener was the information that Köppen provided on the distribution of inversions in the atmosphere up to 1,500 meters. Köppen had calculated the percentage of inversion, that is, the percentage of increases in ambient temperature at altitude, as a percentage of all raw measurements. Köppen's figures showed frequency maxima for inversions between 700 and 900 meters (2,297–2,953 feet) and between 1,200 and 1,500 meters (3,937–4,921 feet). Wegener wrote to Köppen,

> Of greatest interest to me were the investigations of inversions . . . here again these are typical upper surfaces of the cloud layers, namely those at 800 and 1500 m. I don't know whether this is just my own idiosyncrasy, but everywhere I turn I stumble over these typical boundary surfaces! I believe additionally that an even greater maximum percentage will be recorded at the mean altitude of 39–42 hectometers (the level of a-cu [altocumulus]) when kite investigations eventually reach that altitude! . . . I've been going into this rather deeply and this idea seems to me to open an extremely useful perspective. In any case I have a great sack of [research] plans a few of which may hopefully be carried out by sufficiently ardent doctoral students.[76]

Wegener was indeed seeing these "typical layer boundaries" everywhere, as were many others at exactly this time. Wegener, however, was of his contemporaries the most convinced that these held the key to the structure of the atmosphere.

With permissions obtained, and once more with the prospect of an expedition, Wegener happily departed Marburg, but there were maddening delays. The original departure date of his steamship, 17 November, was postponed for ten days, and he found himself cooling his heels in Berlin with his parents. When he finally departed, the freighter dawdled on from Bremerhaven to Antwerp, where it was still sitting in port on 7 December, one week into the planned experiment. From Antwerp he wrote a laconic plea to the Danmark Committee to extend his contract forward yet another month.[77]

Finally under way and out of Europe, the freighter made port in Madeira on 15 December, and a very unhappy Alfred Wegener took the opportunity to write Christmas and New Year's greetings to his parents. Things were not going well. There was too much wind for the balloons, and though they sent up one or two a day, they could follow them for only a short time before they vanished from sight. As his technical correspondence (before the expedition began) shows, Wegener had thought from the beginning that the balloons were much too small to be followed easily, and this turned out to be the case. "I have written Hergesell about it. There you are—this is how we find out!"[78] He also added sardonically, "I haven't been seasick, though not from lack of opportunity."[79]

As if the delays, the bad weather, and the failure of the observation program were not enough, he found out once under way that the ship would lay over in Buenos Aires for three weeks. "When I'll get back, only the gods know . . . if I can't find another ship, I can scarcely be back before mid-March. Since the end of the semester is March 2, I can begin to see that my lectures are for the present on hold."[80]

The remainder of the trip—to Montevideo, to Buenos Aires, and the return—was a complete loss from the standpoint of scientific observation. There was, however, a good deal of time to think and write, and Wegener turned again to the problem (and the idea) of some sort of physical layer in the atmosphere, mentioned ever so briefly in his inaugural lecture at Marburg, a layer that would reside at an altitude of about 70 kilometers (43 miles). The general theme and intuition (long with him) that the atmosphere was riddled with poorly understood layers, and was not a thermodynamic continuum but a *discontinuum*, more and more occupied him as he stirred restlessly in his bunk on the returning freighter.

At least his worst fears were not realized: he managed to find another ship and was able to return to Marburg by mid-February, but with essentially nothing to show for three months of effort. Moreover, in his absence, the world (as was its wont) had moved along without him. Aßmann could not wait for Wegener's return to solve his own staffing problems and had appointed Otto Tetens (1865–1945) as Berson's replacement. Tetens was an astronomer and a photographer of repute (on the model of Adolf Marcuse, Wegener's teacher at Berlin). He had been Kurt Wegener's predecessor at Samoa and was working as an astronomer at the observatory at Kiel at the time of his appointment. He was in no sense a theoretician, but he was an experienced and skilled observer who, like Aßmann, was very interested in instrumentation and error correction. Tetens's career shows "what might have been" for Wegener. Tetens was appointed to Lindenberg in 1909, by December 1910 he was awarded the title of professor, and from 1911 until Aßmann's retirement in 1915 he was acting director of the observatory at Lindenberg. Unbeknownst to Wegener, Aßmann had been offering him a greater gift: not just that he should replace Berson, but that he should replace Aßmann himself![81]

So Wegener was not to go to Lindenberg; he would remain in Marburg and teach—perhaps. Because of the lateness of his return from South America (two weeks before the end of the semester that had begun in November), no one had signed up for his course on position finding, so he conferred with students and informed Richarz that he would offer the course again in the winter of 1910–1911.[82]

Wegener's Theory of Atmospheric Structure

With no teaching duties in the offing until summer semester (15 April), when he would offer a course of lectures in physics of the atmosphere, Wegener was free to pursue his increasingly comprehensive researches into this very topic. The central point of all his work was the layering of the atmosphere. It was to his intuition of discontinuity (as a key to structure) that he returned again and again. From this intuition, rapidly maturing into a conviction, Wegener made an increasingly bold series of speculations, based on slender but real empirical data, about the full height of the atmosphere and its behavior.

Near the end of February 1910, Wegener sent off to Aßmann a manuscript extending and developing the idea that somewhere between 70 and 80 kilometers was another major atmospheric discontinuity. This was an idea broached in his inaugural lectures at Marburg the previous May and developed in the published version of that lecture the previous November (1908). Wegener had worked on this in his ship cabin on the return from Buenos Aires, having no data to work up from this fruitless voyage. He entitled the paper "Über eine neue fundamentale Schichtgrenze der Erdatmosphäre" (On a new fundamental layer boundary in Earth's atmosphere), certainly the most ambitious claim he had made thus far.[83] It is also a remarkably novel and comprehensive approach to an old problem concerning the optical properties of the atmosphere, in which Wegener came very close to a major discovery.

It had been long established that there was a rather sharp boundary, the "twilight limit," between that part of the atmosphere capable of reflecting diffuse light and the layer above it, no longer optically active, and estimated by many observers to occur at about 70 kilometers. Yet because the transition from light to shadow is continuous, it would seem impossible to say with any certainty that a specific layer boundary existed, but only some sort of transition zone. On the latter side of the argument was the evidence of the "noctilucent clouds" injected into the atmosphere by Krakatoa in 1883 to an altitude measured at 83 kilometers (52 miles). In this case, light was reflected back to Earth from a height 13 kilometers (8 miles) greater than the current estimate of the "twilight boundary" layer.

Wegener had argued in his inaugural lecture at Marburg that the height of the Krakatoa clouds was precisely evidence of a sharp boundary layer—in this case, one broken through by the volcanic dust cloud, in the same way that a cumulus cloud breaks through an atmospheric layer and causes a local inversion. More importantly, however, Wegener believed himself in possession of evidence providing "an astonishing confirmation" of what had been only a supposition. The evidence in question was that provided by a consideration of the chemical composition of the atmosphere at different altitudes: "in what follows arguments will be set out which will indeed make this view [of a discontinuity at 70 kilometers] so probable that, I believe, scarcely any ground for doubt will remain."[84]

Here the pattern—imaginative and rhetorical—governing his work is in bold relief once again. He has been reading the literature and seeing ample evidence here and there (but never in one place!), allowing him to move from mere empirical published data, distributions of temperatures and other meteorological elements, to an organizing hypothesis and even to confirming evidence. He had done this with the cloud layers, and again in the seasonal variation of the altitude of the tropopause (the upper inversion), each time extracting a law-like conclusion from some set of them.

"One finds," he wrote, "for this supposition [of a boundary layer at 70 kilometers] . . . a completely unexpected confirmation when one calculates the composition of the atmosphere in different altitudes." If one examines the matter from this angle, he continued, "one comes to the astonishing conclusion that at precisely the altitude of the supposed layer boundary, the composition of the atmosphere changes with extraordinary rapidity, so much so that one is justified to characterize the area below the boundary as the *Nitrogen Atmosphere* and the layer above it as the *Hydrogen Atmosphere*."[85]

Wegener was here arguing that the optical data had been confirmed by chemical data, and for him this was the more exciting point, since the latter notion (of the chemical differentiation of the atmosphere at altitude) was so widely accepted. What Wegener was doing here was not providing new data but once again integrating uncorrelated ideas into a unified picture capable of producing an explanation that others might and ought to have seen but had not.

When this article went to press in 1910, Wegener added the following footnote to his title: "Humphreys [William J. Humphreys (1862–1949), U.S. atmospheric physicist] had already carried out an investigation in 1909, Distribution of Gases in the Atmosphere in the Bulletin of the Mountain Weather Observatory II, which came into my hands only when this article was already in press, with the same calculations of the volume percentages of gases for the different altitudes and even employing the same graphical presentation used here, without however, drawing any broader conclusions from it."[86] For Wegener, priority and originality lay not in the realm of data but in the ability "to draw broader conclusions from it." One mentions this because much of Wegener's work ended up with divided priority, usually because he himself reported the parallel publication.

The data in question on the changing chemical composition of the atmosphere with altitude had been in the literature since 1903. Julius Hann had published an article in *Meteorologische Zeitschrift* in 1903, when the investigations of Teisserenc de Bort and others were making it starkly clear that the "Weather Zone" of the atmosphere, where turbulent mixing took place, was the limit of vertical circulation, and that cyclonic and anticyclonic systems were trapped between the surface of Earth and the tropopause a scant 10 kilometers above it.

Hann had immediately seen both thermodynamic and chemical conclusions to be drawn from the existence of the upper inversion, conclusions drawn directly from Dalton's law of partial pressures, whereby each gas in a mixture of gases behaves independently of the others, based on its temperature. If one knows the mean temperature at a given altitude, one can calculate the partial pressure of any gas at that altitude. Hann had estimated the mean temperature of the atmosphere at 0, 10, 20, 50, and 100 kilometers and calculated the percentage (by volume) of nitrogen, oxygen, argon, hydrogen, and helium, for each of these altitudes. This produced the astonishing conclusion that while the percentage of nitrogen in the atmosphere is virtually the same at 50 kilometers as at the surface, at 100 kilometers it had fallen to 0.1 percent, while hydrogen, composing only 0.01 percent of the atmosphere at Earth's surface, at 100 kilometers is 99.4 percent of the atmosphere.

Wegener (and apparently W. J. Humphreys in the United States at the same time) had interpolated temperature data for 10-kilometer intervals from the surface to 200 kilometers and produced both tabular and graphical data showing the very rapid decrease in the percentage of nitrogen with altitude from 79.7 percent at 50 kilometers to 57.9 percent at 60 kilometers, 24.3 percent at 70 kilometers, and 6.6 percent at 80 kilometers.

Once the turbulent mixing of gases at the surface was no longer a factor, the atmosphere would necessarily, Wegener reasoned, begin to layer out based on specific gravity, producing a very rapid decrease in the presence of nitrogen and in the disappearance of oxygen in all but trace amounts above 60 kilometers. In order to introduce his diagram, Wegener gave the following introduction:

> In the graphical presentation one can see with great clarity the marked boundary between the nitrogen and hydrogen atmospheres. The cause of this extraordinarily sharp separation is the gross difference in a specific gravity of N and H. It is clear that the transition must continue longer the smaller this difference is. On the other hand, the gases would have a tendency to separate from one another and form layers in direct proportion to the difference of their specific gravities. What is of interest in this connection is that the sudden change in composition occurs just at the altitude at which we, on entirely separate grounds, have hypothesized a boundary surface.[87]

Alfred Wegener's reasoning on chemical grounds for a sharp boundary in the atmosphere at 70 kilometers was bold, original, fascinating, and almost entirely incorrect. There is an important change in the atmosphere in this region between 60 and 100 kilometers, as Hann had shown. But current conceptions (those of the early twenty-first century) argue for turbulent mixing and therefore constant composition of the atmosphere up to about 100 kilometers, with that surface taken as actually the outer boundary of the atmosphere, with about 99.5 percent of the atmosphere's mass below that level (figures that Wegener and Hann also knew).[88]

The point, however, is not whether Wegener was correct, but whether his arguments were well founded at the time he made them, which they were. There is a significant boundary in the atmosphere near where he put it, now known as the "mesopause." The atmosphere above 100 kilometers *is* mostly hydrogen. The analysis of the spectrum of the aurora *does* shed light on the composition of the atmosphere, just as Wegener said it would. The aurora *does* indeed extend from altitudes of about 500 kilometers down to the surface of Earth, though there are frequency maxima near 200 kilometers and 60 kilometers, with the former showing similarities to hydrogen spectra and the latter to nitrogen spectra. All this is in Wegener's paper.[89] Further, the absence of oxygen between 150 and 100 kilometers, where most shooting stars appear, *does* mean, as Wegener claimed it did, that they are not burning but vaporizing, and that their spectra as they glow *do* allow analysis of the composition of the atmosphere at the altitude observed.

It is a slight irony that where Wegener was most certain that he was right he was most certainly wrong, and where he showed caution and tentativeness in his conclusions he was more nearly correct. One hastens to add that our current conclusions concerning the character of the upper atmosphere belong to the era of rocketry and investigation of the "ionosphere," 1945–1960.

Near the end of the paper, Wegener ventured some highly interesting if guarded speculations concerning the outermost layers of the atmosphere. The idea of a hydro-

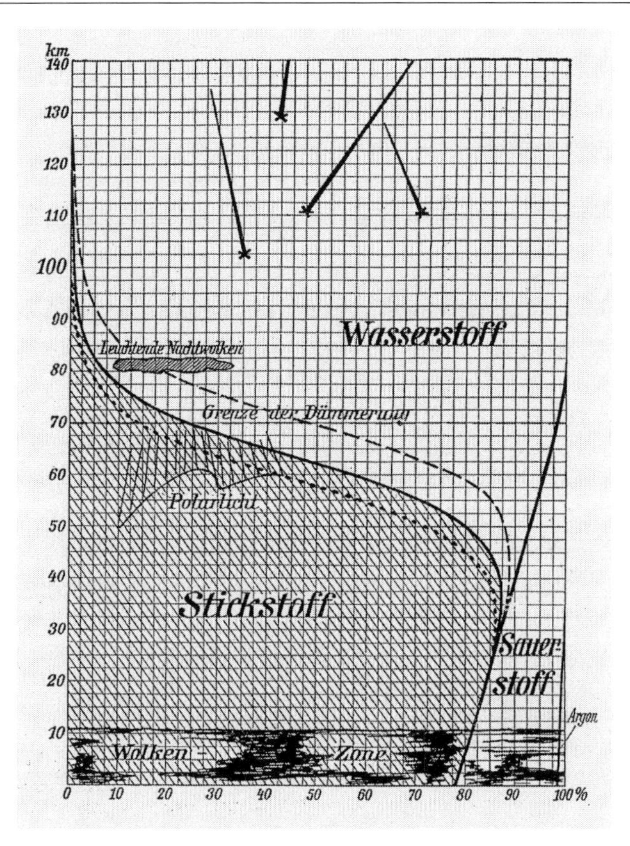

Wegener's picture of the layering of the atmosphere based on chemical composition. The horizontal axis is the percentage of a given atmospheric component; the vertical axis is altitude. *Sauerstoff*=oxygen, *Stickstoff*=nitrogen, *Wasserstoff*=hydrogen, *Wolken-Zone*=cloud zone, *Porlarlicht*=aurora, *Grenze der Dämmerung*=twilight limit, and *Leuchtende Nachtwolken*=noctilucent clouds. From Alfred Wegener, "Zur Schichtung der Atmosphäre," *Beiträge zur Physik der freien Atmosphäre* 3 (1910): 33.

gen atmosphere with a sharp limit, he wrote, offers an interesting analogy to the Sun, where (in the chromosphere) a strongly delimited hydrogen sphere could be detected. "Perhaps one could in the future push this analogy a little further and above the Hydrogen atmosphere find another atmosphere made of a still lighter gas corresponding to the Coronium of the Solar Corona within which the highest polar lights play and which, just as Coronium, gives a green spectral line at 532 Å units producing the known line in the northern lights at 557 Å units."[90]

Wegener did not, out of the blue, hypothesize the presence in Earth's atmosphere of something very like an element "Coronium." Julius Scheiner, the lone stellar astronomer at Berlin during Wegener's time there, had in his book on the spectra of stars (1898) suggested that such a rare, light gas might be found in the upper reaches of Earth's atmosphere. The spectrum of the aurora (*Polarlicht*) was intensively investigated in the first decade of the twentieth century, and Wegener was well read in this work.

Moreover, suggestion of new and rare elements was a common subject of speculation then as now. Helium had been discovered as an element in the analysis of the solar spectrum during the eclipse of 1868, and in 1869 a new spectral line near 530 Å had been suggested as the signature of an extremely light gas: coronium. This element was still under consideration as late as the 1930s. Dmitri Mendeleev, using the same logic that led to his discovery of germanium, had, in 1904, proposed an extremely light element of atomic weight 0.4 and suggested that this element was identical with the "Coronium" detected in the Sun's atmosphere.

Wegener's interest in these auroral spectra, on which he depended for his most important conclusions about the composition of the atmosphere, had several sources. First, auroras were a favorite subject of his old Berlin professor, Förster, and in Wegener's November 1908 article for the *Mathematisch-Naturwissenschaftliche Blätter*, the house organ of the Berlin astronomy department, he had written a semipopular article about the auroral phenomena in Greenland and how they cried out for explanation. Like the composition of stars, the character of the aurora had moved from "things we shall never know" to "things we know well" through the development of spectroscopy.[91]

Second, Wegener's Greenland work, including the time spent at Pustervig, had shown him that auroras, contrary to what he had read, do indeed descend nearly to the surface of Earth. This had bearing on studies of atmospheric electricity and magnetism, then being vigorously pursued everywhere and indeed locally by his professor at Marburg, Richarz. Finally, the interest marks him as a "cosmic physicist" since he was willing and even anxious to explore the analogy between the atmospheres of Earth and the Sun and had always in mind that Earth was a planetary body in orbit about a star and in constant interaction with the rest of the universe: elementally, gravitationally, magnetically, and electrically.

The picture of Earth's atmosphere as a series of well-defined concentric shells marked by sharp surfaces of discontinuity, each with different physical properties, and (above 70 kilometers) different chemical properties as well, emerges here as the kernel for Wegener's dream—as yet more a dream than a plan—to produce a comprehensive physics of the atmosphere. It is important to remember once again that Wegener was not part of the group (Bjerknes, Margules, and others) working on the dynamics and hydrodynamics of the "Weather Zone," the troposphere, and therefore investigating that branch of study now called *dynamical* meteorology. Wegener's atmospheric physics was what we now call *physical* meteorology, the thermodynamic, optical, and acoustic properties of the atmosphere at every altitude. For Wegener, the atmosphere was not the thin shell of ambient, breathable air, but all the regions above Earth that had any thermal, optical, or acoustic signature. Given the height of the aurora, this meant the envelope extending to 500 kilometers above Earth.

Wegener sent this paper off to Aßmann with considerable satisfaction and then turned to an analysis of his lot. He was back in Marburg, he had no students, and he had no money. He was actually reduced to begging. On 15 February 1910, he wrote to the Danmark Committee, asking them to please count the full month as having been work on the expedition results even though he had only worked half a month. He made a good show of work, pushing Lüdeling forward and asking for forbearance from the committee, assuring them that he was as anxious as they were to see the work finished.[92]

He had to find some way to make a living. With the Danmark work winding down and no students, he had no source of income. With his father, Richard Wegener, retired—and a man of modest means who was not much of a money manager—there was no hope or expectation that his father could be much help. Had he been chosen to join the Filchner expedition to Antarctica, he might have expected a new salary stream to begin with the late fall of 1910, when the expedition was scheduled to go to

Spitzbergen for training, an income that would have continued for several years. Filchner, as we have noted, had amassed a great deal of money, and the members of the expedition were to be very generously paid. But as Wegener wrote to Peter Freuchen in March of 1910, this was looking more and more doubtful, to the point where Wegener no longer expected to be chosen, an outcome to which he was now more or less reconciled. He would now have to turn his attention to teaching and preparing a comprehensive course of lectures on physics of the atmosphere.[93] His exploring work would be confined to his study in Marburg, and the closest he would get to the Arctic for the foreseeable future was the polar bear rug at his feet.

8

The Atmospheric Physicist (2)

MARBURG, 1910

> A theory is the more impressive the greater the simplicity of its premises, the more varied the kinds of things that it relates, and the more extended the area of its applicability. Therefore classical thermodynamics has made a deep impression upon me. It is the only physical theory of universal content which I am convinced, within the area of the applicability of its basic concepts, will never be overthrown.
>
> ALBERT EINSTEIN, *Autobiographical Notes* (1946)

Wegener's decision to offer a course of lectures in the summer semester of 1910 at Marburg entitled "Physics of the Atmosphere" was an obvious choice under the circumstances. He had, after all, been brought to Marburg as a physicist; his expertise, though it did extend into astronomy, was in the physics of the atmosphere. On the way back from South America, he had conceived a plan (while working to extend his studies of atmospheric discontinuities) to begin the writing of a complete physics of the atmosphere. The transformation of meteorology into a physics of the atmosphere was spoken of everywhere, and a number of Wegener's closest contemporaries were at work on it. Richard Aßmann's strong desire to have a theoretician to replace Berson as *Observator* (and that is what Berson was, an observer), as well as his desire that Wegener should fill that role, had already, as we have noted, pushed Wegener further in the direction of theory.

Albert Einstein's remark about the impregnability of classical thermodynamics, which serves as the epigraph at the head of this chapter, was made in 1949. Einstein had many times said that the best a theory might hope for was that it should become a limiting case of a theory of still higher generality. His characterization of classical thermodynamics seems to indicate that this theory, and this theory alone, would never suffer such a fate "within the area of applicability of its basic concepts." Since these basic concepts include the law of conservation of energy and the second law of thermodynamics (which says that in an isolated physical system, differences in temperature, pressure, and density tend to even out), it is clear that Einstein was being characteristically droll. The area of applicability of these basic concepts is the universe as a whole. In fact, the third law of thermodynamics expresses the consequences of applying the first and second laws to the entire universe.

Within physics, the unifying power and impregnability of thermodynamics give it a special status that is not often noted in treatments directed at the lay public. Popular works on physics are almost invariably concerned with gravitation, electromagnetism, relativity, and quantum mechanics, and almost never with thermodynamics. The search for a unified theory of physics, whether this is found (as is the case in the early twenty-first century) in string theory or in loop quantum gravity, means the unification of gravitational, electromagnetic, and nuclear forces. In a larger sense, however, the unification of physics is already achieved in thermodynamics, which is universal. Thermodynamics covers everything from the very small to the very large, applies over any time interval, works with both living and dead matter, and applies to the quantum mechanical, relativistic, and classical realms.

It was obvious to Wegener that he should begin a general physics of the atmosphere with a study of the thermodynamics of the atmosphere. We may recall that in his early studies in Berlin he had taken Max Planck's course in thermodynamics, in which Planck avoided reference to microphysical entities and instead developed the first and second laws of thermodynamics out of consideration of bulk properties of the world: temperature, pressure, and volume; he had even given special attention to the atmosphere. The latter reflects the very close attention given to such problems in Berlin by Hermann von Helmholtz (1821–1894) and Heinrich Hertz (1857–1894). In the 1860s Helmholtz, simultaneously with Julius von Hann (1839–1921) in Austria, was concerned with the problem of temperature changes in moist air as it rose and descended adiabatically.[1] In 1884 Heinrich Hertz (then still at Berlin) had introduced a graphical method for determining adiabatic changes in the state of moist air, and this was further developed in 1886 and thereafter by Bezold in his "First Report on the Thermodynamics of the Atmosphere" and subsequent publications.[2]

Wegener was Bezold's student, and Bezold's collected papers (to the year 1900), published in 1906, were mostly in the area of atmospheric thermodynamics. Yet so much had been learned in the period 1900–1908 that Wegener had argued (in his inaugural lecture at Marburg in March 1909) that a full-length book would be required to do justice to this new material; this included a huge bulk of ideas concerning atmospheric temperature distributions. Wegener had further argued in that lecture that the principal function of thermodynamics, applied to the study of atmospheric structure, was to demonstrate the *unintelligibility* of all existing theories of cyclonic and anticyclonic storms. Wegener had also said that the mean thickness of the various layers of the atmosphere, their variation with altitude, and the associated changes in the meteorological elements within them all had to be understood before one could determine how cyclones and anticyclones would behave, especially since these turbulent disturbances were associated with breaches in the integrity of atmospheric layers.[3] Thus, from Wegener's standpoint, the updating of the presentation of basic thermodynamic concepts applied to the atmosphere (to reflect the results of the new aerology) was a fundamental precondition for the understanding of dynamic meteorology.

Bezold had extended his notion of atmospheric thermodynamics to include studies of föhn winds, of water vapor in the atmosphere, of the formation of precipitation, of atmospheric dust, of cloud forms and the macrophysics of clouds, and of cloud elements—the microphysics of clouds. To this he had also added the study of the formation of raindrops and snow crystals. This broad version of the thermodynamics of the atmosphere was pursued by everyone at Berlin: Helmholtz, Hertz, Aßmann, Berson, and Bezold himself, over the span of a half century, had combined theoretical development, laboratory experiment, and field observation to this end.

So it could hardly count as a surprise that Wegener, fetching about for a way to approach the physics of the atmosphere, decided to begin his course of lectures with the material most assiduously pursued by his own graduate school professor. As an added impetus to this very obvious course of action, the pursuit of this material would require some exciting and significant modifications. Bezold's collected papers did not address the profound consequences of the discovery of the *tropopause* and the *stratosphere*. These phenomena, which everyone was learning to call by these new names (since Teisserenc de Bort had announced them the previous September), had changed almost everything in theoretical meteorology.

If Wegener's approach to the physics of the atmosphere was, as the theorists say, "overdetermined" by Wegener's graduate education, it was further reinforced by his very powerful physical and intellectual experience of Greenland. Greenland in winter was a purely thermodynamic universe, the full realization in physical form of theoretical principles. When the ocean was frozen and the shore covered with snow, the world was nothing but the pressure, density, temperature, and volumes of water, with all of water's phases simultaneously present and covering the entire world—ice, water, water vapor, flowing and changing, ascending and descending, transforming one into another, condensing, vaporizing, sublimating, all this physics constantly before one's eyes. What the Tower of Pisa was to Galileo, the Greenland ice cap was to Wegener: the physical embodiment of universal law in action.

There is an important reservation here in referring to thermodynamics as composed of "laws," a reservation of which Wegener was well aware and in full agreement. These "laws" (as we noted earlier) are descriptions, not explanations, and are expressed in "system and boundary" equations. The "laws" of thermodynamics are not derived; they are just written down. This distinction is better observed in German than in English. In Helmholtz's and Planck's writings, as well as in Wegener's, what are referred to in English as "laws" are described as *Hauptsatz*: fundamental principles. "Laws" are something found in nature, whereas "principles" are something found in science. The philosophical sophistication of German university education worked together constantly to remind Alfred Wegener and all other Berlin physics graduates that science was a human construct, not a direct apperception of the natural world.

Wegener's approach through this sort of thermodynamic thinking reflected a desire to bring simplification and conceptual clarity to an atmospheric science that had become enormously complex. In much the spirit of Planck at Berlin, he wished to sever the fundamentals of atmospheric thermodynamics from their applications to theories of atmospheric motion.

The question at issue here was the theoretical turmoil concerning the formation of midlatitude cyclonic storms. Meteorologists in both Europe and North America, from the 1860s on, had worked diligently to adapt physical theory, and especially thermodynamics, to meet meteorological problems.[4] Foremost among all the problems of meteorology was the understanding of the formation and propagation of storm systems.

The complexity and conceptual confusion that Wegener hoped to avoid in his own teaching of thermodynamics came out of the evolution of the thermal theory (also called the "convective theory") of cyclones in the latter part of the nineteenth century. The idea of the conservation of energy had brought much benefit to atmospheric physics in the 1860s and 1870s as an explanation of adiabatic processes. A good deal of work in the 1870s led to the conclusion that the primary source of kinetic energy in storms came from the release of latent heat of condensation in rising air.

The fundamental idea was that a rising column of moist air released heat, and then localized heating and vertical expansion caused the lowering of pressure at the ground center of a storm and potentiated the inflow of air from all sides. Cyclonic motion was the result of the deflection of these winds by Earth's rotation. Constant low pressure at the surface (it was assumed) was maintained by divergence and outflow of air at the top of the cyclone, which would therefore move in the direction of greatest moisture, that is, in the direction of the heat source.

In the 1890s aerological observations challenged this simple view, and by the early 1900s it was evident that both thermodynamic and hydrodynamic considerations would have to be deployed to avoid gross oversimplification. In spite of this, many meteorological theoreticians continue to think of storm systems as thermal entities, and of the thermodynamics of the atmosphere as something that should include consideration of the energetics of storms.[5]

In addition to his desire to untangle atmospheric thermodynamics from the theory of midlatitude cyclonic storms, Wegener wanted to counter a trend, going back to Carnot and Poisson, of developing the subject with great mathematical complexity and elegance, on a foundation of integral calculus. Wegener's strong preference was for the statement of concepts in ordinary language, without reference, whenever possible, to mathematical formalism. In 1911 he wrote to Köppen and referenced approvingly the following quotation from the German edition of Sir George Darwin's book *Tides and Kindred Phenomena in the Solar System* (1898):

> A mathematical argument is, viewed in this light, only a means to organize ordinary human understanding, and it is good when scientific men rather than constantly wrapping their scientific work in a veil of technical terminology such that it is available only to the few, rather from time to time uncover and expose to a broader public the train of thought concealed beneath their mathematical formulas. (From Darwin: Ebb and Flow). . . .[6]
>
> I myself hold the crass and probably exaggerated point of view that such mathematical treatises as I cannot understand (for instance in those works where one can no longer see the train of thought glimmering through the mathematics—it's often still possible to follow the train of thought without working through the formulas) are either wrong or incomprehensible. It is not necessary always to think that one bears the entire responsibility when one does not understand the printed or written word: "for just when ideas fail, a word comes in to save the situation" so it is that where the logic of an argument falters one can usually fill the gap with a few formulas.[7]

The quoted reference within the above passage is from Goethe's *Faust*, where Mephistopheles is telling a student not to think about what he's learning but just to memorize the words: Wegener means this as a critique of memorized, copied, and repeated formulas.

Wegener found complex mathematical presentations in physics distasteful not because he couldn't do the math but because of the tendency of a fully mathematical presentation to make physical subject matters more difficult than they needed to be, and because he considered the superfluous, as opposed to the essential, use of advanced mathematics as a kind of veil, thrown by an author over a subject to cover gaps in his understanding. Wegener's objections in this regard appear to be well founded, as diagrams in nineteenth-century theoretical treatments of paths of airflow in cyclonic storms often appear to owe more to the graphical form (with a particular fondness for hyperbolae) of sequences of partial differential equations than they do to the flow of real air within a real storm. Wegener was insistent that purely theoretical laws had to be modified in favor of the empirical results obtained, in the actual atmosphere, with real and well-calibrated physical instruments.

In addition to his principled opposition to the use of more mathematics than absolutely necessary for the doing of physics, Wegener had another reason for wanting a plain, clear, and straightforward treatment of the subject of thermodynamics, as he prepared his lectures for the summer semester in Marburg. Most of his students, never very numerous, were training to be not university professors but *Oberlehrer,* high school teachers, who would have to re-present even simpler versions of this material to students in their late teens.

Wegener's clear presentation of atmospheric physics in his first course of lectures at Marburg received an enthusiastic testimonial from one of his first students, Johannes Georgi (1888–1972), later himself an atmospheric physicist who did notable work on the jet stream and in the scientific investigation of Greenland.[8] Georgi, then twenty-one and in his last year of university study, arrived in Marburg to complete his physics degree in April 1910 and recalled finding "a notice on the board of the Physical Institute, in a clear and attractive handwriting, announcing that Privatdozent Dr. Wegener was to give some meteorological lectures and demonstrations." Georgi remembered Wegener as firm but modest and reserved: "only here and there could we catch a glimpse of the lion behind the lamb-like manner."[9]

> His lecture on the thermodynamics of the atmosphere was remarkable for the ease of his delivery, which was in complete contrast to the difficulty of the subject. Numerous examples were taken from his recent observations in Greenland; and here for the first time the attempt was made to relate the bulk of measurements from the free atmosphere during the last dozen years to general physical rules for the explanation of the manifold phenomena such as the different atmospheric layers (only eight years had passed since the discovery of the stratosphere!) and the various types of cloud formation. Whoever in those days had the opportunity of following the lectures and practical work of famous scholars would have had to admit that Dr. Wegener's lectures bore no professorial stamp at all. On the contrary, the tutor came down to the level of his audience and developed with them the theme which he had just set down. . . . It is true that the final result still had to be formulated mathematically, but neither before nor since have I had the experience of hearing a tutor state quite simply: "this derivation is not mine; you will find it in the physics textbook by . . . on page. . . ." . . . He always took the greatest pains even in his most specialized work to be as intelligible as possible and not write only for his fellow experts . . . an outstanding trait of his character was his frankness even towards his students. He had an unusually high degree of integrity in such a natural and unpretentious way that one had the impression that he was exempt from the common human temptation of occasionally making oneself appear a little more important than one actually is.
>
> I am sure that young people in particular feel this immediately; and the simplicity of his lectures and demonstrations, obviously based on experience and achievement, always won him the hearts of his audience. At the end of the lectures in Marburg he used to bring out a number of photographs to illustrate the subject he had been discussing.[10]

Wegener seems to have often inspired this sort of loyalty, in which admiration for his scientific ability and understanding was combined with admiration for his honesty and integrity and for his strength of character. This admiration was not limited to his students but also shared by Köppen, Aßmann, Richarz, and in fact every academic supervisor and senior who ever worked directly with him.

Wegener, working without any other duties or distractions, devoted himself completely to the development of his lectures on the physics of the atmosphere. There were only four students actually registered for the course, but the audience was often much larger, depending on the topic, with the rest made up of other instructors and even some of the senior academics who had heard about the clarity and accessibility of Wegener's presentation of these exciting new results in physics. Among his most frequent hearers was the physicist Karl Stuchtey (1880–1950), also an enthusiastic balloonist, and very interested, as was Wegener, in the physical form and characteristics of clouds.[11]

Atmospheric Physics in 1910

Wegener was committed to writing a thermodynamics of the atmosphere and using his preparations for this course in order to advance that project. But a course in the full physics of the atmosphere could not stop with thermodynamics, and Wegener had to go ahead and treat the rest of the physical topics within the subject. In his view, atmospheric physics divided neatly into the same subsections as physics in general: thermodynamics, mechanics, radiation, electricity, optics, and acoustics. For most of these there were textbooks available out of which he could develop his lectures, generally the older meteorological textbooks. Only in acoustics was there a lack of a general treatment.[12]

While Wegener had access to works that could reflect modern developments in atmospheric optics, atmospheric electricity, and radiation, mechanics was something of a problem. The last extensive treatment of the mechanics of the atmosphere (its kinematics and dynamics) had been published in 1885 by Adolf Sprung (1848–1909): *Lehrbuch der Meteorologie*. Sprung, who had worked with Köppen (with whom he was close friends) at the Hamburg Marine Observatory, had later moved to the Meteorological Institute in Berlin to work with Bezold.[13]

Even if Sprung's treatment of the mechanics of the atmosphere grew out of the tradition in which Wegener himself had been trained at Berlin, it was, as Wegener noted, "unfortunately already quite out of date."[14] What had made it out of date more than anything else was the work of Vilhelm Bjerknes, who in 1898, while a professor of mathematical physics at the University of Stockholm, had published a paper on the general circulation of the atmosphere and the oceans, in which he argued that the application of classical hydrodynamic equations to the atmosphere could never advantage practical meteorology (weather forecasting) until the hydrodynamics had been integrated with a thermodynamic treatment. Sprung had tried to treat dynamics and thermodynamics together, but there were severe limitations to his approach. Since Bjerknes was also a product of the Berlin group, as well as a student of Heinrich Hertz, this was for Wegener a case of one Berliner replacing another.[15]

Bjerknes's treatment of the subject had the additional attraction that it covered the circulation (the mechanics) of both the atmosphere and the oceans using the same hydrodynamic ideas. Science is supposed to be "the view from nowhere," but this is rarely if ever the case. Even in theories of great scope, the practical needs of the national and local communities, the personal histories of those who make them up, and the people that they know and whom they trust often have a decisive influence on the form theory takes, especially in its early stages.

Bjerknes had been trained in Berlin and, had Hertz not died, would probably have stayed to work with Hertz in Berlin on electrodynamics. With his return to Scandinavia (a collection of fishing nations sharing a long coastline), it was not surprising that his work on the behavior of continuous media (fields of electromagnetic force) should find encouragement for its extension into problems of the ocean and atmosphere, and this is exactly what happened. Opposed to a vision of the atmosphere as an extension of the kinetic theory of gases, an immense aggregate of particles having elastic collisions with one another and velocities proportional to the temperature, Bjerknes approached the atmosphere as a continuum, as a fluid. We should recall that "the ether" had not yet been banished from theoretical physics, and that theories of electromagnetism were also based on just such a notion of the distribution of some quality in a continuum.

The connection between the atmosphere and the ocean also made Bjerknes's treatment conceptually interesting. A study of the ocean currents and their tracks and the study of vertical temperature distribution in deep ocean basins had led in the mid-nineteenth century (through the work of Emil von Lenz [1804–1865], who predicted a lapse rate of temperature with depth in the oceans) to a general circulation model of the oceans. Cold water was sinking at the poles and converging and rising at the equator, with masses of warm surface water flowing back to the poles. Like George Hadley's (1685–1768) atmospheric circulation, the idea was influential, though clearly not complex enough to account for what was known by nineteenth-century oceanographers about the nonsymmetric arrangement of water masses.

"Water masses" were surface bodies of homogeneous salinity and temperature, and these masses of water would or could sink to a particular depth and travel hundreds or thousands of miles coherently. Hence, patterns of temperature and salinity measurements could give a distribution of these masses and a rough picture of ocean circulation. In Scandinavia the desire to improve the fishery, along with the knowledge that certain species preferred water of a given temperature and salinity, led to a strong research program of "synoptic oceanography," but a theoretical treatment was also required.[16]

Bjerknes's work provided a qualitative treatment, with a sketch of the kind of mathematics required for a quantitative treatment of circulation in both the ocean and the atmosphere. Bjerknes's focus was on the pressure gradient, and he gave striking illustrations of upwelling and downwelling in ocean basins caused by circulation associated with layers marked by sharp density gradients. This, of course, was impressive and congenial to Wegener, who could see in these "oceanographic elements" the same characteristic distributions governed by sharp surfaces of discontinuity with which he was so concerned in the atmosphere. The heaviness of water compared to air and the relatively narrow range of temperatures for ocean water compared to those for atmospheric air made it logical that isobar surfaces (lines of equal pressure) should govern the ocean in the way that lines of equal temperature governed discontinuity surfaces in the atmosphere.

There were other choices in the monograph literature available to Wegener to update Sprung's treatment of atmospheric mechanics. Principal among these was the work of Max Margules (1856–1920) in Vienna. Here the emphasis was on the energetics of storm systems, but discontinuity surfaces still played an important part. Much nineteenth-century work on the motion of storms had to do with the vertical circulation of heat and the generation of thermal energy to drive the storms. But as aerological research indicated and Margules's theoretical treatments elaborated, this was not really the answer. The nineteenth-century notion of a cyclonic storm was that a storm was a self-subsistent entity within a boundary created only by its outer surface with still air. As Margules and others showed, the existence of storms depended on a hydrostatic relationship: they were trapped between the surface of Earth and the tropopause, where, with broad penetrations, generally their convective activity ceased. This was a profound change in the view of storms in air: from full atmospheric circulation to the restricted circulation within the sharply bounded discontinuities between which they lived.

Moreover, Margules was concerned, as were Helmholtz and Köppen before him, with sharp *horizontal* surfaces of discontinuity, of the kind observed in squall lines of thunderstorms with either sharp drops in temperature or sharp rises in temperature.[17] Margules's work differed from Bjerknes's in that Margules was interested in changes in kinetic energy which might drive storms, resulting from the interaction of contiguous air masses, whereas Bjerknes was concerned initially with the question of global circulation; the treatments were complementary but not identical in their content and aim.[18]

Wegener could appreciate the important work being done at the research front by Bjerknes and Margules with attention to the importance of both vertical and horizontal discontinuities in the atmosphere, but the direction his researches were taking had to do with the *permanent* structural features that constrained the movements of such air and water masses, discontinuities that created the conditions for the existence and behavior of these masses, not the reverse. Air masses in the atmosphere, as well as water masses in the ocean, had no constructive role in establishing discontinuities in the vertical dimension.

Wegener's Approach to Thermodynamics

As he worked his way more deeply into the subject in the late spring and early summer of 1910, the impetus to write a book on atmospheric thermodynamics grew stronger with his discovery that the literature was filled with errors, and this included many topics in Bezold's thermodynamic papers published only four years before. Moreover, the treatment of topics seemed chaotic; there wasn't any order or systematic presentation of matters as fundamentally important as the size of cloud elements (such as water drops and ice crystals), or even a systematic treatment of the ice phase of water vapor in the atmosphere, with its various forms of snow, graupel, hail, and ice crystals.

Wegener published two papers in *Meteorologische Zeitschrift* in the summer of 1910 which show the interesting character of the work style he was developing.[19] As usual, a sort of apologia accompanied the papers. In this case it was not his formulaic apology that the "hypothesis might be premature," as in some of his work on layering where the preliminary hypothesis he had formulated seemed to find "striking confirmation" in the scant available data. Here the apology was for publishing scientific communications that contained no novelty in their elements but rather took their novelty from the understanding generated by the systematic presentation of material widely distributed in the literature.[20]

The subjects of these two papers were topics of core significance for meteorology: the constitution and therefore the microphysics of clouds, and a systematic treatment of the different morphologies of ice formation from water vapor in the atmosphere. These were things about which all meteorologists cared and were central to the thermodynamics study of a moist atmosphere. The main line of development was once again Helmholtz, then Hertz, then Aßmann and Bezold, with Bezold making his last significant contribution in 1899. Wegener was again in the position of updating general ideas, based on results from balloon flights and aerological investigations conducted in the first decade of the twentieth century.

Hertz had developed, and von Bezold extended, a picture of an atmosphere with essentially four stages or states. The first of these was a dry atmosphere, where there was only a little water vapor. Then came the rain stage, where saturated water vapor and liquid water coexisted. Third, there was the condition of the atmosphere in which hail could develop—near the triple point of water, where water vapor, water, and ice were all present at the same time. Finally, there was the snow stage, where there was only "ice vapor" and crystalline ice. Hertz had the view that an ascending mass of moist air would cool adiabatically as it expanded and that the water content would pass through these four stages successively. Having reached the snow stage, the ascending cloud would begin to descend.[21]

In twenty-first-century meteorology, the process under discussion is known as the "precipitation staircase." One begins with air with increasing moisture content. Submicroscopic particles, called cloud condensation nuclei, accompany the air as it ascends. These might be salt crystals from extremely small water droplets, created in the burst of tiny bubbles on the surface of the ocean, or extremely small particles condensing from gas. As the air ascends, it expands; as it expands, it cools. As it cools, it humidifies, eventually reaching saturation, when it can hold no more water vapor. At this point, depending on the altitude and temperature and pressure conditions, it will condense into one or another form of precipitation. As some of the precipitation elements grow at the expense of others, the larger drops begin to fall. Sometimes the whole cloud descends, and this begins the "down staircase" where the descending air is compressed, warmed, and dried, finally with all the water vapor evaporating away.[22]

Aerological research, as well as laboratory research in the twenty years before 1910, had shown that it was possible for water in the vapor phase to exist in a supercooled form, where it was well below the normal freezing point of water but not yet formed into ice for lack of something on which to nucleate and grow. Bezold had gone into this topic, and the part of his explanation that Wegener was most drawn to was that moment when, in a cloud of supercooled water, the oversaturation was released and precipitation drops began to form.[23] Bezold had imagined a mechanism whereby a sudden change in the pressure and temperature regime in a cloud would force the condensation and solidification of supercooled water. The mechanism was complicated, and we will not go into it here, but Wegener found it to be physically impossible.[24]

Wegener had a different idea for how water might pass from the supercooled phase, in an oversaturated cloud, into ice crystals. He saw that the simultaneous existence of supercooled water and ice particles would establish a variation in the vapor pressure between the water droplets on one hand and the ice particles on the other. Because the saturation vapor pressure is greater in the immediate vicinity of the water droplets than in that of the ice particles, over time, the vapor pressure gradient would draw water molecules away from the supercooled water drops and toward the ice crystals. These ice crystals would grow until they became heavy enough to fall from the cloud. Depending on the temperature gradient, they would reach the surface as either snowflakes or raindrops.[25]

Wegener imagined he was making a physical correction to an accepted process, without significant novelty. However, in this "correction" of Bezold's work on supercooled water and clouds, Wegener had in fact made one of his most memorable discoveries. The process of migration of water molecules from supercooled droplets to ice crystals in cold clouds is today known as the Wegener–Bergeron–Findeisen (WBF) theory of precipitation in cold clouds. Developed initially in Wegener's paper and expanded in *Thermodynamik der Atmosphäre*, it provided inspiration for the Swedish meteorologist Tor Bergeron (1891–1977) and later for the German meteorologist Walter Findeisen (1909–1945). While the modifications made by both of these other scientists were substantial, Bergeron was quite insistent that the inspiration for this development was clearly laid out in Wegener's work, and indeed one can see it specifically formulated in Wegener's article published in *Meteorologische Zeitschrift* in 1910.[26]

In addition to this clarifying and surprisingly novel work on the ice phase of water vapor in the atmosphere, Wegener also reorganized thinking on the size of the particles making up clouds. William Thompson, Lord Kelvin (1824–1907), had long since argued that given a certain partial pressure of vapor, a water drop must be of a particular radius: drops that are slightly larger will grow by condensation, and drops that are slightly smaller will evaporate. This equilibrium between vapor and small drops is not simple and actually involves drops of different sizes and concentrations. Kelvin had been followed by Joseph Perntner (1848–1908), Julius Hann (1839–1921), and more contemporaneously Victor Conrad (1876–1962). Wegener pointed out, in his parallel paper entitled "Die Größe der Wolkenelemente" (The sizes of cloud elements), that all of these celebrated attempts to study the effect of surface tension on the equilibrium radius of a drop at a relative humidity of 100 percent had failed because they had not considered supercooling. In supercooled clouds, the relative humidity was regularly higher than 100 percent, sometimes 110 or 120 percent, and (rarely) even as much as 400 percent. Under such conditions, the radius of the droplets would be substantially different, and the equilibrium, as the water moved from the supercooled aqueous phase into the ice crystals, would be very unstable. If the volume of a raindrop at 100 percent humidity were 4.2×10^{-9} cubic centimeters, at 400 percent humidity it might be as small as 1.7×10^{-21} cubic centimeters.[27]

The issue of the size of the cloud elements in the supercooled cloud is important because the very small droplet sizes in the exaggerated relative humidity of a supercooled cloud give conditions under which "hygroscopic kernels" (as Wegener described them) would have a significant effect on formation of ice crystals. It would allow not just the very small salt crystals that his friend Lüdeling had observed in sea air to function in this way as condensation nuclei, but even individual (nitrous) gas molecules. This is the so-called ionic theory of nucleation, and it plays an important role in meteorology even today.[28]

The question of condensation nuclei had been in meteorology for a long time, and his mentor and supervisor at Lindenberg, Aßmann, had made some famous observations in 1884 on the mountain outside Berlin called the Brocken, to determine (a) whether the microstructure of clouds and mists was bubbles or drops, and (b) whether or not these drops formed around minute dust particles called "Aitken nuclei," after the Scottish meteorologist John Aitken (1839–1919), who had done some experiments with condensation of steam in dusty and clean air. Aßmann walked up and down the mountain slope with a 400-power microscope, capturing droplets on microscope slides and examining them with a micrometer under oblique illumination to determine their diameters and composition.[29] Aßmann did not observe nuclei, even though he could have seen particles down to 0.0005 millimeters, and therefore assumed they played no role. Wegener's work indicated that actual "Aitken nuclei" might have diameters of only a few millionths of a meter. Thus, Wegener's work on droplet size and "hygroscopic kernels" extended and corrected Aßmann as well as Bezold, Kelvin, Conrad, and Perntner.

Wegener's invention and novelty were not limited to these two efforts, of course. If one reviews Wegener's twenty or so scientific papers published from 1908 through 1910, other than results of the Danmark Expedition, they are all on some aspect of atmospheric physics. In the latter topic he was consciously working to establish himself as the leading theorist of the "Berlin school" of meteorology and atmospheric physics. To do so, he made a special point to employ and feature data collected by the Berlin group, though supplemented with other data sets. He also made an effort to update and correct papers written by both Aßmann and Bezold going back a number of years. He was extremely polite and cautious in these corrections, never saying that "Bezold was wrong" or that "Aßmann got the wrong answer." Rather, his approach was always to say that new data provided by aerology had made thus and such a conclusion no longer tenable, or that such and such an idea needed to be modified.

In addition to his corrections of those senior academics involved in his training and supervision at Berlin, and thus the advancement of their work as well as his own, he was very much involved in bringing together and synthesizing the conclusions of papers by a variety of different investigators, written in the time when he was away in Greenland. Here he disclaimed any novelty in the uncovering of basic empirical facts, casting his contributions as possible avenues of theoretical advance, falling out of the work of keeping some subject matter up to date. Even in his most exhilarating papers on the chemical differentiation of the atmosphere and the new fundamental layer boundary, his stance was always that this material was available in the literature for anyone to see, though no one (save himself) had seen it yet.

One recalls that in his probationary lecture at Marburg the previous year he had spoken of the alternating periods in meteorology between the collection of data and creation of new instruments on the one hand and the theoretical integration in advance of the subject on the other. He had characterized the first decade of the century, following the lead of Köppen, as a time of the collection of new facts—and most of these facts concerned, of course, aerology.

Yet in 1910, as he was increasingly aware, theoretical forward motion of atmospheric physics was everywhere evident—*except* at Berlin and Lindenberg. Aßmann had used up an enormous amount of money building the station at Lindenberg, with little to show for it.[30] He had tried (unsuccessfully) to get Wegener to join him again at Lindenberg and produce theoretical results that Lindenberg could claim as its own.

In donning the theoretician's mantle, Wegener was aware that there were others simultaneously vying for the prize. The work of Ficker, as well as that of Margules at Vienna, was built on the same scale as his own: global atmospheric structure and atmospheric motion. Wegener's historical sense placed him self-consciously in that generational group that would carry out the next set of theoretical advances. His choice of thermodynamics kept him close to his own empirical work and to the greatest scientific successes of the Berlin school.

Wegener was positioning himself in his published work and in his lectures to build a career leading to a professorship at the earliest possible moment. His reasons for wishing this were as much financial as anything else; he was virtually penniless. His lecture fees were almost nonexistent, and his parents could provide almost nothing toward his support. With the help of Richarz, he made a successful appeal in April, May, and June to the Ministry of Education for a supplemental stipend of 1,500 marks per year; there was some humiliation involved, as he had to submit his father's tax returns to show that his parents were unable to support him.[31] In return for this stipend,

Wegener took on some "additional duties" in the supervision of Marburg's Astronomical Observatory atop the Physical Institute, with its modest refracting telescope.

The stipend he arranged in June 1910 from the Ministry of Education was a great help, even though it was less than half of the salary that Aßmann had offered to entice him to Lindenberg. He still needed more money, as evidenced by the correspondence back and forth with the Danmark Committee requesting the use of lantern slides and permission to give paid public lectures, for which he had to obtain permission each time under the stringent conditions set down by the now-deceased Mylius-Erichsen. He was giving so many of these lectures in June and July that he asked if he might hold on to the slides rather than mail them back each time.[32] He fretted at the loss of time these lectures entailed, but he needed the money.

Along with his stipend and his fees for public lectures, he wrote a few popular articles for which he picked up honoraria.[33] His final source of income, as he tried to stitch together a living to support his life as a researcher, was the remainder of his Danmark Expedition salary, now down to a trickle, and constantly in need of extension. In order to keep the money flowing, he had to make continuous progress on the expedition reports. There was the remainder of the Danmark data to publish, and in the summer of 1910 he worked continuously at the next full volume on the station observations from the terminal at Danmarkshavn, on which he had to use extreme care because they were the reference values for all his other comparative measurements and would be the data set most eagerly sought by other investigators, other than the aerology. This work was dreary and time-consuming, very much like his dissertation at Berlin—the reduction of values taken under very different conditions, and requiring absolute vigilance in making sure, as the fragments or sections of this manuscript traveled back and forth to the press in Denmark, that the values were set in type correctly.

Wegener could not concentrate wholeheartedly on the thermodynamics of the atmosphere in summer 1910 because he was constantly pulled away from this work by moneymaking schemes. He obtained some relief from this with a series of balloon flights in association with his newfound friend, the physicist Karl Stuchtey, who had been sitting in on his lectures on atmospheric physics. They began to sketch out a plan to build an instrument to be used aloft to measure the albedo (the percentage of sunlight reflected rather than absorbed by the surface) of Earth and of cloud surfaces, as a way of computing the thermal regime in the middle of the troposphere. They built a prototype that summer and used it on several flights. The instrument allowed them to observe which wavelengths of light were being reflected or radiated back from a cloud and thus make inferences as to the composition of the cloud elements. In aerology, as in astronomy, much work was done by indirect means: the shape of a shadow, the color of light; all these optical phenomena and many others were employed to squeeze out every bit of information concerning the physics of the clouds.[34]

As the summer semester came to an end, he worked on the meteorological records for Danmarkshavn and on his thermodynamics, stealing whatever time he could from the former to devote to the latter. In doing so, he put himself in conflict with his own deadline for completing the expedition work. He wrote to the Danmark Committee on 3 September saying that his hopes to have the manuscript of the station measurements completed by the end of September were now dashed by the pressure of work, but that he would come to Copenhagen toward the end of the month with as much of the manuscript as he had been able to complete, and would send the rest as rapidly as he could.[35]

He would, of course, easily have finished the reduction of the observations had he not been trying to write a book on thermodynamics at the same time. By the arrival of the equinox he had all but the last part of the measurements in print-ready form and had also completed a very rough first draft of his thermodynamics text. He wrote to Eric Henius (1863–1926), now head of the Danmark Expedition Committee, on 24 September announcing that he would arrive on 2 October to help supervise the typesetting for the tabular material.[36]

On 26 September he wrote a postal card to Köppen and asked if he might forward his manuscript on the thermodynamics of the atmosphere so that Köppen could look through it, and then he asked if he might stop by and visit him in Hamburg on his way to Copenhagen, in order to hear his opinion. He planned to stay in a hotel and take the tram to the Köppen house in Großborstel.[37] Köppen wrote back immediately and asked that Wegener send the manuscript along, and he invited him to stay at the Köppen house for an extended conversation on his way to Copenhagen, arriving on Friday and staying over until early Sunday.[38]

Wegener arrived in Hamburg in the later afternoon of 30 September, and Köppen greeted him warmly; the men immediately sat down and began to work. They worked through the manuscript (about 100 sheets at that time) page by page, Wegener defending and explaining his positions and Köppen constantly rising from the chair to bring in offprints of publications that bore on the questions in Wegener's book; Köppen provided as well a stream of alternative viewpoints and interpretations. Else Köppen, who two years before had gone to her first formal dinner as Wegener's partner, was at the time of this meeting between Wegener and Köppen eighteen years old and in her last year of school. She recalled this visit some years later.[39] She remembered sitting across the room from them, trying with half her attention to write a school essay and with the other half to hear and see everything Wegener was doing; she couldn't stop looking at his eyes. Wegener's skin was still very dark; he spent as much time outdoors as his work would allow, and the balloon excursions were an excellent way to become sunburned. His eyes were a pale blue, the same piercingly gray-blue eyes she had noticed two years before. She said that it was interesting to watch him express his thought processes through his eyes. When he began to develop a line of thought, she said, he would look calmly into the distance as he talked, but then at some point his eyes would begin suddenly to flash from side to side as if he were looking around the room for whatever idea had just occurred to him. As he looked around the room, his eyes would occasionally land on her, causing him to pause in his thinking and his face to dissolve into a bemused smile.[40]

At the family meal times, in which Wegener was invited to join, he sat near Else and bantered with her and teased her. She said he was in an infectious good humor, but it took all her concentration just to keep up with him as he launched sally after sally in her direction, flirting with good humor.[41] After dinner on both Friday and Saturday night, the men repaired to the worktable in the front room and continued their conversation far into the night. They were congenial companions in this way. Köppen was a night owl, and Wegener typically slept little, often working in his rooms at Marburg far into the night until eyestrain and exhaustion overtook him.

The meeting was decisive and life altering for both men, and from this point forward in the story of Wegener's life there is a sense in which almost every major work by Wegener should be considered to have been produced with the aid of Köppen's advice or collaboration. Here the theoretical daring, physical intuition, and fierce energy of the young man met the sagacity, vast knowledge, and experience of a man already active in every part of meteorology for more than forty years. There was no development Köppen did not know about, no one he had not met, and no information resource of which he was unaware.

They entered into an absorbing correspondence following this meeting which over the next year or so dominated the intellectual life of the younger man and much occupied the older. Köppen's knowledge of the literature would keep Wegener from making many mistakes and deepen and broaden his grasp of the problems he had so brashly undertaken to write definitively about. At Marburg, Wegener had had good access to physics books and the more recent monograph literature in meteorology. Köppen gave him access to a much larger universe of thinking. Between Köppen's private library, which contained several thousand volumes, and the library at the German Marine Observatory in Hamburg, Wegener suddenly had access to vast resources of hard-to-get publications that he could not otherwise have obtained; without Köppen's guidance, he would never have even encountered them.

Wegener returned to Marburg and worked furiously to finish the temperature data from Danmarkshavn. By 12 October he was done, allowing him to transmit the manuscript to Copenhagen.[42] This task completed, he could turn again to the thermodynamics. As he began to work through the notes he had taken in Hamburg with Köppen, all kinds of new questions occurred to him, and he wrote to Köppen on 28 October asking for more assistance. Köppen once again invited him to Hamburg, an invitation he accepted with alacrity, with Wegener arriving on 4 November, spending a late night with Köppen going over the manuscript once more, and returning to Marburg the next day.[43] Köppen loaded him down with books and reprints and promised more, as they should be needed.

Wegener was anxious to proceed with the revisions to his ideas, but he still had the endless correspondence to deal with regarding the expedition results. The expedition committee was becoming restive over the lack of progress on the magnetic and electrical observations that Wegener had passed on to others—coworkers whom he had not been able to contact in the long vacation of late summer when all the scientists were in the field. Perhaps to assuage the impatience of the Danmark Committee, he announced to them that he had enlisted a high school teacher in Marburg, Walther Brand (1880–1968), as a doctoral student, with the understanding that his doctoral work would consist of an analysis of the measurement of atmospheric pressure at Danmarkshavn.[44]

He worked on his correspondence with his various coauthors all that day and the next, and by Sunday evening, 6 November, he was able to write to Köppen: "Esteemed professor Now I'm finally back on the tracks having finished with my correspondence and having already smoked 10 cigars, and tomorrow morning early I will begin to work through the rich harvest which my 'expedition to Northeast Hamburg' has brought back to my home."[45] He had additional news for Köppen: he had a provisional promise of publication from a good house, Johann Ambrosius Barth in Leipzig, a contract that would allow him at least sixty figures in the text and thirty tables. Wegener continued, "Certainly I will in short order be pestering you with new questions. But I must pause to thank you, esteemed professor, and your family from the bottom of my heart for all your wonderful hospitality for me and for all the trouble and work that you have had on my behalf. I have a terribly guilty conscience that I shall never be able to repay you for all the wonderful friendship flowing in my direction."[46]

Wegener's life at this time, viewed at the distance of a century, is not so different from the lot of any university instructor or assistant professor at the beginning of his career, especially one like Wegener in a small school, with few students and insufficient library resources. Köppen's friendship and offer of aid were godsends. Köppen was a generous and large soul, and his nickname, the "Nestor" of meteorology, was here proven apposite, with his unlimited appetite for conversation, collaboration, and intellectual adventure. Someday perhaps a biography will be written of this remarkable man.[47]

Having returned to Marburg with every intention of beginning work on the revisions to his thermodynamics, Wegener found himself overwhelmed with the projects already started. At this point in his life he sometimes appears to be that apocryphal young man who mounted his horse and rode off in all directions at once. On 14 November Wegener wrote to Köppen, "Well it's all chaos here and I've come to a dead stop with the thermodynamics. Once again I had to put it aside and devote myself solely to other work."[48]

The other work in question was a lecture he agreed to give at Göttingen titled "Critical Investigations into the Nature of the Atmosphere above 70 km Altitude," which he had yet to write. In order to complete it, with the required books not in his own library, he would have to write letters to Potsdam, Copenhagen, Göttingen, and Leipzig, as well as to Hamburg. He was very keen to give this lecture and had already been to Göttingen to visit and lunch with the geophysicist Emil Wiechert (1861–1928), who had a role in selecting the team that was to be in the University of Göttingen's expedition to Lapland and Spitzbergen, to coincide with Filchner's expedition to the South Pole. Wegener had gone to Wiechert to sketch out the scientific plan that he and Stuchtey would pursue if chosen, investigating the spectra of the aurora at different altitudes. "If only I had 20,000 marks!"[49]

Unable to actually work on the book, Wegener wrote long letters to Köppen over the next few weeks, so that he might at least discuss a variety of questions with someone, and these letters make interesting reading. From no other period of Wegener's life does there survive such detailed scientific correspondence, with so many hints of the sources of his inspiration. When we considered his work on cloud elements and the ice phase of water vapor in the atmosphere, which led to the theory of precipitation in cold clouds which today bears his name, it was not evident what exactly was the kernel of the idea for this work, although he did say that "fundamentally it was a problem in molecular physics." Only in his letter to Köppen on 14 November do we discover that the detailed explanation of the phenomenon in question he had learned, in preparation for his own lectures at Marburg, from the work of Otto Lehmann (1855–1922), who had been Hertz's successor at Karlsruhe. Lehmann had worked on the state of matter called "liquid crystals" and would be for this work continuously under consideration for the Nobel Prize from 1913 until his death nine years later. He was also a pioneer in microphotography of growing crystals, and he had done some concentrated work on the growth of snowflakes and the physics thereof. Wegener sketched out Lehmann's ideas for Köppen and noted that while physicists were quite familiar with this work, he had yet to see it referenced in any meteorological literature. He asked Köppen, "Can this be entirely unknown to meteorologists?"[50]

Only three days later, on 17 November, Wegener wrote another long letter, remarking that the thermodynamics still languished in a corner but that he had now written twenty-five pages of the paper he had not yet begun on the fourteenth. This letter also contained his speculation that the outermost layer of Earth's atmosphere was almost entirely a new element, geocoronium, an idea that had already occurred to him but to which he now gave a name. He solicited Köppen's opinion about this notion, with the idea that the separation of the gases in the atmosphere, with the heaviest closest to Earth and the lightest at the outer reaches, was exactly the sort of phenomenon one observed when one centrifuged atmospheric air at high speeds, with the heaviest gases being thrown the farthest.[51]

While most of his correspondence, at least the part in 1910, was Wegener asking for and receiving assistance from Köppen, he was occasionally able to do something for the older man. Responding to a letter from Köppen, Wegener wrote on 28 November to advise him on the perils and prospects of polar meteorology. Filchner had asked Köppen what sort of atmospheric investigations might best be pursued in Antarctica, and Köppen wrote to Wegener to solicit his advice. Wegener responded in great detail. He counseled Köppen to tell Filchner that winching up and down kites by hand just wasn't worth the time and effort it took to get the results: "This was the hardest work I have ever done and ever expect to do, and after all the work you have done you find out generally that the drive mechanism has frozen or that the pen has become stuck or did not write or that the record has been erased by driving snow." Filchner would be much better off making a judicious and restricted use of rubber balloons, as the parchment balloons were too fragile. Wegener's disgust with the amount of work he had performed for such meager results in Greenland was matched only by his despair at the time consumed in editing and publishing it.[52]

In this period of their correspondence the dialogue between Wegener and Köppen had a consistent structure. Wegener claimed to have found a new element in Earth's atmosphere and thought about calling it "Geo-coronium," and he asked Köppen for his advice. Köppen asked him, "Where's the spectral data for the isolation of atmospheric gases at different altitudes?" Wegener went out and found it and dutifully responded. Wegener said that he thought that there was a layer boundary at 225 kilometers (140 miles) just as there was at 70 kilometers (43 miles). Once again Köppen wanted to see the evidence. Wegener said that he had uncovered the mechanism for precipitation in cold clouds, connected with the way ice crystals form in oversaturated environments, and Köppen pointed out to him that this had been under discussion for a long time and sent him another set of papers to read on the subject.[53]

Of the many directions in which Wegener was riding off all at once, the direction he seems to have favored in October and November of 1910 was his attempt to build a cross section of the atmosphere from the surface up to 500 kilometers (311 miles). He was by now convinced that the outermost portion of Earth's atmosphere consisted almost entirely of a new element, his conjectural geocoronium. Dmitri Mendeleev, using the same logic that led to the discovery of germanium, had in 1904 proposed an extremely light element of atomic weight 0.4 and suggested that this element was identical to the coronium detected in the Sun's atmosphere.[54] The question remained open whether what Wegener called geocoronium was a terrestrial appearance of the same gas already seen in the Sun's atmosphere. While there were many persuasive reasons to accept this identity (for example, comets passed through the coronium and meteors through the geocoronium without burning), still the spectral lines were distinct near 530 Å (530.2 Å, 530.4 Å, 531.5 Å, and so on) for coronium and 557 Å for geocoronium.

On 1 December Wegener wrote to Köppen again.[55] He told him that he had just given a colloquium in Marburg on this subject of a novel atmospheric gas shell, and that Emanuel Kayser, distinguished geologist and dean of the university, had mentioned after the lecture that geologists knew that a lot of meteors contain hydrogen and therefore he, Wegener, would have to make a stronger demonstration that the combustion of a meteor could describe the composition of the atmosphere at any given altitude, especially at the higher altitudes. Wegener said he would have to go back into the literature and find some confirmation; he was nevertheless certain that the meteors could not contain enough molecular hydrogen to create all the trails that were observed.

He was thoroughly convinced that there was a new element revealed in the upper atmosphere by these spectral lines. He noted that the spectrum of the upper reaches of the polar lights, at the "twilight limit" (the blue light limit at 225 kilometers), was monochromatic and gave a hydrogen signature—as did about half of the shooting stars. Yet at much greater altitude there was this 557 Å line, which was not exactly the same spectral line as the Sun (although it was impossible to rule out some modification of the spectral signature of coronium). Wegener knew that many gases had very different spectra depending on temperature and pressure relations and positive or negative electricity. It was even possible that the hydrogen trail of meteors was actually a spectroscopically recorded interaction with the green polar light. He wanted, nevertheless, to have geocoronium accepted as a "working hypothesis" until further spectroscopic studies could be conducted.

Here, as in many of his papers in these years, he took everything one step further than his peers and formulated a law, a relation, or a structure, in each case as a "working hypothesis" pending further research.[56] These speculations functioned as a series of moves that might establish his priority in some discovery, or at least a share in it. Wegener's impulse was to determine what was qualitatively plausible and to propose a measurement protocol or test that would lead to either certainty or rejection.

Wegener wanted to publish this material in a professional journal, as much as anything else to determine its suitability to be included in his book on thermodynamics of the atmosphere, which he had guiltily ignored finishing in order to pursue these other lines of investigation. The notion that these speculations about the outermost shell of Earth's atmosphere might have a prominent place in such a text was important to him. As he wrote to Köppen, "I put myself in the position of drowning in work, but it's also time for me to get ready finally to work on the long-promised thermodynamics. I have promised the press to deliver the manuscript before the Christmas vacation. My contract specifies that. Over Christmas I will also work on my own things."[57] One of the reasons that he wanted to incorporate this work on the outer reaches of the atmosphere into his textbook was that he saw this text not as "his own things" but a collection, principally, of other people's work, and he thirsted for some novelty.

There was much time in December for Wegener to work on the thermodynamics. He had completed the manuscript on the chemical differentiation of the atmosphere, especially the upper atmosphere, and including his hypothetical geocoronium, and sent it to Emil Wiechert at Göttingen, who was then editing *Physikalische Zeitschrift*. He had also sent a version of the same paper to *Meteorologische Zeitschrift* and was prepared to publish shorter versions in various other periodicals in order to publicize his find. But for now, the idea in its several forms was off his desk.

The only course he was teaching that term was a practicum on astronomical position finding on expeditions, something that he had done many times and could perform by rote without new work. The weather at that time of year was not conducive to ballooning, and he had a momentary respite from the lecture circuit. Under these advantageous and unusual circumstances he threw himself into the composition and revision of the remaining portions of the book. On the solstice he wrote to Köppen, "I have put a lot of time in on the design of the physical part of the work which I hope was not in vain, but it is such that the size of the book has now grown in consequence, and there are many chapters which you would hardly recognize any longer . . . while I'm in Berlin I plan to finish it and send the manuscript to the publisher."[58]

With his manuscript nearly complete and only a few references to chase down, he was ready for a vacation. He was to travel to Berlin to spend Christmas with his parents. Before he could leave, however, there was something he still had to do. He had a present for the Köppen family and took great pains to pack it with care. Köppen's daughter Else remembers that a few days before Christmas 1910 a very large crate arrived with the legend "Stop!! Don't be curious! Don't open till Christmas Eve." The crate had come from Marburg and within it was a lot of excelsior surrounding a carton, and inside the carton, packed in cotton, was a still smaller box of the kind used for photographic slides. It was sealed all around with tape that bore the legend "open only by lamplight!" Else said she could not bear the suspense and immediately pulled off the tape and opened the box. At the bottom of the box there was a scrap of paper that said, "I told you that you shouldn't be so curious!" She said that she and her brother laughed until they cried. She was almost sorry that she had done it, but the next day another package arrived addressed to her and the rest of the family, containing the German translation of Achton Friis's book about the Danmark Expedition, in which, as we have already seen, Wegener figured prominently. Else was overjoyed: "I buried myself in the adventure in Greenland and for a long time saw and heard nothing else."[59]

Wegener's Thermodynamics of the Atmosphere

By the time he had sent his package to Hamburg, the book manuscript had reached the final form, and this is therefore an apposite point for a discussion of its structure and contents, as well as its intellectual implications for both Wegener and meteorology.

As an entryway to discussion of Wegener's thermodynamics of the atmosphere, let us consider first where, in that period of time, such a textbook might have fit into the state of the science. To put this question in context, let us consider a remark made by the dynamic meteorologist Edward Lorenz (1917–2008) when asked in 1995 to provide an essay on the "evolution of dynamic meteorology" for the seventy-fifth anniversary of the American Meteorological Society: "What constitutes the *state* of an evolving scientific discipline—dynamic meteorology or something else—at a particular moment in history? Does it encompass the ideas taking shape in the minds of the most forward-looking scientists? Does it include only those ideas that have found their way into the refereed literature? Is it the knowledge that is regularly imparted in the classroom in institutions of higher learning, available to all who have the opportunity to enroll? Is it limited to the material appearing in textbooks, available to a still greater audience?"[60]

Each scientific culture has its own answer to this question, but our concern here will be with Germany in the early twentieth century. In bound-book publication in the sciences in Germany at that time there was a quadripartite hierarchy of certainty, and at the top were the *Handbücher*, the reference volumes, containing material deemed certain, such as physical constants and well-established calculation techniques. Slightly less solid, but still generally dependable throughout, were the *Lehrbücher*, or textbooks, a designation given to freestanding introductions to a subject, whether this be a scientific theory or instructions in technical practice for beginning students. Closer to the research front and farther from absolute certainty were *Vorlesungen*, or "lectures," associated with the lectures given in a specific course of study, delivered by a specific professor, and deemed to be the course guide for students enrolled in his course in that subject in that year. Finally, there was just *ein Buch*, a monograph, with no specific rank with regard to certainty.

Wegener's thermodynamics of the atmosphere was a hybrid product that fell between these genres. It had some of the characteristics of a textbook but also read like a series of lectures. It contained well-confirmed results suitable to a textbook for introductory students, as well as frontline results in the form of lectures incorporating summaries of recent monograph literature. The book also contained previously unpublished and therefore, from the standpoint of scientific culture, unverified material. Wegener was quite aware of this hybrid character and the extent to which the book was a *resumé* of themes he had absorbed in graduate school and during his time at Lindenberg, melded with his experiences in Greenland and his published papers in aerology since his return from Greenland. His specification of his desired audience (in the foreword) included active meteorologists whom he hoped to inspire, physicists who might learn about an exciting new field of investigation, aeronauts who might benefit from the work on clouds, and even the "educated lay public," with the latter group able to find, he hoped, "certainly not a popular but rather an easily comprehensible presentation of the most recent branches of meteorology."[61] Neither the book's hybrid character nor its various and probably immiscible audiences can convey the freshness and idiosyncratic character of this book, in so many ways the external realization of his own scientific imagination and the perpetually ingenuous enthusiasm that drove it.

Wegener's chosen standpoint was always that of cosmic physics. This book obviously was not a complete cosmic physics of Earth, which, on the model of Svante Arrhenius's *Lehrbuch der kosmischen Physik*, would have covered the physics of everything from the stars down to the core of the solid Earth and would have weighed in at 2 kilograms (4 pounds), with 1,000 pages of text.[62] Nevertheless, Wegener, the astronomer and cosmic physicist, started on page 1 characterizing Earth's atmosphere as only a single example of a planetary atmosphere within the solar system. Of the eight planets then known, seven had atmospheres, and Earth belonged to a family that included Venus and Mars, but not the gas giants Jupiter and Saturn, nor Uranus or Neptune. Yet even before he can get around to characterizing Earth's atmosphere and its composition, Wegener is already discussing coronium and geocoronium as the most likely gases to be found in interplanetary space. Then attention turns to Venus, Earth, and Mars, which make a nice trio with regard to cloud cover and albedo: Venus entirely cloud covered, Earth about half cloud covered, and Mars with no cloud cover, but all with detectable atmospheres. Here again, by pages 3 and 4 we are involved with work he was currently doing with Stuchtey on measuring Earth's albedo. On pages 6 and 7, we see Wegener's theory of the chemical differentiation of the atmosphere featured in his initial presentation of atmospheric structure: "The researches of the author and others partition the atmosphere in three main sections: a nitrogen sphere from Earth's surface to about 70 km, a hydrogen sphere between 70 and about 200 km, and the still hypothetical sphere of an unknown extraordinarily light gas for which the name 'Geocoronium' has been proposed between 200 and more than 500 km altitude."[63]

Just as one begins to feel that this is not a textbook on the thermodynamics of the atmosphere but a *resumé* of the opinions of Alfred Wegener, the text veers away sharply

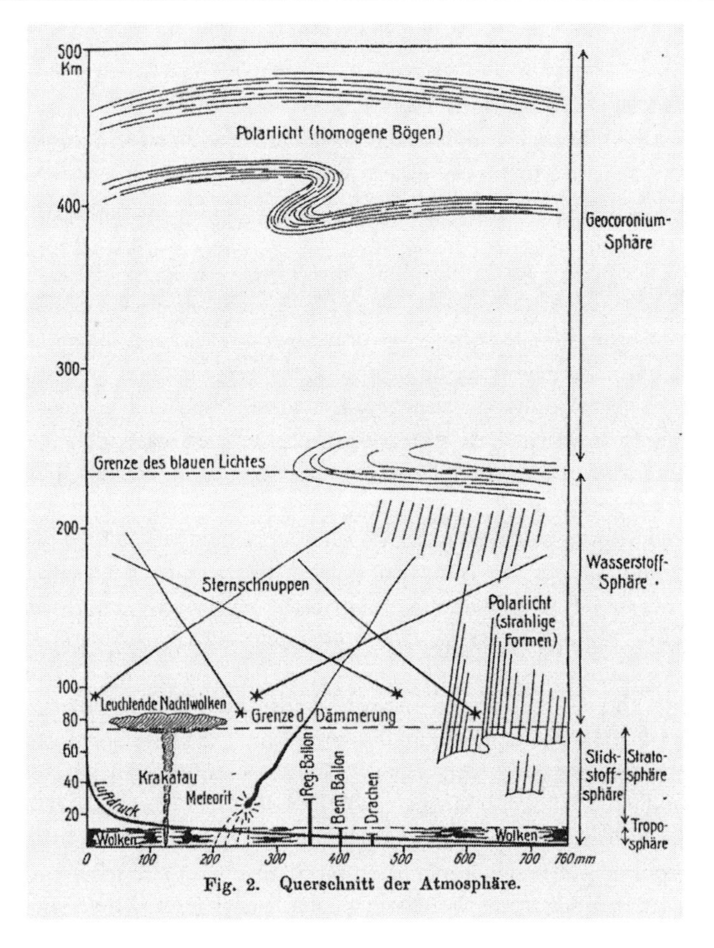

Wegener's cross section of the atmosphere, treating his conjectural geocoronium layer and his hypothesis of the chemical differentiation of the atmosphere as established facts. From Alfred Wegener, *Thermodynamik der Atmosphäre* (Leipzig: J. A. Barth, 1911).

from simple declarative statements and begins a twenty-page explanation of the physical reasoning behind the conclusions given in the first six or seven pages.[64] With great clarity, and with reference to the way the data were collected, how the data have been interpreted, and how we know anything at all about the upper atmosphere, he moved through what we can discern from the aurora, what we can discover from meteors and their trails, what we find out about the twilight limit, what the zodiacal light tells us, what we learned from noctilucent cloud studies after the volcanic eruption at Krakatoa, and finally what we learn from aerology. The chapter ends with an analysis of the gaseous composition of the atmosphere and the way measurements were carried out to determine the fractional volumes of the different gases.

When Wegener wrote to Köppen saying that he had put a lot of energy into the physical part of the book and that this had made it grow larger, he was speaking of this complete revision of the manuscript away from a didactic summary of received results, along with the mathematical formulas for obtaining them, and toward a reasoned and inviting conversation about how it is that physicists interpret evidence in order to characterize things that they cannot see and touch directly. This procedure—giving a result, explaining how the result was obtained, by whom, when, how, and with what degree of certainty—is a view of science from the ground level, a human view, not that of the "eye of God."

This treatment becomes quite striking when he moves on from this introductory chapter to the thermodynamics of gases. In two successive chapters he winnows his focus to the actual world. Beginning with the chapter on the general thermodynamics of ideal gases and the laws by which they are understood, he moves on to the general thermodynamics of real gases: how they flow, how we map them with isobars and isotherms, how we obtain temperatures and pressures, and how condensation takes place. There is room here for his theory of ionic condensation, for condensation nuclei, for all the work of Lehmann on snow crystals, accompanied by beautiful microphotographs. To this he added a clear physical consideration of supercooling and the phase diagram for water including its triple point. In every case there is direct reference to the literature from which he has taken it and an evaluation of the degree of certainty of the results.[65]

The evaluation of evidence becomes even more detailed as he moves from the general thermodynamics of gases, ideal or real, to the special thermodynamics of adiabatic processes. It is only at this point in the text, nearly one-third of the way through, that we begin to encounter thermodynamic reasoning: the concepts of latent and specific heat and the mechanical equivalent of heat, the principle of the conservation of energy, and a discussion of energy and entropy. The mathematics are sparse, nothing more than is absolutely required to understand the relations, with an absolute minimum of subscripts and superscripts. There are a few differentials, but not an integral sign in the full 331 pages. He covers all the essential topics, again referring to the original literature back to the middle of the previous century: convective equilibrium in the atmosphere, potential temperature, mechanisms for the achievement of equilibrium in a water vapor atmosphere, and adiabatic changes in air with condensation. Each of these topics is treated qualitatively wherever possible, with easy-to-read tabular data and very simple diagrams. This section concludes with a very extensive discussion of the mean partition of temperature in the vertical (with a discussion of föhn winds), a fifty-page discussion of all the different kinds of inversions and the implications of inversions for the vertical structure of the atmosphere and its discontinuities, and finally a consideration of the contrast between the turbulent phenomena of the troposphere and the undisturbed laminar flow of the stratosphere.[66]

The diagrams that accompany the text are not what we would see in a modern textbook, in which sophisticated graphics would be used to draw a picture of the atmosphere, perhaps with contrasting colors or different textures and sidebars, among other things. What we generally have are graphic presentations of real data: temperature data from real balloon flights, collated data of mean cloud heights in different geographical latitudes, and the oscillation of relative humidity at different altitudes over Berlin and its correlation with different classes of clouds from stratus at the bottom to cirrostratus 8 kilometers (5 miles) higher. The overall message here is that this is not a picture of the world itself; rather, it is a picture of what we think the world is based on the measurements we take with the instruments at our disposal, and known to a certain degree of accuracy but no more.

The final third of the book, section 5, is the longest and uses all the information previously developed about atmospheric structure, the behavior of gases, and the special thermodynamics of adiabatic processes, in order to elaborate a detailed picture, in three successive chapters, of the physics of clouds. First, a general morphology of clouds is given, and then the physical reasoning behind Wegener's theory of precipitation, with respect to both the special structure of water clouds and finally the special structure of ice clouds. This part of the book he illustrated with very good photographs of cloud forms of every kind, supplementing them with simple line drawings that schematize the contents of photographs.[67] It is in this last section of the book that Wegener goes deeply into his own theory of precipitation, the formation of raindrops, the speed of their fall, a reconsideration of ionic condensation, and the role of supercooling in forming large ice crystals, as a result of differential vapor pressure over supercooled water droplets and ice nuclei. Even here there is a long and sophisticated discussion of alternative hypotheses for droplet formation. The book ends with jarring abruptness after a consideration of optical phenomena in ice clouds: one senses that the pressure to get the manuscript to the press did not even leave time for a conclusion, as there is none.[68]

As striking as what is in the book is what is *not* in the book. It is the only book one is ever likely to see that has the word "thermodynamics" in the title but does not have any pictures of the Carnot cycle or Otto engine cycle, or of isothermal exchanges, or of pistons moving back and forth in rigid cylinders, or discussions of reversible and irreversible processes. The book is not about the atmosphere as a heat engine; it is about the atmosphere as an atmosphere, on planet Earth, without any reference to human contrivances, other than the instruments used to discover the physical values on which the behavioral relations of pressure, temperature, and volume are based. It is not the atmosphere explained thermodynamically with reference to human concerns and machines (though the information on clouds is very much directed to balloonists); it is the atmosphere explained thermodynamically with reference to what the atmosphere *does*, as it has been observed doing by real human beings, looking at real gases, at real elevations, employing real instruments, and employing still other instruments to arrive at sound inferences about things happening at elevations that humans have not yet reached. The only theoretical entities here are inferred molecular processes of otherwise unobservable entities, and these are (throughout the book) restricted entirely to phenomena of condensation. Even these theoretical discussions are premised on and referenced to experimental literature.

The tentative and idiosyncratic character of *Thermodynamik der Atmosphäre* notwithstanding, the meteorology it contained and the structure and the arrangement of its topics were the result of a close collaboration with Köppen and could therefore be expected to be reliable and useful. Wegener's book was certainly the most abundantly illustrated treatise on cloud forms which had yet appeared in a textbook of any kind, and it contained much durable, novel physics. The book, unaltered, would remain continuously in print for the next seventeen years and would give Wegener a permanent place in the history of meteorology. It was his first full-length book, and its completion gave him an immense sense of relaxation, satisfaction, and release. The release, however, was conditional: it released him only to deliver him back to the rapidly accumulating backlog of work on his very long (and growing) list of commitments.

9

At a Crossroads

MARBURG, 1911

Dans les champs de l'observation, le hasard ne favorise que les esprits préparés. (In matters of observation, chance favors only the prepared mind.)

LOUIS PASTEUR, Lecture, University of Lille, 7 December 1854

When Galvani's laboratory attendant saw the leg muscles of the sensitized frog quiver, he had discovered a fact; Galvani himself had not noticed it at all; but when this great scientist was told of the fact, there flashed through his brain a brilliantly intellectual thought, something altogether different from the gaping astonishment of the attendant or the unknown current passed along the frog's leg; to him with his scientific training was revealed the vision of extensive connections with all kinds of known and still unknown facts and this spurred him on to endless experiments and variously adapted theories.

H. S. CHAMBERLAIN, *Foundations of the Nineteenth Century* (1899)

A Christmas Gift

Following the New Year with his parents and Tony in Berlin, Wegener returned to Marburg, with a few days of rest before he resumed his grueling schedule of doing what he must do to make a living, and finding what time he could for his "own work." He was, nevertheless, in a refreshed and relaxed frame of mind, with the satisfaction that he had submitted a book manuscript (well, most of it) to the publisher by the agreed-upon deadline. He was no stranger to the "afterlife" of a manuscript and the endless proof corrections, modifications, and numbering and renumbering of figures and diagrams, but the conceptual part of the book was indeed finished.

Stopping by the Physical Institute on his return to Marburg, he encountered his friend Emil Take (1879–1925), the other young physical assistant besides himself and Stuchtey. Take invited Wegener into his office (their offices were side by side) to show him his Christmas present. It was the Jubilee edition of Karl Andree's *Allgemeiner Handatlas*.[1] Wegener wrote to Else Köppen a few days later:

> My next-door neighbor Dr. Take received the large-format atlas by Andree for Christmas. For hours on end we stared admiringly at the stunning maps. While we were doing so, an idea occurred to me. Take a look (please) at a map of the world. Doesn't the East Coast of South America fit exactly into the West Coast of Africa, as if they had formerly been continuous? It tallies even better if you look at the depth chart of the Atlantic Ocean and look, not at the current continental margins, but rather compare the margins of the continental shelves where they plunge into the abyssal ocean. I'm going to have to pursue this.[2]

This was the experience around which Wegener ordered his memories of the conceptual development that drew so much attention to him and to his career. It was a nearly solitary and qualitative moment of discovery in a time of relaxation and reflection, outside his own field of endeavor, and within a year of his return from a trip across the Atlantic from Europe to Argentina and back. This is the first part of the story he told when asked how the conception of continental drift had first occurred to him.

The point here is not how Wegener discovered continental displacements as much as how they discovered him—how he was fashioned by inclination, training, university education, and the contingent placement of his life in a certain stream of events to be the person through whom continental drift came to be. This entailed not just the raw intuition of similar coastlines but a robust and well-developed theory. If we are really to represent life as humans live it and not construct mythically comforting fables of the complete creative autonomy of individuals, we must attend to the fact that our biographical subjects' lives happen to them, just as our own lives happen to us.

The obvious place to begin our discussion of this "eureka moment" is with an examination of the map that Wegener found so striking. In Andree's *Atlas*, there is a world map toward the middle of the volume which takes up both pages, with two hemispheres conjoined in such a way that the Atlantic Ocean virtually disappears. The left-hand or American hemisphere has South America at its right-hand margin. The right-hand or Eurasian hemisphere has Africa at its left-hand margin.

Wegener was far from the first to notice this parallelism in the fit of the coastlines. It was known to Humboldt and figured in a variety of nineteenth-century theories of Earth, of which Wegener would become aware only much later. There are very few people in the world over the age of twelve who have had access to a globe or a decent school atlas and have not noticed this parallelism of the coastlines. So it was hardly this that was the kernel of his intuition.

What Alfred Wegener found striking here is contained in what he said in his letter to Else Köppen. To match continents at their sea level margins is trivial. The shape of a continental margin is an artifact of sea level, and sea level has varied by hundreds of meters over the past quarter-million years. This subject had been intensively studied since the middle of the nineteenth century in the context of the theory of ice ages, which required the periodic withdrawing of ocean water to produce ice caps and the rerelease of these melting ice caps back into the oceans.

What excited Wegener's interest and imagination was not the margins of the continents at sea level but the oceanic depth contours circumscribing the continental maps of South America and Africa. Andree's *Atlas* was the first atlas in Germany (in fact, the first atlas anywhere) to incorporate the full bathymetry of the Challenger Expedition into a map and to make this information widely available. The Challenger Expedition, under the command of the Arctic explorer George Nares (1831–1915) and the scientific supervision of Wyville Thompson (1830–1882), had spent the years from 1873 to 1876 dredging, sampling, and measuring the floor of the Atlantic Ocean. As often happens with such expeditions, the publication of results takes not years but decades. The scientists aboard *Challenger* were, above all, zoologists and most interested in their biological specimens; they worked this material up first. The depth data, collected by ship's officers as additional duties superimposed on their regular duties of running the ship, therefore languished. The thousands of depth measurements compiled over several years sat uninterpreted and unmapped as raw data throughout the 1890s. Systematic English-language publication of this material did not appear until

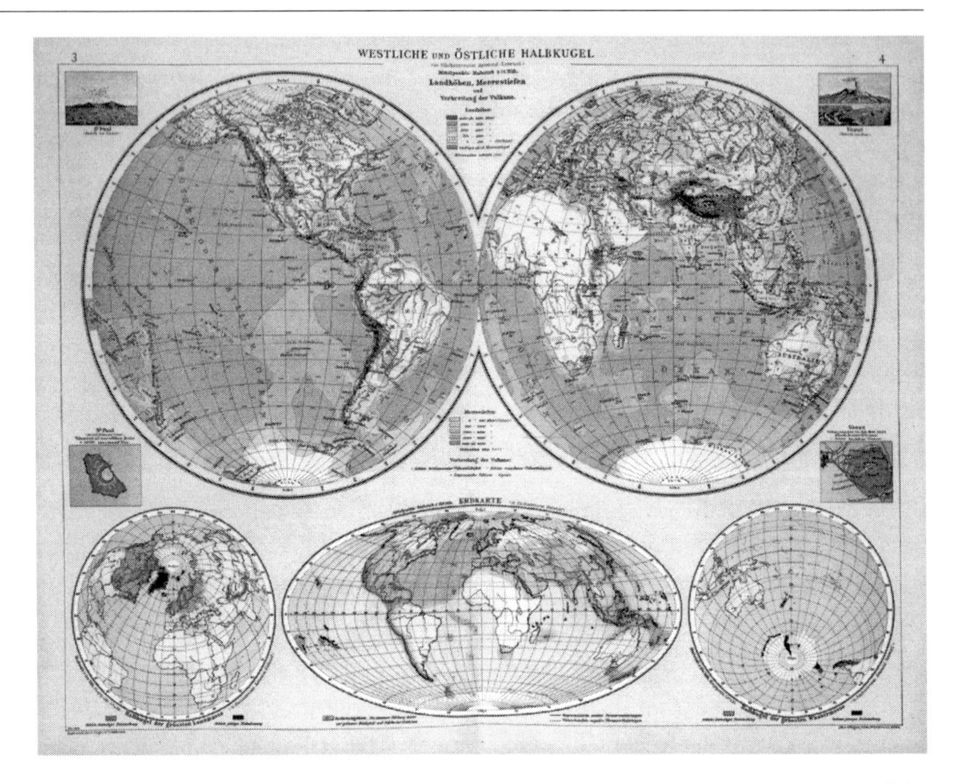

Plate from Andree's *Handatlas* which first brought to Wegener's attention the congruence of the coasts of South America and Africa. The projection is an "equal-area" map with two hemispheres, with the Atlantic cut out on the inner margins, bringing South America and Africa closer together. From Richard Andree and Albert Scobel, *Andrees Allgemeiner Handatlas in 139 Haupt- und 161 Nebenkarten; nebst vollständigem alphabetischem Namensverzeichnis*, 5th revised and expanded; Jubilee ed. (Bielefeld: Velhagen & Klassing, 1907). Photo courtesy of Botany Libraries, Harvard University.

1912, and much of the working up, cartographic and otherwise, was done by German scientists in the interim and made available to the publishers of Andree's *Atlas*.

These depth contours of the Atlantic continental margins of South America and Africa appear in the *Atlas* at depths of 200 meters (656 feet), 2,000 meters (6,562 feet), and 4,000 meters (13,123 feet), the last being contiguous with the floor of the abyssal ocean. Their outlines, as you may see in the map in the adjacent figure, are parallel and nearly identical in shape to the sea level margin of the continents: Africa has the same shape 200 meters, 2 kilometers (1.2 miles), and 4 kilometers (2.5 miles) below the surface of the ocean. This continuity of shape below the temporary position of current mean sea level meant to Wegener that these continental forms were not variable artifacts but structural features of Earth. It was this that he was determined to look into.

As determined as Wegener might have been to look into this matter of the parallelism of the coast of Africa and South America, his experience of the first week of January was, instead, the more or less customary tsunami of correspondence, proof correction, and composition which was his normal routine. The insight was not forgotten; Wegener did not forget plans for research. But he was an expert at postponing them in favor of still other plans and still other research.

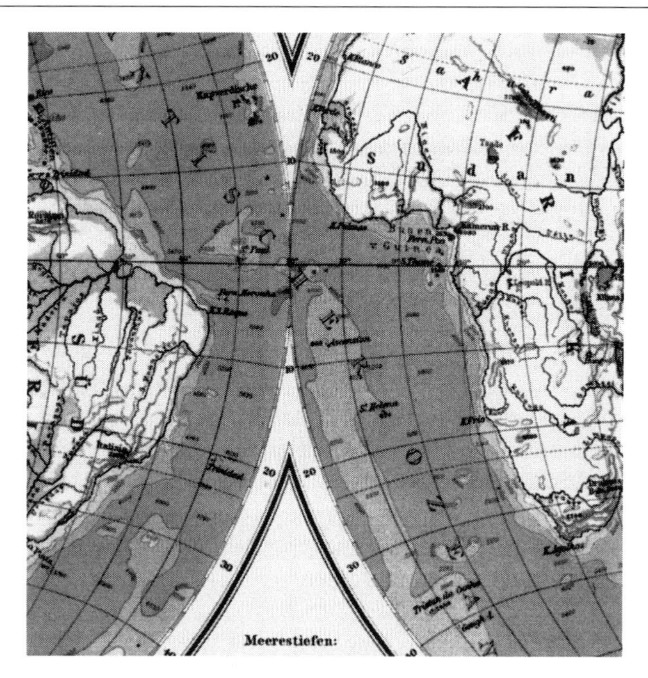

Close-up of the Atlantic margins of South America and Africa in Andree's Atlas, showing the depth contours at sea level, 200 meters, 2,000 meters, and 4,000 meters. The parallelism of these contours indicated to Wegener that the close match of the coastlines was not an artifact of sea level but a structural feature of the continents. He was especially struck by the fit of Cape Sao Roque (*K.S. Roque*) in Brazil and the Gulf of Guineas (*Busen von Guinea*) in Africa. From Andree, *Allgemeiner Handatlas*. Photo courtesy of Botany Libraries, Harvard University.

It is thus with many supposed "eureka moments" recorded in the history of science. The intuition was there, the seed was planted, and Wegener's mind was certainly ready for it. However, as a detailed examination of most such "eureka moments" reveals, there was a much longer process to follow in making something of this intuition. It is good to have the idea, but scientific success comes not just from having an idea but also from working out its detailed consequences in a convincing manner.

So Wegener turned away from what would later make him famous, and for which he had yet accomplished nothing, to face the work at hand. On 6 January 1911 he wrote to Köppen with a variety of news. There was always the need to thank Köppen, for every letter from him arrived with a parcel of papers, books, references, and photographs. In this case Wegener found some heavy volumes and tornado photographs that he hoped to use in his thermodynamics book, and he was delighted. He could also report to Köppen, in response to a query, that he indeed had compared stellar spectra with the spectrum of the northern lights and had found no overlap. The question here was whether the light of the northern lights was actually a chemical consequence of the altitude and the gaseous composition of the atmosphere at that altitude or starlight refracted through Earth's magnetic field.

What may have also diverted him from his plan to look into this matter of the continents was some unexpectedly depressing news from Professor Wiechert in Gottingen. He had written to inform Wegener that the planned polar station at Spitzbergen, to be occupied in parallel with the Filchner expedition to Antarctica, and of which Wegener and Stuchtey had planned to be part, was now not to happen. "Why, he doesn't want to say in print."[3] This probably indicates that expedition and funding politics had killed it off. So he would not be able to go to Spitzbergen at the end of 1911, and he would not get a chance to measure the northern lights spectra to determine further the chemical composition of the atmosphere; he had urged that this research be undertaken in his papers on the chemical differentiation of the atmosphere, and he was more than willing to do it himself if he had the opportunity. Wiechert also told him that he had not yet decided whether to publish Wegener's submission of his investigations into the outermost layers of the atmosphere. The problem was not, as before, that he could not find the manuscript among his papers; he had located it. But the submission was so long that it might be necessary to publish it as a "separate."

The correspondence between Wegener and Köppen, with the turn of the year, began to take on a less formal tone. Köppen had become a trusted older friend and adviser, as well as an intellectual guide through the labyrinthine meteorological literature bearing on the thermodynamics of the atmosphere. Wegener was increasingly at ease with Köppen and felt free to unburden himself to the older man: "Have you seen [Arthur] Wagner's note criticizing my work [on the vertical partition of the atmosphere]? You would think he would have a better use for his time, and he has shown that he can make better use of it. Unfortunately, I don't have time to answer him, as I'm so busy with the details of my manuscript . . . my residence is now a post office and I am constantly overwhelmed with correspondence."[4]

Part of this new informality, and the attendant increase in Köppen's solicitude for Wegener, had to do not only with his admiration for the work of the young man but with something decidedly more personal. Since the Christmas holidays, Alfred and Else Köppen had been in frequent correspondence. She had written to thank him for Friis's book containing his adventures in Greenland; he had replied. In early February she wrote to him to tell him that she had passed her teacher's examination and was scheduled after 1 April to begin as a teacher at her old school. Köppen was at this time revising his climate map of the world, and Else, who had considerable skill in drafting, was helping him by drawing it. She wrote to Wegener about this. He replied to her on 4 March:

> I am so happy that you're helping your father with this necessarily tiring work. For me there is little agreeable to report. I think I'm going to drop dead. Wednesday I had to give a lecture on cloud formation and had no manuscript so I had to dictate on that day to a typist. But the heavy lifting is still to come: on the 13th I give a lecture here in Marburg on "piloting airships," and one the 17th in Cologne on the "Danmark Expedition" with color photographs and special attention to the kite flying, on the 18th once more without the "special attention," on the 20th in Kassel on "travels in a free balloon." Will I live through this? It feels like that everything that could possibly happen has descended upon me all at once: the proofs of my articles for MZ and PZ, from Ambrosius Barth [the *Thermodynamics*] and from Copenhagen, then streaming sunshine, so we have to be outside a lot measuring with the albedo meter, and on Monday I hope "if there's good weather" to make a short little flight . . . soon my brother will return, he should probably have already left Samoa. Where he is stuck only the gods know . . . my parents wrote me if I would please come home to them for Easter but I don't think I'm going to be able after all to do that. With my own work I haven't moved an inch but I shouldn't complain to you so much . . . my thermodynamics is going well, the first 100 pages are at the printer and I hope to have those

proofs soon. Now it's happily already two o'clock in the morning so you will get this letter at the earliest on Monday. Thus, good night. With my heartiest greeting your Alfred Wegener.[5]

By the end of March Wegener was, in spite of his fears, still alive. It had been a wonderfully productive month. He and Stuchtey had collected enough measurements to finish their paper on the albedo of clouds and of Earth using their novel albedometer, and they had sent it off to publication.[6] He had received a commission, with a sizable honorarium, to write a summary of recent advances in atmospheric physics for a new journal edited by the chemist Emil Abderhalden (1877–1950), *Fortschritte der Naturwissenschaftlichen Forschung* (Progress in natural scientific research). Abderhalden was one of the first to see the need for the kind of review-article journal that is quite common today, something like "Annual Review of Geophysics" or "Progress in Microbiology," containing review articles by active researchers sketching out the main line of development of a field. Most of the readers of Abderhalden's journal were medical doctors, and most of the topics were medical and chemical, but the editor was branching out. The new journal was, from its inception, very well received and widely read, and *Nature* reviewed the first annual volume (1910) with the following comment: "The plan of this new publication is to furnish summaries of recent results and select the departments of knowledge in which some degree of settlement and certainty has already been reached. This policy will avoid any risk of wasting time on raw speculations, and, under the able guidance of a man of Dr. Abderhalden's experience and prodigious industry, the series promises to be useful and judiciously chosen."[7]

This commission to write a long article (the finished version was seventy pages) on his area of research was very auspicious. He wrote to Köppen on 3 April about being unable to decide whether he should call it "Physik der Atmosphäre" or "Neue Forschungen auf dem Gebiete der atmosphärischen Physik," but he knew what he would talk about: the new results in aerology in which he played a role, and the new results in the investigation of the outer reaches of the atmosphere, with which he was also vitally concerned. It would end up being, in many respects, a reduced version of his thermodynamics of the atmosphere, with the purely physical chapters left out, and without mathematical apparatus, but with all the conclusions of which he was so proud. It was an excellent opportunity to get his ideas before the public and to proclaim that a very large number of new researches in the area of atmospheric physics, at least the most important ones, were the work of Alfred Wegener.[8] It would also be very nice to have his own particular and original view of the structure of the atmosphere characterized, in *Nature*'s terms, as one in which "certainty has already been reached."[9]

The article for Abderhalden was due on 1 May, but there was more work to come. Wegener's broad responsibilities for the scientific work of the Danmark Expedition included the aerology, the station meteorological observations, the magnetism, the atmospheric electricity, and the marine observations during the voyages to and from Greenland. These also included, as coauthor with his friend and comrade Johan Peter Koch, the glaciological observations of the expedition. There were not a lot of data to be processed for this last publication: photographs of the margins of the Inland Ice, the *Randzone*, with its discolored ice, laminar structure, and complex folding and flow banding. Most of these results were from the ten-day reconnaissance in the spring of 1908 to Dronning Louise Land, that remarkable area of ice-free mountains, Inselbergen, surrounded by outwash gravels and glacial till, which Koch, Wegener, and the others had discovered some tens of kilometers into the Inland Ice cap. Wegener had loved the place and recorded in his diary how frustrated he had been that Koch had kept him constantly on the move and would not allow time for study, poking around, and rumination.

Wegener would never be free of the Danmark Expedition until all the material had been published, and he was pushing ahead as fast and as hard as he could in every way to make this happen; he pushed Lüdeling, he pushed Brand. Koch was determined, apparently, to help him out, by volunteering to come to Marburg for a few days, so that they might talk over the remaining problems with their scant (seventy-seven pages when published) manuscript on the glaciology of the expedition.[10] Wegener was delighted. Other than his conversations with Richarz, or with Stuchtey and Take, he had no social network in Marburg, and it would be good to see his old friend.

In Nansen's Footsteps

Koch arrived on 29 March and stayed until 2 April. He could not afford much travel time out of the country; only a captain in the Danish army, he was one of Denmark's leading cartographers and was also poised to play a leading role in the construction of Denmark's air force. Since time was short, it was fortunate that there was not much that the two men needed to do other than provide a bare description of where they had been and what they had seen. They had not really been prepared to carry out a systematic program of glaciological research. They had taken a few ice borings, but these were contextless and essentially meaningless. Yet the two men had much to talk about. How could they not? Those who share the experiences of the sort that they shared find a unique solace in the company of others who knew exactly where they had been, what it had cost, and how close to death they had been on many occasions.

Reviewing their researches in Dronning Louise Land, they realized how much more they might have accomplished with more time. The death of Mylius-Erichsen had thrown the entire schedule of the last year of the expedition into chaos. Part of the expedition plan had always been a traverse of the Inland Ice from east to west. This would have been the first traverse in thirty years—the first since the pioneering effort of Fridtjof Nansen (1861–1930) in 1888. The scientific program proposed by Nansen after his trek across the Inland Ice had been the inspiration for the Danmark Expedition plan to venture there, but it still remained completely undeveloped and unexploited.

Koch proposed to Wegener that they should go back to East Greenland, as soon as they could raise the money. Even as they were talking, Ejnar Mikkelsen (1880–1971) and a party of Danes were in Northeast Greenland searching for the remains of Mylius-Erichsen and Høeg-Hagen, to recover their lost diaries.[11] Koch knew that these men would be traveling in his own footsteps, as well as the route that Wegener and Thostrup had taken over the edge of the Inland Ice: Mikkelsen's expedition plan included a reconnaissance onto the Inland Ice past Dronning Louise Land. This could only be a prelude to an attempted crossing. De Quervain and others were known to be planning traverses of the ice cap, but this task and achievement, Koch felt, belonged to him and to Wegener. It is what they had talked about and dreamed of in their bunks at their "Villaen" while on the Danmark Expedition. Wegener had confided, several times in his diary, how much he despised the organizational scheme of the Danmark Expedition, in which everyone had a maddening series of everyday duties in addition to their scientific work, and which over time, and in the polar dark, had whittled away at the residual energy to think and do science.

What was needed was a small, compact, highly mobile expedition, with Wegener, Koch, and their old friend and companion the botanist Andreas Lundager as the scientists, along with one nonscientist helper. Koch excitedly sketched out a plan. It was Nansen who provided the template: he had done all the pioneering work and made all the mistakes. His account of the first crossing of the Greenland ice cap, east to west, is one of the most heroic and most comic stories in all of polar exploration. It was dangerous, it was grueling, but it wasn't tragic, and no one died.

Nansen was the great polar explorer of the turn of the century, before the poles were achieved. His book *The First Crossing of Greenland* (1890)—in the original Norwegian, *Paa Ski over Grønland* (By Skis across Greenland), and in German *Auf Schneeschuhen durch Grönland* (1897)—was a huge best seller.[12] It was famous, not least of all, for introducing skiing to the world outside Norway. A good part of the book is given up to the praise of the wonders of skiing and skiing technique: we are speaking of cross-country here, not downhill. It is difficult from our vantage point to imagine that scarcely a century ago almost no one outside Norway ever considered strapping long narrow boards onto his or her feet as a means of traversing snowy countryside, but this is very much the case. Nansen was a great apostle of skiing, which sport he had begun at age four in 1865, and also a great ice skater and a good shot. He was a zoologist who dropped out of school in 1882 to go to sea as a seal shooter, and his ship had been caught in the ice off East Greenland. It was at this time that he got the idea to cross the ice cap from east to west.

A previous attempt to cross the ice cap had been made in 1883 by A. E. Nordenskjold (1832–1901), the great Finnish/Swedish polar explorer whose voyage of the *Vega*, with his navigation of the Northeast Passage from Norway to the Bering Strait from 1878 to 1880, was also one of the great classics of polar exploration. Nordenskjold had failed in his own attempt to cross the ice from west to east, but he had encouraged Nansen to attempt the crossing, and he also encouraged him to take Lapps with him because of their experience in the ice and snow environment.

Nansen had picked a small party: himself; Otto Sverdrup (1854–1930), who was later captain of the *Fram*; Oluf Dietrichson (1856–1942), a lieutenant in the Norwegian army who was there to do the meteorology and cartography; Kristian Kristiansen Trana (1865–?), a lumberjack and deckhand who had worked for Sverdrup's father; and two Lapps, Samuel Balto (twenty-seven years old) and Ole Ravna (forty-six).

Technologically, in the history of polar exploration, the trip was a transition between a British boat-sledge haul and later sled-sledge expeditions. Nansen used no dogs or ponies, with an emphasis on lightweight and fast movement. Nansen and his companions learned a number of painful lessons.

They learned that traveling in July and August was a big mistake, because the snow toward the coast was melting and knee-deep, while the high-ice snow was a wind-driven substance the consistency of beach sand and no easier to ski on or to draw sleds over. They had no ski wax, and they had iron-shod sleds. Every step was a tremendous effort from one side of Greenland to the other. They learned that it would have been useful to have a thermometer that would register below −40°C (−40°F), to take clothes that would keep snow out, and to take a tent that would do the same. They learned that you have to take a lot of fat. They were starved for fat and wolfed down their weekly butter ration.[13] They learned that you need to carry enough fuel to melt snow so that you are not constantly dehydrated, and they learned that without this additional fuel you must wait to eat until your beard thaws, because it freezes so hard that you can't open your mouth.[14] It is for the latter reason that most polar explorers learned to stay clean-shaven or to have very short beards.

Nansen's crossing of Greenland was a major step in the exploration of the interior. From the 1860s onward, beginning with the work of Johannes Rink (1819–1893), there was an attempt to use Greenland as a laboratory for determining the reality and character of continental glaciations and therefore of ice ages. Nansen had been very keen to study the latter problem, especially the calculation of the ice volume disgorged from glaciers.[15] Nansen's account of his crossing of Greenland ended with a brief scientific appendix, containing an agenda of problems concerning the Inland Ice of Greenland, which occupies polar scientists even today.

The program that Nansen had laid out was the program that Koch and Wegener proposed to follow. There were the geographical questions: What is the altitude and extent of the ice surface? Does the ice cap contain substantial outcrops of rocky material, or is the latter mostly absent? Are there more rocky islands like Dronning Louise Land, *nunatak* oases in the desert of ice? Were there any transverse channels connecting the East and the West? Then there were also glaciological questions: What was the profile of the ice surface? What was the nature of the crevasses and fissures, what was its rate of motion, what was the glaciological topography, what was the composition of the ice and snow, and what changes in its temperature and structure with depth? Then there were the meteorological questions: What were the temperature and pressure regimes in the interior? Was it too cold to snow there? Was there a permanent glacial anticyclone? (The winds appeared to blow always toward the coast.)[16]

These were all good scientific questions, questions that might be answered in the course of the traverse they were contemplating. They would have a fighting chance: they would start farther north than Nansen had begun and cross a much longer diagonal from northeast to southwest of perhaps 700 kilometers (435 miles). This traverse would take them right across the geographical center of this gigantic island, so that if there were anything "in the middle of Greenland" they would be likely to find it.

What was more exciting about following in Nansen's footsteps was that he was not only a great polar traveler but also a real scientist with real scientific plans. Nansen's agenda for Greenland was not limited to just the questions given above but included some engrossing "higher order" questions. Greenland contained the last remnant of the continental ice sheets of the last Northern Hemisphere glaciation. It also contained fossils—not merely of temperate plants and coal deposits, but identifiable fossils of tropical plants. Somewhere in Greenland, Nansen was convinced, was the solution to the riddle of the ice ages.

Nansen had thought a great deal about the cause of ice ages, and he knew the wide variety of geographical theories. He was also familiar with the astronomical theories: both the theory of long-term cooling of Earth from an earlier warm state and James Croll's theory that the changes in the eccentricity of Earth's orbit would lead to variations in solar radiation and thus alternate warming and cooling of the Northern and Southern Hemispheres. Nansen rejected all these explanations for a variety of very good reasons, which we need not go into here,[17] in favor of a much more attractive and persuasive alternative: the migration of Earth's pole of rotation. "The easiest method of explaining a glacial epoch as well as the occurrence of warmer climates at one latitude or another is to imagine a slight change in the geographical position of Earth's axis. If, for instance, we could move the North Pole down to some point near the West

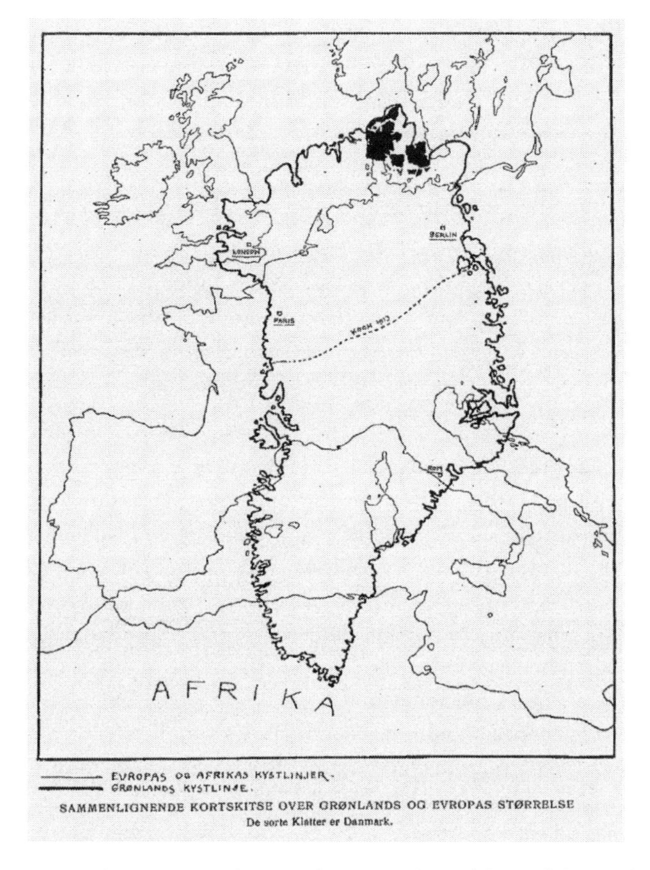

Greenland superimposed on Europe, showing the route planned (in 1911) by Koch and Wegener across Greenland for 1912–1913. The blackened area at the top of the map is Denmark. From J. P. Koch, *Gennem den Hvide Ørken: Den danske Forskningsrejse tvaers over Nordgrønland 1912/13* (Kjøbenhavn: Gyldendalske Boghandel Nordisk Forlag, 1913).

Coast of Greenland between 60° and 65° north latitude we could no doubt produce a glacial period in both Europe and America."[18]

The Arctic explorer and paleobotanist Alfred Nathorst (1850–1921) had documented the existence of floras in Japan (cold-weather plants) and Greenland (tropical plants) at exactly the same time in the Tertiary: this would be impossible with Earth's axis in its present position, and Nathorst had proposed a position for the pole of Earth at 70° north and 120° east, 20° nearer Siberia than at present.[19] These findings supplemented the publications of the great Swiss paleobotanist Oswald Heer (1809–1883), who had done extensive work on tertiary fossils of North Greenland and come to the conclusion that the mean temperature for these regions in part of the Tertiary must have been between 21°C (70°F) and 22°C (72°F).[20] Heer had created a sensation with his *Urwelt der Schweiz* (1865), with its striking engravings of tropical plants growing in Switzerland in earlier periods.[21]

Nansen continued, "That an actual movement of the axis does take place seems to have been established by the fact that observations of several German observatories, Berlin, Potsdam, Strasburg, and Prague agree remarkably in showing alteration of more than half a second in the course of six months."[22] Indeed, the observations taken at Pulkova, Greenwich, Washington, Milan, and Naples seemed to indicate a motion of the pole beyond the simple nutation, which additional motion could displace Earth's axis as much as one second per year.

> My idea, therefore is that, if it be allowed that the axis of the earth can admit of considerable changes of position, there is nothing to be said against the hypothesis that several of these changes may have taken place within the bounds of Tertiary time . . . but even if it be granted that the pole had a different position in the Tertiary period, there still remains the fact that it must have moved twice over towards Greenland, and twice back again, to account for the two different glacial periods. Such movements would be, no doubt, somewhat extravagant and no sufficient reasons for their occurrence are forthcoming.[23]

Whether such displacements of Earth's pole were "extravagant" or not, Nansen continued to support this hypothesis over that of the astronomical theory of James Croll, which, he thought, had the defect that while "it will furnish us with glacial periods returning at indefinite intervals, it cannot account for the recurrence of conditions so favorable as to explain the existence of subtropical climates, such as Greenland, for instance must once have had. Thus we see that from whatever point of view we regard the phenomena we cannot find any satisfactory explanation that covers them all. This must be the work of the future."[24]

Koch and Wegener planned to take up Nansen's challenge: their work would be "the work of the future" foretold by Nansen. They sketched a plan to arrive in Greenland, in the vicinity of Danmarkshavn, in the autumn of 1912 and to erect a prefabricated hut of the kind they had stayed in from 1906 to 1908, wintering over on the margin of the Inland Ice, so that all their supplies would actually be on the ice cap before the attempted traverse the following spring. They might, under other circumstances, have made use of the buildings, including the Villaen, at Danmarkshavn, except that it was possible that Mikkelsen's party would still be in the vicinity as late as 1912.

After the overwintering, they would make the traverse of the ice cap in April, May, and June of 1913. They would need a lot of equipment and a lot of fuel, and the initial estimate was some 20,000 kilograms (44,092 pounds). This meant that both dogs and man-hauled sledges were out of the question: a fully loaded dogsled was only a payload of about 100 kilograms (220 pounds), and this would require 200 trips and many more dogs than four men could handle. Koch proposed that they use Icelandic ponies, *Hyster* or (in Danish) *Hestene*. He had used them in his cartographic work in Iceland, and although they made traveling slower than with dogsleds, their pulling power was enormously greater than a span of dogs. Horse-drawn sleds, equipped with sails when the wind direction was right, could provide the motive power needed. Indeed, Ernest Shackleton (1874–1922) was at that time (1908–1909) using them in Antarctica.

Koch would go back to Denmark to try to raise the money. The Carlsberg Foundation could be counted on for perhaps 30,000 kroner, about half of what they would need, but Koch would have to do some aggressive fund-raising to make up the rest. If Wegener was to be a full partner in the expedition (*selbstständig*), he would have to raise a good deal of money too. He could appeal to the German government for a stipend, common practice for expedition members whose work was deemed important and for the good of the country. It was a long shot, but it was something that they both wanted. Wegener committed to go, and he saw Koch off at the train and then returned to work—not "his work," but just work. He had a bare month to finish the scaled-down version of his thermodynamics book for Abderhalden, plus the endless data reduction of the weather station observations at Danmarkshavn. Wegener was fortunate that, although he had no great mathematical gift, very much like Johannes Kepler, he took an aesthetic pleasure in calculating and reducing numerical data; it put his mind at rest. This appetite for data reduction and calculation was something that he shared with Köppen, who enjoyed working far into the night with numerical data bearing on whatever problem with which he was currently occupied, liking the quiet of the house once everyone else had gone to sleep.

A Dilemma

No sooner had Koch departed than Köppen arrived, having told Wegener that he wanted to visit him for a day or two. Wegener was delighted and spoke excitedly about Koch's visit and the return home of his brother Kurt.[25] The documentary evidence is very incomplete here, but working backward from letters between Wegener and Köppen later in 1911, it appears that Köppen had come to Marburg to offer Wegener a position as an atmospheric physicist at the German Marine Observatory in Hamburg, where they might be colleagues. Köppen was one of several German scientists approached by Hugo Hergesell in 1909 and 1910 to take a role within Germany, generously funded by the Carnegie Institution of Washington, D.C., to turn meteorology into an atmospheric physics. This was an explicit program of the Carnegie Institution in coordination with the U.S. Weather Bureau—although funded almost entirely by Carnegie—to make resources available to investigators in America, Britain, and Germany that they might order, modernize, and scientize the understanding of atmospheric motions. Vilhelm Bjerknes was to be part of this effort, as was Aßmann.[26]

Köppen seems to have pressed Wegener very strongly to take this job. The amount of money was quite substantial, a $2,000 U.S. grant per year to Köppen and Wegener for three years, to work on that aspect of the international network of observing stations being built in preparation for large-scale air travel. Although currency conversion isn't an exact science, $2,000 U.S. in 1912 had the purchasing power of about $45,000 today. Köppen proposed that Wegener should have the lion's share of this money, should he come to Hamburg. Köppen already had an adequate salary and was well settled, and Bjerknes and others had used their grants immediately in order to hire assistants.[27]

Writing to Köppen on 2 May, Wegener said, "You surely know how to pour hot coals on my head."[28] Köppen, who was a respected meteorologist but best known as a climatologist, had said to Wegener that he thought of him as a climatologist and a geographer as well. Wegener replied that he heard this quite a bit about himself, and that if everybody said this of him he might come eventually to believe it, but that fundamentally it wasn't true. He wanted to take meteorology in the direction of physics. "It's likely," he said, "that I could produce climatological and geographical work, if I applied myself to it with a great deal of energy and I could come somewhat closer to the ideas of these sciences . . . but would it work?"[29]

Wegener had told Köppen of the planned Greenland expedition, and Köppen had made his reservations evident. Perhaps in an attempt to put his mind at ease, Wegener told him that "the Greenland expedition is quite up in the air especially the way it is to be organized. Probably it will again be a Danish expedition in which I will be a member."[30]

At the beginning of May, and probably on the same day that he wrote to Wladimir Köppen, he wrote to Else: he was in parallel correspondence with the father and the daughter. He had delighted her earlier in April with a description of a hair-raising crash and destruction of the balloon "Marburg," while he was piloting it. This was the balloon that he, Richarz, and Stuchtey had used for a variety of scientific purposes. The balloon, on the day of the disaster, had not been sufficiently loaded with ballast, and in consequence it was slightly underinflated. It failed to clear a ridge and had been caught in and shredded by tall trees. Wegener was knocked about but not badly injured. With that disaster behind him, he invited Else to come to Marburg for a balloon ride, perhaps a questionable invitation given the context. Nevertheless, she was thrilled. She would have to provide 100 marks as her share of the cost of the ride. He told her he had asked a married couple that he knew (probably *Oberlehrer* Brand and his wife) to go for a balloon flight and needed a "fourth man." He would be delighted if Else would come to Marburg and fly with him. As it turned out, Brand's wife fell ill, and the flight had to be canceled. Else was severely disappointed, but Wegener told her that he would come to see her in Hamburg at the beginning of June instead.[31]

The first ten days of June that year were the Whitsun vacation marking the midpoint of the summer semester; Wegener was teaching but a single course: "Optical Phenomena of the Atmosphere." He had contracted to give two lectures in Hamburg: one on developments in aerology, and the other to the *Luftschiffverein*. There were enthusiastic clubs of aeronauts in every major city, as well as in many small ones, including Marburg. Wegener's world record for time aloft in a free balloon, established with Kurt in 1906, still stood. He was an experienced pilot, with many flights and at least two spectacular crashes, and his devotion to photography of cloud forms during these flights made his illustrated lectures popular and accessible. He was, like many other aerologists at this time, spreading the gospel of the intimate connection between the collection of physical information about the behavior of the atmosphere and the future of air travel generally.

In addition to his lectures, there was the obligatory but much-anticipated visit to the household of the Köppen family in Großborstel. There is much that Wegener had to discuss with Köppen about atmospheric optics (his current fascination) and the next module of the atmospheric physics he hoped to produce. He also had questions and some trepidation about accepting the offer to work in Hamburg. He was in a very difficult position, as Köppen had been of tremendous help to him and was making his career move along rapidly. Köppen's request that he move to Hamburg and become his collaborator threatened to attach him to a project of several years' duration that would be time-consuming (if remunerative) and might restrict his ability to do his "own work." The problem here and *always* was that Wegener was an enthusiast; his mental conformation and imaginative style were such that he would rush from topic to topic as new ideas occurred to him. He had been able, against the background of the burdensome work on the Danmark Expedition results, to do some highly original work in thermodynamics, atmospheric layering, and atmospheric chemistry. What would happen to all of this, were his life to be consumed by another series of routine scientific tasks superimposed on the Danmark work? What would be left for him?

Alfred and Else

There was, in addition to the lectures and the need to talk to Köppen, another and more urgent mission in Hamburg. Wegener arrived at the Köppen house in the afternoon of Saturday, 4 June 1911, dined with the family, and asked Else if she would care to go for a walk after dinner and take a look at the Moon. It was a waxing gibbous Moon, rising just before sunset, and hanging low in the eastern sky as the Sun was setting, very bright against the water haze from the harbor. It was during this walk that he proposed marriage to her, and she accepted him immediately. They had known each other for three years. She had made no secret of her great interest in him during the recent few months in which they had been in regular correspondence. It is quite likely that he had intended to propose marriage to her in Marburg in April, during the forestalled balloon flight on which he had invited her.[32]

Wegener suggested that Else should spend the Whitsun vacation with his own family at Zechlinerhütte. Kurt, finally back from Samoa via his trips across China and then via the trans-Siberian railway to Europe, joined them on the train from Berlin to Rheinsberg along the way. Else was struck by Kurt and Alfred's complete devotion to one another and their unconcealed delight in each other's company. But nearly as striking was how absolutely different they were. Kurt was narrow in the shoulders and very tall, towering over Alfred. Else remembered that Kurt gave an animated account of his travels, interspersed with acid judgments (that she found harsh) about colleagues and conditions. He was nervous in his manner, constantly reaching into his pocket to make sure he still had his rail ticket. Where Kurt was nervous, Alfred was calm; where Kurt was thin, Alfred was barrel chested with powerful shoulders and broad hands—"though these were now white and smooth from his labor at his desk."[33] She liked Alfred's manner of speech, very deliberate, never hasty, and never a word against anyone, even in retailing a story of a colleague's lack of skill, always trying to find and emphasize the positive aspects of people and things.[34]

They arrived at the station in Rheinsberg and were met by sister Tony driving the wagon. Zechlinerhütte was scarcely easier to get to in 1910 than it had been in 1890, and it was a long ride along the sandy path through the forest and then along the shore of the lake. This was the same path Alfred had walked as a child, while the baggage wagon had gone on ahead. Tony, with short-cropped gray hair and a taciturn manner, completely intimidated Else, speaking not a word to her throughout the entire trip. Else's heart sank momentarily. Here she was, nineteen years old, but in the presence of all these adults ten or fifteen years older than she; what was she doing here? "Then," she wrote later, "the wagon stopped under the giant old Linden trees in front of the broad-fronted house, and I looked into the beautiful eyes of my mother-in-law (Anna Wegener). All my anxiety vanished. These were Alfred's eyes!" Richard Wegener, her future father-in-law, greeted her heartily as the bride-to-be of his youngest son, and he did not conceal his delight at their planned marriage: Kurt and Tony were unmarried and confirmed in their bachelorhood.[35]

Else was relieved to find the family as gregarious as her own: the Wegener house was as full of visitors at all times as had been her parents' home in Hamburg. She was entranced by everything: the long table in the dining room was constantly adorned with roses, and the lilacs were still blooming in the huge parkland and garden behind the house. After dinner, in the evenings, Alfred would row her out onto the lake so that she could see the deer come down to the water to drink. She was completely happy and at peace.[36]

When she returned home to Hamburg, flushed with happiness, her father wrote a note of thanks to Wegener's parents for the warm welcome shown to his daughter. He said that he had long admired Alfred as a scientist and found him now transformed in no time into a "beloved son." The impending marriage, about which Köppen was as pleased as he was surprised, nevertheless created difficulties for him and for Wegener. He wrote to the Wegeners senior:

> There is a problem, though probably not too serious a one: I expect great things of Alfred, who is for me extremely important for the future of my science. He is one of the most industrious and productive of the younger generation of meteorologists, and is warmly praised everywhere; I only saw in him the meteorologist but it seems now that I must also speak of him as my future son-in-law. I have therefore written to professors in Berlin and Vienna along with the engagement announcement and explanation of the state of affairs, in order to prevent any misunderstandings. Alfred is certainly no lower in my esteem that he will make my daughter happy and I hope for even greater things from him than I have expected up to now![37]

Köppen was taking pains to make it widely known that he had not been promoting Wegener's career so vigorously, nor offered him the job with the Carnegie Institution money, on the expectation that he was supporting a future son-in-law. Now that this had turned out to be the case, Köppen had worked to clarify, with both Hergesell in Berlin and Julius Hann in Vienna, that his judgment of Wegener as a meteorologist had been proffered without any knowledge of this recently arrived personal relationship. He was working to make sure that Wegener did not suffer professionally by having this personal connection with Köppen which might allow Köppen's patronage to be misinterpreted.

Wegener next wrote to Köppen at the very end of June, with a new salutation that encapsulated his dilemma: "dear father-in-law."[38] He was now bound to Köppen in a different way, one in which professional and personal would not only exist side by side but also overlap and bump into one another. He was under tremendous obligation; Köppen was his true scientific patron, and he knew it. Moreover, the Carnegie money would provide the means for his marriage. His stipend from the Ministry of Education, supplemented by his stipend from the university for his teaching, tutoring, and management of the observatory, amounted together to around 3,000 marks. Thus, the Carnegie salary was four times as great as his university stipend.

Köppen had proposed a published book of about 600 pages which they should produce together uniting aerological research and synoptic meteorology—the latter being the simultaneous production of weather information at different stations and its collation to produce weather maps. Wegener had reservations, however, particularly that the book should be limited to the "unification of aerology with synoptic meteorology. I know," he continued, "you are indicating here weather forecasting. But could not one rather say: 'with the other branches of meteorology?' Why should the global circulation, the outer regions of the atmosphere, the investigation of thunderstorms, and the study of cloud forms, be ruled out?"[39]

Köppen, designing the work plan and getting ready to draw up the contract, also proposed that Wegener take three-quarters of the money, or $1,500 a year. Wegener protested: "But why should you get less than me? There's absolutely no basis for the opinion that I will be doing the lion share of the work and certainly against that must be counterpoised the notion that your portion should be judged to be more important than mine. I protest against this complete 'inversion' and want very much to find the means that you will receive as much as I. If you write Hergesell and say only this, you will see that he will completely agree."[40]

It is painful to read Wegener here, trying so hard to say "yes" and to say "no" at the same time. As his letter continues on, the objections and reservations mount. Köppen must understand that if the Greenland trip should go forward (even though the chances were looking less promising at the moment than before), the contract would have to be abrogated. Further, Wegener lacked any inspiration for the book: "I wish I had more ideas for this project! For every 10 I have I only need one of them to be any good and here I don't even have 10. Maybe something will occur to me when the project is right in front of us." Still further, he had spoken with Richarz, who had been very pleased at Köppen's visit and took the opportunity to say to Wegener, not for the first time, that he was prepared personally to provide the financial means that would allow Wegener to stay in Marburg. Wegener added parenthetically, "He is extremely wealthy." He told Köppen, "I told him that such an effort would put him in a struggle with you and that you wanted me in Hamburg."[41]

A week later, late at night, with his work for the day finished, he wrote again to Köppen, ostensibly to tell him that the first bound exempla of the thermodynamics book had arrived from the publisher and that he had reserved a copy for him. He finally had some ideas about the proposed collaborative book for the Carnegie Institution: some material on the stratosphere. He had been working on the wind speed in the stratosphere, one of several contemporary projects including the beginnings of some ideas on tornadoes. He had already revised some of his ideas presented in the yet-to-appear thermodynamics book, and he announced cheerfully, "Well at least I can say that here is something for the second edition!"[42]

Eventually, on the final page of the letter, he got around to the actual reason for writing. Köppen was, he knew, in the final stages of negotiating with the Carnegie Institution over their plan; Hergesell had written to Wegener asking him what the status of his Greenland plans was relative to his plan to collaborate with Köppen. However unlikely it was that the Greenland trip would come off, Wegener could not give it up. He knew that Köppen was quite opposed to his plan to go to Greenland and thought of it as an interruption to Wegener's career as an atmospheric physicist. For Wegener, it appeared to be a last chance at polar exploration. "By the way, I note that you generally view the matter pessimistically; perhaps it will work out that we can do one after the other, if not in one sequence, then in another. Currently it's for me a matter of stoicism, and I fear that your hope [that it will fall through] will be completely realized!"[43]

Now every time Wegener thought about the move to Hamburg and the Carnegie project with Köppen, his anxiety level rose and his reservations accumulated. Was it really such a good idea to give up his academic status as a *Privatdozent* in Marburg, without acquiring the same status in Hamburg? Was it really wise to give up his stipends? What would his colleagues at Marburg think, what would the Ministry of Education think, if he just "up and left" without advancing in rank?

There were other issues as well. Köppen had urged him to devote "all his energy" to their upcoming collaboration, but Wegener said in reply, "This can't possibly mean that we can't both of us do anything else, that we couldn't publish scientific articles; I would like of course additionally to lecture away to my hearts content, and similarly to pursue any chance for an ordinary professorship." He was feeling pressure not only from Wladimir Köppen but from Else as well: she was urging him to complete the Carnegie contract as quickly as possible so that they might advance the date of their marriage.[44]

With all the uncertainty in his life—where he would live, what he would work on, when he would marry, whether or not he would go to Greenland—there was one constant and dependable item that summer: his knowledge that in all of August and all of September he would be back in the military and on maneuvers. Germany's state of military readiness was accelerating at this time, not in response to any crisis, but continuing the modifications of the Schlieffen plan under the leadership of the new general staff. Wegener had only a few weeks to work before he had to return to the headquarters of the Queen Elisabeth Grenadier Guard Regiment at Charlottenburg. In an undated letter, but probably from early September 1911, he wrote to Köppen, "Dear father-in-law, all my patriotism notwithstanding, I have today set aside instructions on the treatment of Hoof and Mouth disease and have instead been writing meteorology on office memo paper."[45] He continued with news that none of his projects were going forward very well. The Danmark meteorological observations were at 110 pages and growing, seemingly interminable. He had not received his free copies of his thermodynamics book or the free separate copies of his article for Abderhalden. Since Koch's visit, he had heard nothing about the glaciological work except that it had been handed over to the press: "this is really sour pickle time [the 'silly season']."[46]

Wegener returned from eight weeks of military training to Marburg in time to begin the winter semester (early October 1911), in much better physical condition than he had left, once again tanned, and his soft hands once again hardened by work, and with something of his habitual good humor restored. He had been able to visit his prospective in-laws in Hamburg on the way back from maneuvers on 17 and 18 September and had urged the Köppens senior and their family to visit and meet his parents in Zechlinerhütte, something to which Köppen had readily agreed, feeling that the change of air would do himself and the rest of the family good.[47]

We have already noted that Wegener had a tendency to ride off in all directions at once, pursuing every enthusiasm of the moment. If before he departed for the military he had not even had ten ideas from which to choose one good one, now his mind was overflowing. He wrote enthusiastically and at great length to Köppen about his plan to build a new differential psychrometer. This instrument is a kind of hygrometer used to measure relative humidity by comparing the temperatures between a "dry bulb" thermometer and a "wet bulb" thermometer, the latter encased in wet muslin, with the apparatus ventilated in such a way that the latent heat of evaporation on the wet bulb yields a temperature difference between the two thermometers which, by reference to a table, may be used to calculate the relative humidity. Wegener's design (rather too complicated to discuss here) would allow the psychrometer to record continuously, thus providing an opportunity to send it aloft. In addition to the wind speed in the stratosphere (on which he published a paper in 1911), he was also interested in whether the stratosphere was at all turbulent or thoroughly laminar in its airflow, and he wanted to measure the shift in relative humidity within this broad stratum of air.[48]

His correspondence with Köppen indicates that he was also still working on the spectral characteristics of meteors and thus on the elemental composition of the upper atmosphere. As if this were not enough, he had decided, partly at Köppen's suggestion, to try to understand the relationship between the horizontal and vertical components of turbulence in cumulus clouds, and he was planning to try to do some microphotography of cloud elements, having been excited by those photographs he had seen already, and counting on his own skill as a photographer to meet or exceed these standards.[49] Beyond this, as he wrote to Köppen on 21 November, he was trying to understand now the relationship between rectilinear and turbulent motions in fluid media more generally. He thought that the best chance to understand these things in the atmosphere was to take the work done on them already in oceanography, and therefore he was looking again into George Darwin's book on the tides. He wanted to put a footnote in a manuscript in progress on "turbulent motion of the atmosphere" on the relationship between rectilinear and turbulent motions at the surface of the sea, and he asked Köppen, "There must be something about this in Krümmel's *Oceanography* isn't there? I feel like showing a little local patriotism perhaps I can just put in a footnote: compare Krümmel's *Handbook of Oceanography* volume 2 . . . 1911 page." The reference is to Otto Krümmel's massive two-volume *Handbuch der Ozeanographie*, the second volume of which had only just appeared and was not yet in the Marburg University Library.[50]

Whatever his day-to-day activities, the long-term uncertainty about his future had now returned, if not with a vengeance, at least with an insistence he could not avoid. Köppen had written him a letter shortly after 21 November offering a range of options about when, and under what circumstances, they should begin their collaborative work for the Carnegie Institution on advances in aerology. With this range of options of dates, Köppen had also included a draft contract for Wegener to sign, so that he might transmit this back to Carnegie for their approval. Wegener responded with the longest letter of their correspondence thus far.

The wild card here was his on-again, off-again expedition to Greenland with Koch. He wrote to Köppen,

> I am reluctant to return my already completed contract until I get Koch's answer to my question of how things now stand. You know that his expedition with only four men depends entirely on who they are, and I cannot behave so contemptibly as to leave him in the lurch . . . if he gives up, or if there is no end in sight I will send you my contract immediately. . . . I have gone to a good deal of trouble to sort through the various different proposals that you have offered. Finally, I have cut the knot and in the end decided: the beginning for the two of us should be 1 October 1912.[51]

Köppen was on the brink of retirement from the Deutsche Seewarte (German Marine Observatory) and had wanted his collaboration with Wegener to coincide with that date. Moving their collaboration forward to a fixed date the following year, as Wegener saw it, would allow Köppen to retire whenever he wished, as well as give them both a lot of time to map out the work. He was particularly concerned that the contract should contain a stipulation that if one or the other of them was forced by other commitments to reduce the scope of his labor for the Carnegie Institution (in other words, should the Greenland trip actually happen), then that person could get out of the contract and have someone else complete it. His suffering with the Danmark Expedition and its aftermath had taught him a good deal about what to ask for.

"For me," Wegener continued,

> the most pressing question is when I should come to Hamburg. Today I spoke with a number of acquaintances, and with their help have clarified matters. They have one and all warned me against leaving here without having a similar academic position waiting for me there. If I did, I would then be between two stools. A move from here without "rehabilitating" is equivalent to surrendering any chance that I might gain the title "professor." . . . I would have given up one position without having gained

> another . . . it is important that I remain here, above all, in case I get a "call" [to a professorship] or something else like this happens and this plan also has the advantage that I will be able in this period of time not just to read my way through but to learn an excellent little reference library of physics, chemistry, and geophysics, and I need to have this in place for the remaining volumes of my physics of the atmosphere.[52]

"I don't have to tell you," Wegener added, "that a joint work would be every bit as stimulating for me as it would be for you. And you've made the point that when we are together everything goes forward with greater clarity. Surely you can still remember the amateurish bungling of the first draft of my thermodynamics."[53]

There were many arguments against the move to Hamburg, one argument for; then, upon returning to arguments against once more, he added an interesting twist: "In spite of the scientific facilities that Hamburg offers, I prefer Marburg as a refuge from the big city, for which my nerves, like those of my siblings, simply are not 'calibrated.' It is a great defect [of my character] but I have such an overwhelming need for quiet—maybe will happen that I can also, in Hamburg, find a quiet little place."[54]

Else Köppen had contrasted Kurt's nervousness with Alfred's calm, but this calm was an exterior calm with which he could conceal an inner turmoil. In saying that had "an overwhelming need for quiet," he was talking about not simply what he needed to work but what he needed to live comfortably. Wegener's emotional repertoire was certainly limited. He wrote to Köppen, "*Ich habe immer Furcht vor dem Pathos*" (I have always been afraid of excessive shows of emotion). Apparently there did not have to be much emotion for it to be too much for Wegener. He was perhaps not an unusual specimen in this regard.

He was physically strong and courageous, but he had lived in a world composed solely of professional relationships and "comradeship," with the exception of his immediate family. Like many highly intellectual men, faced with a situation or a crisis that actually called for some expression of feelings, he was likely to respond either with verbosity (as in this longest letter ever to Köppen) or with depression. He was trying, in his letter to Köppen, to resolve his anxiety in his divided feelings by multiplying the reasons for staying at Marburg. The truth is, he was not just reluctant to leave but somewhat afraid. Of what? Perhaps he was afraid to be swallowed up by Köppen's concerns and his world, afraid to lose his freedom of action and thought, afraid to abandon his academic career, afraid to give up his chance at another polar exploration, and probably afraid of his impending marriage to some extent as well.

> I believe that we'll have to leave the time of my move to Hamburg open for a while. For the summer semester [of 1912], I have proposed [a course titled] "Dynamic Behavior of the Atmosphere" (vol. II of Phys. Der A.!) and I believe it will be very useful to me when I carried it through, if you would also read it. By October 1, when my lectures [for the winter semester] begin, we may be able to see more clearly what is going to happen . . . you wrote to me that Else will probably remain home until Christmas 1912, and could probably get married by Easter [1913]. That is just over 5/4 of a year but perhaps it can't be any sooner.[55]

So there was the dénouement: he would not sign the contract, he would not move, and he would not marry until the spring of 1913. This last decision was important to him, as he had committed himself to Koch on the understanding that the polar expedition must happen in 1912/1913 or not at all.

Chance and a Prepared Mind

With all of this resting on whether or not the Greenland trip would come off or not, Alfred had to assume that it would, however low the chances seemed at any moment. If it were to happen, he would already have to be scientifically prepared to investigate the glaciology and geology they would encounter. Of course, nothing scientific had ever been done with regard to Northeast Greenland, save by himself and other members of the Danmark Expedition. So he focused his attention instead on Iceland. Koch had mapped Iceland, including its southern ice cap Vatnajøkul. It was their plan, should they obtain funding, to train themselves in driving the *Hestene* and the sleds, and working with *Hestene* on ice and snow, by spending the summer of 1912 on this ice cap in Iceland. In order to prepare himself for this part of the journey, Wegener began a survey of the relevant literature. He began with materials closest to home, including the brand new journal *Geologische Rundschau* (then in its second year), founded and edited by his dean at Marburg, Emanuel Kayser.

In the July–August 1911 combined summer issue of *Geologische Rundschau* he discovered an article by Hans Reck (1886–1937), a geologist and paleontologist at Berlin: "Die Geologie Islands und ihrer Bedeutung fur Fragen der allgemeinen Geologie" (The geology of Iceland, and its significance for questions of general geology). It was a review article, a regular feature of this new journal, trying to bring various topics up to date, with the list of references featured at the very beginning of the article rather than at its end. Reck dealt with the new knowledge of the geology of Iceland obtained in the previous decade, with special attention to glaciation.[56]

Only a few pages further on in the same issue was a review article by Erich Krenkel (1880–1964) addressing what was currently known about the development of the Cretaceous formations of the African continent: "Die Entwicklung der Kreideformation auf dem afrikanischen Kontinente."[57] The list of references was very extensive, more than 100, and although this article had nothing to do with the research project at hand (the geology of Iceland), Wegener's curiosity was easily aroused; perhaps the word "continent" in the title framed up his interest. In any case, he decided to take a look. A few pages into the text, as he flipped through, he came across the following passage:

> Less clear are the relationships on the west coast of Africa. The oldest segments of the Cretaceous are Albien. Before this time it appears that there were no [Marine] transgressions onto the African-Brazilian continent . . . the African-Brazilian landmass still existed in the Cretaceous at least in the lower [Cretaceous], perhaps only on the equator. Everything points to faunal "echoes" such as the relationship of the Trigonians [a genus of clams] of the Uitenhag Formation of the Cape [South Africa] and the same in South America. In the Albien there followed a great Marine Transgression over the current West Coast [of Africa], and it is here that the European and Indian forms begin to appear.[58]

Let us consider the import of this passage. In a perfectly matter-of-fact way this geological author speaks of a continental mass composed of Africa and Brazil together. Toward the end of the Cretaceous this African-Brazilian continental mass was somewhat attenuated, but it still existed, confirmed by the geological reasoning that fossils of identical species of clams (shallow-water organisms that cannot cross abyssal oceans) were simultaneously deposited in (what is now) southern Africa and (what is now)

South America. The appearance of the so-called Albien limestone, subsequent to the deposition of these fossils, was a consequence of an ocean spreading, or, as geologists say, "transgressing" all along the current coastline of all of West Africa. Not only that, but subsequent to this appearance of an ocean separating Africa and South America, the paleontological connections with South America were broken, and the species along the West African coast began increasingly to have similarities to species found in Europe and India in this same geological time period.

Wegener was completely overtaken by surprise. Here was geological evidence to support, in the most forceful way, his intuition that Africa and South America had once fit together as a single continent—the idea he had had at the very end of 1910 and subsequently failed to follow up on. He later wrote about this moment,

> The first notion of the displacement of continents came to me in 1910 when, on studying the map of the world, I was impressed by the congruency of both sides of the Atlantic coasts, but I disregarded it at the time because I did not consider it probable. In the autumn of 1911 I became acquainted (through a collection of references, which came into my hands by accident) with the paleontological evidence of the former land connection between Brazil and Africa, of which I had not previously known. This induced me to undertake a hasty analysis of the results of research in this direction in the spheres of geology and paleontology, whereby such important confirmations were yielded, that I was convinced of the fundamental correctness of my idea.[59]

That this particular article by Krenkel was the source of this reawakening of his interest was first suggested by geologist Aart Brouwer in 1983, though Brouwer did not speculate about the circumstances under which Wegener came across it.[60]

Within the next few days, Wegener's interest was further stimulated in the course of reading an article by Konrad Keilhack (1858–1944) entitled "Alte Eiszeiten der Erde" (Former ice ages of Earth). Keilhack had done work in Iceland in the 1880s on the splitting apart of sections of southern Iceland by "fault troughs." He had also written on the subject of marine transgressions in Iceland and the correspondence between beds of shells in Iceland and those found in the island of Spitzbergen north of Norway.[61] But the article at hand focused not on recent ice ages but on a much older ice age, in the Carboniferous period. By the 1870s most European geologists had become convinced that indeed the Northern Hemisphere had, in the recent geological past, experienced a great series of ice ages. Once European geologists learned to recognize glacial deposits, geologists working in South Africa and elsewhere came upon deposits of unmistakably glacial origin in much earlier periods of time.

Keilhack's article on the ice ages appeared in a popular publication, *Himmel und Erde*, for which Wegener had also written. It roughly corresponded to a magazine today such as *National Geographic*, *Endeavour*, *Smithsonian*, *Natural History*, or *New Scientist*. Keilhack had written the article in 1895, to describe to the general reading public work done in the 1880s by geologists in Africa, Australia, India, and elsewhere on a cosmopolitan flora of ferns known today as the "*Glossopteris* flora." That identical species of land plants appeared at the same time in continental masses now widely separated led inevitably to a conclusion of former land connections between these continents and also explained how an ice age could have taken place that left deposits on all of these continents.

The material was not recondite, nor was it new; it simply belonged to a field of discourse about which Wegener knew nothing. The little geology he had ever known was from practical field experience in hand sample collecting in Greenland, as well as knowledge resulting from a single course during the summer in his undergraduate years. While he was well read in geophysics, geology and geophysics were entirely separate disciplines. He was discovering that geological theorizing throughout the later nineteenth century and well into the twentieth century contained frequent references to former continental masses connecting now-separated continents.[62]

The passage that most excited Wegener appears toward the end of Keilhack's article and was reiterated in a letter of 6 December Wegener wrote to Köppen. "Just now," he wrote, "an article by Keilhack has fallen into my hands: Alte Eiszeiten der Erde (Himmel und Erde 1895, s.249) showing that South Africa, India and West Australia, and also South America, during the Carboniferous, had to have had a simultaneous ice age, with the same flora and so on." He then went on to quote it extensively:

> The extraordinary correspondence of flora and of geological deposits in the broad region between South Africa, the southern borders of Afghanistan, and that of Western Australia make it quite certain that here in the Carboniferous Period there extended a huge continent, which continent certainly at least between India and Africa persisted coherently up until the Tertiary, though the Australian segment detached itself somewhat earlier. Today the greater part of this older continental surface lies in the depths of the Indian Ocean. Whether the newest finds in South America will compel us to connect this southern continent with the others, we cannot yet say. In any case, the existence of this great continental massif, Lemuria, has long been established by a broad range of animal- and plant geographic studies, and geology serves here only to provide a new line of confirmatory evidence, though completely independent. . . .
>
> There has been an attempt to assert a different position for the axis of the earth and to group the region of the Glossopteris flora around a new South Pole. However even when one leaves a South America completely out of the picture and only make the attempt to think of a pole position in which India, Africa, and Australia lie in the most likely position, so would such a pole end up somewhere in between Western Australia and Madagascar, and the location of the sites of the Carboniferous glacial beds farthest from the pole would then only be 30 or 35° from the new equator.[63]

Let us put this long quotation into the context of the discussion between Wegener and Köppen. It appears that Wegener had written to Köppen sometime in late November or early December; that letter does not survive. Neither is it referenced by Else, who remarks only that in the context of Wegener's letter to her on 6 January 1911, first proposing the idea of a continuous continent containing both South America and Africa, her father subsequently "had warned Alfred that he should give up such 'sidelines,' and that it would be enough [work] to explain what was happening in meteorology."[64] Perhaps the letter to Köppen of 6 December 1911 merely picks up the conversation where it had been dropped the previous January, but likely not, since Wegener begins the letter, "Dear father, I must answer your detailed letter immediately."[65] Since the entire contents of this letter pertain to the question of former continents, we may assume that Wegener was answering Köppen's objections in that letter.

Wegener laid out his argument thus:

> I believe that you consider my "Original Continent" [*Urkontinent*] to be more of a fantasy than it is, and you still don't see, that this is merely a question of the interpretation

> of the observational material. While it is true that I only came to the idea by noticing the fit of the contours of the coastlines, naturally the argument must be based on the observational results of geology. These results compel us to accept a land connection for example between South America and Africa that was [later] severed at some specific point in time. One can imagine this course of events in two different ways: 1) through the sinking of a connecting continent "Archhelenis" or 2) through their pulling apart from one another via some huge fault/fracture. Heretofore, because of the *assumed* inalterability of the position of all the land surfaces, most have taken into account 1) and ignored 2). Nevertheless 1) is directly contradicted by the modern doctrine of Isostasy, and by our physical conceptions generally. Thus we are forced once more to take 2) into consideration. And if this results in a series of astonishing simplifications, if it shows us the meaning and allows us to understand the entire history of the geological development of the earth, why should we hesitate to toss the old views overboard? Why should one refrain from expressing these ideas 10 or for that matter 30 years? x) [footnote mark in the original]. Is this, perhaps, revolutionary?[66]

Wegener added a footnote here: "I believe that the older views don't have 10 years to live. Thus far the theory of isostasy is not completely worked out. When this happens the contradictions will become evident and the older view will be corrected."[67]

There is more to this letter that concerns us, but before going on to discuss it in detail, it is important that we stop here to consider what it reflects about Wegener's thought processes and imaginative style. Once again we see this characteristic pattern evident since his work on atmospheric layering. An article "falls into his hands" which contains information on a problem with which he is already concerned, but which does not contain the conclusion that he would draw from the same data. A quick survey of the relevant literature at hand provides "surprising or astonishing simplifications and corroborations." A conviction of the "fundamental correctness" of his intuition forms in his mind well in advance of a detailed survey of the evidence and sends him off on an enthusiastic pursuit of further confirmatory evidence, but not before leading him to assert a bold new working hypothesis that reorganizes the data in a novel way, to produce an argument about the structure of some geophysical entity, dependent to a large extent on analysis of surfaces of discontinuity.

Wegener was decidedly an "enthusiast." His interest, curiosity, and even ardor were easily aroused. Enthusiasm is a double-edged sword and carries both favorable and unfavorable connotations. It may mean someone vitally animated and interested, especially in novelty, or it may mean someone whose judgment is subordinate to his excitement. Wegener was, more often than not, an enthusiast of the former kind. In spite of Köppen's fear that he would be diverted from his core task in atmospheric physics by taking on a "sideline" like the study of former continents, Wegener embraced new enthusiasms without abandoning old ones, by partitioning his interest and increasing his workload. It was this immense capacity for concentrated work that, in part, kept him from superficiality and dilettantism. Once interested in a topic, he pursued it relentlessly.

There is a sense in which the word "enthusiasm" is in danger of trivializing the affective component of scientific investigation, in the case of someone like Wegener. There is a sense in which what we are talking about is more like "love" than "enthusiasm." Sometimes, there is an intermingling of actual romantic love and intellectual productivity, and Erwin Schrödinger is a famous case of this, with bursts of scientific creativity following upon the formation of new romantic attachments.[68] There may be some link like that here. Else Köppen was the first person and for a long time the *only* person whom Wegener told (so far as we know) of his intuition concerning former continents, and he did so in the first stages of a romance that would result in an engagement within a few months and his marriage within a few years.

Yet there is a more firmly anchored and thoroughly intellectual aspect to his pursuit of the idea of continental displacements at the end of 1911 and in the beginning of 1912. Wegener was very much a "theoretical physicist" who, though he had collected data in Greenland in 1906–1908, did most of his work by linking together the data produced by others. Theoretical physics, in the tradition in which he was trained by Planck at Berlin, as well as by Bezold, had a very definite answer to a pressing question that faces all scientists. That question is, when will it be time to gather up everything that we know and make a coherent and synthetic presentation of it in the form of a unified theory? The answer for Planck and for Wegener was that that time is always *now*. For Wegener, a theory was simply an imaginative construct used to order the data in our possession at any given moment. It is an architectural assembly of component elements into a unified picture with the aim of clarifying our thought and guiding further investigation.

For Wegener, a theory was a promising working hypothesis that would lead to new understanding, and *nothing more*. Thus, when he spoke of his idea of former continents, he spoke in terms of not demonstrated *fact* but the "fundamental correctness," and therefore heuristic value, of an *idea*. It was this that led Wegener to say to Köppen, "Why should one refrain from expressing these ideas for 10 or for that matter 30 years?" A theoretical idea (or a hypothesis) is apposite not when it is "confirmed" but when it is likely to lead science in a productive direction. Under such circumstances, hesitation is nothing but procrastination in the face of intellectual urgency.

In terms of the two epigraphs at the beginning of this chapter, we are more concerned with the second, Houston Stewart Chamberlain's characterization of Galvani, and the speed with which exposure to a new physical fact could lead him to make "extensive connections with all kinds of known and still unknown facts and this spurred him on to endless experiments and variously adapted theories." We may recall that Chamberlain's work was part of Wegener's regular inspirational reading while a university student.

Wegener, at the time he came upon these articles by Krenkel and Keilhack, was preparing to return to Greenland and thinking about a scientific program, inherited from Nansen, that could help solve the riddle of the ice ages. Nansen, as we have seen, had entertained the idea of displacements of Earth's pole of rotation as a way to map Northern Hemisphere glacial deposits within some reasonable compass, but he felt in the end that this strategy was insufficient because it could not explain, with continents and landmasses in their present position, how the distribution of flora and fauna in the Northern Hemisphere could be made compatible with any given pole position; he was especially concerned about the inability to have a pole position that could simultaneously produce a semitropical flora in Greenland and a cold-weather flora in Japan.

The extensive quotation from Keilhack, which Wegener sent on to Köppen, dealt with exactly this problem. For his Carboniferous glaciation, Keilhack was unable to come up with any pole position that covered all the glacial deposits without putting some of these regions (showing clear marks of glaciation) within 30°–35° of the paleoequator—the latitude of Buenos Aries, the Kalahari Desert, and Sydney, Australia. A similar distance from the equator in the Northern Hemisphere, with the continents in their current positions, would have required glacial deposits in Los Angeles, Algiers, Baghdad, and Shanghai. Wegener said, at the conclusion of his quotation of Keilhack's article, "What my implication is here, you can see for yourself."[69]

What he meant for Köppen to "see for himself" was that the ice age in the Southern Hemisphere and the distribution of deposits were absolutely confirmed by geology and paleontology. At the same time, no pole position, with continents in their current location, could account for the distribution of glacial deposits, without bringing the ice cap absurdly close to the equator. Thus, the most plausible hypothesis was that the relocation of the pole of rotation of Earth would have to be considered *in tandem* with the relative displacement of continental masses from some former configuration. It is for this reason that Wegener insisted to Köppen that he should consider the idea of an "*Urkontinent*" not as an unfounded or fantastic speculation but as a matter of accounting for all the available data in a way that made things simpler and more plausible. The appeal to simplicity, of course, marks him as a physicist; no geologist of his time would have made a similar claim.

The other essential datum or idea that drove him to consider continental displacements was his astonished discovery that geologists and paleontologists were willing to consider that large stretches of the current abyssal ocean floor were former continental surfaces that had somehow sunk to the bottom of the ocean. The principle of isostasy, to which Wegener refers in this regard, and which we shall consider in more detail in the next chapter, was a way of accounting for many thousands of well-confirmed gravity observations, of the kind he had studied as a graduate student with Helmert at Berlin. These could only be explained by assuming that the crust of Earth floated on the interior of Earth, which must, because of the increase of temperature with depth, be deprived of strength at some point. Astronomical calculations of the mass of Earth, universally accepted and many times reconfirmed, absolutely required that a large portion of its interior must have a much higher specific gravity than any rocks available at the surface, and therefore that this molten, or at least yielding, interior be more dense than the solid rock at the surface. Under these conditions, the idea that huge blocks of Earth's surface could sink into the interior was as physically impossible as the idea that an ice cube should sink to the bottom of a glass of water.

The driving force of Wegener's rapid elaboration of the idea of continental displacements in the next two months, along with his immediate publication of a theory of the origin of continents and oceans, was his recognition that the universally accepted and undeniable geological and paleontological evidence of former connections between continents now separated by thousands of miles of ocean was explained by these very geologists and paleontologists in terms of a theory that appeared to be *physically impossible*. Therefore, the idea of continental displacements seemed to him not a wild or fantastic postulation but simply a means of putting together well-confirmed geology and well-confirmed geophysics into a single working theory.

He closed his letter to Köppen with the following remark: "It is eight days until I begin really to 'collect' [evidence for my hypothesis]. Till then I don't have the time. But I don't intend to take 30 years to do it!"[70] He intended this as a rapid foray into a new field leading to a rapid publication. Instead, the work he embarked on that December would take him the rest of his life, dominate his career and posterity, and overshadow everything he thought of as his "own work."

The Theorist of Continental Drift (1)

MARBURG, DECEMBER 1911–FEBRUARY 1912

> I refer to this new principle [of horizontal displacement of continents], in spite of its broad foundation, as a working hypothesis, and would like to see it treated as such, at least until the persistence of these horizontal displacements in the present shall have been demonstrated, by means of astronomical position fixing, with a precision that eliminates all doubt.
>
> ALFRED WEGENER, 1912

A Copernican Revolution

Wegener had asked Köppen, concerning his hypothesis of continental displacements, "Is this, perhaps, revolutionary?"[1] It is a commonplace that the notion of "scientific revolutions" guides much of the writing of the history of science and has for at least a century. Scientists of the first rank are those who are seen to have precipitated revolutions. Thus, Copernicus, Kepler, Galileo, Newton, Lavoisier, Darwin, and Einstein are each written about as the authors of revolutions. The first modern, and therefore the *archetypical,* scientific revolution was that precipitated by Nicholas Copernicus (these other towering figures notwithstanding). It is no accident that Thomas S. Kuhn's (1922–1996) book *The Structure of Scientific Revolutions* (1962), certainly the best-known book in the recent history of science, was written by a man whose PhD dissertation was published in 1957 as *The Copernican Revolution.*[2]

We celebrate the "Copernican Revolution" as the beginning of not only our current view of the world but also our approach to doing science. This Copernican Revolution, which displaced Earth from the center of the universe and radically simplified the picture of the cosmos, had other revolutionary consequences. In the hands of Johannes Kepler, the idea of perfect circular orbits, an image of cosmic perfection inherited from Plato, gave way to very slightly elliptical orbits. More importantly, Kepler gave up these beloved circles reluctantly, but he did so because he believed absolutely in the accuracy of the numerical data collected by Tycho Brahe, upon which Kepler based his orbital calculations. Historians of science emphasize this event as the first important instance in which a beloved scientific theory was rejected because it conflicted with reliable data, rather than the data being adjusted to fit the treasured theory.

Copernicus's new theory might have been more immediately persuasive had it been better at predicting planetary positions than the old system, but it was not more accurate. It retained, in addition, residual elements of the old theory, which were to be abandoned by subsequent investigators. It contradicted both biblical and classical authority. It advanced itself in terms of complicated mathematical arguments; then and now these are accessible only to the fully numerate (Copernicus himself scornfully rejected the objections of "idle talkers" who would take it upon themselves to pronounce judgment on the theory, though wholly ignorant of mathematics). The theory failed a crucial empirical test—the inability of supporters and opponents alike to observe any stellar parallax. If Earth were actually moving around the Sun at a great distance, it seemed that the relative position of the stars should be different on one side of the Sun than on the other, but no such relative shift was visible. The ad hoc explanation offered by Copernicus, that this was because the stars were so far away, sounded ludicrously improbable given their large apparent diameters viewed (of necessity before 1610) with the unaided eye.

Copernicus's theory nevertheless had from the outset a great virtue to recommend it: with a single comprehensive principle it explained a great number of facts either left unexplained in the old theory or explained in an ad hoc, heterogeneous, and metaphysical way—the retrograde motion of the planets, their varying brightness, the permanent stations of Venus and Mercury close to the Sun. Copernicus solved these problems by centering all motion in the solar system on (or very near) the Sun, now supposed to stand still. This is to say, it made a virtue of simplicity and unity of explanation and, in so doing, established a criterion of adequacy for physical theories—both aesthetic and cognitive—in which unity and simplicity came to have a central place.

By the later eighteenth century a "Copernican Revolution" was already a trope, and Immanuel Kant (1724–1804), in the introduction to his *Critique of Pure Reason*, could celebrate, with the confidence of being universally understood, the "Copernican Revolution" brought about in his thinking by reading the philosophy of David Hume (1711–1776).

Now let us turn these remarks and ideas in the direction of Alfred Wegener. Wegener was himself not only an astronomer by training but also a published author in the history of astronomy. He understood in great detail the sequence from Ptolemy to Copernicus, from Copernicus to Kepler, from Kepler to Newton, and from Newton to Laplace, Lagrange, and Gauss. He had, after all, written a "Copernican" PhD dissertation on planetary tables, specifically to transform a geocentric system written in sexagesimal notation to a heliocentric system written in decimal notation, so that modern astronomers might use the data. Along the way he mathematically modernized and simplified these planetary tables, throwing out a number of spurious "corrections" introduced in the fourteenth century by astronomers desiring to bring observations into harmony with theory.[3]

In considering Wegener's concept of a "revolutionary" development, it is quite sensible to assume that he was speaking of such a "Copernican" move in geophysics. We may recall that Wegener had had university training in the psychology of perception and had studied what we now call "Gestalt reversals," in which the eye rapidly switches from one visual configuration of an image to another, radically reorganizing the data into a new form. He had written to Köppen, in early December 1911, that his hypothesis of continental displacements was not an imaginative creation or a fantasy but a consequence of just such a radical reorganization of existing observational data into a new picture, producing simplification and coordination in the place of previous complexity and contradiction.[4]

Let us then consider Wegener's idea of continental displacements in Copernican terms. In the standard hypothesis of Wegener's time, the geological and paleontological continuity across deep oceans was explained by the sinking of large continental fragments to form ocean bottoms, while what we call "the continents" were the surviving remnants of formerly much more extensive continents, remnants that remained fixed in place. Wegener's Copernican rewriting explained these continuities not by the sinking of vast continental fragments, creating new oceans in their subsidence, but by continents splitting and drifting apart across the face of Earth and creating ocean basins by their lateral motions.

Wegener's hypothesis was Copernican both in its form and in its revolutionary intent; in consequence, it shared nearly every single problem that had beleaguered the new theory of Copernicus. Wegener's hypothesis was certainly no better than the old hypothesis at explaining the continental positions. It contradicted classical geological authority. It defied both aspects of physical theory (especially a leading version of solid mechanics) and common sense. It was based on complicated, numerically framed arguments in a broad range of sciences—some of them, like radioactivity and seismology, very new and unfamiliar to all but a few pioneering specialists. Like the Copernican theory, it failed a crucial observational test—no one could produce compelling, confirming evidence of continental motions continuing in the present, a problem very like Copernicus's problem of stellar parallax. If there were lateral continental motions, then not only certain consequences of these motions but the motions themselves should be observable today.

Wegener's hypothesis of continental displacements nevertheless had from the outset the same great virtue to recommend it as had the theory of Copernicus. It explained with a single comprehensive principle a great number of facts either left unexplained by the old hypothesis or explained in an ad hoc, heterogeneous, and metaphysical way: the jigsaw puzzle geometry of the continental shelves, the different densities of continental and ocean floor rock, the existence of similar life-forms on continents separated by abyssal oceans, the continuations of geological structures and sedimentary sequences on both sides of the Atlantic, the odd dispersal of the Southern Hemisphere remnants of the Carboniferous glaciation, the appearance of tropical fossil biota in high latitudes and temperate fossil biota at the equator, the existence and locations of mountain ranges, the location of volcanoes at continental margins, and a great many similar questions.

Wegener's Sources

Wegener began to work on the question of continental displacements on 14 December 1911, and by 6 January 1912 he was prepared to give a public lecture in Frankfurt, entitled "Die Herausbildung der Großformen der Erdrinde (Kontinente und Ozeane) auf geophysikalischer Grundlage" (The geophysical foundations of the development of the large features of Earth's crust [continents and oceans]). The audience was a plenary session of the German Geological Association, founded two years before by Emanuel Kayser, Wegener's dean at Marburg, who was both chair of the Geological Association and editor of its journal, *Geologische Rundschau,* from 1910 to 1920. It was Kayser who arranged for Wegener to give the lecture. This initial presentation of Wegener's hypothesis was followed four days later by a lecture in Marburg before the Gesellschaft zur Beförderung der gesamten Naturwissenschaften, under the title "Horizontalverschiebung der Kontinente" (Horizontal displacement of the continents).

By the seventeenth of January Wegener had a manuscript of fifty pages, and by the twenty-ninth he had put it through so much revision that he could barely read the first twenty-five pages.[5] A month later, on 24 February, he wrote to Köppen that he had sent off the final draft of his article on continental displacements (now sixty-nine typed pages long) to Paul Langhans (1867–1952), the editor of *Petermanns Geographische Mitteilungen,* the leading journal of geography, and the clearinghouse within the German-speaking world for everything to do with exploration. Wegener's entrée to this journal was dual: the journal welcomed highly speculative and theoretical work on major issues in geography (the origin of continents was certainly such), and Wegener was well known as a polar traveler—his prospective 1912 expedition with Koch to Greenland had already been announced in *Petermanns Mitteilungen*. Wegener sent another copy of the completed manuscript to Emanuel Kayser, who promised publication of an abridged and geologically oriented version of the hypothesis, in *Geologische Rundschau*.[6]

The extremely rapid pace of Wegener's research, presentation, and publication raises an interesting question: How did a thirty-one-year-old atmospheric physicist, who had only known for a few weeks that most geologists believed that Brazil and Africa were once connected, end up producing, in slightly over two months, a seventy-page manuscript that collated an enormous range of geological, paleontological, and geophysical evidence into an entirely novel argument about the origin of continents and oceans?

The answer lies, to a certain extent, in a pattern of scientific publication that no longer exists. If Wegener had tried to piece together his argument from individual geological, paleontological, and geophysical journal articles, his labor would have lasted not a few weeks but many years. However, the pattern of publication in the Germany of his day, along with the corresponding attitude toward new theory, came to his aid. In his world, though no longer in ours, senior scientists felt both the right and the obligation to spend the latter part of their careers producing works that integrated enormous quantities of existing material into coherent syntheses. Wegener was able to avail himself of a series of significant, often multivolume *synthetic* treatises in geology, oceanography, and geophysics, all published within a year or two of his own attempt to write about continents and oceans.

Without these books his initial formulation of his hypothesis would have been an impossible task. Wegener's initial papers on continents and oceans and all of his subsequent publications on continental displacements were made possible by such syntheses. His initial seventy-page paper was, therefore, a higher-order synthesis and distillation of work already highly organized and integrated by senior scientists for whom it was the work of *decades*. When we say that Wegener "discovered" or "invented" the modern theory of continental displacements, it is appropriate to recognize that though the germ of the conception was his own, the evidentiary foundation consisted in predigested syntheses of many thousands of individual journal articles written by many hundreds of scientists in the course of the previous half century, brought together by a few leading figures in geology, geophysics, and oceanography, all at the end of their careers and very near the end of their lives.

The work that, above all, made Wegener's hypothesis possible may be found in the following very short list of very long books. The first, and probably the most important for the Copernican part of Wegener's effort, was *Das Antlitz der Erde* (*The Face of the Earth*) by the Austrian geologist Eduard Sueß (1831–1914). This multivolume work took Sueß twenty-one years to complete (1883–1904).[7] A second important work, which helped Wegener frame his approach to the question of continents and oceans, was the *Handbuch der Ozeanographie* (Handbook of oceanography) by Otto Krümmel (1854–1912), in two large volumes (1907, 1911).[8] In matters of intercontinental correlation of flora and fauna, Wegener relied quite heavily on *Die Entwicklung der Kontinente und Ihrer Lebewelt* (The evolutionary development of the continents and their life-forms), published in 1907 by Theodor Arldt (1878–1960).[9] A fourth significant and in every respect indispensable work was *Physik der Erde* (*Physics of the Earth*) by Maurycy Rudzki (1862–1916), in a 1911 German translation of the 1909 Polish original (*Fizika ziemi*).[10] Finally, as a kind of universal handbook for the study of the intercontinental correlation of sedimentary rocks, the mineralogy of the continents, and data on earthquakes and other dislocating motions of the crust, there was *Lehrbuch der allgemeinen Geologie* (Textbook of general geology) by Emanuel Kayser (1845–1927), particularly the third (1909) edition of this massive, two-volume, frequently updated work.[11]

We should learn something about each of these, to get a sense of what they provided Wegener. The work of Edward Sueß is nearly unique in the history of geology, a science that certainly has had an abundance of towering and comprehensive thinkers. Sent to prison for his participation in the Revolution of 1848, Sueß was barred from university admission and academic employment but began a career as a field geologist in the Alps, taking a job at the central Paleontological Museum in Vienna. Twenty-four years later, having risen to the directorship of that museum, he published a highly influential short book, *Die Entstehung der Alpen* (1875), in which he proposed to interpret the folding of the Alps neither as transverse compression nor as uplift but as the overthrust of the entire mountain system to the north by a one-sided push from the south. His qualitative interpretation of the dynamic cause of this movement was the unequal radial contraction of the cooling Earth.[12]

Sueß represents, as well as serves as a convenient benchmark for, the assimilation of stratigraphical and paleontological work into a dynamic history of the motions of Earth's crust. The thematic introduction of this approach to geology in 1875 was followed thirteen years later by the first volume of his massive treatise *The Face of the Earth*, which in its German original eventually filled three large volumes and more than 2,000 pages. This was not an introductory text for students but a mature summary of geology written by a senior professional for his senior professional colleagues and peers.

Sueß, in his massive synthesis, was deliberately moving beyond the conception of "uniformitarianism" proposed by the great British geologist Charles Lyell (1797–1875), whose book *Principles of Geology* was Darwin's constant companion on his voyage on the *Beagle*. Lyell's doctrine was an argument against the catastrophism (cataclysmic upheavals, collapses, floods, fires, etc.) of earlier versions of Earth history, arguing instead that all the phenomena of geology have been and are produced at the same rate and in the same way that we see them today: "the present is the key to the past." Sueß aimed, in contrast, to enlarge the meaning of "uniform" change to include some phenomena not currently visible, having in mind the tremendous dislocations evident in Alpine mountain systems. *The Face of the Earth* was an attempt to turn geologists away from a too great fascination with miniscule causes (the raindrop hollowing out the stone, the coral polyp building up the reef) and to turn their faces toward the very large dislocations caused by the episodic collapse of large sections of Earth's crust, the subsequent massive overthrust in great mountain belts, and the creation of ocean basins by the downfaulting and subsidence of great blocks of former continental crust.

Sueß was a brilliant writer. *The Face of the Earth* differs from other theoretical treatises because it is not a series of didactic statements or principles that are then evidenced by examples, but quite the reverse: a spellbinding narrative of the genesis and structure of the great forms of Earth's crust seen from an aerial vantage point, imagined as they might be viewed by a visitor from space. Those of us accustomed to seeing pictures of Earth taken by astronauts returning from the Moon can hardly imagine the effect of Sueß's visualization and powers of description on successive generations of geologists into the twentieth century. In its sheer beauty and the rapid pace of its dynamic prose, it is without doubt the most excitingly written book in the history of geology.

No one has ever claimed that Sueß was not devoted to the theory of a contracting Earth, but it is perhaps not often enough emphasized that *The Face of the Earth* was designed as an integrated defense of the contraction hypothesis over against Charles Lyell's notion of oscillating continents. Lyell's vision of the majestic, slow rising and sinking of the continental platforms through geologic time dominated thinking in the English-speaking world well into the mid-twentieth century. Sueß, however, succeeded in converting a very large plurality of the rest of the world's geologists to his views, and it is important to recognize the extent of this success if we are to understand the crisis within the world geological community when Sueß's central conception of Earth contraction began to be challenged at the beginning of the twentieth century by increasingly sophisticated work in geophysics. Edward Sueß's *The Face of the Earth* would play Ptolemy to Wegener's Copernican conception of continental displacements. Everything in Sueß's work is driven by subsidence, by sinking, and it was precisely this subsidence, as the universal motor of geological dynamics, that would be the theory's eventual downfall.[13]

If Edward Sueß's *The Face of the Earth* would provide the structure against which Wegener could push back, Otto Krümmel's *Handbook of Oceanography* not only was an invaluable source of information about the physical characteristics of the oceans but also provided a great deal of information about the topography of the ocean floors. Krümmel was a great follower and admirer of Sueß, and he wanted to treat the ocean floors as submarine dynamical geological regions, dynamic Earth surfaces that happened to be covered with water. Of course, if one believed that ocean floors were former continental surfaces, then their submarine topography should reflect their former life above the surface of the water. Krümmel used the same scheme to differentiate different kinds of ocean floor which Sueß had used to discriminate different kinds of continental surfaces.

Krümmel's book contained an extensive section on physical oceanography, with a discussion of waves, tides, currents, and ocean circulation, and a great deal of information about layering and surfaces of discontinuity in the ocean. There was an extended and highly sophisticated discussion, based on research up through the first decade of the twentieth century, on the behavior of water masses, those coherent bodies of water with a salinity (chemical composition) that demarcated them from the ocean in general. Such density and salinity differences caused these water masses to sink a certain distance and then travel through the ocean; these were used by oceanographers to track global ocean circulation. Wegener was once again reminded, by this approach to the topic, of the very strong analogies between the atmosphere and the ocean.

At the very beginning of the first volume of his work, Krümmel had given an interesting and extended discussion of the various ways in which one might decide on a classificatory scheme for the world's oceans. He reviewed six different possible ways to organize his material. Oceans could be classified by their location and arrangement, by their size, by their form, by their chemical composition and contents, by their dynamic behavior in motion, or, last of all, by their origins, their *Entstehung*.[14] "Of all the classifications of the objects of the dry surface of Earth," he wrote, "it is the genetic classification [*Entstehung*], based on the origin of the forms, which ranks highest. For who explores the intrinsic organization and modes of generation of different forms, encounters, as Alfred Hettner says, the real nature of things."[15]

It was Krümmel who convinced Wegener that his approach to continents and oceans should be framed in terms of their *origins*, rather than some of the other categorical approaches. These other approaches were available, of course, and particularly attractive to Wegener was the approach through consideration of *chemical* differentiation of Earth. In Sueß's account, Earth was seen to be composed of three concentric shells of different density and of different chemical composition. The outermost shell, as well as the thinnest, Sueß had called "Sal," to indicate that its principal constituents were silica and aluminum. The name "Sal" was confusing, because this is the Latin word for "salt," and the term was soon adapted to "Sial." For the mantle of Earth, the shell below the silica and aluminum crust, Sueß used the acronym "Sima," an indication that its principal constituents were now silica and magnesium. Finally, there was the metallic core, composed of "Nife," nickel and iron. Wegener knew he could trust the latter aspect of the conceptual scheme because Sueß had derived it from calculations of Earth's density by Wiechert, at Göttingen.[16]

The chemical differentiation of discrete Earth shells, so attractive to Wegener, would be an important piece of supporting data. The main line of attack, however, would be through the *origin* of the major forms of Earth's surface. It was clear to Wegener that even though Krümmel was correct (that the best approach to such questions was that of "origins"), nevertheless Krümmel's own view of the origins of oceans was as fundamentally mistaken as Sueß's view of the origins and histories of the continents. Krümmel had adopted wholesale Sueß's version of the contraction theory and applied it to the oceans, in detail. Wegener knew that this answer was wrong and that he was in a position to correct it.

Nearly as valuable as either of these two sources was the comparative treatment of the development of life-forms on the continents throughout geological history provided by Thedor Arldt, in his *Evolutionary Development of the Continents and Their Life Forms*. Arldt was working in a discipline well known today, but it was sufficiently novel in 1907 that in the opening pages of his 730-page treatise he felt compelled to define for his audience the word "Paleogeography." Arldt's methodology was interesting and comprehensive. He had read and assimilated the work of most of the major paleontologists and paleogeographers of his day, but rather than using a few "index fossils" to map connections between continents, he produced statistical summaries of identical species on continents now separated by deep oceans (such as South America and Africa). He did this for each different period of geologic time and expressed his results in terms of percentages of species shared at any given time and the likelihood of an actual connection.[17]

As helpful as Arldt's statistical summaries of species' abundance may have been, Wegener did not have to read all 730 pages to get the point. The "business end" of this remarkable book was the ensemble of twenty-three global Mercator maps inserted as sequential plates tipped in at the end of the text. These displayed, in multicolored format, the conclusions laboriously arrived at by Arldt in reading every existing account of the distribution of mammals, of reptiles, and of amphibians and the Dipnoi (freshwater lungfishes, species that Arldt found particularly decisive for establishing continental connections). There were also maps of mountain ranges of Earth, showing continuations of the mountains of Scandinavia in Labrador, of the mountains of Europe (especially the Alps) in the Appalachians of the United States, and of the mountains

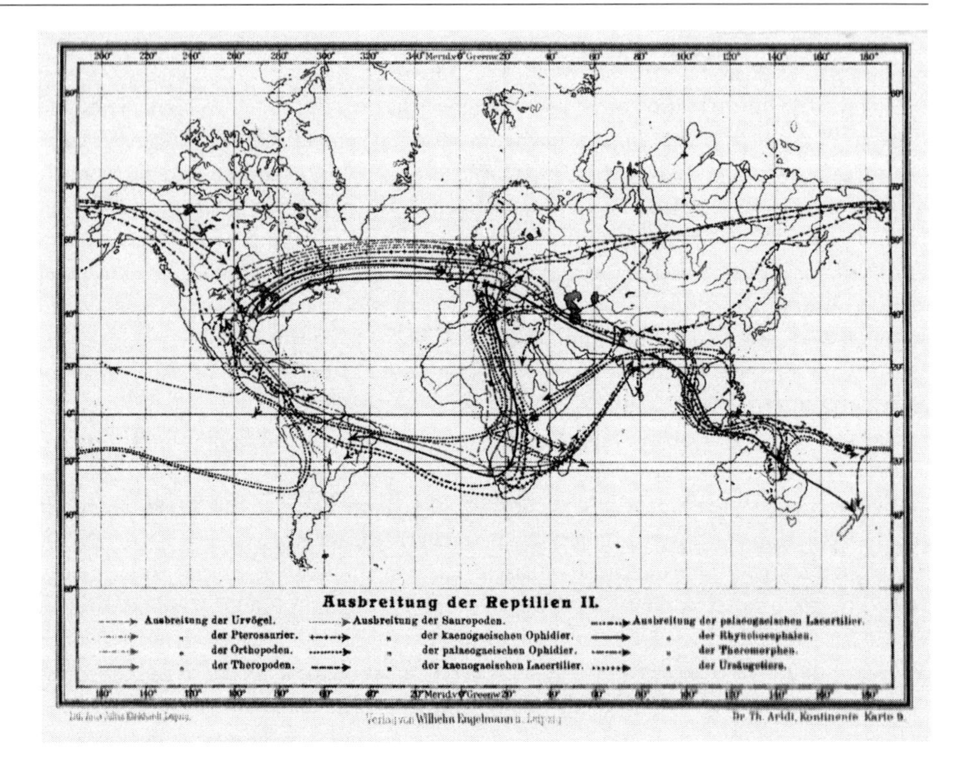

Theodor Arldt's map of the "Diffusion of the Reptilia" (including dinosaurs). Note the main migration routes "across the Atlantic," indicating matching fossil species on both sets of continents (Europe–America and Africa–South America). This and many similar charts in Arldt's book made a persuasive argument for a former continuity of continents now separated by abyssal oceans. From Theodor Arldt, *Die Entwicklung der Kontinente und ihrer Lebewelt* (1907), in *Handbuch der Palaeogeographie*, 2 vols. (Leipzig: Gebrüder Borntraeger, 1919–1922), 1647.

of North Africa in the geology of the northernmost part of South America. Following these were maps of former continental surfaces superimposed on the geographic distribution of the current continents, for the Cambrian, Silurian, Devonian, Carboniferous, Triassic, Jurassic, Cretaceous, early Tertiary, and later Tertiary periods, and finally the glacial age. These paleogeographic reconstructions were derived from already-existing maps that Arldt had found in the literature, which seemed to him to best represent the percentage distributions of fossil species he had calculated on his own.[18]

From Wegener's standpoint, these maps painted an overwhelmingly convincing picture of former geological and, more importantly, paleontological continuities throughout all of Earth history since the Cambrian. This continuity, particularly notable across the North Atlantic and South Atlantic continents, began to attenuate just where Keilhack had said it would: somewhat restricted connections in the Cretaceous, further reduced in the early Tertiary, completely broken in the later Tertiary except for a small land bridge between Greenland and Europe through Iceland, and finally pinched off toward the end of the Quaternary glacial age, or, as Arldt termed it, the "Diluvialzeit."[19]

All of this was empirically reconfirmed and reinforced by Kayser's geological textbook, with its particular attention to intercontinental correlation of stratigraphical sequences and fossil species, in which the material presented by Arldt in a series of maps found a more comprehensive tabular presentation.[20] The message that Wegener got from these geological and paleontological works was that, within the German-speaking world at least, and to some extent the French-speaking world as well, there was nearly universal consensus on the existence of great paleocontinents of some sort, even if the reconstructions varied. Geological orthodoxy seems to have been that, through much of geological history, there was at least some connection across the Atlantic in both the Northern and Southern Hemispheres. Subsequently, for whatever reason (accounts varied from geologist to geologist), this had been progressively attenuated and disrupted in the Tertiary and finally severed within the past few hundred thousand years.

In terms of influence and importance in framing Wegener's physical arguments, it would be difficult to overestimate the influence of Rudzki's *Physics of the Earth*. In chapter after chapter Rudzki gave a dispassionate, calculation-based, comparative treatment of all the chief physical hypotheses still in play in the first decade of the twentieth century to explain the major features of Earth's surface and interior. In Wegener's first two published papers on continental displacements, he made free use of Rudzki's critique of the contraction theory, as well as his critique of Sir George Darwin's hypothesis that the Pacific Ocean had been created in the separation of the Moon from Earth (at a time when Earth had supposedly become an unstable ellipsoid of rotation). Rudzki's division of geophysics into topical sections, his histories of how their results had been achieved, and his conception of their proportional weighting in assessing our conception of the actual state of Earth's interior were all adopted wholesale by Wegener.

It was from Rudzki that Wegener became aware of recent advances in the study of Earth's gravity field and of seismology, as well as how these together provided the first real quantitative data on the state and structure of Earth's interior. From Rudzki he also learned of the implications of the distribution of radioactive materials in the crust for the contraction theory and for geology more generally, especially for its likely decisive resolution of questions about the age of Earth. In the matter of displacements of Earth's pole of rotation, Wegener found in Rudzki data on possible permanent movements of Earth's pole representing translocations, not merely axial wobble or precession. It was here in this volume as well that Wegener found a sophisticated treatment of the possible physical states of matter at different depths in Earth's interior. With these came a thorough discussion of the difference between strength and rigidity in solid materials under high confining pressures and high temperatures, supplementing what Wegener had studied on this topic while writing about the various phases of ice for his book on the thermodynamics of the atmosphere. Not only did Wegener follow Rudzki's arguments and critiques in each of these areas, but he often repeated them verbatim in his published work.

The fundamental task for Wegener was not just to assimilate this new work and recommunicate it but to create an intellectual framework in which he could unite the very solid results of intercontinental correlations of rocks and fossil animals produced by geologists and paleontologists with a sound geophysical theory of Earth's constitution and therefore its dynamic behavior. This geophysical theory would replace a variety of unsound, unphysical, and impossible theories that he found not just embedded in but used as framing concepts for the large geological, oceanographic, and paleontological syntheses he had just encountered and studied. Such a unification of sound geology and sound geophysics would create a milieu in which his hypothesis of continental displacements could receive a fair hearing.

Whose Earth?

There was nothing inherently difficult, recondite, or hard to access in the geophysical model that Wegener was proposing. The question he faced was not so much intellectual as jurisdictional: "who had the right to speak." The latter was a matter of academic specialties and academic communities. We may orient ourselves to the problem that Wegener was beginning to discover by looking at the difference between a geologist's Earth and a physicist's Earth at the time of his writing. Geology was, in the main, a qualitative and historical science, an advanced natural history of rocks, rather than a preliminary approach to material which planned eventually to reduce itself to an aspect of physics and chemistry. There was chemistry, mineralogical chemistry, of course. And there was a mathematical element to the surveying of the globe, measuring the thicknesses of various strata and mapping their lateral extent, considering rates of erosion and sedimentation in a river delta, and such matters. But with the exception of those geologists who received their training in schools of mines and thus had a pressing professional concern with the behavior of solids under load as required for mine engineering, working geologists typically had no formal knowledge of physics whatever and no mathematics past trigonometry.

With respect to conditions in Earth's interior, for geologists, the most useful sort of interior and the most convenient Earth would be a thin strong layer over a weaker layer, that is, an underlayer that should be more pliable and capable of ductile deformation than the surface—but how much more? How weak it should be was not quantified, and geologists did not require that it should be, only that they be allowed to assume it.

As opposed to the hot but qualitative and indistinct interior of Earth preferred by the geologists, physicists offered a mathematical Earth, a homogeneous solid cooling from the center outward by conduction. Cooling by conduction alone was probably not a realistic assumption, nor was the notion that Earth was homogeneous, but these assumptions were adopted because the consideration of convective cooling and of internal inhomogeneity made the calculations too difficult. This physicist's Earth is the one we see in Joseph Fourier's (1768–1830) *Analytical Theory of Heat* (1822), in every way a founding document of modern mathematical physics, and therefore the one we see in William Thompson's (Lord Kelvin's) extensive theoretical development in the last quarter of the nineteenth century of the "solid Earth" so beloved by British physicists.

These preferences for very different Earths could be maintained because into the very late nineteenth century there were no synoptic data about Earth's thermal condition with increasing depth, there was no experimental study of the effect of pressure on the melting curves of rocks and minerals, there was no seismology to reflect waves off boundary surfaces within Earth, and there were limited and theoretically troubled data about Earth's gravity field and the reasons for the departures from "normal" gravity at one place or another on Earth's surface.

Therefore, almost all thermal and dynamic models of Earth's interior before about 1900 were based on neither observation nor experimentation but rather extrapolations from astronomical data. These were still being debated during Wegener's graduate career. Gravity surveys could not provide a criterion by themselves for the condition of Earth's interior. Solid mechanics could as yet provide no real guidance. Until the very end of the nineteenth century there was no consensus on whether rocks generally contract or expand on solidification.

The situation described immediately above, with the inability of these various parts of physics to provide definitive answers to long-asked questions about the interior of Earth, was a nineteenth-century predicament. While a changed situation in "a new century" is often a historian's fiction, it fits the sequence of events in geology and geophysics quite well. In the first decade of the twentieth century, and especially toward the very end of this decade and within a year or two of Wegener's first efforts in this direction, all of this was radically changed.

Wegener's article in *Petermanns Mitteilungen* was the first self-conscious attempt to reconcile the disparate results of the old geology with the new geophysics in the twentieth century, and irrespective of its other (and quite considerable) merits, it would be worthy of our attention for this reason alone. Wegener was always a little more daring than most of his contemporaries, and certainly more daring than most of his senior colleagues. He was always ready to bring about a reorganization and reconsolidation of new data and provide a new picture. But in terms of fundamental novelty, his work in continental displacements was an order of magnitude more significant and more penetrating than anything he had done heretofore.

The Formulation of the Hypothesis of Continental Displacements

Wegener was always under time pressure, but perhaps at no time in his life did he endure it more than in January and February of 1912, when he was assembling the data to support his idea of continental displacements. On the day after his initial address to the Geological Association in Frankfurt (his remarks on continental displacements there on 6 January had evoked, as he wrote to Else, "a storm of indignation"), he wrote to Prof. Warming of the Danmark Expedition Committee to explain his situation. He told Warming of his new hypothesis of continental displacements, on which he had just the day before given his first public address, an idea that had the "strongest possible claims on his attention."[21] He continued, "My publications in the year 1911 amounted to 600–700 pages, and about half of this total belongs to the Danmark Expedition, and at the moment I have six different scientific papers in press." He went on to tell Warming,

> This sweeping hypothesis (on the horizontal displacement of the continents) which I presented publicly, in its development will, I believe, be very interesting to you. This comprehensive work equally concerns geology, geophysics, geodesy, and geography and is not without practical application in the interpretation of the longitude measurements of Danmark Expedition. It will naturally, in the future make very strong claims on my time and I just do not have the heart to leave it alone until I have completely finished my work on the results of the Danmark Expedition . . . it will be a delicate matter to find the right balance.[22]

Warming was predisposed to like Wegener, and he appreciated his candor and honesty in this matter: Wegener knew that he would get a dispensation and be allowed to proceed with his work on continental displacements without being accused of abrogating his expedition contract.

No sooner had Wegener cleared the way with the Danmark Expedition Committee to work on his hypothesis of continental displacements than a massive new time constraint descended upon him. In the first week of February Koch wrote to him to inform him that he had managed finally to raise the 50,000 marks in Copenhagen needed to launch the expedition to cross Greenland in the summer of 1912, and that they were "on." Not only would Wegener have to go immediately to work to raise his share of the money, but he would have to have it in hand by April at the latest.[23] However much Wegener had wanted to go on this trip, it was now also an obligation from which he could not escape. It would interfere with his development of the hypothesis quite severely. Thus, in the published version of "Die Entstehung der Kontinente" in *Petermanns Mitteilungen* (the first of three sections appeared in April) there is a footnote to the title that reads, "Due to my participation in a Danish expedition to Greenland, I am obliged to postpone the detailed treatment I had planned and to publish for the moment only this preliminary communication."[24]

Let us turn our attention now to the "preliminary communication" that Wegener published in *Petermanns Mitteilungen,* beginning in April 1912. On the day he mailed it out to the journal, he had written to Köppen, "The *Urkontinent* (69 typed pages!) is in one copy off to Prof. Langhans (Petermanns Mitt) from whence it will probably soon return with the modest comment: too long!"[25] But Langhans did not cut it, nor did he want it cut, and he considered Wegener's entire argument essential if it was worth publishing at all. Perhaps some of Langhans's enthusiasm stemmed from the extremely careful framing of the argument and the extraordinary efforts that Wegener made at the very beginning to put his ideas into context:

> In what follows, a first rough attempt is made to interpret genetically the large-scale features of our earth's surface—the continental platforms and ocean basins—in terms of a single comprehensive principle: the horizontal mobility of continental blocks. Wherever we have heretofore accepted that ancient land-connections sank to the depths of the oceans, we will now suppose a splitting and drifting apart of continental blocks. The picture we obtain in this way of the nature of our earth's crust is new and, in some respects, paradoxical; yet it does not lack, as will be shown, a physical foundation. On the other hand, in the provisional investigation attempted here, based simply on the major findings of geology and geophysics, so many surprising correlations and simplifications are already revealed that it seems to me, on these grounds alone, justified, indeed even necessary, to replace the old hypothesis of sunken continents with this new, more productive working hypothesis. The demonstrable inadequacy of the former has already been made obvious by the opposing theory of the permanence of oceans. I refer to this new principle, despite its broad foundation, as a working hypothesis, and would like to see it treated as such at least until the persistence of these horizontal displacements in the present shall have been successfully demonstrated by means of astronomical position-fixing, with a precision which illuminates all doubt. It is also not superfluous to point out that this is a first draft. Elaboration of the details will most likely show the hypothesis must be modified in several respects.[26]

On the one hand, we have "a first rough attempt," "provisional investigation," "a working hypothesis," "a first draft"; on the other hand, Wegener asserts "so many surprising correlations and simplifications . . . necessary, to replace the old hypothesis," a "new principle," and a "broad foundation." This is an interesting combination of tentativeness and certainty, of caution and bravado. But, in the end, he was quite clear: it is a working hypothesis, not even a theory, though, like theory, a guide to investigation. It was not a demonstration about the world, not an axiom, law, or truth; it was a heuristic device for combining scientific findings. Wegener claims that the hypothesis is "based simply on the major findings of geology and geophysics." He claims to have discovered no new facts, nor to have invented new justifications. It is his conviction here, as it was with his arguments about the vertical structure of the atmosphere, that the evidentiary foundation for the generalization did not lie in future discoveries but was already amply present in the professional and refereed literature and available for inspection. All that had been necessary was for someone to come forward and say "the obvious."

Wegener began with a brief preview of the argument. He would begin by asking whether, on the basis of general geophysical and geological knowledge, horizontal movements of segments of rigid crust were possible, and if so, how these movements might take place. He would then make a modest attempt to follow the splitting and displacement of continental blocks through geological history and to show the connection of these displacements with the origin of mountain ranges and the displacement of the poles. Finally, there would be a brief attempt to establish the plausibility of contemporary, continuous, measurable displacement of the continents and to offer an explanation of the displacement of the pole of rotation.

Wegener offered one more preliminary: "Before we begin the presentation, a few brief historical remarks are in order."[27] At issue here were his precursors, those who at some point had proposed the splitting and displacement of continents, especially the Atlantic continents, before Wegener had done so. Precursors come in two different varieties: discredited ones from whom one must distance oneself, and credible ones who must be explained away in order to protect one's own claim of originality. Wegener had discovered one of each. (He would later find many others.) The discredited precursor was the American astronomer W. H. Pickering (1858–1938). Pickering had also noticed the parallelism of the Atlantic coasts and had proposed that the Atlantic continents had split and slid apart as a part of the cosmic event when the Moon had been created in "fissipartition" from Earth, leaving the Pacific Ocean as the scar of its removal. Osmond Fisher (1817–1914) had popularized this hypothesis in the English-speaking world, in his *Physics of Earth's Crust* (1881). Wegener had learned from Rudzki that the original calculations for this theory, made by Sir George Darwin (1845–1912) and later advanced by Henri Poincaré (1854–1912), were based on mathematical errors already documented by the Göttingen astronomer and physicist Karl Schwarzschild (1873–1916). Wegener commented concerning Pickering, "Given the removal of the event into an unverifiable grey past, and its admixture with an obviously incorrect hypothesis, the work is at best only of historical interest."[28]

Wegener, however, also had a real precursor to deal with:

> On the other hand, a recent paper by F. B. Taylor [the American geologist Frank Bursley Taylor (1860–1938)] may be regarded as a precursor of the present work, even though both arose in complete independence from one another. Taylor locates, as we do, the horizontal displacements in geologically well-known periods, especially in the Tertiary, and connects them with the great Tertiary fold systems. He particularly emphasizes the breakaway of Greenland from North America based on the parallelism of the coasts. Thus our view concerning a second locale on Earth's surface is already expressed here. . . . Taylor has, I daresay, not realized the enormous range of consequences associated with the assertion of such horizontal displacements of the continents. Since he fails to subject their likelihood to any investigation: despite their contradiction of established conceptions—for the most part his opinions have been greeted with a shake of the head. They could not, as I have stated, have provided any stimulation for the present work as I became acquainted with them too late.[29]

This is a reasonable and fair characterization of Taylor's work. Taylor was a follower of Edward Sueß and was attempting to extend Sueß's theory of the formation of arcs of fold mountains by imagining compressive stresses acting from polar regions toward the equator, pushing continents into plicated arcs. Taylor appears to have believed that Earth was perfectly spherical until some point in the Tertiary, at which time it began to shed fragments of north polar and south polar continents toward the equator, as part of its attempt to become oblate. Taylor's descriptions of these movements, especially of the rifting of Canada from Greenland in the latter part of the Tertiary and of the drifting of South America and Africa away from the Mid-Atlantic Ridge in the Carboniferous, are easy to follow. His physical explanations, on the other hand, are qualitative, confused, and in fairly obvious contradiction of all the celestial mechanics from the mid-eighteenth century down to the time of Lord Kelvin, Sir George Darwin, and Henri Poincaré.[30] One is inclined to agree with Wegener's assessment that these arguments could only be greeted "with a shake of the head."

The Geophysical Argument

When Wegener, leaving precursors and history behind, actually begins his summary of the geophysical arguments against the contraction theory and in favor of his own notion of continental displacements, it is immediately clear how much his powers of exposition had benefited from the two years he had spent writing the undergraduate textbook *Thermodynamics of the Atmosphere*, a volume already popular with students for its lucidity and comprehensibility. Wegener's explanatory skill had further benefited from reducing the scope of that 331-page book down to a 70-page digest for the review article he produced in 1911 for Abderhalden's journal.

Wegener's writing on continents and oceans for *Petermanns Mitteilungen* is concise without being telegraphic, diffident without being casual, exact without being pedantic, and instructive without appearing didactic. He gets to the point immediately, without parenthetical commentary or peripheral excursions. "The problem of how one may explain the platform-like elevation of the continents above the deep-sea floor is an old one."[31] The answer cannot be the forces that produce mountains, for (and here he invokes the authority of Albert Heim [1849–1937], the great Swiss Alpine geologist and theorist of mountain formation) "the continents appear as immense broad pedestals from which, making up only about one five hundredth of their bulk, mountain ranges rise like diminutive ribs. . . . The movements of the crust, which sever the continents and oceans from one another, are thus no doubt different from those which have since wrinkled the crust on the great continental plateaux."[32]

Wegener turned immediately to the initial inspirational idea that led him to this entire line of investigation: the realization that the continents at the 200-meter (656-foot) depth contour have exactly the outline of their sea level surface. He says nothing here about the experience of reading Andree's *Atlas*; instead, in support of his argument, he reproduces the "hypsometric curve" of Earth's crust, the distribution of the elevations and depths of the continents and the oceans, which he had taken from the first volume of Krümmel's *Handbook of Oceanography*. As one can see in the accompanying diagram, the vertical axis is elevation above and depth below sea level, expressed in meters, and the horizontal axis is the area at each of these elevations expressed in millions of square kilometers. The result is a mean continental elevation of 700 meters (2,297 feet) above sea level and a mean oceanic depth of 4,300 meters (14,108 feet) below that level: Earth's distribution of elevations is bimodal. The shape of the curve

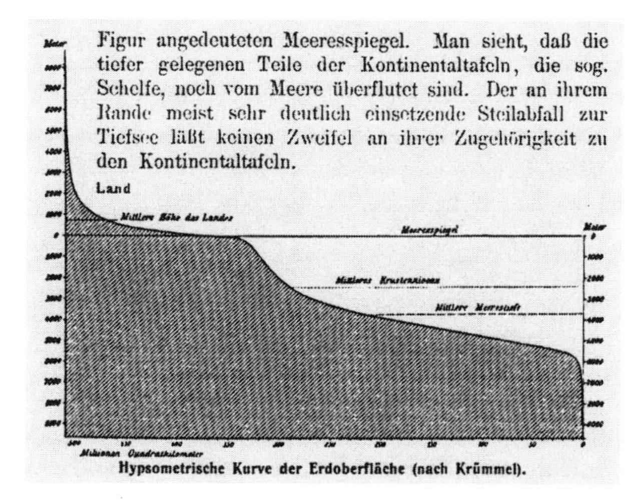

Wegener's "Hypsometric Curve of the Earth's Crust (after Krümmel)." The vertical axis is elevation in meters; the horizontal axis is in millions of square kilometers. The chart shows a bimodal elevation: a mean level for the continents, and a mean depth for the ocean—both of these very far from sea level. Many of Wegener's critics would have great difficulty understanding the significance of this bimodality. From Alfred Wegener, "Die Entstehung der Kontinente," *Petermanns Mitteilungen* 58 (1912).

shows that the continental shelves down to about 200 meters are indeed a part of the continents, and the steep drop thereafter leads to a much gentler curve of distribution of different ocean depths, the abyssal ocean floors.

So the continents "exist." They are not an artifact of sea level but a distinct layer in Earth, one of two preferential modal distributions of altitude. But why are there two distinct layers? "The opinions which exist today," Wegener continued, "concerning the origin of these remarkable tabular elevations of Earth's crust present an extraordinary instance in natural science of self-contradictory befuddlement. Although in this sort of work polemics ought, in principle, to be avoided, we cannot avoid casting a brief critical glance at these prevailing views, in order to see what we forfeit when we replace them with our hypothesis."[33] In 1912 many geologists were still using the analogy of a baked apple coming out of the oven, with the corrugation of mountain ranges compared with the shrinking of the cooling apple skin over the reduced interior. This simple analogy, of course, broke down completely, because the contraction theory was held responsible not only for the origin of mountains but for the origin of continents and oceans, with no analogical explanation or picture as to how first a continent might sink, and then the adjacent ocean basin, and then again a continent, and so on. "However," Wegener wrote, "it is precisely the relentless consistency with which Sueß has worked through these ideas that has already opened the eyes of many to their weaknesses and has thus indirectly opened the way for a more correct interpretation."[34]

The self-contradictions of the contraction theory in its various simple and qualitative analogies were a persuasive argument against it, but the idea was now also faced with two formidable obstacles of quite recent origin. The first of these was the discovery of radioactivity in Earth's crust, which challenged the notion that Earth necessarily had to cool and shrink at all, because radioactivity provided an additional source of internal heat. Not only was Earth not cooling down, but it might possibly be heating up.[35] More pressing and less conjectural than this objection was the great mass of information collected concerning Earth's gravity field: "However, even if all of these arguments against the breakdown of the terrestrial globe did not exist, we would still have to reject the conception, because it contradicts gravity measurements. If the oceanic deeps were nothing more than sunken continents, they would consist of the same material as the latter. Gravity measurements show, however, with inescapable logic, that under the ocean lies rock heavier than that of the continents and not only heavier, but so much every that the difference in elevation is compensated and equilibrium prevails."[36]

While European geologists, in the main, were still in the grip of the hypothesis of universal long-term contraction of Earth, North American geologists had embraced the notion of "isostatic compensation." Extensive gravity measurements carried out by the U.S. Coast and Geodetic Survey indicated that Earth's interior only some scores of miles below the surface must be in a yielding state, as a consequence of heat and pressure, and therefore the surface had to be seen as floating on the interior, because it was composed of lighter material. There was simply no other plausible explanation for these many thousands of gravity measurements.[37] Wegener noted that however correct American geologists might have been in opposing the notion of "foundering continents," and having instead asserted the notion of "the Permanence of Continental Platforms," the American geologists had attached to this well-founded and geophysically well-supported conclusion the "dubious doctrine of the Permanence of Oceans."[38] Of course, hundreds and hundreds of paleontological finds of both terrestrial and shallow-water organisms, confirming unhindered communication back and forth across abyssal oceans that such organisms could not possibly have traversed, contradicted this notion.

So here we come to the crux of Wegener's argument. In Europe we have the notion of foundering continents and newly created oceans, and in North America the notion of permanent continents and permanent oceans: "thus the two conceptions stand in complete opposition to one another. Both parties have good, incontestable arguments, but both to draw inadmissible conclusions from them. I shall try to show that the correct claims of both can be encompassed more simply in the context of the hypothesis of the rifting and horizontal displacement of the continental blocks."[39] Wegener saw his hypothesis of continental displacements as a way to reconcile two different geological communities, each attached to a partial truth and to a demonstrable falsehood. He imagined that both groups would welcome a third way out.

Wegener then pursued an excellent, detailed, nuanced, but extremely technical discussion of the various ways to measure gravity over continents and oceans, and he evaluated the existing survey data on Earth's gravity field, all of which led him to the nearly inescapable conclusion that the material under the oceans was more dense than the material of which the continents were composed; the ocean floors could not have ever been former continental surfaces. The intellectual issue involved was relatively straightforward. Since water is only about half as dense as rock, gravity should be less over the oceans, but measurement shows that it is not. Why? The likely answer is a higher specific gravity of rocks under the ocean.[40]

To this discussion Wegener attached an equally sophisticated and detailed notion of "isostasy." The flotation of light crust on the denser subcrust would lead to near-equilibrium values of gravity over the oceans and continents in spite of the higher specific gravity of the suboceanic rock. This principle, known as isostatic equilibrium, had a history of investigations going back to the middle of the eighteenth century.[41]

Wegener then tried to entice geologists to accept this geophysical argumentation by showing the utility of isostatic equilibrium for a variety of geological problems. For if the continents were in a floating equilibrium with the layer of Earth below them, the loading and unloading of the continents must produce a sinking into this magma and a rising out of it. Such loading of a continental surface, on a continental shelf, could explain how to accommodate long-continued sedimentation in a region, a so-called geosyncline, where the sedimentation surface was always just below the surface of the water, but many tens of thousands of feet of sediment might accumulate over time, by pushing down of the subcrust under the weight of this load.

Isostasy also provided a good explanation of Charles Darwin's celebrated theory of coral atolls that, as they grew and added weight, must cause the underlying island to subside, encouraging the further growth of coral above. Wegener applied isostasy to a number of other well-known geological phenomena, such as the "postglacial rebound" of Scandinavia with the removal of the continental ice sheets: well-documented measurements along the shores of the Baltic had shown the progressive rising of the land over more than 150 years, previously unexplained but now characterized as an "isostatic rebound" some thousands of years after the removal of the ice load. Calculations of the amount of such sinking and rebound, given the relative specific gravities of ice and the Scandinavian rock, even provided a way to estimate the thicknesses of these ice sheets. Wegener was at great pains here to try to show geologists that a little bit of geophysics went a long way to help them pursue their own aims without geophysics trying to conquer them or tell them to solve new and different problems.

In this context, we need not spend much time determining the details of his argumentation on gravity, isostasy, and the impossibility of the ocean floors ever having been continental surfaces. His sources were recent, and he had read them carefully, interpreted them correctly, and presented them fairly. The theoretical situation—or, if you will, the theoretical crisis—to which he thought he was addressing himself in 1912 was actually there, and it was there in precisely the terms he presented it. Subsequent very detailed accounts of this episode by historians of science have confirmed the accuracy of Wegener's historical sense, his physical intuition, and his mastery of the relevant argumentation.[42]

More to the point than his physical sophistication was his political naïveté. What could possibly have led him to imagine that a short piece like this, presenting results already in the literature, written by someone who was not a geologist, would convince the many thousands of busy field-workers and the handful of theoretical workers in the global community of geologists to readily abandon the theories so laboriously erected on a foundation of many decades of fieldwork stretching back to the beginnings of the nineteenth century? The answer is that he was a young man from a field—atmospheric physics—that had just been revolutionized by a series of novel theoretical developments of the very sort he thought he was presenting: new results, all less than a decade old, discovered with new instruments, and many times confirmed, leading to an entirely novel picture of the structure of Earth's atmosphere and the dynamic behavior of the weather.

The situation he saw in geology and geophysics was precisely that which he thought he had just lived through in atmospheric physics. Most of the confirming (or disconfirming) evidence that he was presenting in his paper concerning the inadequacy of the contraction theory, the structure of Earth's interior, and the probable origin of continents and oceans was less than a decade old. Most of the gravity work had been done since the turn of the century, and the most significant and decisive material, that of the North American Survey, had been published only in 1909. In seismology (which we have yet to discuss) most significant work was also less than a decade old, with crucial parts of Wegener's argument underwritten by material first published in 1907 and 1910. The same is true for his work on solid mechanics, plasticity, viscosity, fracture and flow in Earth's interior, and the chemical composition and behavior of the outer shells of Earth.

Wegener assumed that geologists would be persuaded by these new developments in physics, even though they had no particular physical training themselves. The world meteorological community had been persuaded and had actively welcomed the new developments in atmospheric physics as a positive revolution in their science, even though most meteorologists were field observers busily collecting and collating data from a synoptic network of observing stations, extending the "atmospheric map" across the face of Earth with the same relentless avidity of field geologists extending their geological mapping. Wegener assumed that geology would welcome "a new head from physics" because this had happened to him already in meteorology.

To a certain extent, however, Wegener was also encouraged to proceed with this novel hypothesis and spend January and February elaborating it, not because of what he did *not* know about geology, but because of what he *did* know. The geologist he knew best was Emanuel Kayser, in the front rank of the science, president of the national association in Germany, and editor of its newest and most forward-looking journal, in which he had invited Wegener to publish. Kayser was a great champion of intercontinental correlation of sedimentary rocks and paleontological data, and he had doubts about the adequacy of the contraction theory. As a geologist predisposed to welcome physics, closely associated with the geophysicist Richarz at his own institution, and close to the group at Göttingen led by Wiechert, he was clearly inclined to accept the notion of a comprehensive earth science much more amply defined than nineteenth-century "geology." Kayser was perhaps a significantly *misleading* model of a geologist for Wegener.[43]

Wegener also assumed, again somewhat naïvely, that his arguments would carry themselves without any particular reputation of his own to endorse them. His own experience in atmospheric physics may have led him to believe that this was true, and to a certain extent it was. But Wegener, in spite of his gratitude for the patronage he received, may not have understood what it meant for him to have been advanced by Aßmann, praised and supported by Hergesell, aided by Süring, and to have the full support, cooperation, and collaboration of a giant like Wladimir Köppen. What Sueß was to world geology, Köppen (along with Julius Hann) was to meteorology and increasingly to atmospheric physics. As we return to a brief consideration of the rest of Wegener's chief geophysical arguments, we can better understand his fresh and infectious enthusiasm to present this interesting material in a completely ingenuous way, as a consequence of his own previous experiences.

Moving forward from the idea that continents are blocks of crust of specific gravity lighter than that of the ocean floor beneath them, Wegener proposed the question, how thick are these blocks? Once again, gravity measurements provided some guidance. Friedrich Helmert, Wegener's old professor at Berlin, had made an extensive study of this problem, measuring gravity along the coast of the ocean at fifty-one stations and finding, as one moved measuring equipment closer and closer to the coast, that the value of gravity showed "positive anomalies." In other words, "the

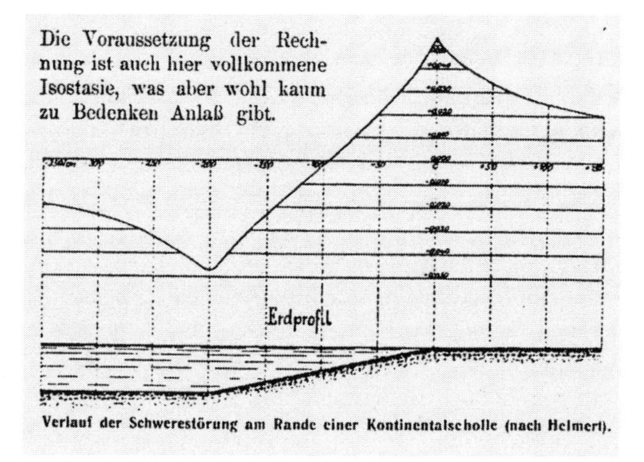

Verlauf der Schwerestörung am Rande einer Kontinentalscholle (nach Helmert).

Friedrich Helmert's gravity profile of a continental margin. The peak of the curve to the right is a higher-than-expected value for gravity on land, the trough to the left a lower-than-expected value at sea. Helmert's (and Wegener's) interpretation was that the anomalies were caused by the adjacency of dense ocean floor and lighter continental material. The dense ocean floor makes the continental value "too high," and the light continental material makes the oceanic value "too low." Farther out to sea, or farther inland, the anomalies vanish. From Wegener, "Die Entstehung der Kontinente."

measured value of gravity was higher than should have been expected." Subsequently, measuring gravity at sea and moving the instruments progressively farther offshore, the values suddenly became "negative," meaning that gravity was less than might have been expected. Still farther out to sea, the anomalies vanished. Helmert had interpreted these data as the result of a sharp coastal boundary between dense suboceanic and light continental material, and he calculated, based on the size of the anomalies, a thickness of *continental* crust of 120 kilometers (75 miles). This was a very encouraging corroboration for Wegener of the North American Gravity Survey, which had estimated the thickness of the "level of isostatic compensation" (and therefore the thickness of the outer layer of Earth) at 114 kilometers (71 miles).[44]

It is characteristic (and habitual) in physical argumentation to use a variety of instrumental sources to investigate the same phenomenon and compare the results. A result achieved by one method alone is not to be trusted. Thus, Wegener turned from gravity data to the corroboration available from earthquake wave studies pioneered by his senior colleague, Emil Wiechert, at Göttingen. These studies, following the oscillation of fixed waves that could only exist in a thin and elastic crust, arrived at a value for the thickness of the outer shell of Earth of around 100 kilometers (62 miles). Wegener also invoked very recent and very interesting work on earthquake travel times by the Croatian meteorologist Andrija Mohorovicic (1857–1936), suggesting the lower boundary of the solid material of the crust, in places, at depths of no more than 50 kilometers (31 miles). Although Wegener did not know Mohorovicic personally, they had a broad range of shared interests, including tornadoes, the lapse rate in the atmosphere, and the character of clouds; Mohorovicic's work had been published in a report from a meteorological observatory, so that Wegener, as a meteorologist, came upon it before many geophysicists were aware of it.[45] Finally, Kayser had produced a comparative study of the depth of focus (the hypocenter) of eleven major earthquakes around the world since 1873, producing a mean depth (which Wegener took to be accurate only to an order of magnitude) of about 100 kilometers.[46]

Wegener had tried to establish the thickness of the continental blocks using first gravity data and isostasy and then seismology. He then turned to a third line of evidence: the material of which the continents were made, to establish the reality of the differences of specific gravity between the continental blocks and the ocean floors. Wegener knew that geologists were overwhelmingly concerned with sedimentary rocks and their fossil contents; this was the only sort of geology he himself had ever done. Wegener thought he needed to establish, in the minds of geologists, the importance of considering *not* the detailed, differentiated mineralogy of sedimentary rocks, as well as the fine discriminations of their specific gravities, but the bulk undifferentiated density of the continental massifs as a whole.

> It is simple to calculate that with the removal of the water in the oceans, the continental platforms would plunge downward an additional 1500 m in the dense magma and the current 5 km difference in elevation [produced by the weight of sediment] would thus be reduced to 3 1/2 km. But while the thickness of the sediment is also of the same order of magnitude as this difference in elevation it entirely disappears when compared to the total continental thickness of 100 km; and one first sees this clearly, when one takes isostasy here also into consideration. If we were, of course, to have removed sedimentary cover from the whole earth, the continents would rise to the level of the old surface almost everywhere, and the relief of Earth would be only slightly altered. From this it is obvious that the continental platforms are forms of a higher order, compared to which the processes of erosion and sedimentation play only the role of secondary, superficial phenomena. The continental material is made of an archaen rock, the ubiquity of which, in spite of many objections, is not to be denied. If we, in order to establish these ideas, stick to the principal representative, then we can say: the continental blocks are made of Gneiss.[47]

It is clear what Wegener means: from a geophysical standpoint one is concerned not with the details of rock type (a geological matter) but merely with the mean specific gravity of an entire continental mass. A thermodynamic system is only concerned with the bulk properties of matter, not its manifold complexities: just pressure, density, volume, and temperature. Wegener is still in the midst of his geophysical argument and has not yet moved on to make the argument for continental displacements from the standpoint of geology, but it was impolitic, to say the least, to suggest that "the processes of erosion and sedimentation play only the role of secondary, superficial phenomena," since the results of these processes absorbed the working life and intellectual energy of most geologists.

Pursuing this line of argument, Wegener was ready to undertake a calculation, the results of which can be seen from inspection of his simple diagram. Wegener imagined the continental block to be 100 kilometers thick, with a mean density of 2.8, and in isostatic equilibrium with (that is, floating on) the material making up the ocean floor. He took 4.3 kilometers (2.7 miles) as the mean depth of the ocean and 1.03 as the specific gravity of seawater. This left 95 kilometers (59 miles) of subcrust of unknown density, from the level of the sea floor down to the bottom of the floating continental block. It is a matter of simple multiplication to produce $x=2.9$ as the density of the subcrust within the first hundred kilometers or so, "which harmonizes quite satisfactorily with the assumption that the material is essentially identical to the Sima."[48]

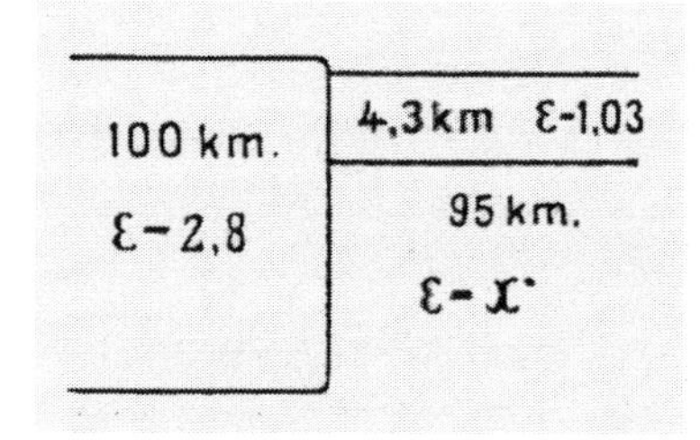

Wegener's sketch of a floating continental block 100 kilometers thick, density 2.8, adjacent to an ocean with mean depth of 4.3 kilometers and density of 1.3. This yielded a density of 2.9 for the 95 kilometers of rock beneath the ocean, close to the estimate for Earth's shell of Sima, thus supporting Wegener's supposition of continental flotation in the denser material. From Wegener, "Die Entstehung der Kontinente."

It is near this point in the argument that Wegener refers his readers to a plate at the end of the volume, which contains the accompanying figure. Wegener has this to say about his cross section:

> As a further illustration of the dimensions [of the continental blocks], in Figure 2 of Plates 36 a cross-section (along a great circle) of Earth between South America and Africa is presented in true vertical scale. The irregularities of the crust and even the great hollow of the Atlantic Ocean are so slight that they occur within the thickness of the line representing Earth's surface. In contrast, the continental blocks are readily identifiable. For comparison, the figure also contains the Wiechert Iron-Core, and the principal atmospheric layers: the Nitrogen Sphere, the Hydrogen Sphere, and the hypothetical Geocoronium Sphere. The zone of clouds (the Troposphere) is not thick enough to be represented.[49]

Wegener was at this juncture so anxious to move along to a discussion of the material out of which Earth is made and its physical behavior (we shall return to this in a moment) that he had nothing more to say about this diagram, yet it is certainly worth our attention. Notable here is that the ocean of the oceanographer, Earth known to the geologist, and the troposphere of the dynamic meteorologist have all vanished. What we have instead is the chemically differentiated Earth and its chemically differentiated atmosphere, beginning at the outermost Earth shell with Wegener's hypothetical geocoronium, and proceeding through layers of increasing density down to the core of Earth. Here is the essence of Wegener's picture of Earth, a series of spherical shells with sharp surfaces of discontinuity and sudden marked change in physical properties at these boundaries. It is an Earth stripped of almost everything that most "earth scientists" would find interesting, but it accentuates the point that Wegener most needs to make: that the continents are higher-order features of Earth's constitution and form a distinct Earth shell.

While Wegener needed no encouragement to think in terms of spherical shells separated by sharp surfaces of discontinuity, it is extraordinary how much support he obtained from every quarter in emphasizing this particular structural model. Seismology, in the work of the Göttingen group and Wiechert, was founded on the existence of surfaces of discontinuity, as well as differences in wave propagation at such surfaces. Geologists appeared universally to be entirely comfortable with a chemically differentiated Earth in three large segments: crust, mantle, and core (Sal, Sima, Nife). All discussions

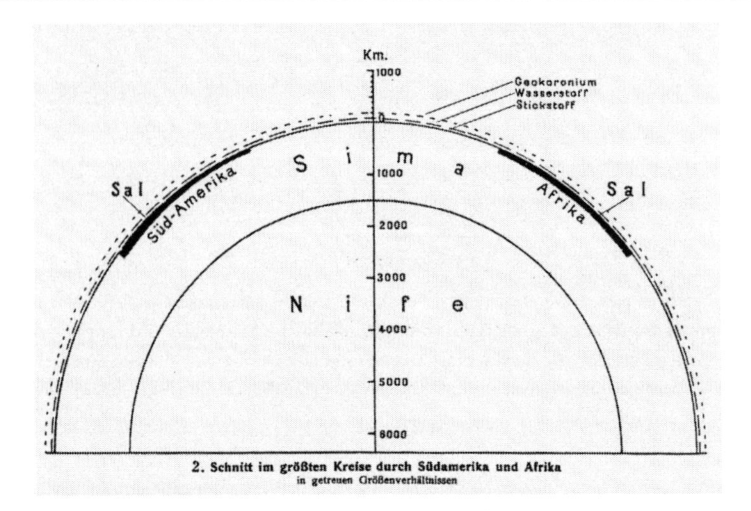

Wegener's cross section of Earth, showing the floating continents (Sal), the subcrust in which they float (Sima), and the nickel-iron core (Nife), as well as the major layers of the atmosphere, including his "geocoronium layer." The relief of the continents (mountains) and the oceans are too small to be seen at this scale, underlining Wegener's contention that they are features of "secondary importance." From Wegener, "Die Entstehung der Kontinente."

of Earth's gravity field contained inferences of the existence of "surfaces of compensation," which were either perfectly spherical or penetrated by bulges, in the case of the idea that mountains had deficits of gravity because they had developed lighter "roots." Within oceanography, the existence of a "mixed layer," at the surface of the ocean, with a sharp discontinuity surface separating it from the deeper ocean water, carried this theme into a lighter and faster-moving medium, and of course we need not elaborate further on Wegener's own conviction about the chemical differentiation of the atmosphere and the importance of discontinuities within it.

In this context, once again Emanuel Kayser emerges as Wegener's great geological guide. In the introduction to all of his textbooks, Kayser emphasized this viewpoint. He spoke of the surface of Earth in terms of an atmosphere, a hydrosphere, and a biosphere (the organic world) and noted that the subject matter of geology, the solid crust of Earth, had recently come to be spoken of as the "Lithosphere, in the context of these other spheres."[50]

Wegener now moved along to what were unquestionably the most difficult and taxing parts of his explanation, as well as the area of greatest contention and misunderstanding concerning the hypothesis of continental displacements. How is it possible for a continental block made of solid rock to displace horizontally through equally solid material that responds to the tidal pull of the Moon as if it were as rigid as steel? The intuition that such motions are absurd would pursue Wegener and his hypothesis throughout his entire career. Nowhere was the geologists' lack of training in physics more of an obstacle than in this matter.

The essential distinction here is between two characteristic aspects of "solid" matter, never entirely grasped by most geologists during Wegener's lifetime. These are "strength" and "rigidity." As late as 1940, the distinguished Canadian geologist Reginald A. Daly (1871–1957) was moved to write a textbook with the title *Strength and Structure of the Earth*, with the specific aim of teaching geologists to discriminate these two physical quantities.[51] The many editions of Harold Jeffreys's (1891–1989) influential geophysical textbook *The Earth*, from 1929 onward, devoted much attention in each edition to the establishment of fundamental physical quantities: strength, rigidity, viscosity, elasticity, and so on. Jeffreys remarked in every edition of his work that "rigidity and strength are quite distinct properties, but are habitually confused in the geological literature."[52] Wegener's task was made no easier by the circumstance that a discussion of Earth's interior was the most absorbing and interesting area for the establishment of the meaning of these physical terms throughout the first half of the twentieth century. Thus, any debate about the actual behavior of Earth always involved a more fundamental debate about the meaning of the terms of the argument.[53]

If we leave the technical discussion aside, we can accomplish much in a small space by considering the difference between strength and rigidity using an example given by Wegener himself. "Pitch offers an extreme example: if one lets a piece of it lie for a long time, it begins to flow under its own weight, and tiny lead balls sink into it in the course of time; yet under the blow of a hammer it shatters like glass."[54] This is to say that *strength* is the property of matter which represents its response to a long-continued stress, while *rigidity* is that property of matter which represents its response to an instantaneous stress.

The critical quantity here is the time of application of the stress. The behavior of Earth under a short stress anywhere from three seconds to about four hours (an earthquake wave) is that of an elastic solid with a strength limit. For intermediate stresses anywhere from three years to about 15,000 years (Chandler wobble, Earth tides, faulting), it is an elastic body recovering its shape slowly after long-continued stress, but still rigid for shorter-term stresses. For longer intervals, from 15,000 to 100,000,000 years and longer, Earth has no rigidity and no strength at all, but a high viscosity, and can be plastically deformed indefinitely. These long-continued stresses would include such processes as mountain building, postglacial rebound, polar wandering, and continental drift.[55]

Wegener termed this section of his argument "Plasticity." Thus, it was clear that he had in mind the long-interval stresses. Crucial to his argument about the ability of continental blocks to displace horizontally within the Sima was the notion that the melting point of the Sal, the continental rock, should be 200–300°C (392–572°F) higher than that of the Sima. Thus, at an identical depth, say, 100 kilometers, the Sima, already virtually at its melting point, under long-continued stress would behave more like a fluid (lacking both strength and rigidity), while the continental block, several hundred degrees below its melting point, would behave more like a solid, though still capable of plastic deformation. Using a variety of experimental data, Wegener inferred (a highly conjectural inference) that from the standpoint of long-continued stresses, the entire 1,500-kilometer-thick (932-mile-thick) layer of the Sima ought to be treated as a viscous fluid, in which floated continental blocks that were plastic solids.

Wegener's confidence concerning the behavior of the Sima came from his reading of Rudzki, especially the latter's remarks on the physical characteristics of ice. Rudzki had argued that what we learned about the behavior of flowing glacial ice, assumed to be at its melting point at the undersurface of the glacier, was generalizable to the behavior of all materials close to their melting point, with glacier ice providing an example of viscous flow of a crystalline solid on a timescale that could easily be observed. "On the basis of our experience with the plasticity of ice we can form a view about the plasticity of solid rock in Earth's interior."[56]

After a considerably technical discussion of the rigidity of Earth at various densities and various depths and the interplay of pressure and temperature at depth, Wegener was able to conclude to his own satisfaction that "all of these phenomena thus in consequence point to the idea that the Sima represents a plastic but in no sense completely mobile material, and that the Salic crust has a considerably greater strength, yet is not because of it completely devoid of plasticity. We will thus, on this account, have no reason to dispute the possibility of extraordinarily slow but nevertheless great horizontal displacements of the continents, acting in the same way, unaltered, throughout geologic time."[57]

Wegener's final sentence in this argument points to an attitude toward physical hypotheses: what is one permitted to imagine and to assert? In physical argumentation, then and now, there seem to be two clearly defined schools. On one side we find physicists like Wegener, for whom it is only necessary to establish consonance with the laws of nature and plausibility with regard to known observed and experimental values in order to imagine and sketch out a state of affairs, in this case the character of the interior of Earth. In this school, one may imagine and propose almost anything for which there is not direct contradictory physical evidence to prevent it. On the other side, perhaps more common in the Anglo-American world than in Germany, is the sense that some sort of demonstration of existence is necessary, in addition to the characterization of plausibility. This divide was perhaps more pronounced in the early twentieth century because there was nothing in the Anglo-American world like the theoretical physics appearing in Germany. Additionally, the mathematician David Hilbert (1862–1943) at Göttingen had only recently legitimized the notion of "existence" proofs as opposed to "constructive" proofs in mathematics.

One has the sense that in the Anglo-American world, then and now, a "theory" is something about which truth claims may be made. There is nothing like that in what Wegener is saying here. While there can be little doubt that he believed he was describing the actual state of Earth, science was, for him, not a matter of belief but a matter of evidence. Wegener was arguing that the evidence from geology and geophysics taken together was more consistent with the idea of lateral continental motions than with the idea of a shrinking Earth.

This large-scale framework of theory, or of a working hypothesis, had still to make contact with phenomena appearing at the surface of Earth. Even if erosion, sedimentation, and mountain building were second-order phenomena compared to the great motions of the continents plowing through the subcrust, it remained to demonstrate the relationship of these geological objects to the geological theory under consideration.

Wegener attempted to connect his large-scale hypothesis to important phenomena of geological dynamics, especially mountain building and volcanic activity. The theory of mountain building, or "orogeny," was one of the most important and theoretically developed aspects of geology, and the theory of mobile continents should have something to say about it. This was especially true because geologists at all interested in the dynamical behavior of Earth's crust were likely to have a special interest in the origin of mountain ranges and to want to know how the motions of the subcrust would affect the great fault and fold structures at Earth's surface. Similarly, if the continents were rifting and pulling apart, as Wegener indicated they were, one might plausibly expect volcanic eruptions to be preferentially located on the trailing edge of such continents, where newly exposed high-temperature subcrust finds itself on the ocean floor. That this was not the case, at least with regard to the Atlantic, the heart of Wegener's hypothesis, clearly required explanation.

Wegener's rapidly sketched and qualitative discussion of these matters looks something like this. The great fold structures seen in the Alps and elsewhere, assumed to be the largest motions on the surface of Earth, represent, relatively speaking, only the most superficial aspect of mountain building. Underneath the great fold structures or "nappes" of the Alps and similar ranges (including the Himalayas) are flow structures, which absorb compression by the tight folding and then downward flowing of large masses of older rock beneath the sediments. Wegener here also adopted the recently expressed views of Albert Heim, among others, that the amount of compression in the Alps amounted to 4–8 times the current width of the range. Thus, with the Alps being currently 150 kilometers (93 miles) wide, a region between 600 and 1,200 kilometers (373–746 miles) would have been thrust together. "A consequence, that before the lateral compression the continental platform must have had a significantly different outline, has, in my opinion not yet been sufficiently appreciated. If, for example the chains of the Himalayas consist of thrust together landmasses of a corresponding width, where would the southern tip of the Indian subcontinent have had to lie? In particular does any room remain at all for a sunken Lemuria?"[58]

Wegener imagined a history of Earth in which mountain building causes a continuous thickening of the continental blocks at the expense of their horizontal extent. He imagined that once a salic (or "sialic") crust, roughly 30 kilometers (19 miles) thick, had covered the entire Earth, covered in its turn by a "Panthalassa," a universal ocean about 3 kilometers (1.9 miles) deep. This vision is the exact opposite of that put forward by Edward Sueß in *The Face of the Earth*. Sueß imagined the "breaking up of the terrestrial globe," in which the continents would eventually sink entirely beneath just such a universal ocean.[59] Rather than the destruction of continental masses and their transformation into oceanic deeps as a consequence of ongoing Earth contraction, Wegener imagines the successive emergence of thicker continental blocks separated by broader oceans. Once again the Copernican motif, in consonance with his view of the direction of the evolution of the crust, emerges in high relief.

Wegener's treatment of the question of volcanism, like his treatment of mountain building, considered the topic from a geophysical rather than geological standpoint. "Up to now," Wegener wrote, "we have completely disregarded an obvious objection. With every displacement of continental platforms, the underlying high temperature Sima will be laid bare under the oceans. Must this not lead to catastrophic events?"[60] Wegener's answer is that it need not. If the Sima is denser than the Sal of the crust, it cannot rise into an open fissure and cause a volcanic eruption, unless it is under pressure; this is because of isostasy. If a continent should rift into two fragments, these fragments would still float at the same depth in the subcrust, and there would be nothing to force the hot material below upward. Isostasy predicts that the crust and the subcrust are in equilibrium, and there is no reason on the physical side for the hot matter exposed by this continental drift to surge upward. Rather, Wegener argued, one should expect volcanoes to be concentrated in regions of compressive stress (such as the Pacific margin of South America), rather than regions of tension (the Atlantic margin of South America). Of more interest to Wegener was the episodic character of volcanic outbreaks, which seemed to him—or at least he allowed himself to imagine—to correspond precisely to those periods of continental displacement: one in the Carboniferous and Permian, and another great phase in the Tertiary. These would correspond directly to Wegener's notions of the breakup of the continents first in the Southern Hemisphere and later in the Northern Hemisphere.

This was important material for anyone wishing to talk to geologists, and one might have expected a fuller discussion. Yet one's overall impression is that Wegener handled it in a hurried and cursory fashion. One can sense the pressure he was under to complete this manuscript in the short window of time allotted him before his departure for Copenhagen and then on to Iceland and Greenland. The haste, incomplete character, and compression of argument are evident in his brief and somewhat apologetic remarks at the end of the first section of his paper, when he turns to "Remarks on the Causes of the Displacements":

> The question concerning which forces cause the horizontal displacement of the continents we have advocated is so obvious that I cannot completely overlook it, although I am of the opinion that it is premature. It is unquestionably necessary first to establish exactly the reality and the type of the displacements before one can hope to fathom their causes. It can essentially only be a question of forestalling false interpretations, and, to a lesser extent, to present such interpretations that might already be able to make a claim to correctness . . . an obvious candidate to bring forward as a cause of variation of the pole of rotation.[61]

Wegener was not, however, at this point predisposed to accept variation in the position or inclination of the pole of rotation as a cause of continental displacement, but rather to imagine that displacement of the pole would be a *consequence* of redistribution of continental masses. The displacement of the pole must then itself have a cause, and Wegener's hurriedly sketched hypothesis was for lunar tides in the body of Earth leading to meridional fractures, in combination with "fortuitous streaming in the body of Earth," as the cause of displacement following these initial meridional fractures. One hardly knows what he meant here (and it is not clear that he knew either), except that the continents, especially in the Southern Hemisphere, have long tapering forms with their long axes along meridians of longitude rather than parallels of latitude. The vagueness of these suggestions and the impatient, naïve enthusiasm they betray seem to have struck even Wegener, who completed this section with a firm declaration that "the time seems, however, as I said, not yet right for these questions."[62]

Having established the qualitative plausibility of continental displacements from the standpoint of geophysics, Wegener's task in the subsequent section of the paper—on the specifically *geological* arguments—was to bring to the forefront of discussion those structural features of Earth's crust left completely unexplained, or only partly explained within the context of the contraction theory, in the form given it by Sueß. He aimed to show that the theory of continental displacements did all the epistemic "geological work" of the contraction theory and a good deal of additional geological work that the contraction theory was unable to perform. It explained all the same geological phenomena that the contraction theory explained, as well as a variety of features and phenomena it did not explain. His arguments on these crucial matters are laid out in the next chapter.

11

The Theorist of Continental Drift (2)

MARBURG, FEBRUARY–APRIL 1912

> Before we undertake to pursue the processes of rifting and compression of continental blocks in the history of Earth, discussed in the first part of this work, it should be pointed out once more that such an initial, tentative investigation must of necessity prove incomplete in many respects, and in other respects perhaps incorrect. All the same, it must be ventured. For once the major considerations have been established, it will not be difficult, through detailed investigations, to eliminate errors.
>
> ALFRED WEGENER, 1912

Geological Arguments and Measurement

Wegener had worked hard to establish the plausibility of his hypothesis of continental motion and the corresponding implausibility of the theory of Earth contraction. His was a hypothesis of very great generality: continents in general, mountains in general, volcanoes in general. But geology is a very concrete science. If physics and physicists took pleasure in simplicity and unity, geologists seemed to glory in variety and heterogeneity. A geologist's brain is a gigantic lumber room of Platonic forms, not a sparse laboratory containing a few essential shapes. Fortunately, Wegener seems to have had an inkling of this and the necessity of making his case apply to specific kinds of structures and specific sets of geological data.

His reading of Sueß and of Kayser had given him a broad overview of the major problems of geology seen from the standpoint of geologists. His line of geological attack, as suggested at the end of the previous chapter, was to select major problems discussed by Sueß and others but not solved by them. He wished to show that his theory could not only reorganize existing geological knowledge but also contribute to the solution of outstanding problems.

He therefore constructed his argument around specific large-scale features of Earth's surface, features left only partly explained in the existing geological literature. Here are the major geological features of Earth's crust which, in Wegener's estimation, had been left unexplained by the contraction theory:

1. The character of large fault troughs (*Grabens*).
2. The relationship of the Atlantic to the Andes.
3. The extent and location of the paleocontinent Gondwanaland.
4. The area and extent of the Permian Ice Age.
5. The difference between Atlantic and Pacific sides of Earth.
6. The question of the displacement of the poles.[1]

Wegener first considered the matter of the fault troughs or *Grabens*. These structures, appearing across a broad range of scales, had a very important function in Sueß's theory of the formation of the continents. For Sueß, the continents were "Horsts," giant fault blocks flanked by oceanic "Grabens," which he imagined to be downfaulted sections of former continental crust. Wegener had already argued against the plausibility of this model, based on density data for the continental surfaces and the ocean floors. Gravity surveys taken at the continental margins suggested that gravity anomalies in this region could best be explained by assuming the ocean floor to be more dense than the continental material adjacent to it, and therefore that it was a distinct Earth shell and not a former continental surface. Now Wegener wanted to turn from such general *geophysical* considerations to specific *geological* examples of these phenomena.

The first of Wegener's geological examples was the Rhine Graben, a North German fault trough well known to his audience. It had been drilled to considerable depth and the sediment cores from the borholes minutely analyzed. These core samples showed that the material at the bottom of the trough was indeed the same sedimentary material, in the same sequence, as that of the remaining "horsts" flanking it. Gravity data, however, indicated no negative gravity anomalies within the graben. Even though very light sedimentary material was supposed to have "sunk" (which should reduce the value of gravity in this location), the specific gravity of the material deep in the trough was higher than in the flanking hills. Wegener took the opportunity provided by these data to infer that the graben was not a coherent "dropped" fault block but a tensional rift contemporary with the splitting of America from Europe. The equivalence of the sediments in the graben with those in the horsts on both sides was the result, rather, of the slippage of sediments at the margins of the widening rift. The gravity data could be explained by the rising of the Sima into the tensional rift, reestablishing isostatic equilibrium.[2]

Wegener's second example bearing on this question was the rift valleys of East Africa. Sueß had offered no dynamical explanation for why these great rifts should be where they are. On purely morphological grounds, he had asserted that they appeared to be tensional structures.[3] In these African rift valleys, gravity measurements showed, in contrast to the Rhine Graben, negative anomalies (less than expected values of gravity). Wegener's interpretation of this case is that the East African rift valleys were tensional rifts opening from the top down which had not yet cut through the full depth of the continental block to let in the Sima. Since the East African rift valleys were generally agreed to be structurally continued in the Red Sea, to the north, and since the Red Sea was isostatically compensated, Wegener was able to argue that the width of the rift at its northern end had allowed infilling with more dense material from depth, something that had not yet happened farther south within the margin of the African continent.[4]

The case of the Rhine Graben shows gravity data incompatible with a simple down-dropped block of crust. On the other hand, the East African rift valleys, admitted by Sueß to be tensional structures, could now find an explanation where they had none before, and the two sets of phenomena could be united within the framework of the theory of continental displacements. In both cases the argument for rifting and lateral displacements was conformable with the existing data and resolved anomalies in the previous explanation.

The next geological example Wegener chose, "Atlantic and Andes," took up the most famous aspect of his displacement hypothesis: "The large-scale parallelism of the Atlantic coasts is not to be underestimated as an argument in favor of the supposition that these constitute the margins of an enormously widened rift. A single glance at the map suffices to establish that where mountains lie in the East, such are also found in the West, and where such are absent here, they are absent there as well."[5] Wegener's approach, highly schematic, began with the circumpolar regions: where there are mountains in Greenland, we find them in Scandinavia. Farther south, where there are long sections of North American coast with no mountains, the same condition prevails in Europe. Farther south still, the great concave shapes of the Caribbean, as well as its European counterpart the Mediterranean, form one long, transverse sea-lane. The great South American tableland matches the great African tableland, and where the coast straightens out farther south in South America, it does so in Africa as well.[6]

In the North Atlantic, in "the parts of the world best known to us, namely Europe and North America, there prevails almost perfect agreement regarding the details as well."[7] Eduard Sueß's great work had pursued these continuities in relentless detail right down to the mineralogy of the gneissic rocks on both sides of the Atlantic, matching those of the Lofoten Islands and Hebrides in the East with those of Greenland in the West. Wegener was following Sueß down the Atlantic coast, and one can sense his excitement in his point-for-point substitution of his interpretation for that of the acknowledged master of this material.[8]

From Wegener's point of view the most striking data were maps, published in a number of geological memoirs and books since the later 1880s, showing the continuation of individual and distinct mountain chains across the Atlantic Basin. Different authors had given different names to these ranges—Caledonian, Hercynian, Armorican, Alpine, Altaide—but most observers agreed that between the mountains of Scotland and those of Nova Scotia, between those of northern Europe, Great Britain, and Ireland and the American Alleghenies, and between the mountains of North Africa and those of the northernmost tip of South America, there was demonstrable continuity in structure and mineralogy. European geologists had concluded that these mountain segments were once continuous.

There was already a very strong argument against the idea that large sections of these mountains had subsided along with the rest of some former Atlantic continent. Albrecht Penck, the great glacial geologist, had pointed out that the missing segments of such mountain ranges would then be thousands of miles longer than their remaining segments, and a former continuity across the abyss would render them the longest mountain ranges on Earth.[9] Viewed in this way, Wegener's notion of an Atlantic rift gave mountain ranges of plausible extent while employing the same geological evidence.

One may note that the map of the Atlantic in Andreé's *Atlas* which had so inspired Wegener in 1910 did not show him what he might have seen in a different presentation: that the depth data for the Atlantic floor of the *Challenger* showed no transverse structures, while clearly showing the north-south extent of the Mid-Atlantic Ridge. This casts further light on the question of priority between the American Frank B. Taylor and Wegener in the matter of continental displacements. Taylor's map of the Atlantic floor, using data from John Murray, gave no hint of suboceanic mountain ranges east and west across the Atlantic floor. Had Wegener come upon Taylor's work while still composing his paper, he would likely have employed this important piece of data in his argument for displacements.[10]

More abundant than this structural evidence of former continuity across the Atlantic abyss was the massive compilation of paleontological data suggesting such connections. A former continuity of fossil forms, followed by discontinuity at some later time, might have several explanations: "It is easy to see that these questions are completely independent of whether one accepts the horizontal displacement of continental blocks or believes in a subsidence of land bridges. . . . Portions of one and the same continental platform can also become faunally and floral distinct through shallow transgressions, and the decision will often be difficult as to whether we are viewing a rifting, or a separation via transgressing seas."[11]

Wegener's consideration of the paleontological data concentrated, as one might expect, on the area of his original inspiration: the former Africano-Brazilian continent, sometimes known as "Archhelenis." His treatment of the evidence here suggests his strong dependence on Theodor Arldt and Emanuel Kayser, and through their work he asserted, in a few brief paragraphs, the existence of a general consensus on a former land connection across a broad front (and thus not a narrow "land bridge" or "isthmian link") between Brazil and Africa throughout the Mesozoic. There was also general (though not unanimous) agreement that this connection had ruptured in the Tertiary, at the end of the Eocene.

When Wegener turns his attention to the paleontological evidence of the North Atlantic, one begins to see his larger view of how the separation of the Atlantic had taken place. Wegener concluded that Atlantic rupture opened from south to north, appearing first in southern South America and southern Africa, with the fission of the North Atlantic somewhat later than the South Atlantic: "According to our view, North America, Greenland, and Europe would have constituted a continuous block up into the ice age, and the ice cap would have had a much smaller perimeter than one was heretofore obliged to accept."[12] This geological story was "reconfirmed" by paleontology, with species persisting across the North Atlantic continents while divergence of faunas was already evident between Africa and South America.

Moving quite briskly, Wegener turned to his consideration of the Andes. These mountains had been left unexplained by Sueß, who had, as we noted above, a complicated theory of the origin of fold mountains in the "overthrust of a foreland subsidence," a theory that the phenomena of western South America could not support. Wegener's interpretation is exactly what one might expect:

> As the upfolding of the Andes is essentially contemporaneous with the opening of the Atlantic Ocean, then the concept of a causal connection between them is provided from the start. The American blocks, in their drift to the West, would have encountered opposition from the certainly very old and no longer very plastic floor of the Pacific, whereby the vast shell, which once formed the western margin of the continental block, with its thick sediments was thrust together into fold mountains . . . the folding of the Andes in no way need be equivalent to the full breadth of the Atlantic (about 4000 km).[13]

Wegener then added quickly in a footnote, "I wish to point out emphatically that this presentation is in many respects and of necessity rather schematic. For instance, in North America only the western ranges of the Cordillera are of Tertiary origin, while those in the East are older, and more so the farther east they lie. Naturally only the Tertiary folds can be linked to the separation from Europe."[14] This note, added as an afterthought, is an attempt to resolve an apparent contradiction in his argument: if North America drifted away from Europe later than South America from Africa, the mountains on its western margins should be younger, rather than older, than the Andes. Yet most of the Rocky Mountains of the United States and Canada are in fact substantially older, as are portions of the Basin and Range Province of the western United States, and both are hundreds of miles inland from the Pacific coast.

In turning to the next case, that of "Gondwanaland," Wegener continues his argument concerning the Tertiary separation of great continental fragments. "If we also apply the views derived above concerning the correlation of folding with horizontal displacement to the Tertiary folds of the Himalayas, we arrive at a range of astonishing relationships."[15] The argument here is dense but can be summarized rather easily: India, which Sueß had described as a "fragment all the way around," may now be interpreted as a northward-moving fragment of a very extensive Southern Hemisphere paleocontinent (*Urkontinent*), with the Himalayas representing the compression of about half of its former extent in its northern motion. If the Himalayas could be unfolded, they would then produce an India, the southern point of which would be adjacent to South Africa. Following his reading of Keilhack and others, Wegener rapidly sketched out the former extent of this continent, to include Africa, Madagascar, India, Australia, New Zealand, and Antarctica.

Wegener's argument now embraced the following picture: there were once *two* great paleocontinents. The Southern Hemisphere continent, consisting of South America, Africa, India, Australia, New Zealand, and Antarctica, began to break up in the Mesozoic, probably in the Triassic. As Africa and India split apart, they both severed their connection with Australia, though the latter remained for some time in contact with the southernmost tip of South America and with Antarctica. In the Jurassic, the Antarctic block slid away from South Africa in the direction of the Pacific. In Wegener's mind this would make the mountains of Graham Land and Victoria Land in Antarctica the results of the same kind of leading-edge crumpling that created the Andes. India, after its separation from Africa, moved northward, progressively crumpling itself against the southern margins of the Eurasian continent in the Tertiary and forming the Himalayas. South America began to break away from Africa at the end of the Mesozoic, splitting from south to north, with the Africano-Brazilian connection also being broken in the Tertiary period. It was at this same time that the Andes were created by the crumpling of the leading edge of the west-drifting continent.[16]

The Northern Hemisphere continent also began to split from south to north in the Tertiary, and with the opening of the Atlantic the progress of severing of the land connections was finally completed in the Quaternary ice age, with northern Europe, Greenland, and northern North America still close enough together at this point to create a comprehensive land surface for a more compactly situated Northern Hemisphere ice cap. Thus, the ensemble of former paleontological connections, later broken, and the worldwide appearance of great fold mountains in the Tertiary were to be united under the aegis of the displacement theory as complementary aspects of the same dynamic process.

Wegener's reassembly and then separation of a Southern Hemisphere continent was constructed out of a paleontological record of former continuities now severed. Such a severing of connections, as he himself pointed out, could have come about from oceanic transgressions onto the continental surfaces at relatively shallow depth and need not have resulted from continental rifting and drifting. However, the question of the Southern Hemisphere ice age at the end of the Carboniferous and the beginning of the Permian presented an opportunity (of which Wegener had been made aware by Keilhack) that could tip the balance away from "transgressions" and toward splitting and drifting of continents. Incontrovertible glacial deposits, landforms, and other glacial phenomena occurred in southern India and Africa and northern Australia, as well as Madagascar. Even imagining Earth's pole of rotation to have been displaced to a more likely location for the focal point of a great southern ice cap, such a move to a more plausible site left that pole in the middle of the ocean between western Australia and Madagascar and still required, Wegener noted, glacial deposits within 30°–35° of the equator.[17]

> The Permian ice age has created up to the present a hopeless enigma for paleogeography. For these unmistakable ground moraines of an extensive inland ice cap, lying on a typically striated basement, are located in Australia, South Africa, South America, and above all in the Eastern India. Koken [Ernst Koken (1860–1912), author of *Die Vorwelt* (1893)] has indicated in a special essay, and illustrated with a map, that, given the present disposition of the landmasses, such a great extension of the polar ice cap is absolutely impossible. For even if one dismisses the South American findings as untrustworthy, which by now should be hardly permitted anymore, and were to place the pole in the most advantageous position imaginable, namely in the middle of the Indian Ocean, the furthest regions of the inland ice would still be assigned a geographical latitude of about 30–35°. In the face of such an extent of ice cover, scarcely a single part of Earth's surface could have remained free of glacial phenomena. And in that situation the North Pole would have been in Mexico, where not a single trace of Permian glaciation is known. The South American remains would, moreover, come to rest almost on the equator.[18]

The question of the Southern Hemisphere ice age was, as Wegener noted, impossible to solve with the continents and the pole in their present positions. Even displacing the pole of rotation, without also displacing the continents, produced an ice cover so massive that no region of Earth could have remained unglaciated. The contradiction was so blatant and so severe that even an otherwise rather conservative geologist like Albrecht Penck had been moved to remark that the current dispositions of the Southern Hemisphere continents with regard to this glaciation made so little sense that "the motion of Earth's crust in a horizontal sense be envisaged as a working hypothesis to be seriously taken into consideration."[19]

Beyond this summary of the geological and paleontological evidence, there remained two additional large-scale data sets that Wegener wished to address: the differences between the Atlantic and Pacific sides of Earth, and the anomalous distribution of tropical, semitropical, and temperate species of fossils in high latitudes of the Northern Hemisphere. The difference between the two sides of Earth was something noticed by the relentless Eduard Sueß, who had left it unexplained, therefore providing an opportunity for Wegener to move against the contraction theory on yet another front. Sueß had characterized the Atlantic side of Earth as bordered by indented, rough, and broken coastlines, with fractured margins, while describing the Pacific side as ringed uniformly by fold mountains parallel to the shore. The Pacific was deeper than the Atlantic, its lavas had a somewhat different composition, and it was more heavily sedimented in its abyssal depths with radiolarian ooze and red deep-sea clay.[20]

Wegener had already dismissed (at the beginning of the paper) the notion that the Pacific was a scar left by the departure of the Moon: Schwarzschild had mathematically refuted George Darwin's calculations supporting this hypothesis. Wegener argued instead that the Pacific was deeper simply because it was older, more isostatically compensated, cooler, and more rigid—thus its ability to crumple the leading edge of moving continents. The Atlantic margins, on the other hand, show the scars of their more recent rifting, and the Atlantic floor, having been uncovered more recently, was warmer and less dense than that of the Pacific; the Atlantic was therefore shallower and showed less extensive deep-sea sedimentation.[21]

The last of Wegener's six geological arguments, and perhaps the most interesting, was that concerning the displacement of the pole of rotation. Here he abandoned all remaining tentativeness: "Notwithstanding the great and justifiable caution with which one approaches in geology all hypotheses concerning the displacement of the poles, yet so much material has been produced lately from that point of view, that a great displacement of the poles may in any case be regarded as proven [*nachgewiesen*]. In the course of the Tertiary Period, the North Pole shifted from the vicinity of the Bering Strait over towards Greenland; the South Pole from South Africa toward the Pacific side of Earth."[22]

It may seem that Wegener is here suddenly dropping geological argument and returning to geophysical argument by discussing the displacement of Earth's pole rotation. This would be a source of confusion to Wegener's readers in 1912 and thereafter: he consistently characterized the notion of the displacement of Earth's pole rotation as a *geological*, rather than a *geophysical*, argument. It is hard to imagine something more geophysically significant than the displacement of Earth's pole, yet from Wegener's standpoint it was, and would remain, a *geological* hypothesis. Wegener classified his arguments not according to their topical focus but according to their *evidentiary basis*. The geophysical arguments for continental displacements were those based on data concerning Earth's gravity field, the propagation of seismic waves, and the laboratory determination of the physics of the solid state under high temperatures and confining pressures: these were all geophysical data. The geological arguments for continental displacements (and for displacements of the pole of rotation) are those that are founded on geological data: petrology and mineralogy, paleontology, surface morphology, glacial deposits, and mountain ranges.

The evidence that appeared decisive for Wegener in the matter of pole shifts came from studies of successive layers of fossil plants in the Tertiary of Europe and the Arctic, with the shift back and forth among subtropical, temperate, and cold-climate plant remains. The study of very recent plant fossils had the immense advantage that one knew from their living near-relatives, in an entirely unambiguous way, what sort of climates they inhabited. This gave great force to the latitude-zone reconstructions of the geographical distribution of these same plants in the more distant past. Almost all of the authorities and sources quoted by Wegener in this section of his paper he obtained from Nansen's scientific appendix to *Auf Schneeschuhen durch Grönland*, the same book he was reading and rereading in preparation for Greenland.[23] Most notable among these paleobotanists was the work of Alfred Nathorst (1850–1921). Nathorst had plotted pole positions for the ice age and most recent parts of the Tertiary, and Nansen had highlighted and endorsed Nathorst's work, as well as the puzzles it presented, as a presumptive reason for scientific expeditions to Greenland, of exactly the kind on which Wegener was about to embark. We may recall that Nansen, following Nathorst, had agreed that this sequence of plant fossils could not be explained by the normal rhythm of glacial and interglacial intervals within an ice age: tropical plants were simply too far north.[24]

Wegener imagined that in the course of the Tertiary, the North Pole shifted more than 30° from its current position and lay then in the Bering Strait. The pole of rotation then traveled back toward Greenland, overshooting its current position by perhaps 10°, and had moved back to its present position only at the end of the ice age.

A Tertiary position of the North Pole in the Bering Strait would put the corresponding South Pole about 25° south of the Cape of Good Hope. Thus, the absence of glacial remains in the Southern Hemisphere in the Tertiary would be explained by the fact that this South Pole was at that time, if not at sea, then on the margin of the former South Pole continent, with much of the polar ice at that time being sea ice—as is characteristic of the North Pole regions in our own time. Similarly, in explaining the Permo-Carboniferous remains of the Southern Hemisphere, the placement of the Permian South Pole as much as 50° away from its current position (dictated to give a rational distribution of faunal remains) would have put the Permian North Pole beyond the Bering Strait in the Pacific, thus leaving no Permian glaciation in the Northern Hemisphere, which is consistent with what is observed.

> Of the greatest importance for the understanding of the whole phenomenon is however the circumstance that the great displacements of the pole clearly are contemporary with the great displacements of the Continental platforms. Especially evident is the temporal coincidence of the best-verified pole displacement in the Tertiary, with the opening of the Atlantic Ocean. One may also perhaps be able to connect the (relatively insignificant) return journey of the pole since the ice age with the separation of Greenland and Australia. It seems in this connection as if the great continental displacements are the underlying cause of the displacement of the pole. The pole of rotation must in any case follow the pole of inertia. If this is altered by the displacement of the continents, then the pole of rotation must travel with it.[25]

Wegener's hypothesis in this version includes both continental displacements and true polar wander. The displacements of the continents redistributed large masses across the surface of Earth and changed Earth's pole of inertia. The pole of inertia is the pole (the axis of figure) defined by Earth's center of mass, and if the surface masses redistribute, this pole will move to reflect this redistribution of masses and Earth's new center of mass. The so-called Chandler wobble, or nutation, is the oscillation of the pole of inertia around the rotational pole, caused by Earth tides, atmospheric motions, and other variables. Wegener imagined, superimposed on this well-known phenomenon, a larger shift of the pole of inertia caused by the redistribution of very large continental masses, with the astronomical (rotational) pole following. Wegener did not imagine the whole "lithosphere" sliding over Earth's interior, but individual continental fragments moving in different directions: South America, North America, and Greenland were moving to the west, while India and Australia were moving to the north.

Wegener's argument here is quite dense, difficult to follow even using a good map, and typically baffling to readers not conversant with the relevant geography. What is important is that Wegener was creating a spatial/temporal grid of Earth, integrating continental positions, pole positions, and continental glaciations throughout geologic history, as an index to the distribution of latitude zones with appropriate vegetation. He was, however, trying to say in a scientific paper the sort of things usually said in a good-sized book, consequently with somewhat indifferent results.

The abandon with which Wegener pushed Earth's pole this way and that across many degrees of latitude and longitude may seem reckless, but Wegener noted that the planetary astronomer Giovanni Schiaparelli (1835–1910) had in 1889 concluded that in a plastic Earth (the sort of Earth that Wegener had considered most likely: rigid for short-term stresses, but indefinitely deformable by long-term stresses) the pole of rotation might be displaced by any amount. Schiaparelli considered three cases: a completely rigid Earth, an Earth with "plastic" adjustment but some lag in its response, and "an earth sufficiently plastic not to lag appreciably behind."[26] Wegener favored the third of these alternatives, and he found these investigations extremely interesting, not least because Schiaparelli had argued that "in order to obtain a displacement of several degrees it would probably be necessary to move beyond those phenomena already revealed by the study of Earth's crust."[27] By this he meant a resort to the sort of geophysical data produced many years after his argument—exactly the data on which Wegener depended. Schiaparelli had also asserted that "geological actions, sufficiently prolonged," could at any time destroy the stability of the geographic poles and give rise to rapid movements of the pole of rotation: "once admitted, this will open new horizons for the study of the great physical revolutions that the crust of Earth has undergone."[28]

Schiaparelli was an astronomer of note, and the quotation given above is indicative of just how many scientists—geologists, geophysicists, and astronomers—were willing at the turn of the twentieth century to consider displacements of Earth's pole over rather large distances as an entirely plausible hypothesis to explain geological and paleontological phenomena. From Wegener's standpoint, his own originality consisted largely in showing that pole displacements alone, however extensive, could not provide a sufficient explanation for the existing paleontological evidence; these had to be joined with actual displacement of the continents.

Attempts to calculate pole displacements caused by observed mass displacement had, up to the time of Wegener's writing, concentrated on "only the very slight displacements that could be noted in earthquakes, for example," and therefore "the conclusion was invariably reached that the produced pole displacements had to be immeasurably slight."[29] Hayford, whose North American gravity survey was so important to Wegener in establishing the thickness of the continents, had calculated that the shift of the crust along the San Andreas Fault during the San Francisco earthquake of 1906 could have shifted the pole of inertia only about 2 millimeters (0.08 inches), through the dislocation of a 40,000-square-kilometer (15,444-square-mile) block 118 kilometers (73 miles) thick, 3 meters (10 feet) to the north.[30] Wegener noted that the displacements of continents with which he was concerned involved masses at least 100 times larger than those considered in this example and might cause pole shifts as large as 1° in 360,000 years, an order of magnitude that could shift the pole enough to produce the paleontological transitions from warm to cold climates and the reverse in the allotted time.[31] Wegener estimated that it had taken roughly 10,000,000 years to open the Atlantic with this amount of annual displacement.

Measuring Contemporary Displacements

Wegener's estimates for the rates at which the continents were moving apart are enormous by current standards. He estimated that the westward separation of Greenland from Scandinavia, in 1912, was proceeding at a rate between 14 and 28 meters (46–92 feet) per year. He obtained this astonishing rate by assuming the separation of the two continental blocks from one another to have taken place between 50,000 and 100,000 years ago. This would have been after the "great" ice age but before the last ice age, according to the geochronology of his time, and was an estimate based jointly on preliminary radiometric dates and paleontological data supplied by Arldt and others. The distance of 1,400 kilometers (870 miles) between Scandinavia and Greenland in 1912 and the assumption that they had been separating for between 50,000 and 100,000 years accounted for the spreading rate of 14–28 meters per year.

Wegener did not obtain this number by any geophysical calculation having to do with any quantity concerning plasticity, fracture and flow, isostatic adjustments, or any other geophysical considerations. He got his spreading rates the way we get ours: he took the total distance and the total time and divided the distance in meters by the time in years, to get meters (or fractions of meters) per year. In spite of a general impression to the contrary, our current plate tectonic spreading rates have yet to be unambiguously measured. Scientists, since the 1990s, using very long baseline interferometry, satellite laser ranging, and GPS systems, continue to claim to have demonstrated such motions. Without exception, their claims repeatedly confuse the precision of their measurements with the accuracy of their results, which have uncertainties of 40 percent and greater. The uncertainties are largely driven by the need to incorporate into their calculations a theory of Earth's nutation and a crucial term for atmospheric refraction, the latter depending on the amount of water vapor in the atmosphere, a quantity that changes hour by hour. The measurement results agree with models, and no one doubts that the plates are moving, but accurate measurements of plate motions are still, at this writing, beyond the limits of resolution of the best technologies, regardless of many strident claims to the contrary.

With the current estimate of the end of the Cretaceous as the time for the beginning of the opening of the Atlantic (this was also Wegener's timing of this event, again following Keilhack), this gives a time of around 57 million years ago for the original separation. Current estimates for the opening of the Atlantic are on the order of 2 centimeters (0.8 inches) per year, with the notion that earlier spreading rates may have been as high as 5 centimeters (2.0 inches) per year. The spreading rate is obtained by the ratio of the distance (3,500 kilometers [2,175 miles]) and the time. This is identical to Wegener's method, with his much-attenuated time span producing a reciprocally increased spreading rate.

Wegener wanted to test his estimate against measurements of the longitude of East Greenland, but he had only three sets of measurements with which to deal. There were those made by Edward Sabine in 1823, those made by Börgen and Copeland on the Germania Expedition in 1869/1870, and finally those made by Koch (and Wegener) on the Danmark Expedition in 1907–1908. The technique of measurement was an old one, that of the observation of the Moon against the background of the fixed stars. Each of these three expeditions used a different version of this technique. Sabine used lunar distances (distance between the Moon and a fixed star), Börgen and Copeland used lunar culminations and stellar occultations, and Koch and Wegener had used lunar azimuths. None of these observations were taken at exactly the same place, and all were subject to very taxing and minute corrections to yield any accuracy at all. In spite of the possibility of errors in the range of hundreds of meters for the location of these observing stations, Wegener insisted that the difference between Sabine's measurements and those of Koch and himself, after a lot of calculating, showed a westward drift of Greenland of 950 meters (3,117 feet) over the span of eighty-four years. This gave a rate of drift of 11 meters (36 feet) per year, within his predicted range, and he chose to believe that this figure was not the summation of large individual errors.

> It is well known that in the measurement of longitude with the help of the moon, the attainable accuracy is poor. One may estimate the probable error of a single series of measurements as quite easily several, and perhaps many hundreds of meters. The difference of 950 meters, however, that has here become apparent over the course of

> time, seems to me to be somewhat too great already to be justifiably set aside as merely an unfortunate summation of individual errors. It might rather be quite considerably more likely that it has been caused by an actual displacement of the continents of the order of magnitude given.[32]

Understanding (and then discounting!) the fragility of the inferences from such measurements, Wegener then offered as evidence of continental displacements a series of measurements made between 1866 and 1892 comparing the longitudes of positions in North America with those at points in England and France. These measurements used the (then) recently laid submarine telegraph cables to exchange exact time signals of the culmination of celestial bodies above established observatories. At the precise instant of culmination the observer struck a telegraph key that sent a telegraph time signal across the Atlantic, to be recorded on a revolving drum on an instrument at the other observatory. The difference in the times of culmination of the same celestial body between two stations on opposite sides of the Atlantic represented the difference in longitude between the stations.

Such measurements strove for great accuracy, and observers traveled back and forth across the Atlantic carrying their recording equipment, so that the U.S. observer and his telegraphic equipment would then be used in England, and the English observer and his telegraph equipment would be used in the United States. The resulting difference in longitude as measured was 0.23 seconds of longitude over twenty-six years, or somewhere around 1/100 of a second of time per year, equivalent to a continental displacement of about 0.27 meters (0.89 feet) per year.[33]

These measurements were, of course, not carried out to measure continental displacement but were part of an ongoing international program to correct the accuracy of geodetic networks and also of stellar ephemerides. The possibility of exchanging instantaneous time signals was something that Wegener was familiar with through his own astronomical work at the Berlin and Potsdam observatories, and of course something in which he was keenly interested as a student and author of tables of planetary and lunar positions. At the very time Wegener was working on his article on continental displacements, an international conference was preparing to convene (in Paris) to establish a system of universal time, to be based on time signals transmitted telegraphically from the Royal Greenwich Observatory in Great Britain. Thus, the 0° of longitude would also become the home of the reference time signal and make highly accurate determinations possible according to a single international standard.

Wegener had to admit, however, that with regard to measurements of continental displacement "these numbers are not only for the time being still completely uncertain, but also that they are, up to now, scarcely to be regarded as at all sufficient to establish the reality of displacement. The observed time difference amounts to only 0.23 s and is thus still so small, that it, if need be, could be explained by the greater imprecision of the earlier measurements. If however some new longitude measurement—twenty years have now already passed since the last—again should likewise yield to change, the reality of the displacements could no longer be doubted."[34]

Wegener also considered that latitude measurements could be employed to measure displacement, both to measure the displacement of India and Australia, which he believed to be moving northward, and through measurements of the position of the North Pole, by the International Latitude Service. This international organization, established in 1899, with six observatories located at latitude 39°08″ north, was a joint effort of the U.S. Coast and Geodetic Survey, the Prussian Geodetic Service, and other organizations. These were a continuation, using zenith telescopes, of the observations initiated by the Potsdam Observatory in Germany, under the direction of Wegener's professor, Friedrich Helmert, and the observations of Chandler himself in the United States, to measure the wobble of Earth's axis of rotation. Wegener was again intimately familiar with these measurements; indeed, they were the inspiration for his doctoral dissertation.

While no such measurements had yet detected a permanent offset of the pole of rotation such as might be produced by a displacement of the pole of inertia, Wegener believed that a longer time series would also soon show this displacement and thus serve as an argument for the displacement of the continents as a cause of the displacement of the poles. "Naturally, these views will be considered correct only when it will have been possible to support them through an exact mathematical treatment. At present I do not have time to undertake such an investigation. Perhaps this brief note has, however, succeeded in calling attention to this problem, which through its connection with the hypothesis of the horizontal displacement of continents, seems to me, in a manner undreamed of previously, to be thrust into the foreground of interest."[35]

With this laconic observation, Alfred Wegener brought to an end his first extensive presentation of his theory of continental displacements. The abrupt termination of the argument, without summation or conclusion, reminds one of the last lines of Wegener's *Thermodynamik der Atmosphäre*, published the year before. As in that case, there was more to say but no time to say it: the publishing deadline loomed, and other tasks were at hand.

When we examine the lines of geophysical and geological evidence he brought to bear in hypothesizing continental displacement, it is astonishing how many of the principal authors he employed in his arguments were either his professors (like Helmert) or senior colleagues with whom he was closely associated (like Wiechert and Kayser). He had firsthand knowledge and training in astronomical position fixing, geodesy, gravity measurement, solid mechanics, thermodynamics, and physical chemistry. In this regard, his work in continental displacements was similar to his work on atmospheric physics, where his confidence to handle the material stemmed not least from his personal contact with the leading practitioners of that science and his direct training and experience in what appears, in retrospect, to be an astonishingly broad range of subfields of physical science.

In consequence, one should not be misled by later characterizations of Wegener as a rebel, an outsider, a loner, a maverick, or an isolated genius figure. Wegener was a well-trained, experienced, highly competent physicist who saw all his work, continental displacements included, as legitimate extensions of solid and well-accepted conclusions concerning the physics of Earth. He was professionally well placed, well known, and widely published.

It may, nevertheless, seem a great overreaching for a theorist of the atmosphere to develop and advance publicly (within months of its first formulation) a comprehensive theory of the origin of continents and oceans. Perhaps it was. From Wegener's standpoint, however, the unification of material parted out to different disciplines within Germany—geology, oceanography, meteorology, geophysics, geography, and astronomy—was precisely the task of cosmic physics, already recognized as a discipline meriting professors and professorships in Scandinavia and as a subject matter in German secondary schools. Wegener made no secret of his ambition, expressed to his

Alfred Wegener in a photo taken in 1911 or 1912. His demeanor suggests the confident and expansive mood that characterized his work on the origin of continents and oceans and the thermodynamics of the atmosphere. Photo courtesy of the Heimatmuseum, Neuruppin.

brother, his parents, his wife-to-be, his future father-in-law and collaborator Wladimir Köppen, and his superiors at Marburg, Richarz and Kayser, that he should become one of the first university professors of cosmic physics in Germany. His confidence that he would achieve this was as absolute as his confidence that he would mathematically confirm continental displacements within a few years. In both cases it was merely a matter of time.

Greenland

When he had learned in early February 1912 that his Greenland expedition—his and Koch's—was now imminent, Wegener knew he had to raise 15,000 marks in a little less than two months and additionally had to face another difficult truth: almost everyone he knew and loved opposed the trip. Köppen thought it an unnecessary diversion from the opportunity, offered by the Carnegie grant, for Wegener to move to Hamburg and begin their closer collaboration. Else was unhappy with the prospect of the danger and the postponement of the marriage. Wegener's parents, Richard and Anna, were aghast that at this critical juncture in his professional life, on the cusp of being a candidate for a real professorship, he would venture off on another one of his vagabond larks.

The timing was certainly inconvenient for him professionally as well. It would force him into hurrying and compressing his revisions to the paper on continental displacements, which so obviously showed its unevenness and the haste of its composition, especially in the final section. Clearly, however, he wanted to go back to Greenland. There was his private ambition to be known as a polar explorer. Moreover, with his work on the Danmark Expedition science complete, he had consequently run out of any data of his own which he might offer as an original scientific contribution. He could certainly make his mark as a theorist, but it was also necessary to do field science if one wished to be listened to in a world as data driven as that of meteorology and geophysics.

There was another significant dimension to this expedition plan. Wegener had a score to settle with the way that the Danmark Expedition had been run. It had been sprawling, wasteful, and disorganized and had sacrificed science to everything else, from Mylius-Erichsen's unquenchable ambition to the insistence that all the scientific staff perform the work of ordinary deckhands. The motorcar that was the only source of power for the kite winch, the loss of which cost him so many hours of painful labor, had been sacrificed in a joyride. This and dozens of other irritations he painstakingly recorded in his expedition diaries from those years. While his loyalty to and friendship with Koch might have carried him again to Greenland anyway, this expedition was very much the fruit of their many conversations about how things should have been done in 1906–1908; it was a chance to do things "the right way."

This new expedition would be a study in contrast to the previous one. It would be well organized, well planned, compact (if 20,000 kilograms [44,092 pounds] of equipment and supplies can be called compact), and, best of all, collaborative on an equal footing. Wegener, Koch, and the botanist Andreas Lundager would be equal scientific partners; Vigfus Sigurdsson would see to the horses. Lundager had been their companion in the "Villa" on the Danmark Expedition, along with the painter Aage Bertelsen, and was a proven quantity, as was Vigfus, well known to Koch. There would be none of the personal struggles and petty squabbles of the Danmark years.

Wegener had to make the case to his family, friends, allies, and colleagues that he should, instead of pursuing his career, finding a professorial post, defending his radical theoretical claims in both meteorology and geophysics, getting married, settling down, and entering into some stable and long-term professional relationships, take off on an expedition to traverse the middle of Greenland, simply because it had never yet been done. The expedition had a scientific program, but it was not extensive, and neither the time they would spend in winter quarters on the east coast of Greenland nor the pace they would set in order to make it across the ice cap in the summer months would permit a concentrated program of scientific work anything like what had been accomplished between 1906 and 1908. In polar science, the sheer physical challenges to survival were immense, and the chances of failure in every scientific undertaking were very high. For anyone who knew anything about Arctic science, this was "a long shot."

Persuasive help in making his case arrived from an unlikely quarter. Wegener's *Thermodynamik* had been rapidly reviewed and received positive notices in the *Physikalische Zeitschrift* and elsewhere. In December of 1911, however, it had received a terse and dismissive review in the one place it could hurt most: *Meteorologische Zeitschrift*.[36] The reviewer was Felix Maria Exner (1876–1930), who was only four years Wegener's senior but professionally very highly placed. He had been, since 1907, the secretary of the Central Institute for Meteorology and Geomagnetism in Vienna. As a scion of the Exner family's academic dynasty, extending back into the nineteenth century, he had been named professor of cosmic physics at the University of Innsbruck in 1910. His uncle, Franz Serafin Exner (1849–1926), a celebrated physicist, had become rector of the University of Vienna in 1908.[37]

In February of 1912, shortly after Alfred broke the news to his parents that Koch had obtained the money from the Carlsberg Foundation and that he would therefore be leaving for Greenland in a few months, Kurt Wegener wrote to Richard and Anna to explain the situation. The context makes it clear that Wegener's parents had asked Kurt to intercede with Alfred and to induce him to abandon the Greenland plans. Here is what Kurt wrote:

> The [Greenland] plan is scientifically significant . . . the possibility of Alfred suddenly becoming a professor has, in consequence of the extremely unfriendly review of his beautiful book (Thermodynamics) by Exner in the Meteorologische Zeitschrift, for the time being completely vanished, since externally or to all appearances the opinion of the Meteorologische Zeitschrift passes as the standard measure for the profession, though the true professional opinion which one has heard everywhere speaks of Exner's careless and cursory review, and this must be given time to slowly penetrate. So that the expedition year signifies thus from this viewpoint no loss, especially since in that year the first reviews of Alfred's many completed papers (including those concerning his "Urkontinent") will appear and have time to work. Finally, it seems to me for his later finding a permanent situation, it will be advantageous for him to actually carry through an already announced expedition so that he doesn't develop a reputation as a blowhard and puffer. You will probably be very angry with me but I can't for all the above reasons fail to believe that this is true and above all for the good of the expedition cannot dissuade Alfred from this undertaking. And also his health makes me think that a "sport year" seems to be desirable. This constant bent-over sitting makes him seem like he'll become just as slight as I am.[38]

Kurt's reference to Alfred "suddenly becoming a professor" may sound odd, with Alfred barely having enough money to live on and even considering leaving academia altogether to go to work in Hamburg on the Carnegie project with his future father-in-law. It turns out that Alfred had been on the short list of nominees, in the summer of 1911, for the newly created professorship of geophysics being established in Leipzig.[39] This prestigious post had gone instead to Vilhelm Bjerknes (those working at Leipzig already had a strong preference for his highly mathematical style of work), but that Alfred was nominated for the job is an indication of the extent to which he was considered a rising star in the meteorological community at this point in his career.[40] Indeed, the great Russian climatologist Aleksandr Voeikov (1842–1916) had used those exact words—"a rising star"—in his review of Wegener's thermodynamics in Russia, an assessment that he passed along to Köppen in a letter.[41]

Kurt's assessment was that it would take at least a year for the damage done by Exner to fade and for Alfred's thermodynamics to be seen in the context of other published work. Kurt was persuaded that the work on continental displacement would by itself make Wegener famous and, in the context of his work in atmospheric physics, would certainly propel him into the front rank of the profession.

Alfred had himself told his parents, on 6 February, that he had promised Koch that he would make the trip, including both the overwintering and the crossing, but only if it could happen in the year 1912–1913. He told them that he and Else were going to go forward with their marriage plans and had set the date for 1 November 1913. He also told them that all his scientific and professional work would be completed and delivered before the expedition began, especially the work on the continents. In addition, he assured them (and he was proud of this) that his doctoral student at Marburg, Walther Brand, was completing the last remaining part of the Danmark Expedition work.[42]

These statements were all true, but they were not as persuasive as Kurt's announcement that the review by Exner had for the time being dashed Alfred's hopes for immediate professional advancement and a professorship. Kurt's letter also played on Anna Wegener's feeling that her children had to get out in the countryside because they were so "pale and listless," an opinion expressed in anguish after Käte Wegener's death so many years before in Berlin.

Alfred, of course, saw Exner's review when it appeared and had discussed it at length in a letter to Köppen on 17 January 1912. "Exner's review is actually only his opinion," Wegener wrote, "not an objective review, and I think the tone of the whole thing was a misstep. The advertisement of the publisher, which appears in the same issue as the review, provides a fitting complement to it, since the things that Exner made an especial point to say were *not* covered in the book, can at least be found in the advertisement among the titles of the chapters."[43] Wegener also told Köppen that the definition of thermodynamics, given by Exner in the review, was fundamentally mistaken and confused: "And what's with the reference to the steam engine? The atmosphere is not a steam engine, quite the contrary. If you look, for example, at Planck's Thermodynamics, you will look in vain to find a steam engine."[44]

It was certainly generous of Wegener to stick to the intellectual substance of the review, especially since Exner, rather than characterizing it as his own opinions and then disagreeing with Wegener's standpoint, had instead intimated that Wegener was a careless autodidact who did not understand the subject he was writing about. "Reading this book," Exner wrote, "gave me the impression that this could only have been written just now and not published a year from now . . . it is a collection of his work up to now . . . the selection of topics and the way they are treated is very uneven . . . what the author has studied in the literature and published in the last few years is here, often not very accurately presented or clearly brought together."[45]

This sort of personal attack was characteristic of Exner's manner of reviewing works of broad scope and theoretical ambition, especially those that might challenge his own theoretical commitments and orientations. Some years later, when Lewis Fry Richardson published his epochal *Weather Prediction by Numerical Process* (1922), today recognized as the foundation of all weather and climate modeling, Exner dismissed Richardson's work as unlikely to bear fruit and too burdened with detail.[46] He was even more ill disposed toward Vilhelm Bjerknes, and Bjerknes's biographer, Robert Marc Friedman, has described Exner's response to Bjerknes as "troublesome" and even "ugly."[47]

Exner's personal nastiness aside, however, there were real scientific issues being discussed here. Wegener took thermodynamics to be what we now call "physical meteorology," largely concerned with temperature distributions in the vertical and their effect on rising and descending parcels of air, as well as the formation of clouds and precipitation. He thought, as written in his book, that the empirical determination of the layering of the atmosphere and its temperature regime was a necessary precondition to a dynamic theory of cyclonic storms. Exner, on the other hand, associated atmospheric thermodynamics entirely with the theory of cyclonic storms—a heritage of the nineteenth-century "thermal theory of cyclones." He criticized Wegener sharply for not taking up the questions posed by cyclonic motions or discussing Margules's solutions to them.[48]

Wegener disagreed with this whole approach and had written so to Köppen in January, in the letter quoted just above. The things that so excited Exner in Margules's work on the energy of storms were not about thermodynamics, Wegener wrote, but "belong to mechanics." He reiterated his position that thermodynamics was about the distribution of temperatures in the vertical. As for Margules himself, Wegener went on at length about the enormous theoretical difficulties left unsolved by Margules. Not the least of these, from Wegener's standpoint, was that Margules's combination of thermodynamics and hydrodynamics (the same problem that Bjerknes was working on) had not solved most of the major problems of this synthesis, especially since work on hydrodynamics dealt with an incompressible fluid (water), while the work on the atmosphere dealt with a compressible fluid (air). Finally, Wegener thought that Exner had a too exalted view of the role of theory, a view that Wegener found illusory. "The foundations of a science are not constructed through such theoretical works, but through individual, empirical investigations. That's my view anyway."[49]

The vision of atmospheric thermodynamics as independent of atmospheric dynamics, rather than as the handmaid of the latter, pitted the institutional and theoretical aims of the group at Berlin and their students distributed throughout Germany against the group in Vienna and their network throughout the Austro-Hungarian Empire. Moreover, the Viennese, and Austrians in general, were committed to mountain meteorology, while the Berlin group argued vociferously that mountain meteorology had to give way to aerology. This was also a source of friction between them.

When we see what Exner and Margules were doing in Vienna, we see that Exner's review of Wegener's book could hardly have been otherwise. The Viennese work was highly theoretical and highly mathematical. Wegener's disinclination for abstract mathematical formulations, especially in an elementary textbook, led Exner and indeed others to believe that Wegener was not capable of frontline work. One should say, parenthetically, that the reason that Margules's fundamental work became recognized only decades later was precisely because it was so theoretically difficult and so mathematically dense.

Even though meteorologists have come around to accepting Wegener's view of the appropriate scope for the subject matter of the thermodynamics of the atmosphere, in 1911–1912 almost all the intellectual energy in meteorology was being thrown into dynamics. Köppen's plan for his collaboration with Wegener, under the auspices of the Carnegie Institution, emphasized dynamic meteorology and weather prediction as proximate goals. Wegener had objected to this strenuously in their correspondence in November 1911, especially to the notion that the atmosphere could be referred to, without further qualification, as that portion of the atmosphere below the tropopause, the "zone of weather." Wegener understood the financial and intellectual pressures at work, but he did what he could to counteract them, in the service of understanding the behavior of the entire atmosphere.

In any event, his involvement with the theory of continental displacements was so deep and so all-consuming that thinking about the atmosphere seemed almost like a relic of a former life. When he wrote to Köppen on 29 January, after a brief discussion of the kind of observational network that would be required for the Carnegie project, he turned immediately to a long discussion of Earth's pole of rotation. "It seems to me ridiculous," he wrote, "that it should be so easy to come up with an explanation [of the cause of the displacement of Earth's pole] which those who are actually specialists in this area have been unable to find so far despite their most astute mathematical analysis." Then he added, almost as an afterthought, "Lately I've had a few letters about the Thermodynamics from Trabert, Aßmann and—Exner. You would find Aßmann's letter interesting. . . . What should I do with Exner, answer him?"[50]

Along with his obsession with continental displacements, there was another consuming occupation able to replace atmospheric physics in his mind: raising enough money to finance his part of the expedition. Along with the letter congratulating him on the publication of his *Thermodynamics*, Aßmann had sent a check in the amount of 500 marks to support Alfred's expedition to Greenland. Additionally, Wegener had received a large donation made privately by the chemist and historian of science Ludwig Darmstädter (1846–1927), a gift arranged by the meteorologist Gustav Hellmann (1854–1939), Bezold's successor at Berlin.[51] Wegener needed 15,000 marks and was writing many letters of appeal for funds each day. He was able, in February and March of 1912, to obtain a large grant from the Prussian Interior Ministry, one from the Prussian Ministry of Culture, and another from the Berlin Academy of Sciences, totaling more than 10,000 marks.[52] With this money in hand and confident that he could obtain the remainder, he wrote in early April to the (new) dean of the philosophical faculty at Marburg, Ernst Elstner (1860–1940), asking to be excused from his summer semester lectures, already announced, and additionally to be granted leave until the fall semester of 1913.[53]

Whatever the career troubles for Wegener that Exner had generated with his review, they were not echoing loudly in Marburg, where Wegener was currently something of a celebrity. Not only did the dean immediately grant the requested leave, a request that was accompanied by a strong endorsement from Richarz, but also the faculty obtained from the Ministry of Education permission to continue his stipend during his absence on the expedition. The faculty also passed a resolution endorsing the scientific importance of his Greenland expedition.[54] Richarz himself provided "very opulent terms" for Wegener's continuation on the faculty of Marburg. He told Wegener that he was taking steps, beginning with the vote of the faculty, to have the Ministry of Education create a more secure place for him at Marburg, perhaps an extraordinary professorship (the equivalent of an American associate professor position, something short of the creation of a new *Lehrstuhl*, or full professorship). Wegener was very touched by this outpouring of support and wrote about it excitedly to Köppen.[55]

In the midst of this, he also had a personal life, albeit one confined to letters. He and his fiancée, Else, were in near-constant communication. He had invited her to come to Marburg when her teaching year ended in April. She would stay with his student Walther Brand and his wife, allowing them time together, appropriately chaperoned. There was to be an annular eclipse of the Sun on 17 April, which they would view from Marburg. After that would be the big surprise: he had arranged a balloon flight for her, to replace the one canceled the previous year. They would travel by train together to Göttingen, where they would meet Kurt and Tony, and the four would make the flight together.[56] Else had told him that making this flight was "her dearest wish."[57]

Even before their reunion in April, Alfred and Else had come up with an interesting plan. Wegener wrote to Köppen in February, "Together with Else, I have had a stupendous idea, but won't know for sure about it until tomorrow evening; I won't say anything about it now."[58] Here was the idea. Else had been deeply disappointed by the postponement of their wedding plans and felt in limbo in her parents' house in Hamburg. Wegener had broached the idea with Vilhelm Bjerknes that while he, Wegener, was in Greenland, Else would travel to Oslo and live with the Bjerknes family for a year, more or less as a daughter of the house. While Bjerknes had taken up the Leipzig professorship in 1912, he was not quite ready to move his family. The Bjerkneses had four sons, the oldest of whom, Jacob, was 15. The plan was that Else would teach the boys German in preparation for their *Gymnasium* course in Germany in 1913, while simultaneously attending classes at the university in Oslo, where she would learn Norwegian and Danish. While Wegener and Bjerknes differed in age (Bjerknes was twenty years older), they were made comrades by Exner's hostility to both of them, by their Berlin connection, and by their mutual knowledge that they had been competitors for the same professorial chair in Leipzig.

It was a poignant time for Alfred and Else. Though their marriage would be postponed a year, Alfred was doing everything he could to create an aura of honeymoon and married life for Else before he departed for Greenland. In this spirit, her trip to Marburg in April was a resounding success. Alfred was relaxed and happy, he had the money in hand, the last of his manuscripts had been sent out, his doctoral students were set to defend in June, and he was confident that they would finish successfully.[59] It really was a vacation. Else was thrilled to be treated as an adult woman. Here she was with her fiancé, having a picnic with the Brands; Walther was Alfred's student, but they were both thirty-one years old, two couples at a garden party with a solar eclipse as entertainment.

If the Marburg visit was a success, they had even better luck with the balloon flight. The eighteenth of April dawned clear with light wind. They lifted off without incident, and Else marveled at the silence and the smoothness of the flight. "No shaking, no sound of a motor disturbed us. Now and then the barking of a dog in a village below would break our silence."[60] They traveled for hours to the northwest, finally settling down—Kurt made a flawlessly smooth landing—in an open field near the train station in Leer, near Bremen, about 250 kilometers (155 miles) from their start.[61] After a festive dinner, with the balloon packed up and shipped back to Göttingen, Else traveled to Hamburg, while Alfred, with Kurt and Tony, went to *die Hütte*, so that Alfred might take leave of his parents.

Departure of the expedition for Iceland and then Greenland was drawing close, but there was still much to do. After leaving his parents, Alfred traveled to Hamburg and spent several days at the Köppen household in Großborstel. Then, at the beginning of May, he and Else traveled to Copenhagen. Once again, Else thrilled to have a taste of adult life with her husband-to-be. They were guests of Koch and his wife, and Else had a chance to take part in the preparations. Alfred wanted her to have a sense of where he was going and what he was doing. As they packed the crates with the 20,000 kilograms of equipment and supplies, she made lists of the contents. An experienced meteorological assistant in spite of her age, she was able to help calibrate and pack the meteorological instruments for shipment.[62]

The most interesting part of the preparations in Copenhagen was perhaps the assembly and then disassembly of the prefabricated house in which they would overwinter. It had been built to specification for Koch by the Danish military. It consisted of a series of bolted-together panels of plywood, fabricated such that there was an air

Balloon flight arranged for Else Köppen in April 1912 by Wegener. *Left to right*: Alfred Wegener, Kurt Wegener, Else Köppen, and Tony Wegener. Photo courtesy of the Heimatmuseum, Neuruppin.

space between two layers of plywood in each panel. The central *Stube* for sleeping and work, with bunks, worktables, a stove, and a photographic darkroom, was flanked to the outside by parallel rooms, one a store room and the other the stalls for the Icelandic ponies. Measuring 6.6 meters (21.7 feet) × 5 meters (16.4 feet), the interior area of the building was 33 square meters (356 square feet). While not palatial—though Frau Koch christened it "Borg" (the Castle)—neither was it claustrophobic, and it was certainly as well designed and comfortable as anything taken into polar regions by an Arctic expedition to that date.[63] The panels were much smaller than those for the prefabricated "Villaen" that Koch, Wegener, Lundager, and Bertelsen had occupied in the winters of 1906 and 1907. The larger panels of that building, lashed down on the deck of the "Danmark," had warped in transit and had to be completely disassembled before the building could be raised. The panels of this newer design were small enough that each Icelandic pony could carry two at a time strapped to the packsaddles; their smaller area also lessened the chance that they would warp in transit.

Koch, Wegener, and Lundager numbered the panels sequentially as they assembled them. They bestowed similar care on the crating and labeling of their supplies. Everything was designed either to be carried on an Icelandic pony packsaddle or to sit flush to the rails of one of the horse-drawn sleds. The boxes had to be strong enough to withstand battering but light enough that a man could lift them onto a horse pack and down a number of times each day.

The expedition was scheduled to depart around 1 June and to travel to Iceland. Here they would meet up with Vigfus Sigurdsson and his Icelandic ponies. Koch had worked with the horses before, but Wegener and Lundager had not. Koch had done considerable cartographic work in Iceland, including the Vatnajøkul ice cap, which he proposed to use as a training ground in the management of the Icelandic ponies in snow and ice, for the benefit of Wegener and Lundager. They would practice riding, leading pack trains of horses, loading and unloading the packsaddles, feeding and grooming, and learning to be comfortable with and earning the trust of the horses. There were a number of experiments to try. Each man was to have an impressive array of footwear: leather climbing boots, Eskimo kamiks, finnesko (reindeer-skin boots stuffed with soft grass), gum boots, wooden clogs embedded with ice spikes, and woven hemp slippers, all for different tasks in different weather conditions, each of which had to be tried out.

The plan of the expedition was simple; the execution would be difficult. They would sail on their chartered sealer the *Godthaab*, with their ponies and equipment, with the intention of being put ashore at Danmarkshavn; this would allow them to make use of their knowledge of the terrain. Then would come the hard part: transporting 20,000 kilograms of baggage 150 kilometers (93 miles) inland to Dronning Louise Land, to establish an overwintering station on the Inland Ice. In the following spring, they would attempt a 1,200-kilometer (746-mile) traverse from the northeast to the southwest, hoping to emerge near the depot at Upernavik on the west coast of Greenland, laid down in anticipation of this expedition, in 1911, by Lundager.

If it proceeded as planned, the expedition would take slightly more than a year: the summer for training in Iceland, a late summer and early fall arrival in Greenland, and an overwintering from October to April. This would be followed by a concerted push diagonally southwest across the ice cap, hoping to emerge in the early summer, before the ice melted and made travel difficult, as it had thirty years before for Nansen. The group would then have a few months to pack up, find a ship, and return to Denmark. They could arrive home no later than the end of October, for Wegener had an appointment he could not miss: his marriage to Else, which under no circumstances was to take place later than 1 November 1913. This was a promise he intended to keep.

12

The Arctic Explorer (2)

GREENLAND, 1912–1913

Kamerater:	Comrades:
hen over Sne og Is	There over snow and ice
vor Vej mod Vest fører hjem	To the west lies the way home.
Vi vover Livet mod højest Pris	We wager our lives for the highest prize
at bringe i Sollyset frem	To bring to the light of day
vor Saga af Rimtaagens Dis.	Our saga, out of the ice and fog.

JOHAN KOCH, "On the Inland Ice," 19 May 1913

The expedition's departure from Copenhagen was smooth, although there was the usual flurry of activity at the end. Since the expedition was under "Royal Protection," there were the obligatory speeches, as well as a public send-off to be arranged. The press had taken some interest in the assembly and disassembly of the prefabricated "Borg," but neither Koch, nor Wegener, nor Lundager had an appetite for publicity, nor facility in arranging it, so they were entirely spared the circus atmosphere that had accompanied their departure from Copenhagen in 1906. On the very day that Wegener departed for Iceland, Else departed for Oslo, to begin her eleven-month residence with the Bjerknes family: her own "expedition."

Also in sharp contrast to their departure in 1906 was the rapidity of their progress once at sea. In 1906, with breakdowns, engine trouble, loading and off-loading of stores, and other distractions, the trip from Copenhagen to Iceland had taken almost three weeks. In 1912, on a commercial steamer bound for Aukureyri, it took only twelve days, including even a stopover in Bergen, Norway, and three days beset by fog. On 12 June 1912, the three arriving members met up in Iceland with their fourth, Vigfus, and the expedition could be considered to be really under way.[1]

It was shrewd of Koch to have planned so much time in Iceland in June. It was not just a matter of learning to drive the horses. Wegener and Lundager were soft from years of urban life and needed the physical exercise and training; Wegener had done no real physical work since his military training the previous summer. Iceland provided good versions of the kinds of terrain over which they would have to drive the horses. They could use the extensive lava fields as proxies for the stony shoreline around Danmarkshavn, along which the horses would pack their fodder and much other equipment. On the Vatnajøkul ice cap, even in June the climate conditions were those of the Greenland interior: fog, snow, and freezing temperatures.

Two days after their arrival, on 14 June, they set out to work the horses in the volcanic terrain, stopping to climb a volcano and look at the activity of the sulfur springs. Wegener needed the exercise, and he commented on the astonishing strength of Vigfus in a letter to his parents. "We will," he wrote, "have use for strong legs in Greenland."[2] On the nineteenth, with fewer horses and this time unguided, they took an eight-day trip onto the Vatnajøkul ice cap. This was valuable practice: they enjoyed a full suite of inclement weather, including wind, rain, and fog. Though it was high summer, when the sky cleared the temperature at night was below freezing, and at the northern margin of the ice cap there was fresh snow. The terrain was difficult, with much gain and loss of elevation and a precipitous and uneven ice surface at the margins of the ice cap.[3]

After traversing the ice cap, they encountered more volcanic terrain, which was extremely difficult but good practice in finding a way that would not damage the horse's feet, always a concern on such sharp rocky surfaces. It was also a good place for learning to select a line of advance and an angle of slope which would not cause the horses' shoes to slip on the smooth rock. It was exhausting but necessary, and Wegener wrote to his parents on 1 July, the day of their return to Aukureyri: "Today at noon we returned, healthy and lively from our trip and found the *Godthaab* here, on which we will travel to Greenland."[4]

Wegener himself may have returned from this trip feeling "healthy and lively," but not so Andreas Lundager. Sometime between their return to Aukureyri and 4 July, he announced to his companions that he was going to withdraw from the expedition. He was too old, he said, at forty-four, and had found the training expedition to be so physically taxing that he doubted he could perform up to standard on the full expedition in Greenland. "Both of them [i.e., Koch and Lundager] are so reasonable, to draw the necessary conclusions, without caring how it looks outside."[5] In other words, they would not put a bold face on it and try to muddle along; rather, they would replace him immediately. After some discussion, Koch's choice was one of the mates on the *Gothaab*, Lars Larsen (1886–1978), who was extremely strong and mechanically adept. Wegener noted in his letter to his parents on 4 July that they had to have a fourth man, not for the crossing of the Greenland ice cap, for which three people would be better, but because all the rations and the work plans were carefully divided four ways, especially with regard to getting up onto the ice cap. Wegener fretted about this, having calculated that the advantage of a fourth person to move supplies onto the ice cap would be very nearly canceled out by the extra weight of supplies required to maintain him. Regardless, they would have a fourth, and it would be Larsen.[6]

Into the Ice

Shortly after the return to the north shore of Iceland from the training expedition, Wegener packed up his sea trunk with mementos and presents (he had bought Else a filigreed knife in the Faroe Islands, to be given to her at Christmas). He also sent along a very carefully written (one gauges the status of the recipient by the quality of the penmanship) letter to Ernst Elstner, the dean at Marburg, thanking him for his support, announcing their departure within a couple of days, and estimating that they would be in the Greenland pack ice "at the most in eight days."[7] In fact, they arrived much sooner. Even though they sailed far to the northeast, passing to the east of Jan Mayen Land (a full degree of longitude east of Iceland) in order to obtain a high northing before reaching the pack ice, they ran into the pack in the late evening of 10 July, only four days out. Fortunately, they broke through the pack to open water on the morning of the eleventh at latitude 74° north and sailed in open water at the pack ice margin another full day, turning and sailing directly west into the pack on the twelfth.

The pack ice off Greenland, which they entered on 13 July, was thick and difficult to navigate. This was always hair-raising, though the *Godthaab* had a stronger engine than the *Danmark*, and they had great confidence in their helmsman, Wegener's old sledding companion from 1907, Gustav Thostrup, who served as ice pilot. The ice forced them to the south and then far to the east, and it was only on the fifteenth that they were able to turn back to the west. They were forced farther south, moving away from their destination at Danmarkshavn, and not until the eighteenth were they able to proceed west again. Finally, around midday on the nineteenth Thostrup spied an open lead to the northwest, and within forty-eight hours, in reasonably open water, they found themselves making landfall at Danmarkshavn.

As the ship steamed through the open lead, Alfred hurriedly completed a letter to his parents in his cabin and enclosed his German translation of Koch's account of the trip to Vatnajøkul for publication in *Petermanns Mitteilungen*.[8] Always the photographer, he had taken a number of photos of the sea trip, which he enclosed with greetings to all, writing, "Now comes the most difficult part of our program, the transport trip towards Dronning-Louise-Land. I am certain however, that everything will go smoothly."[9] This final phrase deserves an honorable mention in an anthology of "Famous Last Words."

Their landfall at Danmarkshavn was emotional to a degree that surprised them all. They arrived on a beautiful day, with a cloudless sky. Koch, not normally expressive, was quite overcome. He said that he and Wegener just stood there staring from the lookout atop the mast. Every ravine, every stone, every curve in the shoreline of the fjord seemed to welcome them like an old friend.[10] Wegener, Koch, Thostrup, and the captain went ashore with Vigfus and walked up the slope to the old "Villa." The inside of the hut, which they had left in such good order four years before, was a shambles and filthy. Ejnar Mikkelsen and his companion Iverson, who had made an expedition in the intervening time to try to find the lost diaries of Mylius-Erichsen, had stopped here on their return from the north, and the Villa gave every evidence of their desperation and depression.[11] Wegener and the others could not bear it; not only did they leave, but they asked the captain to take the ship as far west as he could go. Navigating inshore, they were able to gain 10 kilometers (6 miles) into Dove Bugt, a place called Stormkap, and here they began to unload the ship. They were delighted to be able to penetrate to this well-known anchorage, which saved them at least 10 kilometers of hard ground.[12]

The elation at penetrating into the fjord was short-lived, however. The next morning, 23 July, a powerful föhn wind began to blow and break up the pack ice in the fjord; it threatened to trap the ship, and the unloading had to be suspended while the ship returned to Danmarkshavn, which was reliably ice-free. They hurried to unload the rest of the supplies at Danmarkshavn. The ship had a good hoist and canvas slings fitted to the horses. One by one they were lowered into their barge, *Schachtel* (the Box), and ferried to the shore. This barge, 10 meters (33 feet) in length, was the sort of beamy towboat one sees everywhere in the canals and quays of North Europe, particularly in Denmark, even today. They had a motorboat, 6.5 meters (21 feet) in length and with a 4-horsepower engine, to pull it.

After off-loading, they led the horses to a mossy knoll, where there was fresh grass, thinking that the horses might like to graze, after a steady diet of fodder on the sea voyage, and that perhaps they needed to move around a bit. Move around they did. When Vigfus and Wegener went to check on the horses several hours later, they were gone. They were nowhere in sight, except for three standing nearby, which Vigfus (with some foresight) had hobbled; the other thirteen had completely disappeared. Vigfus

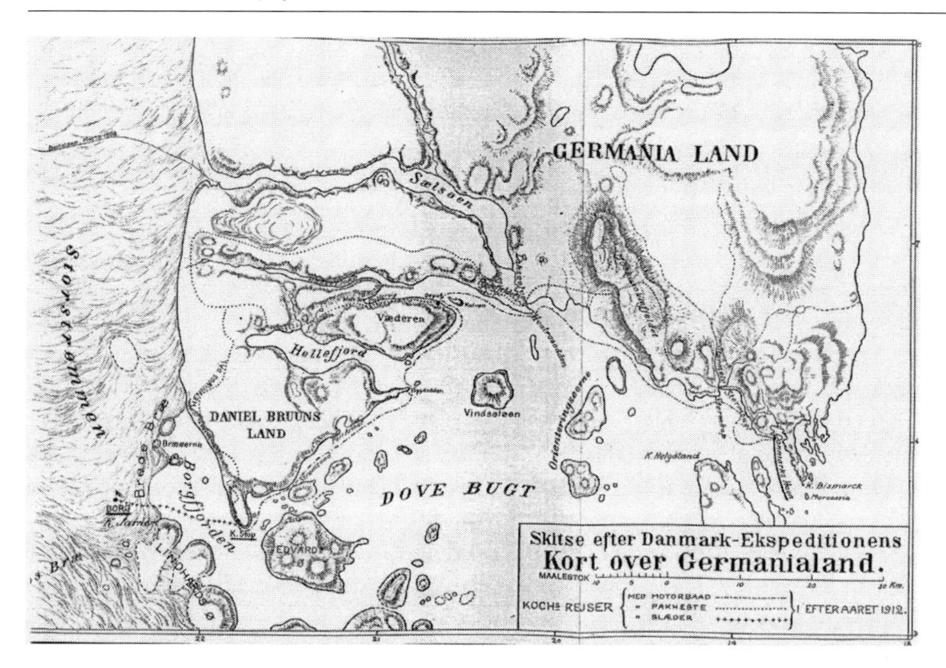

Map of the transport routes (land and water) in the fall of 1912 between Danmarkshavn and Kap Stop, and from there to winter quarters on Storestrommen Glacier (marked "Borg"). The legend indicates motorboat (dot-dashed line), pack horses (dotted line), and horse-drawn sleds (plus signs). Note the long roundabout route to the north taken by Sigurdsson and Wegener to skirt open water. From J. P. Koch, *Gennem den Hvide Ørken: Den danske Forskningsrejse tvaers over Nordgrønland 1912/13* (Kjøbenhavn: Gyldendalske Boghandel Nordisk Forlag, 1913).

and Wegener hurriedly assembled sleeping bags, food, and guns and rode off in pursuit. On the stony ground the horses left no trail. They found three of them by a small lake later that afternoon, and the next day they found three more back at the harbor. Rounding up the rest of the horses took six full days, and this was just a taste of what was to come. The best that Wegener could say about this entire excursion is found in a brief note from the twenty-seventh: "Last night I slept very well, as the wind drove away the swarms of mosquitoes."[13]

No sooner had Wegener and Vigfus taken themselves off in pursuit of the delinquent horses than the engine in the motorboat died. Fortunately, the ship had not left yet, and they were able to hoist it from the water, so the *Godthaab*'s machinist could work on it. While the boat was being repaired, Koch took the opportunity to go back to the "Villa." He was not sentimental by nature, but the state of their old house ate at him. He couldn't stand the filth and disorder in which Mikkelsen left the place, and he returned there in the afternoon with Larsen and spent several hours cleaning and reestablishing it until it once again felt like "home," even though it would not be their home.[14]

Here is the gist of the transport plan, which one can follow on the accompanying map. They had devised two parallel transport systems. The first would move baggage on the backs of the horses along the shore; this was to be accomplished by Vigfus and Wegener. The second transport system was the motorboat and barge, and this was the responsibility of Koch and Larsen. The redundancy was essential for several reasons. The wooden barrels containing petroleum, as well as the sleds to be drawn by the horses, could not easily be transported by the horses, as there was no possibility of dragging heavily laden sledges across the rocky summer shoreline. Even the lighter sledges, drawn empty across the rock, would (and later did) suffer damage to the runners and their fastenings; there was no question of loading these either. The horses, on the land route, would carry their own fodder compressed into large "hay sacks." This was no small task, as fully half of the 20,000 kilograms (44,092 pounds) of expedition baggage was horse fodder: 6,000 kilograms (13,228 pounds) of hay, and 4,000 kilograms (8,818 pounds) of a kind of "horse pemmican" devised for Koch by a professor of agricultural sciences in Denmark, Harold Goldschmidt. It was an interesting mix of corn, rye, wheat, molasses, and fat, and it was heavy.[15]

Since there was a tremendous mechanical advantage to moving things by barge as opposed to overland, the plan was to move the expedition's supplies to a series of appointed rendezvous along the shore; the idea was that Koch and Larsen would move successive loads of supplies (the petroleum barrels, boxes, and sleds) as far to the west as possible and off-load them, before returning to Danmarkshavn for more. Reference to the accompanying map shows that the horses were forced, by open water, at what was roughly the halfway point of the boat trip, to a "long way round" to the west and then to the south to Kap Stop. This was the final destination of the bifurcated transport system, as the expedition members assumed that by the time they had ferried and packed their material by both boat and horses this far to the west they would be forced to rely on the boats alone over the last stretch of open water, which they christened Borgfjord. This was the plan.

This conception of how things should go was a realistic concession to their expected difficulties and the realities of the terrain. In practice, its execution was much more difficult than they had imagined, from beginning to end. Wegener's journal entries and Koch's diaries, later collated in book form, tell a story of frustration, mishap, bad luck, miscalculation, and unforeseen obstacles. Their tale was at once typical of polar travel and yet especially poignant because these experienced polar travelers had worked so hard to devise a system that would avoid the obvious missteps made by other expeditions with which they were acquainted, including their own previous expedition to this part of Northeast Greenland.

Rather than tell this story, day by grim day, week by grim week, and month by grim month, a digest of the principal difficulties will do. Koch divided his account of this part of the expedition into two sections: motorboat trips and land transport. They might well have been entitled instead "Struggles at Sea" and "Struggles on Land."[16]

Let us begin with the struggles at sea. The trouble began on the first day of the voyages and did not cease until the abandonment of the boat and barge on 4 September. The route they followed Koch and Wegener had been over many times, and Koch had mapped it not just as a reconnaissance but also as part of the survey of the Danmark Expedition. Route finding was therefore not a problem. The problem was Greenland. At the far western edge of Dove Bugt, along whose shores they traveled, was the great Storestrømmen Glacier that flowed north to south. Here it met another great glacier tongue, Bistrups Brae, flowing in the opposite direction. Their confluence formed an ice tongue at 90° to both flows, which calved icebergs at Breda Brae, near the final depot at Kap Stop. The wind here blew either from the west, off the ice cap, or from the east and the ocean. If it blew from the west, it pushed the surface water with it, and the underpowered motorboat and its overladen barge had to struggle to make any headway. The tidal ebb and flood in this area was anywhere from 0.5 to 2 meters (1.6–6.6 feet) for an ordinary diurnal tide, which meant that some combination of west wind and ebbing tide could easily overpower the forward motion of the boats. On the other hand, a strong easterly wind, which helped propel the boats inland with a favoring tide, also drove pack ice from the mouth of Dove Bugt into the narrow channel by the shore; the ice could stack up so rapidly from the west shore outward that forward motion into the ice became impossible.

If wind and tide were the first obstacles, the boats themselves provided their own resistance. The motor proved extremely unreliable, and though Koch and Larsen were able to fuss with and repair it, the real mechanic on the expedition, Vigfus, was always away to the northeast pursuing his primary mission—management of the packhorses. The mechanical difficulties both with the working of the engine itself and with the propeller and rudder were unrelenting. Moreover, motoring coastwise in uncharted waters with uncertain tide conditions, they frequently ran aground, scraped and nicked their keel, and damaged the propeller and its shaft. In addition to the problems with the motorboat, the high gunwales of the barge, which served to increase the carrying capacity and to keep the cargo from wetting in the chop, made it extraordinarily difficult, when the barge was beached (virtually the only way they could load it), to move heavy cargo in and out; the sleds and petroleum canisters made this evident the very first day.

Then, of course, there was the ice—treacherous, grinding, unpredictable, moving as if it were alive, and capable, as were the sharp rocks along the shoreline, barely submerged, of scraping and puncturing both boats. These were often damaged in this way and repaired only with great difficulty. Moreover, in order to repair them, they had to ground them, and when heavily laden, the keels were often overstressed and precariously balanced on rocky surfaces. It was also all too easy to become icebound in such a small boat, much easier than it was with a large ship (and that was easy enough). At one point, they were held ice-fast, with the ice constantly grinding against the hulls of their boats, for twelve days.[17] That single round trip from Storm Kap to Kap Stop and back took them twenty-three days.

If the weather was clear, the temperature fell at night and new ice formed, and the winds tended to blow briskly from the west. If it was cloudy and rainy—and Koch had never seen so much rain—they were miserable. In between, if it was foggy, they could not navigate and find their way. All of these uncertainties made chaos of their preplanned rendezvous with the shore party, to compare notes and plan the next steps. Koch and Larsen on the one hand and Vigfus and Wegener on the other saw each other almost never, communicating by written messages at the depots.

Now let us consider the struggles on land. When one reads Wegener's diaries from this period of time, July through September, it does not take many entries to grasp what's going on. At a depot and at each stop along the way, the horses would be bridled, saddled, and brought one by one to a pile of gear where the hay sacks and the few provision and equipment boxes would be tied to the saddles. With the pack train assembled, the two wranglers, Vigfus and Wegener, would set out with the pack train, one before and one behind, mounted on their horses. Moving along the shore in relatively uninterrupted terrain, already well known to Wegener, who had traversed it many times (alone and in company) on foot and by dogsled in 1906–1908, they estimated that they should have been able to make about 15 kilometers (9 miles) per day (about half of what they would have done with dogs over snow), which would mean roughly a month of travel, with a little cushion at each end. They would bring a load of hay sacks half the distance, turn around and go back to Storm Kap for another load, and return. With three round trips, of perhaps 150 kilometers (93 miles) each, they estimated that it would take them at most a month. In reality, it took more than six weeks.

The typical day of their journey sounded something like this.[18] They would leave their tent at five or even four o'clock in the morning and (without coffee!), in snow and rain usually, set out to find the horses. They had to leave the horses unhobbled in the evening and night, so that the horses could feed themselves on the available grass, in order to conserve fodder, and so that they could flee a bear if menaced by one. It usually took one and a half to two hours to find the horses, a half hour to feed them, and sometimes as long as four hours for the two men to bridle, saddle, and then pack them. The hay sacks were cumbersome, and it took all of their strength to cinch them down securely enough on the packsaddles that they would not slip off during the day. By the time this work was done, Wegener's arms were so tired that he could barely lift them.[19]

Having risen in the early morning, it was sometimes early afternoon before they could set out, and they tended to travel through the afternoon and evening, but there were always miscellaneous disasters to hold them up. Rain and snow were constant companions, along with the occasional fog. The way along the shore turned out to be narrow and constrained, and they sometimes had to move on the rocky surface of the intertidal zone at low tide in order to avoid overhanging rocks—wave-cut benches that tended to knock the hay sacks awry on the saddles and loosen the straps. Their prized puppy, Gloë, who proved herself again and again by alerting them to the presence of bears and finding game to hunt (everyone agreed that the dog was the most valuable member of the expedition), would sometimes flush an ermine or a rabbit and take off after it; then they would have to wait for her to return.

The biggest obstacle was geographical and required a good deal of forethought. About a third of the way along, directly opposite a spit of land called Hvalrosødden, where Wegener had first learned to drive dogsleds back in 1906 under the instruction of Mylius-Erichsen, there was an outflow stream from a dammed-up lake immediately to the north, which in 1906 they had named Lakxelv. In August 1912, this outflow stream had turned into a broad lake more than 90 meters (295 feet) wide, too deep to ford, and too broad and dangerous for the laden horses to swim. The level rose and fell with each tide, but even at full ebb tide it was still too deep to cross. Rather than venture across it, they decided to create a depot on its eastern margin.

By 6 August they had transported the hay, sixty-three sacks in all, to their depot just this side of the outflow stream, now a slow-moving rising and falling arm of the sea. Wegener wrote in his diary on 8 August, "We finished with the pack straps early, about noon, and after having something to eat we were enjoying a welcome smoke of our pipes. Then, as often happens when you're really in need, a good idea comes to you. I suggested to Vigfus that we make out of two empty petroleum barrels, and our waterproof provision box, floats for a raft."[20] They had three long boards intended for the prefabricated house and were able to build a triangular raft, with sections of packing cases as the decking. They tried it and found that it would indeed support a man. They had 30 meters (98 feet) of rope, which was just able to stretch across the lake; they were then able to construct a ferryboat of sorts (their raft) whereby, over a period of three days, they transported all their gear, one sack and one box at a time. At the end Vigfus was able to entice the horses to swim across.

Now past the watercourse, they could begin to wind their way west and south to the final depot at Kap Stop, with a side journey to Wegener's old meteorological station at Pustervig, to pick up material from a nearby depot left by Koch and Larsen.

The same scenes and dramas were repeated day after day: runaway horses, the difficulty of packing, stopping and hunting for food, fording rushing streams, being stalked by bears, incessant rain while sleeping in the open. Planned meetings with Koch and Larsen rarely came off. The latter were struggling constantly with the engine, the tide, and the ice, and all four travelers were struggling with wet reindeer-skin sleeping bags, matches that wouldn't light, and tobacco that wouldn't burn; they were working themselves to near exhaustion each day.

It was not until 29 August that they found themselves together at the depot near Pustervig. The meeting was almost accidental. Wegener and Vigfus were there with the horses: they had moved all the hay and fodder to the final depot at Kap Stop and had turned back to pick up a miscellaneous load at the Pustervig depot. Koch and Larsen, who might have pursued a reconnaissance farther into the ice-choked Borg Fjord, could not force the boats any further along in the thickening ice and decided to stop at Pustervig for the night. They were all exhausted, with the possible exception of Vigfus, who was indomitable, but they took time that evening for a celebration. They broke out a bottle of port, a bottle of champagne, strawberry preserves, and dried apricots. The only thing they had to drink the champagne out of was their 1-liter (0.3-gallon) enamel mugs, out of which they ate and drank everything, and not wanting to combine the precious port with champagne, they eventually decided to drink the port out of empty cartridge cases "the size of one's finger."[21]

Polar exploration, like the practice of medicine and conduct of war, alternates periods of frantic activity with extended periods of boredom. So it was with the expedition in the first two weeks of September. Having pushed themselves to the limits of their strength (and beyond) to arrive at Kap Stop at the earliest possible moment, they now found themselves unable to proceed.

The ice tongue of the Storestrømmen Glacier merged with that of Bistrups Brae, at a location 15 kilometers to the west of their camp at Kap Stop. The flow of the glacier off the ice cap was so rapid that the ice tongue pushed its way halfway across Borg Fjord and calved icebergs so rapidly and in such profusion that the entire mouth of the fjord was jammed with a chaotic mélange of icebergs and sea ice, excepting only a narrow open lead of water, about 100 meters (328 feet) in width, adjacent to Kap Stop, a lead that opened and closed in every tide cycle.

Koch had imagined that once they had arrived at Kap Stop, the others would organize the equipment and provisions, while he motored in the boat back and forth along the 30-kilometer (19-mile) range of coastline to the west of them, looking for the best point of access onto the Inland Ice and the best road into Dronning Louise Land. He was so convinced that this is how things should unfold that it took him three days of struggling with Larsen to move the boat forward even a kilometer to realize that this would not happen, that the nautical part of this expedition was definitively over, and that the boats must be abandoned.

They now had to wait for the new ice to form in the fjord, before they could begin their reconnaissance. There was, now, no other way forward. They occupied themselves by moving their camp slightly to the north, where they could see clearly across the fjord. They turned their attention to mastering the various complicated harness arrangements for hitching the horses to the sleds. Wegener wrote in his journal, "It's a complete science, and compared with it, understanding the origin of continents is quite trivial. I made an attempt to enter into these mysteries, studying it energetically, but had completely to give it up and surrender it to the other three."[22]

In the midst of the beauty of their surroundings and the fantastic light both on the new ice in the fjord and reflected from the ice front to the west, there was the insistent cruelty of this mode of life. Vigfus had analyzed the health of their horses. Two had already gone missing in the trip from Danmarkshavn, and the remaining fourteen had clearly begun to lose weight. The calculation of the horses' rations had been inadequate. This was something Wegener had experienced in his own food supply between 1906 and 1908. Just as he had starved on his trip to the north with Mylius-Erichsen, these horses were starving in the persistent cold and overexertion. On 10 September, they shot the four weakest animals, one of which was Wegener's mount, a horse he had ridden in Iceland and that had carried him for the past six weeks. He noted in his diary, "I was in a black mood all day long."[23]

On 15 September, as soon as a two-day storm blew past, Wegener and Koch set out with a man-hauled sledge to try the same reconnaissance over the thickening ice: without the weight of the horses and the large sled, the ice might hold them. The account of the next day or so in their diaries is remarkable only for its tale of ordinary misery. In polar exploration, even when things go well, compared to everyday life in cosmopolitan Europe, they go very badly. They dragged the sled across broken and uneven ice. They were wearing one of their novel forms of footwear: wooden shoes studded with nails. These gave good purchase on the slick surface but were clumsy and difficult when climbing the frequent ridges and protuberances of sea ice. They were there to investigate and photograph the ice front on the other side of the fjord, to locate a place where they might ascend. It was exhausting work at every step, with much backtracking.[24]

The next morning they decided they could move faster without the sled, and they eventually found a place that, although broken and rough, would allow them to ascend onto the main body of the glacier, avoiding the sheer 70-meter (230-foot) ice wall that dominated most of the western margin of the fjord. Wegener took a number of photographs—he photographed at every opportunity—and then strapped the camera and tripod on his back for the return trip. Along the way they encountered a bear with a freshly killed seal and had to take a circuitous route around it.

On their way around an ice hummock to avoid the bear, Wegener lost his footing on the slick sea ice and flipped over backward, landing on the tripod and camera, which punched through his parka and into his back, just above his pelvis on the left side. He tried to rise to see whether his apparatus was intact (he had, in fact, broken the lens plate in three places), but he fell over bellowing with pain and could not rise again. He passed out and then regained consciousness. He tried to rise again, but he had no strength and could not move his left arm. Then he fainted again.[25]

When Wegener regained consciousness, Koch gave him two *Kolapastillen*. These stimulant tablets had as their active ingredient Kola (*cola acuminata*), the original and principal ingredient of those beverages called "colas." It was a cardiac stimulant that, in quantity, also increased alertness and physical energy, elevated the mood, and suppressed appetite. That it also increased body temperature, blood pressure, and respiratory rate made it an excellent emergency drug for polar explorers, and Koch was a deep believer in its potency.

Wegener was, stimulated or not, unable to get up, and Koch had to go for the sled. They had only one gun between them, and there was a bear a few hundred meters away. Koch left the gun with Wegener and set off as fast as he could go to retrieve the sled. He was as close to panic as he had ever been in his life. He was a man of immense physical strength, patience, and courage, but he was so terrified that Wegener might die that he could barely concentrate. It would take two hours to get the sled. What if Wegener lost consciousness? What if he froze to death on the ice? What if the bear came? What if the bear comes after me while he's got the gun? Would he be able to fend a bear off with only his Alpenstock?[26]

After an endless hour, Koch found the sled and loaded it with essentials: sleeping bags, anoraks, and food. Then came the most difficult part: finding Wegener again. It is impossible to exaggerate how difficult this was: in icy terrain there is no scale. Everything is the same color as everything else: that block of ice might be 5 meters (16 feet) tall and 100 meters away, or 20 meters (66 feet) tall and 0.5 kilometers (0.3 miles) away. Without Koch's experience as a cartographer and way finder, without his automatic, subconscious registry of the location of each landmark and his ability to triangulate automatically in his head, Wegener would have perished that afternoon.

Another hour later, after some anxious searching, Koch found Wegener. In the two hours he had been gone, Wegener had been able to pull himself up and, bent over, with his useless arm resting on his left knee and a ski pole in his other hand, had managed to drag himself 2 kilometers (1.2 miles) across the ice. He kept going via frequent pauses, more kola pills, and a good deal of loud singing, but he was exhausted, retching with pain and exertion, and nearly incoherent when Koch found him.[27]

Koch bound him to the sledge and dragged him back to their campsite, but the ordeal was far from over. Koch knew he could never drag Wegener over the last of the broken ice between their campsite and Kap Stop. He would have to leave Wegener alone in the tent and go for help. That night Wegener at least showed some appetite for food, and even more for tobacco. The latter encouraged Koch mightily, as he knew Wegener's passion for smoking and welcomed the calming effect of nicotine on Wegener's system.

The next morning Koch left Wegener alone and headed back for the encampment at Kap Stop. This time he took the gun with him. Wegener lay all morning in the tent and in midmorning had a good scare, which he recorded in his diary. He had seen that the ice tongue coming off the glacier was actually plunging into the fjord, beneath the new and rapidly thickening smooth ice on which he was camped. He had seen, several times in the previous day, the startling spectacle of an iceberg calving from the submerged tongue of the glacier and bursting up through the newly formed ice, sending radial cracks in every direction with thunderous reports. While he was waiting for Koch to return, such an event took place, with one of the radiating cracks passing within 10 meters of his tent and opening a lead in the ice more than a meter wide.[28] Further fissures could radiate from that spot at any moment and drop him into the fjord in his sleeping bag, too weak to clamber back on the ice—certain death.

Rescued by Koch and Vigfus that afternoon and taken back to Kap Stop, Wegener was still in a bad way. The next morning, 18 September, Koch dressed the wound in Wegener's back, first with Mercurochrome and then with a coal tar plaster in a tight bandage. At the very least the muscles were badly torn, and more likely he had snapped his ribs close to the spine. Koch wrote in his diary, "Whether there's a trace of sense than anything I did, only God knows."[29] Wegener noted that even though he could not lie on his back or stay comfortable in any position for long, he could at least keep the stove burning, cook the food, and do various housekeeping chores while the others prepared to drag their 20,000 kilograms of gear across the fjord to the ice road (which he and Koch had discovered) at the base of the great glacier.[30]

Wegener convalescing after the fall that tore his back and possibly broke a rib. From Koch, *Gennem den Hvide Ørken.*

Transporting their supplies and equipment across the fjord took eleven days, experimenting with different arrangements of sleds and loads for the extremely difficult and rapidly changing ice conditions. Wegener wrote on the twenty-third, "Half our baggage is now on the inland ice. . . . That the work has gone so smoothly is entirely due to Koch's energy and initiative, without pushing the others or driving them unreasonably."[31] By the twenty-eighth of September, Koch wrote in his journal, "For the first time since we arrived in Greenland, we have all our gear in one place."[32] That place was an ice tongue in a 100-meter-wide ravine in the face of the Breda Brae ice front—younger glacier ice flowing between two rocky "headlands" surmounted by older, stable portions of the glacier. It was so sharply inclined at the upper end that they would have to cut stairs for themselves and the horses, but it was the only ingress in what was otherwise a sheer wall of ice. Here they piled their baggage and built a stall out of packing crates to keep the horses out of the snow and wind. This would be their staging area for the traverse to Dronning Louise Land, visible 20 kilometers (12 miles) away from the summit of this narrow road of ice.

In spite of his satisfaction that they had all the baggage assembled, Koch was worried about their situation. He had written in his journal on 22 September, "The margin of the glacier is not a secure place for a depot. Catastrophe threatens, something that could undo our expedition. But there's nothing to be done: we have to proceed from this place forward or we won't get to Dronning Louise Land this winter."[33]

Koch's fears were well placed. About three o'clock in the morning on 30 September, they were awakened in their tents by a series of loud reports, tremendous sounds of cracking and groaning. Wegener later wrote, "I crawled out of my sleeping bag as fast as my back pain would let me."[34] As he tied on his kamiks, he felt the ice beneath him bob up and down and the tent tilt over. He crawled out into the moonlight: "What a sight greeted my eyes around our camp. There, where our pathway up from the sea ice had been—30 meters from the tent and 20 meters from the stable—a colossal menacing black ice block now thrust upwards into the moonlight! (About 15 meters tall)."[35] The ice tongue had split in half and calved a great iceberg, and the entire stretch of ice seaward from their tents—about 150 meters (492 feet)—had overturned. The remainder, including their depot, stable, and tenting place, was a chaotic jumble of rifted ice blocks. Half their baggage had fallen into a newly opened rift in the ice, which continued to crack, groan, and move. They scrambled up the ice road in the ravine, only to discover that they were now cut off by a new crevasse too wide to cross. One wall of the stable had collapsed, though neither affording a way for the terrified horses to escape nor falling in on them.

The cacophony of grinding ice continued as they tried to assess their situation. They were all mostly unharmed, as were the horses. Vigfus had bloodied his feet, running out barefoot onto the broken ice, but sustained no other injuries. They looked at each other in astonishment. "How is it possible," wrote Wegener, "that with this incredible destruction proceeding all around us, that we have not lost our lives, and seem only to have lost a single piece of baggage?"[36]

Their painfully prepared ice road up the ravine was now once again a jumble of great broken ice blocks and would have to be rebuilt. Their position was precarious in the extreme. The ice tongue was unstable and in constant motion with the tide; it could calve again at any moment, and if it did, they would not escape so lightly. Wegener was still in such pain that he could barely stand erect, so he could not carry anything. He continued to cook and do what he could while the others worked furiously to move their provisions and gear onto the main stream of the Storestrømmen Glacier, nearly a kilometer beyond the ravine. As they looked seaward, they could see that the former ocean surface was now nothing but a jumble of icebergs and fragments of bergs: they counted seventeen new icebergs in the half square kilometer of ocean just to the east of them. They repeatedly heard the great crashing and tearing as new bergs formed, and the sound encouraged them to move faster. In the first two days after the catastrophe, they moved 14,000 kilograms (30,865 pounds) of baggage onto the Inland Ice—thirty-nine loads of 400 kilograms (882 pounds) each on the fourth of October alone—and they hoped to get the rest there in two more days.[37]

By 4 October they were 800 meters (2,625 feet) away from the calving front of the glacier, but they could not now pause, if they were to make it to their destination before deep winter. The very next morning they started off with six loaded sleds and a packhorse. It took them three and a half hours to cover a single kilometer of ice hummocks, and immediately they broke the runners on one of the lighter "Nansen" sleds. It was clear that these sleds, required for their traverse in the following spring, would be destroyed in the process of moving this final 20 kilometers inland. That meant that only the larger two-horse sleds could be used, and Wegener did the math. While the big sleds were supposed to carry a load of 3,500 kilograms (7,716 pounds), they had so far never done better than 2,000 kilograms (4,409 pounds). On this uneven ice, the most they could expect the horses to pull was 400 kilograms. With five days for each round trip, it would take them fifty days. They would never make it.

"The die is cast," wrote Wegener on 7 October; "We can't reach Dronning-Louise Land with our baggage. In spite of the whole summer we spent scanning the horizon, searching in the distance for that place our house would stand, we now must finally completely abandon all hope of reaching it. It is a real blow to us, but we tell ourselves: it is our future."[38]

They would move a kilometer or two farther in, to the middle of the glacier, away from any chance of further calving. Wegener put a brave face on it: "If we stop here and build our house on the inland ice, we'll have enough light to make a reconnaissance to Dronning Louise Land, which is much better now than trying it in the winter night."[39] If they stopped short of Dronning Louise Land, they could explore routes, lay down some depots, and start doing some science. If they tried to force the plan into late November and early December, they would exhaust themselves and accomplish nothing. There were plenty of crevasses nearby for Koch's glaciology, and they would be sufficiently on the Inland Ice, a kilometer or two farther inland, to get meteorological measurements that could be said to have been made on the ice cap and not on its margin.

The Wintering

They were committed to staying where they were; therefore, they had to get the house assembled as quickly as possible. The temperature was well below freezing each day, with more and more blowing snow, making the work difficult and painful. Finally, on the ninth of October, it was clear, though very cold, and they quickly assembled the house; by the next day they were sleeping under its roof.[40]

There were various problems to be solved. The house was too snug, and the petroleum stove smoked, was finicky, and threatened several times to asphyxiate them; they noticed this when the lamps went out. Someone would yell, "Everybody into the next room!" It became a standing joke with them, when things went wrong through the next year, for someone to yell, no matter what the circumstances, "Everybody into the next room!"[41] Dampness was a persistent problem, and the walls dripped water all during the winter. The photographic darkroom always had a wet floor, a problem they were unable to solve. With the horses, it was even more difficult. They had imagined they could manage without a floor for the horse stalls, but it turned out that the horse urine melted the ice and the horses kept sinking into the slush, so Larsen had to build floorboards out of packing crates. They had neglected to provide sufficient ventilation, and the methane from the horse dung created nauseating gas. Finally, ventilation pipes in the main room and the horse stalls were rigged in a way that increased the airflow sufficiently to keep them all, the horses included, from suffocating.

By the end of October, things were somewhat better. They had solved every problem in the house except the persistent dampness. Their books were on the shelves, and they had hung the pictures they had brought with them of loved ones and home, though Wegener fretted because he could not find two pictures of Else, which had gone missing on the disembarkation from the ship and hadn't turned up since.[42]

With the coming of the winter night, Wegener's mood darkened. He had already noted in his diary that he feared the depressed, enervated state (*Energielapsus*) that had so compromised him in the winter of 1906. In the volatility of his mood toward the end of October and the beginning of November 1912, one can see his struggle with winter depression beginning anew. Every day brought a new crisis of sorts. The iron runner of one of their sleds broke transversely and was irreparable. Their mercury barometer, though nested in a packing case of its own, shattered on the last leg of the transport from the final depot to Borg. Their only galvanometer was useless. Material and mental casualties mounted in the midst of their own physical recovery from the wounds of the trek from the coast.[43]

There was further reason for Wegener's pessimism and incipient despair. He, Koch, and Vigfus attempted a brief reconnaissance on the twenty-ninth and thirtieth

Wegener at his worktable in "Borg," with the meteorograph recording temperature and humidity. Note the dampness of his hair; it proved impossible to keep the moisture level low and the house warm at the same time. From Koch, *Gennem den Hvide Ørken.*

of October, one of many attempts that winter to locate some way forward onto the main ice cap. They sank to their knees in the freshly fallen snow and had continuously to draw their sleds through empty watercourses atop the glacier and over ice hummocks the size of small hills. They estimated their forward progress on the twenty-ninth at 7 kilometers (4 miles), but they could not be sure because an accident broke their hodometer, the wheel attached by a strut behind one of the sleds, which measured the distance traversed in its revolutions and recorded their number. The second day they traveled barely 2.5 kilometers (1.6 miles): "at this rate," wrote Koch in his journal, "it would take us more than a year to cross Greenland."[44]

Wegener was already in a dark depression by the end of the second day of travel. He wrote in his diary on the twenty-ninth (remembering Mylius), "The chances for our expedition are now not very good either. What lies in store for us? But these are ideas prompted by the winter darkness." He then added, "The sunlight of spring will paint with different colors."[45]

There was as yet little time for dark musing, in any case, as in addition to setting up the house they had to get the scientific program under way. Wegener worked to set up his meteorological station, with the precipitation gauge, the anemometer, the barograph, and the recording thermometer, and to get the line of poles sunk in the snow between the thermometer hut and Borg, to guide him back and forth in the winter night.

The loss of the mercury barometer, shattered in transport, was a very serious matter. They had also an aneroid barometer but now no means of calibrating it against the mercury barometer, and on the traverse across the ice cap the following spring, the measurement of altitude was to have been done by comparison of the reading on the aneroid barometer with that of a "boiling point hypsometer." The difference between the boiling point of water at any given altitude and the boiling point at sea level gave a means for calculating the absolute altitude. One read the barometric pressure off the aneroid barometer and then discovered the boiling point of water by lighting a spirit lamp underneath a thermometer encased in a jacket of water. When steam was ejected from a pinhole vent at the top, one read off the temperature; one could then use this datum to determine the correction factor needed to reduce the aneroid barometric pressure to sea level, thus providing a measure of absolute altitude above sea level. With the mercury barometer gone, there was now no way to calibrate the aneroid barometer or the barograph.

Wegener's meteorological program, other than station records, was a continuation of several of his current enthusiasms. He planned to continue his photography of ice crystals, as a part of his study of their formation under different temperature conditions. He had located a flat-topped knoll 5 kilometers (3 miles) to the west of Borg, which he could photograph to show the character of inversions in the first few meters of air above the surface; he christened this "Gundahl's Knoll." Finally, he planned to use their theodolite, their sole position-finding instrument, in order to calculate the altitude of the aurora. If all had gone as planned back in Europe, his brother Kurt would now be in Spitsbergen, ready to make observations of the aurora. They had agreed that on any night in which the aurora was persistent, they would measure its altitude. Knowing the exact time and date of these observations and their longitudinal differences would allow them to calculate the altitude of the aurora (in kilometers) to a high degree of accuracy across this long baseline and, by observation of its color, make inferences about the composition of the upper atmosphere.

The glaciological part of the expedition was a combination of the geography of the glacial ice with studies of the morphology of glaciers, including phenomena of fracture and flow and the interpretation of the "blue bands" in the ice. The latter, the alternation of bands of clear bright blue ice with bands of white ice, was not well understood. Were these bands surfaces of deformation, seasonal laminations, or infilling of fractures with meltwater? A good deal of this work was to be photographic documentation of the appearance of these various structures, principally by Wegener. In addition, Koch was very keen to study the laminar structure and the depth of the firn. Firn is snow that is left over and compacted after a full season of (summer) melting and ablation. The hexagonal crystals of newly fallen snow melt and then refreeze into a granular snow called névé. If this granular snow survives the melt season of one year, under the weight of newly fallen snow in the next season it is gradually compacted into firn. How long this process took was one of the most interesting things to be discovered about the ice cap; they had plans to dig deeply into it.

As a way of getting this scientific program under way, in the failing light of 5 November Koch set off with Larsen back in the direction of the last depot (where the iceberg had calved) to make a determination of their height above sea level. Larsen went first, probing the snow ahead of them to locate the crevasses with his alpenstock; there were several they knew they had to traverse. The abundant snow of late October and early November had begun to bridge the crevasses, and Larsen probed the snow bridges, as Koch followed in his footsteps.

They were most of the way to their destination when Larsen crossed a snow bridge and waited for Koch, who was carrying the theodolite, to follow him. The snow was deep and soft, and as Koch reached the middle of the bridge over the crevasse, it gave way completely and he fell, disappearing. The crevasse was not quite 2 meters wide, and he bounced off its walls as he fell, managing finally to stop his fall by grabbing hold of a ledge 12 meters (~40 feet) below the surface. He pulled himself onto the ledge and sat down in the snow-covered shelf. He was bleeding badly from his forehead, and his right foot was sharply painful and felt hot. He noticed the contrast because he could feel his other foot begin to freeze almost immediately.[46] He was still conscious, and he told Larsen to go back for help and to bring rope and their rope ladder.

Now the roles were reversed, and it was Wegener's turn to be filled with panicked thoughts as he and Vigfus skied toward the crevasse carrying the rope and rope ladder; Wegener had also thought to bring his novel electric flashlight. It had taken Larsen most of an hour to return to Borg, and they had to be careful themselves as they skied, not to fall as Koch had done. Wegener thought, "How will we find him . . . the entire expedition hangs on the strength of Koch's personality. Is it possible that a badly injured man can lay motionless for two hours in this cold, without freezing to death?"[47]

Arriving at the crevasse, Wegener threw himself to the ground and yelled "hallo" as loudly as he could into the crevasse. To his intense relief, Koch was able to answer and told him that he had likely only broken his foot and that he was not yet frostbitten. They lowered the rope ladder down and tied the electric lamp to the rope, lowering it as well, to give Koch a chance to locate the theodolite. The latter, however, had vanished into the crevasse and was nowhere to be seen. Koch pulled himself up the rope ladder with both hands, bracing himself with his left foot, his right hanging useless. His strength gave out 3 meters (10 feet) from the top, and they pulled him out the rest of the way.

Hours later, back at Borg, Wegener gloomily assessed their situation. "Has then, everything turned against us? There lies Koch, with a sprained ankle—that is if it's not broken—with a swollen and bloody forehead and a fever, and we fear that he may have even more serious [internal] injuries not yet manifest."[48] Moreover, the loss of the theodolite was an unmitigated disaster: uncharacteristically, Koch had not brought a backup instrument. "We are going to have to go back down into that crevasse," Wegener wrote, "because the theodolite is absolutely irreplaceable for our trip across the inland ice."[49]

On 6 November, Koch was slightly better, but it was snowing heavily, and they could not leave to search for the theodolite: "Heaven knows where all this is going to end," wrote Wegener.[50] When the weather finally broke, they made two attempts to recover the theodolite, but to no avail. The snow-covered ledge on which Koch had landed was now under several meters of drifted snow, and the theodolite was still further down in the narrowest part of the crevasse. On the eleventh, Wegener finally surrendered to the inevitable: "Now we are completely out of luck . . . the theodolite is irretrievably lost, and it was not only our sole instrument for exact angular measurements, but I'd counted on for my meteorological work. Determination of time in winter (the proposed simultaneous auroral measurements with Kurt!), cartography, measurements of glacial movement, navigation on our great traverse, measurements of Sun rings and mirages—all of these will now be inaccurate, though at this point we are not yet completely resigned."[51]

Much to everyone's relief, Koch had not sustained any more severe injuries, but he was incapacitated, and their reconnaissance was over. Throughout most of the rest of November they settled into winter quarters and began to work on solving the problem of the lost theodolite. In the end, they constructed three instruments.[52]

All of these instruments show the mark of Wegener's training in the history of astronomy, without which they would likely have had to abandon all hope of serious measurement. The first and easiest to construct was a nontelescopic "meridian instrument."

Only a few days before the loss of the theodolite they had taken a sighting of the passage of a bright star from a point about 10 meters away from the hut, with the meridian telescope focused on a point just above the tip of the flagpole that was mounted on one end of the building. They had been able to synchronize their three chronometers and found only a three-second difference between them. To replace the theodolite, they mounted a meter-long staff with a metal disk with a hole punched through it, and they placed it in the spot where the theodolite observations had been made. Peering through the hole at the top of the staff, and sighting at the top of the flagpole, they were able to record the passage of a known star across the meridian. Using this technique, they would be able to photograph the northern lights while taking astronomical time measurements with the meridian instrument.

They then constructed (or, rather, Vigfus constructed according to Wegener's instructions) both a sextant and a quadrant instrument, the latter of which they christened a "Jacob's Staff," though it did not have the movable crossbar of the instrument of that name used throughout the history of pretelescopic astronomy. The former instrument, a *Senkelquadranten*, consisted of a square of wood with carefully scribed lines at right angles to one another, allowing the construction of a meridional arc marked off in degrees at the bottom of the instrument. The principle is more or less that of an astrolabe. A small pendulum was suspended from the vertex opposite that of the meridian arc.

The design of the sextant was good, but it was hard to hold it steady in even a relatively light wind or to keep the pendulum still long enough to take an accurate reading. They elected instead to depend on the "Jacob's Staff." This was a rectilinear wooden frame, not quite a meter long in its horizontal dimension. The two vertical struts had holes bored in them near the top, with metal disks, also with holes bored through them, set into the frame. The top horizontal piece was there to confer rigidity, but along the bottom horizontal piece and up the sides of the verticals about halfway, they carefully glued a flexible celluloid ruler. Also mounted in the face of the bottom horizontal piece was a spirit level that Vigfus had fabricated out of the spirit-filled gauge of the broken galvanometer—a round one that allowed the bubble to be centered in the bull's-eye and thus assured the leveling of the instrument both in its long axis and perpendicular to it.

Koch was delighted with this instrument. He wrote, "Our primitive Jacob's staff proves to be a splendid little instrument. It gave the Sun's altitude to an accuracy of 1 to 2 minutes [of arc]—and of course we needed no greater accuracy than that—and had other valuable characteristics. It did not need to be packed and unpacked, and wind, blowing snow, and hoarfrost had no effect on it; it took almost no time to set it up, it weighed practically nothing, and could be used without taking one's hands out of one's gloves."[53]

It was remarkably easy to use. One simply set it in the snow, leveled it, and aimed it at the Sun. The Sun's rays, passing through the small aperture at the top of the upright portion, made a tiny oval of light on the celluloid ruler. If one measured both ends of that tiny oval and took their mean, that gave a measure of the Sun's altitude. After taking the Sun's altitude, one reversed the instrument 180° and took the altitude again. The average of these readings canceled out any error of squareness in the instrument and gave a reading of the latitude sufficient for their purposes, as would any noon sighting with the more sophisticated sextant. While it was insufficient for any of their more sophisticated scientific applications, it would get them across the ice cap in the spring,

This navigational problem solved, they settled comfortably into their winter quarters. They worked out a rotating schedule in which they took turns cooking, caring for the stove, and feeding the horses and cleaning their stalls. This work took usually only half the day even under the most severe circumstances, leaving a good deal of time for reading and scientific conversation, most of the latter, of course, between Koch and Wegener. Wegener now had time to complete the setup of his meteorological instruments and their calibration and to finish developing all the photographs on the expedition, as well as time for microphotography of snow crystals and at least some qualitative observations of the northern lights. They began the first of a number of boreholes into the glacial ice, one of which would eventually reach a depth of more than 25 meters (82 feet).[54]

Wegener later recalled this intense scientific collaboration in the winter of 1912 and 1913 as one of the most wonderful periods of his life.[55] During his convalescence, Koch read through Wegener's thermodynamics and made a number of suggestions that Wegener found important. They discussed both meteorological and glaciological topics on a daily basis: as any history of the subject will show, glaciology, at the beginning of the twentieth century, was still in its infancy, and Wegener noted that Hans Heß's *Die Gletscher*, which he had brought along in order to prep himself in the subject, contained many more questions than it did answers.[56] In the evenings, Koch and Wegener played chess, and by Wegener's account he never won a single game from Koch the entire winter.[57]

The regularity of their work, the ease of their circumstances, the stimulation of their conversation, and their ability to pursue their scientific program, however limited by the damage to the instruments, warded off the depression of the winter darkness. As Wegener later said, "There had probably never been an overwintering of this sort in the Polar Regions. The obviously progressing scientific work amidst a daily stimulating exchange of ideas, provided great satisfaction; its rays banished the shadows of the winter night and even warmed our comrades Vigfus and Larsen right through: such that even though they often did not understand what we were doing, they could sense that it was succeeding and that the aim and object of the expedition was daily drawing closer to fulfillment."[58]

As on all expeditions, the Christmas festivities were duly noted in their diaries—extra food, sweets, champagne, and port. The festive Christmas in Borg also marked the halfway point of their trip in time: they should be on the west coast sometime in June. The solstice, Christmas, and New Year's Day marked the darkest days they would see, and for Wegener the opening of gifts, carefully stored away against the event, contained a special treat: as her Christmas gift Else had carefully packaged two color photographs of herself, and Wegener was overjoyed to have these to replace the ones he had lost.[59]

January and February were nothing but a waiting game. Wegener's diary is extremely sparse, and in the published accounts of the expedition the months of January and February take up a bare two pages of text. On 10 January the temperature hit −50°C (−59°F), very close to the coldest temperature ever recorded on an Arctic expedition. In spite of the return of the Sun on the seventeenth of February, which prompted panegyrics of joy in Wegener's journal, February remained bitterly cold, the temperature never getting above −40°C (−40°F).

The return of the light broke the routine and allowed Wegener to pursue science outside of books, and beyond the time-consuming and engrossing but rather limited task of boring holes in the glacier to examine the layering of the ice and take temperatures with depth. He could begin once again his microphotography of snow crystals and continue the series of photographic studies of Gundahl's Knoll, the flat-topped hill 5 kilometers to the west, to document the complex layering of the atmosphere near the ground, in the form of complex mirages.

The Crossing of Greenland

The first half of March was devoted to tentative attempts at venturing outside. On the ninth they made a reconnaissance to Gundahl's Knoll, a 10-kilometer round-trip that took them ten hours, and which left Wegener's nose and cheeks badly frostbitten, but allowed them a view of their "beloved Dronning Luise Land," their gateway to the west. A week later they tried again, spending a night in a tent—the first night outside their hut since October—and they practiced digging into the firn, as the horses would need to be out of the wind each night on the traverse, or they would freeze to death. It was a matter of exercising the horses and themselves: all of them had soft muscles from their winter's inactivity.[60]

The second half of March and the first half of April were devoted to creating a series of depots to move their supplies into Dronning Louise Land. The weather was terrible. They were anxious to get going, but the snow was deep and loose and the crust of the snow so fragile that the horses could not pull the large sleds, and they had to be satisfied with hitching the horses to the lighter "Nansen" sleds. By 10 April they had succeeded in moving the material basis for their expedition only 20 kilometers inland, to a narrow neck of ice between two mountains that appeared to be their only access to the Inland Ice.

As it warmed from −40°C to −30°C (−22°F), the clear and bitter cold gave way to spring snow, which made the exhausting and stupefying work of hauling their supplies through the deep snow even more exasperating and demoralizing. What was worse, they were soon to discover that what they imagined to be their last depot before entering the Inland Ice was still well within Dronning Louise Land. They now paid a bitter price for their inability to complete the reconnaissance the previous fall. As the accompanying map shows, they were headed in the wrong direction: had they gone slightly north, they would have mounted the Inland Ice and cleared the zone of crevasses 60 or 70 kilometers (37–43 miles) sooner than they actually did. Their commitment to a southwesterly route from the beginning gave them an additional month on fissured and uneven ice.

On 20 April, they committed to leaving Borg and headed west. The weather was terrible, and they were immediately pinned down. They wished and hoped for things to get better, but things got worse. In early May, when they had hoped to be on the Inland Ice, they were still trapped in their tent by weather and had not begun their crossing. Wegener confided to his journal on 3 May, a date when the wind was so strong and the cold so bitter that they could still not think of traveling, "Are we ever going to see the good weather that is supposed to be characteristic of the inland ice? Are we just unlucky? This is a dog's life we are living!"[61] He was furious and frustrated, confessing to outbursts of anger at Koch's leadership, or lack of it, in getting them under way. They were stuck in their tents with nothing to do, in wet sleeping bags, utterly miserable.

Perhaps because of Wegener's urging, but also perhaps because of the invariably bad weather, Koch decided that they had to move. They departed on 5 May, traveling 45 kilometers (28 miles) the first day![62] But it was still back and forth, back and forth

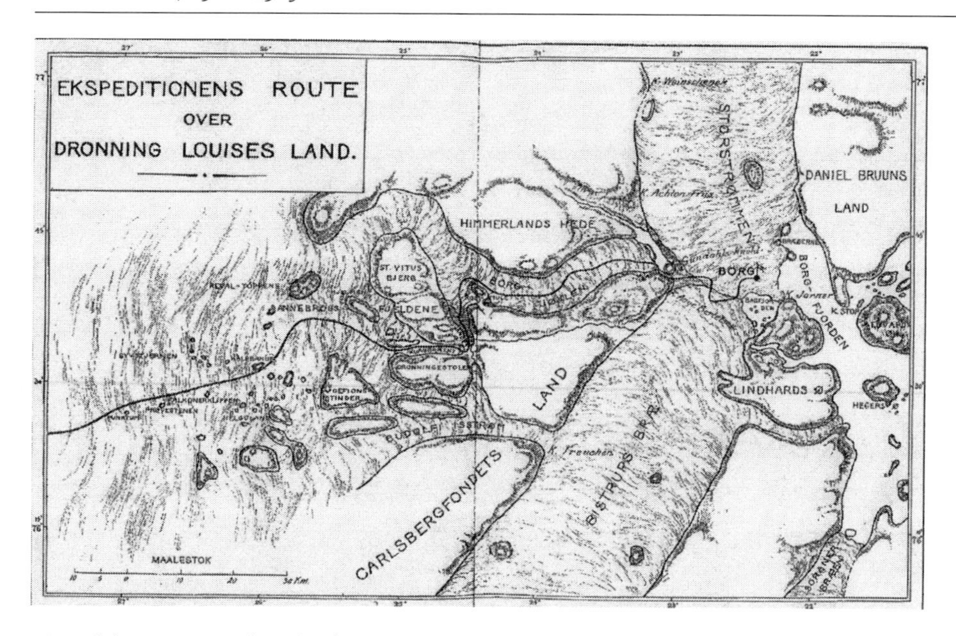

Map of the route onto the Inland Ice, April–May 2013, through Dronning Louise Land, showing the consequences of having chosen the southwesterly route. A more northerly course around the area would have saved them weeks of travel. From Koch, *Gennem den Hvide Ørken*.

to the depots: in spite of all their attempts to cut back, they were still hauling 1,800 kilograms (3,968 pounds) of gear on five sleds. There was nothing but clothing, fuel, food, and their journals and photographic negatives, but still there was too much.[63] Conditions were much worse than anything that Nansen had faced.

The horses, after a month in the cold, were already exhausted from the Sisyphean task of pulling their own fodder across the snow; by the middle of May two of them had gone snow-blind and were so weak that even a half-loaded sled was too much for them. They had to divide each day's travel into two parts, with a long midday pause to rest the horses. Wegener, who had chafed at the delays in travel scant weeks before, liked this no better: "May 13. Our altitude, from my calculation is now 2287 m. Latitude, longitude, and altitude, these are the three most interesting things that there are here. Beyond them there's only the blue of the sky and the white of the snow." Larsen took the mirror out of his private kit and passed it around so they could see themselves. The skin on Wegener's frostbitten nose hung in black tatters, and his chin was also blackened with frost burn. "It's a good thing that it will be a long while before anyone in the civilized world sees me again. Anyone would think I was a leper."[64]

Things got very bad, very quickly in the middle of May. On the fifteenth Wegener wrote, "A terrible day. We had to shoot 'Polaris.'" The horse, exhausted and snow-blind, could not pull the load. They repacked on the four remaining sleds and were about to leave when a huge snowstorm blew up, trapping them in their tent for an additional four days. Wegener tried to think of scientific problems but found his mind drifting constantly into fantasy. All he could think about was what it would be like to be living with Else in their own home, and what they would make for dinner. He reckoned that he had spent so much time thinking about these things in this four-day period that, had he the energy, he could have written a treatise on each theme, "compared to which the Origin of Continents would have been a sixth form essay."[65]

Crossing the Inland Ice. Wegener and Koch are finding their latitude during the noon break by "shooting the Sun" with the homemade "Jacob's Staff" built by Vigfus to replace their lost sextant. Note the bright Sun and wind, both constant adversaries. From Koch, *Gennem den Hvide Ørken.*

The chronicle of their suffering is such that it suffices to summarize it. Its character was just as uniform as it was awful. The men were continually wet, exhausted, cold, and frostbitten. The horses, snow-blind, starved, and overworked, failed one after another. They shot them as they failed, then reconsolidated the loads on the remaining sleds, and went on. They maintained their course using the Jacob's Staff, shooting the Sun at every opportunity and on every pause in the march. On 2 June it was still 30° below zero, and Wegener noted that the one bit of good news was that the boiling point barometer showed that they were not gaining much altitude, so that they were near the halfway point: the summit of the Inland Ice. They found their way by altitude and latitude alone: the hodometer, which measured the mileage, had shattered irreparably early in May, and their one compass, in which the needle floated in alcohol, was always frozen, leaving only celestial navigation, with an instrument fashioned from pieces of a packing crate, to find their way across a trackless flat expanse of ice and snow.

On 5 June they took a day of so-called rest. The horses were exhausted. The four men took turns and carved and dug a 6-meter (20-foot) hole in the snow to photograph the layering and to measure the temperature with depth. Wegener's diary for this day shows a brief eruption of good humor with the possibility of doing any science at all, though to balance this good humor he noted that he had smoked his last pipe and was now out of tobacco, except for a few cigars held in reserve. That week it turned colder,

Koch's drawing of the "7-meter hole" they dug on 12 June 1913 to investigate the structure of the "firn" layers—the multiyear recrystallized snow on its way to becoming ice. Wegener recorded this as one of the best days of science on the whole expedition. From Koch, *Gennem den Hvide Ørken.*

and as they reached 2,853 meters (9,360 feet), they were still deep in the grip of winter. On the eleventh of June they shot "Roter," the third of their five horses, leaving only "Grauni," their favorite, and "Fuchs." The need to increase the horses' rations, to get them to pull at all, left them with barely enough food for a single horse.[66]

On their next rest day, the twelfth of June, they dug a 7-meter (23-foot) hole, and Wegener was extraordinarily pleased with the beautiful temperature curve they obtained, revealing to them that the cold they had experienced—roughly −30°C—was indeed the annual mean temperature, measured in layer after layer of snow down more than 6 meters.[67] That night, one of their two remaining horses, Fuchs, ran away. Koch wrote in his journal on the fourteenth, "Yesterday we saw the tracks of 'Fuchs' and it started our fantasy in motion: is there truly, west from us, some snow-free area in the middle of the ice?"[68] This is a reference to the hope of every expedition venturing onto the Inland Ice since the end of the eighteenth century: ever since the missionary pastor Hans Egede had imagined that he would find the lost Norse colony hidden away in an ice-free, verdant valley in the middle of the ice cap, expeditions had dreamed of such a land, with or without Vikings. No such land existed, as their steady, wearying trip to the west soon revealed. They never saw the runaway horse again.

Wegener took an altitude on 12 June with great care and found that they had reached 2,937 meters (9,636 feet). They had been, for several days, very close to that altitude, and he was sure they would cross Greenland and had reached the highest elevation, though more than three-quarters of their food was gone. Each day on the march they went through their belongings and put together a sack to leave behind. With the wind now blowing from the south and mostly behind them rather than in their faces, they could raise the sail every day to help move the sled. Grauni could no longer pull alone; he wanted to pull but lacked the strength. So the men took turns harnessing themselves to the sled along with the horse.

At this point, the expedition took a remarkable psychological turn. From this (halfway) point forward, all of their will to live seems to have become focused on one task and one task alone: saving this last horse, Grauni. "Tomorrow," Wegener had written on

12 June, "we will test our new method of travel with the last horse and all of us pulling just one sled. We estimate now we have only two more days of rest [two weeks of travel] on the inland ice and hope that we will be able to save Grauni."[69] They had plenty of trouble to record in their journals: Vigfus ill, probably with pneumonia, everything soaked by walking through the melt puddles now appearing atop the snow in midday, the breaking of their tent poles and their attempt to use the sled to make a lean-to, the depletion of the rations. Yet none of this concerned them so much as Grauni's condition. They pulled, and the horse followed the sled on a tether.

When the horse stopped, they stopped.[70] They decided to cut their own rations in order to supplement the horse's, by soaking their allotted bread in water and feeding it to the horse, supplementing this in turn with their rapidly diminishing supply of ground peameal and even pemmican with chocolate.[71] Each day the horse got weaker; each day they paused longer. By 26 June their measurement of the Sun's altitude at noon showed that they were only 100 kilometers (62 miles) from the depot on the west coast: "a stones throw for a healthy strong horse—but will we be able to get Grauni there with us?"[72]

On 28 June the horse could go no further, and the men looked at each other and came to a decision: they picked up the horse and put it on the sled, binding it gently so it would not fall off, and then put themselves in its harness and began to pull him toward the coast. They were within 60 kilometers of the depot and could have reached it within a single day had they not had the horse, but they would not give him up.[73]

It took them five more days to reach the depot, and on the last day, 4 July, they left their tent, the sled, and Grauni and skied downhill the last few kilometers to the depot. "It was a strange feeling," wrote Wegener, "after all that snow, to feel earth, real earth under your feet, to see flowers swaying in the wind, bumblebees and butterflies darting about, to hear the twittering of birds."[74] They opened a case of provisions and quickly wolfed down a cold breakfast. A "council of war" ensued, and Vigfus and Koch left to hike the 30 kilometers further on (according to their map) to the head of Laxelv Fjord to see if there were any signs of Eskimos. Wegener and Larsen quickly packed up bread, to take back to the last tent place and feed Grauni.

When Wegener and Larsen arrived at the tent site with food for the horse, they found Grauni on his side, barely breathing, clearly dying. They had to use their last cartridge to put him out of his misery.[75] After everything they had done, harnessed together, struggling along, resting when he needed to rest, feeding him their own food, pulling him for a week on the sled mile after mile, he was dead. "And he died a mile from the depot! So this wasteland gathered in its sacrifice in the last hour while we thought we had already brought him to safety."[76]

With the horse went their hope for anything out of the expedition but their own survival, and this was by no means certain. They had a few days of food in the depot, otherwise nothing, and it was many miles over rough terrain to the settlement at Prøven; they would not arrive there without help, as it lay on an island. There was no time to grieve, only to haul the remains of their camp down to the depot. Wegener and Larsen struggled all that day (4 July) and all the next, in rain and fog over terrible uneven ice: "in an hour's hard work, we would make 300 m, a distance I could've covered twice in 5 min. just walking."[77]

Koch and Vigfus did not return until 6 July. They had traveled 90 kilometers (56 miles) in fifty-eight hours over difficult terrain in terrible weather, with almost no food. They had found an abandoned but quite recent Eskimo encampment on the shores of the fjord and left a note there, with some expectation that the Eskimos would return in a few days.[78] They were completely exhausted (it was four more days before Koch could even write in his diary). The news was not promising: not only were there no Eskimos, but they would have to cross the fjord; they would have to fashion a boat out of their sled and tent, a sort of homemade coracle known to polar explorers as a "Leffingwell," after the American Ernest DeKoven Leffingwell (1875–1971), who had first described the technique of lashing a sled inside a tent to make a canvas boat.

While Koch and Vigfus tried to recover their strength, Wegener and Larsen moved the rest of their equipment from the edge of the Inland Ice to the depot and pared it down to the essentials. They would be traveling across bare land and could not pull a sled: everything would go on their backs. Wegener packed all the journals, diaries, scientific results, photographs, and photographic negatives, including those from the traverse of the ice which he had not yet developed, into their medicine chest. They would also haul the sled, the mast and sail (to raise as a signal flag), as much food as they could reasonably carry, and their tools, scientific instruments, and camera. They would leave their sleeping bags behind, heavy and soaked as they were, with most of the hair gone off the reindeer hides; their tent would travel only as far as the other side of the fjord. Still, the gear to be packed totaled more than 90 kilograms (200 pounds): Vigfus would carry the sled, the others rucksacks with 23-kilogram (50-pound) loads on their backs and the remaining poles in their hands.[79]

They left for the coast on 8 July, four men and their dog, heavily overloaded. Disaster struck almost immediately. While using the sled as a bridge over a fast-rushing stream, whoever was carrying the medicine chest, with all the expedition scientific work in it, lost control of it, and the box fell into the stream. If Vigfus had not immediately plunged into the water and grabbed it, it would have been swept away and lost in the lake into which the stream emptied a few meters away. Everything in the chest was soaked, and they feared that the photographic negatives of their traverse might be irreparably damaged. (In the end they lost about two dozen photos.) They had to spend the day spreading their journals, the photos, and the rest of their scientific gear on rocks in the meadow and wait for the Sun to dry them.[80] Wegener could think of nothing but going home.

Koch wrote in his journal that all that he and Vigfus had learned on their reconnaissance to the fjord was that they would not be able to retrace their steps via that route carrying a heavy load: they would have to find another path up and over the mountainous arm of the headland behind which their depot had been situated.[81] Every step was an agony. Their feet, unaccustomed to rough terrain and knocked about by rocks, began to swell. Larsen had injured his foot and was limping, and in the afternoon of the second day Wegener developed leg cramps and couldn't go on. In all, it took them two days to go 10 kilometers, wet, cold, and bitten ferociously by huge swarms of mosquitoes that they could not escape. Arriving on the shore of the fjord at three o'clock in the morning on the eleventh, they set immediately to building their boat and fashioning kayak paddles. It took them eleven hours to ferry their gear across in the leaky boat: two men to paddle and 25 kilograms (55 pounds) of gear per trip. At the end of each trip the canvas of the tent would be soaked and the boat would need to be disassembled and the tent dried in the wind.[82]

Once across the fjord, they made a cache of their equipment, scientific records, and instruments, in a place on the rocky shore easily visible from the fjord, and covered it with the remains of the tent. Each man now had a load of only 9 kilograms (20 pounds) or so—food, matches, but little else. "We made no distinction from then on between day and night. We traveled for 50 min. and rested for 10 and when we had been going on in this way for four hours we cooked some food, a long and difficult process without our stove" (they boiled food in open tins on a campfire of gas-soaked heather twigs).[83]

Moving forward in this way over the next twenty-four hours, they covered nearly 22 kilometers (13 miles), and the next day another 18 kilometers (11 miles). Toward the end of this day they had to climb up and over the arm of the mountainous promontory named Kangek and down to the coast again, which should have brought them in sight of the settlement at Prøven; their map, however, was faulty, and they had completely misjudged their location: the settlement was still 11–13 kilometers (7–8 miles) distant and invisible behind a boggy peninsula that they could not traverse.[84] They were exhausted and starved and nearly at the end of their food, and whatever courage they might have summoned at this point was drowned in a cold rain that soon turned to snow. They had to find shelter, quickly, and found a rocky overhang that they could block up with stones and moss. Koch, Vigfus, and Larsen were so exhausted that they couldn't move and stumbled in out of the rain. Koch later wrote that Wegener was the only one of them at this point with any remaining energy. While they lay exhausted under the overhang, Wegener searched here and there under the scattered boulders to find dry heather twigs, and he boiled the last of their milk and soaked the last of their bread in it.[85] The snow intensified, and fog rolled in, making it impossible for them to find a way forward; for the next thirty-two hours they lay, as Koch said, "in a semiconscious stupor."[86]

They no longer knew whether it was day or night, but when the weather cleared slightly on the fifteenth, they knew they had to move forward or die. Koch was at this point hallucinating, and they were all, with the exception of Wegener, using smelling salts to keep from blacking out. They started out to the west carrying their few remaining belongings, in wet snow that made it seem "as if each foot had a 10 pound weight tied to it."[87] After five minutes of travel Koch was so out of breath that he stopped, sat down in his tracks, and then fainted. As he later wrote, "I experienced this blackout as a foretaste of death." They were in the final stages of physical and mental exhaustion and had traveled only 30 meters from their miserable hut. There was no going forward; they had to go back to shelter. Wegener remembered thinking at this point, "Are we going to die like animals, here at the end of this long and danger filled trek, scarcely 2 miles away from the colony [Prøven, 6 miles distant], in the month of July? Where is a trace of logic in that?" He also recalled thinking, "Everything in me rebelled against this, and all my mental energy was focused in a single powerful thought: I will live, I will reach Proven, even if the sky falls."[88]

They had not eaten in thirty-seven hours, and they had one hope to survive and took it. They killed Gloë, their faithful and loyal companion in every extremity, built a fire, and cooked their dog. "It was our sole remaining hope," Wegener wrote flatly, "that a big meal would allow us to continue to walk." They were cutting up the half-raw dog meat into portions to eat it when Wegener jumped up and pointed toward the fog-shrouded fjord. Wegener saw something that he had initially mistaken for an iceberg, but it was moving too fast. A quick look through the telescope confirmed that it was a sailboat. Their exhaustion forgotten, they jumped and yelled wildly, and Larsen waved the sail back and forth, but it was their yells that brought the boat to them. Along with the crew of Eskimos was the Danish pastor, Chemnitz, from the mission at the

Wegener and Koch at the home of the Danish trader in Prøven, on 17 July 1913, a day after their rescue by Pastor Chemnitz and a group of Inuit traveling with him. They are wearing borrowed clean clothing and enjoying tobacco for the first time in many days. From Koch, *Gennem den Hvide Ørken.*

Upernavik, who had come to Prøven for a confirmation class and, knowing of the expected arrival of their expedition, had set out to see if he could find them.[89]

Within a few *hours* they were sitting in the European-style house of the Danish trader at Prøven: fed, dry, warm, and smoking their pipes *indoors*. Two days later, they were wearing new, clean clothes purchased at the trading post and waiting for a festive dinner with speeches praising their heroism. "It would be difficult to imagine a starker contrast than this," wrote Wegener on the seventeenth of July, "from the furthest limits of desperate need, to the full comforts of civilization."[90]

Wegener and Koch spent most of the day on the seventeenth re-creating their travels from the time of their departure from the depot on 9 July, eight days and a universe away, before the memories faded into unreality—so great was the contrast between their safe haven in Prøven and their near demise a few days before. On the nineteenth Vigfus, Larsen, and twelve Eskimos took a boat back up the fjord to recover the scientific records and equipment cached on the eleventh. The trip took a single day to cover a distance that had taken them five days and nights. The pastor, Chemnitz, had returned to Upernavik and brought back the motorboat that the expedition had shipped there the previous year. A supply ship reached Prøven on the nineteenth, and Wegener was able to send news home of his safe arrival. Else, who had finished her year in Bergen with the Bjerknes family and was waiting at Alfred's parents' house for news of the expedition, received a telegram from Copenhagen on the thirteenth and a letter directly from Alfred on the fourteenth. Her joy at the news was moderated by Alfred's announcement that though all he wanted to do was come home, he was committed to stay in Greenland until the autumn.[91]

However astonishing and anticlimactic it may seem, their work was not over, and Wegener was indeed committed to carrying out measurements with Koch of the

Jacobshavn Isbrae, some 386 kilometers (240 miles) to the south of Prøven on the west coast of Greenland, opposite Disko Island. This storied glacier, with an iceberg-calving front some tens of miles across, moved at speeds that defied imagination: 15–30 meters (49–98 feet) *per day*, roughly a meter per hour: fast enough to actually see.[92] It moved so fast that most of the early measurements had been discounted, but these speeds, as Wegener confirmed, were actually the case.

As was his practice, Wegener abandoned his daily journal as soon as the traverse of the ice was complete: in spite of the necessity of remaining in Greenland, the expedition was "over." It was a "crossing" expedition, and they had crossed. His sketchy notes on the outflow of the glacier at Jacobshavn filled a few pages at the end of his last journal, but they began from the back of the notebook and formed a separate entity. These scant notes contain no speculations on his theory of continental displacements, but it is interesting to think of the impact on his imagination of the outflow velocity of this immense glacier, considering how clearly he had made the analogy between flowing rock and flowing glaciers and between floating continents and floating glaciers. He had hypothesized that Greenland was moving west at 10 meters per year: here was a glacier moving almost 1,000 times faster. That such a velocity could be achieved by something in the solid state, something this massive, would not have escaped his attention.

As interesting as this may have been, his mind was reaching beyond the expedition, already collating material for his preliminary reports of their scientific findings, on the model of the articles he had written in 1908, and for the same semipopular journals, *Umschau* and *Himmel und Erde*. For all their effort, and nearly at the cost of their lives, they had produced rather little in the form of science, although their traverse was notable for its length and the great altitude reached at the center of the ice cap—nearly 3,000 meters (9,843 feet)—as well as the temperature data concerning the astonishing perpetual cold at the center of Greenland.

For all Wegener's dissatisfaction with aspects of the Danmark Expedition, it had made his reputation and allowed him to produce volume after volume of scientific work. In sharp contrast, he had since then set out on two expeditions of his own, each a failure in its own way. His trip to Argentina in 1910 had failed in making upper atmospheric wind measurements. He had now failed in Greenland in 1912–1913 to amass a significant scientific output, especially for the problem he cared most about: auroral measurements to compare with those made by Kurt simultaneously in Spitsbergen, and their implications for understanding the composition of the upper atmosphere. He had some photos of ice crystals, some photos of mirages, and some borehole measurements of ice temperatures. He had some photographic data on the lamination of the ice cap, its deformation, and the blue bands in glaciers, though his explanations were conjectural. That was it. Neither he nor Koch mentioned it, but their deepest intention, to show everyone in the polar science community the "right way to do polar science," had been less than a complete success.

Though anxious to return home, he was happy and excited. Both Else and Wladimir Köppen had sent letters to him in Greenland on the first ship leaving Copenhagen in the spring.[93] Köppen's letter had some encouraging news. In Wegener's absence, his patron at Marburg, Richarz, and his new dean, Ernst Elstner, had bombarded the Ministry of Education in Berlin with appeals to establish an extraordinary professorship for Wegener at Marburg, with the hope of keeping him there. The Ministry of Education had ruled, in September of 1912, that such a decision must wait for Wegener's return from Greenland and his return to Marburg.[94] This, from Wegener's standpoint, was excellent news. He had wanted to return to Marburg and felt that the setting was ideal for him there: he had lived with the conviction, for more than a year, that with persistence, such a position in physics, or even (his deeper wish) in cosmic physics, must open for him there. The ministry had not said yes, but they had also not said no. Writing from Jacobshavn to his future mother-in-law in August, he said, "The best news, after hearing that all of you are well, is the news of the planned position for me in Marburg. I think we [he and Else] can marry as soon as I get an official communication concerning the position. In any case as soon as possible. I have written the same to Else and my parents."[95]

He had every reason to be happy. His reputation as a polar traveler and polar scientist was now completely secure. His persistence and hard work in building a position for himself at Marburg appeared to be paying off, and he would now be able to both marry Else and avoid moving to Hamburg, for the Carnegie Institution and its money had moved along without him. He would have a few responsibilities in working up the expedition's findings, but since the meteorological portion of the work, including studies of the aurora, had largely failed, most of the data reduction and all of the cartography would be the responsibility of Koch. Wegener would not find himself, as he had in 1908, buried in the aftermath and afterlife of an expedition. He might have wished for a bigger harvest of data certainly, but although he thought of himself as a practical field scientist, it was increasingly clear to him, and to those who read his work, that his greatest successes were coming from his ability to collate, integrate, and reinterpret the work of others, and that his work in the field was a source of inspiration as much as or more than a source of data. He would return to Germany without debt, without obligations, and ready to pursue a course of life he had mapped out himself. All he had to do now was find a ship home.

13

The Soldier

MARBURG AND "THE FIELD," 1913–1915

> I don't have anything to do with science now, and have hung it up until the war is over. A man can't serve two masters and if I'm going to have to be a soldier, I will do my best, and do it wholeheartedly.
>
> WEGENER TO KÖPPEN, September 1914

Wegener's ship from Greenland docked in Copenhagen on 18 October 1913, and Else was there to meet him. They were thrilled to see one another, but there was bad news: three weeks earlier, while Wegener was at sea, his mother had suffered a stroke; she could now not speak and was paralyzed on one side. Wegener wanted to leave immediately but had to remain until at least the following day, when he and Koch were to be knighted by the king of Denmark.[1] There was to be a formal dinner afterward to honor the expedition, but Wegener begged off, and he and Else left for Zechlinerhütte immediately after the official ceremony.[2] Traveling via Hamburg, they arrived to find his mother still in poor condition, and they decided to stay on until their wedding, scheduled for the middle of the following month. To pass the time, they walked each day in the autumn woods, where Alfred found the colors delightful after the monochrome world of the ice cap.

During the waiting period at Zechlinerhütte, Alfred made a trip to Berlin to obtain further information about the progress of the appeal for a better appointment at Marburg. Here also, the news was not good. He wrote to Köppen that even though his dean had personally traveled to Berlin to talk to the ministry, the application was likely to be refused. Under the circumstances, he and Else would still marry as planned, return to Marburg, and accept the current stipend, while waiting for another round of applications. Between his stipend and the continuation of his expedition salary while he worked up his scientific results, they could, with care and economy, last a couple of years: by then he would know either that he had a position or that it was time to "take off."[3] Wegener's reluctance to give up an academic career and move to Hamburg was as strong as ever, and it is clear that Köppen's desire to bring him there was still intact as well.

In late October, the couple visited the marriage license bureau. In Hamburg they had been startled by the requirement that Alfred produce (from the Greenland government) an attestation that he had not married while there; Else was asked to provide a similar document from the Norwegian government in Oslo. Of course, this late in the year, there was no chance that such a document could be procured from Greenland even if one could have been obtained from Norway, and it appeared that their marriage would thus have to be postponed another year. Fortunately for them, things were less formal in rural Rheinsberg than in Hamburg, and the marriage registrar, a local farmer, decided to waive this requirement, announcing to the couple, "I know that you have been in Greenland, Herr Dr., because I read about you in the newspapers, and I can't believe that you would have married an Eskimo girl when your bride was here waiting for you."[4]

Their wedding was a modest affair, in Hamburg in mid-November. From Alfred's side there was only Kurt, already in Hamburg at the Marine Observatory with Köppen; Anna Wegener was improving but too ill to travel, and sister Tony was needed in Zechlinerhütte as caretaker to Anna and Richard both. Alfred had written to Marburg asking that his winter term lectures be put off until 1 December, and the newlywed couple spent the next two weeks packing and shipping household goods and furniture—the crates of linens, chairs, and dishes a humorous contrast to the boxes they had packed in Copenhagen in the spring of 1912.

Arriving in Marburg, the couple moved into their newlywed lodgings, a spacious floor-through apartment, in a corner building on the Biegenstraße. This was a brand new commercial and residential district that had, in the past few years, sprung up between the old town and the train station. When the city fathers had decided to restrict the train station to a spot some kilometers out of town and across the river Lahn, they had no concept of the urban sprawl they would engender, as businesses and residents alike chose easy access to modern transport, via the construction of a new neighborhood that brought the city to the train station.

Else agreed with the local opinion that these faux Bavarian buildings along the Biegenstraße, with their obviously fake external timbering and their decorative onion domes, were "hideous" and spoiled the view of the river from the Oberstadt. On the other hand, the new apartment was large and airy and provided a good view, from the rear windows, of the palace, the university, and the observatory. From the street-side windows they could see the river and the wooded hills beyond. If they needed additional money (which they might), they had extra rooms to let to students. Else found the old town enchanting, and she would bypass the markets in her neighborhood to walk a mile and climb the many stairs to the vegetable stalls in the old market square near the town hall, where the mechanical rooster in the clock tower crowed the hours. For Else, Marburg was rural, rustic, authentic. The food and air were fresh; she was thrilled by her new life.

There was a sense of ritual commencement in the way that Alfred and Else set up their apartment. The greater part of their attention went to Alfred's workroom: arranging the desk with good light over his left shoulder from the large windows facing the street, unrolling the polar bear rug from his first trip to Greenland. They would "work together." Else had learned to weave while in Norway, and she set up her small loom in the workroom to keep him company as he wrote. If she had not done so, she might never have seen him; she knew enough of his habits to know that unless she established herself early, this would certainly be the case. She set about as well to build a card index of his hundreds of offprints of scholarly articles. Alfred instructed her, "When you dust around here don't even think of moving anything on the desk."[5] Visions of early twentieth-century bourgeois life notwithstanding, they had no servants: she would cook and clean, and laundry would be sent out. Else later wrote, "But that was part of my upbringing, I had learned long before from my mother how to take care of a writing table overflowing with papers without moving anything out of order."[6] The unpacking and reordering of books, boxes, and papers, the beginnings of domestic routine, gave an external order to their life, as did the resumption of university lectures.

A photo taken of Alfred in Copenhagen a day after the ship had docked from Greenland had shown him serene and happy—so much so that Achton Friis decided to use it as the basis for his portrait sketch of Wegener in Koch's account of the trip. But even as Alfred set to work, he felt an interior sense of dislocation and disquiet. He had difficulty concentrating. The inessential and the essential were, uncharacteristically, all mixed together in his mind. He fretted about the cost of their move from Hamburg to Marburg, which had used up almost 10 percent of his annual stipend. He could not find the latest issues of journals published just before he had departed from Greenland, and this worried him. He had tried to buy a copy of Krümmel's *Ozeanographie* in Berlin, but he had not been able to obtain one. He began, just before Christmas, an article on the motion of vortices (*Wirbelbewegung*) but suddenly found that he could make no progress.[7]

The situation was very different from his return from Greenland in 1908. Then he had a rich harvest of his own data and two years of station records to work up, material that he was sure held the signature of important developments. There was none of that in 1913. He had come back full of ideas but bereft of data. Greenland had once again fired his speculative imagination: the fracture and flow of the Inland Ice, the layering in the firn atop the ice cap, the speed of the Jacobshavn Glacier, the behavior of the aurora, and his photographs of ice crystals and of complex layered inversions a few meters above the ground all suggested lines of advance. Yet none of these allowed him to expand or extend his principal line of work: the theory of the overall structure of the atmosphere.

Else could sense the problem. Her year in Norway had been very instructive for her, not least scientifically. Living with the Bjerknes family, she was in a scientific milieu she understood: the paterfamilias made an effort to explain his work, as clearly as he could, to his family. This was just as it had been at the Köppen household. But Bjerknes's work was new to her. It was purely theoretical, and Bjerknes measured the progress of meteorology in his ability to translate atmospheric motions into the differential equations of theoretical physics; he had scarcely any need to come into direct contact with nature. Very much in contrast, her father had devoted his life to building a network of meteorological stations in North Germany, and his scientific work consisted in attempting to extract statistical regularities from this mass of data and to turn these regularities into the laws of meteorology and climatology. Alfred's method of work differed from both of these, as Else saw it, and sprang from the necessity that Alfred felt to work in direct and unmediated contact with nature, whether in his balloon flights or through his life on Greenland's ice cap. She saw the wellspring of his invention in his emotional connection with—his love for—the natural objects he studied.[8] Where Bjerknes generated no data himself and Köppen worked with summaries of data collected by others, Wegener depended for his inspiration on having his own data. Generally speaking, he preferred data of a kind not before collected by anyone, or otherwise scanty, scattered, or new.

Back in Marburg, but without a harvest of novel data generated in "the midst of nature," Wegener found himself forestalled. Trying to focus his mind on work, he did everything he could to minimize distraction. He reduced his lectures in astronomy from two hours a week to one. He went to the university only for his lectures, for meetings of the physics colloquium, and for the sessions of the Marburg Natural Science Society. The rest of the time he spent at his desk. He had, of course, plenty of work to do. He had several popular manuscripts to finish, retailing his Greenland adventures, and he had voluminous expedition correspondence. He could proceed (using his photographic evidence) to think about questions of polarization; moving on from atmospheric thermodynamics to optics had always been part of his general plan for a full physics of the atmosphere. He had still to transcribe and edit his Greenland diaries, and he had a moneymaking project to begin, organized by Köppen. Curt Thesing, a well-known popular author and producer of scientific works in encyclopedic format, had solicited an encyclopedia of the atmosphere from Köppen, who had agreed to take it on, with the notion that he would farm out the work, as well as the proceeds from it, to both Kurt and Alfred. It was the sort of systematic and synthetic work at which Alfred excelled, but there seemed to be nothing new in it; it was bread labor.

In the midst of this scientific impasse, Alfred tried to turn back to the study of vortex motions in the atmosphere. In the summer of 1912 he had made a number of observations of "dust devils" on the volcanic landscape of Iceland, while preparing for Greenland. It was clear to him, as it had been to most meteorologists (it had been published long since in Hann's *Handbook of Meteorology*), that these were thermally driven and not the same sort of phenomena as tornadoes.[9] He was far from the only meteorologist in 1912 to be interested in vortex motion: his sharpest critic, Exner, had performed a number of experiments with rotating tubs, using colored dye to chart the development of turbulent eddies.

Wegener had worked on turbulent motion in the atmosphere at the very beginning of his career, back in 1906. He had written on "Helmholtz waves" at the boundary between two atmospheric layers of different density, as well as about the formation of cumulus mammatus. His interest in these topics had been reawakened at the Munich meeting of the German Meteorological Association in November of 1911. Köppen had given a paper discussing the interaction of "circulation layers" and "boundary layers" in the atmosphere, and it occurred to Wegener in the course of the discussion of Köppen's paper that this provided a new way to understand vertical convection in the atmosphere, particularly the way in which it brings about the fall of temperature with altitude. If one considered, Wegener had speculated, that parcels of air in a circulation layer were cut off from interaction with the surface of Earth by the action of a boundary layer, then thermal convection via the contact of the air with the (warmed) surface of Earth could not be the cause for the vertical convection, which would have to be explained, on the contrary, as a mechanically forced mixing, which was above all a product of the velocity of the flow, in combination with the internal friction of the air mass.[10]

His reflections on atmospheric turbulence, published in early 1912, contained two generalizations that, taken together, constitute one of his more notable contributions to meteorology. The first of these was the general principle that "the troposphere is the zone of turbulent, the stratosphere of laminar motion." The isothermy of the stratosphere was, in fact, a consequence of its laminar motion, while the troposphere was a zone of constant or nearly constant turbulence. He declared, "I believe, that this new or at least heretofore unremarked difference of the two principal layers of our atmosphere will be of the utmost importance in the future; for through no other principle will one be able in such a simple manner to explain the sudden sharp boundary between temperature lapse and isothermy, between the constancy of potential and the constancy of actual temperature, and between constant entropy and constant energy."[11]

The succeeding generalization contained, in a single sentence, the theoretical shift from the thermal theory of cyclonic motion to the mechanical: "If the above representations of the situation are correct, then the strongest vertical circulation should be expected not at the time of maximum heating, but at the time of the strongest winds. It is very interesting that, in fact, we [all] came to this conclusion some time ago, but without being able to provide a useful explanation of it."[12]

Whatever progress was being made in the theory of storm systems, empirical work had demonstrated that while storms were not thermally driven, and therefore theoretically outside the province of thermodynamics, on the other hand the theoretical understanding of the mechanical basis of these (turbulent) phenomena still lay well in the future. The barriers to theoretical understanding, Wegener indicated, were well laid out in encyclopedic textbook treatments by A. E. H. Love (1863–1940), Wien, Horace Lamb (1839–1934), and Rudzki, if anyone cared to pursue them.[13] This direction forward for meteorology—resolving the problems of using hydrodynamic models, problems outlined in Wegener's paper with regard to the study of the movement of the atmosphere as a whole—proved indeed to be the main line of advance in the next two decades, especially in the work of Bjerknes, Tor Bergeron (1891–1977), Wilhelm Schmidt (1883–1936), Harald Sverdrup (1888–1957), and, above all, Lewis Fry Richardson (1881–1953).[14]

Wegener's qualitative grasp of this problem back in the autumn of 1911 had appeared in his use of homely analogies. The difference between laminar and turbulent motion in the vertical could be seen, he indicated, in the smoke rising from a cigar, which turns from a smooth flow to a turbulent billow within the first few centimeters. The same phenomenon might be seen in the difference between a cleanly burning and a flickering candle, or in the blackening of the chimney of a kerosene lamp.[15] On a larger scale, and in the horizontal dimension, Wegener sketched the smoke from a factory chimney in the early morning in the presence of a temperature inversion. Smoke from the chimney makes a smooth line: laminar flow (a thin, straight, condensed horizontal plume) under the influence of light prevailing winds, as the inversion capped the vertical ascent of the air. Within a few hours, however, one saw turbulence with a widening amplitude, with the smooth plume of smoke from the factory turning into a widening billow. This argument was typical of Wegener: the qualitative evidence for the phenomenon was all around but had not been aggregated into a single picture or *properly* quantified.[16]

At the beginning of 1914, Wegener still had an inclination to pursue the study of turbulence, but during his absence in the preceding year the character and content of such work had become ever more mathematical. He saw this as a serious problem: in his view, a good deal of this mathematical work was just plain *wrong*. He wrote to Köppen in January 1914, "The mathematical work is simply scandalous! Nine out of ten people think that when they calculate with a formula according to some rule they memorized, that they have discovered something, while completely missing the nonsense in their arguments. . . . Most of them either don't know or refuse to acknowledge that in mathematics, the formula is either exactly correct, or it is wrong, and so they blithely wade into deeper and deeper waters, having accepted the first [formula] they came across, without any reasoned examination."[17] He then went on to point out an error in a recent issue of *Meteorologische Zeitschrift* (a dropped exponent in a differential equation) that made nonsense of the entire working formula presented: "Pity the poor reader!"[18]

Given this state of affairs, Wegener was at a loss in early 1914 on how to proceed with his atmospheric work. He had nothing new to say about thermodynamics on a large scale. His attempt to do something with atmospheric motion led him to attempt to write about the mechanics of vortices, but he found himself here with neither sufficient new empirical (numerical) material nor any adequate conceptual treatment to apply to it. He thus abandoned this work on vortex motions and its connection to atmospheric turbulence in February 1914, almost as soon he had picked it up. His work in atmospheric optics was in its beginning stages, nowhere near ready for publication, and his "bread labor," the encyclopedic project for Thesing, took him over familiar ground. He was mentally "stuck." With this encyclopedia project facing him (as had been the case with the proposed 1912 work for Carnegie), working on something for which he had no appetite had caused his imaginative resources to desert him.

In the midst of this confusion and disappointment there was at least one bright spot. Richard Aßmann had solicited (as Wegener mentioned in a letter to Köppen on 4 January 1914) a fuller version of his article on the origin of continents, for a series of popular-science books that he edited for Friedrich Vieweg & Sohn.[19] Vieweg was a well-known scientific publisher, and the series for which Aßmann had recruited the book was the "Sammlung Vieweg," which featured short (circa 100 pages) titles on novel themes. The motto of the series reflects this: "Current Issues [*Tagesfragen*] in Science and Technology." Aßmann's offer of publication was a statement of support; Wegener was still, for Aßmann, his young "theorist." What he had seen in Wegener's publication on continents and oceans (in *Petermanns Mitteilungen*) was the same boldness of conception which he admired in Wegener's atmospheric physics.

Aßmann was not alone in finding something to admire in Wegener's work on the origin of continents and oceans. The previous September (1912), while Alfred had been struggling to balance hay bales on the backs of Icelandic ponies in Greenland, the Seventeenth Conference of the Geodetic Association had taken place in Hamburg, with the participation of (among many others) Friedrich Helmert, as well as Wegener's old astronomy professor from Berlin, Wilhelm Förster, and also Carl Albrecht (1843–1915), who, with Förster, had been a leader in the development of the International Latitude Service.

Wladimir Köppen was at the meeting and had mentioned to these old friends and colleagues (in a break between sessions) Alfred's hypothesis of continental displacements. Köppen was clearly testing their reactions, worried about what they might think. Much to Köppen's surprise, Carl Albrecht said that he thought that Wegener's proposal of continental displacements was a really interesting idea, and it probably wouldn't take more than a year or two of measurements to find out whether it was true. Förster and Helmert agreed completely.[20] Albrecht's interest in Chandler wobble and the motion of Earth's pole had been the driving force behind the program at Potsdam in the 1890s to establish the reality and extent of such polar motion. Moreover, this interest and the program were some of the main reasons that Wilhelm Förster had directed Wegener to do a thesis on the Alfonsine Tables. In fact, Wegener's instructor in geodesy at Berlin, Marcuse, was Albrecht's student and assistant, and it was Albrecht who had arranged with the U.S. Coast and Geodetic Survey to send Marcuse to Honolulu, Hawaii, in 1890.[21]

While Albrecht's name is mostly associated with his work in the determination of latitude, he had also been deeply concerned with exact measurements of longitude. He had been a pioneer in the electrotelegraphic measurement of longitude distances, as well as an observer in the 1890s in the measurement of the longitude arcs joining Paris, Warsaw, Bucharest, Pulkovo (St. Petersburg), Greenwich, and Horta (Azores). This work was preliminary to the unification of global geodesy and time reckoning, finally cemented in place in the year 1912, when all astronomical measurements and time measurements henceforth were to be calculated from the 0° of longitude at Greenwich.

Albrecht wanted very much to use Wegener's hypothesis as an excuse for a test of longitude measurements using radio time signals, then still called, quaintly, "Hertzian telegraphy," but this proved impractical. Nevertheless, Albrecht went immediately to work to establish a protocol to measure the longitude arcs from Borkum (in the North Sea) to Horta and from there to New York, using the German telegraph cables connecting these places. Albrecht was anxious to use the station at Horta because he had established its longitude himself in the 1890s and reconfirmed it many times in the succeeding years. Albrecht was so excited by the idea of testing for shifts in longitude that he determined that he himself, at the age of 71, should go to the Azores and take the measurements that were to get under way in July 1914.[22]

Wegener was, of course, pleased to hear of this test. He was confident that his hypothesis of continental displacements was correct, and he was certain that an appropriate measurement program would confirm it in short order. He even referred to the longitude measurements as an *experimentum crucis*, employing Isaac Newton's term for a definitive experiment that must remove from consideration every hypothesis but one.[23] In this context, one might have expected him to reinvolve himself with the topic of continental displacements, especially since the rush to prepare for Greenland (in 1912) had constrained the detailed presentation of evidence for his argument. Yet there is no hint in his correspondence from this time (early 1914) of any enthusiasm for the book project or any impetus to work on the idea of displacements. Wegener appears, at this juncture, to have been content to await the results of the planned measurement program. This would make sense: if the measurements determined no movement, there would be no reason for such a book.

Even had Wegener been willing and able to go forward with further research on continental motions, or atmospheric turbulence, or atmospheric layering, or optics, or any of the choices before him, he had no opportunity. He barely had time to finish the magazine articles on Greenland and to recover his journals (they had been in Berlin being transcribed) before facing the reality of two months of intensive military training at Charlottenburg in Berlin, in March and April of 1914. Else would accompany him to Berlin, and they would be allowed to live together in civilian quarters, but his time would be taken up all day, every day.[24]

In the meantime, there were additional distractions. He and Koch had agreed to give, in January 1914, a joint lecture in Berlin at Urania, the observatory where Wegener had apprenticed; this was associated with the piece Wegener wrote for *Himmel und Erde*.[25] In February Koch and his wife paid a visit to the newlyweds in Marburg, which delighted Else. While there, Koch repeated with Wegener before the university community the Berlin lecture, after which they had to suffer the "dignity" of being crowned with beribboned laurel wreaths. Back at the apartment, Else and Frau Koch laughingly unwound the ribbons and reserved the laurel leaves for kitchen use. The ribbons would come in handy for children's clothing, Else wrote, and this was much on her mind, for she had just discovered that she was pregnant, with the child most likely due in the latter part of August 1914.[26] This was wonderful news for the young couple, but it must have increased Wegener's sense of financial anxiety, as money was a constant and nagging worry.

During the time that Wegener had been in Greenland, Richarz (as Wegener's professor and director of the Physics Institute) had worked with successive deans at Marburg to try to improve Wegener's financial situation. Their joint failure to create an "extraordinary" professorship for Wegener bothered Richarz very much. Using a combination of his own and Aßmann's influence, he had named Alfred, in January 1914, as Germany's representative to the impending conference of the International Commission for Polar Aeronautics, scheduled to meet in Copenhagen in February. The "Marburgers" were using every means at their disposal to raise Alfred's profile, nationally and internationally, in hope of inducing the Minister of Education to create a professorship for their distinguished younger colleague. The rector of the university joined in and issued a proclamation, published both at the university and in the city of Marburg, reporting Kaiser Wilhelm's official letter of congratulation to Wegener for his knighthood in Denmark.[27] It appeared now that the best chance for a professorship for Wegener lay in accentuating his celebrity as one of Germany's rising polar scientists and using the fame derived from his (still standing) world record for time aloft in a balloon back in 1906.

Wegener accepted Richarz's invitation and dutifully went to Copenhagen to participate in the conference, though (notably) he begged off, while he was there, from giving a lecture before the Greenland Scientific Commission on the details of his ice cap traverse, citing the press of work.[28] Upon his return to Marburg, he announced to Richarz that his participation in this aeronautical effort was redundant and that he probably need not attend again. Wegener was here and elsewhere almost willfully blind to the sorts of work that one did within the social system of science in order to position oneself for professional success. He seems to have pinned his hopes for advancement on his scientific work alone, and while accepting Köppen's arrangement of paid scientific writing, he shied away from any kind of administrative role in polar science or meteorology. In this matter, he was completely consistent. He had refused to give up the Marburg instructorship in 1912 when Köppen had pressed him to move to Hamburg and accept a position at the Marine Observatory. He had earlier declined Aßmann's very generous terms to return to Lindenberg as a theoretical scientist without observational duties, again sure that he could make a go of it in Marburg.

In early March, Alfred and Else moved to Berlin, to small and temporary lodgings in Charlottenburg, close to his regimental headquarters. She remarked (years later) that Alfred seemed exhausted throughout the entire time in Berlin and had difficulty concentrating. He tried to work, but without his books and papers and the calm of his workroom in Marburg, he seemed unable to keep his ideas clearly in mind, and he spoke of this to her. Even traveling by streetcar from Charlottenburg to the Meteorological Institute, in order to obtain books, left him depressed and unable to concentrate. He found the noise of ordinary street life in Berlin almost unbearable. His "nervous constitution" and inability to tolerate noise (of which he had spoken to Köppen two years before) were certainly a reality here, yet Else found herself surprised by it; it was astonishing, she wrote, that someone with the physical strength and courage to do what he had done in Greenland, and who had survived the solitude of the overwintering so well, should have his nervous constitution undone by the crowds and hubbub of Berlin's city streets.[29]

Preparing (in April) to return to Marburg for the summer semester, at the end of his stint of military training, Wegener informed the dean that he had decided to change his lectures in thermodynamics and mechanics of the atmosphere to a shorter course entitled "Optical Phenomena of the Atmosphere."[30] Just before a Whitsun (June) visit to Marburg by the Köppen family, Wegener reported to Köppen that he had begun to make progress in the work on the meteorological encyclopedia for Thesing. He had begun by writing the articles on meteorological optics and acoustics. His chapter on synoptic meteorology for the planned encyclopedia allowed him to work through this topic historically: the development of weather maps, a history of the idea of low-pressure systems, the ideas of Bjerknes on squall lines (the emerging concepts of frontal weather), and other topics.[31] These broad reflections were stimulating and encouraging. His ability to focus and work seemed to be returning to him, and his episodes of anxiety and depression were becoming more rare.

With the departure of the Köppen family for Hamburg after their Marburg visit, Alfred and Else turned resolutely in June and July of 1914 to their own work; few letters from this period survive. Alfred seems to have been consumed by the encyclopedia volume; this now had intellectual content for him and no longer felt so much like bread labor.

While he was consumed by his meteorological work, Else was busy translating Koch's expedition account of the ice cap traverse into German. Koch had spent the last two months in Greenland, while waiting for the ship home, collating his and Wegener's journals into a single expedition record. Koch, who had still not worked up all of the cartographic materials from 1906–1908, nevertheless completed this popular book in record time; it was published before the end of 1913. It was a beautiful volume under the title *Gennem den Hvide Ørken* (Through the white desert), published in large format, with striking halftone photographs on almost every page, as well as an abundance of maps and diagrams.[32] Wegener reasoned that he had essentially written half the book published by Koch: all but a few of the photographs were his, and they told the story as he had seen it; his diary entries additionally recorded his participation and his thoughts. He had no desire or impulse to write a "Greenland book" of his own. However, when it became clear (in early 1914) that the Danish edition had sold well, Alfred and Else decided that she should undertake a German edition, translating Koch's journals and the Danish parts of Wegener's. She had learned Danish and Norwegian during her time in Bergen the previous year, with the intention that she could "help with his scientific work." This would be the first major help of that kind, both scientifically and financially.

By early July 1914, Wegener was at his desk eight to ten hours a day, sometimes more. Else admired his capacity for work. "He scarcely went out, and when he did he missed his books and the quiet of his study. Sitting in the house at his writing desk he could concentrate for hours at a time writing; from time to time contemplating the blue smoke rising from his cigar, occasionally exchanging a few words with me, or reading me a passage, asking me whether it was clear, and then plunging back into his writing, most often remaining at work deep into the night."[33]

His output of encyclopedia articles during June and July must have been prodigious. On 23 June, Köppen wrote him in mock despair: "It is raining manuscripts from you over here [in Hamburg]!"[34] Even with Köppen's fabled capacity for work, he could not keep up with the flood of paper that Wegener was sending. Alfred was back in stride now, doing what he did best. Köppen estimated that at this pace they might be done with the work on the encyclopedia by 1 August, and at the latest by 15 August.

With the work on the encyclopedia nearly complete, Alfred and Else plunged into the reorganization and cataloging of his (by now) huge collection of reprints. The new file boxes for these had arrived, and the young couple worked out a library system to provide Alfred with easy access to documents, as he needed them. The scale of the synthesizing work he envisioned (in the completion of his physics of the atmosphere) was such that merely stacking reprints on the writing table, so that they could be conveniently "to hand," was no longer practicable.

The work of organizing meteorology in an encyclopedic form and of considering the organizational scheme for his various reprints in meteorology, astronomy, geophysics, geodesy, and (now increasingly) geology gave Wegener further opportunity and cause to reflect on his place in the world of science. His superiors at Marburg had gone forward to the Ministry of Education with the plan that he should have an extraordinary professorship in astronomy and meteorology. This initiative had failed. Looking at his own interests and the needs of his students, he began to think that the position might better be reconfigured (in the next round of applications) as an extraordinary professorship in cosmic physics—an idea he had long cherished.

Wegener mused aloud to Köppen, in a letter on 23 July, that cosmic physics perhaps had stronger claims as a university subject than meteorology, especially since high school teachers already had to demonstrate a proficiency in the subject matter in order to pass their examinations. Meteorology would for a long time remain a special subject within physics, but cosmic physics promised to be more pragmatically comprehensive with regard to his own interests and "the emerging scientific worldview," as he saw it.[35]

Crossing disciplinary boundaries was, for Wegener, an expression of an interest in *problems*, rather than an interest in *topics* or subject matters. Wegener had been, from his earliest days in meteorology, interested in the manifold problems generated by sharp surfaces of discontinuity. These surfaces had led him to work on atmospheric layering, to work on the chemical composition of the atmosphere, and eventually to thinking about surfaces of discontinuity within the solid body of Earth. In physics, then as now, solving a problem in one area leads one to think about other areas in which the achieved solution might apply. Wegener's inspection of the depth contours of the Atlantic had led him to imagine that surfaces of discontinuity played a role. This had led, in turn, directly to the idea of continental displacements.

As he moved forward in the summer of 1914 into atmospheric optics and acoustics, it was clear to him that these areas of physics would also yield important problems that might be solved with reference to sharp surfaces of discontinuity. Mirages, which he had photographed extensively while in Greenland in 1912 and 1913 (and had thought about as early as 1906), were the result of discontinuities in the atmosphere close to the ground and of inversions. His work in atmospheric acoustics, not yet well formulated, would also have to do with reflection and refraction of sound waves at the contact surfaces, the "layer boundaries," of media of different density. To respect the disciplinary partitions of Earth into geology, geophysics, oceanography, meteorology, and atmospheric physics would be to miss the invariant elements consistent across these different sections of a single subject matter: Earth as a physical planetary body in a solar system.

To understand this attitude toward problems and solutions is to see why he had been reluctant to move ahead with a book-length treatment on continental displacements at this time. From Wegener's standpoint, the physical work of formulating a hypothesis and producing a "crucial experiment" to determine its validity had already been done. His only interest in the topic of continental displacements in the summer of 1914 was the possibility of strengthening the measurement-based argument for the validity of the hypothesis.

With the measurement protocol in place to establish the drifting apart of North America from Europe using the German cable across the Atlantic, Alfred wrote in June to his old professor at Berlin, Julius Bauschinger, to secure the exact latitude of those Australian astronomical observatories with which the Berlin astronomers correlated their astronomical observations.[36] Since, in Wegener's vision of continental motion, there were major *latitude* shifts (India and Australia moving to the north) as well as major *longitude* shifts (the Americas moving to the west), cable communications across the Pacific with Australia might be able to produce a record, in the Southern Hemisphere, of the contemporary and continuous latitude shifts of India and Australia to the north. The telegraph cable between Brisbane, Australia, and Vancouver, British Columbia, on the western coast of North America had been completed in 1902; Australia also had cable connections to the Cocos and Keeling Islands and to India. In conjunction with the observations of the International Latitude Service, these would be three independent sets of measurements directed to the establishment of the reality and extent of lateral motions of continental fragments across Earth's surface at the present time.

Wegener's War

Completely immersed in their own affairs in a world where news accounts were often days behind actual events, Alfred and Else were largely unaware of the international crisis in late July and early August of 1914. Not until 31 July did the seriousness of the situation become apparent to Alfred, and he wrote to his father-in-law that he feared mobilization and the outbreak of war.[37] His mobilization order arrived the next day. Despite his frequent military training, the notion that a war would ever actually happen had been beyond his imagining. He owned a field uniform, but for several years now he and Kurt had shared a pair of binoculars, a revolver, and a field pack with associated kit, since their military training never happened at the same time.[38] Alfred was still in his Guards regiment, while Kurt had transferred to the Fliegertruppen des deutschen Kaiserreiches in 1910, as one of Germany's first pilots of fixed-wing aircraft. The mobilization, therefore, found Alfred materially as well as mentally unready, and he had to borrow equipment from friends. After seeing him off at the station in Marburg, Else was able to follow his progress for a day or two, but after 3 August, all communications ceased.

Wegener opened the war in the Second Guards Reserve Division of the X Reserve Corps commanded by Gen. von Kirchbach; this unit was part of the German Second Army under the command of Gen. von Bülow, advancing into Belgium. Just as Wegener's old military textbook from 1903—*Das Volk in Waffen*—had predicted, everything went wrong from the beginning. The Belgians put up unexpectedly fierce resistance, and with German forces divided in order to protect their southern flank against a counterattack, the Second Army had a difficult time taking the Belgian fortifications, even though they outnumbered the Belgian troops by about 3 to 1.

Alfred was wounded in his very first battle, when the Guard Reserve Corps of the German Second Army was thrown into the line in the battle for Namur, one of the Belgian forts blocking the line of advance into France. In the early-morning hours of 23 August, while taking part in a massed infantry assault on the fortifications, he was shot cleanly through the forearm. Treated for a short time in a military hospital, the stiffness in his thumb indicated that the tendon might have been cut, and he was sent home to be x-rayed, arriving on 2 September in Marburg. There he discovered he was the father of a little girl, Hilde, born four days before on 29 August. Both Else and the child were doing well.[39] It might seem strange that he should be sent away from the front back into Germany for such a slight wound, but the reality was the reverse of the appearance: the Germans had taken such terrible casualties in the assault on Belgium that the field hospitals were overloaded, and most with minor wounds were sent home in the expectation that the war would in any case be over in a matter of weeks.

After two weeks at home, he was recovered enough to be ordered to Charlottenburg on 15 September, where a hastily assembled replacement battalion was ready to be shipped back to the front; he joined them as the head of an infantry company composed of men he had never met. So much for the long years of being a "father" to men that he "knew intimately," in an atmosphere of mutual trust and dependence, the foundation of Colmar von der Goltz's Darwinian theory of military solidarity he had studied in his cadet years. This composite battalion rejoined the Guard Corps in the freshly dug trenches a few kilometers outside the fortified city of Reims, where the German armies had fallen back after their loss at the First Battle of the Marne.

From his bunker outside Reims, he could see the cathedral, with its Red Cross flag flying. He put his replacement troops to work deepening the trenches and strengthening the bunkers, and just in time. About three o'clock in the afternoon of his second day at the front, he heard the hiss and the crash of the first shell of his first artillery barrage, and then he heard the explosion and felt the blast wave—first one shell, then another, and then forty in rapid succession. He wrote to Else that the noise was deafening, even though he had plugged his ears with cotton, and that the pressure waves of the explosions were shocking to the nervous system. The French artillery was positioned only about 1,500 meters (4,921 feet) away, just out of rifle shot, and had the range of the German positions exactly. Wegener said that the only thing that kept him from panic was the calmness of everyone around him, continuing to fire their weapons even as their comrades a few feet to the right and left were blown apart by artillery.[40]

During a break in the bombardment, Wegener had occasion to converse (while seeing to his own wounded) with the regimental chief medical officer, only to discover that this surgeon was a "graduate" of the Schindler Orphanage and someone he had known in childhood. Learning that Wegener had already been wounded once, the surgeon conducted a brief examination. Wegener had told him that he was experiencing a recurrence of an irregular heartbeat that had plagued him during the initial march into Belgium. The doctor asked him to make an appointment for a full workup and diagnosis of the ailment, and Wegener did so, but the appointment was never kept, as that very night (30 September) his unit was pulled from the line to take part in what has become known as the "Race to the Sea."[41]

The German forces, having been stymied by stiff resistance and unable to turn the French flank, were putting their troops farther to the west, with the French pacing them and foiling the maneuver. Wegener's unit was sent by rail to Cambrai and then had a 28-kilometer (17-mile) forced march through the night, arriving at the front on 1 October.[42] They were thrown into the line after two hours of rest in the assault against the French Tenth Army. Here Wegener's Reserve Corps rejoined their original "brother" regiment, the Kaiser Alexander Grenadier Guards. There was a lull in the fighting for two days, and on 3 October the Germans went on the attack again. The Kaiser Alexander Guards were in the first wave, Wegener's unit of the Guard Reserve Corps in the second, with a third wave behind them.

Wegener's description of the battle of 4 October in the village of Puisieux is a hair-raising account of the hellishness of this sort of combat. The wave in front of them was pinned down and unable to proceed; they were diverted to the left and charged forward, without artillery support. The French artillery found its range almost immediately and inflicted terrible casualties on the advancing line. The German artillery, trying desperately to silence the French guns, could not find the range and landed its first barrage on the German troops, blowing to bits the infantry companies to Wegener's immediate left. Finally, the German guns found the range and laid down a furious bombardment. Wegener found himself listening to the sound made by the different kinds of projectiles, some making a "pfft," some a melodic singing sound that changed, as the shells got closer to the ear, to a whistle. As he lay on the ground, he was calculating, by the change in pitch, the Doppler effect of the projectiles, and he was rapidly learning to distinguish those incoming from those that had gone past.

As the German barrage lifted, Wegener and his troops rushed forward. Being to the left of the village, now they found themselves able to advance without entering the barbed wire that defended the main road into town. Rushing pell-mell with no obstacles or resistance, they soon realized that they had gone entirely past the village, and they heard a loud hurrah from the rear, signaling that the Alexander Guards had cleared the barbed wire and entered the town. Coming through a hedgerow, Wegener saw that he was facing a firing line of French troops which had fallen back several hundred meters from the village, and he was, all of a sudden, being fired on by these troops in front of him and by snipers from hedgerows on both sides.

Wegener later reported that his men were so maddened with bloodlust and fear—they were being shot at from three sides and there was no way to advance—that he had to keep them from firing at German troops approaching them from behind. In the middle of this perilous situation, Wegener had an episode of severe heart palpitations and had to stop, too dizzy to go on. These subsided briefly and he dashed forward, revolver in hand, firing at the French in the hedgerows. His heartbeat became wildly irregular a second time, and he had to stop again.

He continued to fire and could hear the sound of the French bullets from snipers in the hedgerows getting closer and closer; he was out in the open and a stationary target. All of a sudden he felt a sharp blow to his throat which spun him entirely around. He clapped his hand to his neck and took it away: no blood! Had it been a grazing shot, which had hit his collar? He turned to the man next to him and asked him if he could see anything wrong, and the man reached up and pulled out the bullet lodged in his neck: a spent shot fired from a great distance away. It was only then that he began to bleed profusely, and he applied a field dressing. Going to the rear under artillery fire, he happened on his divisional commander, who noticed the flowing blood and asked about the circumstances of his wounding. When Wegener recounted his company's charge beyond the village, the officer clapped him admiringly on the shoulder and wished him a good recovery.[43] With the field hospitals still overflowing with critically wounded, Wegener was sent to the rear for medical treatment and recuperation with a wound that would not even have taken him out of the trenches in 1917 or 1918. In transit to Germany, Wegener learned that Else and the baby were in Hamburg with her parents, and that is where he went to recuperate. Else recalled that in Hamburg in the first days, even though he was away from the front, the front was still with him. For a week or so he was disoriented and could not find his way about in the house in the early morning. While his mental condition soon improved, his neck wound, on the contrary, became badly infected and was a long time healing.[44]

By the middle of November he was sufficiently healed that he could be ordered back to Charlottenburg, where he was put to work training a replacement battalion. He was a valuable asset in this regard, as he was a senior veteran with years of service and had actual combat experience, including leading the assault that had led to his wounding. Wegener found he could perform these training duties without too much strain while the troops were still in barracks, but he discovered, during field maneuvers, that climbing even a modest slope left him breathless and started the heart palpitations once more.[45]

In the middle of December he was examined to determine the cause of his ailment; this was the much-postponed evaluation recommended to him by the surgeon in the field. A cardiologist, Dr. Friedlander, examined him and provided the following medical certificate: "Lieut. Herr Dr. Wegener of the III Queen Elizabeth Grenadier Guards Regiment suffers from a chronic heart ailment that was diagnosed some years ago. It has, however, been apparently aggravated by military service. Lieut. Wegener is permanently unfit for field and garrison duty. An official medical certificate will follow in due course."[46] He was to be sent home, not returned to combat, with an initial recuperation leave of six months.

Wegener returned to Marburg with his wife and child in early January 1915 to begin his life anew, once again. One says this without irony: he was a man of many beginnings and endings, of constant excursions and expeditions. This particular return was filled with relief and a simple gratitude just to be alive. He could relax in his sense of duty performed. He had been twice wounded and was now a decorated combat veteran: he had received the Iron Cross 2nd Class. He had acquitted himself honorably and was free, both officially and emotionally, to return to his original master: his scientific work.

While he was still convalescing in Hamburg in October and early November, Köppen had urged him to produce a reprise of recent advances in meteorology, similar to the one he had published in 1911. This was Köppen's way of pulling him mentally and spiritually away from the war and back to civilian life. He could guarantee publication in the *Annalen der Hydrographie und Maritimen Meteorologie*, the monthly periodical produced under the auspices of the German Marine Observatory in Hamburg and thus a publication over which Köppen had complete control.

Wegener undertook this project and completed it in early January on his return to Marburg. It was much shorter than his very ambitious review of 1911 and has the feel of an effort designed to reorient him to areas in which he would continue to work. This was telegraphed by the title he chose: "New Researches from the Field of Meteorology and Geophysics" (treating the two subjects as one), rather than just "New Researches in Meteorology." In addition to short summaries of recent work in aerology, studies of the upper layers of the atmosphere, and studies of the aurora, he added, somewhat surprisingly, a brief section titled "Displacement of the Continents."[47]

Wegener had typically approached such reviews with the brash assurance that wherever he was headed, meteorology must follow. This had been evident in his inaugural lecture in Marburg in 1908, evident in the review he produced in 1911 for Abderhalden, and detected by Exner in the *Thermodynamik der Atmosphäre*, arousing the latter's antagonism. As Wegener reflected in early 1915 on his career, he could see that this brusque confidence and bravado had been an expensive strategy for him. Exner had proclaimed him self-involved and overconfident. The assembled geologists at the Geological Association meeting in Frankfurt in 1912 had responded with scorn and outrage to his inaugural lecture on the displacement of continents. These responses had been humorous then; they were not funny now.

The tone of this new review of "recent developments" was, in contrast to his previous efforts, notably subdued. It contained no marching orders for others and no particular focus on his own work. In the initial section on the advances in aerology, all emphasis was on the differentiation of the troposphere and stratosphere, and he directed attention to the work of his predecessors: Aßmann, Teisserenc de Bort, Köppen, Hann, and Hergesell. He included reference to North American observers and ended with an approving summary of the method of approach of Vilhelm Bjerknes, who was attempting to work from synoptic charts to calculate the weather.[48]

In reviewing work on the upper atmosphere, especially the idea of the chemical partition of the different layers above the troposphere, he was even more tentative and generous. He spoke of how long it had taken to understand, by indirect methods, the characteristics of the different gases in the different layers. Here he mentioned laboratory work, observations of meteors, noctilucent clouds, and the explosion of Krakatoa. He reported the speculations of others that the "outer zone of audibility" (a problem that had begun to interest him) suggested the possibility that atmospheric layers of different chemical composition might reflect or absorb sound in different ways and thus dictate the distance traveled by sound waves. Of additional interest was the change in color of large meteors, which might indicate not the spectrum of the elements composing them but the gaseous spectrum of the layer through which they were passing.[49]

Emblematic of this new cautiousness on his part was his presentation of the status of his hypothetical element: geocoronium. In 1911, in his textbook of thermodynamics, he had noted the hypothetical character of this outermost layer of the atmosphere, but then he had gone on to describe it in great detail and treat it as an accomplished fact awaiting only final confirmation. Now, in presenting his own work in a review, he took the following line: "An even greater level of uncertainty attends the hypothesis, which I have proposed, that outside the Hydrogen shell of the atmosphere there is a zone composed of an unknown even lighter gas: Geocoronium."[50] He noted that the spectral line of this gas was close to that of the Sun and that it already had a place in Mendeleev's periodic table, but that its existence was by no means certain.

At the close of this ten-page summary of recent work he turned to a discussion of his own work on the displacement of continents. Here the tone was positively self-deprecatory: "The theory of the displacement of continents proposed by Alfred Wegener in 1912 has awakened increasing interest, though one notes the majority of scientists are today divided between those who view it unfavorably, and those who (at best) are withholding their judgment."[51]

The Origin of Continents and Oceans

Wegener's reprise of his hypothesis of continental displacements and his rueful assessment of its reception contained as well the (bad) news that the measurements of longitude differences between Europe and North America, which were to have put the hypothesis of displacement on a sound empirical footing, had come to an end with the outbreak of the war.[52] The Geodetic Institute in Potsdam had pledged 10,000 marks toward these measurements at the beginning of 1914, with the expectation that this amount would be matched by the U.S. Coast and Geodetic Survey. The measurements, begun in July 1914 on the European side (in the Azores) by Carl Albrecht himself, ended abruptly in early August when cable communication between Germany and North America was severed.[53]

This suspension was a severe blow to Wegener's hopes for the hypothesis. Measurement of longitude using "lunars" was woefully inexact and could never substitute for electrical signals transmitted along a cable.[54] His hypothesis of continental displacements postulated motions of Greenland to the west of between 14 and 28 meters (46–92 feet) per year. The average of the existing lunar measurements, under the best interpretation, gave a maximum shift to the west of 11 meters (36 feet) per year, clearly insufficient to demonstrate the hypothesis. He needed more accurate measurements, and these were not to be had. Since there was no way to obtain Greenland measurements via cable, the planned measurement of longitude shifts between Europe and North America had been his best chance for an empirical test.

Facing the problem head-on, Wegener turned in January 1915 to the hypothesis of continental displacements. He had deferred, in the spring of 1914, the offer of a book in the series edited by Richard Aßmann, in which he might have elaborated his idea and the evidence for it. Now, with the measurement protocol abandoned, he had either to strengthen his claims for his hypothesis or to abandon it until the end of the war and the resumption of measurements. He had written nothing on the topic, nor worked systematically on it, for nearly three years. While a few geologists had spoken against it, no one but himself had spoken for it. If the idea were to survive (let alone gain support), he would have to respond to the criticisms against it and to supplement his original data with additional arguments. The offer of a book-length treatment, still open, provided the necessary opportunity.

Were it not for the outbreak of the war and the suspension of the measurement program, it is quite possible he would never have re-engaged the hypothesis of continental displacements at all. He had pinned all his hopes on telegraphic time signal measurements, expecting them to yield rates for the westward drift of North America of about 4 meters (13 feet) per year. This figure was close to the existing but inexact lunar measurements available in 1915, but it is *200 times greater* than estimates current at the beginning of the twenty-first century; modern spreading rates are in the range of a few centimeters per year. Thus, no motions within two orders of magnitude of his prediction would have been detected in 1915 and 1916, and Wegener would have witnessed the (apparent) empirical disconfirmation of his hypothesis. He would then perhaps have concentrated for the rest of his professional life on working in the realm of atmospheric physics and pursuing polar meteorology. Had that been the case, he would be known today only to meteorologists interested in surfaces of discontinuity and the theory of precipitation, as well as to historians of Arctic exploration.

In returning to the question of continental displacements, Wegener was aware that geologists had found his hypothesis not so much wrong as repugnant. Why should this have been? He had found geodesists and geophysicists willing to consider the idea seriously, but not, other than Emanuel Kayser (the only geologist he seems to have known personally), the geologists. He needed a geological informant who could help him out of his impasse. Fortunately for Wegener, Emanuel Kayser had recently brought to Marburg, as his protégé, a young geologist named Hans Cloos (1885–1951). Cloos had taken his degree at the University of Freiburg in 1910 and had spent the past few years abroad. He had worked for two years in Java, looking for oil on behalf of the American

Petroleum Company.[55] Traveling to and from Indonesia, he had been through the Suez Canal and to the Red Sea. He had also circumnavigated Africa and spent some time in German South-West Africa (today Namibia).[56]

Cloos had not intended to begin an academic career; he was an exploration geologist and traveler stranded by the war in Marburg, awaiting military assignment. Cloos remembered his first encounter with Wegener. He said that Wegener had just "shown up" at his office one day and announced to him that he needed a guide to the geological literature, someone who would be willing to provide geological information and concepts for him. Cloos remembered being struck by his pleasant countenance and especially by his piercing gray-blue eyes. Wegener then sketched for him an "extraordinarily strange train of thought concerning the structure of the earth" and said that, as a physicist, he needed help in exploring it. Cloos took an instant liking to Wegener and said that he found Wegener likable and his ideas strange in about equal measure.[57]

This initial encounter was the beginning of an intense two-month intellectual relationship, lasting throughout the rest of January, February, and early March 1915; it was also the beginning of a friendship.[58] Wegener was grateful for Cloos's guidance through the complexities of the geological literature, but the heart of their collaboration was the stimulating daily argument and conversation. It was here that Wegener learned for the first time how geologists thought.[59] Cloos experienced their interchanges as a spirited exchange of views, in which he would raise geological objections to various aspects of Wegener's theory, and Wegener would respond with physical arguments. Cloos was sympathetic, although not convinced—a stance he would remain in for more than twenty years. Meanwhile, he saw that Wegener's principal achievement, already evident in the first version of the theory, was that he had "placed on a solid scientific foundation an easily comprehended, sensationally provocative intellectual construct."[60]

Cloos, in his debates with Wegener, provided the latter with valuable information about how geologists would view the various aspects of the hypothesis, outlining problems more evident to a geologist than to a geophysicist. Many parts of the resulting short book, the "first edition" of *Die Entstehung der Kontinente und Ozeane*, show the direct influence of Cloos's travels and field experience: the problem of the Red Sea, the need to address the folded mountain chains encircling the Pacific, and especially the question of the island arcs arrayed like festoons from the Kurile Islands south through Japan, the Philippines, and Indonesia. The latter were very fresh in Cloos's mind and experience, close on his return from the Dutch East Indies.

A few overarching generalizations can guide a closer examination of the book that resulted from this intense collaboration. Perhaps most striking is the extent to which Wegener's morphological descriptions of continents and oceans remained two-dimensional. This must sound self-contradictory. The hypothesis was, after all, one of floating continental blocks in a viscous fluid substratum; a number of diagrams in both the 1912 and 1915 versions are schematic cross sections of Earth's interior. Yet in this new version of the idea, Wegener's principal concern was to explain the various surface elements (relief and structure alike) as a consequence of lateral displacement of continental blocks, via coastline matching where it was evident, and via explanation of the lack of such matching where it could not be seen. There was no detailed concern (other than the existence of mountains) with the morphology of continental surfaces or their undersides.

Equally worthy of note is the way in which Wegener reorganized his materials in the 1915 version. In 1912 Wegener had divided up his presentation in terms of the disciplinary source of his arguments: "geophysical arguments" and "geological arguments," the latter including discussions of paleontology. In 1915 the twelve chapters of this short (ninety-eight-page) book were focused on specific geological and geophysical *questions* tied to specific *regions* of Earth. That is to say, Wegener was shifting his argument into a format (with the help of Cloos) that would be intuitively easier for geologists to grasp. Moreover, in service of this same aim, the terseness of the 1912 version was abandoned in favor of a more colloquial and patient (if still spirited) advocacy.

Finally, the decision to reorganize the materials into geologically comprehensible sections and topical divisions led to a good deal of repetition, in which the same geological and geophysical arguments ended up under different chapter headings, some being repeated as many as four or five times. The book reads as if it had been written not so much as a single extended argument but as a series of essays addressing specific problems and concerns within the ambit of the hypothesis. This seems to be a residue of the way that Wegener and Cloos went at their discussions: "What about island arcs?" "What about the Red Sea?" "What about the question of New Guinea and Australia?"

Let us now turn to some of the details of Wegener's argument, emphasizing the things that were new in 1915. Immediately, the first chapter announced in its title a modified stance: "The Displacement Theory as Mediator [*Vermittelung*] between the Doctrine of Sunken Land Bridges, and the Doctrine of Permanent Oceans."[61] Wegener now placed his argument within the history of geology, rather than outside it, by noting that the dispute over continuities of flora and fauna on opposite sides of the deep ocean went back to Charles Lyell. Rather than criticizing the contraction theory (the theory of Sueß) directly, he invoked paleontologist Emil Böse (1868–1927)—who worked both in Europe and in North America—to declare that while the contraction theory was no longer fully accepted (as adequate), no theory had come forward to take its place and explain everything that it had explained.[62]

Wegener no longer spoke of sunken continents, but of sunken land bridges, alerted by Cloos that this was a more acceptable formulation. Addressing the difficulties faced by both the contraction theory and the theory of the permanence of oceans, he now acknowledged rhetorically the radical character of his proposed hypothesis: "These difficulties [with the two theories] will disappear completely if one will allow oneself to take but a single *momentous* step: if one accepts this that the continental platforms can undergo lateral displacements across the face of the earth."[63]

In the second chapter, Wegener took up the subject of isostasy and the flotation of the continents. Most of the documentation was the same as in 1912, as were most of the references, but more prominently featured were the arguments of Joseph Lukashevich (1863–1928). Lukashevich had published in 1911, in St. Petersburg, a small book entitled *Sur le méchanisme de l'écorce terrrestre et l'origine des continents*. Wegener had adopted in 1912 both Lukashevich's explanation and diagrammatic representation of floating continents and his formulation that on planetary bodies of a certain size "the molar forces prevail over the molecular forces."[64] Wegener changed that formulation slightly here, to say that for geological structures of a certain size mass forces (gravity) would prevail over molecular forces (solidity); for large dimensions Earth was plastic, whereas at small dimensions it was stably solid.[65] This was perhaps an unfortunate choice of words. Plasticity implies the tendency of the material to suffer permanent deformation. That is not really an issue where isostasy is concerned: the issue is the displacement of the viscous substratum of the Sima by large blocks of Sal which float on it, according to their differences in specific gravity. Indeed, the entire explanation of isostasy here was dense and difficult to follow even if one knows the material, which his audience probably did not.

As if sensing he needed to make another attempt to explain these difficult concepts, in the third chapter, "Salic Continents and Simatic Ocean Floors," Wegener undertook to discuss the mineralogical differentiation of salic (crust) continents and simatic (mantle) ocean floors. Here he reintroduced (from 1912) as evidence of this differentiation the bimodal elevation of Earth's surface, the flotation of continental blocks, the difference in specific gravity of continental platforms and ocean floors, the evidence for this gravitational difference measured at continental margins, and most of the other arguments from the 1912 paper—including every single one of the cross-sectional illustrations in the earlier publication. In retrospect, it seems that it would have made more logical sense for this chapter (on material differentiation based on density differences) to have come before the consideration of continental flotation presented in the preceding chapter.

Wegener approached the same material *yet again* in the fourth chapter, entitled "Plasticity of the Sal and Viscosity of the Sima." In 1912 Wegener had argued that because of the difference in the melt temperature of Sal and that of Sima, at a depth of some tens of kilometers, the Sal, with its higher melting point, would still be solid, whereas the Sima would already have lost its rigidity (*Riegheit*) and would be indefinitely deformable. In the 1915 version, Wegener no longer speaks of the lessened *rigidity* of the Sima, but of its greater *viscosity* (*Zähflüssigkeit*) and tendency to flow. Conversely, in discussing the continents as *plastic*, he is indicating that as the continents move through the Sima, they are capable of accumulating deformation (fold mountains on the leading margins), while the Sima, though more dense than the Sal, is already at a temperature where its fluidity prevents it from maintaining its deformation: it simply flows.

It had taken Wegener four chapters to work his way through the geophysical material: flotation, gravitational differentiation based on density, and differential deformation and flow as a consequence of differences in melting points at depth. Attempting to reconfigure this material in a way that might prove convincing to geologists had turned out to be very difficult for him; understanding it would be a daunting problem for geologists as well. Other than a few textbook writers and theorists working at the highest level of generality, geologists in 1915 rarely thought on a continental scale. They were rarely interested in what went on deep within Earth. In 1915 only about one-eighth of the continental surfaces had been geologically mapped, and almost all of the workaday effort of geologists was involved in extending this work. Moreover, Eduard Sueß had declared in the final volume of his great synthesis that he did not even believe in isostasy; for those who nevertheless "believed in it," the American version of isostasy differed from the British, and both of these differed in important respects from French and German work on the same subject. The ease that Wegener had in understanding and in making up his mind was confronted here by a daunting blend (in his geological audience) of ignorance, agnosticism, and indifference to physical theory.

Wegener was now more than a third of the way through the book and had yet to address a new question or say anything much that was not already in the version of 1912. Only in chapter 5, "Mountain Ranges, Island Arcs, and Deep-Sea Trenches," do we see new perspectives and new material, and here (probably more than anywhere else in the book) we see the influence of Hans Cloos. Just back from the Dutch East Indies (Indonesia), Cloos had much to say about the complex relations of the mountain ranges along the coast of Asia and the associated convex archipelagos running parallel to the coast, with deep-sea trenches on their outer sides.

The mountains of Asia were of great interest to European geologists but had barely been reconnoitered. One of the most absorbing parts of Edward Sueß's compendium and treatise had been his stirring descriptions of the complex folding of these huge and often sinuous ranges. The American geologist Bailey Willis (1857–1949) had undertaken an extensive reconnaissance in China in 1903–1904, reporting (in English) for the first time on such great ranges as the T'ien Shan.[66] These observations supplemented those of the great German geographer Ferdinand von Richthofen (1833–1905), who had been active in China, as well as in Java, Burma, and Japan, in the later 1860s and early 1870s, and who had developed a theory of these coastal ranges in China and of the associated island arcs.

Using Cloos's knowledge of the region, Wegener was emboldened to pose a novel hypothesis for these unusual structural elements, in the context of his general picture of continental displacements. It was well known that the coastal mountains of Asia exhibit compressional folding parallel to the coast, and that these compressional folds also appear to have been compressed in their long (north-south) axes, as if subsequently the folding had rotated 90° from its original direction. Wegener proposed that the folded mountain ranges of the coast dated from the time (in the Tertiary) when Asia had been moving *toward* the Pacific. Subsequently, the continental motion had shifted, this time pushing Asia from north to south, and subjecting the existing mountain ranges to a pressure that forced them into sinuous arcs. When the direction of displacement changed again, with the Asian continent moving once more to the west, the outermost components of the ranges had detached themselves. Wegener used the analogy of what happens when you squeeze a deck of playing cards between the thumb and forefinger: as the pressure increases, eventually the outermost card will fly away, forming a convex bulge.

There were, of course, as many unexplained as explained elements in this construction. Wegener had no idea why the inner, continent-facing concave sides of these archipelagoes should be ringed with volcanoes, while their outer sides were composed of steeply tilted Tertiary sediments, except as a logical consequence of tensional rifting; the Tertiary sediments on the Pacific side of the islands would bespeak their continental origins. As for the ocean deeps parallel to the outer side of the islands, Wegener had no explanation or even conjecture. That he mentioned these structures at all was due to Cloos: the young geologist reaffirmed and strengthened Wegener's own instinct that if you can see a problem that your theory cannot explain, it is much better for you to mention it than not mention it, that is, not wait for an opponent to point it out.[67]

When we turn to the next chapter, "The Mechanics of Displacement," it becomes clear that everything to this point has been a "backstory" to prepare the reader to consider the large-scale regularities of the displacement theory. It was an uphill struggle to have proceeded in this way, to have to teach his audience the basic conceptual vocabulary and physical foundation for his hypothesis in addition to the hypothesis itself. Now the laborious work of didactic presentation of physical concepts could give way to the broad sweep of how the entire process would work, if it were true of the world.

Wegener noted with rueful humor that the one aspect of his theory that critics had found impossible to oppose was the direct connection between the lateral compression of Earth's crust and the origin of mountain ranges. No other question dominated thinking about Earth's crust more than the question of mountain ranges; their undeniably compressive structure had been the greatest argument in favor of the theory of a contracting Earth, a theory that Wegener characterized at every opportunity as physically impossible.

Of all the mountain ranges on Earth, the most intensely studied by European geologists were those in their own backyard, the component ranges of the great Alpine system running east to west across the middle of Europe. Their manifold complexities had absorbed the ingenuity and analytical energy of European geologists for 150 years by the time of Wegener's writing. It was therefore a monument of good sense, as well as a great tribute to the sagacity of Hans Cloos, that Wegener brought up the question of the Alps, only to announce that "on account of their great complexity, it would not be possible to go into the [Alpine] question further in this context."[68] Rather, he urged, it is a much better test of the theory to see what happens in the simpler case: what happens when the Indian peninsula was compressed against Asia, producing the Himalayas, or the smaller case of Spain colliding with Europe and producing the Spanish Pyrenees. This was, of course, the physicist's approach: pattern first, anomalies and difficulties later. Abstract from the particular phenomena to find the underlying rule. The first apparent rule was that, in general, peninsulas (e.g., Baja California, Spain, Arabia, and the southern half of Greenland) are more compressed than the rest of the continental blocks from which they have become separated.[69]

The theme of the origin of mountains by compression was here for Wegener the through line uniting all aspects of the displacement theory, just as it had been the through line uniting all aspects of the contraction theory. And here, without any special announcement, Wegener undertook the most dramatic expansion of his theory over the 1912 version. His argument proceeds thus. In considering the characteristic phenomena associated with the free displacement of continental blocks, it is best to think of them as if they were oceanic islands; this is as it was in the first version. Their motions, however, can now be divided into two groups: one in which displacement through or over the Sima actually takes place, and one in which a block is passively carried along by a current flowing through the Sima (*Simaströmung*). The larger blocks—those the size of South America, for instance—are more likely to travel against the Sima, producing a range of characteristic phenomena. Smaller blocks, such as Madagascar and India, are carried passively by these currents in the Sima. Moreover, these currents flowing in the Sima are geologically active even when not carrying continental blocks. For instance, the compressional folding of most of the coast of East Asia is the result of the underthrusting of the continental margin by a current of mantle material (Sima) flowing to the north.

The dynamism and fluidity of the displacement theory in this version (1915) stand in marked contrast to Wegener's earlier and later pictures of continental displacement. Once again, one sees the powerful influence of Hans Cloos in urging Wegener to consider the idea of streaming in the Sima as a component of the theory.[70] The notion of such streams in the subcrust had been hypothesized by an Austrian geologist, Otto Ampferer (1875–1947), in 1906, as an important part of the folding of the Alps.[71] The American geologist Bailey Willis had proposed that the marginal ranges of Asia had been created by the underthrusting of the continent by the mobile ocean floor moving toward the continent in a process he described as "gravitational creep."[72] What Wegener was proposing was quite different from both Ampferer and Willis in its scope and details, but the inclusion of the idea, in general, was clearly designed to bring the displacement hypothesis closer to current geological conceptions and to smooth the way for its serious consideration.

On this highly mobile globe, the larger continents move across and through the subcrust, while fragments of former continental masses are carried by a variety of currents moving in this same hot and fluid substrate; all the characteristic phenomena of the continental outlines and margins are generated from the motions themselves. On the leading edge of the largest continental blocks, mountains are thrown upward, because the plastic crust can hold the deformation caused by the resistance of the viscous fluid Sima, while the latter material cannot permanently deform. As the continents move, the subcrust plows underneath them, emerging on the other side. Wegener's favorite example of this ensemble of phenomena was the Drake Passage between South America and Australia. The fragmentary archipelagoes of Tierra del Fuego on the one hand and Graham Land on the other are pieces of the leading coast of South America and Antarctica torn away by the resistance of the Sima. The South Orkney Islands and South Georgia and the South Sandwich Islands are fragments torn away from Antarctica; the Falkland Islands, on the east coast of South America, represent former sections of the west coast of South America torn away and carried in the Sima stream completely underneath the continental tip, only to emerge at the surface again on the other side.[73]

Wegener concluded his discussion of the mechanics of displacement by noting that all questions of displacement had to be considered relatively. It was, for instance, not absolutely certain that South America was drifting and pushing against the Sima of the Pacific; it was equally possible that the Pacific Sima was streaming against the continent, although not capable of pushing it along, as might have happened with a smaller continental fragment. Moreover, the morphology of continental margins, some of which have mountains on more than one side, indicates that there must have been, in Earth's history, back-and-forth motions of continental blocks, such that the long-term development of the continental crust is not a matter of one-sided motion, with a single leading edge always susceptible to deformation. Rather, the long-term result of continental motions was the emergence of ever-thicker continental blocks as Earth's surface moved from a time of universal oceanic cover to the emergence of higher and higher continental platforms; most of this elevation was concealed as deep "continental roots" caused by the lateral pressure of the fluid but still viscous Sima against the stronger but still plastically deformable continental blocks.

The choice of the word "mechanics" to describe all this was technically correct, but it was bound to be misleading to the audience and thus unfortunate. By "mechanics" Wegener means only what any physicist would mean, a comprehensive description of the interaction of the parts of the system. But "mechanics," or "mechanism," is often an abbreviation in geological discourse for "causal mechanism," that which brings something about. Thus, in modern geotectonics, the mechanism of continental displacement is the motion of tectonic plates, and the mechanism of plate motion is supposed to be convection in the mantle (what Wegener would have called Sima).

When encountering Wegener's succeeding chapter entitled "Likely Causes of Displacement," it becomes evident why he made this choice to speak of mechanics in one chapter and causes in another. He considered it in 1915, as he did in 1912, too early to talk about the causes of displacement.[74] Yet because several of his critics had identified the lack of a presumptive cause (as he reports here) as a drawback in the presentation of his theory, he reluctantly decided to remove this objection by suggesting several possible presumptive causes without choosing among them. By separating out mechanics from causes, making these two things distinct, he hoped to blunt criticism of the theory. He mentioned here in passing the possibility of tides within Earth, some magnetic action because of the discrepancy between the magnetic pole and the pole of rotation, and the possibility of a so-called pole-fleeing force in which masses close to the poles would, by Earth's rotation, be forced toward the equator because their center of buoyancy was higher than their center of gravity. None of these were demonstrated or demonstrable.

He then added a few highly interesting speculations that the cause of such motions on Earth might be "cosmic" since observational astronomy permits us to see phenomena on the surfaces of Mars and Jupiter which suggest (and which our observations of sunspots show us) that planetary surfaces and even the surface of the Sun are capable of motion. For planets above a certain size—perhaps Mars, and certainly Jupiter—displacement of surface elements by currents beneath the surface might be a real possibility.[75] He concluded by pointing out (and the distinction is important) that we might be able to establish the reality of continental motions by measuring them astronomically, and yet we might *never* know their cause.[76]

Notably, one well-documented motion on Earth which Wegener refused to invoke as a cause of continental displacement was the wandering of Earth's pole of rotation, which Wegener took to be independently established by paleontological evidence. Wegener argued, in this edition as in 1912, that displacements of the pole had to be invoked to explain the distribution of the Permo-Carboniferous glaciation in the Southern Hemisphere, as well as the fossil record of Tertiary land plants in the Northern Hemisphere. The causal connection between continental displacement and polar wandering (if there were any) must run the other direction: the displacement of large continental fragments, as well as the significant transfers of mass across the surface of Earth, would be most likely causes of pole wander, as the axis of rotation moved to correspond to the axis of inertia.[77]

For all this attention to geophysics, mechanics, possible causes, and polar wander, Wegener still believed that the fundamental evidence for the existence (as opposed to the plausibility and possibility) of continental displacements was geological and paleontological. The matching geological formations across thousands of miles of abyssal oceans, the fossils of closely related shallow-water and terrestrial species on different sides of the Atlantic, and the provenance of the Gondwanaland flora in widely separated Southern Hemisphere continents (along with evidence there of massive continental glaciation in the Carboniferous and Permian) were the facts that pointed directly to continental displacement. He announced, at the beginning of his chapter entitled "The Atlantic Ocean," that "this and the following chapter [on Gondwanaland] contain the most important evidence for the displacement theory."[78]

Cloos had been helpful at every stage in the recasting and expansion of the argument and increasing Wegener's sensitivity to addressing major anomalies, in hopes of winning a geological audience. But it was in the areas of geology and paleontology that Wegener had sought out his guidance in the first place. It was precisely the most recent literature produced since 1912 and the literature on regions of the world where Cloos had worked (or had plans to work) that figured in the revision of the chapters on transatlantic continuities and on the reconstruction of Gondwanaland. From Cloos he learned of very recent work, presented at the International Geological Congress in

Toronto in 1914 and not yet published, establishing very close connections between the Sierra south of Buenos Aires and the mountains of the cape in South Africa.[79]

Cloos also introduced Wegener to a geological resource of which he had been previously unaware, the *Handbuch der Regionalen Geologie*, an extensive international collaboration that had just begun to appear in many large volumes and would appear continuously for another decade. In these works, all the stratigraphic, paleontological, economic geology and petrology, as well as the geological structure and morphology of all regions of the world, were to be gathered together under the editorship of recognized experts from every major geological community in Europe and North America.

Of most immediate use for Wegener's theory were the volumes by Paul Lemoine on West Africa (1913) and Madagascar (1911) and those by Patrick Marshall on New Zealand (1911, 1912).[80] These were very helpful to Wegener in strengthening his argument for the former connection of Africa to South America, the connection of Madagascar to the coast of East Africa, and the bewildering geological connections stretching from Indonesia to New Guinea and Australia and south to New Zealand. The material on Australia, New Zealand, and New Guinea, which constitutes the major addition to Wegener's consideration of Gondwanaland, was provided directly by Cloos, and his knowledge of these areas contained much information not yet published, and some known to Cloos only from his field experience and conversations with other workers in this region.

It was also Cloos who urged Wegener to be cautious about his advocacy here, even if he could not blunt Wegener's editorializing entirely. After completing an extended argument about the former unity of Australasia, Wegener concluded, "The uncertainty of such reconstructions cannot be eliminated by the citation of geological sources."[81] Nevertheless, he continued, not only does the bulk of the material harmonize with the displacement theory, but also it harmonizes so much better with the displacement theory than with the theory of sunken land bridges that it provides the most striking evidence for the former.[82]

Throughout the process of revision and extension of his theory, Wegener was in good spirits and surprised by the speed with which he could work. In early February, one month into the project, he had written to Köppen, "With the continental displacements I am making good and unexpected progress. This is unbelievably fascinating work. I already have a draft of about 50 pages, though this is only about half [of what I plan]."[83] In the same letter, he praised Cloos for his help and also passed along another piece of information: "The geologist Dacqué (Munich) has sent me the page proofs of his book on paleogeography, in which he takes a stand in favor of my theory."[84]

The unexpected support of Munich paleontologist Edgar Dacqué (1878–1945) had much to do with Wegener's elevated mood. Dacqué had taken his PhD with the great paleontologist Karl von Zittel and had the previous year (1914) been named extraordinary (i.e., associate) professor of paleontology at the University of Munich, in conjunction with his appointment to superintend the paleontological collections of the Bavarian State Museum. The proofs he sent to Wegener were those of his major work: *Grundlagen und Methoden der Paläogeographie* (Fundamentals and methods of paleogeography).[85] Dacqué was very explicit in the introduction that this was a treatise not *in* paleogeography but *on* paleogeography.[86]

Like Wegener's work, Dacqué's was theoretically innovative and comprehensive, addressing directly such questions as continental motion, pole wander, and "continental permanence." It was a massive and beautifully produced book of more than

500 pages, with attractive and well-produced graphics and maps and an exhaustive index. Dacqué was at this time a rising star in his field. As had Wegener, he had written a PhD thesis with a historical slant: a history of evolutionary theory from antiquity down to the present.[87]

Dacqué had read Wegener's 1912 paper and been immediately persuaded by the power of this "profound synthesis."[88] He provided in his book extensive summaries of various aspects of Wegener's hypothesis and found Wegener's synthesis the best existing approach to the question of faunal and floral distributions, especially for Gondwanaland, more so than those of the great geologists of the previous generation: Sueß, Neumayer, and others.[89] In addressing the question of continental and oceanic permanence, he made a novel interpretation of Wegener's findings. If it was clear that the continents were not permanent (in place, outline, or size), Wegener's ideas gave a new meaning to the concept of oceanic permanence: the deep sea was always there because it was a distinct Earth shell, whether any given part of it happened to be covered by continental fragments at any one time.[90]

Dacqué's book placed Wegener on an equal footing in the citation cascade with Sueß, Émile Haug (1861–1927), Koken, Rudzki, Neumayer, Krümmel, Bailey Willis, and Albrecht Penck. It treated his ideas with the utmost seriousness and made a point of their superiority to other dynamic theories of crustal motion, especially the novel and influential ideas of Damian Kreichgauer (1859–1940). Kreichgauer, in *Die Äquatorfrage in der Geologie* (1902), had produced paleogeographic maps for every geological period, mapping (with climate and fossil evidence) the motion of the equator across the face of Earth throughout geologic history, and demonstrating that polar motion was the best way to account for all the faunal and floral distributions in each period.[91] His qualitative theory of how this took place was based on the notion that the fluid Earth produced an upward pressure on the outer crust, accounting for volcanic activity on the one hand and making the crust highly mobile on the other. It was evident to Dacqué, as it had been to Wegener, that this was in complete contradiction of the principle of isostasy, a concept with which Kreichgauer seems to have been unfamiliar.[92] There was much for Wegener to absorb in Dacqué's book, too much for the time he had allotted to the writing of this short book. The impact was there, however, even though Wegener had time to make only a few references to it in his manuscript.

If the work of writing the book had been fascinating, the results were less than satisfying. Wegener noted in the preface, written in early March 1915, that the "crucial experiment" of longitude measurements had been abandoned with the outbreak of war, and that he had undertaken this revision during a convalescent leave after being wounded at the front. Because of the disadvantageous circumstances in which it was written, he remarked, "the presentation is marred by many imperfections, that might have been avoided in more peaceful times."[93] Nevertheless, he was able to bring together some new evidence and also to give a more detailed formulation of the number of points "that in the first presentation [1912] were misunderstood largely as a consequence of their brevity."[94]

Given that in March only half of his six-month convalescent leave was over, we might inquire why he did not spend more time polishing and improving the work if indeed he found the results so unsatisfactory. The answer to this lies in the simple fact that continental displacements, for all their fascination, were a sideline to his own work in atmospheric physics, work that he had been pursuing concurrently with the writing of this book. Even before he had begun drafting the new version of his argument concerning displacements, he had begun working extensively on the problem of lunar tides in the atmosphere.

The concurrent work Wegener pursued in the spring of 1915 on lunar tides was quite different from the qualitative work on displacements and refreshingly complicated. He wrote to Köppen on 11 January that he had begun working on the problem of lunar tides they had discussed during his convalescence in Hamburg the previous October and November. He wrote again the very next day, to say that he was really getting deeply into the problem.[95] Wegener often wrote a note to Köppen merely to share his excitement at the way his work was going; this feeling about scientific work was a strong bond between the two men.

The problem had to do with inconsistent observations of lunar atmospheric tides taken at different latitudes at different times. At low latitudes, near the equator, there was a clear pattern of semidiurnal atmospheric tides matching (in phase with) the oceanic tides at those locations. Such tides in the atmosphere could be measured in wavelike deviations of some hundredths of a millimeter in barometer readings. This semidiurnal tidal fluctuation was commonly observed in barometers on board ships, for ships must rise and fall above the surface of the solid Earth as the tidal bulge lifts the water beneath them and allows it to recede. Yet, interestingly, such fluctuations could also be observed on tropical ocean islands and on continental surfaces at low latitudes, and such fluctuations could not be attributed to tidal bulging of the solid Earth; they had to record actual tidal action on the atmosphere itself. This was all fine, consistent, easily understood. The problem was that at higher latitudes such as Hamburg and Berlin, the semidiurnal fluctuations were sometimes present, sometimes offset from the ocean tides by some length of time, and sometimes entirely absent, leaving no clear barometric signature.

Wegener wrote on 21 January to Köppen that he was making progress on lunar tides, although he said, "It is moving in a somewhat different direction than I had planned."[96] He had analyzed the Moon's motion in relation to atmospheric tides in a series of published measurements from the 1870s and 1880s, taken at both high and low latitudes. He considered carefully the relationship of the atmospheric tides to the oceanic tides and the variations brought about when the Moon was closest to Earth (perigee) and when it was farthest from Earth (apogee). None of these details completely resolved the anomalous offsets.

The answer, he found, lay in the question of the Moon's declination. This refers to the altitude above the horizon of the Moon when it reaches its culmination, its highest point in the sky. Because of the tilt of Earth's axis at 23.5°, the Moon will reach a maximum and minimum altitude above the local horizon two weeks apart. But because the Moon's axis is *also* tilted at 5° away from the orbital plane of Earth around the Sun, in a long and complicated cycle this variation in declination can be as small as 18.5° or as large as 28.5°. Wegener observed that the long-term variation in the Moon's declination could explain the apparent anomalies at high latitudes and the absence of such apparent variation at very low latitudes.[97] He was extremely pleased with this result, and it was an outstanding version of his ability to find in published data the signature of some physical phenomenon not noticed by the authors. By late February he had checked his result by extending the time series back into the middle of the nineteenth century, and he put the paper into publishable form.[98]

As if this were not enough, he was working on another problem as well. Wegener's old boss and his constant champion, Richard Aßmann, at seventy not in the best of health, had retired from Lindenberg in 1914. His students and colleagues had quickly organized a celebratory volume to be published as a supplement to *Das Wetter*, a journal he had helped to found. Everyone from "the Berlin school" was to be in it, including Köppen, Richarz, Kurt, Hergesell, and Berson. This, of course, meant that Wegener had to produce something, and he turned to a topic he had taken up in his review of recent work in meteorology and geophysics, the attempt to verify the chemical differentiation of the upper atmosphere at different altitudes by the change in color of large meteors.

He wrote to Köppen in late February that he had finished his contribution and gotten a nice result although "it had not cost much work."[99] Once again Wegener was mining data available in the meteorological literature of the nineteenth century and applying it to a novel aim. Observations of the color of large meteors taken in the nineteenth century had not been directed at the spectroscopic analysis of the layers of the upper atmosphere but were clearly useful for such purposes. Where the meteor and its color had been observed by more than one observer at distant points from one another, there was a possibility of determining the altitude by knowing the time. Wegener's interpretation of this phenomenon was that the color of the meteor trail was determined not by the composition of the meteor itself but by the atmospheric gases through which it was passing and which it was igniting in its passage. We may recall that Wegener believed the atmosphere to be chemically differentiated with sharp boundaries between an oxygen atmosphere, a nitrogen atmosphere, a hydrogen atmosphere, and eventually a geocoronium atmosphere.[100]

All this writing and thinking in the spring of 1915 transpired in Marburg, in Alfred and Else's apartment, with their new child, Hilde, and, fortunately for them, the assistance of a hired girl, their first household help. Money was a problem as always. Wegener had no announced lectures and thus no lecture fees, although these were never a large part of their income. In October 1914 his stipend of 1,500 marks per year had been discontinued.[101] Now that he was back in Marburg, the stipend was to be renewed, on the condition that he fill the role of "assistant" in the Physical Institute. This was something arranged for him by Richarz, knowing his financial difficulties. Wegener begrudged the time, 9:00 a.m. to 1:00 p.m. and 3:00 to 6:00 p.m. each day, but had not been able to say no to Richarz, who had lost both of his actual assistants to military service.[102] The institute work turned out not to be nearly as burdensome as he had imagined, and in early April there was a visit from Vilhelm and Honoria Bjerknes.[103] While technically a neutral (Norwegian), Bjerknes could not help but be involved in war work. The German navy had always been deeply involved in meteorology, but now there was an air service, and with the army preparing for large-scale use of poison gas, knowledge of prevailing winds would be a matter of life and death. Indeed, Germany would make its first major gas attack on Allied positions later that month, in the Battle of Ypres. Leipzig's (and Bjerknes's) meteorological institute became a center of military meteorology.[104]

However different their work, Bjerknes and Wegener saw eye to eye: both considered dynamics and thermodynamics parallel but separate disciplines within meteorology. Both were deeply focused on surfaces of discontinuity: Wegener in the vertical, Bjerknes in the horizontal. During the Marburg visit, they discussed the "squall line theory" (*Böentheorie*) of Wilhelm Schmidt (1883–1936). A squall line refers to what we would now call a cold front, and Schmidt had done laboratory experiments on the behavior of adjacent masses of different temperatures, with the idea that the thermal difference was a cause of the instability. Wegener had seen in Greenland squalls without thermal difference, as well as huge thermal differences that did not produce squalls. If there was a relationship, it was not causal. Bjerknes was grateful for Wegener's critique and for knowledge of Schmidt's work, for which Wegener provided references.[105] During the visit, Else was not feeling well. Hilde had been feverish for more than a week, and Else became feverish herself and ultimately suffered a miscarriage—she was in the first or second month of her second pregnancy. The Bjerkneses moved to a hotel and came over for meals—Honoria Bjerknes was able to provide advice and solace for Else, as she herself had miscarried earlier in her marriage.[106] While Honoria spent time with Else, Alfred and Vilhelm went for walks or retired to Alfred's study.

Both men were deeply involved in scientific aeronautics, and with the advent of war, suddenly knowledge of winds aloft was of pressing interest: in January of 1915, Germany had begun bombing raids on England, employing zeppelins—large and navigable rigid airships filled with hydrogen. In these first raids, the zeppelins encountered unexpected upper-level winds and were blown far off course. As winter turned to spring, the kaiser wished to extend and expand these raids, but he had ruled historic buildings in the heart of London off-limits, restricting the raids to the docks on the Thames.

It is therefore not surprising that Alfred was requested, in early May 1915, to cut short his military leave and depart immediately for Brussels, where he was to make a circuit of the aerodromes for the zeppelin bombers and teach the pilots and crew aerial navigation. Night raids made it harder to shoot at the zeppelins, and Wegener's navigation technique for astronomical position finding in balloons using the sextant that he had designed back in 1906 was the only well-known information on the subject.[107] Wegener's doctor objected strenuously to the trip, arguing that the rapid changes in elevation in an airship could affect his heart, but Wegener agreed to go. Kurt, after all, had been in the newly created air force at the outbreak of war, and in April he had suddenly been made adjutant to Hugo Hergesell (1859–1938), head of the newly created Field Weather Service; doubtless the appeal had come directly from them, and Wegener spent the month of May in and around Brussels, teaching aerial navigation to zeppelin captains.[108]

Returning to Marburg in early June, he found that the doctor's reservations had been merited, as he began to experience palpitations when reclining.[109] Else's health was also not good. It had been decided that her miscarriage had not merited "an operation," but even before Alfred left for Brussels she had experienced chest pain diagnosed as angina. One possibility for her diagnosis is listeria, a food-borne illness and a frequent cause of miscarriage in early-term pregnancies; listeria is a bacterial infection that proceeds in some cases to endocarditis—with fever and chest pain.

Alfred decided that they needed a "cure" and took Else on a two-week vacation including a cruise down the Rhine and a walking tour in the Taunus, the low mountain range inside the state of Hesse; they were not far from Marburg but worlds away from their recent cares and fears.[110] By mid-June, when they returned home, she was feeling fit and refreshed, but Alfred's military physical in Berlin on 19 June (at the end of his official six-month leave) again revealed a pronounced "click" in his heart sound, which, together with the palpitations, indicated a prolapsed mitral valve. The battalion medical officer, Friedlander, declared him permanently unfit for combat but available for "temporary garrison duty."[111]

On 19 July, Alfred was reexamined by Dr. Hildebrand (his physician in Marburg), who found his symptoms diminished and pronounced him ready for full garrison duty within about four weeks.[112] Even before his July medical examination, he had written

Alfred, Else, and Hilde Wegener in June 1915, shortly before Alfred's return to active service in the German Army. Photo courtesy of the Alfred Wegener Institute, Bremerhaven.

to Köppen that he daily expected orders to return to his regiment in Berlin. The realization that he would return to active duty had hit him hard, and once again he had difficulty working. "The last month has been very unproductive for me scientifically."[113] He was unwilling to start work on the optical phenomena from Greenland—or anything else—only to be interrupted by military service.

There was more to Wegener's dispirited disinclination to work than just his impending return to military service. At the end of June Köppen had written him a long letter about the book on the displacement theory, having read through the proofs. Wegener was pleased, even delighted, that Köppen, who had always treated the work on displacement theory as a distracting sideline, had warmly approved it. But Wegener's percipient guide and mentor had also raised a number of pointed questions about the manuscript, not least the inconsistent, confusing, and (finally) inconclusive treatment of oceanic deeps. Wegener had been persuaded by his discussions with Cloos to include something about these interesting features on the outside of the great island arcs of the western Pacific, even though he could not give a convincing explanation of them. Now that Köppen had asked for details, he found that he could give none. Moreover, the existence of the Peru-Chile Trench, pointed out by Köppen, 8 kilometers (5 miles) deep and 5,000 kilometers (3,107 miles) long, on the leading edge of the South American continent made nonsense of the explanation that such trenches were somehow connected with the creation of island arcs springing away from the *trailing* edge of a continent. Why had he not seen this before?

Wegener held onto Köppen's letter for several weeks before replying. He expressed delight at the approval and thanked him for his comments, and he admitted his inability to explain ocean trenches in any greater detail ("the whole thing is still not clear to me"). His reflection went much deeper than this, however: "I think, by the way, that it is a weakness of all my work that I am forever going into too many side issues. I managed to cut out many of them from this book, but I should have limited myself to an even greater extent to the straightforward establishment of the displacements. Now we'll just have to wait to see how it is received. We can scarcely expect much discussion as long as the war lasts."[114]

Wegener's rueful self-assessment in this letter strikes home. His book on thermodynamics and now this book on continental displacements had a tendency to lose central focus in an overly meticulous treatment of peripheral issues. By the time he had sent the proofs (in 1911) of the book on thermodynamics to Köppen, he had felt the need to apologize for incorporating a large number of physical speculations that they had not discussed. It was exactly these speculations and this miscellaneous character that Exner had savaged in his review of the book. Now he could see clearly that this same tendency had come to the fore again. This motif in his work was closely tied to the source of his enthusiasm, his thirst for novelty, and his pleasure in discovery. Moreover, it could also be linked to something drummed into him by his father in childhood: that all intellectual work, all creative work should be infused with what Schiller had called *Spieltrieb*, the spirit of play.

Wegener believed that theoretical work should incorporate some sense of the novel play of ideas, but he also believed in hard evidence and that novelty comes not so much from theoretical speculation as from careful empirical work. He had said so to Köppen in responding to Exner's review, noting that he thought that Exner overvalued theory as a tool of discovery. Moreover, in his inaugural lecture in Marburg back in 1909, he had argued that in all sciences, periods of theoretical elaboration must alternate with periods of renewed empirical work and the gathering of data. Certainly his work on continental displacements was a theoretical elaboration of a great quantity of empirical data, and as certainly it was infused with the spirit of creative play of ideas, both in 1912 and to a much larger extent in 1915. It was now, perhaps, time to move to the other extreme.

Also at work here was his sense of science as a vocation, very like a call to the clergy. This was strong, though not exaggerated, in him, and it was not uncommon in men of his generation. Two of the most famous essays by Max Weber, written at just about this time, are entitled "Politics as a Vocation" and "Science as a Vocation." Perhaps because Wegener's own father was a clergyman, he easily adopted clerical vocabulary to describe his intellectual life. He had already referred to the deepest part of himself as his *Allerheiligstes*, his "inner sanctum," and he referred to the science-interested but not professional public as *Die Laien*, "laymen." This is common in English as well, as in "layman's language," but in German it has a somewhat stronger sense. With his overreaching and speculative forays in his work on displacements now evident to him, he felt something close to a sense of sin, and certainly a sense of remorse.

Two weeks after his confession to Köppen, at the end of July, these strands of thought seem to have come together in the form of a plan. He wrote to Köppen that he had turned again to his work on *Tromben*, a generic term for whirlwinds of all kinds, but which in 1911 he had used as a synonym for "tornado," a distinct rotating column of air usually occurring along with a thunderstorm and spinning off away from a squall line. He had not been satisfied with the work and found it poorly substantiated, and now he had turned to original descriptions of tornadoes and had come to the conclusion that "it would be very useful to write a book on this topic, perhaps in the 'Science' series of Vieweg [his publisher on continental displacements]." He went on to say (in a long and enthusiastic letter) that it would be very important to report all the descriptions of tornadoes "*word for word*" (his italics) from the original and present them together with the original illustrations. Information about tornadoes was scattered throughout the literature, often in small notes and reviews tucked away between articles. In the few collections of descriptions which existed, the descriptions had been edited and the figures adapted in ways that appeared to serve the theory of the formation of tornadoes which went with that book.[115]

Köppen feared a repetition of the same sort of theoretical excursion as that on continental displacements, a work that would run off in all directions at once, as Wegener's work so often did. Therefore, he urged Wegener to consider instead a theoretical article or supplement to a scholarly journal like *Meteorologische Zeitschrift*, and Wegener immediately replied that he did not want to write something narrow for meteorologists or something theoretical. This was to be just a compilation and not an investigation. While professional meteorologists would certainly find it interesting, the audience he had in mind was a lay audience—for the general reader, for meteorological observers (who manned weather stations and sent data as part of meteorological networks), and especially for sailors—and his aim was not just to inform them of what was known but to give them, as it were, a template to encourage new descriptions.[116]

So Wegener's book would be limited to descriptions, not theoretical speculation, and further limited to description of these phenomena only in Europe. He would consolidate the descriptions into lists of various kinds, try to clarify terminology, and for each verifiable occurrence of a tornado or waterspout give the year, day, hour, and place. In addition to these lists, he would organize the material in other ways: the direction and speed of travel, descriptions of the origin, and descriptions of the disappearance.

He would analyze meteorological elements: cloud conditions, precipitation, the direction of rotation, the diameter, the height, and further specification of its effect on trees, crops, buildings, and ships, the better to assess the energy involved. As he wrote to Köppen at the end of August, "I have come with much more clarity to the ways in which I will limit my work on tornadoes."[117] These limitations—scrupulously avoiding all theorizing, not even paraphrasing, not even redrawing original illustrations—would be one of those efforts he described as necessary in 1909 in his inaugural lecture at Marburg: the accumulation of data preparatory to theory. At that time he had spoken of meteorology as a science that alternated accumulation and assimilation as creative moments. It also underlined his feeling, in this case in some sense a remorseful feeling, that empirical description should be kept separate from theory, and both of these separate from advocacy.

There are additional echoes here with his earlier work. He was constructing, or was about to construct, a "time series" of observations of tornadoes. His dissertation in astronomy had been an extension of the time series of observations of the motion of the Moon back to the thirteenth century. His hypothesis of continental displacements depended on establishing a time series of longitudes for Greenland and North America and of latitudes for India. His contribution to Aßmann's *Festschrift* had pointed out that the historical record of the observation of the change in color of large meteors would be a valuable source of data for understanding the chemical differentiation of layers in the upper atmosphere. Modern approaches to these observations were of such recent vintage that pursuing a variety of clever stratagems for gaining information indirectly, from earlier documents, was an important scientific strategy, and one that came repeatedly to Wegener's mind when faced with a difficult problem.

While planning this work in July and August, Wegener was making hurried preparations for his return to the war; he was now scheduled to report for duty in Berlin on 1 September. He had taken Else, Hilde (or "Hildemops," as she was now known), and the hired girl to visit his parents and Tony at Zechlinerhütte. He was pleased with the health of his parents and sister, and they were enjoying their visit; things were going well.[118] He wondered whether Else's mother might come and stay with her and the baby in Marburg now that his orders to Berlin had come through.[119] When the Wegeners returned to Marburg in early August, Alfred reported to his father-in-law that he and Else worked every afternoon in their garden, and that Else was constantly cooking and putting up vegetables as a wartime food reserve.[120] Rationing was already in place, and some foodstuffs were beginning to be in short supply.

Along with her strenuous domestic routine, Else was also pursuing her scientific work: she was translating an article by Koch, on his Greenland glaciological work, for the *Zeitschrift für Gletscherkunde*. The work had been passed her way by the journal's editor, the great Austrian geographer and climatologist Eduard Brückner (1862–1927), who just happened to be her godfather.[121] The young couple was, in every way, squeezing as much normal life as they might into the few remaining weeks and days—both work and play. But time was very short, and they knew it.

14

The Meteorologist

"IN THE FIELD," 1916–1918

> Such people will refuse any reorientation of their ideas. If they had already learned the displacement theory when they were in school, they would defend it for their whole lives with the same lack of understanding, and with the same sort of incorrect information, they now use to defend the sinking of the continents. The best thing is just to wait until they die off.
>
> WEGENER TO WLADIMIR KÖPPEN, 25 December 1917

Wegener was called back to service in Berlin in mid-September 1915. October found him at the registration office (*Melde-Amt*) at German headquarters in occupied Antwerp, Belgium, where he briefly served as an adjutant.[1] By early November, he was in Tongeren, in eastern Belgium, and well behind the front lines, where the Germans were building a major railhead to supply the front, and which also seems to have had a military airfield.[2] It is not clear how long Wegener remained there, but by January 1916, 1st Lt. Wegener was in the Field Weather Service of the German army, as commanding officer of Field Weather Station 12, in Mülhausen, in the far south of Alsace. Nestled in the triangle between the French, German, and Swiss borders, it was both an administrative center and the southern anchor of the German lines on the western front.[3]

Mülhausen

Wegener's duties as a field weather officer in Alsace throughout 1916 were much like his duties at Lindenberg in 1905–1906. Field Weather Station 12 was one of many such stations in a synoptic network extending throughout Germany, occupied Belgium, and France—and into Poland as the Germans continued to drive the Russian armies back in the East. The cadre of officers under his command made observations at fixed times each day, including kite and captive balloon ascents, and they also sent up pilot balloons to measure the winds at higher altitudes.[4] The latter were especially important as they determined the operational range of aircraft flying into the face of prevailing westerly winds. Bombing and reconnaissance aircraft could, in 1916, fly at altitudes of 2–3 kilometers (1.2–1.9 miles), somewhat above the usual range of kites and captive balloons. The staff officers (and Wegener) telegraphed daily weather observations to central weather stations, to be combined into synoptic weather maps of the kind developed by Bjerknes at Leipzig, in order to forecast prevailing winds, temperatures, precipitation, and especially cloud cover, the last because all airborne navigation was by landmark.

The relative safety of Wegener's assignment and his consequent lack of strain, the regularity of his duties, and the presence of assistants and subordinates, including an orderly, left him with time and inclination to pursue his own meteorological research. He devoted these (many) free hours throughout the first seven months of 1916 almost exclusively to his book on *Tromben* (whirlwinds). His time in Mülhausen was vastly more comfortable and more congenial to scientific research than either of the expeditions to Greenland. He missed his home and family, but his sense of being divided between duty and science had vanished: now his military duty *was* science. He was quartered with his officers in a villa abandoned by its owner, and the only source of strife seems to have been keeping the peace between the cook, who had stayed on, and his junior officers.[5]

Wegener's most regular (surviving) correspondence with Wladimir Köppen is from this period of time, and these letters show a sense of intellectual excitement, energy, and an appetite for work. Most have to do with the difficulties, failures, and successes of his book-length catalog of all European tornadoes since the 1500s. Wegener had access to the library at University of Freiburg, some 48 kilometers (30 miles) to the northeast of Mülhausen and only a short trip by rail.[6] This university access was vital in obtaining periodical series and monographs in which he could find accounts of tornadoes. His proximity to Freiburg also allowed him regularly to visit with and converse with Kurt, who was in 1916 commander of a bomber squadron based there, and who was writing, between bombing and reconnaissance sorties, a book on the future *scientific* use of fixed-wing aircraft.[7]

Köppen was his usual helpful self, obtaining records of tornadoes from his collection of reprints, from the library at the German Marine Observatory in Hamburg, and from Julius Hann in Vienna, and Gustav Hellmann in Berlin also rendered assistance and provided materials that were difficult to find.[8] Frequent rail service, as well as the importance of Mülhausen as administrative center for the German military on the western front, sped the post and ensured the safe arrival of parcels of books and papers.

By early February things had settled into a routine and Wegener felt that his work was going well, though the material on whirlwinds was "hard to knead into shape."[9] This is not surprising: he was drawing on descriptions of tornadoes from a variety of sources, written with different levels of scientific understanding, with widely varying detail, and in different languages, across a span of 400 years. Moreover, he was (for once) not guided by a theoretical idea and was not *interpreting* a data set. Rather, he was committed to assembling one. He not only lacked a guiding theory but also refused, as a principle of organization, to entertain one; selecting data to fit a guiding theory of tornadoes was precisely the flaw he found in all previous compendia on the subject.

Tornadoes and Waterspouts

In spite of these difficulties, by early March he could write to Köppen that he was fully committed to going forward with the book. He felt compelled to emphasize once again that "it is not so much an investigation, but rather more a compilation, and this will make it much easier to read."[10] By this he certainly meant that there would be little or no complicated mathematical apparatus, and no hydrodynamic theorizing requiring a specialized knowledge of this area of research. The pace and character of the project suited his circumstances, as it was the kind of work which could be picked up and put down without the necessity of reestablishing a complicated train of thought.

His principal concern was straightforward: how to organize the material and how to manage the details of publication.[11] Part of the problem of organization was solved by the quality of the various descriptions. He had a relatively small number of detailed descriptions by naturalists and scientists who had observed and carefully recorded individual instances of tornadoes and waterspouts, descriptions that had been published as whole scientific articles. He would present these, perhaps a dozen or so, in an introductory chapter. Then he could proceed with a chronological catalog of all known descriptions in the scientific literature; this looked to be 200 or 300 items. These could be listed by date, place of occurrence, author, publication, and whether or not the text was accompanied by an illustration.[12] This list of descriptions he could then "data mine" for statistical analysis and for grouping under various topical headings: size, direction, length of path, weather at the time of occurrence, and as many other categories as he could think of.

In early April he was expanding his search for tornado records, by contacting observatories to obtain weather maps for the days on which tornadoes occurred, and even obtaining simple lists of "weather events" from the Marine Observatory in Hamburg, hoping to uncover in them records of tornadoes not otherwise reported. Despite his promise not to theorize, he could not help seeing patterns. His early statistical analysis showed that in the United States most tornadoes occurred in the spring, with very few in the winter, while the reverse was true in Europe. How to explain this? Could it be because in America the ocean was to the east, while in Europe it was to the west? Such ideas would find their way into the final book in isolated sentences, not organizing principles.[13]

Continental Displacements, 1916

April 1916 also brought to him the first reviews of his book on continental displacements, and the news was surprisingly good. He had sent a copy of the book to M. P. Rudzki (on whom he depended so much for geophysical data), hoping that Rudzki might review it, and indeed he did, in *Die Naturwissenschaften*. In the review, Rudzki praised both the "clarity and elegance of the style, that makes the reading of Wegener's book very pleasant," and Wegener's command of the literature supporting his hypothesis, which was "completely free of inaccurate representations [of the work of others] or misunderstandings [of their arguments]." Rudzki cautioned that Wegener had probably overstated the accuracy of gravity measurements over the ocean and the certainty of large-scale gravity differences between the ocean floor and the continental surface. He also noted that longitude measurements could shift as a result of the warping of the continental surface caused by local deformation and mass displacements, and that quite large latitude and longitude differences would have to be measured and confirmed before actual alteration of the distance between two continents could be assumed. He concluded, "Wegener shows that his hypothesis explains very well a variety of observations concerning continental morphology, geology, zoogeography, phytogeography, and so on."[14]

Wegener's ideas were also finding their way into standard textbooks even before the initial cycle of reviews had run its course. For example, Alexander Tornquist (1868–1944), professor of mineralogy and geology at the University of Graz in Austria, gave Wegener an enthusiastic notice in the introduction and concluding sections of the second edition of his *Grundzüge der allgemeinen Geologie* (1916). He described Wegener's "bold" hypothesis as something "far from our current conceptions," but nevertheless an idea of great interest and promise, worthy of further development and demonstration.[15] Such enthusiastic endorsements appeared, naturally, in parallel with doubts and criticisms expressed by other (and equally established) writers. Franz Kossmat

(1871–1938), director of the Geological and Paleontological Institute at the University of Leipzig, in the second (1916) edition of his *Paläogeographie*, wrote (like Tornquist) of the "bold hypothesis of Wegener" but found Wegener's notion of the dispersal of the southern continents "beset with tectonic difficulties."[16]

The reception of Wegener's ideas in this area of science was actually somewhat warmer than the reception of his textbook of thermodynamics in his own specialty. No reviewer mentioned or referenced, in any of the reviews and notices in 1916, Wegener's scientific specialty (i.e., that he was *not* a geologist), and all referred to him simply as Wegener, A. Wegener, or Dr. Wegener. Established geologists, geophysicists, and paleontologists do not seem to have viewed him as an interloper or a beginner. His own self-identification on the title page of the book had been "Instructor in Meteorology, Practical Astronomy, and Cosmic Physics in the University of Marburg." He also mentioned the measurement protocol between North America and Germany (and its abandonment because of the war), and he declared that he had only undertaken the scientific task of expanding the argument to pass the time during his rehabilitation leave, subsequent to being wounded in battle.[17] He was glad of the reception of the work and felt that he was now part of a dialogue with a range of coworkers, sponsors, and colleagues working full-time in geology and geophysics.

As exciting as this was, he had pressing concerns that diverted his interest away from continental displacements. He wrote at this time to Köppen that he would finally be eligible for home leave at Easter (at the end of April); he had not been home for seven months.[18] There had been a few hurried meetings with Else, in train stations when he knew his route (often to Berlin), or returning to the field, when he had managed to send her a telegram in time. Else wrote, in her memoir of this period, that this was the way they saw each other throughout most of the war between 1915 and 1918. Sometimes they found each other; sometimes they missed connection. Else learned to consult the registration books in the train stations, as all officers were required to sign in at the moment of their arrival, so she would know whether Alfred had not yet arrived, was still there, or had come and gone.[19] Now, at Easter, they would actually spend a span of days together for the first time in more than half a year. Arriving in Marburg in mid-April for his two-week furlough, Alfred was overjoyed to be home, to see his wife and child, and to be among familiar surroundings.

The Treysa Meteor

Upon his arrival in Marburg, he learned of an astonishing astronomical event earlier that month. On 3 April, at about 3:30 p.m., a huge fireball/meteor had streaked across the sky and detonated and crashed somewhere in Hesse. Richarz was wildly excited, as was Kayser, and they were desperate to locate it, as it had clearly landed in "their backyard" very near Marburg. Knowing that Alfred had written about color changes in large meteors, they sought his advice on the matter, and Wegener not only agreed to help them but told them that he would find the meteorite for them.

He was glad to help his patrons in their search, as he felt himself deeply in their debt. In his absence, Richarz (and Aßmann, from Berlin) had bombarded the Ministry of Education with appeals to establish the special examination in cosmic physics which Wegener had requested, the better to position him for a professorship in this topic at Marburg. They had also, with the concurrence of the dean of the university, urged that Wegener—not only a promising young scientist but also a wounded war hero—be made a titular professor for the duration of the war. At the forefront of this calculation was the unspoken desire to provide his wife and child with a steady income. Unable to offer courses and unable to work in the astronomical observatory, Alfred's stipend would always be in danger, and a professorial title could stabilize his finances and situation.[20]

This debt to his patrons notwithstanding, the search for the meteor was clearly a labor of love, and even a sort of fascinating game. The work that came out of it was a minor part of his scientific legacy, yet it is one of the most polished and thorough pieces of work he ever did.[21] Wegener was well prepared to write about the topic; he had just considered the color changes in large meteors, in his contribution to Aßmann's retirement volume. He had been at work for many months assembling newspaper and periodical accounts of the occurrence of tornadoes and had developed a strong interest in determining the accuracy and usefulness of (lay) accounts of unusual meteorological phenomena; this was certainly one such event. Finally, he was a trained astronomer, with a degree from a department that specialized in the tracking of near-Earth objects: asteroids, comets, and meteors.

Wegener began his search by writing to major newspapers in Frankfurt, Cologne, and Magdeburg, to assemble press reports of the meteor. The descriptions in these news reports allowed him to determine that it had indeed fallen somewhere in Hesse, and that the limits of visibility approximated the Rhine River to the west and the Main River to the south. Armed with this information, he wrote to fourteen additional local newspapers. From their reports he was able to divide the recorded observations into those that had seen the fireball and those that had both seen the fireball and heard the explosion. He could also plot on a map (with small arrows) the direction of travel of the meteor as seen from the standpoint of each recorded observation. From these descriptions he was able to inscribe a circle on the map, with a radius of roughly 125 kilometers (78 miles), within which all the sightings appeared. He located its center by the directionality of the arrows of the observed meteor path, placing it in the area of the greatest density of those observers who had seen the fireball and also heard the detonation.[22] This put the likely impact site near the village of Treysa.

He requested and obtained an extension of his leave in order to complete the investigation, and between the ninth and twelfth of May he traveled to the vicinity of Treysa (two days with one of his colleagues, and two days with Else), where he went house to house and farm to farm collecting testimonies of what people had seen and heard. This pleasant and diverting foot tour brought the number of observational accounts to more than 100.[23] Wegener asked his informants to sketch the shape of the fireball, and for those who had seen the meteor in the last few seconds before it hit the ground to indicate not just the direction of travel but the angle and the appearance of the smoke plume. The persistence of a smoke plume (for more than a minute) allowed a better sense of the direction and angle of incidence. He asked them about the sound and about the color, and he got wonderful descriptions. It was described variously as white, gleaming silver, yellow like the Sun, lemon yellow, blue like an acetylene torch, blue like the fire of hot coal, black with red streamers, and a red core surrounded by a green halo. This was, of course, what he expected—that the color should change based on the altitude of the meteor at the time of observation.[24]

The exercise of interviewing ordinary people about a scientific event was highly instructive. Mixed in with sober and clear observations were all sorts of fantastic apparitions. One observer claimed to have seen a fiery cloud in which the face of the kaiser appeared.[25] The most interesting phenomenon was perhaps the invention of

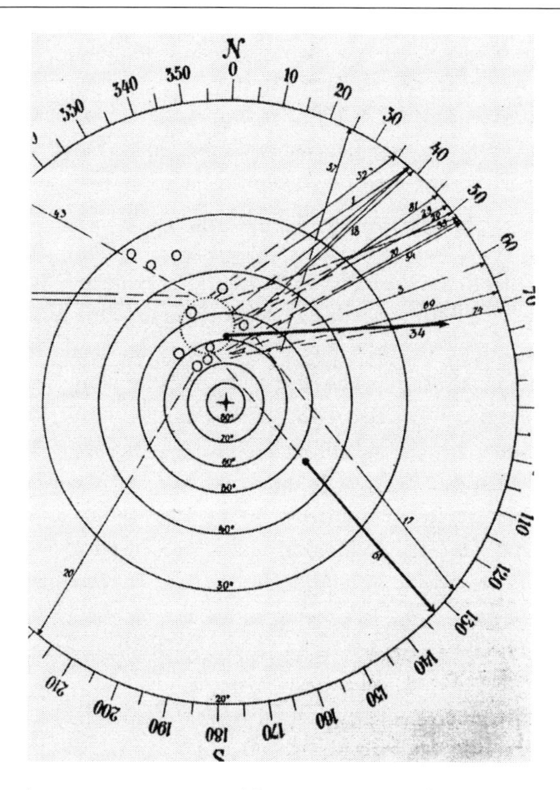

Rosette of meteor paths Wegener constructed from interviews in the late spring of 1916, producing a likely impact site near the village of Treysa. (The meteorite was eventually discovered just outside Wegener's target circle.) From Alfred Wegener, *Das detonierende Meteor vom 3 April 1916 3 1/2 Uhr nachmittags in Kurhessen*, vol. 14, *Schriften der Gesellschaft zur Beförderung der gesamten Naturwissenschaften zu Marburg* (1917).

sounds—especially whispering sounds—by those who had seen the fireball and claimed to have heard it simultaneously. This was, of course, impossible for anyone not within 500 meters (1,640 feet) of the impact site; for most observers the sound would have arrived several minutes later.[26]

At the other end of the spectrum from apparitions and imagined whistling and whispering sounds, there were observations of great value. One such was a newspaper account of observations by a painter named Fritz Behr, who lived in Römhild, more than 100 kilometers (62 miles) from the estimated impact site, and reported hearing an explosion about five or six minutes after the fireball passed overhead. Another observer, perhaps 10 kilometers (6 miles) distant from Behr, had also heard the explosion "some minutes" after the fireball had passed. There were no other claims of having heard the explosion outside the circle roughly 50 kilometers (31 miles) in diameter around the village of Treysa.[27]

These reports of audible explosions more than 100 kilometers from the impact site indicated to Wegener that the meteor impact had created an "outer zone of audibility." He was already attuned to this phenomenon because while at the front in Belgium in October 1914 he had heard multiple versions of the same explosions and had begun collecting accounts of such multiplications of sound. The exact nature of this phenomenon was not well understood, but it had to do with the refraction or reflection of sound from layers in the upper atmosphere. Indeed, improbable as it may seem, some residents of Marburg had reported (throughout early 1916) hearing faint artillery fire. Wegener had urged his PhD student Walther Brand, now a high school teacher in Marburg, to pursue these reports, and Brand had determined that they were hearing the furious artillery fire from the Battle of Verdun, some 360 kilometers (224 miles) to the west. "Outer zones of audibility" from Verdun had appeared at 225, 275, 325, and 360 kilometers (140, 171, 202, and 224 miles, respectively) away from the source.[28]

The analysis of all this meteor data would take some time; Wegener would take his observations back with him to Mülhausen at the end of his leave. Before he left on 10 May, he was able to give an address to the Marburg Natural History Society with preliminary conclusions. Based on the observations, and using the algorithm devised by his professor at Berlin, Julius Bauschinger, he was able to determine the velocity of the meteor (37.5 kilometers [23 miles] per second) and its angle of descent (55°), the latter achieved by reducing a variety of qualitative descriptions to quantitative estimates. The calculations were extremely laborious, but he was used to it and enjoyed it.[29]

One of the most interesting calculations he presented to the Natural History Society was the time of impact, which he determined by reference to railroad timetables. In Germany it was standard meteorological practice in the early twentieth century to determine the time of weather events—including tornadoes and severe thunderstorms—from observations made by passengers on trains, referenced to the train schedules. Wegener was able to calculate that the fireball had streaked across the sky at 3:25 p.m. on 3 April 1916. He knew this because a reliable account had been given by a passenger on the train between Cassel and Marburg, a train that had arrived on time in Marburg at 4:08 p.m. The observer had recorded that he saw the fireball while the train was between Neustadt and Allendorf. The train schedule showed that an on-time train would have been between these two villages at 3:25 p.m. on that day.[30]

Unable to stretch his home leave any further, Wegener returned to Field Weather Station 12 in mid-May 1916. To his surprise and pleasure he found that things had gone efficiently and productively in his absence. He had an unusually competent staff of junior officers, even if most of them were not meteorologists. In fact, the only other meteorologist on staff was Erich Kuhlbrodt (1891–1972). But among Wegener's officers were some outstanding scientific intellectuals. For instance, there was Walter Porstmann (1886–1959), who had been an assistant to Wilhelm Ostwald before the war and was just finishing a book with the provisional title *Normenlehre*, on the history and theory of fundamental measurement units and their relationship. It contained an ambitious plan to reform and reorganize the entire system of measurement and time units which Wegener much admired.[31]

Also among the junior officers was Ulrich Hellmann, a specialist in German literature. Wegener had had a standard classical literary education but did not know much about modern German literature, and Hellmann was his guide here, especially to literature in a lighter vein. Hellmann procured a volume of the nonsense verse of Christian Morgenstern (1871–1914), a much-beloved author and the German counterpart of Edmund Lear, Ogden Nash, and Dr. Seuss. Else said that Alfred memorized a number of these nonsense poems (from Morgenstern's *Galgenlieder* [Songs from the Gallows, 1905]) and drove her crazy with them by reciting them when they hiked together—until she screamed at him to stop.[32]

Wegener was touched by the dedication of his junior officers and impressed by their ability to learn the science. It gave him a deep satisfaction that they were—in the midst of this terrible war—doing good science, and he had pleasure in his command: it was the first time he'd ever been in charge of anything. Previously, there had always been a superior in close proximity: Mylius, Aßmann, Richarz, even his good friend Koch. Now, in Mülhausen, he was actually in charge of a scientific institution, albeit a modest one. It brought him out of his shell, made him more social, less likely to retreat at every opportunity to the isolation of his study.

Wind- und Wasserhosen

The industriousness of his junior officers allowed Wegener to make great progress in May and June on his book on *Tromben*. He was in something of a quandary about terminology. *Tromben* was derived from the Italian *tromba*, which meant variously "trumpet," "trunk" (i.e., as of an elephant), and "water pump." It was the broadest generic designation for a rotating column of air. It was also confusing, as the word "trombe" was sometimes used to designate only waterspouts. There was the additional problem that the Americans referred to these phenomena as "cyclones" and also as "tornadoes." What Americans said about them was important because America had many more and somewhat larger tornadoes than Europe, and tornado research in the United States was ahead of that in Europe.

He eventually solved this problem by giving primacy to considerations of shape, scale, and size: once again he had found a morphological solution to a problem. He noted that whirling vortexes of air could be partitioned into classes by their size. It was a nice problem. Tropical cyclones and barometric depressions (we would call the latter low-pressure systems, or "lows"), he noted, were anywhere from 300 kilometers (186 miles) in diameter (in the case of tropical cyclones) to 3,000 kilometers (1,864 miles) in diameter (in the case of low-pressure systems). At the other end of the size spectrum were *Staubwirbel* ("dust devils"), at most 200 or 300 meters (656–984 feet) in diameter. In between were the phenomena he was actually interested in cataloging: tornadoes and waterspouts. For these he chose descriptive terms already in use: *Windhose* and *Wasserhose*. The series of whirling vortices thus fell neatly into successive orders of magnitude, with the largest of the "dust devils" 100 times smaller than tornadoes, and with the largest tornadoes 100 times smaller than the smallest tropical depressions.

Wegener accepted Hann's distinction between these classes and extended his discrimination into a formal definition. *Windhosen* and *Wasserhosen* were *Großtromben*. They were large rotating whirlwinds with the vertical axis extending from a cumulonimbus cloud to the ground, rotating fast enough to cool and condense the air and make them visible as a cone, funnel, hose, or pillar, narrower at the bottom than at the top. These were the result of mechanical action, with thermodynamic consequences (the condensation of water vapor in the rapidly whirling column). On the other hand, *Staubwirbel* (dust devils) were thermally driven, not rotating fast enough in most cases to cause serious destruction, short-lived, and visible only because of the dust particles picked up at their base, since they were not able to cool the air enough to condense it; neither did they have the marked thinning of the air so characteristic in the cores of tornadoes.[33] These smaller, thermally driven objects (*Staubwirbel*) fell into the class of *Kleintromben*.

So his book would be about the *Großtromben*, and since this was a classification he himself was inventing or reinventing, he eventually settled on the book title *Wind- und Wasserhosen in Europa*. The book he produced in these months is exactly what he set out to make: a historical and descriptive catalogue of all known tornadoes in Europe since the sixteenth century, in which these were analyzed and characterized by their size, shape, duration, velocity, speed of rotation, altitude, frequency, length of destructive path and extent of destruction, seasonality, characteristic geographic location, mode of formation and dissipation, relationship to squall lines and to individual large cumulonimbus clouds, and other weather phenomena. Each of the items in the above list had a full chapter of its own in the finished book.

Wegener worked steadily on the book in late May and all of June. By the end of the first week in July he estimated that he had only a week's worth of writing left.[34] Things were going well for him and seemed to be going well for Germany. He wrote proudly to Köppen that he had been promoted to captain, something so unexpected that he found himself without the appropriate uniform, and he had had to write to his student and friend Brand in Marburg and borrow one from him; Brand was a captain in the home guard. The promotion came with a very substantial raise, which embarrassed him, and he expressed to Köppen his hope "that the war will end before I become a millionaire."[35]

In early July, in spite of Wegener's hopes for an early end to the war, things suddenly took a turn for the worse. The British offensive on the Somme River had begun on 1 July. The beginning of this battle is notorious for the terrifying casualties taken by the British: 60,000 killed and wounded on the first day. In the English-speaking world this catastrophe tends to overshadow what was happening on the German side. The Germans had been caught unawares by the attack and took terrible casualties all during July, even though they managed to stop Allied advance after only about 10 kilometers. Kurt Wegener wrote Alfred that since 3 July things had gone very badly. He had been shot down and was in no-man's-land for several days before being rescued. Returning to his aerodrome, he found that it had been bombed by the British. In anticipation of further Allied advances, Wegener's entire staff had been ordered to a staging area closer to the front, should they be needed to shore up the line. Wegener hoped they would get no further toward the action, but he was worried.[36]

Things at home were little better. Else had money, but prices for food were inflating wildly, and there was little to buy in the market in Marburg. Indeed, that summer there was rioting in Vienna over the price of bread, and serious unrest and unease gripped the home front in Germany and Austria. Mail was slowing down between his forward station and home, and he was having difficulty finding out any news. He asked Köppen to find out some information. Else had help and support from the Brands, as well as from Cloos and his wife, but they had no more access to food than she. That summer and autumn she and Hilde would travel to Hamburg and to Zechlinerhütte; food and fruit were plentiful in the older Wegeners' garden.[37]

By the third week in July the British and French advances had been definitively stopped, although the cost was horrifying. Stabilization of the front lines allowed Wegener to return to Mülhausen and resume his routine. With a full staff and complement of officers, he could once again proceed with his scientific work, and within a few days he was able to send the manuscript of his book to Köppen to look through, before he sent it along to the press.[38]

The completed book was just over 300 pages long. There are several things worth noting about it. The first of these is the meticulous, even exhaustive, detail. Every description is worked over again and again to extract every piece of information. Wegener had numbered each of the tornado descriptions and referred to them by number whenever he needed them in a specific context. They seem almost to emerge as individual personalities, so repeatedly are they named and discussed.

As an example of the meticulous detail, we can consider the material in chapter 9, "Rotation," appearing as pages 176–185 of the finished book. In this chapter Wegener tried to determine one fundamental question: "is the rotation [of tornadoes and waterspouts in Europe] cyclonic or anticyclonic?" Of the 255 records of tornadoes he assembled, he determined that only 25 could, with absolute certainty, give the direction of rotation. Of these, he wrote, "18 (72%) are cyclonic and 7 (28%) give an anticyclonic rotation."[39] He then briefly listed them by number, giving a short description (two or three sentences) of each, in the form of a direct quotation from the original source. One recalls that he had committed, in the very early stages of this work, to stay with exact descriptions taken verbatim from the original texts. He then went on: "Four of these *Tromben*, namely number 176, 219, 198 and 205 allow us to determine the direction of travel; the first two traveled towards the SW, the last 2 towards the W since these are the same directions of travel of all *Tromben* generally considered there can be no ground for holding that anticyclonic rotating *Tromben* are governed by a particular direction of travel."[40]

This was a theoretical question of some importance, but the theoretical importance is implicit, and what is explicit is the description. At stake theoretically is whether the rotation of Earth governs the direction of rotation of these phenomena, these *Großtromben*, in the same way that it governs cyclonic and anticyclonic circulation. In the Northern Hemisphere, cyclones rotate counterclockwise and anticyclones clockwise; in the Southern Hemisphere, the reverse is true. Wegener makes no determination, pointing out merely that there can be no doubt that there are anticyclonic tornadoes in the Northern Hemisphere, even if the majority are cyclonic. It may be different in the Southern Hemisphere, but a determination of this would require more than the three existing descriptions from that half of the world.[41]

One discovers such theoretical questions embedded throughout the text, with one or more in every chapter made evident in bold type. For instance, here is his evaluation of a description of a waterspout with a "double funnel" (a funnel descending from a cloud, getting narrower but then having its base embedded midair in a large bulge that is the top of a second funnel descending to the water surface). Wegener gave a graphical and theoretical analysis of the difference in wind speed and lines of rotation within such paired funnel clouds, as a preface to the following statement: "We will in this context suggest another question which is of fundamental significance for the explanation of Tromben, **namely whether Tromben continue on further into the inside of the cloud, or whether they terminate at the cloud floor.**"[42] This is less a form of covert theorization than it is sagacity and understanding of the problem and of what sort of information is likely to be of interest and of use to investigators of these phenomena. It also poses a question successfully answered only late in the twentieth century: they actually do continue to the inside of the cloud as "supercells."

Wegener came to explicit consideration of, as he put it, "Views on the Origin of Tromben" (*Ansichten über die Entstehung der Tromben*) only in the final chapter of the book (chap. 15). Quickly dismissing the volcanic theory, the downpour theory, the electrical theory, and the wave theory in a few pages, he noted that "there remain then only two theories, the mechanical and the thermodynamic." In rejecting the thermodynamic theory, he had the same response that had occurred to him so many times in evaluating atmospheric turbulence: that the principal argument against a thermodynamic theory was that in rising columns of air, whether it be the smoke from a cigar, the air above a lamp, a volcanic plume, or simply a cumulus cloud, there is no rotation about a vertical axis.[43]

This left only the mechanical theory, whether strictly mechanical or hydrodynamic; he included them both under the same heading. He noted that of all the theories of tornadoes this was the oldest, and he verified this with an extensive quotation from Lucretius, a description he had been happy to find some months before. This is delightful in its own way, as it links the oldest-known hypothesis for the formation of *Tromben* to the most recent, which Wegener considered for four or five pages at the very end of the book, before concluding with a mild comment that the phenomena required much further investigation.[44]

The completeness of the catalog, the sagacity in selecting analytical categories, the clear delineation of questions arising from the descriptions themselves, the tentative endorsement of a mechanical/hydrodynamic theory as the only one consistent with the phenomena, and the demarcation of interesting points and rhetorical questions from the declarative portions of the text all contributed to the lasting power of this work. Of all Wegener's publications, it is virtually the only one that remains actively cited in the twenty-first century, not in a historical context or as background, but as data pertinent to the investigation today of the phenomena it considers.[45]

With the book sent to the press, Wegener was again somewhat at loose ends. He had started a paper on the "outer zone of audibility" but found he had not enough data, so he confined himself to theoretical speculations.[46] Already in early August he was impatient, having little scientific work of his own to do.[47] He thought about working up the weather observations at his field weather station for publication, especially the pilot balloon observations, which he thought consistent and interesting, and he wrote to his commanding officer to see if this would be possible.[48]

It was characteristic of Wegener, when he had no scientific work available, to become both nervous and impatient. In mid-August he wrote to Köppen and complained of Vieweg's glacial pace in the production of proofs: "If they continue the dawdling tempo of the last few weeks the book will not be ready for Christmas."[49] He hoped that he would have leave at Christmas, but as both of them knew, Wegener had planned to finish the book to present it to Köppen on 25 September 1916, his seventieth birthday. Of course, Vieweg was not moving slowly because he wanted to, but because a scarcity of workers and a shortage of paper were slowing the entire intellectual apparatus of this great scientific civilization to a crawl. Wegener was not immediately aware of this at the front, but at home shortages of all kinds were becoming routine, and frustrations and delays at every level in every activity were the order of the day.

One suspects, in his complaining and fretting, his fear of a recurrent bout with one of his transient, recurrent depressions. On 21 August he sent a brief note to Köppen apologizing for not answering anyone's letters, noting that he was depending on Else to transmit news both to the Köppen household in Hamburg and to his own family at Zechlinerhütte.[50]

In any case, in September he received permission for a brief trip to Hamburg to celebrate his father-in-law's seventieth birthday with the whole family. Returning to Mülhausen, he found a completely unexpected present of his own: an official letter from his dean and a copy of the "Patent as Professor," a citation from the Ministry of Education announcing that "in recognition of his scientific achievements Dr. Alfred

Wegener is awarded the title: *Professor*."[51] This was not, to be sure, the permanent chair in cosmic physics he had hoped for, but rather appointment as an "Extraordinary Professor" (*Extraordinarius*), something akin to an associate professor: salary, title, but no tenure. Wegener's superiors had finally achieved a part of their aim, at least temporarily, and moved a step closer to finding Wegener a permanent academic home at Marburg.

Wegener was delighted with news of this appointment and immediately wrote to Hamburg to tell Else that "he had a late birthday present" and that she "was now a Frau Professor."[52] All of this put him in an expansive mood. He remarked that the "Patent as Professor" was so pompous in its wording that it could have provided a good theme for a Christian Morgenstern poem. "If only these happy events [his salary as a captain and the title of Professor] could outlast the war."[53] And title was most of what it was. It was a ghost professorship, even though it came with a title and a stipend. At Marburg, as the catalogs for each new semester were published, his lectures were regularly announced and just as regularly canceled on account of his war service.[54]

Beginning in October 1916, and for many months thereafter, Wegener's scientific correspondence was very scant. He had no major project under way. Such involvements had always meant steady correspondence with Köppen, but in the period from October 1916 to May 1917 Wegener wrote to his father-in-law only three times. Else Wegener, in her memoir of Alfred's life, skips over the entire period from October 1916 to June 1917. We know that in the fall of 1916 Wegener was still negotiating to try to raise money to support publication of Else's translation of Koch's book about the 1912–1913 expedition, that he was still working on his ever-lengthening account of the Treysa meteorite and its context, and that he continued to receive interested communications concerning his work on tornadoes from colleagues in Germany, Austria, and even Turkey.[55]

Forecast Meteorology

The one area in which Wegener could "keep going" as a scientist was forecast meteorology. He had not been able to get permission to publish the results of his field weather station's aerology—a naïve hope at best, as such information would certainly have been classified. Still, he was thinking more about weather forecasting (and doing more of it) than at any previous time in his life. His career to this point had been that of an atmospheric physicist. He had used the tools of meteorology but had never before used them in order to find out the weather the next day or the next week. He had used daily weather the way he used meteors: as a means to find out about the structure of the atmosphere. Now, for the first time, he began to pursue synoptic meteorology as a regimen and as a scientific interest.

One aspect of this pursuit of forecast meteorology was his review of E. Neuhaus's *Die Wolken in Form, Färbung und Lage als locale Wetterprognose* (1914).[56] Book reviewing was never a mainstay of his scientific work by any means; in a career stretching back to 1906, he had written only five book reviews. For Wegener to review the books of others almost always indicated an interest in pursuing a line of investigation which touched the subject of the book in question. Neuhaus's book was a slender volume (forty-eight pages) prepared as a handbook for local weather forecasting in the Swiss Alps. It gave instruction on how to use the form, color, and altitude of cloud layers to forecast local weather, an aim furthered by beautiful cloud photography of the kind that Wegener much admired and had practiced. In the review, he compared Neuhaus's work with that of Ernst Mylius (1846–1929), a pharmaceutical chemist and talented watercolorist specializing in paintings of clouds. In 1914 Mylius had written a charming and useful book, *Wetterkunde für den Wassersport*; Neuhaus was doing for mountain weather what Mylius had done for clouds over the ocean. Wegener's review was genial and supportive of both efforts.[57]

Wegener had written to Köppen in August 1916 and noted in passing that the Swiss meteorologist Alfred de Quervain (his contemporary in both meteorology and Greenland exploration) was planning a new cloud atlas. Köppen had himself coauthored a major cloud atlas in 1890 with Hugo Hildebransson and Georg Neumayer, and since then, there had been an international cloud atlas that had standardized the representation of cloud forms.

Somewhere around this time (August–October 1916) Wegener began to consider producing a cloud atlas of his own. He was a strongly visual thinker, as we have already noted; photography has always been a major part of his approach to science. Between the Danmark Expedition and his traverse of the ice cap in 1912–1913, he had produced thousands of photographs; their value as documents in the history of exploration has far outlasted the scientific results of either of these undertakings. His *Thermodynamics of the Atmosphere* illustrated everything from snow crystals to cloud forms, using his own (and others') photographs. His work on mirages depended crucially on photographic evidence.

Yet whatever the value of photographic documentation, it had its limitations. Color photography was still in its infancy, and cloud colors are an important part of the information clouds provide about weather. The thirty-one color plates of the *International Cloud Atlas* had employed both color photographs and paintings. While a marvel of technology at the time it was published (in the 1890s), this atlas's color photographs had an important drawback: a photograph can capture only those features of a natural object available at an instant. Wegener was therefore attracted to the idea of a cloud atlas that would rely on colored paintings of clouds, in preference to photographs. That possible project lay somewhere in the future.

There is every indication that Wegener sank—in the autumn and winter of 1916–1917—into one of the winter depressions that had afflicted him in both Greenland trips. When he finally reestablished contact with Köppen on 20 January 1917, his letter contained much news, but it was for the most part flat and listless. Köppen had inquired earlier in the month about whether meteorological stations in Germany should be equipped with "masts" containing meteorological instruments. Wegener replied that he didn't know and couldn't think of what they might be used for, or whether it would be worth it, or what the cost would be. Köppen had asked him to visit Hamburg on his next leave, but he declined. He said he had to stay at the weather station in the interest of *Kameradschaft*, having skipped the Christmas festivities. The latter admission is another indication of his depression; he had found such celebrations unbearable in Greenland. Now he had to put in an appearance at the kaiser's birthday celebration at the headquarters of the Field Weather Service and then write his monthly report.[58]

There were numerous reasons for his depression. He was far from his family and had only spent a few months with his only child, who was now three and was growing up without him. His mother was no better (if no worse) after her stroke and still paralyzed. With food rationing and scarcity, Else and Hilde had gone to stay at *die Hütte*, where the garden and orchard provided plentiful food, much more than

Anna, Richard, and Tony required. While Alfred's wife and child enjoyed the bucolic scenes of his childhood, he superintended increasingly rancorous daily disputes between his officers and the cook at the weather station. Moreover, after his appointment in September as "extraordinary" professor, there had been no more news from Marburg, thus indicating that the Marburgers had had no success in making such an appointment permanent. This also depressed him. He had now been a *Privatdozent* for eight years, and many colleagues junior to him had already found professorial appointments.

He was beginning to give up the possibility of a permanent professorial post at Marburg and starting to look elsewhere. It appears that he and Kurt had discussed a plan that they would both relocate (after the war) to the Meteorological Institute at the University of Straßburg. Otto Stoll (1885–1923) was already there. Stoll had replaced Kurt at the Spitsbergen Observatory in 1913 and was working with Kurt during the war; Kurt had an appointment at Straßburg, and it seemed to all of them that the Central Meteorological Institute at Straßburg would be both the center of meteorological research in Germany and the center for polar exploration, with which they were all deeply involved. It also seemed about to become a major center for geophysics: Beno Gutenberg (1889–1960), a student of Wiechert at Göttingen, was a *Privatdozent* there, and in the Army Weather Service as well. Alfred went into great detail about the range of scientific facilities and its suitability for Else. The institute where he would work was outside the city, and Else wanted no more of city life than she had to have. This was, of course, partly because of Alfred Wegener himself; he did very poorly in large and loud urban areas.

"Sixty-Seven Topics"

In a letter of 20 January 1917 to Köppen, outlining plans for Straßburg, Wegener closed with an interesting comment: "My Meteor is now finished and sent off, and I'm looking around for new work, so I'll spend a few days thinking about which of my 67 [i.e., numerous] topics I will work on now. Probably it will be [the question of] Color Change of Meteors (once again, this time with expanded material)."[59] The choice of this topic is not surprising, and indeed he did proceed with it. It is the comment about the "67 topics" that draws one's attention. It is an indication that his depression was lifting: he was once again scientifically engaged. Wegener picked up new topics with enthusiasm, almost by chance, and then held these in reserve until something else pushed him in the direction of one or another of them. At any one time there were more things he wanted to work on than he had time or resources to pursue.

We should also be aware, though, that however many topics Wegener pursued, they were always in the service of the same scientific problem. Since receiving his PhD in 1906, Wegener had investigated and published on the following topics: Helmholtz waves, mirages, atmospheric polarization, the ice phase of water vapor in the atmosphere, sun dogs [*Nebensonnen*], noctilucent clouds, twilight layers, color changes in meteors, blue lines in glacier ice, temperature changes in snow layers, atmospheric inversions, atmospheric layering (generally considered), föhn winds, atmospheric zones of intermittent audibility, atmospheric turbulence, squall lines and surfaces of convergence and divergence, tornadoes and waterspouts, and, finally—in his hypothesis of continental displacements—the physical properties of the solid Earth. Every one of these "sixty-seven" topics is about the *same physical problem*: sharp surfaces of discontinuity in otherwise continuous media, leading to dynamic changes in behavior, usually the result of a jump in temperature, pressure, density, or chemical composition.

Wegener was aware of this tendency in his thought and had remarked on it as early as 1909.

In this instance, the push he needed to move him to more work on meteors was not long in coming. In early March 1917, Wegener received an electrifying telegram from Richarz in Marburg: "We have it! Pure iron!"[60] They had found the Treysa meteorite—that is, it had been discovered in the woods outside of Treysa by a hunter (who gladly claimed the reward of 300 marks offered by Richarz). The site of the impact was not quite 730 meters (2,395 feet) from Wegener's predicted point. The discovery of this meteorite created quite a stir, as well as creating something of a problem for Wegener, as his booklet on the meteor was already in page proof. Should he put something about the discovery in the book? He decided merely to insert a note after the title page saying that the meteor had been discovered while this publication was in press. Wegener added that it "seemed more interesting to leave the text completely unaltered, especially in those portions which were not confirmed by the manner and circumstances in which the meteorite was discovered."[61] In other words, he wanted to show how right he had been, by showing those (few) places in which he had been wrong.

Just one month later, in early April 1917, Wegener got another uplifting and stimulating communication, this time from his old companion in Greenland, Andreas Lundager. Wegener had sent a copy of his 1915 book on continents and oceans to Lundager but had heard nothing for two years. Wegener now learned that Lundager had taken upon himself to write a long review in *Geographisk Tidskrift* (Copenhagen). He included a synopsis in the letter: it was quite positive and contained the judgment that "to this day we still work with the old contraction theory. Its days, however, seem numbered. It must be assumed that Dr. Alfred Wegener's displacement theory will take its place."[62]

With this letter from Lundager, Wegener now knew he had exactly two supporters: Lundager and Edgar Dacqué. The latter had devoted considerable space in his *Grundlagen und Methoden der Paläographie* (Principles and methods of paleogeography, 1915) to explaining and promoting Wegener's theory. It was therefore a pleasant surprise and an interesting coincidence that almost exactly a month after Lundager's letter (early May 1917) Wegener received a letter from Dacqué asking him if he had seen the most recent "anti-continental displacement publications (Soergel, Diener, Semper)" and asking if he planned to respond to these or had already done so. In any case, he wanted to discuss them with Wegener.[63]

Dacqué, also in "the field" in the German army, was trying to keep his scientific work alive and to defend (against detractors) positions he had taken "at home." This was a significant problem for younger scientists throughout the war, a problem rarely addressed by historians though amply documented: the extent to which ongoing debates in all the sciences were disproportionately influenced by the writings of senior scientists too old for military service, still teaching in their university positions and training yet another generation of undergraduates without the "balance" that might have been provided by access to younger instructors, most of whom were in uniform. The slow pace of publication, paper rationing, and the difficulty of getting scientific periodicals at the front also slowed down scientific discussion and debate, especially after 1916–1917, extending into the early and middle 1920s. By 1918 the pace of publication had slowed to about half of what it was before the war; it would not regain its normal expansion until 1926.[64]

Wegener had not seen any of these "anti-continental displacement" publications, let alone replied to them. Characteristically, Wegener did not respond directly to criticisms of his work, except by revision of a successive scientific publication on the same topic when he found that the criticism merited it. We do not know what he responded to Dacqué, but he did not undertake to obtain and read any of these criticisms.

In any case, his mind was elsewhere. He had spent the spring reviewing his "67 topics" and come to something of an impasse. He had thought again about turbulence in the atmosphere and had come to the conclusion that he should take the issue off his list.[65] He was now trying to decide between the idea of the cloud atlas with (the artist) Ernst Mylius and his work on meteors. He was very enthusiastic about the latter but was coming to realize that his scientific interests were falling into cracks between disciplines and not attracting the audience that he wanted. He was a cosmic physicist, but there were no journals of cosmic physics, and there was no university discipline of cosmic physics. His problem was captured beautifully by his desire to work on meteors. He imagined a series of three publications: one on color change in meteors (already complete), another on the direction of travel and speed of meteors, and a third on the propagation of sound from meteor detonations. There was a problem, though, as seen in his plaintive note to Köppen:

> But where could I publish it? Meteorologists should be very interested in these questions, but astronomers seem to have a monopoly on the study of meteors, and they naturally ignore the atmosphere. All of these are part of the "investigation of the outermost layers of the atmosphere," but I think we need now a new word for this, something that will allow us to reorganize the phenomena. . . . I'm not going to bury them [his new publications] in the Proceedings of the Marburg Natural History Society, and the MZ [*Meteorologische Zeitschrift*] doesn't see enough interest for a 22 page article [the manuscript on color change in meteors], as they, in the manner of [Felix] Exner, understand meteorology as limited to the study of the weather.[66]

The passing reference to Exner was not just an aside. The editor of the *Annals of Hydrography and Maritime Meteorology* had sent Wegener Exner's recently published *Dynamische Meteorologie* for review. Wegener had agreed to review it, writing to Köppen that "he liked the book, and this has produced a charitable compulsion to read carefully. 'Charitable' because it helps overcome my aversion to his very abstract treatment."[67] If he read Exner's book with a certain benevolence, that benevolence did not spill over into the review, which damned with faint praise and praised with faint condemnations. In 1911 Exner had accused Wegener of producing a book (*Thermodynamics of the Atmosphere*) that was incomplete, of mixed quality, seemingly composed of whatever the author had been reading at the moment, and in any case something that would not last more than a year or two. Here Wegener returned Exner's favor, albeit without the haughty demeanor and biting tone.

Wegener's review of Exner's book began by announcing that most of the major topics covered and major findings had already appeared in a 1912 encyclopedia article in the *Encyclopedia of Mathematical Sciences* which had not reached its intended audience because reading it required the purchase of the entire volume. In the existing literature, Wegener continued, Exner's book most closely resembled the work of William Ferrel and Adolf Sprung, mathematically updated, with special attention to Exner's own work and "even more attention to the work of [Max] Margules, which in the words of the author [Exner] 'is woven through the book like a red thread.'"[68] So, suggested Wegener, the book mostly resembles textbooks of weather science characteristic of the later nineteenth century, with updated mathematics: an antique approach to the subject, with some new mathematics, fronting the work of the author, but depending throughout on another scholar's conception of the subject.

The book, Wegener continued, was very much like the work of Bjerknes: a purely theoretical treatment of the subject matter, not based in observations, in sharp contrast to the work of Teisserenc de Bort and Hildebrandsson's *Les bases de la météorologie dynamique*, which took these foundations not to be mathematical principles but observations, and which was organized historically rather than theoretically. While there is some treatment of thermodynamics in the works of Exner and Bjerknes, said Wegener, it stands in sharp contrast to his own view of thermodynamics, where thermodynamic principles are introduced to explain the observations and not the other way around. In both Exner and Bjerknes, the relationship is rather like that of experimental physics to theoretical physics: the only observations of the atmosphere they discuss are those they can explain by current thermodynamic principles. In a more complete treatment of the phenomena, there would also be mechanics, acoustics, radiation, electricity, and the optics of the atmosphere. This sort of work would be an applied physics, "while Bjerknes and Exner provide instead applied mathematics, the first in the form of a single large theory, the latter more referential in form as a collection of important mathematical-theoretical investigations."[69]

The great utility of Exner's book, from Wegener's standpoint, was that it demonstrated (largely through the author's admission in the later chapters) that the problem of atmospheric dynamics was very far from a solution. Exner's clear presentation of the main lines of mathematical theorizing showed the difficulty of trying to harness meteorology to preexisting mathematical structures of theoretical physics or laboratory investigations alone. This, of course, was Wegener's standpoint from the beginning of his career in meteorology. In his book on *Thermodynamics of the Atmosphere* he had insisted that we move from the treatment of thermodynamic processes in ideal gases to those processes in a real atmosphere. He had insisted that the state curves of the atmosphere predicted by theory surrender to the state curves demonstrated by observation.

We can see now that Wegener was removing the study of turbulence from his list of "67 topics" because the approach taken in the literature of meteorology was moving away from the empirical materials necessary to build the atmospheric physics that would be the bridge between pure theory and pure observation. Exner's book, the predilections of the editors at *Meteorologische Zeitschrift*, the theoretical work of his good friend Bjerknes, and even the orientation of his father-in-law and closest collaborator all pointed to meteorology as the science of atmospheric dynamics in the troposphere, aimed first and foremost at understanding the genesis and history of midlatitude cyclonic storms. Wegener, on the other hand, was devoted to a much more phenomenological viewpoint: "meteorology" was the study of everything that happened in the atmosphere from Earth's surface to the outermost layers, and meteorological phenomena were any phenomena that helped to understand the physical characteristics of that atmosphere, irrespective of the discipline to which such phenomena seemed to belong.

Wegener's conception of meteorology was not quite as broad as that of Aristotle, but nearly so. If we consider that he was also actively at work in geology, geophysics, and climatology, through his work on continental displacements, then in some sense his science was equivalent to Aristotle's "meteorology," which held all these things within its ambit. This created a very serious problem for the kind of work he wanted to do.

The partitioning of meteorological research in the early twentieth century, and indeed of research in most fields of science, is closely documented in the multiplication of journals. Within meteorology in Germany there was *Das Wetter*, largely anecdotal reporting of interesting weather phenomena and qualitative discussion of their patterns; there was *Meteorologische Zeitschrift*, with its predilection for study of weather from both a practical and a theoretical standpoint; and there was Aßmann's *Beiträge zur Physik der freien Atmosphäre*, invented specifically to provide an outlet for studies of the atmosphere not pertinent directly to weather.

While *Meteorologische Zeitschrift* occasionally permitted Wegener and others to publish short notes on studies of the upper atmosphere, a review of Wegener's publication history (up to 1917) shows that he was repeatedly driven to physical and chemical journals not typically read by meteorologists in order to report the results of his investigations. In 1917, he could see that meteorology was moving farther away from and not closer to his broad conception of the subject. He was increasingly aware of the extent to which he was being forced to "bury" his research (his own word) in marginal publications, and he was increasingly unwilling to put energy into topics that would not find an audience.

Continental Displacements, 1917

In the midst of all this discussion of the future of meteorology, as well as Wegener's own research and his publication plans, one might well ask, Where are continental displacements in all of this? The answer is, effectively, nowhere. At the end of a long letter to Köppen in mid-May Wegener closed with the following:

> Prof. Andrée, the geologist at Königsberg (earlier a Priv. Doz. [instructor] in Marburg) sent me a long article from *Petermanns Mitteilungen*, a review of the displacement theory, where he takes an intermediary position. He reviews the contrary evidence put forth by Prof. Diener (Vienna), which made a scant impression on me, because he [Diener] has for the most part misunderstood me. In one place he has mistaken a position on the globe by 90°! Articles by Soergel, Semper and others have appeared in the same vein—all unknown to me—and Dacqué, the author of a substantial paleogeography, has written to me that he will fire off a review of the various recent "anti-Wegener." Best regards, Alfred.[70]

The author of this article, Karl Andrée (1880–1959), was an acquaintance of Wegener from Marburg and a protégé of Emanuel Kayser (as was Hans Cloos). Andrée, no relation to the celebrated cartographer, had taken his PhD in Göttingen in 1904, and in 1917 he was a professor (a wartime "*Extraordinarius*," like Wegener) of geology and paleontology at Königsberg. The full title of his review was "Alfred Wegener's Hypothesis of the Horizontal Displacement of Continental Blocks, and the Permanence-Problem in Light of Paleogeography and Dynamical Geology."[71]

That Wegener showed no inclination to study or respond to this article in 1917 does not make it any less significant a part of his life story. It was the first extensive review of his work in a leading earth science periodical which was not an outright attack or an outright defense. It was influential in framing the terms of the debate in

Germany on Wegener's work for some time to come. The issues Andrée chose to address, as well as those he did not address, would turn out to be of some importance for Wegener at a later date. This is to say that if Wegener had a strong influence on Andrée in the conventional sense that we speak of one thinker being influenced by another, Andrée influenced Wegener by helping to determine the way in which he was read and understood by German earth scientists. We will return in a moment to a consideration of Andrée's critique of Wegener's hypothesis of continental displacements, but first we must consider Wegener's situation in May and June of 1917.

Even had Wegener wanted to engage with the question of displacements in spring 1917 (and he clearly did not), there was no time. He was immersed in plans to produce a cloud atlas with Mylius, who was so excited about the project that he had crated and shipped 400 watercolors of cloud forms to Marburg, so that Wegener might go through them with Köppen—should Wegener obtain leave.[72] Alfred had also promised Kurt to look through the latter's manuscript "Vom Fliegen" on the prospect for scientific research after the war using fixed-wing aircraft.[73] He was also still running Field Weather Station 12; his regular duties as station commander, though not onerous, were real and continuing even with the stabilization of the western front.

Jüterbog

On top of all this, Wegener learned in June that he would be ordered home to Germany in early July 1917 to take over a reorganization of the Domestic Weather Service. The central office, the *Hauptwetterwarte der Heimat*, was headquartered at Jüterbog, a picturesque old town 65 kilometers (40 miles) southwest of Berlin. Gustav Hellmann, head of the Field Weather Service, had engineered this transfer and promotion. Hellmann had assisted Wegener in finding accounts of tornadoes and waterspouts for his book and had tried to find money to publish Else's translation of the account of the crossing of Greenland; Hellmann admired Wegener as an excellent officer with outstanding organizational skills. Their collegial relationship and coinciding interests made Wegener a natural choice for this job, and Hellmann's choice reflects Wegener's growing reputation and importance as a meteorologist.

This pressure to reorganize Germany's *domestic* weather service in time of war may seem odd, but it was driven by the need to accommodate military and civilian aviation, which had grown rapidly since the beginning of the war. Kurt and Alfred had often discussed the need for accurate forecasts of the wind for the use of pilots; forecasting wind speed and direction had been one of the principal tasks at Field Weather Station 12. Alfred had known since his Greenland observations in 1906 that the direction and velocity of the wind even 10 meters (33 feet) above the surface could be very different from the surface wind, and that the prevailing wind half a kilometer aloft would be different still.

This need for wind forecasts in a reorganized weather service makes sense of the request from Köppen in January 1917 that Wegener advise him about the height of meteorological masts for weather stations. Now Wegener was in charge of deciding this for all of Germany: masts with anemometers mounted on them should be the same height above ground at each station, and similarly the precipitation gauges, barometers, and thermometers should all be identically shielded, located at standard distances from surrounding obstacles, and established at fixed heights above ground. All of this had to be written down and integrated with protocols for making observations and calibrating the instruments. Because much of the work was to be done by volunteer weather observers (as is the case in many parts of the world today), it was essential that the instructions be clear, simple, and direct.

All of this pointed to Wegener's suitability for the task: he had turned ordinary college professors into research meteorologists on the western front, and he had turned deckhands and field scientists in other disciplines into weather observers while in Greenland. He understood colloquial descriptions of weather phenomena and was expert in the calibration of simple instruments. He wrote fluidly and well, and he had an enormous appetite for work.

Wegener would remain at this task all of July and part of August 1917, during which period he barely had time to eat and sleep, let alone respond to comments about a hypothesis on continents and oceans he had offered in 1912 and (somewhat reluctantly) recast in 1915. Nevertheless, the displacement hypothesis was "out there," and on the home front it was attracting increasing attention, not least because of the serious consideration given by writers such as Dacqué, Andrée, Tornquist, and Koßmatt. While Wegener was at work in Jüterbog on forecast meteorology, his ideas were at work in a broad and ongoing discussion of major topics in geological theory. It is appropriate, then, to turn to a consideration of the wartime geological debate in which Wegener's ideas, but not Wegener himself, took part.

The Permanence Problem

In 1912 Wegener had framed his argument about continental displacements as a way out of an impasse. According to Wegener, European geologists, in order to explain continuities in flora and fauna on widely separated continents, had opted for a version of the contraction theory in which the deep oceans were relatively recent geological phenomena created by the collapse of huge blocks of Earth's crust. North American geologists, however, who had eagerly embraced the principle of isostasy, asserted that the deep oceans, like the continental blocks themselves, were permanent and primordial features of Earth's crust. The continents had always been continents, and the oceans had always been oceans. Wegener concluded that the Europeans had a sound version of paleontology allied with a deficient geophysics, whereas the North Americans had a sound version of geophysics allied with a deficient paleontology that ignored strong evidence of former continuity in geology and paleontology. Wegener characterized his own work as a way to put sound paleontology and sound geophysics together.

This framing of the question injected him prominently into a major debate in the earth sciences over the permanence or impermanence of the continents and the age of the oceans. While Wegener had certainly grasped the logical importance of "the permanence problem," he was not initially aware of the extent and contentiousness of the debate. In 1915, Edgar Dacqué had said, "Wegener himself has so far only alluded to the idea that his theory also throws new light on the 'question of permanence' but has not yet explained how."[74] Dacqué actually went to some trouble in 1915 to reframe Wegener's contribution to this debate.

While the two positions Wegener had sketched out as the "European" and the "North American" positions did exist, it was not so much that Wegener mediated them, Dacqué wrote, as that he had created a third position, or a third solution to the problem of permanence. That this should be the case was because there was also a *fourth* position. In this last solution, the continents have always been continents, and the oceans have always been oceans. Shallow seas have sometimes covered the continents, and in general in the past the oceans were shallower than they are now. Through time, weighted down by sediment from continental interiors, the continental shelves had been pressed down and area lost permanently to the ocean. Only Australia and Antarctica were once part of a larger continent; the others had always been as they are, where they are. No attention was given to geophysics in this solution, which remained attached to the contraction theory, albeit a contraction that only had the effect of gradually creating abyssal oceans slowly through geologic time. This deepening of the oceans swamped the relatively unimportant (geologically speaking) but paleontologically vital land bridges and island arcs and archipelagoes that explained floral and faunal connections that could not otherwise be explained away.[75]

One can now see that the debate in the teens of the twentieth century was essentially a four-way conjugation of two variables: geophysical evidence and paleontological evidence. North Americans paid maximal attention to geophysics and minimized intercontinental flora and fauna—minimal attention to paleontology. A certain group of Europeans paid minimal attention to geophysics and maximal attention to intercontinental correlation of species of animals and plants. These were the two positions that Wegener saw himself reconciling. Actually, however, Wegener's position was maximal attention to geophysics, combined with maximal attention to intercontinental correlation of animals and plants—maximal paleontology. As Dacqué pointed out, there was a fourth position that was opposite Wegener: this was minimal attention to geophysics, combined with an absolute minimum of intercontinental correlation of animal and plant species. All four of these alternatives, in Dacqué's view, could be concentrated in a single point: "One can reduce the entire permanence-impermanence debate to a single question: At what point did the deep-sea come into existence?"[76] Wegener's strikingly original view was that the continental surfaces had always been continents (allowing shallow marine incursions) and the oceans had always been deep oceans—*but not always in the same configuration or in the same location on Earth's surface*. The continents themselves were the necessary "land bridges," interrupted by rifting and drifting, not by sinking either of narrow isthmian links or of broad conjectural surfaces over the bed of current oceans.

How did Wegener's solution to this problem fare in the hands of Karl Andrée? In 1917 Andrée, working parallel with Dacqué, wanted to simplify the permanence debate by giving prominence to Wegener's ideas on geophysics. On the other hand, he expressed a view shared by almost all geologists who had any sympathy for Wegener, that Wegener represented an extreme position, and that he was overstating a good case. Andrée nonetheless began his discussion of Wegener with a forceful advocacy. Whatever our reservations and criticisms of his views on paleogeography, Andrée wrote, all geologists must be grateful to Wegener for his presentation of geophysics, a presentation that clearly describes, more vividly than anyone has ever described them before, the actual geophysical conditions and forces governing the largest features of Earth's crust: continents, mountains, and oceans.[77]

Wegener has made it clear, Andrée insisted, that in order to explain the great surface features of Earth, we must turn away from molecular forces and toward the "cosmic body forces," understanding that the surface shapes of the lithosphere (the outer crust) are governed not simply by fracture at the surface but by flow processes deep within Earth. Wegener, said Andrée, was following in the footsteps of reputable geological theorists, including Otto Ampferer and Bailey Willis (and Andrée himself), in emphasizing the importance of flow for any hypothesis of continent and mountain building. If further research would be required to determine the extent to which geologists must follow Wegener, "taking Wegener's hypothesis totally *ad acta* [that is, shelving it or filing it away] would mean turning a blind eye to all the advances of modern geophysics."[78] Andrée went on to say that even if we do not go so far as to accept broad horizontal migrations of continental blocks throughout their full thickness, some amount of drift and displacement by some mechanism, at least in earlier periods of Earth history, is clearly required to reconcile the evidence at our disposal.[79]

Having given this ringing endorsement of Wegener's geophysics, Andrée turned to Wegener's paleogeography and paleontology and proceeded to take issue with every major conclusion in Wegener's book. The proposed drifts were probably "impossible."[80] South America and Africa had not been connected up to the Tertiary, and if there had been any drift, it had to be before the Triassic. South America and North America could not have drifted away at different times. The account of the creation of the Indian Ocean by India's movement northward contradicted the fossil evidence. Neither the Andes nor the Himalayas could have been created by folding up the leading edge of a drifting continent or by collision between two continental masses.[81] Pole wander as an explanation for Southern Hemisphere glaciation does not fit the evidence for the rest of Africa and Australia.[82] Finally, a drift of North America away from Europe could not be the reason for the creation of the Atlantic Ocean, because this would presuppose a land bridge between Alaska and Siberia that, until the Tertiary, would have stretched over 35° of longitude.[83]

Most of the issues Andrée raised had already been raised the previous year by the Viennese geologist and paleontologist Carl Diener (1862–1928), and most of the criticisms of Wegener laid out in Andrée's paper are taken directly from that work.[84] Perhaps Andrée felt he needed some heavier artillery in dealing with Wegener's confident paleogeography. Diener had taken a doctorate in 1883 and was a tireless field worker. Throughout the later 1880s and the 1890s he traveled across Europe and Asia, as well as to the Rocky Mountains in North America, the Himalayas, Spitsbergen, the Urals, the Caucasus, Siberia, Hawaii, and much of Canada. This extensive field experience gave him great credibility, as did his position (since 1906) as the professor of paleontology at the University of Vienna. He was internationally well connected, a member of many scientific associations, and an influential teacher and writer.

However well established Diener was, his (and Andrée's) criticisms of Wegener consist mostly of assertions of opinion and obiter dicta, referenced to those paleontologists with whom Diener happened to agree, and referenced to the geology of Sueß, who had been, throughout Wegener's work, a principal foil for his own views—his "Ptolemy." Diener was simply reasserting the minimalist version of correlation against those paleogeographers, cited by Wegener, who held the maximal position (such as Theodor Arldt). Moreover, though accepting some notion of lateral displacement of continents at an early stage of Earth history (read: "wherever he needed it to fill a hole in the story"), Diener gave no clear mechanism for such drift—at the same time that he criticized Wegener for lacking a mechanism for his own drift. Moreover (and this spoke against the coherence of Andrée's paper as well), Diener was thoroughly resistant to taking geophysics seriously, which was the one point of agreement between Andrée and Wegener.

Nevertheless, if this "minimalist position" had certain contradictions, it at least took Wegener seriously and as worthy of engagement. As Dacqué had pointed out, Wegener's position was not halfway between two existing positions (North American and European) but one of four possible positions conjugating the two variables of geophysics and paleontology; Wegener's opposition, as it turned out, had already existed before his early papers had been written; it was, in some paradoxical sense, waiting for him to appear.

To Diener's and Andrée's seriousness and intellectual engagement, we may contrast another early opponent of Wegener: Max Semper (1870–1954), a geologist and paleogeographer at the University of Aachen, and a student of Karl von Zittel (1835–1904). It is from Semper that descends the polemical approach to Wegener as a "scientific outsider," "ignorant of geology," a "reckless" mischief-maker ignorant of geological methods, given to wild speculations, absurdly unprepared to discuss geological matters. Semper's 1917 attack on Wegener, entitled "What Is a Working Hypothesis?," was one of the most intemperate pieces ever written about continental displacement.[85]

Semper was what in the modern world is known as a "methodologist," a scientist who believes that progress—or lack of progress—in science is largely a matter of adherence to good methods (on the one hand) or failure to follow strict inductive procedures (on the other). Semper had developed, before he encountered Wegener's work, the sense that geology was not making progress as a science because of its poor grasp of scientific methods. He had developed a sequential view of stages of accumulation of data in which a working hypothesis was something just above the level of empirical data, somewhat below the level of a "synthesis," and far below the level of a general theory. To a large extent these views were extrapolated from the published opinions of Zittel, whose history of geology and paleontology followed the ideas of the British geologist Archibald Geikie. Geikie believed that German geology had stalled in the eighteenth century because it had let grand theory overwhelm empirical evidence. Certainly there was a case to be made that extensive fieldwork would yield more results concerning the history of Earth than theorizing following the intensive examination of a single locale. However, Semper went far beyond this and attempted to formalize a hierarchy of theory building.

Semper had already applied his theoretical ideas to what he saw as a lack of progress in the study of ancient climates (especially the anomalous distributions of Northern Hemisphere animal and plant remains in the Tertiary) and in geology generally, in papers published in 1910 and 1911. Writing in 1917, he took exception to the work of Wegener as exactly the sort of approach to the subject which would cause grave trouble for geology and had caused it trouble in the past. It is not that Semper's views on this matter were atypical; most early twentieth-century geologists were much more attuned to the accumulation of data than to its interpretation on a large scale, and it is not at all clear that they would even have understood the process that physicists call "data reduction." Indeed, North American geologists, led by the Chicago geologist Thomas Chamberlin (1843–1928), spent a lot of time talking about a method of "multiple working hypotheses," whereby one evaluated ensembles of fieldwork according to several different explanatory theories, supposedly reserving judgment. It was a method more preached than practiced, mostly used as a rhetorical device to attack anyone who had very definite theoretical ideas different from those of the advocates of "multiple working hypotheses."

Thus far, many geologists might have followed Semper in his strictures on Wegener, had he not appended to them a series of absurd and contradictory views: that we didn't know what the ocean floors were made of, that they might be made of iron, that it didn't matter whether Greenland was drifting or not, that geophysics was irrelevant, that synthetic treatments of geological data were a terrible way to go at theory building, that a working hypothesis that involved synthetic data from different geologists was an illegitimate approach to the subject. Finally, at the end of an increasingly personal and nearly hysterical attack, he had written, "Geologists must never forget to inscribe over their doorways: O Holy St. Florian, protect this house and burn another."[86] This apparently obscure phrase was a popular saying, the figurative meaning of which is "not in my backyard!" The implication, quite retrograde, was that geophysicists should have no role in geological theory. If this sounds initially absurd, it is not. The creation of geology as a separate science in the nineteenth century demarcated this study from astronomy and physics, to create a separate space for a historical science, descriptive and inductive in character. At the end of the nineteenth century, however, with geology well established everywhere, the demarcation served less to legitimize geology as a science than to absolve geologists of the necessity to consider physical, astronomical, and geophysical approaches and theories, all of which were increasingly necessary to interpret the data at hand.

The essential theoretical question here was clearly not whether geophysics had a role in geology; much more at issue in 1917 was the question of continental permanence. After, for instance, Wolfgang Soergel (1887–1946) gave his inaugural lecture at the University of Tübingen in 1916, he expanded his already-published critique of Wegener's notion of the splitting and drifting of continents into a book entitled *The Problem of the Permanence of the Oceans and Continents*.[87] Soergel was one of the original three "anti-Wegener" writers named by Dacqué in his letter to Wegener in May 1917.[88] Dacqué had planned to respond to all three of these: Semper, Diener, and Soergel; in the end, he responded only to the last named. Semper had disgraced himself to a certain extent with his intemperate fury, and Diener and Soergel were exponents of the same "minimizing" theme, which attempted to rule out further consideration of Wegener's hypothesis by chipping away at the paleontological evidence for former continuity, to the point where Wegener's solution became an isolated and unwelcome attempt to solve a nonexistent problem.

When Dacqué reviewed Soergel's "anti-Wegener" book in 1917 (wartime paper shortages kept the review on hold until 1918), he seemed less a proponent of Wegener than someone with a judicious interest in clarifying the "permanence problem." Nevertheless, he insisted that Wegener could not be dismissed and that the problems he had raised would not go away by themselves.

It simply would no longer do to wave away isostasy and geophysics, reassert the contraction theory, insert land bridges and island arcs when needed, and make them disappear when they were no longer necessary, though both Diener and Soergel had done just this. Moreover, Wegener, in Dacqué's view, had solved an associated problem that neither the sunken continents theory nor the minimal land bridge theory dealt with: the *Wasserfrage*, or the "water question." Where, in the land bridge theories of recently created oceans or of gradually deepening oceans, did all the abyssal ocean water come from? Only two theories seemed to have a reasonable answer: the North American theory of extreme permanence, and Wegener's theory. The problem "went away" in the North American version because there had always been deep oceans. The problem went away in Wegener's theory because Earth was originally covered with a global ocean and the continents grew over time, split, and moved apart, simply rearranging the geometry of continents and oceans and revealing more of the simatic floor. Of these two theories, the "extreme" version of the permanence theory and the theory of continental displacements, Wegener's was still the more reasonable once paleontology was seriously considered. Dacqué thus arrived at a position like that of Andrée: Wegener's theory might push things too far, but those who opposed it would still have to "get serious" about the geophysical problems facing their own alternatives and deal realistically with the remaining paleontological data.[89]

It would be wrong to see the long debate over Wegener's displacement theory as one of continental "mobilism," as opposed to another stance called "fixism"; these terms were introduced in the 1920s by Émile Argand (1879–1940). This division is true up to a point: Wegener insisted on the lateral mobility of whole continents, while other thinkers insisted with equal force that the continents remain locked in their current positions relative to one another on the surface of the globe.

Yet it is now clear that this "mobile" versus "fixed" argument was originally only one aspect of a debate (with multiple dimensions) about the *permanence* of continents and oceans, irrespective of their motion or fixity. All of Wegener's early opponents accepted some version of polar wandering as a solution to otherwise anomalous distributions of animals and plants in the geological past. Large-scale displacement of Earth's pole of rotation—again and again through geological time—is clearly a "mobilist" idea.

Some of Wegener's opponents also accepted a primordial lateral displacement of continents, most often in the context of the hypothetical partition of the Moon from Earth and the sliding of remaining continents toward the "hole" created by the Moon's departure. Some others, including those later most strongly opposed to Wegener, allowed for contemporary or recent displacement and deformation of continents, albeit local and modest in extent, by flowing currents of molten rock in the subcrust. Still others, also opposed to Wegener, and vigorous in their defense of continents fixed in place, allowed for folding and overthrusting within the margins of continents, making way for massive lateral displacements—hundreds and even thousands of kilometers of *intracontinental* mobility.

The early twentieth-century debate over the question of permanence of continents and oceans not only contained this broad range of ideas about mobility and fixity in the horizontal sense but included an equally great variety of ideas of vertical mobility: up and down movements of the continents in place, progressive and permanent sinking of large blocks of former continents to make abyssal oceans, permanent gradual deepening of the oceans through time via contraction of Earth, and intermittent vertical elevation of mountain ranges caused by Earth contraction.

While large numbers of geologists, geographers, and paleontologists subscribed to various versions of the "mobility" of one or another aspect of Earth, they did so with reluctance and out of necessity. What made Wegener different was that he enjoyed the idea of mobility, and his theory contained mobility of every sort, all of it continuous throughout history, and much of it so rapid as to be inconsistent with the majestic cadences of geological imagination: the coral building up the reef, the raindrop hollowing out the stone. Wegener had Earth's pole of rotation in constant motion in response to the splitting and drifting of the continents, which also proceeded continuously, as did the gradual increase in continental elevation and the gradual decrease in lateral extent of continental surfaces.

The range of opinion concerning Wegener's theory in 1917, from Semper at one extreme to Dacqué at the other, was about how far one had to move toward "mobilism" to solve geological problems that would not go away otherwise. At Semper's end of the spectrum, geologists were willing to throw out huge accumulations of paleontological data suggesting former connections between continents, in order to avoid most or all forms of lateral mobility. At Dacqué's end of the spectrum, there was a hopeful (if cautious) embrace of the possibility of continental displacements as a unified solution to a large array of geological and geophysical problems for which all other existing solutions appeared inadequate.

Here, in 1917, is where the matter rested: four different "solutions" to the problem of the permanence of continents, of which Wegener's was one, and the only one to which the name of a single theorist was attached. To attack Wegener was, therefore, not always to engage the details of his theory, or even to try to understand him, as much as to express animosity toward the notion of extensive, varied, and, above all, *rapid* lateral displacements with which Wegener's name was already becoming associated. An attack on Wegener was sometimes also an attempt to head off a geology anchored in the new geophysics and governed by large-scale correlations rather than local details. Even an ardent exponent of the new geophysics (and a professor of paleontology) like Andrée measured the situation thus: it was not "how much of Wegener" one should accept, but how little, while still achieving a workable resolution of the problem of the "permanence of continents and oceans" to which Wegener had offered a challenging new solution.[90]

While his defenders tussled with his detractors on the issue of continental permanence, Wegener was working long hours to reorganize the domestic weather service. He was exhausted and chronically undernourished because he could not remember to walk to the dining hall (some distance from his office) in time for the meal service in the afternoon and evening. Bread rations were small, and he typically ran out of bread by the middle of the week. Overwork and lack of proper nourishment sapped his health, and within a few weeks of his arrival in Jüterbog he contracted dysentery, and in his weakened condition he could not get well. He continued to work; this was not his first experience with exhaustion and illness, but he began to lose weight and strength.[91]

While Wegener was languishing in Jüterbog in July of 1917, Else and Hilde had left Marburg and gone to stay with Wegener's parents, as they had the previous year. Food was increasingly scarce and expensive in all German cities, and the kitchen garden and preserved food at Zechlinerhütte were, once again, a powerful attraction. Else found her mother-in-law unchanged: still paralyzed on one side, barely able to speak, and spending most of her days sitting in silence. Sister-in-law Tony was still tirelessly caring for her and for Richard, the more so since illness had struck the family again: Alfred's father was dying of esophageal cancer. He had been hale and hearty the previous fall, but in the spring of 1917 his illness had progressed rapidly. He was thin, could barely eat, and had difficulty staying warm even while sitting in the summer Sun.[92]

Sofia, Bulgaria

Wegener wrote to Köppen in early August (his first letter to him since May) to tell him that he had been reassigned as commanding officer of the Central Meteorological Station in Sofia, Bulgaria, an appointment that would begin in September. Bulgaria had entered the war in 1915 on the German side and had used the alliance to pry territory away from Serbia (in Macedonia) and Greece. German and Bulgarian troops had also made significant inroads into Romania. The Central Meteorological Station in Sofia coordinated all weather data from stations in Turkey, Albania, the Black Sea, and the annexed portion of Romania, and it worked together with the Bulgarian Meteorological Institute to provide weather forecasts throughout the Balkans.

Alfred was alternately resigned and enthused about the transfer; the press of administrative work in Jüterbog had prevented him from doing any science at all, and he hoped that he might be able to do some real scientific work in Sofia. His junior officer at Field Weather Station 12, Porstmann, had loaned him a volume of "Sunday Sermons," by the secular humanist philosopher and Nobel Prize–winning chemist Wilhelm Ostwald. Ostwald urged an "energetic imperative," using the second law of thermodynamics as a guide to personal conduct, maximizing energy and productivity. Wegener had found these "Sunday sermons" a little bit bland and colorless, if well meaning, and noted sardonically that, as director in Sofia, he would probably maximize "energy wasting." He hoped that he would be able to off-load as many administrative duties as possible to his junior officers in order to make some time for his own work. He had already determined to have all the weather stations in the Balkans make measurements of atmospheric polarization, and he asked Köppen whether he had any suggestions for other systematic measurements. Here as elsewhere in his military service, he tried to do as much science as was possible, or rather not to miss a chance to do science in the context of the war work.[93]

In mid-August he was rotated out of his command at Jüterbog and given home leave. It was not much of a vacation: Else and the family physician put Alfred on a ten-day diet of nothing but cottage cheese in order to cure his dysentery (it worked). His sickness in July and his rapid recovery in August made it clear to him and to Else how much he needed her care and solace, as well as how much he missed her. They talked about this in Marburg and determined that as soon as he could process the paperwork with his superiors she should come to Sofia and stay for as long as possible; Hilde would go to her maternal grandparents in Hamburg. The pretext was a sponsored program: "the advancement of friendship between the Bulgarian and German peoples."[94]

Wegener traveled to Sofia at the end of August and was immediately buried in work, learning the duties he would take up officially on 1 October. He was also charged to make a tour of his component stations, including those in Constantinople and on the Black Sea. Soon after his arrival, he realized that money would be a problem once again. He had to pay for his own lodgings, and a single room (even without breakfast coffee) cost 100 marks per month, larger than his monthly rent in Marburg for a full apartment. Though Marburg University had paid him two different stipends in the summer of 1917 and he had his captain's salary, even with a 300-mark-per-month subsistence allowance, things were still bad.[95]

If food was expensive, it was at least plentiful, and he told Köppen that he would soon be sending food to him in Germany, even though this was forbidden and treated as a black market activity. In fact, throughout his time in Sofia, Wegener superintended a very large network to obtain food and to send it in small packets, unlikely to be intercepted by the authorities, to the families of his subordinates in Germany. He would eventually turn an unused storeroom in the Meteorological Institute in Sofia into a sort of henhouse, where chickens and geese could be held and fed until someone was going back to Germany on leave, at which point the fowl could be slaughtered and travel as baggage with the furloughed officer. In the meantime, the chickens provided a steady supply of eggs, very scarce in Germany. Flour and fat (shortening, lard) were also prime ingredients in the packets sent home.[96]

Wegener was not rebellious in temperament; he justified his illegal activity as a mandate of his military training to act as a father to his subordinates and to care for them. In this case, given the time-wasting, pointless activity of the Sofia weather station (the hoped-for chance for real research did not materialize), he lived as soldiers came to live everywhere in that war, following those regulations that made sense and willfully ignoring those that did not.[97]

He continued to work on science whenever possible; if there was no possibility of actually generating meteorological results from the Sofia station and its satellite stations, he could proceed with "book work." He generated manuscripts, both reviews and short articles intended for *Meteorologische Zeitschrift*. He worked on wind speed and direction and, inspired by Porstmann's book on scientific measurement standards, attempted to sort out and rectify the use of the concept of "friction" in meteorology. He determined that it was being used in at least four distinct ways, and he wrote a short note attempting clarification. He had it in mind to propose a new meteorological element: "*Gatterung*" in place of *Reibung*, or friction. The former would be a new unit measuring the severity of turbulence at the border between two air masses, or at the contact surface of two different atmospheric layers. None of this ever saw the light of day. *Meteorologische Zeitschrift* was fearfully behind in publication, paper was in short supply, and every activity in Germany not directly related to the war was grinding to a halt.[98]

In October he received permission from his commanding officer that Else could visit him in November in Sofia. Hilde, as planned, would stay with her grandparents in Hamburg. Arriving in Hamburg with Hilde, Else learned that Richard Wegener had succumbed to his esophageal cancer on 12 October. Alfred obtained a brief compassionate leave to bury his father, and then, together with Else, he traveled to Sofia. November was a good month for them; they traveled widely—she accompanied him on his inspection tours of outlying stations. She pitched in to help with the wrapping of the food parcels to be sent back to Germany. She made the acquaintance of most of the German scientists working at the Bulgarian Meteorological Institute; many of these were colleagues of her father, and they were able to socialize in a way that seemed almost like peacetime.[99]

Else returned to Hamburg in December and stayed with her parents. As the financial difficulties increased for their small family, with Alfred's expenses outrunning his income and allowance, it appeared that they would have to give up their Marburg apartment (convenient to both the train station and the university) and move to cheaper lodgings on the outskirts of town, at war's end or even sooner. Meanwhile, Else and Hilde could stay in Hamburg. Lack of food was still a consideration, as it was easier to send food parcels directly to Hamburg than try to send them to Hamburg and Marburg both.

In Sofia, things seem to have been static throughout December and into the New Year. With the success of the revolution in Russia, Lenin and Trotsky had decided to end their war with Germany, and although the Treaty of Brest-Litovsk was not signed until the spring of 1918, there was an armistice on the eastern front. While this decreased the urgency of the meteorological work in Sofia, it produced no letup in the station work itself. Wegener wrote to Köppen in January to say that he was constantly on the move and had no time for work: work here, as always, meant his own scientific work, not his military duties.[100]

It appears that there was to be, in early February, a meteorological conference in Brussels bringing together meteorologists from all of the Central powers and perhaps some neutrals such as Bjerknes. Wegener became quite excited about this and wrote to Köppen about a scheme he was developing for a new way to represent wind direction and speed. He wanted to use hash marks of different lengths at boundaries of air masses, thinking of a way to represent the strength and direction of wind within an air mass and thus the likelihood of the development of zones of convergence and divergence.[101] This scheme (which was not adopted) he presented in Brussels in early February 1918. His failure to gain assent notwithstanding, it indicates his continued enthusiasm for dealing with the emergence of turbulent phenomena in the lower atmosphere, and it shows that in spite of his determination to strike this off his list of "67 topics," he could not help but think about it. It also indicates that the war had succeeded, in no small measure, in turning him from an atmospheric physicist into a "meteorologist."

Continental Displacements, 1918

Just before Christmas, Wegener received a letter from Köppen which included a letter from the Hamburg zoologist Georg Pfeffer (1854–1931). Pfeffer had given a lecture on 12 December 1917 in which he had taken a minimalist position with regard to correlation of species across the Atlantic Ocean. He had read Diener and Andrée, and his criticisms are interesting, because they show the extent to which a specialist in one field may feel entitled to ignore results from another field. Pfeffer said that it was immaterial to him whether or how formerly connected continents came to be separated. He was interested in the distribution of animals and of plants, and neither the geology nor the geophysics of the situation was his responsibility. He would stick to the facts and not worry about hypotheses.

Apparently, in the question session following the lecture, Köppen had asked Pfeffer whether he was aware of Wegener's work and of isostasy. Pfeffer was out of time and told Köppen that he would write to him. Pfeffer wrote to Köppen on 14 December and said that he was indeed aware of Wegener and that Wegener's work on gravity and on Sial and Sima was "theoretically interesting but of no practical significance."[102] This is another way of saying (somewhat more politely than Semper's "O Holy St. Florian, protect this house and burn another") that he felt that it was not his responsibility, he didn't need it, and he didn't want to talk about it.

Wegener's response was interesting and vigorous. "The letter from Pfeffer is typical. He <u>will</u> not listen to anyone. He belongs to that crowd of people who boast about standing on a solid ground of facts and having nothing to do with hypotheses, never realizing that their solid ground of facts has embedded in it a completely false hypothesis!" Wegener nevertheless wanted to hold on to the letter if Köppen no longer needed it, though he added,

> I don't think it is worth the time and effort to deal with people like Pfeffer. His letter lacks any hint that he has the penetration to get to the bottom of things, and expresses only his joy in being able to find fault with other men's views. Such people will refuse any reorientation of their ideas. If they had already learned the displacement theory when they were in school, they would defend it for their whole lives with the same lack of understanding, and with the same sort of incorrect information, they now use to defend the sinking of the continents. The best thing is just to wait until they die off.[103]

Wegener's response is revealing in another way: it was clear that even his own father-in-law, with whom he had many times discussed these ideas, fundamentally misunderstood his views on the displacement of Earth's pole and on the rising and sinking of the continents under load. Here, however, there was an expectation that he was speaking to someone whose mind could be changed. "I have a completely different take on the question of pole wandering than you." He went on to explain that in opposition to the older view of Kreichgauer (that the entire outer crust of Earth would rotate over Earth's interior as the result of pressure and motion in the subcrust), he held the more modern view that redistribution of mass in the outer crust, caused by the lateral displacement of continents, must force the relocation of the axis of Earth, as Earth assumes a new ellipsoidal shape in consequence of the motion of the surface masses. The interior of Earth did not have to be in motion; all it had to be was plastic enough that the pole and therefore the equator, 90° away from the former, would move to positions that reflected the distribution of masses at the surface. The technical way of saying this is that "the pole of rotation would move to coincide with the pole of inertia." This was the view of Schiaparelli and of Lord Kelvin which "Kreichgauer and especially the older geologists know nothing about."[104]

On the question of the rising and sinking of continents and the related question of isostasy, Köppen seems also to have been confused. Hence, Wegener reverted to the question of glacial isostasy: the well-established fact that tide markers in the Baltic Sea were continuing to rise, which exponents of the theory of continental glaciation had explained with regard to isostasy. With the load of thick continental ice sheets removed, the land beneath the ice had risen again. The Sima, the mobile subcrust beneath the continents, played no dynamic role: if the continent were to be pressed down under load, the Sima would flow away laterally; if the load were removed (the melting of the ice), the Sima would flow back underneath, thus lifting the continents. Wegener's explanation here is matter-of-fact but very simple, slow, and painstaking, almost as if talking to a child.[105]

This exchange brought home to Wegener the extent to which his efforts in 1915 to explain geophysics to geologists had been a failure. Even his own father-in-law had not understood what he was talking about. The disconnect between geology, geography, and paleontology on the one hand and geophysics on the other was much greater than he had imagined. Moreover, with his comments about "waiting for people to die" rather than trying to convince them, as well as talking about "the older geologists," there was a dawning realization that the difference was generational and might not be resolved by empirical evidence or reasoned debate. Younger paleontologists like Andrée and Dacqué had training in geophysics, geodesy, and map projections. They might not agree with all of Wegener's arguments, but they knew he was correct in the matter of geophysics and its necessity for any general theory of earth sciences. Older geologists had, in the main, neither the training nor the will to understand the arguments, and they simply ignored them.

Nevertheless, the exchange seems to have galvanized something in Wegener. He had it in mind, since his arrival in Sofia, to establish a wartime colloquium, where the various officers under his command (not all of them scientists) and their Bulgarian counterparts might present the results of their research to one another. He also had access to real scientific institutions—especially the Geographical Institute—with a library and facilities to support such an effort.

During the first week in January 1918, he inaugurated the colloquium with two lectures on his displacement theory, on two successive days. He addressed his audience at the Geographical Institute, in a room with large wall maps. Reporting this colloquium to Wladimir Köppen in a letter on 9 January 1918, he also responded to new questions from the latter, sent in the interim, concerning the displacement theory. In addition to not understanding isostasy or Wegener's version of displacement of the poles, Köppen also could not see how the displacement theory could account for the sinking of the land surface. Colleagues in Hamburg had told him that while the Baltic tide markers might be rising, parts of the North Sea were simultaneously subsiding. Köppen then wondered how Alfred could account for a larger subsiding of the land, of the kind that might allow a marine transgression.

Wegener responded to these new questions with the same calm demeanor that was his habit, even though one suspects he was experiencing considerable frustration. He had, in effect, already explained to his father-in-law that with the removal of the ice cover from Scandinavia, the land must rise as the Sima flowed back underneath the rising continent. This would mean (implicitly—Wegener had not stated it because it was so obvious) that the Sima would flow back in from the oceanic areas surrounding, and that their floors must necessarily then subside. In this letter, however, he took a different approach and explained to Köppen a more general and hypothetical case.

When a continental mass begins to rift, he wrote, the rift does not immediately open from top to bottom and separate completely. Whatever is pulling the continent apart will at first stretch it, which will produce plastic deformation on the underside of the block, with a certain amount of thinning. On the exposed Earth's surface, the opening of the rift would be signaled by a complicated system of faults. Progressively, as the continent was pulled apart, sections of this fault system would drop down and eventually form a large rift valley, with some blocks remaining elevated. When such a rift and fault system reached a shoreline, a marine transgression could proceed even before the rift was complete. Wegener said he believed that this process would explain the appearance of the Aegean Sea (a large map of which he had just been studying at the Geographical Institute), and he interpreted the archipelagoes of the Aegean as the few remaining elevated blocks of a complicated fault system. On the other hand, Wegener wrote, this need not always be the case, and he noted that the split of Africa from South America seems to have proceeded more smoothly and more "of a piece."[106]

This is the sort of back-and-forth interchange which he had had with Hans Cloos in Marburg, where Cloos would offer a geological objection and Wegener would answer with a physically plausible hypothesis, but it was exactly the sort of "picking away" at his theory that came from not understanding the larger geophysical principles involved. Wegener had assumed from the beginning that these general principles, once introduced and understood, would allow his audience to answer their own questions; obviously, this was not the case even for his closest collaborator and patron, his own father-in-law.

Wegener had included in this letter (of 9 January 1918) a compilation of all the mentions he could find of his displacement theory. This list of references has been lost, but it likely contained almost all references to the theory in German and French geological and geographical periodicals, as well as Dutch periodicals; all these would have been available to him at the Geographical Institute in Sofia. The list cannot have been very long, as exhaustive literature reviews made in the early 1920s turned up about twenty items prior to the end of the war, and many of these were short acknowledgments of the existence of the hypothesis, rather than scientific engagements with its content.

The list certainly would have featured Rudzki, Dacqué, Andrée, Lundager, Semper, Diener, and Soergel, as well as perhaps Koßmat and Tornquist. All of these authors' publications would have been available to Köppen in Hamburg, and all of them could have been read in a day or two. They would have immediately shown that, whatever reservations there might be about Wegener's overall displacement theory, there were strong supporters of his geophysical approach, and that there were geologists who favored both displacement of the poles and displacement of the continents as necessary in any general theory of the history of Earth.

It took almost three months for Wegener to clarify these arguments sufficiently that Köppen began to understand them. They exchanged letters about once a month in March, April, and May of 1918.[107] One would not say that Wegener struggled to explain this to Köppen, but he certainly labored. He spent a lot of time correcting the misunderstandings of Semper and Diener, who were under the misapprehension that the evidence for isostasy was somehow equivocal, or that there were other interpretations of gravity data than that the floor of the ocean was more dense than the continental surfaces. Wegener patiently explained that while it was true that Rudzki in his 1916 review had thought that Wegener might have overstated the size of this difference (something that Wegener was willing to admit), this did not change the overall situation. He had to explain to Köppen that the notion of a "mass defect" or a "positive or negative gravity anomaly" did not mean that something was wrong with the mass of Earth, or that gravity in some spot was somehow incorrect, but only that gravity measured in some place was either less than expected or greater than expected given the estimated distribution of mass. It was a rhetorical figure to speak of a "defect."

Along with the gravity data, Wegener patiently explained, one had to consider geological data at the same time. If there were many possible distributions of mass that would account for the gravity observations, most of these made little geological sense. Wegener continued this line of argument to talk about the continuity of mountain ranges across the Atlantic. When two mountain ranges on opposite sides of the Atlantic had almost identical mineralogy (something accepted by geologists), the assumption was that they were the same mountain range, now separated by an ocean, either because a former Atlantic continent had sunk or because the continents had drifted apart. But, said Wegener, let us imagine that there is some measurable chance, say, 10 percent, that the mineralogy is just a coincidence. Thus, when you have two mountain ranges, the chance of coincidence in their identical mineralogy would be 1 in 100; with four mountain ranges, 1 in 10,000; and so on. With a small chance of so many geological coincidences, as well as geophysically compelling evidence that the sinking of a former Atlantic continent was impossible, the displacement theory then gained force.[108]

Home Leave, 1918

Eventually, Wegener just had to give up, as he had with him none of his books out of which he might have explained the numerical values for various gravity corrections and scenarios. In despair, he wrote that since Köppen could not understand explanations of these phenomena given by Helmert and Rudzki in their published work, and since he could not seem to get maps with the kind of projections that could show him the proper continental reconstructions, he would just have to wait until his planned leave in May and June in Marburg. He hoped that Köppen could come, both to discuss this and to see their new apartment. They would be moving to the outskirts of town to a high-ceilinged, floor-through apartment on the Gisselbergerstraße, facing a wooded park that ran along the bank of the river Lahn.[109] He would have three weeks of leave plus a week of travel days.

Wegener was anxious to go home, naturally, but with special reason this time. He had not had leave in almost a year, and he had not seen Else since the previous November. Moreover, she had conceived a child during his last home leave in the late summer of 1917, and in early March 1918 she had given birth to a second daughter, Sophie Käte. The child was named for Alfred's sister (who had died in childhood in 1884), and Alfred had never seen her. Moreover, Hilde was nearly four years old and barely knew her father.[110]

The Wegeners used the home leave to dismantle their apartment on the Biegenstraße near the train station and move to their new home approximately 3 kilometers (2 miles) to the southwest. Alfred also spent part of the leave working with Else on the finishing touches to her translation of Koch's book on the Greenland traverse; in May the Prussian Ministry of Culture had finally awarded Wegener 1,500 marks to support publication of this book.[111] It was an odd time, and their life was in every way the negative image of their life before the war. Back then Alfred had gone away for several weeks of military training each year and been home, very much at home, the rest of the time. Now he was constantly away, except for these few weeks of annual leave.

While in Marburg, Wegener went to the university to confer with Richarz and review his situation. In May the university had petitioned the Ministry of Education once again for a professorship for Wegener in cosmic physics, but there had been no response.[112] Wegener's financial situation was even more precarious in 1918 than usual because of the extra expense of living in Sofia, and he told Richarz of his difficulties. Richarz had offered, as early as 1912, to help Wegener make ends meet out of his own private resources. Wegener was unwilling to accept charity, but Richarz, who had his own foundation, the Bernd-Richarz Stiftung, reconfigured the offer in June as a prize honoring Wegener's scientific work, in the amount of 800 marks, to be paid out in late July. This would barely allow Else to maintain the apartment home, when combined with the stipend as extraordinary professor, but it was something.[113]

Wegener had only just returned to Sofia in early July when the entire staff of the Central Weather Bureau in Sofia was ordered to the western front. The Germans launched repeated offensives in the spring and summer of 1918 and had broken through the Allied lines in a number of places, although they were unable to hold all of their gains. From July to September Alfred was on the western front helping to reestablish a network of field weather stations immediately behind the ground gained by the German army in northern France, just inside the Belgian border. He was constantly on the move and often under artillery fire. The Germans were making extensive use of gas warfare in these offensives, and weather stations had to be numerous and closely packed (parallel to the front) in order to determine those areas in which the prevailing wind would send the gas toward the enemy and not toward their own lines; the weather increasingly determined the timing of attack and retreat.[114]

Dorpat, Estonia

Already in July 1918 it was clear to most officers on the western front that Germany would lose the war and would have to sue for peace. When this would happen was not clear, but it would happen. Wegener's time on the western front was the most dangerous work he had done since the advance into Belgium in 1914. It was very far from the "garrison duty" to which he was supposed to be restricted; the German army in 1918 was far from observing such niceties.

At some time during the summer of 1918, Wegener began to make plans for the end of the war and his career afterward. It was increasingly likely that Germany would have to give up Alsace, and this meant that the plan that he and Kurt had formulated to relocate to the University of Straßburg was no longer tenable: it would again be a French university (renamed Strasbourg), as it had been before 1871.

It is not clear how or why he settled on the idea of moving to Dorpat (Tartu), Estonia. Ulrich Wutzke, Germany's leading expert on Wegener's career and publications, looked carefully into this matter in the 1990s but was unable to determine exactly how the "call" was arranged.[115] Dorpat was deep inside Estonia and had been progressively Russianized since the end of the nineteenth century. While most of the urban bourgeoisie, and thus the civic culture, of this town was dominated by German speakers, the old and distinguished university had been transformed from a joint German/Russian-language institution to a Russian-only university in 1906.

In 1917 Germany had occupied all three of the Baltic nations (Estonia, Latvia, and Lithuania), and although the treaty of Brest-Litovsk (March 1918) had ruled these nations to be independent, at the end of hostilities on the eastern front all three were still occupied by a large German army garrison. In the summer of 1918, someone in the high command of the German army determined that it would be a good idea to occupy these troops by reopening the University of Dorpat as a German institution and allowing the university students among the garrison to continue their education. This reopening as a German institution was to commence in September 1918.

Once again, it was clear that Köppen played a hand in advancing his son-in-law's career. Köppen, who had been born and raised in Russia and was a fluent speaker of Russian, had received an honorary degree from the University of Dorpat in 1901.[116] In August 1918 he had written to his colleague Boris Ismailovich Sreznevskij (1857–1934), professor of meteorology at Dorpat, presumably to arrange a "call" for Wegener. This appeal was apparently a success, as Wegener received a letter from Köppen in September 1918 announcing that he would receive such a call.[117] Two weeks later, Wegener wrote to Carl Becker (1876–1933), who would later become the Prussian minister of culture, to confirm his acceptance of the offer that he should go to Dorpat and teach meteorology to German garrison troops pursuing university study.[118]

Wegener was still at the front in late September 1918, at Marle, in Picardy, well inside France. He was to be reassigned to the University of Dorpat in October and would proceed there after a short home leave in Marburg. Wegener settled in quickly in Dorpat in October, having found lodgings near the astronomical observatory. He wrote to Köppen on his own birthday, 1 November 1918, with news of his colleagues, his plans for a colloquium, the need to plan lectures for the winter and spring semesters, and his plans for purchasing equipment to update the meteorological observatory—since the equipment that he had was "somewhat primitive but certainly still useful."[119] The most useful part of this equipment turned out to be the typewriter in the Army Weather Service office, on which he was writing up his lectures on the investigation of the upper layers of the atmosphere, lectures he planned to give in the spring of 1919. Everything seemed to be moving along, and Kurt had confirmed that Straßburg would not remain German.[120] The choice of Dorpat seemed plausible, as its atmosphere was thoroughly German, and there would be a professorship of meteorology, which might extend for him into the peace.

All of these plans and dreams concerning Dorpat came to a crashing halt as soon as they had begun. Even though it was clear that the war was ending, the German Admiralty ordered the fleet to sea at the end of October 1918, and the sailors at Wilhelmshaven mutinied. Their revolt quickly spread to Berlin, where the workers declared a republic and forced the abdication of Kaiser Wilhelm, who relinquished the throne on 9 November. Two days later, an armistice was signed on the western front. Events moved quickly after that. On the thirteenth of November Russia annulled the Treaty of Brest-Litovsk. The German evacuation of the Baltic republics began almost immediately, and the German command declared the University of Dorpat closed. Wegener was ordered to Berlin, where he arrived at the end of November, and where he was demobilized in early December. He returned to Marburg on 5 December 1918.[121]

In Marburg he found his wife and his two daughters living in a single room in their apartment, on Gisselbergerstraße, trying to stay warm using wood from the university's forest to fuel their coal stove. Wegener was thirty-eight years old, and he was once again an unsalaried instructor preparing to announce courses of lectures for the winter semester 1918/1919—in general meteorology and in general astronomy. His extraordinary professorship had expired with his captain's commission in December; he now had no salary at all. His entire publication record from 1913 through 1918 contained fewer items (and many fewer pages) than he had produced in the academic year 1911–1912. He had no data to work with, no manuscripts in process, no prospects of publication, no students, no support, and no clear research program. He was cold, exhausted, and broke. But he was home, and his war was over.

15

The Geophysicist

HAMBURG, 1919–1920

> Because of the extremely intricate paths that scientific ideas follow in the course of their development, it is often very difficult to trace their origin and growth. It is therefore not surprising that erroneous accounts abound, particularly in the works of younger investigators. Because our findings increase in value when we pursue their origins, and because this pursuit is itself extremely interesting, it is well worth the effort, now and again, to take such a retrospective view . . . questions of priority are very difficult to decide; therefore we may say with Mephisto: "who can say what is foolish, what is brilliant, if he has not contemplated the past?" There are of course original ideas, but they enter the stage so occasionally, and in such varied costume, and receive at first so little notice, that they generally originate and evolve in many minds, independently of one another.
>
> WLADIMIR KÖPPEN (1920)

As soon as the Christmas vacation was over, Wegener returned to teaching in the (now) unheated lecture halls of the University of Marburg. There was no coal, although there was gas (or kerosene) in small quantities for illumination a few hours of each day. His lecture courses were titled "General Meteorology" and "Introduction to Astronomy with Illustrations and Demonstrations"—the latter a short course from February to April, in a semester calendar designed especially for returning veterans.[1] Almost as soon as he returned to this academic career (many times ended and restarted), he learned that it was—at Marburg—finally really over for him. The administration had made one last attempt to procure a professorship for him, and the ministry had replied, in no uncertain terms, not only that they would not create a professorship for Wegener but also that they would forbid his reappointment as an extraordinary (associate) professor any longer, in expectation of such a future appointment. He could, of course, stay on as an instructor, but without a guaranteed salary.[2] There was no animus toward Wegener, but Germany was defeated and financially staggered and had lost at least two universities: Dorpat and Straßburg. Moreover, nationalization of universities in the newly independent nation-states of Hungary, Romania, and Poland, among others, was forcing repatriation of dozens of German "professors in rank," now without professorships or a means of making a living.

Fortunately, for Wegener, there was a way out. In November 1918, shortly after the armistice, Köppen had received a visit in Hamburg from Ernst Kohlschütter (1870–1942), whose post of *Admiralitätsrat* was roughly equivalent to an "Undersecretary of the Navy" in the United States. Kohlschütter stayed for two or three weeks, and he was in Hamburg to plan the future, in the postwar period, of the Deutsche Seewarte (German Naval Observatory). Kohlschütter was an astronomer and mathematician who had participated in an expedition to East Africa to make gravity measurements over the East African rift valleys; it was this expedition that had produced evidence of the negative gravity anomalies (less gravity than expected) that had so impressed Wegener and had figured in his account of splitting and rifting of the continents.[3] Kohlschütter was an expert in astronomical position finding and nautical astronomy and, like Wegener, an enthusiast of improved and modified geodetic instruments.

The ostensible occasion for Kohlschütter's visit was to sound out Köppen about a replacement for Louis Großman (1855–1917), who had headed the Weather Forecast Division (*Abteil M.*) at the observatory until his death the previous year. Köppen brought the conversation around to his own replacement. He had turned seventy-two in September 1918 and would soon celebrate his fortieth year at the observatory; he was anxious to retire and pursue his scientific work. He suggested to Kohlschütter that he consider Kurt Wegener, but by Köppen's account in a letter to Else of 2 December 1918, Kohlschütter asked instead about Alfred. Köppen told him that Alfred wanted to pursue an academic rather than government career, and he passed along three names, with Kurt heading the list.[4]

Kohlschütter was, nevertheless, determined to approach Alfred, and he wrote to Köppen about it. He pointed out that were Alfred to accept the position at the Seewarte, there would be several places where Alfred might give lectures in Hamburg: the Colonial Institute, the Maritime Academy, and especially the university, just then opening its doors for the first time since 1895.

Köppen in turn wrote to Else and asked her to intercede with Alfred: "What I mean is, Alfred should grab it—it won't be easy to find an offer as good. It is very uncertain that professorships in cosmic physics or meteorology will be offered in the foreseeable future in German universities." He went on to point out that if Alfred should take the job in Hamburg, there was nothing to stop him from accepting a call to a professorship elsewhere if such should appear.[5]

Neumayer had created Köppen's position, meteorologist of the observatory, for him in 1879. Previous to that he had been head of Division III, in charge of assembling weather forecasts, but the new position was "Department Head without Department" (*Abteilungsvorstand ohne Abteilung*), meaning that he had no real official duties and was free to pursue scientific work. This wonderful freedom persisted until Neumayer's retirement in 1903, when Neumayer's naval codirector, Captain (later Admiral) Herz, won a bureaucratic struggle to make it less a scientific institute and more an administrative clearinghouse for naval affairs and took over the directorship of the observatory. He had apparently been unpopular in the navy, an achievement he duplicated at the observatory.[6]

By Köppen's account, the new director after 1903 had no understanding of scientific work and took his task at the observatory to be the bringing of order and system to its affairs. This meant much filling out of cumbersome forms, clock punching, and deadline meeting, none of it, as Köppen said, very compatible with scientific work. Moreover, the admiral seemed not to care what sort of work was done as long as the forms were filled out and the deadlines met. Köppen managed to get his scientific work done anyway because he had finally obtained funds for his meteorological kite and balloon station at Großborstel. It was at this time that Köppen moved his family out of Hamburg to the big house on the Violastraße. He was able to convince the admiral to allow him to work at the station two days a week and come in to Hamburg and the observatory on four days.[7]

However much Alfred hated administrative work—and he did—he would at least have two weekdays, his Sunday, and his evenings for his research, as had Köppen. He did not deliberate long. Already, by the first week of January 1919, he had decided. He wrote to Köppen that he had been to Berlin, to meet with various officials and discuss his future. "With Marburg," he wrote, "it is now a definite no. At present it is impossible to create any new positions. . . . B. and K. advised me, off the record, to take the position at the Observatory."[8] "B and K" were, respectively, Carl Becker and Kohlschütter. Becker was the official at the Prussian Ministry of Culture who had approved the short-lived professorship for Wegener at Dorpat. It appears that Kohlschütter, during the Berlin meeting, virtually assured Wegener that he would appoint brother Kurt to be the head of the Weather Forecast Division (Division III) at the observatory.

So, wrote Else, "for us the die was cast."[9] Wegener accepted the appointment as meteorologist of the observatory, effective 1 April 1919 (though not due until September 1919). He was touched by the regret of his Marburg colleagues: "I had the feeling, for the first time this time, that they really wanted to keep me at the University and just were unable to do so. The Faculty seems really to have put their shoulders to the wheel."[10] He would remain at the university in Marburg until the end of the winter term, and then he, Else, and the children would move to Hamburg in the summer. Else was delighted; she had wanted this as early as 1912 and in some sense had never wanted to leave home at all, but only to bring her husband to her family home. This she had now almost accomplished.

The Köppens offered to give up their house entirely and find a new, smaller lodging in the neighborhood, but Alfred and Else refused; it was just too much to ask. They would expand the house, and in the meantime they would take the ground floor, while the Köppens senior moved upstairs. This would be somewhat inconvenient, if only because Alfred required a dedicated workroom away from children, but they could not count on having enough money to expand the house in any case; Kurt predicted that the Central Bank in Berlin would fail as early as March. Berlin was in the midst of the Communist revolution known as the *Spartakus* uprising, and there was street fighting between the Communist revolutionaries, led by Rosa Luxemburg, and the anti-Communist *Freicorps*. Alfred began to think he should get his family to Hamburg soon, in case the predicted collapse of the bank turned their currency and savings into waste paper.

Alfred's previous reluctance to move to Hamburg had vanished. He was not, as he would have been in 1912, a temporary, part-time assistant writing for hire; he was now to be the successor (*Nachfolger*) to one of the greatest meteorologists of the era, as well as a salaried and pensioned senior scientist at a major and respected research institution. He would live in a house that was a pilgrimage site for every meteorologist who came to Germany, in which he had been an awed guest in 1908. He sketched out enthusiastic plans. Kurt would come and head the Weather Division; Alfred was certain that Kurt was capable of making the kite observations himself even though he had not pursued forecast meteorology since early in the war.[11] Alfred envisioned expanding the kite station to include a runway for fixed-wing aircraft, as Kurt was convinced that this mode of observation was the wave of the future. "How strange it would be," he mused, "if the brothers Wegener were to come together once more now, as they were in Lindenberg."[12]

Meteors and the Moon

Before Hamburg could be a reality, there was still the winter semester in Marburg, to which he was already committed. He was also extremely anxious to restart his scientific career, so many times interrupted in the war years. In December 1918 and January 1919 he was able to discover that the last of his publications from the war years had finally appeared. These included a digest in the *Astronomische Nachrichten* of his successful plan to find the Treysa meteorite, a similar short summary of his book on tornadoes and waterspouts in the *Meteorologische Zeitschrift*, and a highly mathematical treatment of mirages, on which he had worked intermittently over the past three years, in the *Annalen der Physik*.[13] As he had promised himself and Köppen, he was determined not to bury his work any longer in obscure journals. This was very much the case for his greatly expanded treatment of color change in meteors, which he published in the *Acta* of the Royal German Academy of Sciences.[14]

When he finally saw the published work on color changes of large meteors, he was very pleased with the way his ample (thirty-four-page) treatment had turned out. It was the initial article in volume 104 of the *Acta* and thus very likely to be read. He had followed the conservative and cautious template he had established with his work on tornadoes and waterspouts, providing a detailed descriptive catalogue of all reliable accounts of the color changes in large meteorites, going back through the nineteenth century.

His catalog recorded the measured elevation, angle of descent, color of the fireball, and color (if any) of the *Rauchschweif* (smoke trail), especially in those rare instances confirmed by spectrographic analysis. He had then aggregated these data, to show that the color change was almost invariably from green to red, and that this shift could be explained by reference to the composition of the atmospheric layer through which the meteor was passing. He then illustrated this with his own diagram (first published in his *Thermodynamik der Atmosphäre*) of the elemental segregation of the atmosphere in the sequence hydrogen→nitrogen→oxygen and correlated the shift in color with the passage from a predominantly hydrogen to a predominantly nitrogen atmosphere. He put forward the opinion, largely confirmed by later observers, that the color of the incandescent trail of a meteor passing through the atmosphere is mostly the result of the ionization of the air and only slightly the result of the meteor's own composition.

As often happened in reviewing his own published work, new ideas occurred to him, and aspects of the topic emerged that were not at first apparent. He saw that there was something more to say about air resistance. He had calculated the probable diameters of meteors producing green and red spectral signatures. He knew that iron meteorites tended to burn up more rapidly in the atmosphere, but he was curious why more large stony meteorites were not discovered, given the large size (more than 100 meters [328 feet] in diameter) estimated for a number of his catalogued objects. One notes that objects in this size range are today denominated as asteroids, with meteoroids restricted to smaller diameters; this discrimination was not yet established in the early twentieth century.

In thinking about air resistance, which caused the frictional heating of the face of the descending body, he had to consider the extent to which the material on the leading edge would not only be heated but also be compressed until the resistance of the air overcame the strength of the stony material and caused it to disintegrate. This would explain the repeated fireball puffs in a descending smoke trail, as the leading face of the meteoroid underwent a resistance greater than its strength and disintegrated. This repeated process of heating and disintegration in the atmosphere would explain why so few meteorites were discovered, and it also explained the survival of his Treysa iron meteorite from 1917, which had landed intact as a 63-kilogram (139-pound) mass. The cohesive strength of iron is much greater than that of rock, and an iron meteorite impacting a solid rock surface could pulverize that surface without itself being destroyed. Such considerations of solid mechanics had been important for him in his work on continental displacements, and his phrase "molar forces overcome molecular forces" here had a parallel, in which the resistance of the air overcame the molecular cohesion of the heated body.[15]

This study of meteors, their trails, and their occasional impacts on Earth's surface led him to consider another related problem: meteor impacts on the Moon. A planetary body with no atmosphere would be subject to bombardment by large meteors that would impact at very high velocities (up to 80 kilometers per second [~18,000 miles per hour]) and likely excavate large craters. There were various hypotheses about the origin of lunar craters: that they were the remnants of exploding bubbles of gas rising in an original magma, that they were the result of "tidal overflow" of lunar magma through fissures in the surface, that they were volcanoes, and finally that they were the result of meteoric impacts. While both the bubble hypothesis and the tidal hypothesis had had strong advocates, the leading candidate was the volcanic hypothesis, with the infall (*Aufsturz*) or impact hypothesis a distant fourth.[16]

The Moon was constantly reappearing in his work. Each of the major geophysical theories about motion of continents (proceeding his own) had also contained a hypothesis of the origin of the Moon.[17] He had just finished (in the fall of 1918) reviewing the lunar observations of Koch in Greenland during 1906–1908 (finally published in 1917), measurements that he had hoped would provide evidence for the westward movement of Greenland. Additionally, he had completed and published, within the previous two years, an important and widely read paper on lunar tides in the atmosphere. Lunar tides had also figured prominently as one of the several speculative mechanisms for lateral displacement of the continents in both his papers of 1912 and his 1915 book on the origin of continents and oceans.

As if these considerations were not enough, interest in the Moon had burgeoned in the first two decades of the twentieth century. "Selenology" was the name given to the study of the origin of the Moon, the mineralogy of its surface, and the interpretation of its surface morphology. In the nineteenth century these considerations were the province of "selenographers"—those telescopic observers who devoted their careers to exactly describing and recording the features of the Moon's surface. By the early twentieth century, however, the most prominent figures among selenologists were geologists. Leading the way to a fuller consideration of the surface features of the Moon, geologists developed the analogy between the lunar craters and those forms on Earth which were most like them in shape: volcanoes. The development of volcanology in the early twentieth century reinforced the long-standing interest in the analogy between lunar craters and terrestrial volcanoes; numerous investigations of volcanoes and volcanic phenomena continually added new empirical content to an old impression of similarity. By 1914, volcano observers had established the range of sizes of volcanic craters on Earth—altitudes, diameters, and depths—and it was also possible to estimate the number of active and extinct volcanoes on Earth's surface.

In 1909, in the final volume of *The Face of the Earth*, Eduard Sueß had devoted a chapter to the Moon, in which he endorsed the ideas that the Moon had separated from Earth and that the lunar craters were volcanoes. It is not incidental that Sueß, in the very same chapter, vigorously rejected the idea of isostasy and the idea of a floating crust as incompatible with both his notions of the strength of rocks and his fundamental commitment to the contraction hypothesis. To the extent that Wegener was rewriting Sueß on the face of Earth, he might also rewrite him concerning the face of the Moon; this he had not yet done.

As we have already seen, Wegener took great intellectual pleasure in the process of sorting out areas of study he considered unnecessarily muddled. He repeatedly characterized his own creative work as the reorganization of existing elements into more systematic and meaningful arrays—reorganizations that he had the pleasure to carry out, but that might well have been carried out by others. Typically these began and ended within the confines of the existing literature and the realm of calculation.

His consideration of the infall or impact hypothesis of lunar craters began as such a reordering of existing literature, but it did not end there. Writing about the work he did in the winter of 1918/1919, he remarked that

> [in] the literature we can see the gross uncertainty afflicting [previous] conceptualization of the details of the impact process, an uncertainty that is the cause of the many missteps and conflicting opinions on the subject. These follow from our nearly complete lack of experience with the sequence of events an impact sets in motion, and the forms resulting from it. The few experiments that have been conducted to date in this area are completely unsystematic, and hardly serve to support the hypothesis. It was to correct this lamentable state of affairs that I performed in the winter of 1918/1919 at the Physical Institute in Marburg, a systematic series of experiments with impact craters.[18]

Wegener wanted to pursue a *systematic* experimental approach to lunar cratering. The principal failure, in his mind, of all the previous attempts to simulate lunar craters in the laboratory was twofold. The first failure was that none of the previous experimenters, back into the beginning of the nineteenth century, had any clear sense of what their experimental simulations were meant to demonstrate beyond the simple indication that something that looked like a lunar crater could be created by throwing or dropping some projectile into a yielding medium. The second failure was the lack of any attempt to quantify the results in physical terms to see whether the results obtained could "scale up" from centimeter and millimeter dimensions to tens or hundreds of kilometers, the latter being the dimensions of the largest known craters on the Moon.

The best example of this failure, the more notable since it was a favored hypothesis, was the idea that lunar craters had been formed by gas explosions in hot, viscous magma, with the ring mountains (the crater rims) as the remnants of exploded bubbles. Part of the appeal of the bubble hypothesis was the ease with which a laboratory simulation could be made with a variety of substances (wax, clay, and pastes of gypsum, lime, or sulfur) in which a viscous mass was heated to boiling and then cooled, or heated and aerated as it cooled; most supporters of the idea had undertaken just such simulations.

In parallel with his experiments on impacts in 1918/1919, Wegener repeated experiments in the recorded literature on the bubble hypothesis and found that the resemblance between the small ringwalls produced in this way and the rims of actual lunar craters was quite superficial. Even more importantly for Wegener, who would return to this point again and again in evaluating the candidate hypotheses, the numerical proportions of craters produced in this way did not approximate those compiled (by careful selenographic mapping) for the craters of the Moon.[19]

Though it was easy to demonstrate that the bubble crater experiments did not produce a crater morphology like that of the Moon, it was easier still, from Wegener's standpoint, to show that "the hypothesis is based on a fallacy."[20] In attacking the bubble hypothesis on physical grounds, Wegener understood—in a way he had not in 1915—that while the concepts involved in scaling physical forces in models were easily grasped, they were not part of the working vocabulary of most geologists or astronomers of the time. The hostile, baffled reception of his book on continental drift and the failure of his geological critics to understand isostasy, or to even understand the nature of geophysical arguments, had alerted him to the necessity of spelling out the reasoning behind what was, for him, an obvious physical point.

In the laboratory setting, and on the centimeter scale, one can produce a variety of effects that depend on the action of what Wegener generically called "molecular forces": electricity, magnetism, cohesive strength, and so on. In working with these forces, it is possible in most cases to ignore the "mass forces" (other than weight); in the laboratory, the gravitational attraction between two objects can scarcely be measured even with sensitive instruments. The point Wegener pressed home was that on a cosmic scale, the situation is reversed. When one is dealing with planetary-sized bodies, molecular forces are completely overshadowed. The larger the masses involved, the more powerful are the mass forces, while the molecular forces do not increase at all. The result is that on a cosmic scale, laboratory experiments based on molecular forces do not carry over.[21]

In the case of the bubble hypothesis, the physical phenomenon involved was that of surface tension (*Oberflächenspannung*), and though "it is possible to form bubbles of different sizes . . . always within definite limits . . . there cannot be, nor could there ever have been larger bubbles on the Moon. . . . Those who would explain immense lunar features hundreds of kilometers in diameter as burst bubbles commit a fallacy as outrageous as someone attempting to ascribe the flotation of ocean liners to the surface tension of water, by analogy with water beetles or water striders, or the fact that a needle can float."[22]

While the tidal and the bubble hypotheses of lunar craters could be dismissed easily on physical grounds, the volcano hypothesis, which was physically plausible but could not be convincingly modeled in the laboratory, required a much more extensive critique. A range of geologists who were experts on terrestrial volcanoes had long supported the idea that the craters of the Moon were volcanoes. Wegener's argument against the volcano hypothesis may be summarized thus: though lunar craters may seem to look more like terrestrial volcanoes than like other known geological phenomena, they don't look much like them. When the two sets of forms are systematically compared, they can be seen to differ in shape, to sit differently with respect to their surroundings, to manifest a very different range of sizes, to be arranged differently relative to one another, and to be distributed quite dissimilarly on the planetary surface.[23]

> The Moon's surface is covered by so many craters that it is doubtful whether there is a single point on it that has not been at some time in the interior of a crater. The volcanoes on the earth, on the other hand, cover an infinitesimal portion of the surface, and their craters are so small that were there astronomers on the Moon, even using the most powerful telescopes they would be able to detect them with great difficulty

> or not at all—who knows if such lunar astronomers would even be aware of their existence? On the Moon, on the other hand, the crater form is represented at every size up to the gigantic dimensions of the Mare, and are not only the most typical and common surface forms, but, as one can see at a glance, practically the only forms.
>
> I do not understand how anyone comparing the Moon to the terrestrial globe can come to any other conclusion than this: the forms are fundamentally different, therefore their origins must be different as well. The contrast between the Earth and the Moon is so glaring that the next generation will surely laugh at our frantic attempts to establish an identity between them.[24]

Turning to the impact hypothesis, Wegener noted that such experiments had been performed with a wide variety of materials. Experimenters had thrown canister shot into a viscous paste of lime, gypsum, and cement. Others had spilled knife tips of dextrin (a fine powder used as sizing and adhesive) onto a surface of the same powder. Still others had used lycopodium powder (the extremely fine spores of ground pine, used in pyrotechnics and to pack burn injuries) as the impacting surface and balls of rubber or yarn as the impactors. All of these had produced satisfying imitations of lunar craters, often including the characteristic central peaks these forms exhibit, but all had some manifest implausibility, physical or geological, mixed in. The canister shot experiment (and that with a rubber ball in lycopodium powder) had used very hard but low-velocity impactors, together with a very yielding surface; the materials were too dissimilar. The experimenter with dextrin had the right scaling, but its proponent thought he had demonstrated that the Moon was covered with dust or sand and had been bombarded with dust or sand.[25]

Wegener's experiments, on the contrary, are notable for their strict protocol, careful scaling, accurate measurement, and systematic variation. Above all, they are interesting because they represent the only *blind* experimental trial in the whole literature of impact craters to compare laboratory craters and real lunar craters, numerically and exactly. Wegener decided, in the course of his experimental design, that he would do a blind comparison with measurements of real lunar craters compiled by Hermann Ebert (1861–1913) in 1890 from a multitude of crater measurements by previous observers. Ebert's measurement data were well known and well respected; Grove Karl Gilbert (1843–1918) had discussed them in a review article on the history of impact cratering experiments in 1892, an article Wegener had much admired and used to guide his own experimental plans.[26]

Ebert had looked at ninety-two craters with diameters from 13 to 161 kilometers (8–100 miles). He had measured the diameter of the craters, the altitudes of the ringwalls above the impact surfaces, the depths of the crater floors relative to the impact surfaces, the altitudes of the central peaks relative to the impact surfaces, the ratios of the crater depths to the ringwall heights, the ratios of the diameters of the craters to the crater depths, the ratios of the crater depths to the heights of the central peaks, and so on.[27] Wegener made note of these various kinds of measurements and ratios but did not examine the actual numerical results of Ebert's measurements before performing his own experiments.

Wegener used cement powder for both the target surface and the impacting mass. As he explained, "The reason for using such a weakly coherent powder is that since the impacts were produced by hand power alone, the kinetic energy of the impacting bodies was also extraordinarily reduced compared to those striking the Moon. If one wishes to achieve similar results in spite of the reduced kinetic energy, one must proportionately reduce the molecular cohesive strength of the rocks of the target surface, since this is overcome within the limit defined by the borders of the lunar craters: thus we arrive at this kind of powder, with a very small degree of cohesion."[28]

This did not, of course, mean that the surface of the Moon was covered with powder, but rather that the powder in the experiment corresponds to the rocky surface of the Moon. Cement was also advantageous because he was able to fix the produced craters by spraying their surfaces carefully with water and then the next day, when they had hardened, soaking the entire mass in water.

After several attempts, he settled on a smooth surface of cement powder in a shallow paper box. He pressed down a centimeter or so of cement powder with a sheet of paper and, removing the paper, sprinkled the surface with about a half centimeter of loose cement. He had discovered that when the surface was too tightly packed, no craters could be produced, and that he had to vary the depth of the impact surface in order to produce central peaks. He produced the craters by throwing a half tablespoon of cement powder, which typically produced craters of 4 or 5 centimeters (1.6–2.0 inches). When satisfied by the general morphology of his craters, he produced eighteen with central peaks. He then measured the diameter of each crater, the altitudes of their ringwalls, the crater floor, and the central peak relative to the undisturbed impact surface. From these values he generated measures of the total crater depth and the total height of the central peak, and using these, he generated additional ratios. He presented these in tabular form, dividing the data into two groups of nine craters each.[29]

These eighteen craters provided the basis for the comparison with the measurements of real lunar craters by Ebert. While the ratio of diameter to crater depth was off by a good bit from the data of Ebert, Wegener's ratios of crater depth to ringwall height and of crater depth to central peak height corresponded almost exactly. Wegener subsequently argued that his largest crater, which was 12.2 centimeters (4.8 inches) in diameter, corresponded to a crater approximately 121 kilometers (75 miles) in diameter on the Moon. Ebert had a set of craters up to 161 kilometers (100 miles) in diameter; thus, the loading of Ebert's data with large craters must make the ratios differ. Wegener was therefore able to argue that he achieved his main objective: the demonstration that by the use of appropriately scaled materials one could produce experimental lunar craters that quite precisely match the dimensional ratios for real lunar craters, and moreover one could do this without introducing selective bias into the experimental trials.[30]

Outside of this test with preexisting data for the actual Moon, Wegener produced systematic variations in his protocol, changing the material of both the impactor and impact surface (viscous cement pulp) to produce impacts exhibiting double ringwall structures, oblique impacts, crater rows, and furrows, and finally he created a system of rays, some longer than 2 meters (7 feet) in length, by dumping a half tablespoon of gypsum powder on a piece of black pasteboard—thus imitating similar features in the lunar maria.[31]

In one series of experiments, Wegener used cement powder for the impacted surface and gypsum powder for the impactor, and then when the concrete had set, he cut a cross section through his experimental crater and demonstrated that the impact debris from the impactor spreads all the way to the crater wall, a question of some interest in analyzing the form of the craters.

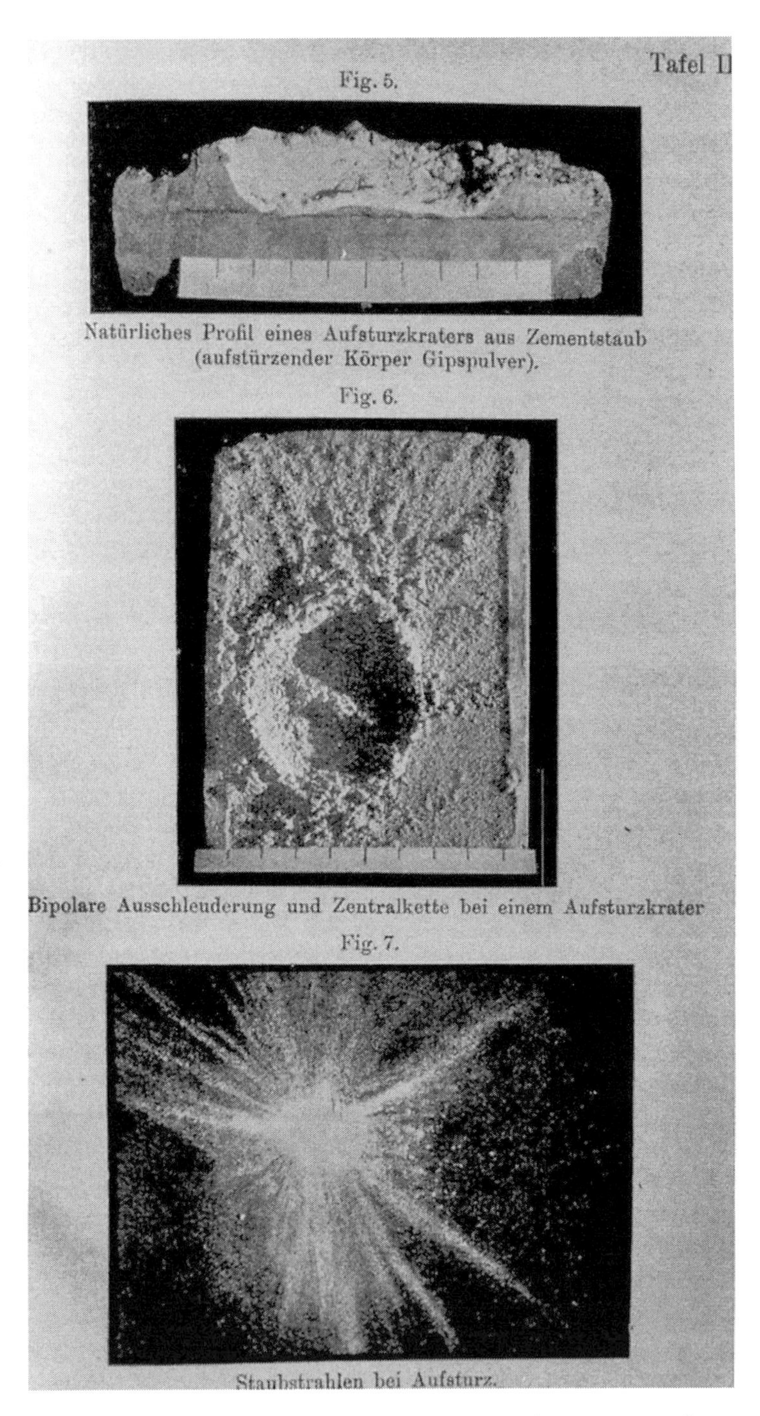

Wegener's crater experiments. At the top is a crater produced with an impactor of gypsum onto a surface of cement powder, to show that the debris of the disintegrated impactor would be blown out to help construct the ringwall of the crater. In the middle is a picture of "bipolar ejection" of impact debris, creating a transverse ridge like those observed on the Moon. The bottom photo shows the impact dust streams created by tipping cement onto black paper, creating the structures observed in the lunar maria. From Alfred Wegener, *Die Entstehung der Mondkrater* (Braunschweig: Friedrich Vieweg & Sohn, 1921).

Wegener emphasized that he had "arrived, following a purely morphological-empirical path, at the result that typical lunar craters are best explained as impact craters."[32] Both qualifications are important. Wegener's experiments explain only the typical crater form and not the great crater seas and maria, and he was "arguing to the best hypothesis" based on available evidence, restricting the latter to the "purely" empirical (i.e., quantitative) study of crater shapes.

Here as always, Wegener was a strongly visual thinker, and his papers on meteors and the Moon are amply illustrated with drawings and photographs. Objects were, for him, physical processes frozen in time, mathematical law made manifest in three-dimensional, measurable reality—out of the noumenon, into the phenomenon. Wegener took the theory of lunar craters in a direction familiar in his other work, especially that on the origin of continents and oceans: the determination of the origin of a series of forms in terms of a "single genetic principle."

Wegener gave a talk on his work on air resistance and impacts before the Marburg Natural History Society in February 1919 and published a brief note in their proceedings, which appeared at the end of March.[33] He then began work on a longer manuscript for submission to the Royal Academy of Sciences, making his work on impact craters a companion piece to his paper on color change in large meteors which had already appeared in this distinguished journal. He had decided to dedicate this manuscript to Richarz, as a gesture of thanks for all his support over many years.

Hamburg

Even though the winter term at Marburg was not yet over and Wegener was not due to take up his position in Hamburg until September, the political and social instability of Germany in the spring of 1919 forced a change in his plans, and he decided to move as soon as possible. Travel by rail could be accomplished only with police permits for each traveler, and the political situation in March 1919 seemed to be deteriorating rather than getting better. Since the road to Hamburg led through Berlin, time was of the essence. They packed hurriedly and in late March set out for Hamburg, having to carry their own food and an alcohol stove with which to cook it on board the train. It took two days to reach Hamburg, with frequent stops and police searches of the train. Wegener saw them as far as Hannover, the last leg of the trip, and then caught a train back to Marburg; he would live for the rest of the semester in a room at the Physical Institute.[34]

Even before the move, Alfred had decided not just to get his family to Großborstel but also to take up his position at the Deutsche Seewarte immediately—five months early. While he could not take up his job as Köppen's replacement until September, he could step in as head of the Weather Forecast Division, replacing Großman, and he indeed took this title (between April and September 1919) and the welcome salary. While Germany was not yet in the grip of the hyperinflation that would beset it between 1921 and 1923, the value of the German mark in 1919 was only half of what it had been before the war; Alfred deemed it unwise and unnecessary for them to expend their slim savings in Marburg while waiting for an appointment to begin in September.

The Köppen-Wegener household in Großborstel, roughly 10 kilometers (6 miles) north of the observatory, was luxurious in comparison with any place that Alfred and Else had ever lived. Surrounded by trees and very close to open country, it had a large fenced yard with berry bushes and fruit trees, as well as an ample kitchen garden. The streetcar line stopped not far from the Violastraße, and from there it was only a few kilometers to the rail station on the main line into Hamburg, and a second streetcar ride to the Seewarte.[35] Kurt Wegener made the move to Hamburg at the same time, and Else found him lodgings near the Köppen-Wegener household. Still unmarried, Kurt preferred living in "rooms" rather than a full house of his own.

Beginning in April 1919, Alfred (and Kurt as well) commuted to the Naval Observatory four days a week. The observatory had a number of divisions, but fundamentally the scientific output was split between oceanography/hydrography on one side and meteorology on the other, with this joint preoccupation expressed in the title of the institution's fifty-year-old publication: *Annals of Hydrography and Maritime Meteorology: Journal of Navigation and Oceanography.*

The oceanography and hydrography section proceeded with its work much the same way as before the war, though now under the direction of [Carl] Wilhelm Brennecke (1875–1924), who took over general editorship of the Seewarte's monthly journal in 1919. Brennecke had been at the observatory since 1904 and had been a participant in the German Antarctic Expedition of 1911–1912. Brennecke, Kurt, and Alfred were very close in age and represented the new generation in both meteorology and oceanography. This is exactly what Kohlschütter had wanted: to revitalize the entire institution under the direction of younger men.

Kurt and Alfred were again a team, working together doing practical meteorology. Kurt had been at the University of Straßburg just before the war, in a setting that was pioneering new forecast methods. Alfred had been, as we have seen, a full-time forecast meteorologist since 1916 and had played a role in the modernization of the Domestic Weather Service that year. In the spring of 1919 Kurt and Alfred were able to assemble a staff of men contemporary with themselves, or slightly younger, all able and eager. Kurt recruited Paul Pummerer (1884–1957), trained in both astronomy and meteorology, and who had lost his academic position when the University of Straßburg reverted to French control in November 1918; he was part of that flood of talented men displaced by the settlements of the peace. Alfred brought in his wartime subordinate and favorite coworker Erich Kuhlbrodt (1891–1972), and they were joined by Wegener's former student from Marburg, Johannes Georgi (1888–1972).

The brothers Wegener decided to begin their tenure at the observatory by completely revising and modernizing all the protocols for collecting, compiling, and disseminating weather data. Forecast meteorology in Germany had suffered in the war years from an inability to get weather data from England and France, those countries immediately to the west and in the path of prevailing weather that would arrive in Germany a day or two later. The observatory had tried to make up for the lack of foreign data by increasing the number of upper-altitude observations; the leading edges of frontal systems (as we would now describe them) are visible first at higher altitudes. This, however, did little to increase the completeness and accuracy of wartime forecasting. Moreover, telegraphic communication between the observatory and its satellite stations was often interrupted—by physical interruption and by censorship during the war, and by strikes and civil unrest during the peace; weather telegrams often arrived in Hamburg too late to have any role in the forecasts.[36]

One of the first technical changes they introduced at the observatory was a shift away from wire-telegraphic communication to *Funkentelegraphie* (radio telegraphy), in which messages were transmitted in Morse code on radio frequencies. Not only did this speed up communication, but even while Germany remained out of scientific contact—let alone cooperation—with England, France, and Belgium and had not yet established or reestablished ties with Holland and Denmark, the rapid shift in all of these nations to the transmission of weather data by radio telegraphy in 1919 allowed the Germans to intercept and process the "three-day forecasts" that were the standard in these countries and to employ these data in the preparation of weather forecasts at the observatory. They also engineered a transition in the way forecast data were to be assembled and altered the format and content of the observatory's weather maps.[37]

Alfred and Kurt understood that this was an ideal opportunity to reestablish the German Marine Observatory, and especially the aerological station at Großborstel, as a major center of meteorological research and innovation, returning it to the prominence it had enjoyed when Köppen had founded it at the beginning of the century. Köppen's aerological kite station, constructed in 1903, had burned to the ground in 1913. Though it had been rebuilt shortly thereafter, the recent erection of electrical power lines in the vicinity had made regular kite flying almost impossible; the danger that the kite cables would cut the power lines was simply too great to risk. The station, therefore, was reduced to a daily pilot balloon ascent. Alfred and Kurt would have to rethink the entire plan of observations at Hamburg.[38]

In May and June of 1919 Alfred and Kurt drew up extensive plans for a state-of-the-art research institute, to be located a few kilometers to the north of the current kite station and adjacent to the new Hamburg airport, in 1919 already serving civilian domestic aviation. A new kite station, expanded machine shops, a more modern winch, an instrument building, and a repair facility would make this a scaled-down version of Aßmann's plans for Lindenberg twenty years prior. This new aerological station would, as laid out in detail in Kurt's 1918 book *Vom Fliegen*, progressively shift the standard instrument platform away from kites and manned balloons toward fixed-wing aircraft, which could be taken aloft several times a day to record meteorological elements in short flights that could attain readings at altitudes of several kilometers, without the immense expenditure of effort required to obtain the same data with aerological kites.[39]

So, as Bjerknes had his institute in Bergen and Exner had his institute in Vienna, Alfred Wegener would have his institute in Hamburg. He would push ahead with his plans to advance meteorology as a science, reestablish a new center of gravity for that science in Germany (after the loss of Straßburg), and begin to train a new generation of German meteorologists in modern methods of aerological research. He hoped and expected that the new style of weather maps and the new observation protocols would quickly generate the kind of data needed to understand (in much greater detail) the structure of the troposphere. Rather than accepting the bifurcation of meteorology into a theoretical atmospheric physics on the one hand and practical weather forecasting on the other, he would unify these into a theoretical and practical meteorology that could take its place as an equal among the other sciences in the German university system.

In his expectation that he would rebuild German meteorology and train a new generation of meteorologists, Alfred pressed forward with his plan to habilitate at the University of Hamburg, an institution that had reopened its doors in March 1918 and would offer its first full semester of courses in the fall of 1919. Writing from Großborstel in early July 1919, he requested permission from the dean of the mathematical-scientific faculty at the University of Hamburg to be allowed to transfer his habilitation from Marburg to Hamburg. He asked to be named an instructor in meteorology and geophysics and noted that he had been told that, in association with his appointment as the meteorologist of the German Marine Observatory, "there would be no obstacle to a possible instructorship at the University of Hamburg in consequence of his appointment." He also noted that, in expectation of this appointment, he had "recently refused a call to be director of the Baden Central Bureau of Meteorology and Hydrography, which came with the promise of a teaching appointment in Meteorology at the Technical Institute in Karlsruhe."[40] It had been no great loss for Alfred to turn down the call to Baden and Karlsruhe; this state was the new southwestern borderland of postwar Germany now that Alsace had been returned to France, and it was at a disadvantage from the standpoint of observation—shielded from a full range of weather by the Jura Mountains to the southwest.

He sent this letter off at the end of almost three months of intense work, and he was anxious for a vacation. Ordinarily, in July, they would all have gone to *die Hütte*, but rail travel through Berlin was still a problem, and they had other resources nearby. Alfred, Kurt, and Else decided on a hiking tour in Schleswig-Holstein to the north, in an area known as the Holsteinische Schweiz, a forest and lake district very like the Ruppiner Schweiz in the Brandenburg of Alfred's youth. Alfred and Else took their first hike together since the fall of 1918, when she had come to Sofia. Hiking along the shores of the beautiful lake known as the Plöner See, they could see the sailboats in the fresh summer breeze, and this, like the rest of the landscape, reminded them of the pleasures of their youth.[41]

On the trip, enjoying a freedom they had not known for the many years of war and separation, they could actually plan something besides a brief rendezvous, or a way to get more food and fuel. Over the course of several evenings, they (and Kurt) decided that they would buy a sailboat—not a small open boat like the one that Alfred and Kurt had sailed in their childhood, but a 9-meter (30-foot) boat with a full cabin and galley. There were many such boats for sale currently in Hamburg, and they could obtain one at a price they could afford to share; they would be able to sail every weekend if they wished. Hamburg, 100 kilometers (62 miles) inland from the sea, was at the narrow end of the great estuary of the Elbe River which opens to the North Sea at Cuxhaven. When the tides were right, they would be able to ride the flow all the way to the ocean and back again, catching the outgoing tide on a Saturday morning and the incoming tide the following Monday.

Returning to Hamburg from their trip, they immediately set about finding a boat, and by early August they had found a *Kajütkreuzer* (cabin cruiser), a 9-meter gaff-rigged ketch, in wonderful condition. It was in many respects a greatly enlarged version of the 3-meter (10-foot) boat of Alfred's childhood, and thus familiar, but with the wonderful addition of the cabin with berths, a head, a larder, and a galley.

The purchase of the sailboat led to another idea. They had planned to go and see Alfred's mother, Anna, at the end of August, in the break between the summer and fall semesters, and immediately prior to Alfred taking up his position as successor to Wladimir Köppen at the observatory. They decided that it would be fun to take Anna sailing. She had been largely housebound for years and was increasingly lonely and isolated after Richard's death. Still partially paralyzed from the stroke she had suffered in the summer of 1913, she was nevertheless alert. Between 27 August and 6 September they sailed with her through the chain of lakes in Brandenburg which Alfred and Kurt had sailed more than a decade earlier. Else kept a charming record of this voyage, with observations about the weather, the sights, and their passage through the locks. They had taken along Hilde, who was old enough to be a companion to her grandmother and could be trusted not to launch herself out of the boat at an inconvenient moment. The sailing trip turned out to be bittersweet, as Anna died within a few weeks of returning home; it was the last time they saw her alive.[42]

Hamburg University

The death of Alfred's mother in late September 1919 gave a sense of finality to the shift in professional plans and domestic arrangements signaled by the move to Hamburg. *Die Hütte* was now his no longer; it belonged to Tony. They were (of course) always welcome there, but as guests. Alfred's home and his work were now in Hamburg. He took up his duties officially on 14 September, becoming the head of the meteorological work of the observatory and in charge of all its official correspondence, which was extensive and burdensome. Two weeks later, on 27 September, he received the expected appointment as instructor in the University of Hamburg, with his subject area defined as "Geophysics, especially Meteorology."[43]

Wegener had been excused the necessity of a *Habilitationsschrift*, but he still had to give his inaugural lecture (*Antrittsvorlesung*). In keeping with his new title and his academic appointment, he elected to give a lecture entitled "Meteorological Problems in Polar Research," which required little in the way of new preparation and could be expected to be of interest to a broad audience, especially when illustrated amply with lantern slides. He gave his lecture on 11 October, and as there was another lecture scheduled immediately after his—by a recently appointed botanist, Edgar Irmscher (1887–1968)—Alfred and Else decided to stay on, as a courtesy to their new colleague.

As Alfred and Else took their seats in the lecture hall to await the lecture by the young botanist, Else looked up at the front of the room, where an assistant was hanging a series of maps. She elbowed her husband to look at them: they were maps of continental positions in different periods of Earth history, from the standpoint of Wegener's theory of continental displacements.[44] The young botanist, Irmscher, strode to the podium and announced his lecture topic: "The Origin of Continents in Relation to the Diffusion of Plants."[45] The lecture, to Alfred's intense surprise and pleasure, was a full and detailed defense of the displacement theory as the most reasonable hypothesis for explaining the diffusion and distribution of plants on different continents. Irmscher posed the problem as a contest between the hypothesis of continental permanence, in exactly the form it had been given by Soergel, and Wegener's hypothesis of continental displacement.

Moving from map to map over the course of an hour, Irmscher explained that the distribution of certain families of vascular plants required unhindered communication, which had previously been explained by the land bridge theory—Irmscher referred to this theory alternately as the *Brückentheorie* (bridge theory) and *Permanenztheorie* (permanence theory). He gave a detailed exposition of the bridge theory of Soergel as outlined in the 1917 book in which the latter had repeatedly attacked Wegener. Irmscher went on to assert that whatever the utility of the bridge theory to geologists, it was clearly a geophysical impossibility because of isostasy. Soergel was simply *wrong*. He concluded, "The acceptance of polar wandering, and the associated displacement of climate zones is, in any case, essential for the understanding of plant diffusion. Of the views that have been put forth concerning our picture of the origin of the largest features of the earth, the displacement theory of Wegener accords better than any of the others with what we know about the distribution of vascular plants. This material [the distribution of plants] may be considered to offer support for this theory."[46]

Irmscher had no idea that Wegener was even in Hamburg, let alone that they were prospective colleagues at the same university, and Wegener had no idea that he had garnered support and even allegiance from botanists and plant geographers. Heretofore he had mostly employed paleontological data gathered by men hostile to his own theory, with the sole exception of Edgar Dacqué, and yet here was a young botanist and paleobotanist willing not just to consider his displacement theory as a reasonable hypothesis but to judge it the sole adequate candidate proposed to date to explain the distribution of vascular plants. Moreover, this support had come from a man who was soon to head the Institute of General Botany at the University of Hamburg and could provide counterweight to Hamburg zoologist Georg Pfeffer's woolly dismissiveness. The contrast between the botanist and the zoologist confirmed in Wegener's mind his intuition that a generational shift was necessary for a change in opinion: Pfeffer was sixty-five; Irmscher was thirty-two.

Continental Displacements, 1919–1920

Whether Irmscher's lecture was the inspiration or simply a spur to drive him on, in the fall of 1919 Wegener decided to proceed with an updated and revised edition of his 1915 book *Die Entstehung der Kontinente und Ozeane*. As preparation for this revision, he sent to a bookbinder a copy of that book, to be rebound with blank sheets between each pair of printed pages. On these blank pages he meticulously copied out passages from any work that had appeared between 1916 and 1919 which referred to, or had any import for, his displacement theory; these references and sources he placed directly opposite those points in the book to which they referred.[47]

In the autumn of 1919, now with access to the observatory's extensive library and periodical collections and to his father-in-law's personal library of several thousand volumes, Alfred was able to work his way systematically through this material and to think through the comments and criticisms. He recorded (in his newly rebound version of his 1915 book) those emendations that seemed most apposite and addressed those misunderstandings (and errors) that most needed amplification and correction.

More important for this effort than his proximity to the library and reference materials was his proximity to Köppen himself. In 1911, and again in 1915, Wegener had made major changes to a book manuscript *after* its last review by Köppen; in each case these changes and additions had been magnets for the criticism of his reviewers: from Exner in the case of his 1911 book on thermodynamics, and from Diener, Semper, and Soergel in the case of his book on continents and oceans. Alfred had written to Köppen in late 1915, admitting his disastrous penchant for being deflected by side issues and his inability to keep himself focused on the main line of argument.

In their correspondence in the spring of 1918, Köppen had repeatedly pressed Wegener to explain isostasy, "mass defects," "negative gravity anomalies," and other aspects of his geophysical argument. Köppen had indeed at first misunderstood these ideas, but in the main the fault lay not with Köppen's intellect but with Wegener's tortured exposition. The proof of this is that on 6 March 1918 Köppen had given a lecture before the Hamburg Natural History Society, on the subject of "Isostasy and the Origin of the Continents."[48] A page-and-a-half summary of this lecture which appeared in the society's proceedings gave a clearer picture of isostasy, of the difference in gravity over the continents and over the oceans, and of the density difference between the Sal and Sima than Wegener had been able to produce in the first four chapters of his 1915 book.

Köppen, talking to a nonspecialist audience in March 1918, kept it simple and made it clear: the rocks of the continents are less dense than those of the ocean. Somewhere 40–50 kilometers (25–31 miles) below the surface there is a "Lava Sea," in which the continents, 100 kilometers thick, float. These continents can no more sink further into the material below them than icebergs can sink in the ocean. This is true, he continued, whether one accepts the notion that continents move laterally or the notion that continents are permanent in place. If one wants to believe that former continents were connected by broad land bridges, that is one thing; but to imagine that such continental land bridges can sink when no longer needed is contrary to the known facts.[49]

Georg Pfeffer was in attendance at the lecture, and one may suspect that Köppen was delivering it as a response to Pfeffer's lecture in December 1917 attacking Wegener's hypothesis. Köppen's lecture drew a strong response from Pfeffer: the secretary of the society noted drily in the minutes of the meeting that "Prof. Pfeffer turned the discussion after the lecture into a long oration against the principles underlying Prof. Wegener's views, attempting to invalidate them, using facts from the realms of paleontology and the zoogeography."[50]

What Pfeffer missed is that Köppen *did not* defend that theory but merely proposed it as an alternative to the theory of permanent continents and oceans. In fact, in 1918 Köppen was not yet convinced that his son-in-law was correct; what he had finally understood was that the contraction theory in its current form—that is, the *Brückentheorie* of broad intercontinental connections via now-sunken land bridges—was impossible, and that this impossibility extended to all paleogeographic theories that depended on the existence of such sunken land bridges. It was as if Köppen had grasped for the first time Wegener's formulation in 1912: extensive evidence exists of former connections between the continents. The geophysical theory (the contraction theory) that explained these former connections and their later interruption (with its sunken land bridges) had been shown to be impossible. Only a few choices were available: reject the paleontology, reject the geophysics, or find a way to reconcile the paleontology and the geophysics. What was *not* possible was what Diener, Semper, Soergel, and Pfeffer wanted: to make this all go away. That was the sole point of Köppen's lecture: that everyone who knew anything about geophysics accepted the principle of isostasy; it was not negotiable, and it would not go away.

Köppen completed a revised and expanded version of this lecture in May 1918 and offered it to the *Geographische Zeitschrift*. In this article he demonstrated that the principle of isostasy was not some geophysical oddity useful to a hypothesis about floating continents, but absolutely essential to understanding the postglacial surfaces of northern Europe and Scandinavia. He showed a map of a postglacial lake running the length of Norway and Sweden, long since vanished but with its shorelines still existing, noting that only the deep depression of the rock by the weight of ice in central Scandinavia could have created a basin large enough to hold this amount of water; the documented uplift of Scandinavia would explain the lake's disappearance. He also employed Wegener's map of the reconstructed North Atlantic continent before the separation of North America and Greenland from Europe, showing that Wegener's calculated pole position allowed the farthest extent of the ice in both North America and Europe to occur exactly on the 60° north latitude circle, which matched the disposition of ice age glacial remains (especially the moraines) in the present.[51]

Köppen's choice of a geographical journal was no accident. From the beginning Wegener's hypothesis had been of most use to geographers, paleontologists, and

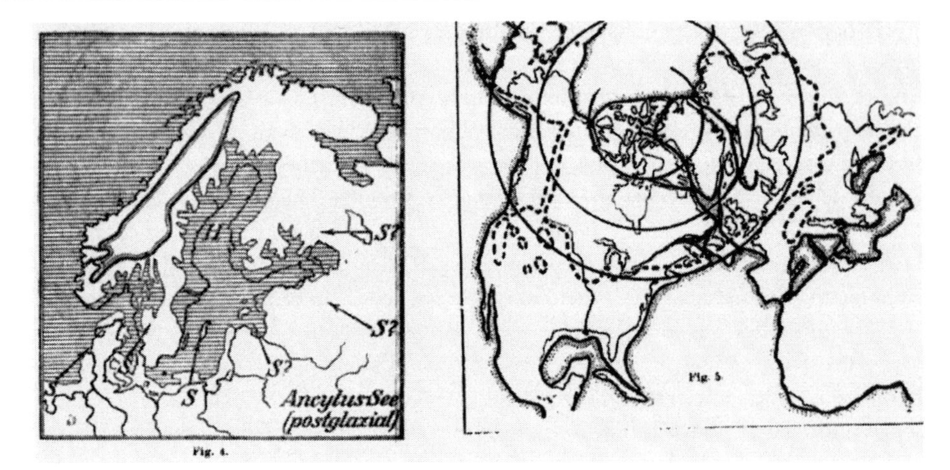

Köppen's figures showing the extent of a former postglacial lake covering most of Sweden, verified by shorelines, which had disappeared by isostatic uplift (*left*), and his use of one of Wegener's figures of a reassembled Northern Hemisphere continent to show how it made sense of the extent of the ice during the Pleistocene glaciations (*right*). From W[ladimir] Köppen, "Über Isostasie und die Entstehung der Kontinente," *Geographische Zeitschrift* 25 (1918); see note 51.

paleogeographers who could employ it to explain puzzling distributions of plants and animals in the fossil record. While this could be done for most of the history of the Northern Hemisphere merely by moving the pole of Earth, it could not be done for the Southern Hemisphere by pole motions alone. Köppen understood the need to find support where it could be found. Wegener's original statement of the hypothesis had, after all, been accepted to a geographical journal; Wegener's early champion Dacqué was a pioneer of paleogeography. Wegener's initial plan to have the hypothesis tested by telegraphic time signals had been endorsed by senior geodesists and geographers such as Carl Albrecht and Albrecht Penck. Köppen used his article in *Geographische Zeitschrift* to keep the terms of the argument in front of this audience.

Now, eighteen months later, Köppen still needed to be convinced, not so much of the scientific plausibility of his son-in-law's hypothesis, but rather of its accuracy with regard to *all* classes of existing data. This was an important point. The adequacy of a global hypothesis of any kind must be a compromise between its scope (how many things it gathers within its ambit) and its precision (how many things it can explain in detail). This is particularly difficult in combining the data of geology and paleontology on one hand and the data of geophysics on the other into a hypothesis of the "whole of Earth."

Once Wegener began the revisions for the second edition, in October 1919, Köppen and Wegener discussed the book almost every day, though the time for this joint work and these conversations was severely limited. Throughout the fall of 1919, Alfred was extremely busy at the Marine Observatory, putting into place the new protocols for gathering weather data. He was also teaching a course at the University of Hamburg that autumn, entitled "Aerology," aimed at training a new generation of German meteorologists from among the many veterans enrolled.[52] The choice of this course was tactical with regard to his time, as the subject matter could best be taught at the kite station near Großborstel, rather than in the city. This saved a few hours of travel each week, but the time pressure was still intense and the commutes long.

The work on continents and oceans was thus restricted to the late afternoon after Alfred returned from the observatory (conversations with Köppen) and the later evening after the children had gone to bed (the actual writing and revision). In spite of Alfred's desire for a separate workroom, lamp oil was in short supply, and there was only enough for one lamp each evening, and thus no illumination in Alfred's study while family life was under way in the main room of their lodgings. Once the children were asleep, Else worked at her sewing, sharing the lamplight with Alfred as he worked at his desk, and in this way it was much as it had been in Marburg in the early years of their marriage.[53]

On Sunday afternoons, often a social time at the joint household, Alfred would retire to his study after the midday meal, leaving Kurt to play with the children. Else made it clear in her memories of this time that Alfred was not in any case very good at playing with the children, and that for him social time and small talk were generally considered time wasted.[54] Johannes Georgi, Wegener's younger colleague at the observatory, noted that when he was a Sunday guest at the Köppen household, Wegener invariably excused himself "to attend to urgent work in his study," something that Georgi very much admired.[55] This "urgent work" was his work on continents and oceans, for which he had to seize every spare hour.

We know fewer of the details of the collaboration between Wegener and Köppen in 1919–1920 than in any previous period; shared residence and daily contact made letters unnecessary. We do know that Köppen became so deeply involved in the book that Wegener would write in the foreword to it that "several chapters grew out of an exchange of ideas with him [Köppen] so close and continuous, that it is no longer possible to partition credit."[56] He went on to say that "just as the first edition was advanced through the generous geological advice and collaboration of Cloos, so is the second characterized by the no less valuable collaboration of a climatologist; its development proceeded in a daily exchange of views with W. Köppen."[57] We know from Else's recollections of this time that "it was an uncommonly productive collaboration. Many days my father waited impatiently for Alfred's return from the Observatory so that he might discuss a newly discovered argument with him. At this time he always carried a small globe of the Earth in his pocket, so that he might at any time check his reflections on pole positions and climate zones."[58]

The small globe that Köppen carried in his pocket at all times in 1919–1920 is a key to a central difficulty in the debate about continental displacements in this period. Paradoxical as it may seem, geologists (Wegener's fiercest opponents) lived and worked on a flat or nearly flat Earth. Geologic mapping proceeded on a much finer scale than geographical reconnaissance; this explains to some extent why geographers were easier to convince than geologists: their sense of adequate work was not so aggressively fine-grained.

Thinking about the whole Earth as a planetary body was relatively new to geology, even in 1900. The first major geology text in the United States to consider the origin of Earth as a *geological* topic was Chamberlin and Salisbury's *Geology*, published between 1905 and 1907.[59] One of the striking novelties of Edward Sueß's *Das Antlitz der Erde* had been his insistence on viewing Earth from space and seeing Earth's surface on a continental scale, emphasizing the continuity of Earth's largest rather than its smallest surface features. This sort of presentation stands in marked contrast to the work of someone like Oswald Heer (1809–1883), who, like Sueß, worked in great detail on the geology of Switzerland. Heer's most popular work, *Die Urwelt der Schweiz*

(1865), contained stunning lithographs of the appearance of Switzerland in different geological eras, complete with tropical plants and animals and sometimes oceans. These scenes are, however, always viewed from the vantage of someone standing at ground level; it is the world seen from the surface of Earth, not seen by a "visitor from space."[60]

Geological fieldwork was so tightly constrained by concern for detail that the curvature of Earth rarely came into play; nevertheless, in the margins of geological and topographical maps there are lines called "Mercator grid ticks," which show the offset of the position on a spherical Earth from the position on a Mercator map of Earth. For the lower and middle latitudes, within 45° of the equator, these differences are rather small, but they increase dramatically at higher latitudes, becoming infinite at the poles. This unreflective commitment by geologists to the Mercator map created real problems for them in visualizing Wegener's argument. It turns out that a number of objections by geologists to Wegener's continental reconstructions, as well as a number of their confusions in understanding his scheme of continental motions and reconstructions, came from their automatic resort, both physically and in memory, to the "Mercator world." To understand this, we have to stop for a moment and discuss the problem of map projections.

To portray the spherical Earth on a plane surface, one struggles to balance the preservation of true angles (*conformal* mapping) with the preservation of the true surface (*equivalence* mapping). These two properties cannot be perfectly attained in the same projection, which is why there are different projections for different purposes. The Mercator map, introduced by Gerardus Mercator (1512–1594) in 1569, was a revolutionary aid to navigation and is the ultimate *conformal* map, with every aspect of projection sacrificed to true angular measurements and thus true direction. It allowed sailors on a spherical Earth to navigate dependably while charting their courses using compass directions on a flat map. To accomplish this, Mercator drew the meridians of longitude so that they would be equally spaced and did not converge toward the poles. Latitude and longitude were then in a rectangular grid. However, this progressively distorted the area of land surfaces at high latitudes, making them many times their true area. This mattered little for most European navigation when it was largely confined to midlatitudes and especially the Atlantic. If you kept to your course on a Mercator map, you would eventually arrive at the destination you chose, but because of the distortion of areas at higher latitudes, you might be some days off in your estimate of when you would arrive.

This Mercator map is even today the most familiar of all projections, and it is also the reason why most children think that Greenland is the same size as South America (which is how it appears on the Mercator map), when actually it is the size of Mexico. The Mercator map's smaller distortions in midlatitudes, in contrast, give a fairly accurate representation of the land surface of the United States, southern Canada, most of Mexico, and most of Europe and the Mediterranean. It can therefore be employed for a number of purposes, including the portrayal of climate zones in the most familiar inhabited areas. When Köppen published his epochal *Versuch einer Klassification der Klimate* in 1901, with the climates of Earth defined in terms of groups of plants and animals found at different latitudes, he used the Mercator map to delineate these latitude zones.[61] Köppen's mental world was also a "Mercator world"—thus his need for constant resort to the globe in his pocket in 1919: he had to learn to see Earth differently, and to do this, he had constantly to look at a physical globe instead of the Mercator map firmly lodged in his consciousness and memory.

Wegener was trained as an astronomer, and the image of Earth that always came first to his mind was a rotating sphere revolving around a star. Because of this, the map of Earth which Wegener carried in his mind and used in thinking about Earth was entirely different from the Mercator projection: Wegener (in 1920 and after) always drew continental and paleocontinetal positions using a very different projection: the "azimuthal equal-area projection" constructed in 1772 by the German philosopher, mathematician, and astronomer Johann Heinrich Lambert (1728–1777). Lambert had been something of a hero to Wegener in his university days. Self-taught in science, Lambert studied comets and planetary astronomy, was interested in meteorology, and delighted in mechanical devices—all interests Wegener shared.[62] Wegener had argued, in his brief history of cosmology, written while still a student, that Lambert should have had a full share of credit for the origin of the nebular hypothesis, most often attributed to Kant and Laplace.[63] Wegener's reasons for choosing Lambert's map projections were, however, entirely scientific and unsentimental.

Once again, to portray the spherical Earth on a plane surface, one struggles to balance the preservation of true angles (conformal mapping) with the preservation of the true surface (equivalence mapping). The Lambert azimuthal equal-area projection that Wegener favored is an *equivalence* map with two outstanding characteristics: all geographic directions from the center of the map are true directions (thus "azimuthal"), and all the area measurements at all latitudes and longitudes on the map correspond to the real surface areas on the spherical Earth (thus "equal area"). The one parameter sacrificed in this projection is true distance: distances are compressed at the limits of the plane circle representing a hemisphere of Earth. Lambert also constructed an "azimuthal equidistant projection" for use when distance was more important than area.

Not only was this the first map that came to Wegener's mind, but it is also the projection he saw in the atlas in 1911 which inspired his idea of continental displacements. The original inspirational pages he had seen in Andree's atlas had *three* versions of this map: the main map was the Equatorial Aspect of the Lambert Azimuthal Equal Area projection centered on longitudes 110° west and 70° east (to put the oceans at the margins and minimize distortion of continental surfaces). Along with this map, in the margins of the same page, Andree had used the Hammer Elliptical (Lambert) Azimuthal Equal Area projection centered on the Greenwich Meridian and two aspects of the Lambert Oblique Azimuthal Equal Area centered on the Greenwich Meridian and 50° north latitude (Berlin) and the facing hemisphere on 180° longitude and 50° south latitude.

Wegener, in both 1912 and 1915, had made a fundamental error in assessing his audience. He had thought that he had to teach them geophysics; what he actually had to teach them was how to read a map properly, as well as how to read the *proper map*. Köppen had been able to demonstrate—in his Hamburg address and in his article in *Geographische Zeitschrift*—that one could appeal to authority with regard to geophysics, rather than try to teach it to geologists who would or could not learn. On the other hand, it was absolutely essential, in order to represent in two dimensions what happened on a three-dimensional Earth, that the directions of motion of the continents be true directions and the mapped area of the surfaces of former continents be their true areas. Köppen became convinced, and in turn convinced Wegener, that this insistence on the proper visualization of the planet would be the key to persuading the audience. Wegener's method, even in his own description, was "morphological-empirical," and it was in this arena that the battle would actually be fought.

Nothing made this clearer than the errors of Carl Diener in his wartime criticism of Wegener's book. Diener had accused Wegener of using a continental reconstruction in the Northern Hemisphere which had created a 35° separation between Russia and Alaska across the Bering Strait, and in another part of his argument Diener had misplaced the North Pole by 90° relative to Wegener's reconstruction of the gathering of continents around the South Pole, before their dispersal in the Carboniferous and Permian. Wegener demonstrated to Köppen, first in a letter and then in person in early 1918, that Diener's visualization of the situation could only have been created by someone looking at a Mercator map, someone who saw Earth distributed on a rectangle, rather than a globe, with consequent distortions of area. It was at this point that Köppen actually understood the significance of map projections when considering continental displacements, so much so that he added a section to his manuscript to be published in *Geographische Zeitschrift* showing that Diener's accusations—that Wegener had made absurd claims about continental motions—were actually themselves embarrassing errors by someone looking at a Mercator map who could not visualize Earth as a globe.[64]

Edgar Dacqué had anticipated these considerations already in 1915, though too late for Wegener to incorporate them into his own book. Chapter 9 of Dacqué's book on paleogeography was titled "The Design of Paleogeographic Maps and Their Details."[65] In the final section, "Principal Objections against [different] Paleogeographic Maps and Their Data," he had remarked, "I direct your attention to the theories of continental displacement treated extensively in Chapter IV. Even if such theories are not demonstrably true, but merely arguable, they show already what intricacies and especially what precautions are involved in the design of paleographic maps."[66]

He went on to spell out his concern that however suitable the Mercator projection for the demonstration of faunal and floral latitude zones, when discussions moved into polar regions, it was preferable to use equal-area maps, and this was of crucial importance when considering theories of displacement of the poles and displacement of the continents. Even here, not all equal-area maps would be equally suitable, and in consequence, "the best procedure is and will remain, especially for educational purposes in lectures and seminars, the simultaneous use and comparison with the globe, always with the realization, naturally, that neither the equator, nor Earth's poles are in the same constant positions throughout Earth's history, and that neither a globe, nor maps that represent the world in the present, can accurately and absolutely represent all paleogeographic details."[67]

This idea was foremost in Wegener's mind in preparing the second edition. When he came to write the introduction to the book in April 1920, after six months of continuous work, he gave this final bit of advice to his readers: "The reader is emphatically advised that a large number of questions—if one does not want to give up one's own judgment [i.e. does not wish to resort to the arguments of others to make up one's mind]—certainly require the use of a terrestrial globe. An atlas is not sufficient because of the distortions caused by the projections used. The criticism of the first edition suffers throughout from a failure to refer to the globe."[68]

The tone of the criticisms of the first edition was not as troublesome to Wegener as that the majority of objections raised by his (intemperate) critics were based on misunderstandings. Now, in 1919, the major problem for Wegener was that these misunderstandings, misreadings, and inaccurate descriptions of his work had been carried over wholesale into a variety of subsequent reviews. This was quite serious, and he found himself accused of gross errors of fact by men who were being treated (by a second round of reviewers) as authoritative critics, when the errors involved were those of the critics themselves, almost all attributable (as Wegener saw it) to their failure to evaluate the arguments using proper paleogeographic maps, or by reference to a globe of Earth.

While Wegener could not make his critics agree with him, he could make the book easier to understand, make the argument easier to follow, and reorganize its elements in a way that might fit the conceptual world of his readers better than the structure he had chosen in 1915. Assuming that a new edition of the book would produce both new supporters and new critics, at the very least he could try to ensure that these critics would have to disagree with what he had actually said about continental motions and pole displacements on a spherical Earth, not on a Mercator map. Thus, at every junction in his argument, he would strive to remind his readers that they should be thinking of lateral displacements on a globe, not on a flat map.

Given the inherent complexity of his argument (much in the spirit of Darwin's *Origin of Species*, the book was "one long argument"), Wegener decided to produce a very detailed analytical table of contents. The previous edition had a table of contents merely twelve lines long, and some of the chapter titles were but a single word: "Splitting," or "Gondwanaland." In the second edition, the argument was laid out in full detail in the table of contents. Nearly every page in the book would have its own subhead title, referenced to the page or pages on which that material appeared. The second edition has 135 pages, and there are 114 titled subheadings in its table of contents. This detail would provide an overview of the work for new readers, but it would also alert his critics (and supporters alike) to the location of the revisions and corrections he had made in light of previous criticisms and suggestions.

To further the aim of making the book's argument transparent and easy to follow, Wegener also decided to include a comprehensive index of authors, concepts, and geological place-names. Of his previous works, only *Thermodynamik der Atmosphäre* had an index; the circumstances of composing *Wind- und Wasserhosen in Europa* and the first edition of *Die Entstehung der Kontinente und Ozeane*—both in wartime—had persuaded him to proceed to print without an index in each case. This had definitely been a mistake, especially in the latter book, in which he sometimes covered the same material in four or five different chapters, leaving the reader to page through the book to try to remember where an argument or concept had appeared before.

In preparing this revision, Wegener also made a radical change in his mode of work. He had corresponded with Köppen about everything he wrote from 1908 onward, but he had worked on the first papers on continental displacements in 1912 almost entirely alone, and he had worked through the first edition of his book in 1915 on the topic with a single collaborator, Hans Cloos, who had given him a firmer footing in the literature of early twentieth-century geology. However, in all the other areas from which he had drawn evidence for his hypothesis, he had depended on his own reading, with mixed results. To remedy this obvious fault, he recruited several scientists to read through his manuscript as he prepared it and to offer criticism in advance of publication.

He first recruited Edgar Irmscher, close by in Hamburg and already a supporter of his hypothesis; Irmscher could review anything having to do with the distribution of plants. Wegener also engaged the cooperation of Ernst Tams (1882–1963), a young seismologist, also at Hamburg, then on his way to becoming one of Germany's leading geophysicists. Tams was pioneering the use of travel times of seismic waves to show the difference between continental and oceanic crust. Additionally, Wegener got the enthusiastic support of Wilhelm Michaelsen (1860–1937), an assistant at the Zoological Museum in Hamburg, about to be promoted to its curator. Michaelsen was an expert on the distribution of earthworms in the present and the past, and he became intrigued by the possibility that Wegener's hypothesis of continental drift could help explain how so many earthworm species or closely related species could exist on widely distributed continents, while not being able (obviously) to traverse abyssal oceans in order to get there.

Wegener also enlisted Karl Andrée, who had depended heavily, in his highly critical 1917 review of Wegener's book, on the opinions of Carl Diener. Wegener could use Andrée as a test case to see whether correcting Diener's misunderstandings could make a convert of a skeptic, and in any case he could have a sound critique by Andrée from the standpoint of paleontology. Finally, throughout the first half of 1920, Wegener corresponded with Hans Cloos, who had just settled into a professorship at the University of Breslau. Wegener depended on Cloos to send him any further notices concerning the first edition of the book, and he hoped to count on his support when the new edition appeared.[69]

Unlike the first edition then, this manuscript would be read through before publication by a geologist, two paleontologists (one for animals and one for plants), a zoologist, a geophysicist, and a climatologist. Under such circumstances, it would be much more difficult for his critics to claim that he had made factual errors, as had happened with Semper, Diener, and Soergel. Now, with his empirical evidence fortified and supported by recognized experts, Wegener could force his critics to admit that many of his supposed "mistakes" were actually differences of expert opinion, with well-respected scientists supporting Wegener's views on these issues. Wegener was learning that in such debates, it was not just what was said but who was saying it that mattered.

Wegener said that this new edition of his book was so different from the first that it should actually be considered a new book. The papers that Wegener had written in 1912 were reports of a novel hypothesis, with special attention to the problems that it solved, or might solve, and a conjugation of several lines of evidence that rendered it plausible. The short book that he published in 1915 amplified some of the arguments and evidence for which there was no room in a journal article, but once again it contained fundamentally no new insight. The book on which he embarked in 1919 and 1920 came into a very different environment from all his previous work on this topic. His hypothesis was now established in the literature as something worth defending; even more importantly, it was something worth attacking.

Not only had the environment of his theoretical ideas changed, but his reputation and role had also undergone a transformation. He was no longer an unknown instructor in physics and astronomy in a small university in the far southwest of Germany, but a senior scientist at the German Marine Observatory. Whatever the controversial character of his work on continents and oceans, he was a recognized authority on the thermodynamics of the atmosphere, and his work on tornadoes and waterspouts had had an effect he had not imagined: it gave him a large international following among forecast meteorologists, not as a theoretician but as a reliable synthesizer of difficult data sets. Not all of these things, of course, would be available for comment or even inspection to his opponents or his supporters in the realm of continents and oceans, but they were available to him, and his self-confidence had undergone considerable revolution.

The new edition would be a different argument, addressed to a different audience. For one thing, Wegener's hypothesis was already of interest to the large public in Germany which read popular scientific magazines. As early as 1916, the geodesist Andreas Galle (1858–1943), an expert on variations in Earth's axis, had written an appreciative article about the attempt (begun by Albrecht in 1914) to measure the separation of Europe and North America in the *Deutsche Revue*, "Are Europe and North American Moving Apart?," and treated the matter as an intriguing and serious idea under active investigation by experts.[70]

By 1919, discussions of Wegener's ideas written for popular audiences were increasingly common. Edgar Dacqué, still Wegener's strongest supporter, wrote a short popular book entitled *Geographie der Vorwelt* (Geography of the prehistoric world) which fully incorporated Wegener's ideas about the displacement of the pole and continents as essential for a correct understanding of Earth's past. Heinrich Schmitthenner (1887–1957), reviewing the book for *Geographische Zeitschrift*, noted that "not every reader will be as willing as the author [Dacqué] to follow Wegener in his brilliant explorations of the dark sea of hypotheses," but that Dacqué had indeed made a good case that the "permanence problem" could be solved by adopting Wegener's ideas.[71]

Wegener's publisher, Vieweg, wanted to move the book in the second edition from the *Sammlung Vieweg* (Vieweg Series) to the *Sammlung Wissenschaft* (Science Series), itself a statement that the book was no longer a speculative suggestion but a serious working hypothesis under active consideration; this was a significant boost. The editor of the Science Series, Eilhard Wiedemann (1852–1928), was a physicist who specialized in thermodynamics and optics, as well as an editor of the distinguished journal *Annalen der Physik*, where Wegener had just published, in 1918, his detailed mathematical treatment of mirages.[72]

Ordinarily the move from a popular to a technical series would have meant that Wegener should have made his argument deeper and more technical, but it was evident from the response to the first edition that the argument was already too technical for many readers. Therefore, he chose not to deepen the argument but to reorganize it in order to make it clearer. This was a delicate matter, as he was both responding to critics and attempting to gain new supporters. The book had to answer criticisms raised of the first edition, but it could not be defensive or entirely reactive; it had to move forward.

The first order of business was to provide what the first edition had so conspicuously lacked, namely, a clear, brief, and direct summary of the hypothesis at the very beginning of the book. Wegener wrote such a summary as his first chapter, entitled "Land Bridges, the Permanence of the Oceans, and Isostasy." It was a reprise of the trajectory of geological theory over the previous fifty years. In a bare twelve pages it moved geological theory in a continuous arc from general satisfaction with the contraction theory (here called the *Schrumpfungshypothese*—the shrinking theory) toward the necessity of the displacement hypothesis. It emphasized that the contraction theory, while still popular with some geologists, had been abandoned by most and entirely abandoned by geophysicists, who rejected the idea that it could be the cause of Earth's surface morphology, and especially not the cause of mountain ranges. The vestigial remnant of the shrinking theory was the land bridge theory, in which huge sections of formerly continuous continents were supposed to have broken down throughout geological history to create abyssal oceans. However, this vestige, Wegener asserted, was no longer seriously advocated by anyone.[73]

Wegener updated the context of his hypothesis so that there were only two alternatives: the permanence theory and the theory of continental displacements. The advocates of the permanence theory and of the displacement theory both accepted the same geophysics; their principal differences lay in explaining the distribution of plant and animal remains. The advocates of permanence tried to explain these remains in terms of transgression and regression of the ocean from broad continental margins, such that, given the migration of Earth's pole of rotation, sometimes broad land bridges were exposed across the Bering Strait, in the North Sea, in the Aegean Sea, in the Bass Strait between Tasmania and Australia, and in many other regions.[74]

In setting up this opposition, Wegener adopted the framing of the problem offered to him by his supporters, most recently by Irmscher in the fall of 1919. Now his task would be to show the superiority of displacement over permanence by showing the difficulties of the permanence hypothesis. The first of these difficulties would be how the land bridges could have risen to the surface of the ocean in recent geologic history. On an earlier Earth, with shallower oceans, this need not be a problem. In the recent past, with a mean difference between continental surfaces and ocean floors of about 5,000 meters (16,404 feet), "this would seem to be extremely difficult to explain."[75] The second difficulty (of which Wegener was made aware by Dacqué) for the permanence hypothesis was explaining where the water of the ocean went when these land bridges were elevated. This was the inverse of the vexing question of where the water *came from* in the land bridge theory of sunken continents. Both of these processes—the elevation of the bridges and the displacement of the oceans—would involve tremendous disturbances of isostasy, for which the permanence hypothesis gave no account.

In this construction, the displacement hypothesis was no longer a middle ground between other theories; it was "the only way out" of a series of fatal contradictions.[76] By accepting this idea of horizontal motion, "we obtain the possibility of reconstructing broad land connections, where today we find deep ocean, and are able to do this without coming into conflict with isostasy and without creating difficulties about the total volume of water on the earth."[77]

There remained now the problem of detail: how much did his displacement theory explain? His geological critics hammered this point in their attacks, and even supporters asked for more specific information. Here Wegener sought to finesse the issue: "It is the task of the following sections to show that the displacement theory produces a range of unexpected simplifications, and that it is capable of bringing together the totality of our contemporary knowledge into a single picture. A document as brief as that before you can, naturally, only produce a sketch, and the filling out [of the hypothesis] would require a very great number of painstakingly detailed studies."[78]

If Wegener's book is a sketch, here we can only give a sketch of that sketch, pointing out the principal lines of attack, especially those that differ from his previous versions of his hypothesis. These lines of attack do not appear in different sections of the book but all mixed together. This is nowhere any more apparent than in the first substantive chapter, "The Nature of the Deep Sea Floor." "The theory of the displacement of continents finds its strongest physical justification," he claimed, "in a new conception of the nature of the floor of the deep-sea, that one might formulate as a principle: 'the floors of the deep-sea are not parts of the lithosphere but composed of the heavier material of the barysphere.'"[79] The crucial evidence for the correctness of this new view, he wrote, was the existence of a bimodal distribution of elevations of the crust. "In the entire realm of geophysics there is scarcely another example which provides as clear a law as that provided by these elevation statistics."[80]

Where, in the previous edition, the difference between the continents and the oceans had rested most strongly on gravity measurements, now he shifted back to morphology. Rudzki, in his 1916 review of the first edition, had suggested that Wegener had overinterpreted the clarity and certainty of gravity measurements. On the other hand, Soergel, in his attack on Wegener in 1917, had concentrated on attacking the bimodal elevation as evidence for displacement. Wegener listened to his supporter, Rudzki, and backed off a bit on gravity measurements; this afforded him the opportunity to respond to Soergel.

Wegener noted that while the bimodal elevation of Earth's surface had been known for a long time, he had been the first to offer an interpretation or an explanation of it. Soergel, Wegener said, was the only one who seems to have understood this, as well as how crucial a piece of evidence it was for displacement; Soergel had therefore attacked it directly "in his polemic." He had argued that such a difference could simply be an accidental outfall of a certain range of elevations and depressions; the disturbances of isostasy would be the same. This argument, Wegener replied, was based on a fundamental lack of understanding of statistics. Any accidental distribution of elevations and depressions would produce a "normal curve," a Gaussian distribution with a single central peak. The only way to get a bimodal distribution was the causal forcing of the situation by some factor other than elevation.[81]

Wegener pressed forward, citing items of evidence not included in either of his previous versions of the hypothesis. The ocean floor has no mountain ranges, which we know from the explorations of the Challenger Expedition. The ocean floor has only volcanic rocks and a very small class of deep-sea sediments, not the kind of rocks that we find on the continents; we know this from a series of dredging operations. The rocks of the ocean floor are differently magnetized than the rocks of the continents, and seismic waves travel at different speeds through the continental surface and through the ocean floor. This last bit of information he got from his newfound ally Ernst Tams.[82]

Now the aim and structure of the book begin to come into focus, and its contrast with the previous versions becomes more evident. The material that follows this exemplary treatment of the floor of the deep ocean has four sections, each with a separate but coordinate function, which reveal together Wegener's new conception of what continental displacements mean for the sciences of Earth, how best to make the case for them, and to whom to make that case.

Gone from this edition is the sense that Wegener, working alone, had come up with a Copernican rewriting of the history of Earth as a counterpoint to Edward Sueß's "Ptolemy." In fact, gone is the idea that continental displacements are something that belongs to the realm of geology, or that the scientists that need to be convinced of their reality are geologists. Wegener's new view is that he, as a geophysicist (his new self-description), is uniting the findings of geophysics with the findings of geography. His aim is to give a comprehensive explanation of the history of the large-scale morphology of the surface of Earth and therefore also an explanation of the distribution of life-forms.

This new description of his profession and the reasons for his book is certainly worth a few moments of reflection. At the end of the opening chapter, Wegener acknowledged that others before him had seen the parallelism of the coastlines and had spoken of the link between the displacement of the continents and the creation of mountain ranges. Yet, he said, "as a *geophysicist* [italics added] I came across these works naturally for the first time only through my examination of the geological literature as I worked out the details of the theory."[83] This is a different picture of the relationship of geology and geophysics than we have seen previously; Wegener now expects his readers to understand that these fields of investigation are so separate that no one who worked in one field might reasonably be expected to master the contents of the literature of the other without undertaking a detailed study for some reason. In this new view of things, he was in 1920—and had been in 1911–1912—a *geophysicist*.

Yet what of his attempt to unify the structure and bulk physical properties of Earth's landmasses, oceans, and atmosphere on the one hand with the details of topography and animal and plant geography on the other? Certainly these seem farther afield from one another than the relationship between geophysics and geology. Of course, one could account for this odd conjugation through his collaboration with Köppen. Wegener had first become acquainted with Köppen as a meteorologist when in 1906 he attempted to borrow or buy some of Köppen's "Hamburg Style" kites to take to Greenland. It was only as their relationship matured that he came to appreciate that his father-in-law's fame stemmed principally from his work as a climatologist, as the inventor of a system of climate classification based on the geography of plants. Köppen's system of latitude zones became the basis of their collaborative work on the migration of Earth's pole of rotation through time. Indeed, Köppen seems to have been won over to Wegener's ideas in 1919–1920 as the "Ariadne's thread through the labyrinth of paleoclimatology" (*der rote Faden im Labyrinth der Paläoklimatologie*).[84]

Earlier (in chap. 9) we approached Wegener's initial intuition of continental displacements by suggesting that in addition to finding out how Wegener may have "discovered" continental displacements, we should perhaps be interested in how "continental displacements had discovered him." We suggested that his training and circumstances prepared him not just to have an intuition but also to develop it into a mature theory. Now, as we move from an examination of the formulation of Wegener's hypothesis in 1911–1912 to its extensive modification in 1919–1920, we may ask not only what Wegener did to revise his hypothesis but also in what ways his changed milieu and audience helped to *revise his hypothesis for him*.

Consider the following. Wegener had published several articles between 1906 and 1912 in *Gerlands Beiträge zur Geophysik*. Georg Gerland (1833–1919), a geographer, had founded this journal with the express intent of reforming geography scientifically. For him, this had meant a fourfold limitation of geography to geophysics, plant geography, animal geography, and sociology. Such an endeavor had no immediate antecedents in geography in Germany, since he was directing geographers to the phenomena of geophysics (the physical interactions of land, water, and air), where they had made little contribution.[85]

Gerland's founding editorial in the first issue had argued that (logically) the study of the forms of relief on Earth's surface should begin with an explanation of the continents, but that after a half century of work in geophysics there was still no explanation of why there should be continents at all. Gerland wanted to push geography toward consideration of high-level generalizations, and he thought of geography as the senior science in which the results of the geophysical sciences—meteorology, seismology, volcanology, terrestrial magnetism, geodesy, and hydrology—all served to help understand Earth. This universal science of geography had the final aim (for Gerland) of understanding how all these processes, interacting in a realm of land, water, and air, produced habitat for plants, animals, and human beings.[86]

Not just Gerland but his student Alfred Hettner (1859–1941), who founded *Geographische Zeitschrift*, remained interested in questions of high generality. Many of the leaders of German geography in this era committed themselves to a study of the origins of the largest landforms—both continents and oceans. This was true of Ferdinand von Richthofen (1833–1905) and Albrecht Penck (1858–1945), who succeeded him at the Institut für Meereskunde (Oceanographic Institute) in Berlin. Penck, like Köppen, was a climatologist and deeply committed to a physical understanding of the origin of ice ages.

In short, just as Wegener was "born into other people's lives" in 1880, so his hypothesis of continental displacements was "born into other people's careers," finding, as it turned out, its natural home in the second and third decades of the twentieth century not in geology, which remained either indifferent or hostile to the idea of the *origin* of continents and oceans, but rather in this comprehensive and uniquely German form of geography represented by those geographers named above.

Wegener's new self-definition as a geophysicist embodied his dawning realization that in Germany the cosmic physics he had sought to found already had a (university) home in the union of geophysics and geography, and this is where his audience resided. As a "geophysicist" he was enrolling himself as a member of this particular German approach to geography pioneered by Gerland, Hettner, von Richthofen, and Penck. This rapidly growing community also included paleogeographers such as Andrée, Irmscher, and Dacqué.

Among geologists, what support he might hope to find would come from the geophysical side, from those investigating the formation and deformation of mountain ranges by huge lateral overthrusts—a not insignificant group, and one growing rapidly in the twentieth century. He could hope to find some understanding from those geologists who believed that some sort of flow of molten material underneath deeply buried sediments could drive such mountain folding.[87]

These hopes for future support notwithstanding, Wegener still had to respond to his current opponents, who had tried to discredit him by pointing to a long list of detailed geological structures and locales that seemed to depart from or contradict his overarching scheme. Some of these critiques were more cogent than others. Carl Diener had quite rightly asked of Wegener how South America could move to the west, throwing up huge mountain ranges and deforming along its western margins, while the delicate structure of the Isthmus of Panama remained undeformed, and he had asked why the leading edge of western North America in the United States had no large coastal mountains like the Andes but was dominated by north-south faulting. Koßmat had questioned the timing of the separation of Madagascar and India and had wanted details concerning the complex geology of East Africa. Even Köppen had inquired how an oceanic trench could appear before the leading edge of the continent, as in the case of South America.

Rather than admitting that *any* phenomena departed from or contradicted his scheme, Wegener asserted (in broad and general terms) that all but the smallest and most local features of Earth were the result (direct or indirect) of the lateral displacement of the continents. He asserted this to be true on every scale, from the continents themselves down to individual mountain ranges and such structures as the submarine canyons at the mouth of the Congo and Hudson Rivers and the fjords of Norway. Continental displacements caused the folding of the Himalayas, Andes, and Alps; they caused the great fault and rift systems of East Africa and the San Andreas Fault in California. They explained everything on every scale *in which isostasy might play a role*—the continents and their major surface features, the continental shelves with all their complexity and variety, and the flat, barren, cold floors of the deep seas.[88]

This first section of the book might well have been titled "on continents, their shelves, and the oceans *in general*." The morphology of all three of these areas of Earth was, he wrote, governed by geophysics, and the geophysical principles in question were the flotation of the crust on the interior and the rifting, splitting, displacement, and lateral compression of continental fragments. Repeatedly, wrote Wegener, portions of the continents are pulled apart. The opening of a fissure *above* first reflects this, with the downfaulting of blocks into the emerging rift, while at the underside of the continental block (many kilometers below) the intensely heated material is stretched and thinned. Eventually, such fissures ruptured and exposed the denser Sima below, creating new ocean floor. As the water poured into these fissures created by departing continents, new oceans appeared.[89] The origin of continents *is* the origin of oceans.

As the continental "skin" was pulled apart in one area, it was shoved together in another, and about 95 percent of such compressive thickening happened below the surface. Wegener found an exact analog here to the formation of pressure ridges in sea ice: when two ice floes collide, a pressure ridge—many meters tall—is created. However impressive these ridges may be, only a small portion of their structure appears at the surface, and most of the ice is thrust down deeply into the water. What appears to be "elevation" is simply the buoyant effect of the huge (invisible) mass below—a consequence of flotation. The way sea ice moves, rifting on one side and piling up by resistance on the other, is also the way that continents are created.[90] It has been suggested a number of times that Wegener's experiences with Greenland ice provided fuel for his ideas about drift; this is made fully explicit here and at many other points in the book: the continents are indeed "continental bergs," and Wegener extensively develops the analogy of Sal and Sima with sea ice and ocean water.

Wegener's discussion of the details of the folding of mountains in this first section shows his increased awareness of the so-called nappe theory of the Alps: that the entire Alpine system consisted of huge and complex sedimentary folds. Geologists had, since 1900, increased their estimates of the amount of shortening of the crust by such folding from hundreds to perhaps thousands of kilometers. In 1920 European geologists were extending this analysis of the Alps into an explanation of the great arc of mountains extending through the Himalayas to China. Wegener approved of this move and noted that, as in the case of sea ice, when mountains are squeezed together, most of the crustal shortening is absorbed below the surface, and that the surface mapping of the strata would only yield minimum values of the amount of crust swallowed in these episodes of folding.[91]

His point was plain: with the contraction hypothesis rejected as the cause of the formation of mountains, "the forces which fold mountains must be the very same which cause also the horizontal displacement of the continents."[92] These forces have most recently and extensively produced the great fold systems of the Tertiary period to which the Alps, the Andes, and the Himalayas belong. These most recent fold systems have a remarkable configuration, namely, a compression toward the equator for the Alps and the Himalayas, and an equally extensive meridional folding system for the Andes. For the Himalayas and Alps, "the principal cause of the continental displacements would seem to be a 'flight from the poles' [*Polflucht*] by the continents."[93] The cause of folding of the Andes is less immediately apparent, though it must be somehow connected with the Atlantic Rift, the Rhine Graben, and the East African Rift System, all of which form a series of meridional parallels. Reserving a full discussion of these forces for a later chapter, he notes with some excitement that "[Damian] Kreichgauer believes he is able to detect Equatorial fold zones also for earlier geological periods, especially the Carboniferous folding, and a placement for the Carboniferous equator that contains the coal deposits of Asia, Europe, and North America."[94]

Moving on to "continental shelves in general" (*Kontinentalränder*), Wegener had considerable work to do. Even within the bounds of his morphological-empirical method, his first edition had been much too schematic. There were details on the scale of hundreds of kilometers he had not addressed, and most of these were on the edges of the continents. Now that he was going "head-to-head" with the permanence theory, which placed so much stress on continental shelves as zones of sedimentary accumulation and deformation and as rising and sinking land bridges, he had to give a much more detailed account. In his earlier work, the edges of continents were vertical or steep angular surfaces that merely indicated that the continent was over and the ocean had begun. He had now to provide a plausible account of the complex phenomena of a region to which he had given absolutely no attention heretofore.

Wegener provided, as was his wont, a cartoon schematic of his new version of continental flotation. Instead of a vertical margin, the continents are now shown in cross section to be convex on their outer sides. Wegener conjectured that in the part of the continent submerged in the Sima, the forces on the outer margin would be up from below (buoyancy) and inward toward the continental margin from the ocean side (resistance). In the topmost few kilometers between the floor of the ocean and the top of the continental block, the forces would be downward (gravity) and outward (spreading against the lighter resistance of the water). The result was that continental margins (between their surface elevation and the floor of the deep sea) would be preferentially subject to slumping, spreading, and fracture, as a result of all the force vectors just named.

The weakened and preferentially fractured marginal slabs might then have several fates: on the trailing edge of a continent they might be left behind as islands, or if longitudinally compressed like the Asian island arcs, they might spring away in echelon from the mainland. Segments might be torn off and left behind as in the Drake Passage between South America and Antarctica, with the shattered, confused archipelagoes on the western side of South America and islands left behind like the Falklands and South Georgia. Another outcome might be sections being stripped off parallel to the coast and entirely pulled away as in the case of Malaya and Sumatra. In a less extreme outcome, there might be meridional faulting along a scarp many hundreds of kilometers long without detachment, as in the case of the San Andreas Fault in California. Everything depended on the vector addition of forces, each of these geological phenomena representing a determinate but unpredictable outcome of the contest of pushes and pulls on the continental mass.

As for the floors of the oceans, these were not merely the outer skins of a static fluid medium with a density slightly greater than the continental bergs that floated in it, but the upper, outer surface of an Earth shell pursuing its own thermodynamic evolution. The three great ocean regions, Pacific, Indian, and Atlantic, were different in depth based on their age—with the Pacific being the oldest, the Atlantic being the youngest, and the Indian Ocean having an intermediate age, less deep near Africa and deeper beyond India. The depth differences were a result of long-term cooling, but even as the outer shell shrank and solidified (deepening the oceans), this did not bring an end to the mobility of the ocean floor. If the Sima was indeed so fluid that it could pass underneath moving continents, it must also be fluid enough to have local and independent turbulent streams and eddies.[95] These currents would then be capable of moving and rotating various pieces of the torn-off crust—islands and other continental fragments—so that Wegener could declare it "ridiculous" to ask the theory of displacements to account for every single local motion, since local idiosyncratic movements of the subcrust were capable of causing them.

All in all, this is a much more confident and well-structured argument than the first edition. Yet it introduces new dangers while eliminating old ones. Particularly at risk is the mixture of solid geophysical results with highly speculative inferences about how some previously unexplained phenomenon might be achieved by such and such a disposition of forces. There is nothing wrong with this approach, except that this was exactly the sort of thing that his (highly positivistic) critics had railed against in their attacks on the first edition. The heady brew of data, established results, and unfounded conjecture (the qualitative notion that continental margins are particularly susceptible to massive rupture) opened a new line of attack for Wegener's critics.

When Wegener moves on to the next section (chap. 4), "The Displacement of the Continental Blocks," there is a decisive shift in the narrative, focusing on a history of Earth with its evidences of former connection, as well as the history of the splitting and drifting. The detail is very dense here, much more so than in the first edition of the book. Moreover, even though Wegener postpones treatment of the "cause" of continental displacements until the end of the book, he constantly repeats the theme of folding, thickening, and splitting apart—over and over again. He refers all geological consequences back to the folding and tearing of the outer shell—a cumulative, one-directional process, constantly raising the mean elevation of the continents relative to the floors of the deep sea.

"For some time," he began, "the lithospheric skin covered the entire globe [*Erdball*]. It could then not have been 100 kilometers thick, but only about 30 kilometers and was entirely covered by a 'Panthalassa,' the mean depth of which Penck estimates at 2.64 kilometers, and from which only a very small part, or perhaps nothing at all of the Earth's crust remained free."[96] This accords, he continued, "with everything we know about the earliest development of life, all forms of which came from the ocean. Before the Carboniferous we have no record of quadrupeds or insects, before the Devonian no land plants, and before the Silurian, no air breathing organisms."[97]

"Then, in response to whatever forces [*durch irgendwelche Kräfte*], this earth shell, still plastic, and capable of sliding over the interior, was torn open, and on the side [of Earth] opposite this tear, compressed. Thus were formed the first fold mountains and, at the same time, the ocean began to divide itself into deep and shallow seas, that were even then marked off from one another by a steep slope."[98] The early continents were not, at first, large blocks standing above the surface of the sea, but more or less like the ribs of a paper lantern, like the island festoons of East Asia today.[99] From the beginning until the present this process of tearing and compression has been continuous, and in the future the mean elevation of the continents will be higher, though their total area on the surface of Earth smaller. The smaller the continents, and the greater their thickness, the more difficult it will be to fold them; this is to say that the rate of the process has not always been the same (and was greater in the past than in the present).

It is remarkable how well Wegener defends this speculation, in contrast with his conjectures about continental shelves in the previous section. When he proposes a "Panthalassa," he uses the (published) authority of Albrecht Penck. When describing what we know about the emergence of life and its relation to the deep sea, he uses the authority of the great paleontologist Gustav Steinmann (1856–1929) and the well-known geologist and paleontologist Johannes Walther (1860–1937). The appeal to Walther carries another message: for he is the author of Walther's law, which is that when we see paleontological remains reflecting different environments stacked above one another, these reflect lateral shifts of the paleoenvironments. Walther probably meant this with regard to the advance and retreat of shallow seas over continental edges, but Wegener was able to adapt and adopt his authority to the idea of continental displacement. These men were all active and living at the time of this writing, and his point clearly is not simply to use their authority *but to try to win them over.*

Finally, now, at the halfway point of this short book, we arrive at detailed reconstructions of the continents. Here Wegener pressed the necessity of viewing Earth as a globe and of using sophisticated map projections. Wegener drew a map of the relative positions of the continents in the Carboniferous. The map is the oblique case of the Lambert Azimuthal Equidistant Projection that he favored, centered on a meridian passing through Russia, East Africa, and the Carboniferous South Pole (for his "land hemisphere") and centered on the open ocean of the Pacific, touching only a section of North Asia through Kamchatka (for his "water hemisphere"). The caption underneath the map has the parenthetical note that it has been drawn "without regard to water coverage," meaning that shallow transgressions of various parts of the continents are not indicated. The stippled areas represent the "shelf seas."

The choice of the Carboniferous was obvious from the standpoint of the reconstruction of the fossil remains of Gondwanaland and of the remains of the Permo-Carboniferous Glaciation, with the single great paleocontinent reassembled around a South Pole. Wegener refers to it later in the book, passingly, as the "Pangäa" of the Carboniferous.[100] The North Pole, in this reconstruction, lies in the Pacific, about 20° of latitude away from the Aleutians and Kamchatka, thus fulfilling his prediction for the absence of remains of this early glaciation in the Northern Hemisphere: there was no land area adjacent to the pole.

Wegener tells his readers, "The map, naturally, has a peculiar appearance, and suggest[s] the presumption that the procedure in drawing it was somewhat arbitrary. This is, however, not the case. This is an exact reconstruction—within given limits—and it should be noted that in multiple independent repetitions there were no significant deviations from the map as published here."[101]

Then, with a sort of naïve enthusiasm, Wegener tells the story of how he produced the map. Working with a globe with a diameter of 0.5 meters (roughly 20 inches), Wegener used tracing paper to produce outlines of the continents, including their shelves, and cut them out. To make them fit better on the sphere, he cut them into smaller sections, and to account for his estimate of the tremendous compression of Eurasia by the Tertiary mountain folding (subsequent to the time this map represents), he increased their lateral extent by 10°–15° of latitude. He then glued these tracings to the globe, assembling them in such a way that South Africa was immediately adjacent to

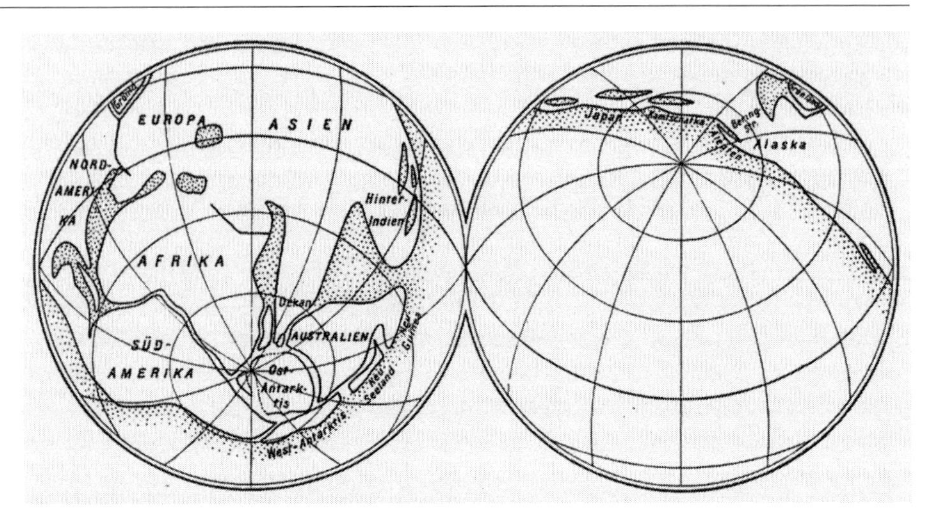

Wegener's drawing of the reconstructed continents of the Permo-Carboniferous period, with the reassembled continents on the left and the Pacific Ocean on the right, illustrating his view (in 1920) of the very early partition of Earth into land and water hemispheres. From Alfred Wegener, *Die Entstehung der Kontinente und Ozeane*, 2nd (completely revised) ed., Die Wissenschaft (Braunschweig: Friedrich Vieweg & Sohn, 1920).

the South Pole and the equator passed through Germany. He used the parallelism of the current continental coasts to guide the reconstruction. This was most secure along what would be (millions of years later) the Atlantic rift and moderately secure in the area of what he called "Dekan," the proto-India that would later be folded up into the Himalayas in the "Lemurian Compression." Least secure, he admitted, was the reconstruction around Antarctica, for which there was as yet little real geological data.[102]

On this Carboniferous Earth, with the continental land occupying one hemisphere and the Pacific Ocean the other, the "westerly migration" meant that the leading edge of the protocontinent would be the stretch of coastline from the Bering Strait to the southern tip of South America, and the trailing edge the margin extending from the southern tip of South America around Antarctica, Australia, Southeast Asia, and Japan to the Aleutian Islands, reaching the Bering Strait by the opposite great circle. Wegener imagined that to this point in Earth history, the entire Pangäa was capable of rotating to the west, and meridional rifting was in the initial stages.

The principal basis for this reconstruction was a compilation by Theodor Arldt, in his recently published *Handbuch der Paläogeographie* (1917).[103] Arldt had made a compilation of land bridges postulated by twenty different paleogeographers for each geological era, in an attempt to develop a consensus view of when, in the geological past, the present continents had been connected and when they had not. The list includes Arldt himself and (notably) Carl Diener, Franz Koßmat, Melchior Neumayr, and the Americans Charles Schuchert (1858–1942) and Bailey Willis, the latter two both strong "permanentists." Arldt's table shows that, of those paleogeographers postulating land bridges in this era, *all* agreed that there were land bridges across the North Atlantic, across the South Atlantic, between Lemuria and Africa, and connecting all the segments of Gondwanaland.[104]

To produce a contrasting view, Wegener then jumped forward in time to the Eocene, an early epoch in the Tertiary period, and again drew a map of the continents,

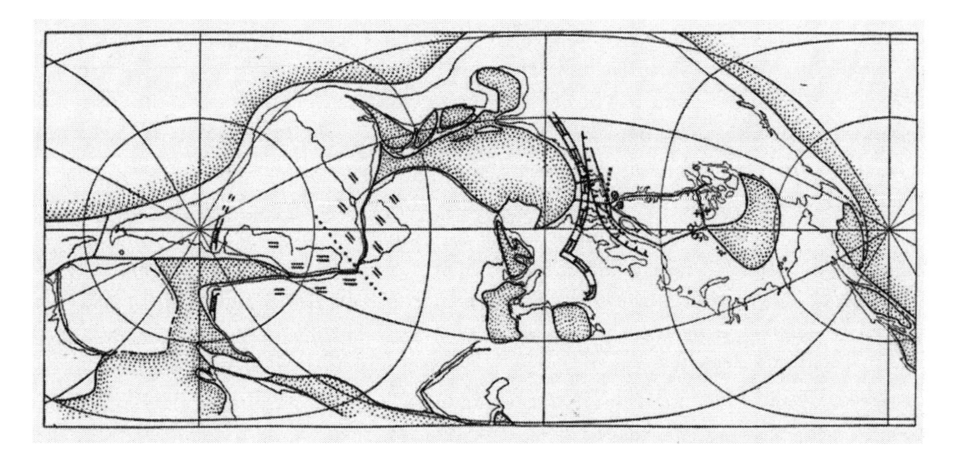

Wegener's reconstruction of the continents in the Eocene (Quaternary) in an unusual projection, a transverse pseudocylindrical projection of the Mollweide equal-area map that had recently been developed to minimize distortion close to the poles. The projection is appropriate to the task, but difficult for the unpracticed eye to interpret. From Wegener, *Die Entstehung der Kontinente und Ozeane* (1920).

using the same tabular data. The map, reproduced here, is in an unusual projection, a transverse pseudocylindrical projection of the Mollweide equal-area map that had recently been developed to minimize distortion close to the poles. The projection is appropriate to the task but difficult for the unpracticed eye to interpret. Wegener used this map to show the geological continuities of mountain ranges, including the Tertiary, between North America and Europe and between South America and Africa. It is notable that even in the Eocene, which Wegener imagined to have begun fifteen million years ago (the current figure is fifty-six million years), the Atlantic split has barely begun and existed only as a shallow sea in its northern half, while Africa and South America were still connected directly.[105]

The next twenty or so pages in the book are taken up by a history of the splitting and drifting of the several portions of the original Pangäa, and they are so dense and so detailed that they must have baffled all but the most expert geographers. Wegener divided his argument into three sections: "The Atlantic Rift," "Lemuria," and "Gondwana-land." The bulk of his argument, fifteen of the twenty pages, he devoted to the Atlantic, at once the most contested of the splits and the one for which he had the most evidence. There is a good deal of arguing here with unseen adversaries (above all, Diener), only occasionally mentioned by name. To read this material successfully, one needs access to a globe, an atlas, a gazetteer of place-names, a paleontological textbook, a geological text or at least some geological maps, and considerable time and patience. There is no doubt of his mastery of the argument, but it is an uneven mix of geological and paleontological data, geophysical theory, and speculative reconstruction which makes few concessions to the audience. The juxtaposition of solid geological data with speculative morphological reconstructions—especially of the complex rotations of various parts of Gondwanaland (Australasia, New Guinea, and New Zealand)—is jarring, unless one decides that Wegener is only trying to sketch what *might be true*, and on this point he is not always clear.

There are other problems as well. The section on Lemuria, the part of his argument about India's northern movement and compression to form the Himalayas, is treated in a few pages, very speculatively, with but a single footnote. This is very odd, given the detailed critique of this part of his hypothesis by Andrée and Diener and others, critiques he had noted in his interleaved volume.[106] The section on Gondwanaland is somewhat more detailed, but Wegener admits that the details of his reconstructions, apart from the connections among South America, Africa, and India around a conjectural Carboniferous South Pole, are almost purely hypothetical.

Wegener was never very comfortable with geological data and argumentation. He was a physicist, and for him geological argument was too cluttered by individual details. This speaks to a point deeper than his own discomfort: the appalling difficulty that geologists face in creating general theories for Earth out of geological argumentation. There is almost no economical way to present a global geological theory based on geological data. It is not that it cannot be done, but that the amount of detail required can be suffocating to the imagination. One can try to economize the empirical base, but beyond a certain threshold of generality, geologists are no longer interested and become radically more difficult to convince. It took the brilliance of someone like Sueß to reduce thousands of individual monographs into a few volumes of synthesis, and this took him more than twenty years. Wegener was overmatched by this particular task, and he had neither the time nor the will to pursue it.

Thus, when he turned to the question of the migration of the pole of rotation of Earth, there was a palpable sense of relief: not only was he back in the world of geophysics, but he was faced with the kind of problem that he knew best and most enjoyed: "As for the question of the positions of the pole in earlier geological periods, there is currently a mess [*Verwirrung*] that, even if it is historically comprehensible, no longer appears to be necessary, given the overall state of our knowledge."[107]

Wegener was never happier than when he imagined he was cleaning up a mess in the published literature. When he worked with geological literature, he began with a mess but generally ended with a slightly improved mess. With geophysics, as in the case of displacements of the pole of rotation, the timeline was generally short, the number of players few, the principles reasonably clear, and authorities easier to come by. Not only that, but in physics there is an expectation that there will be an interplay between the theoretical principles and the collection of data, as well as a series of iterations of the theory showing the impact of these new data as they are collected.

Wegener was beginning to see how this worked, and he said so in his introduction to the section on displacements of Earth's pole of rotation. The data we have, he wrote, are not very good for the Northern Hemisphere or for the Southern Hemisphere. The pole positions we establish in arguments like these are not final and are open to correction. No one is asking for mercy for being wrong, he said; what one wants instead is not to have, as geologists have repeatedly tried to do, a "death sentence" proclaimed on all attempts to map the movement of Earth's pole, simply because the data aren't as good as we would wish.[108]

He was also beginning to understand that men like himself appeared odd to working scientists remaining faithfully and clearly within the designated boundaries of their disciplines. When it came to geophysics, he was in excellent company discussing displacements of Earth's poles: Laplace, Euler, Lord Kelvin, and Giovanni Schiaparelli (1845–1910). However, in correlating these displacements with the history of life, he was in decidedly less august company. His predecessors here were Paul Reibisch (1867–1934), an engineer with an avocational interest in mollusks; Heinrich Simroth (1851–1917), a biological systematist working on slugs and snails; and Damian

Kreichgauer (1859–1940), a physicist turned monk teaching general science in a monastery in the Netherlands.

When we begin to examine the details of Wegener's chapter on *Polwanderungen* (migrations of the pole), we see a simple explanation of a geophysical process tied to a very complex set of data about plant and animal distributions. Wegener intended to use the evidence for shifting latitude zones collected by his predecessor "outsiders"—Reibisch, Simroth, and Kreichgauer—but to throw out their poorly founded geophysical theories. He would then use his own theory of pole migration, forced by continental displacement, and use a modified version of Köppen's scheme of latitude zones to predict the position of the equator, the temperate regions, and the poles in every period from the Carboniferous down to the Quaternary.

This was a clear and concrete plan, but one cannot wonder that geologists (almost universally) took exception to it: it paid almost no attention to geology. Wegener used certain geological formations the way one might use fossils, to make an argument about shifting environments, but he had no intrinsic interest in them. Almost everything that interested geologists—the details of local structures, erosion and sedimentation, geosynclines, the creation of drainage networks and river valleys, ore genesis, mineralogy and petrology, volcanoes, igneous intrusions, metamorphism—received a brisk tangential treatment except where these matters played into Wegener's history of displacements. Wegener even subtracted major aspects of geology and put them into geophysics: the fjords of Norway and the submarine canyons of the St. Lawrence, Hudson, and Congo Rivers were for him no longer erosional features but emergent rifts on the margins of continents. These continental margins did not acquire thick sequences of sediment by the weight of sediment pushing down the shelves. Rather, fracture and downfaulting of the margins of continental blocks (caused by their lateral displacement) created the space and elevation differences, which allowed sedimentation to proceed. Elevation was not itself a geological process, only apparently so, as the squeezing together of the crust and flotation by isostasy jointly took credit for this activity; thus, the erosion subsequent to such "uplift" was also geophysically dictated.

The description of the displacement of Earth's pole is the same here as in both of Wegener's previous attempts to explain it, albeit in an expanded version. Wegener had always followed the argument of Schiaparelli, who was himself working out consequences of a speculation by Lord Kelvin from the 1870s. Wegener quoted Kelvin in 1915 and again here, so it is probably worth repeating. Kelvin had said, "We may not merely admit, but assert as highly probable, that the axis of maximum inertia and the axis of rotation, always very near one another, may have been in ancient times very far from their present geographical position, and may have gradually shifted through 10, 20, 30, 40, or more degrees, without there being at any time any perceptible sudden disturbance of either land or water."[109]

In a completely solid Earth, the pole can never move, other than a minor free oscillation. In a completely fluid Earth, the adjustment of the pole of figure to the pole of inertia would be instantaneous. The interesting case is in a viscous Earth, where the pole of figure can move but lags behind the pole of inertia for some period of time. Following Schiaparelli, Wegener imagines that there is a critical threshold of difference (geographical distance) between the pole of inertia and the pole of figure such that when this figure is exceeded, Earth, in order to reestablish its oblate spheroid to match the new distribution of mass, begins to flow. Once this flow begins, it may continue for quite some time, and if the forces causing the initial redistribution of mass continue, the pole of inertia will continue to move and the pole of figure will run to catch up to it.[110] For small events and short time periods, Earth behaves as a solid. For large events and long time periods, Earth behaves as a viscous fluid.

From the standpoint of explanation, that was the easy part; the difficulty would be marshaling (in a comprehensible way) the thousands of paleontological data points that would indicate the position of the poles and the different climate zones in each geological era, and thus the path of the pole. In order to keep track of this, Wegener began to use the globe of Earth on which he had pasted tracings of the continents, to record the position of the equator and the poles in different periods, using small bits of colored paper attached with pins from his wife's sewing kit.[111]

This was the work on which he and Köppen collaborated so intensely in the winter and spring of 1919/1920. Using the work of Reibisch and Kreichgauer to establish the position of the equator and the pole in the Carboniferous (these were in general agreement), Wegener and Köppen established that period of Earth history as a starting point. It was not, of course, the beginning of geological time, but it was the first period in which they felt reasonably certain of the data.

They chose to assume that Earth, in every period between the Carboniferous and the present, had a varied climate similar to the present. Köppen's scheme for the climates of Earth in the recent past was too complicated for the distant past; they had not enough detailed data. They settled for a four-zone system of latitude bands with distinct fauna and flora. They started with a tropical rain zone between the equator and 20° of latitude, then a dry, desert zone between 20° and 30° of latitude, followed by a temperate rain zone from 30° to 60° of latitude, and finally a polar zone from 60° to the poles.

At the beginning of their investigation they focused on the history of the climate of middle Europe, for which the paleontological data were most dense, and they recreated, using these assemblages of plants and animals which indicated to Köppen distinct and latitude-dependent climates, the shift of latitude of this region of Earth throughout geological time, from just before the Carboniferous to the recent past. This was an engrossing exercise for both, and they were able to generate a curve showing that in the Devonian and Carboniferous through the Permian, central Europe had a tropical climate, in the Permian and the early Triassic it had a desert climate, and in the Jurassic it had been transformed into a temperate rain zone. By the Cretaceous it was passing again through a desert phase, and in the early Eocene it had for a time been (once more) tropical. Throughout the Tertiary there was a rapid and continuous motion of the climate from warm to cold: moving from an Eocene climate compatible with a latitude of 15° away from the equator to a climate in the Quaternary equivalent to latitude 60°+ north, then returning, in the recent past, to a temperate rain zone. They got the idea of graphing this as a continuous curve from Dacqué, though his attempt had been an impressionistic graph for the ups and downs of the climate of the whole Earth from warmer to cooler and back again.[112]

It was a successful effort, but they realized that to accomplish an accurate and verifiable position of the equator, and therefore of the poles, for each period of the geological past, they would have to repeat this labor for every major continental segment; *it was the work of years, not months*. For the first time in Wegener's career he had run out of high-level syntheses that could accomplish the task of collating the individual data he needed. Much of the work in the decade 1900–1910 was still appearing only in

1919–1920 because of publication delays caused by the war. Even an experienced synthesizer and textbook writer such as Emanuel Kayser could barely keep up with the flood of new information, as he noted in the preface to his 1918 edition of his text.[113]

It dawned on Wegener early in the spring of 1920 that he would finish this book, but when it came out it would not be "finished." He had enough data for a rough estimate of the pole positions, and by mapping the movement of the North Pole, he could get an estimate for the location of the South Pole. Knowing the latitude of Germany in different periods, and estimating the amount of its shift in latitude which was due to displacement rather than pole migration, he could sketch a paleoequator, using interpolated data for the other continents as a check. It was an extremely rough pass, and he knew it and said so. On nearly every page of this long chapter he repeated the warning that the results were provisional and subject to revision and that he had depended trustingly on different authorities for different places in different periods.[114]

In spite of the uncertainty attending this technique, Wegener did manage to produce pole positions for every period from the Devonian down to the recent, with a total displacement of about 65° since the Carboniferous, and with several important oscillations in between. Wegener was not able to produce a map with an accurate paleoequator, and his summary chart gives only positions of the North Pole, the South Pole, and Germany—the poles in terms of latitude and longitude; for Germany, latitude only.[115]

From the standpoint of physics research it was a good first pass through the data, but from the standpoint of the way geology and paleontology were "supposed to be done" it was sketchy, conjectural, and premature. This is not to say that it was not well sourced and intricate. Yet it was (and is) even more difficult to follow than the geological data in the chapter on continental displacements. It required familiarity with geographical place-names, geological time divisions, and plant and animal species in the present and the past, and its compression works decidedly against it: Wegener was trying to summarize in about thirty pages of text what others had struggled to compress into books of 400–600 pages. It was just too much.

It is in the final sections of the book that we see the most remarkable transformation of Wegener's argument. These two chapters, on the causes of continental displacement and on their measurement by astronomical position finding, take up only about ten pages, but they are together a radical revision of his entire hypothesis. Not only is the content different, but so is the status of the investigation. It is no longer the solution to a long-standing problem, but very much a long-term investigation and a work in progress which will lead to such a solution in the future.

"Although," he wrote, "on first inspection the movements of the continents present a quite variegated picture, behind this variety one can see the outlines of a great System: the continental blocks move equatorward and they move westward. It is advisable to treat these two components of the movement separately."[116] He titled this chapter "System, Cause, and Effects of Continental Displacements." The displacements form a system of movements, these movements have a cause, and this cause has a series of effects—these should be considered as three separate entities.

In this novel scheme, the principal displacements are in the direction of the equator, and Wegener calls these (a coinage he will attribute to Köppen) "*Polflucht*" (pole-fleeing). Such motions affect large blocks more than small blocks, and their effects are most visible in middle latitudes, as evidenced by the great Eurasian fold systems of the

Tertiary period. Evidences of this are, however, available for every continent, including North America, Australia, Lemuria, and South America, and even, in some periods, Antarctica.[117]

Equally primordial is the secondary component, the "Westward Migration" (*Westwanderung*) of the continents. This is evident even in the time of Pangäa, where the reconstruction of the single protocontinent shows a crumpled (western) leading edge and a torn, trailing (eastern) edge.[118] These westward movements are probably derivative from the *Polflucht* and represent a consequence of Earth's rotation, where here the movement toward the poles is deflected to the west, by analogy with the trade winds in the global atmospheric circulation. It is also possible, though there aren't any data to support this sufficiently, that solar and lunar tides in Earth might play a role; the problem is that for the short time periods involved Earth's behavior is probably purely elastic. In any case, "the question must for the time being remain undecided whether we must choose between these two different causes for an explanation of the westward migration or whether they are both simultaneously effective."[119]

The fundamental point he wishes to drive home before turning to the "cause" of the "system" of movements is that even if "these two components, the *Polflucht* and the *Westwanderung*," cannot explain every individual detail of the displacements, and if such explanation must wait for the future, nevertheless the principal movements (*Hauptbewegungen*) both in the present and in the past are clearly evident. In 1912 Wegener had assigned no causes at all for the displacements, and in 1915 he had mentioned a few suggestions. As Köppen later said of him, this well-considered caution came from his geophysical training, which made him wary of beginning from some plausible geophysical explanation and then proceeding deductively toward the evidence, rather than the reverse, which is to examine the empirical evidence and establish a pattern in the phenomena without specifying a necessary causal mechanism.[120] Even here, in 1920, Wegener was extremely reluctant to specify a cause and devoted not quite two pages to the question. Köppen, he said diffidently, has recently proposed an explanation for displacement on a rotational ellipsoid, which he calls *Polfluchtkraft*, and which he has allowed me to summarize from an article that will appear in *Petermanns Mitteilungen*.

Here was Köppen's idea. In a floating body, the center of gravity (what we should now call the center of mass) is determined by the figure and density of that body under the attraction of gravity. However, for a floating body, the relative densities of the body and the medium in which it floats determine the center of buoyancy. A simple calculation shows that for the relative densities of the continents and the Sima, the center of gravity in a continental block lies on average 2.4 kilometers (1.5 miles) above the center of buoyancy. On a rotating (viscous fluid) body like Earth, the disparity between the center of buoyancy and the center of gravity results in a small but detectable resultant gravitational force. Where the center of buoyancy lies higher than the center of gravity, this resultant will be in the direction of the pole; where the center of gravity lies higher than the center of buoyancy, the resultant will be in the direction of the equator. It is obvious that for objects at the pole or at the equator there will be no resultant force, and that the largest resultant would appear in midlatitudes.[121]

Under the action of such a force, however slight, masses as large as the continents would eventually slide toward the equator from the midlatitudes. Köppen had only described this qualitatively, Wegener said, and had not made any calculations. If we knew the elevation and area of the continents in different periods, he wrote, we could conceivably predict the path of the pole through numerical integration, though this is far beyond our capacity now.

If the causes are conjectural, it is the system—the motions—and not the cause of the motions which produces the effects. The principal effect of continental displacement is the migration of the poles, for reasons already described in an earlier section. The second effect of continental displacement is meridional rifting: if the continents were not moving, the stretching and thinning of the lithosphere which produce rifting would not take place. The third consequence of the motions is oceanic transgression and regression. If displacements of continents force the migration of the pole, the viscous Earth will adjust with a considerable lag time, but the water in the oceans will respond immediately, resulting in a recession of ocean water in front of the moving pole and a transgression of ocean water on the land behind it. Many writers on the displacement of the poles had discussed this, but never in connection with a sound geophysical explanation.[122]

Wegener has shifted quite radically here from a theory emphasizing changes in longitude to one emphasizing changes in latitude. He has also shifted emphasis from the creation of mountain ranges along meridians of longitude to creation of mountains by compression along parallels of latitude, as the continents shift toward the equator.

This theoretical move creates an interesting situation. In both 1912 and 1915 Wegener could assert that the American Frank B. Taylor, who had an idea somewhat similar to his own with regard to the rifting and drifting of the Atlantic continents, had made an offhand suggestion in a long paper mostly concerned with the argument that the great mountain ranges from the Alps through the Himalayas to the ranges of Asia were the result of a compression, as continents moved away from the poles and toward the equator. Similarly, Wegener had barely mentioned Kreichgauer as an interesting proponent of the idea that the shift of the poles and the movement of the lithosphere were connected phenomena.

Now, however, Wegener found himself accepting the same general motions of the Taylor theory (without the highly suspect geophysical basis) and also found himself accepting the basic idea and the basic mode of presentation of Kreichgauer's 1902 book *Die Aequatorfrage in der Geologie*. There were important substitutions here, of course. Kreichgauer had assumed that Earth had been spherical in the Carboniferous and had only assumed its oblate ellipsoid over time, with the centrifugal force driving the sphere toward an ellipsoid offered as the cause of the continents fleeing the poles; this was a geophysical absurdity. Moreover, because he had no lateral displacement of continents, his paleoequators were conjectural and often on the wrong continents. Nevertheless, Wegener now found himself in a position where he had to acknowledge the general tendency of the crust to slide from the poles toward the equator, mentioned by a number of authors, among them notably Kreichgauer and Taylor.[123] In this context the reader is referred to the epigraph at the beginning of this chapter, in which Köppen remarks on the great difficulty of tracing the lineage of ideas, which appear over and again in a variety of forms and in a variety of ways, associated with good ideas and bad such that the idea of priority becomes extraordinarily difficult to establish, and the actual path followed in the development of the theory is not linear or intuitive.

This leads to another interesting perplexity for Wegener: if others have proposed both the equatorward and the westward displacement of continents, and if others have connected these ideas with displacements of the pole of rotation because of shifts of mass on Earth's surface, and if others have documented these sorts of shifts with reference to former geological continuities and the distribution of animals and plants, what exactly is new and different about Wegener's theory of "the origin of continents and oceans"?

Wegener's answer to this very good question is to throw everything into the question of measuring displacement: "Among all the theories with similar far-reaching ambitions, the displacement theory has the advantage that only it is capable of being tested by exact astronomical position finding."[124] Here he had great hope but slender means. His handwritten notes in the rebound copy of his 1915 book show that he had been informed of an error in his own transcriptions of supposed changes in longitude between Europe and North America. In 1912 and again in 1915 Wegener had recorded the change in longitude between Cambridge, Massachusetts, and Greenwich, England, as having been 4 hours, 44 minutes, 31.065 seconds of longitude in 1870, and 4 hours, 44 minutes, 31.12 seconds in 1892. However, the second figure was quite wrong, and the published figure was actually 31.032 seconds, a change 10 times smaller than Wegener had recorded in order to declare a continental displacement of 4 meters (13 feet) per year over twenty-two years. Moreover, Wegener's notes show that the actual measurements carried out in the Azores in 1914 by the Geodetic Institute, before the cable was cut, gave a distance between Cambridge and Greenwich of 31.039 seconds, virtually the same longitude as in 1892. Finally, Wegener had to admit that Galle's 1916 article in the *Deutsche Revue* had noted this discrepancy and had concluded that there had been no longitude shift measured since 1892.[125]

Wegener had clearly made a major error in using these telegraph time signals, and even though he corrected it in this edition, he persisted in another error, of claiming that Koch's longitude measurements in 1906, published in 1917, lent confirmation to the idea that Greenland was drifting to the west. Examination of Koch's published measurements, in the context of Wegener's own published review of the same, shows a much greater uncertainty about the accuracy of these measurements.[126] The reader may refer to the earlier chapter in which these measurements were made (or not made) to connect the longitudes of the German expedition of 1870 to the measurements of the Danmark Expedition, on a trip to Sabine Island in November 1906 which nearly cost both Wegener and Koch their lives, and on which Koch neglected to wind his chronometers, arriving there with no way to determine the longitude at all. Even though Koch had returned to try to make these measurements again, discrepancies of hundreds of meters (almost a kilometer) remained in the measurements.

One important change in his attempt to measure displacement was a shift from the exclusive focus on the opening of the North Atlantic, or the shift of Greenland, to the amount of displacement per year required for different separations: across the Arctic, the North Atlantic, and the South Atlantic. To this he added estimates of the drift of Madagascar from Africa, as well as of Southeast Asia from Madagascar, and finally a single estimate of the displacement between Tasmania and Wilkesland (East Antarctica).

He made it quite clear that he was using recent (1910) estimates of the number of years elapsed since the beginning of different geological periods and dividing these by the number of kilometers of separation since the proposed rifting. This produced, for every region of Earth other than Greenland, Iceland, Norway, Scotland, and Labrador (all supposed to have begun their separations between a half-million and one million years ago), comparatively small annual displacements in centimeters per year, not meters. For instance, across the arc from Buenos Aries to Cape Town, the displacement was only 30 centimeters (12 inches) per year. Between Tasmania and Antarctica it was 36 centimeters (14 inches) per year, and between Brazil and West Africa it was 24 centimeters (9 inches) per year. While he was still proposing recent spreading rates as high as 20 meters (66 feet) per year in the high Arctic, he hoped that these smaller rates would do much to assuage his critics and supporters alike, almost all of whom had proclaimed that his spreading rates were unbelievably high in terms of the physics of Earth.[127]

In this context, given the stakes involved, it is no wonder that Wegener turned to an increasing reliance within his theory on shifts in latitude rather than shifts in longitude. The International Latitude Service had discovered and published shifts of anywhere between 0.17 and 1.51 seconds of latitude in the past half century for a variety of stations around the Northern Hemisphere. Wegener admitted that such shifts could, of course, be either displacements of the continents or displacements of the pole of rotation and were not themselves definitive, but these figures gave a good match to the distance the pole would have had to travel from the Eocene to the time of the ice age, assuming a time gap of about ten million years.[128]

Here, quite characteristically, Wegener ended his book, with no conclusion, no summary, no retrospective judgment, and simply a final sentence. It was April 1920, and he was exhausted. He had worked on this revision every day (every evening) since October 1919 while holding down a very demanding job and teaching at the university. As he signed his name to the introduction and sent off the manuscript to Vieweg, he awoke again to the life of his family, where Else was giving birth to their third daughter, Lotte, born on 16 April.

The book was gone, but the work was not complete. Even as the book was going to press, he and Köppen were working together on two detailed articles for *Petermanns*, one on the relationship of pole migration and continental drift to the history of climate, and a second, more technical article on the causes and effects of continental displacement and migration of the poles.[129] As if to underline the continuity of this commitment, Wegener began, in April 1920, to teach a course of lectures at the University of Hamburg entitled "Climatology."[130] He was preparing to change his profession once again.

16

From Geophysicist to Climatologist

HAMBURG, 1920–1922

> The ways in which WEGENER's aforementioned hypothesis and DARWIN's theory developed are widely different. While DARWIN gathered arguments during a period of about twenty years until his ideas had matured, WEGENER put forward a supposition, which afterwards underwent many alterations. Possibly a dissertation may one day be written about all these alterations. Many of them are highly interesting if one knows the background.
>
> C. E. WEGMANN (1948)

Through the fall of 1919 and into the spring of 1920, Wegener was following two parallel careers. By day he was a senior meteorologist in an oceanographic institution, and in the evenings, on holidays, and on weekends he was a geophysicist working on the history of continents and oceans. It was a contrast of extremely short-term and extremely long-term events, as his work at the observatory was principally concerned with forecast meteorology, on a scale of one to three days. The time horizon of prediction for his work on continents and oceans, on the other hand, was from decades to millions of years.

The intermediate ground between these two different imaginative timescales was the study of climate. Climatology is the study of the long-term pattern underlying day-to-day weather. In English-speaking countries, climatology used to be regarded as merely the statistical aspect of meteorology, in contrast to the French and Russian view that meteorology was a subsection of climatology. From the standpoint of Köppen and most other Germans, it was not so easy, as meteorology referred mostly to the dynamics and physics of the atmosphere and was a theoretical science, while climatology was a descriptive science where one did not abstract very far from the full ensemble of factors acting on a region.

Wegener's attempts, begun in 1911, to convince his father-in-law of the reality of continental displacements had taken hold only in 1918, in their extended correspondence on isostasy. Köppen could see the linkage between geophysics and climate once he had grasped the idea that continental ice sheets were so heavy that they could actually push down large sections of Earth, not by fracturing them, but by forcing the fluid material below to move away laterally. This began to make real sense to him when applied to the postglacial landscapes of Scandinavia, including the long-term uplift of the shoreline of the Baltic, as laterally displaced fluid rock far below the surface flowed back beneath the landscape and elevated it, *thousands of years after the ice melted.*

It was only at that point that Köppen began to believe that continental displacements could be the "Ariadne's thread through the labyrinth of climatology" and could resolve intractable problems in understanding climate zones in the past. As we have seen, since the 1870s the primary strategy for climatologists and paleontologists to explain tropical plant fossils in high latitudes had been to (conjecturally) displace Earth's pole of rotation. Köppen knew, as did all other climatologists, and as did Wegener, that conspicuous absurdities remained if one simply moved the pole; given the resulting pole positions, polar flora and temperate flora, or temperate flora and tropical flora, appeared simultaneously in the same latitude zones in the reconstructions. This was intolerable not only because it made it difficult to understand past climate but also because it threatened all climate schemes based on latitude zones. There would, of course, in any latitude zone, be differences based on the configuration of land and water, since oceans moderate climates and continental interiors are both hotter in the summer and colder in the winter than coastal regions. But the anomalies in the geological record were much too great for this to resolve the differences.

Wegener's and Köppen's interests had grown together in the fall of 1919. The more Köppen learned about the ability of continental displacements to explain past climate, and the more he interrogated Wegener on the connection between continental shifts and climate shifts, the more Wegener was drawn in to seeing his theory as a solution not so much to a geological problem as to a climate problem. In 1920 his situation was very different from that in 1912 or 1915. Wegener's targets in earlier years—the contraction theory and the theory of sunken land bridges—he could now dismiss. The remaining contest was between continental permanence and continental movement. Here the question was no longer (simply) how certain kinds of shallow water and terrestrial creatures could have gotten from one side of the deep ocean to another, but how one could explain (latitude-based) climates over the full Earth in every period of the past. For Wegener, the climates of the past became the key to demonstrating the superiority of his displacement hypothesis, just as, for Köppen, the displacement hypothesis had become a potential key to the understanding of past climate. The outfall of this was not just that Wegener increasingly became a climatologist, but that Köppen increasingly became a geophysicist.

Their joint residence and close relationship were the catalysts, but not the cause, of these changes in their views. Köppen had been working on climate systems since the 1880s, and his interest in climate had begun much earlier, when he was a thirteen-year-old child growing up in Russia. His father, already seventy-six in 1860, found the raw climate of Petersburg increasingly hard to bear and had bought an estate in the Crimea in Karabagh. Köppen, in a memoir written in 1931 (when he was eighty-five), remarked that the transition from north to south in Russia was much more stark than in Western Europe, and that as a child he had marveled, on rail trips from St. Petersburg to the Crimea, at the constantly changing vegetation outside the windows of the train.[1] Near St. Petersburg there was boreal coniferous forest, and in the Crimea a Mediterranean, arid, subtropical climate; in between were a variety of woodland and steppe climates.

Köppen was certainly a founder of the field of climatology. Julius Hann's *Handbook of Climatology* (1883), Aleksandr Ivanovich Voiekov's *Climates of Our Globe and Particularly Russia* (1884), and Alexander Georg Supan's (1847–1920) climate province scheme all appeared at about the same time as Köppen's initial (1884) article in *Meteorologische Zeitschrift*: "The Heat Zones of the Earth Considered with Reference to the Length of Hot, Temperate, and Cold Seasons and the Effect of Heat on the Organic World."[2]

Köppen took as the basis of his system the temperatures in different regions during the warmest months of the year and prepared isothermal maps based on them.

"Isothermal maps" are maps that show lines of equal temperature (isotherms). He showed that the 10°C (50°F) isotherm was the boundary of forest and tundra (the *Baumgrenze* or tree limit), and he classified places where the mean monthly temperature never went below 20°C (68°F) as "tropical."

The system that Köppen chose began from his discovery that the northernmost latitude for the growth of trees is well defined by the temperature of the warmest month. Where the mean (diurnal) temperature is 10°C or less in the warmest month, trees cannot grow: this is the *Baumgrenze*. It actually exists, and it exists independently of knowing *why* it exists. It is a description of a climate phenomenon rather than an explanation of why it occurs. Having decided to proceed in this way, Köppen superimposed his isotherms on existing maps of vegetation and showed that definable suites of vegetation showed up when the mean monthly temperature remained above 10°C for one, four, and twelve months. He developed the idea extensively, mapping temperature distributions by drawing maps for the 10°C isotherm and for the 20°C isotherm, for periods of one, four, and twelve months. This allowed him to develop a latitude zone system of climate representing the whole globe.

In 1901 Köppen republished this in book form: *Versuch einer Klassifikation der Klimate vorzugsweise nach ihren Beziehungen zur Pflanzenwelt* (An attempt at a classification of climate chiefly with respect to its relationship to the world of plants).[3] In this extremely important and influential book, Köppen attempted to meld his work with standard botanical classifications. Some climatologists thought this a step backward from a modern and scientized (read "thermodynamic") approach to climate, which would be much closer to dynamic meteorology, but it was a sound choice. The identifying mark of a climate is its vegetation, seen as an integral response to temperature, rainfall, orientation to the continental interior or the oceans, and soil type. Climate zones were, Köppen argued, differentiated by this full ensemble of factors: not just temperature, but the entire weather picture; not just the amount of precipitation, but its type and its seasonal distribution.

Because Köppen was a founding editor of *Meteorologische Zeitschrift*, he could publish revisions to his scheme almost on a monthly basis, and he interspersed these modifications (at decadal intervals) with bound book publications and summaries. In 1890 he published a cloud atlas, along with Neumayr and Hildebrandsson, and was able to insert his climate scheme into what quickly became a widely used reference book.

By 1918, all the other founders of climatology were either dead or nearly so. Voeikov had died in 1916, Supan would die in 1920, and Hann would die in 1921. Hann had published his last great work, his three-volume *Handbuch der Klimatologie*, in 1908–1911. Hann and Köppen were close colleagues and friends, and Köppen was strongly influenced by Hann's inclusion, in his later work, of geology in the system of climate indicators: the use of coral reefs and limestone, salt and gypsum, peat, coal, and glacial tills, in addition to plants or plant fossils—the latter in mapping the climates of the ice age. In 1918, Köppen had published a new edition of his standard work on climate, *Klimakunde I: Allgemeine Klimalehre* (Climatology I: general theory), incorporating some of this material.[4]

In that same year, he had worked out a simplified five-climate-zone scheme: polar, boreal, temperate rain, desert, and tropical rain; he had included this as a brief addendum in the 1918 edition of his work. Collaborating daily with Wegener on the climates of the past, in order to understand the relationship of continental displacements and displacements of the pole, he had further simplified this five-zone scheme to four zones, consolidating the boreal and polar zones into a single zone.

Köppen had also, from the beginning, been deeply interested in the causes of climate *change*. The theory of continental glaciation, or of the "Great Ice Age," had only come into general acceptance during his university years in the 1870s. By the later 1880s there was incontrovertible evidence of a much earlier Carboniferous/Permian Ice Age in the Southern Hemisphere, as well as suggestions that in that Achaean past (known then generically as the Algonkian) there had also been ice ages. The theory of ice ages went far to explain climate change at different latitudes, even if there was (yet) no explanation of ice ages.

It seemed to Wegener and Köppen that climate change, as well as the resulting long-term changes in vegetation, depended on ice ages, migration of the pole of rotation, continental displacements, and *nothing else*. These explained the facts of climate change as no other theory could, particularly when the distribution of identical animals and plants separated by great oceans was taken into account. If no one knew the cause of ice ages in 1920, there was, with Wegener's hypothesis, a causal story connecting migrations of the poles and the movement of the continents: the latter caused the former, and then the former restarted the latter. The distribution of ice age remains was intimately connected with the dynamic of continental displacement and migration of the poles. This was indeed the Ariadne's thread through the labyrinth.

Köppen and Wegener were excited and pleased with their hypothesis, but at the time they were writing there was no shortage of theories of climate change. Some of these were theories of ice ages alone, and others were more general. There were cosmic theories, of Earth's climate being changed by passing through different gaseous regions of space, or passing through cosmic dust. There were theories that climate change was caused by variation in solar radiation. There were astronomical theories: that ice ages were caused by changes in Earth's orbital parameters. There were theories of variations in Earth's internal heat, variations in elevation, variation in land and sea distribution, variation in ocean currents, changes in the composition of the atmosphere, episodes of volcanic dust, changes in atmospheric circulation, and movements of the poles and displacement of the crust. Some of these theories went back to the beginning of the nineteenth century, but most of them had been proposed or updated since the 1890s, and the majority of them had been developed during or after the First World War.[5]

What separated Köppen and Wegener from this welter of theoretical propositions was that they were actually attempting to demonstrate something, support it geophysically, sequence it in time with fossil evidence, map it geographically, measure it with astronomical position finding, and in general integrate it as fully as possible with the existing literature of the earth sciences. Theories of variations in Earth's orbital parameters were supported by calculation, but none of the other theories were anything but an idea. Changes in ocean currents or atmospheric composition left no signature; there was no global, rhythmic record of volcanic eruptions to make the air dustier or less dusty, let alone a record of cosmic dust; there was no record of long-term variation in incoming solar radiation. While the bulk of these theories emphasized one or another factor, the consensus was that "they all played a role," which is a way of saying either "I don't know" or "We may never know."

Wegener had surrendered the manuscript for the second edition of his book to Vieweg in April, not because it was done but because he could not afford the time to work on it any longer. He had lectures to prepare for his course in climatology at the University of Hamburg; these were given at the Geographical Institute on the west side of the Außen Alster, one of the two great artificial lakes in the center of Hamburg. His daytime job at the observatory was some distance away from this, adjacent to the harbor. The Geological Institute, which held all the publications in modern geology and geophysics which he required for his geophysical work, was on the east side of the Alster. He was constantly running from one place to another, and Else said that one day he came home so exhausted that he had remarked, to her considerable consternation, "Ten years in Hamburg, and I'm done for" (*ich bin erledigt*).[6]

Moreover, he had to turn his mind back to meteorology, not just to the management of the maritime meteorology part of the observatory but to theoretical issues. Bjerknes had announced that he would hold an international conference in Bergen, Norway, in early August. It was a move to standardize the collection and dissemination of weather data and the format of weather maps, something that Wegener was keenly interested in. It was also a plan for the establishment of a circumpolar weather service. Actually, Bjerknes had announced two conferences. One would be held in July for the Allies—Britain, France, and the United States—along with scientists from Iceland, Norway, and Sweden. He then had planned a second conference for the Central Powers—Austria, Finland, and Germany—to meet with Norwegian and Swedish meteorologists.[7]

This pressure from meteorology was real, though Wegener continued work on the maps he had wanted in the book but had not had time to finish. He had started two maps of the changing continental position between the Carboniferous and the Quaternary, as well as maps of the path of the poles in both the Northern and Southern Hemispheres since the Carboniferous. For the book he had had to settle instead for a map from Kreichgauer's book and another borrowed from Dacqué.[8]

Köppen continued on with the work as well. He was preparing two articles for *Petermanns*, summarizing different aspects of their joint work. The choice of venue was important for him; it would be his first public declaration of support for Wegener's hypothesis, and he wanted it to have the greatest possible visibility and reach the widest possible audience. *Petermanns* was, of course, also the place where Wegener's hypothesis had first appeared, as well as the place where the widely read criticisms by Andrée (and, via him, by Diener) had appeared in 1917.

The first of these two articles was the easier one to write: "Polwanderung, Verschiebungen der Kontinente und Klimageschichte" (Migration of the pole, displacement of the continents, and the history of climate).[9] Köppen and Wegener's work on the history of climate now fell into an established tradition, which included Reibisch, Simroth, Kreichgauer, and Karl Löffelholz von Colberg (1840–1917), all of whom explained changes in climate by a shift in latitude zones following a displacement of Earth's pole. Since Reibisch and Kreichgauer were both still alive and still active (Löffelholz had died in 1917, Simroth in 1920), Köppen thought it especially important to have a detailed discussion of the similarities and differences between the work of these predecessors and his own and Wegener's work. There was neither room nor time in the second edition of the book for such an extensive summary, and Wegener had handled the matter in a few dismissive sentences. Köppen's relaxed, ample, and discursive style made a much better job of it and put their work in context, while managing to show, without any sting, that what they had just done was much better than anything that had come before.

Köppen's presentation of their joint theory for *Petermanns* provides an opportunity to reemphasize how much the fate of Wegener's ideas depended, after 1920, on Köppen's collaboration, coauthorship, reputation, and support, all rendered more powerful still via his persuasive, clear, readable prose. Köppen wrote fluidly, well, and often. In the period from 1918 to 1920 (for half of that time fully employed at the observatory) he wrote and published forty-two notes and articles in scientific and popular scientific periodicals.[10] He was as famous as he was relentless; his name and his work were everywhere. He had been publishing scientific papers in refereed periodicals since 1868; he spoke German, Russian, French, and Esperanto (he was a great promoter of the last as an instrument of international scientific cooperation). In addition to these languages, he read English, Italian, and Spanish and kept abreast of scientific developments in all seven languages.

Köppen understood the importance of seeing novelty in context and in clarifying the issues at hand. In the very beginning of this article for *Petermanns* he laid out the "most obvious causes of climate change": changes in solar radiation, changes in the proportion of land and water, and the shift of Earth's pole. Solar radiation could of course change Earth's climate, but because the whole Earth is illuminated by the Sun, the climate of the whole Earth must change everywhere in the same direction if the amount of solar radiation changes; this can't explain regional variation. Similarly, the proportion in each hemisphere of land and water and the possible alternation of land and water must influence climate, but a look at the temperature in the Northern and Southern Hemispheres shows that the mean sea level temperatures measured in them vary by only a few degrees, in spite of the fact that the former has most of the land and the latter has most of the water; the influence is only perhaps a third or a quarter of that of latitude alone. Moreover, no theory of alternation of land and water could explain the fossil remains of deciduous trees above latitude 80° north.

This means, Köppen argued, that we must consider the displacement of the pole of rotation and a shift of latitude zones. The possibility of displacement of Earth's pole, he continued, does not come from geophysics, but from the data of paleontology. Here he quoted Rudzki: "in case the paleontologists ever come to the conclusion that the distribution of climatic zones in one of the past geological epochs points to an axis of rotation totally different from the present axis, there will be nothing left for the geophysicists but to accept this contention."[11]

While Köppen continued to work enthusiastically on that paper and a second paper on the causes and effects of continental drift and polar wandering, the latter had to stay on hold throughout the spring and early summer of 1920. The intuition of a "pole-fleeing force" had been Köppen's, but the mathematics and physical working out were clearly something for Wegener to do, and he was simply overwhelmed; it was all he could do to finish the maps.

With the coming of spring, Wegener made some attempt to return to his family life, if not a more relaxed schedule. Else loved their sailboat and was anxious to get back on the water. She pointed out to Alfred that their newborn, Lotte, could sleep in a basket as they sailed, while the older girls remained at home with their grandparents. So, beginning in May, they spent every weekend they could on the water, often with Kurt. Kurt had sailed the boat almost every weekend all winter long; he loved being in a boat or in an airplane—his restless nature demanded more movement. Else later wrote that in the Hamburg years Kurt's pessimism about the future, as well as his (from her vantage point) pointless longing for the prewar world, left both her and

Alfred often depressed; the solution to this problem was to see him mostly on Saturday afternoons and Sundays when they might sail together, during which his pessimistic attitude evaporated.[12]

With the end of the "Easter term" in July, Alfred and Else were able to take a longer sailing trip, but Alfred's mind was very much on the coming conference in Bergen. While he had been able to catch up with the literature, the events of 1918 and 1919—leaving the army, returning to Marburg, leaving Marburg for Hamburg, plunging into the second edition of his book—had severed him as thoroughly as the war years from active conversation with colleagues; the conference would be his best opportunity in a decade. Moreover, most of the senior scientists of the prewar period were now retired or dead, and Wegener would be in the presence of contemporaries and younger men, with the exception of Vilhelm Bjerknes, who, though twenty years Wegener's senior, was always full of youthful energy and pushing strenuously forward to recruit the next generation to his scheme for streamlining and unifying weather forecasting.

As is the case with most scientific meetings, the real action in Bergen that August took place away from the lecture hall, and here there was much of interest for Wegener. One surprising development was Wegener's meeting with Felix Exner. It was a long time now since Exner's damaging review of Wegener's thermodynamics text, and a few years since Wegener's response in 1917 to Exner's work on dynamic meteorology. They discovered, once they were face to face, that they admired one another's intelligence, that they enjoyed one another's conversation, and that they were probably going to become friends.[13] Exner was still involved in an acrimonious dispute with Bjerknes, as he had his own scheme for weather maps and thought, as did most of the Austrians, that Bjerknes had undervalued their contribution to meteorology.[14] Wegener had no role in this controversy and did not want one. Perhaps, like his contemporary Ficker (trained in Germany and working in Austria), he could see Exner's pain at the collapse of a huge and beautiful imperial scientific establishment and the impoverished and restricted future that appeared to face Austrian science.[15]

For Wegener it was a pleasure to meet and befriend Exner, but there was much else for him here. For the first time he met the young Swedish meteorologist Tor Bergeron (1891–1977). Bergeron, a protégé of Bjerknes, had spent the winter of 1917/1918 reading Wegener's *Thermodynamics of the Atmosphere* and was completely engrossed in it because it was the only source he could find to feed his own fascination with the physics of clouds.[16] Wegener also met Johan Sandström (1874–1947), whose work on the thermodynamics of layered ocean water provided Wegener much to think about in terms of his own work on the atmosphere.[17]

Wegener found in Bergen that he was known, understood, and admired. He also found that while controversy swirled around him, he need not always take sides. The Norwegians and the Swedes were undisputed leaders in the study of ocean circulation and layering, and Bjerknes had long aimed to match that leadership in the realm of meteorology. Both dynamic meteorology and dynamic oceanography were fertile grounds for controversy. The controversy between Exner and Bjerknes was both longstanding and bitter, but there was no need for Wegener to take any role in it, because he was, by his own admission (and everyone else's admission), not a dynamic meteorologist. He enjoyed the same immunity in Hamburg, where he could easily refrain from taking sides in the bitter and protracted dispute between Carl Brennecke, his colleague in Hamburg, and Alfred Merz (1880–1925), Brennecke's Berlin counterpart, in the determination of cause and character of the deep-water circulation in the Atlantic.[18]

These disputes involve theoretical positions so close together that it might seem (in retrospect) that they involved the "pathology of small differences." This would be a mistake: while a controversy is alive, even the smallest details seem important, and only years or decades after the fact, when the question is decided, do these differences become small. In any case, Wegener, in Bergen, was the friend of all and the enemy of none.

Wegener was deeply affected by the sense of goodwill and scientific community he found in Bergen. He had a chance here to "see himself" outside of Germany for the first time in seven years. That he should feel close to the Scandinavians was no surprise, given his work with Koch in Greenland; he was known and respected as a "man of the North," both in meteorology and in exploration. But there was another and more striking change: he was meeting younger men who had been affected by his writing, as well as older men who accepted him as a colleague both because of the quality of his scientific work and because of his position at the observatory in Hamburg. He was not someone who required the approval of others; now, others wanted approval from him.

Wegener and Exner traveled back from Norway to Germany together, and Wegener urged Exner to stay on for a few days at Großborstel. During his stay, Exner gave a colloquium at the observatory, and the response was so positive that Wegener decided to establish a regular geophysical colloquium at Hamburg. He had, both at Mülhausen and at Sofia during the war, organized colloquia that brought stimulation to his young subordinates and local colleagues. Though the geologists at Hamburg had made it clear that they had no interest in Wegener's work, there were physicists and geophysicists at the university (including his helpful colleague, the young seismologist Ernst Tams). He had the education and careers of his own assistants and colleagues at the observatory to think about, and even some students from the university. Karl Frisch (1892–1953) had come to Hamburg from Dorpat to study with Wegener. He was Wegener's first PhD student since Walther Brand—before the war in Marburg—and Frisch was working on atmospheric layering; Wegener eventually assigned him a thesis topic to study the structure of inversion zones in an atmosphere without turbulence, a difficult problem and a good beginning for Frisch's successful career.[19]

Hamburg was a major seaport and a major rail terminus, as well as an easy destination for scientists from both neutral countries and the Central Powers in the First World War (Holland, Denmark, Norway, Sweden, Finland, Austria, Hungary, and Serbia). Alfred, Kurt, and Köppen began regularly to attach invitations to their correspondence, and these quickly bore fruit. Willem van Bemmelen (1868–1941), a great early investigator of geomagnetism, had been the head of the Dutch scientific station in Batavia (Jakarta), and when he returned to Holland in 1920, Kurt got him to come to Hamburg. Köppen got his old friend and colleague Svante Arrhenius (1859–1927) to come and speak; they were both enthusiasts for a single universal language.[20] Alfred's old Greenland companions and supporters in displacement theory, Lundager and Koch, came, as did Sandström and Bergeron.

Vilhelm Bjerknes came several times, and he sent his sons one after another for a semester at Hamburg, during which time they lived at the Köppen-Wegener household. Almost all the invited guests stayed for several days in Großborstel, and Else remembered spirited conversations lasting far into the night: it was reminiscent of the intellectual excitement that characterized this household in the years before the war.[21]

Wegener was excited enough by what had happened in Bergen to shift his attention (temporarily) from climate and from the work on continental displacements back to meteorology. He had published a brief article in January 1920 in *Meteorologische Zeitschrift* (written in the fall of 1919) extending his theory of cold cloud precipitation to the formation of cirrus clouds. He had also written a very interesting short piece on considering the atmosphere as a colloid structure when supercooled droplets were interacting, referencing there the idea of Brownian motion. It would take us too far afield to consider this argument in detail, but it shows that Wegener's physical imagination, even with this deep involvement in geography and climate, was still vital and productive.[22] While he did not begin a new research program in meteorology on returning to the observatory in August, he did decide to shift his teaching at the university, for at least one term, from climatology to "Weather and Weather Bureaus," probably an introduction to the new forecast network that Bjerknes envisioned, using protocols that (Bjerknes hoped) would be adopted by the League of Nations.[23]

For the book on displacements, the summer of 1920 was a time for reading proof, thanking friends and collaborators, and waiting for the finished book to appear. Wegener wrote to Cloos in June, thanking him for his help in locating geological references to the first edition, and again in July, requesting an opinion about the work of Leopold Kober (1883–1970); apparently Wegener was now reading prospective manuscripts for Vieweg on geological topics. He wrote to Cloos again on 1 October 1920 to tell him that the book had come out, expressing his hope that in its new form it would be more widely read and make an impression sooner rather than later. He also added, touchingly, "Thanks for your expert advice about what I can do to speed up the process of the adoption of my ideas, so that I may live to see the result."[24]

Sometime in the period between the sending off of final proofs in July and the appearance of the second edition of *Die Entstehung der Kontinente und Ozeane* in October, Wegener was able to help Köppen work his way through the final draft of the first of the two articles for *Petermanns*. Köppen had constructed his argument to show that the pathway led from paleontology to geophysics and not the other way around. It is important for us to consider this article in more detail, because both Köppen and Wegener declare that they wrote it together, and it expresses both their views; it is therefore a part of Wegener's scientific development, even if it appears under someone else's name.

As we noted above, Köppen had begun the paper with the history of attempts to explain past climates, aiming to show how both the displacement of the poles and the relative displacement of the continents were required to match the history of two data sets. The first was a shift through time of latitude zones based on the fossil record for given locales. The second was Theodor Arldt's compilation of fossil data concerning the connection, or lack of it, between different regions of Earth in different periods.

The core of the article was a discussion to accompany "Maps of the Earth for the Carboniferous and the Quaternary."[25] In introducing the maps, Köppen requested the following of his readers: "In the following journey through the history of the earth, pursuing the question of the changes in the position of the poles and the continents, I would ask you if possible to have a globe ready to hand and not simply make good with a world map; a small globe of about 10 centimeters diameter is sufficient to the task and very convenient. Only," he continued, "by the use of this means is one safe from the sort of glaring errors to which, for instance, Diener (for example) fell prey in an attack on Wegener's first edition whereby he believed that with a South Pole in Natal [South Africa] the corresponding North Pole would be between Florida and Bermuda."[26]

The maps, five in all, appear in the published version in a foldout at the end of the issue. These were the maps Wegener had not had time to complete before sending his

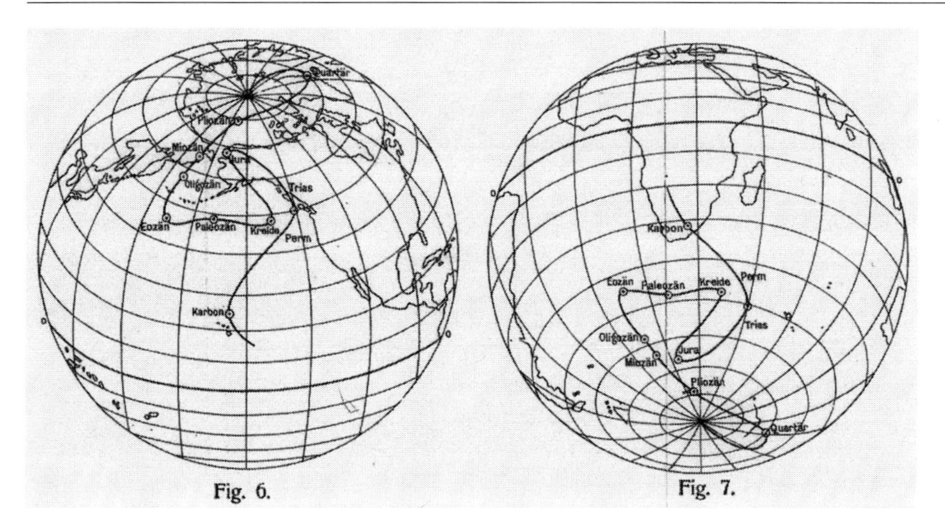

Wegener and Köppen's calculated paths (for both hemispheres) for the displacement of Earth's pole of rotation from the Carboniferous to the present, on a Lambert oblique equal-area projection. The continents are in their present position, and the latitudes and longitudes are calculated from the current position of Africa. From W[ladimir] Köppen, "Polwanderung, Verschiebungen der Kontinente und Klimageschichte," *Petermanns Mitteilungen* 67 (1921).

manuscript to the press. They marked a significant step in his visualization of continental displacements and included new map projections and new conventions of representation. Wegener used two different kinds of equal-area maps. In his figures 4, 6, and 7, showing the movements of the pole, he used the "oblique orthographic Lambert azimuthal equal-area projection." Translated into ordinary English, this is a Lambert equal-area map with true directions away from the center, drawn to show Earth as a sphere and not a flat surface (thus "orthographic), and with the pole not at the top of the map but inclined slightly toward the viewer (thus "oblique"). He had used this projection in his second edition in the map showing the clustering of the various portions of Gondwanaland.

These maps show Earth as a sphere, with lines of latitude and longitude curving as they would on a globe held in the hand. The position of these lines, the *Gradnetz* (graticule), shows Africa in its current latitude and longitude. This was a convention Wegener would maintain ever afterward: *arbitrarily* holding Africa in its current latitude and longitude, and measuring all continental motions relative to the current position of Africa. In his scheme of (absolute) continental motions, Africa had probably moved the least, sliding a bit northward in the Carboniferous, but remaining largely fixed, or oscillating slightly north and south. There were good reasons to do this, but it would be a source of considerable confusion, as many readers then and now have imagined that he thought that Africa had never moved, and never would, and this is not the case. It is also worth noting that Kreichgauer used a similar oblique orthographic projection to show the path of the pole, on the title page of his *Die Äquatorfrage* (1902).[27] The more one looks, the more Wegener and Köppen seem to owe to Kreichgauer.

Wegener's figures 1 and 2 use a different projection, this time borrowed directly from Kreichgauer's use of the same in his work. This is the Mollweide equal-area projection of a sphere on an ellipse, obtained by stretching out the equatorial diameter to

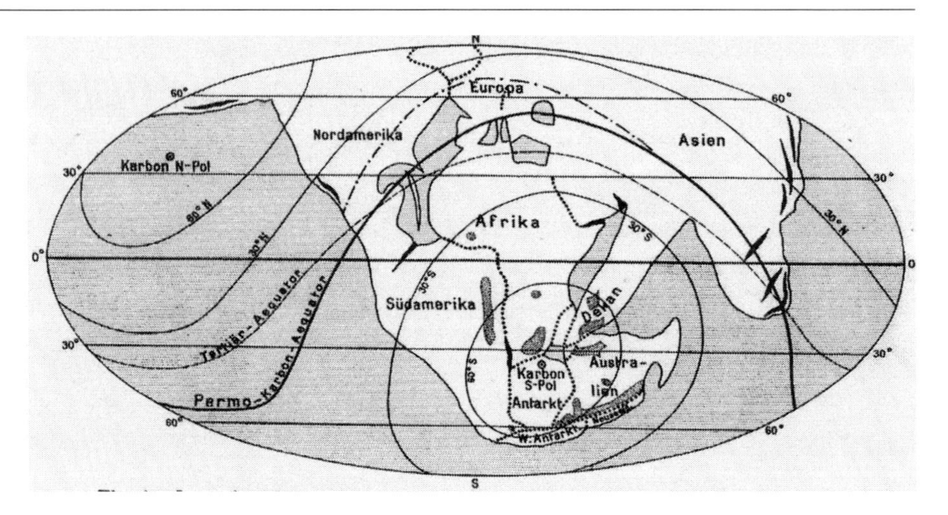

Wegener and Koppen's map, on a Mollweide projection of the whole Earth on an ellipse, showing the position of the continental blocks and the parallels of latitude in the Permo-Carboniferous. Outlines of the present continents are shown with dotted lines, and the paleoequator of the early Carboniferous is shown as a dot-dashed line. The graticule is that of contemporary Africa. From Köppen, "Polwanderung."

be twice as long as the polar diameter. The linear distances are true only on the major axis (the equator), and shapes become more distorted the farther away they are from the center of the map. It was a favorite of geographers and paleogeographers because it showed the whole Earth on one elliptical map. Astronomers liked this map because it allowed them to show the entire celestial sphere, and meteorologists liked it because it is ideal for drawing isobars and isotherms for the whole planet in a single view.[28]

On these maps, Wegener displayed two data sets simultaneously: the difference between the continental positions in the Permo-Carboniferous (his fig. 1) and those in the Quaternary (his fig. 2), and the displacement of the equator through time. Seeing Earth represented thus takes a little getting used to, but one can see very quickly that it allows one to superimpose, as curving lines, the positions of the equator in previous periods and how they pass through different parts of the continents than they do today. Once again, he represented Africa in its current latitude and longitude on the *Gradnetz* in both maps, while arranging the other continents relative to Africa and to one another depending on the time in the geological past. Thus, in the Permo-Carboniferous map (his fig. 1) South America is snugged up against Africa in its current position, and in the Quaternary map (his fig. 2) it has moved far to the west of a "stationary" Africa.

Wegener derived the maps of the displacement of the equator and the maps showing the path taken by the migrating poles of Earth under the influence of the shifting continents from paleontological data about plants and animals at different latitudes at different times. Köppen had contrasted this strategy very strongly with the attempts by Wegener's predecessors, Simroth and Kreichgauer. Simroth had a fixed idea that the poles had swung back and forth like a pendulum, and Köppen criticized him for trying to shoehorn empirical data into this fixation, in the context of a nearly impenetrable book in which he had "ridden the idea to death."[29] Kreichgauer, on the other hand, had been comfortably readable in his presentation and flexible in his attachment to the idea of a shifting crust over the interior of Earth. His problems had been a lack

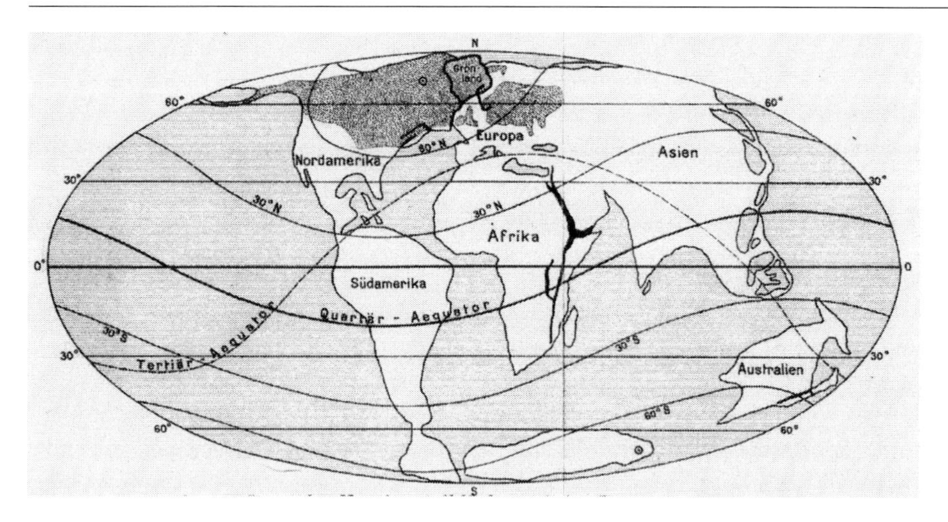

Wegener and Koppen's map, on a Mollweide projection of the whole Earth on an ellipse, showing the position of the continental blocks and the parallels of latitude in the Quaternary. The stippled area covering northern Europe, Greenland, and northern North America is the Quaternary glaciation. The graticule is that of contemporary Africa. A parenthetical at the bottom says, "Epicontinental seas are not represented." From Köppen, "Polwanderung."

of sufficient data for continents outside of Eurasia and a mistaken picture of the physics of Earth.

This approach by Köppen, linking the work to predecessors, positioned Wegener not as a brilliant theorist but as someone in possession of enough adequate data to see the deficiencies of previous schemes and to inductively, not deductively, determine that the continents must move, and the poles must move as well, if we are to make sense of the empirical data at our disposal. That is how Köppen's argument ran and how it concluded in a commentary on these maps, drawn by Wegener and reproduced by *Petermanns* with a craftsmanship and clarity that invoke admiration even today.

The second edition of *Die Entstehung der Kontinente und Ozeane* appeared in October 1920, to Wegener's intense pleasure and excitement, and in November there was another major event: he turned forty. It caused him to reflect. When he had first thought of the idea of continental displacements, he had just turned thirty and was an unmarried instructor in physics and astronomy at Marburg who had liked the pose of the "brilliant theorist," someone who drops in on an existing dilemma and provides a breakthrough idea. Köppen and Aßmann had encouraged him in this view of himself and thought that his ideas would lead to breakthroughs in meteorology, especially in the area of atmospheric layering. Now he was a forty-year-old government scientist with a wife and three daughters, as well as a veteran of the Great War and two expeditions to Greenland.

He could look over his work in meteorology with some satisfaction, but little that he had done had led to major breakthroughs in that scientific world. His work on atmospheric layering and atmospheric thermodynamics had certainly made an impression; his postulation of geocoronium and his notion of an atmosphere vertically segregated in terms of the abundance of elements were still very much in play. Yet the conference in Bergen in August 1920 made it clearer than ever that the next big steps in meteorology would be in dynamics, and that Bjerknes would lead the way. In this area Wegener could be a commentator and a resource, but not a principal player. Exner's bitterness at having lost out to Bjerknes was not something Wegener felt himself, but he had to think about the direction his own career was going. He had no research program under way in meteorology, and other than his work on tornadoes and waterspouts—deliberately written in a way that suppressed his penchant for theorizing ahead of the data—his only original contributions to atmospheric physics in nearly a decade had been his papers on atmospheric optics and his speculations on concentric, intermittent zones of audibility.

His theory of continental displacements was, in 1920, by far the most prominent and interesting science associated with his name. *Thermodynamics of the Atmosphere* had received good and bad reviews, but no one had taken time to write book-length defenses of or attacks on his work in meteorology. In that world, even his most vociferous opponent, Felix Exner, was now on his way to being a close colleague and a personal friend. It was much different with his theory of continental displacements, which had generated a major controversy, and in which supporters and opponents (most of whom he had never met) had devoted considerable time and intellectual resource to defending or opposing his ideas.

He was learning that it was not enough to bring a new idea into the world; for a hypothesis to survive and grow, one must tend it and defend it. The contest is not between ideas; it is between scientists who hold them. Wegener had imagined he could defend his ideas by making them better *on their own*: a new edition, a better presentation. It was Köppen, above everyone else, who was teaching him not only what to do but how to do it. From Köppen he was learning not just to *produce* the work but also to *present* it; the most effective way to do this was to position it as an improvement in a promising line of advance, rather than as something entirely novel. This gave the audience a greater sense of participation in a collective enterprise and made them players rather than mere spectators.

Lunar Craters

There was nothing Wegener could do, immediately, to advance the cause of his book. To write an article, or summarize it in too great detail, could risk hurting the sales; his first wish for the book was that people should buy it and read it. He fetched about for something to do and decided to apply the lessons he was learning from Köppen, about tending and advancing hypotheses, instead to his work on the origin of lunar craters. He had published a paper (dedicated to Richarz) in the *Acta* of the Academy of Sciences earlier that year, as well as a companion piece in the popular astronomy journal *Sirius*.[30] Neither of these efforts had been more than seven or eight pages long; they were the extremely condensed "physics style" papers he had greatest experience in writing.

Therefore, in the fall of 1920, he contracted with Vieweg for a short book on lunar craters, to be published in the "Sammlung Vieweg" series, in which he had first published his book on continental displacements in 1915. Vieweg was happy to do this; the 1915 volume had sold out, and it appeared that the 1920 volume would as well. Wegener was a good author, was meticulous, made few changes in the final proof, and was scrupulous about deadlines.

Wegener, in this work, made no alterations to his hypothesis of lunar craters and conducted no new experiments. He added a few supplementary line drawings comparing the profile of lunar craters with the profile of his experimental craters and also

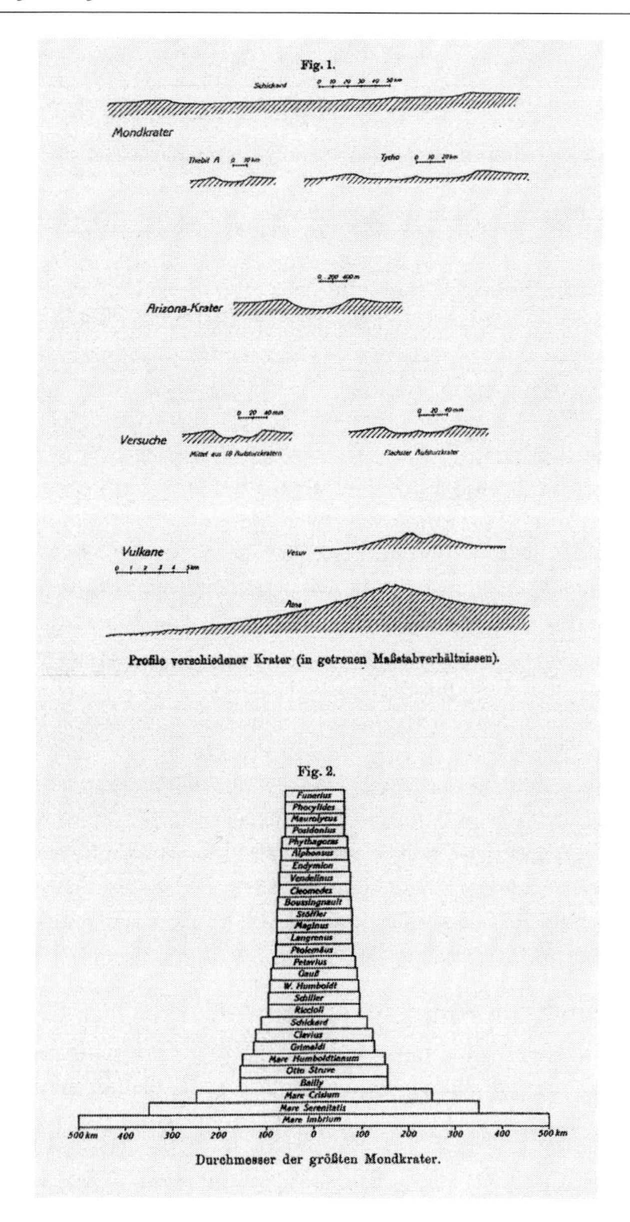

Wegener's supplementary diagrams from his short book on the origin of lunar craters showing the very different morphology of lunar craters and terrestrial volcanoes (top) and the range of the largest crater sizes on the Moon (bottom), indicating how much larger these are than any volcanoes on Earth. From Alfred Wegener, *Die Entstehung der Mondkrater* (Braunschweig: Friedrich Vieweg & Sohn, 1921).

with the profile of Earth volcanoes, to show how different they were in elevation and form.[31] He also drew a figure stacking the largest lunar craters on top of one another, showing that though they were identical in profile to much smaller lunar craters, their diameters ranged from 300 to 1,000 kilometers (186–621 miles); the point here was to reemphasize the difference between crater size on the Moon and volcanoes on Earth, the largest of which were only tens of kilometers across.[32] He produced a chart showing the latitude and longitude on the Moon of the largest craters and showed that they were uniformly distributed over the surface, whereas Earth's volcanoes exist only in certain zones.

More significant than his few novel illustrations was the way he inserted himself historically into a tradition of research. He went deeply into the details of every one of the predecessor theories: earlier infall (impact) theories, the bubble theory, the tidal theory, and the volcano theory, showing how they had developed over time, stressing the care his predecessors had put into the work, and noting the physical and morphological problems they faced, including the issue of scaling. He reduced his sarcasms, though he could not restrain himself from repeating that the bubble hypothesis was an "egregious fallacy."[33] He greatly expanded his historical coverage of the volcano hypothesis and went on a hunt for every conceivable predecessor of the impact hypothesis. He may have solved the problem, but he was part of a great tradition of research, not a rebel against it.[34]

He also chose a new title. He had originally used the term *Aufsturz-Theorie*, but the word *Aufsturz* (infall) is so unusual that it does not occur today even in very large German-English dictionaries. He entitled his new book *Die Entstehung der Mondkrater*, emphasizing its focus on a single genetic principle that explained all the craters on the Moon. This tied the work to his second edition of *Die Entstehung der Kontinente und Ozeane* and stressed that Earth and the Moon, so different in appearance, were so because of the very different *origins* of their surface features.[35]

He better understood now the need to "tend" his hypotheses. He prepared the way for this book (under the new title) by writing a three-page note, "Die Entstehung der Mondkrater," for the scientific weekly *Die Naturwissenschaften*. He also wrote a four-page summary of the book for *Die Umschau* (a scientific monthly very much like *Scientific American*) and titled it "Das Antlitz des Mondes" (The face of the Moon), clearly a reference to Eduard Sueß's great work *Das Antlitz der Erde*. In both of these he presented his work as a new solution to an old problem, succeeding through a new approach, but depending on a tradition of research.[36]

The Berlin Symposium on Continental Displacements, 1921

Wegener may well have planned an even longer book about the Moon; as published, it was less a book than a pamphlet of about fifty pages. He had received in late November or early December 1920 an invitation to present his hypothesis of continental displacements to the Geographical Society in Berlin in February 1921. He was to give a formal lecture outlining his theory, and there would be three commentators, Franz Koßmat (1871–1938), Albrecht Penck (1858–1945), and Ernst Schweydar (1877–1959). Wegener would then be given the opportunity to respond; all the papers would be published together in the society's journal. The invitation probably came directly from Gustav Hellmann (1854–1922), the head of the Prussian Meteorological Institute and professor of meteorology in Berlin; he had worked with Wegener throughout the war, had helped him with his book on tornadoes and waterspouts, and was a close colleague of Köppen. Hellmann was, in 1921, president of the Geographical Society.

It was a tremendous opportunity for Wegener, and his commentators were major figures. Koßmat had been one of the first to comment on Wegener's hypothesis, and while finding "tectonic difficulties" in Wegener's rearrangement of the southern continents, he had referred to the "*kühne Hypothese von Wegener*" (Wegener's bold hypothesis).[37] Koßmat was a mineralogist, geologist, and paleogeographer, head of the Geological Survey of Saxony, and director of the Geological-Paleontological Institute at

Leipzig. In 1920 he had published the first gravity map of Central Europe; he was thus expert in several of the areas drawn together by Wegener's hypothesis. Albrecht Penck was even more famous and was the head of the Berlin Oceanographic Institute. He had written a classic work on the ice ages, *Die Alpen im Eiszeitalter,* and was a world leader in the study of geomorphology and climatology; he had been one of the original supporters of the plan to test continental displacements with telegraph time signals in 1914. The third commentator, Ernst Schweydar, was somewhat less well known, but he was an expert on Earth's gravity field, astronomical position finding, and Earth tides, and he had been an assistant at the Geodetic Institute in Potsdam since 1897; he was now a professor as well.

The Berlin session was scheduled for 21 February 1921 and would be chaired by Ernst Kohlschütter, the undersecretary of the navy who had insisted on having Wegener to replace Köppen at Hamburg; he was also one of Germany's leading experts on Earth's gravity field. It would be a *Fachsitzung* (technical session) rather than a meeting aimed at the general public.

The panel favored Wegener's chances of a good hearing. Kohlschütter and Koßmat were ten years older than Wegener, and Schweydar only three years older. Penck, twenty-two years his senior, was on the other side of a generational divide that Wegener understood very well. Along such lines Wegener could expect the most support from Schweydar. Kohlschütter would not speak but could be expected to be sympathetic and to show it. Koßmat was more traditional and geological but had published geophysical work and understood the arguments. Penck would be the problem. With the exception of Penck, everyone on the panel had published detailed work in geophysics. Penck was not hostile to geophysics, and where he moved away from it he moved in the direction of a morphological-empirical approach, which Wegener also championed. Schweydar would be generous and favorable, Penck opposed but probably not dismissive, and Koßmat somewhere in between. Whatever the speakers said, the most important thing that would come out of the meeting would be that Wegener's ideas were important enough to merit a symposium. It is generally true that in science, as in other areas of endeavor, "the only thing worse than being talked about is not being talked about."

Things appeared to be going his way. Shortly after hearing that he was to take part in the symposium on 21 February, he wrote to the German Meteorological Society and offered to give a lecture on "Climates of the Past," a theoretical approach to paleoclimatology, evaluating the different classes of evidence. They accepted with alacrity, and he and Else traveled to Berlin on the eighteenth to be ready for these two appearances.[38]

The larger challenge was the symposium at the Geographical Society. Wegener knew that for his opening remarks he would have perhaps an hour, and he would not have time for details. Details would be a trap in any case, because his opponents could pounce on any specific detail at the expense of the whole argument. So rather than talk about individual pieces of evidence, Wegener chose to talk about classes of evidence, groups of arguments collected by many scientists working independently of one another over a long time, none of whom had had the idea of continental displacements in mind. He chose to describe these sets of data as each having a sign, pointing either toward his ideas being likely or away from them. The idea was that if the general tendency or preponderance of evidence in some realm of the study of Earth made room for his hypothesis, he could count it as potentially on his side. The more independent classes of evidence he could count as being on his side, the stronger the plausibility of his overall argument, and the greater the likelihood of it being true.

On the day of the event, he began briefly and diffidently, remarking what an odd picture he had drawn of Earth, with all the continents far away from their accustomed positions and all clustered together. Then, through time, there was the spectacle of their moving away from one another, some to the south, some to the north, and some to the west. He said that he had done the best he could to give the foundations for this strange idea in the second edition of his book, and that in a venue like the symposium he could only give a few of the pertinent facts, briefly presented.[39]

He then went through the geophysical evidence we have seen him rehearse before, beginning with the bimodal elevation of Earth's surface: this could not be a matter of chance, but one of necessity, of structure. He quickly indicated the gravity data that suggested that the continents and oceans were made of different stuff, and how the study of geomagnetism supported this idea, as did the study of the propagation of earthquake waves and the dredging up of samples from the ocean bottom; all elements pointed in the same direction: the continents and oceans were two different layers made of different materials. He pointed to the generally accepted notion that Earth's outer layer floated on the layer below it: isostasy. So, Wegener said, you have to admit that most of the geophysical evidence is on my side. "In fact," he continued, "I have never met, *or even heard of* a geophysicist who disagrees with anything that I've just told you."[40]

Then he turned to geology. Here again Wegener ran through the classes of evidence: he pointed to the huge overfolds in the Alps and other great mountain ranges that almost all geologists accepted, and the impossibility that they should be caused by Earth contraction—not his own conclusion but that of a variety of experts. There was the correspondence between the coast of South America and the coast of Africa, both in the contours of the coastline and in the continuity of geological formations across them; together these suggested that they were once connected. His critics, he allowed, have made much of his failure to exactly specify in his reconstructions of the continents where some mountain range in South America connects with another in Africa, but with the distances involved, he mused, did it really matter if it were 100 kilometers (62 miles) south or 100 kilometers north of where he put it on his map across a distance of 6,000 kilometers (3,728 miles) of open ocean? The geological evidence, he judged, ran heavily in his favor.[41]

He then moved on to talk about a completely separate universe of material, from biology and paleontology. This had nothing to do, he said, with geophysics and had nothing to do with mountain ranges or other geological structures. Then he introduced the compilation by Arldt, about former connections between all the continents at different times, inferred by paleontologists to explain the unity or disparity of flora and fauna. Wegener had known he could not possibly present this material in a lecture by showing a table with all the pluses and minuses of twenty different thinkers for thirty different periods of time concerning thirteen different land bridges—as he had published it in his book—so he came up with a graph for the four most important land connections for his theory. These showed, in a way that was immediately visible, where expert opinion judged that various continental segments began to lose connection, and how the timing of these breaks fit with his own reconstructions.[42] Thus, he argued, paleontology and biology count in his favor as well.

There is, he went on, still another class of data, that of the climates of the past. Here he made the case that the discrepancies in latitude zones could only partially be solved by the mobility of the pole. Especially in the Southern Hemisphere, motions of the pole were not enough: the continents needed to be in different positions. Here, he said, his hypothesis extends and completes earlier work explaining these climate shifts as shifts of latitude. Once again, this separate class of evidence pointed in his favor.

So, he asked, when we pile up all of this together into a single picture, all these independent lines of evidence gathered by different people having nothing to do with one another, all of which point to the direction of his hypothesis and none of which point away from it, and all of which, taken together as he had assembled them, seemed to fill most of the holes in most of the proceeding explanations given for them, how are we to judge? "Can a theory be wrong," he said, "in the light of which the features of the face of the earth acquire in such a wonderful way, meaning, and life?" Even so, he said, "The displacement theory is ready to undergo one final, demanding interrogation. It is willing to prove itself true by precise astronomical position finding." He then declared that Koch had provided unambiguous proof of the westward motion of Greenland based on his Danmark measurements.[43]

"Ten years ago," he concluded, "when the theory of continental displacements was published by me for the first time, it stood facing a mountain of difficulties and questions. Today, these have turned into a glittering array of confirmations."[44] Since the publication of the second edition of his book, he continued, he had received numerous letters from scientists telling him that this work had transformed their previous negative attitude into one of acceptance—though still with reservations: "Even as we speak here, I happen to know, a range of plant and animal geographers, geologists, geographers, and geophysicists are working with this new theory with astonishing results. These developments will, I am convinced, not stop until the theory of continental displacements has become the fundamental basis [*Grundannahme*] of our understanding of the evolution of the face of the earth."[45]

Koßmat came next to the podium and began a respectful, detailed, and professional critique of Wegener's overall theory, speaking to him and his assertions from the following standpoint: "The idea of crustal wandering ties together facts from geophysics and Earth history, so Wegener's work should be considered carefully from the geological standpoint."[46] Wegener's ideas, he continued, have many problems, but they also show much promise. These ideas, Koßmat said, work better in the Southern Hemisphere than they do in the Atlantic. He agreed with Wegener that reuniting the continents alone could not solve the problem of the climate history of the Southern Hemisphere in the Permo-Carboniferous, nor could polar motions alone; whatever the eventual solution to this puzzle, it would require some lateral motion of continental surfaces.[47]

That he accepted the need for some motion of continents at some point in the past was not, he hastened to say, for him to accept Wegener's theory; he did not. Like most geologists, his interest in Earth movements concentrated on the vertical, and he liked the idea that mountains were created by "magma injection." Molten rock from deep within Earth pushing up from below elevates mountain ranges in thick sedimentary basins that were once areas of ocean.[48] Continents and oceans were, he insisted, interchangeable, at least in areas like the shallower part of the Indian Ocean and the Mediterranean. Geophysicists, Koßmat said, will have to give way and admit that some amount of sinking of the land is possible, and that the oceans are deepening through time.[49]

Koßmat went on to disagree with a number of Wegener's assertions. He saw the same problems that Cloos had seen: the explanations for the East African rift zone;

the timing and nature of the split between Africa, Madagascar, and India; the timing of the motion of India and the compression of the Himalayas. He tried to constrain Wegener's theory to the few places and times on Earth when it was absolutely required to make sense of the data. He denied that the theory was really in a position to unify all of the earth sciences. He was more skeptical of geophysical evidence than Wegener, but not dismissive of it (one recalls that even Rudzki thought that Wegener had overemphasized the difference between continental and ocean rock). There are many complications, Koßmat said, on the face of Earth, which take this simple picture Wegener has provided and force us—in following out the details of the manifold of appearances—to try to understand all of the forces that have brought them about.[50] In other words, continental displacements were a piece of the puzzle, but not the solution to the puzzle in general.

Next at the podium was Albrecht Penck, and he was condescending and intransigent. As we all know (he began), almost all the evidence we see is of vertical movement. As for this business about continental and ocean crust being different, he did not think it established. It could just be that the continental crust that covers the ocean floor has been cooled by ocean water and shrunk down, thus producing these gravity readings.[51] He did not believe that the continents could plow through the Sima and at the same time be upthrust and folded on their leading edge. This was just intuitively impossible for him to imagine. From his perspective, the second edition of Wegener's book did not solve the problems left by the first edition.

He liked Diener's solution to the problem of getting animals and plants across the North Atlantic by island arcs, and he said that Wegener's hypothesis did not rule this out.[52] As for coastline similarity, that did not mean a former connection, and maybe there *were* mountain ranges transverse across the floor of the Atlantic which we just hadn't found yet.[53] The continental shelf is not a vertical wall but a stepwise descent, so coastline matching at the edge of the continental shelf doesn't really mean anything more than coastline matching at sea level.[54]

He then went on to question the accuracy of the longitude measurements of Koch. Here he moved not in vague generality, as above, but in effective detail, pointing out that Wegener's account of what Koch had said and what Koch had himself said were different, that the uncertainties were gigantic, and that what Koch had proposed as a possibility Wegener had transformed into a revolutionary certainty that did not exist. Penck was on excellent ground here and scored heavily against Wegener, given Wegener's challenge that proof of displacement was a final test his hypothesis had already passed.[55]

However, Penck still had to deal with the problem of the Southern Hemisphere, and here he quickly went from a definitive demolition of Wegener's measurement claims to vague arm waving against the reassembly of the southern continents. This is, he said, "beset with many difficulties," and "our knowledge of the details is sketchy," and "we have to be cautious."[56] Wegener's idea of the horizontal motion of continents is a plausible explanation for a range of quite various and differentiated data, he admitted, and this makes it extremely "seductive," but there is more evidence for sinking than for splitting and drifting. He concluded by raising an entirely different issue and saying that no matter how much evidence one collects in favor of a hypothesis like this, no matter how seductive it is and how well it functions, one really cannot accept it until we know the mechanism.[57]

Penck sat down, and it was time for Schweydar to discuss the geophysical implications of Wegener's hypothesis. His remarks did not seem to start well for Wegener, because he began by pointing out that Wegener's hypothesis depended crucially not just on isostasy but on a particular *interpretation* of gravity differences associated with the British astronomer George Airy, whereby gravity deficits and excesses are removed by assuming that mountains (and in Wegener's case continents) have deep "roots" of light material. He then went on to say that he had looked into this question rather deeply in his own research and had become convinced that the approach taken by Wegener to the interpretation of the gravity data was the right one.[58]

After this brief scare for Wegener, everything else in Schweydar's remarks was smooth sailing. Geodesy supports Wegener's idea of floating continents, Schweydar said, and our knowledge of the strength of materials makes it plausible that they could plow through the Sima. Whether there could be currents in the Sima is another and more difficult matter to establish, but even that cannot be ruled out. As for Penck's claim that the lack of accurate measurement data for the motion of continents counts against Wegener—that Koch or others did not establish this motion—this was not, said Schweydar, the case. The measurements do not count for him, but neither do the lack of same count against him.[59]

Schweydar went on to say that the idea of a pole-fleeing force, or at least the tendency for continental masses on a fluid Earth to move from the poles toward the equator, was perfectly well established, and given that Earth does behave as a fluid on appropriate timescales, both the continents and the poles might move. Not only that, said Schweydar, but "in any case it must be taken as *likely* that the continents, under the influence of a pole fleeing force, suffer displacement towards the equator."[60] Schweydar was uneasy about the linkage of the pole-fleeing force to the westward motion of the continents, and he wondered about the size of forces generated in the movement toward the equator and the velocity of the continents. Finally, he expressed doubts about the ability of the displacement of continents, even large continents moving through several tens of degrees of latitude, to force the pole of Earth more than 2°–3° away from its current position. This, however, was a matter for investigation.[61]

When Schweydar concluded, Wegener was offered, as was customary in such symposia, a final word, and he did not conceal his satisfaction with the outcome for his theory. "Anyone," he began, "who has followed the geological literature on continental displacements from the first heated review by Semper, to the remarks before you by Koßmat, cannot help but see a slow but continuous transformation from complete rejection, to at least partial consent. Such an accommodation I see especially in Koßmat's unreserved affirmation of the necessity of accepting large horizontal displacement of portions of Earth's crust as well as large displacements of the pole. The remaining differences, I believe, we can now eliminate."[62]

He spoke first to Koßmat's arguments and made a respectable claim: that geological opinions contrary to Koßmat's assertions existed in most cases, and particularly that with regard to the connection between the Appalachians and the mountains in Europe Koßmat was swimming upstream against the preponderant geological opinion. What remained was a mild difference of opinion on specific geological facts, and Wegener ended by saying that he thought he had been able to answer most of the objections and did not doubt that Koßmat and his colleagues would come in the course of time to agree that this was the case.[63]

Penck came in for much rougher treatment. He would not, said Wegener, even touch Penck's notion that he can make a free choice between the geophysical data and his alternative explanation for the difference between the continental and the ocean crust. Similarly, Wegener said, he found himself unable to make any sense of Penck's notion that the statement "land bridges, or permanence of oceans, it is just one assertion against another" is somehow a point against my hypothesis: "That was exactly my point!"[64] The displacement hypothesis is what allows us to move beyond these dueling assertions, "as anyone would know who had read the first chapter of my book."[65] When, Wegener continued, Penck cites only negative opinions about former connections across the Atlantic, he could easily give the impression that this represents some sort of geological consensus, but this isn't so. And when he cites Diener's work, on which his rejection of such links depends, he has had to resort to assertions that a simple review of the literature would show have already been stamped as "grotesque misunderstanding" and thoroughly refuted.[66] Wegener went on in this tone and in this vein, pounding Penck on point after point for about twenty minutes, and one is left with the impression of Penck as a senior scholar who thought he could get away with a casual dismissal, instead of a serious argument, and who paid a heavy price.

Wegener's response to Schweydar was but a single paragraph. Wegener said that he was looking forward to Schweydar's future publications on the topic, that he had found Schweydar's comments on the pole-fleeing force extremely interesting, but that nothing he had said had surprised him, as he had never yet met a geophysicist who opposed the possibility of displacements. Wegener ended with a general thanks to the group for all the objections brought against the displacement theory, as the only way forward to clarify and improve it. He made a single reservation: "However, I think that any unprejudiced observer will be able to reconcile any of the facts brought forward by Penck with my presentation of the displacement theory."[67]

One can imagine that Wegener would have been excited and even elated by this outcome. Schweydar, as the representative of geophysics, had supported him in every single assertion in the displacement hypothesis, though urging some revisions in Wegener's characterization of the action of the pole-fleeing force. Schweydar had stated that he intended to work on the problem himself, thus underlining Wegener's contention that a community of investigators was already using the displacement hypothesis as a way to guide their research. Moreover, Wegener had been able to represent Koßmat's remarks not as a thoroughgoing critique of the displacement hypothesis but as a major step forward in the willingness of geologists to accept both large lateral displacements of the continental crust and large displacements of the pole as absolutely necessary in order to explain the data at hand. Finally, he had come up just short of accusing Penck of intellectual dishonesty in representing his own opinion as the consensus of the field and in using an already-discredited publication by Diener to support his claims.

There was something larger at stake than this "victory" over Penck. Winning some war of words with a critic was something that Wegener cared little about, and had he not been goaded by Penck's attitude in the symposium, he might never have responded as he did. Much more important to him was the realization that the status of his ideas about continental displacement was now really as he had described it: a working hypothesis guiding an identifiable and growing community of investigators from a number of different disciplines.

Yet a victory it was. Otto Baschin (1865–1933), a geographer and meteorologist who had been to Greenland with Drygalski, attended the session and wrote a glowing summary of it for *Die Naturwissenschaften*, which appeared in early April. He said that Wegener's presentation had been brilliant and that the large audience was swept away with enthusiasm for the idea and for the astonishing number of facts that Wegener was able to bring together. Wegener's critics, he concluded, had not been able to marshal one single secure fact to challenge the central premises of the theory of continental displacements.[68]

It is a major step forward when a hypothesis becomes severed from the name of its proposer and takes on a life of its own. So it was now with Wegener, with his displacement theory launched into the world, and with a community of investigators moving this hypothesis where it would go, not because he told them to, but because that is what happened *to the hypothesis* as they wrote and published scientific work. He could help inform the community, he could continue to contribute on his own, and his work would always be of prime importance to those attracted to working in terms of the idea, but he was not any longer in charge.

Still, in the mind of the public and the scientific establishment, in the spring of 1921 this was "Wegener's theory." Scarcely a week after Otto Baschin had previewed the proceedings of the Berlin symposium in *Die Naturwissenschaften*, the same magazine devoted an entire issue (other than news and book reviews) to an extensive review of Wegener's hypothesis by Bruno Schulz (1888–1944), an oceanographer at the Deutsche Seewarte in Hamburg. It was less a review than a summary of the hypothesis, but Schulz concluded with the following judgment: "As we look at the elaborate structure of the displacement theory and its manifold connections, it is hard to avoid the impression that the theory possesses an inner core of truth, especially since no one has, as yet, been able to bring forward any convincing counter-evidence."[69]

When Wegener returned to Hamburg, buoyed by his Berlin reception, he could see that he needed to begin *immediately* on the new edition of the book. His strategy in Berlin of avoiding a mixture of hypothesis and fact in his presentation, as well as his decision to group the evidence in a series of modular categories emphasizing their independence from one another, had paid a huge intellectual and rhetorical dividend. He was stung—but convinced—by Penck's remark that the second edition of his book had not solved the problems of the first edition. Wegener had rejected and overcome the specific criticisms that Penck had in mind, but he had to agree with the overall judgment. The strength of the hypothesis would not, in future, lie in a firm integration of every bit of evidence into a seamless whole, but in the production of a summary digest of what different investigative communities, not in direct touch with one another, were producing.

Wegener had known even before the second edition was out that there would be a third edition, and he had already begun to make provision for it in 1920. In December of that year, Wegener had purchased a quadrille-ruled, clothbound notebook, about 13 centimeters × 19 centimeters (5 inches × 7 inches), and had written on the cover "Kontinental-Verschiebungen." Beginning in late November 1920, he had begun to make notes and extracts from books and papers. This is the same process he had followed in preparing the second edition. Then, his materials had been organized in a different way, within the covers of the first edition rebound with interleaved blank sheets. He would not repeat this strategy this time with a copy of the second edition;

perhaps money was tight, and perhaps he realized that he had tied the structure of the second edition too much to the critiques of the first. He was prepared now to move to a new structure for the book in its third incarnation, and thus the notes would find their home in the appropriate sections of the new edition of the book and would not need to be married to any specific portion of the second edition.[70]

There was unexpected time and opportunity to move forward with research on continental displacements in spring 1921: Wegener returned to Hamburg to discover that his ambitious meteorological program had been effectively canceled by Germany's growing budget crisis. The plans for the expansion of the kite and balloon station at Großborstel, long since approved by the ministry, had now been shelved for lack of money. Without the buildings, offices, and machine shops, there could be no large-scale experimentation with new instrument designs and no regular program to fly them.

Wegener would still be able to carry out a modest but steady program investigating ways to improve meteorological instruments; with the promise of meteorological flights using fixed-wing aircraft, Wegener worked hard on the problem of finding ways to keep recording instruments, especially their clockwork, from freezing up: this was important both for polar meteorology and for atmospheric research.[71]

Working with Eric Kuhlbrodt, he was trying to modify the design of theodolites for tracking balloons. This was a problem that dated from his days in Lindenberg and his very first solo publication there on a theodolite designed by Alfred de Quervain. Wegener's seagoing trip in late 1909 and early 1910—to track upper-level winds across the South Atlantic—had failed to produce results for a number of reasons, but especially because the balloons he sent up were too small to be easily followed with a theodolite, and the controls adjusting this telescopic instrument were so fine that once a balloon was lost to view, it was very difficult to locate it again. Wegener and Kuhlbrodt wanted to find a mechanical way to allow very fine adjustments and very large adjustments of altitude on the same instrument.

These plans notwithstanding, the budgetary news was still a catastrophe. Kurt and Alfred's program to turn Hamburg into Germany's new center for atmospheric research had ended, they realized, before it even began. Kurt had worked hard through 1920 to obtain surplus warplanes to convert to meteorological use, and these planes arrived in Hamburg in April 1921. However, by the time they arrived, Kurt had already decided to leave the observatory and move to Berlin, where there were more planes and, more importantly, the gasoline to fly them. Although the Hamburg aircraft did fly throughout 1921, there was only enough gasoline for two flights per month of about one hour duration each.[72]

Alfred would also be able to continue a very limited program of aerology in the context of his university teaching, and he scheduled a course in that subject at the university for the Easter term 1921 (April–July).[73] However, it appeared to him that his days of active data gathering in meteorology and atmospheric physics might be over, and that more and more he would be a teacher and mentor to younger colleagues.

At the observatory he was not a selfish or distant leader, nor did he misuse subordinates. Johannes Georgi, Wegener's former student at Marburg and now his assistant at Hamburg, recalled,

> I observed how cautiously Wegener made use of his intellectual superiority. We were walking along the corridors of the Seewarte one day when Wegener talked about various experimental equipment, and also wanted to hear my suggestions. Although a

> meteorologist, I happened to have read about some hydrodynamic experiments with pulsating balls in water, which had been described in 1876 by the father of the famous meteorologist Wilhelm Bjerknes. Wegener listened with interest, and I was flattering myself that I had told him about something new when, without any unkind intent on his part, he showed in the course of conversation that I had not mentioned all the facts of the matter—in short that he knew far more about these old and rather off-track experiments than I did. That was not the only occasion during the course of the next ten years that I came away red in the face after conversations with him, shamed by his more extensive knowledge and at the same time by his kindheartedness.[74]

In addition to this downsizing of his meteorological work, there were other decisions pending in spring 1921. He and Köppen had to sit down and decide how they were going to proceed together. Köppen's paper "Migrations of the Pole, Displacements of the Continents, and the History of Climate," prepared in January and February 1920, finally appeared a year later in the January/February and the March 1921 issues of *Petermanns*. The delay was frustrating but unavoidable; paper was still scarce, and the journal was much reduced in size, with each monthly number about one-third the size of its prewar counterpart. Köppen's second paper, "Causes and Effect of Continental Displacements and Migrations of the Pole," had been postponed again and again and was now scheduled to be published, also in two parts, in July/August and September 1921. This second paper, worked out with Wegener, and partly written by him, was a mathematical discussion of the pole-fleeing force, in which they had actually attempted a calculation of the size of the force. Using data from Krümmel's *Handbuch der Ozeanographie* plotting the relative area of land and water on Earth's surface for every 5° interval of latitude (in thousands of square kilometers), they tried to estimate the moment of inertia of the amount of continental mass at each latitude, but they found the calculation too hard.[75] Indeed, Wegener had said in the second edition of his book that someone, someday would carry out the numerical integration, but not him, and not now.

Köppen was, in March 1921, reading final proof for the second article for *Petermanns*, and although he and Wegener could not make any major changes, they found that they could add a few paragraphs at the very end, without running over the page limit. Köppen had already written a list of eleven major points in the theory, a summary and conclusion to both papers. To this he and Wegener added a short section entitled "Remaining Questions to Be Clarified."[76]

These questions are of some interest because they show just how far Wegener's theory had moved in a decade from its original inspiration in the outline of the Atlantic continents. The Atlantic was now the most problematic part of the entire picture, and one of the most pressing work points was how the opening of the Atlantic had affected the relocation of Earth's axis. Where did the North Pole move, and what was the effect on the total obliquity? The latter quantity, the obliquity, is the inclination of Earth's axis with respect to the plane of the orbit.[77] The effect of the opening of the Atlantic on the position of Earth's axis was especially important, because of the movement of the Atlantic continents (mostly) toward the west while everything else moved toward the equator; the *Westwanderung* needed a special explanation.

The important question of the Atlantic continents led directly to a more general question. If it happens, as Wegener supposed, that the continents are thickening and elevating through time, what effect does this have on their movement? Does their velocity increase because of an increasing separation of the center of buoyancy from the center of gravity, or do they slow down because their greater bulk makes it harder for them to move through the Sima? Perhaps it could be that the Americas had torn off from Eurasia because the compression of North Asia, moving toward the equator, made that part of the continent thicker and made it harder for it to move through the Sima; thus, did the thinner western section tear off and drift away?

How could the paleontology of the Southern Hemisphere be integrated with what happens in the Tertiary and Quaternary movements of the pole? Wegener and Köppen had interpreted the paleontological data for the Northern Hemisphere to indicate that the North Pole was oscillating back and forth, with swings of 10°–20° in the past 20,000–30,000 years. This they inferred from the rapid movement of climate zones in the Northern Hemisphere. What about the Southern Hemisphere? Were such short-term rapid swings characteristic of any earlier geological period, and were they related in some way to other known astronomical periods of Earth's axis?

Not all of these are questions that they intended to answer themselves, though they would certainly try, and on the physical side this would mostly be work for Wegener. They knew that in any case they needed to build on the momentum of the Berlin symposium. The twentieth meeting of the *Deutscher Geographentag* was to be held in Leipzig at the end of May, and Wegener would go and give a shorter version of his Berlin address. Köppen was preparing a shortened version of his two articles for *Petermanns* for the *Geologische Rundschau*. It was not likely that a contribution from Wegener would be welcome there, but Köppen had a huge reputation, and in any case Hans Cloos had joined the editorial board in 1921 and was most likely the conduit for this article.[78]

They decided on the following plan of action. Wegener would reorganize and rewrite the material of the second edition following the division of his presentation at Berlin, dividing the evidence discipline by discipline. This had the advantage that members of the component disciplinary communities they were trying to reach could find, in one place, the results and remaining issues to be resolved to which these disciplines could uniquely contribute.

Wegener would pursue this work in the two days per week he was spending at the aerological station at Großborstel. Though there was no observational work there other than the releasing of pilot balloons, his employment contract still specified his right to work there two days a week, and he could split his time between his work on displacements and work on the theodolite that he and Kuhlbrodt were trying to modify. In the summer of 1921 they took a prototype of their new instrument on a cruise on the research ship *Poseidon*, in the North Sea, and learned enough to settle on a design they could fabricate at Hamburg in winter 1921/1922.[79]

Wegener and Köppen together would continue their work several evenings a week on pole positions and paleoclimate. Moreover, Köppen had made a decision to update and expand his own textbook of climatology, incorporating his thinking of the past few years into the first major revision of his scheme in more than two decades. His longtime publisher, Goschen, had been merged in 1919 with the publishing house of Walter de Gruyter in an era of tight money, scarce paper, and skyrocketing printing costs. Köppen's 1918 revision of his classic scheme had a short press run and had sold out quickly. De Gruyter had more resources and was willing to consider a new and much larger edition; the new book would be three times the size of the previous one.[80]

We know very little about Wegener's activities from September 1921 through March 1922, other than that during those months he was hard at work researching and writing the third edition of his book on the origin of continents and oceans, doing his job at the observatory, and teaching at the university. Else's memoir of her life in Hamburg gives few details of their life during this year. She has vivid individual anecdotes for the "Hamburg years," but these are notably disjointed and out of sequence when compared with her recollections of earlier and later periods in her life with Wegener. It was a difficult time for them and became progressively more difficult the longer they were in Hamburg, with the shortages of food, fuel, and clothing, the skyrocketing inflation and consequent lack of money, Wegener's endemic fatigue from driving himself too hard, and three small children to care for.

We can derive some information about the process of composing the third edition of his book on continents and oceans by looking at the notebook he began in November 1920. Almost *none* of the material in this notebook can be matched verbatim with what appeared in the published version of the third edition; these are notes and extracts from a variety of sources on a variety of topics. In the first dozen or so pages of this notebook there is a good deal of geology and geophysics, but as Wegener moved along, geology and geophysics gave way rapidly to the collection of citations to paleontology, and these in turn gave way to geological distribution of glacial, desert, and coal deposits. As these notes accumulated, they became longer and longer and tended to become divided by topic, and by the end they were barely short of page drafts for a projected manuscript.[81]

These notes fit with another story. Wegener later wrote that in the process of researching the book it became clear that the material on paleoclimatology was beginning to outweigh the rest of the evidence for displacements. He had wanted to make sure that he devoted nearly equal space to each of the lines of evidence involved in advancing the displacement theory: geophysical, geological, paleontological and biological, paleoclimatic, and geodetic. These were categories he had used in 1912, and they represented distinct groups of scientists publishing in different journals.[82] He wanted to treat their work as equally important and equally worthy of discussion.

Additional evidence for his deeper focus on climatology comes from his teaching. Wegener was nominated for an extraordinary professorship at Hamburg in March 1921, shortly after his Berlin triumph, and this was confirmed in July. This had happened very fast and was very promising: it bode well for his ability to someday become a full professor at Hamburg or elsewhere.[83] The course he elected to teach in the fall of 1921 and spring of 1922 was his course on climatology first offered from April to July 1920. For the Easter term, April–July 1922, he taught his course on weather and weather service but also added a course entitled "Climates of the Past."[84]

The Third Edition of *Die Entstehung*, 1921–1922

Each of Wegener's versions of his theory had a character sharply different from the one that preceded it. The first version, that of 1912, preserved faithfully the way in which the idea occurred to him and the way he filled out his original intuition. It was dense and heavily geophysical: deeply concerned with gravity data, isostasy, solid mechanics, and strength of materials. In this version even geological topics such as mountain building and volcanoes were considered to be part of geophysics. The geological arguments were secondary and brief, and geology, paleontology, the theory of ice ages, pole displacements, and the difference of the Atlantic and the Pacific sides of Earth were all shoved together as "arguments from geology."

The first book-length version, in 1915, pursued and expanded the initial conceit of a "Copernican rewriting" of Eduard Sueß's *Das Antlitz der Erde*. Wegener expanded the treatment of geophysics (though that was already most of the paper of 1912) in a redundant and repetitive way, taking up more than sixty of the ninety-four pages of this brief book; he compressed the geological and paleontological evidence down to about twenty-five pages.

The second edition of the book, in 1920, was driven by Wegener's desire to respond to his critics. Responding in detail to criticisms of the first edition tied the book very closely to its predecessor, now expanded with more geophysical data, some novel cartography, and considerable speculation. Wegener wrote an analytical table of contents and a much clearer introduction. Still, one had to enter his theory through thirty-five very dense pages of *Geophysikalische Erläuterungen* (geophysical elucidations) of the continents, the continental shelves, and the floors of the deep oceans. Here fact, interpretation, and speculation were all mixed together.

The actual narrative history of the displacement of the continents was much improved in the second edition, but it was also extraordinarily dense, with new factual material once again side by side with speculation. The novel material on displacements of the pole, which followed immediately, was partly a defense of the theory and partly an attempt to show the utility of the theory in the study of past climate.

When Albrecht Penck had said that the second edition had not solved the problems of the first, he was talking about not just missing explanations (the mechanism) but the difficulties in the book's structure, which made it so hard to read and understand. Wegener had said that the book was a "sketch," but he also insisted that the book should be studied, not just read, and with a globe and a series of maps and reference books, and even monographic literature ready to hand. He had gone so far as to suggest that his readers might well obtain some tracing paper to make cutouts of the continents and perform his experiments on a globe. It was less a sketch than an epitome, and it embodied objectives in direct competition with one another.

In writing his preface to the third edition, Wegener made it clear how well he understood all this: "The third edition is again completely revised and differs almost as much from the second edition as this did from the first. The reason for this lies not solely in the fact that in the intervening two years an extensive literature has appeared that directly or indirectly concerns the displacement theory, but also and especially that the entire substance of the book has been cast in a new, and I hope, more convincing form, in which the essentials are better separated from the embellishments [*Beiwerk*]."[85]

The third edition is vastly better than any of the versions that preceded it: better structured, clearer, and easier to read. Its tone is cautious, diffident, and accommodating—to the critics of displacement theory, to the proponents of other views, and to the reasonable doubts of the general reader. It is conciliatory with regard to the problems and prospects of the land bridge theory and the theory of continental permanence, and it makes evident how much work still needs to be done to fill out the bare details of the displacement theory.

Wegener divided the book into three parts. The first is titled "The Essential Content of the Displacement Theory." His intention for simplicity and directness is evident from the very first sentence, as is his decision to mobilize and re-create his own original inspiration. Here is how the book begins:

> Anyone who has looked at opposite coasts of the South Atlantic must be aware of the similar shapes of the coastlines of Brazil and Africa. . . . Experiment with a compass on a globe shows that their dimensions agree precisely.

> This striking phenomenon was the starting point of a new conception of the nature of the Earth's crust and the movements that take place in it, which is called the Theory of Continental Displacements, or, for short, Displacement Theory, since its most striking component is the assumption of the great horizontal drifting movements that the continental blocks have undergone in the course of geologic time, and which presumably continue even today.[86]

Wegener then laid out the entire theory in all its aspects in two and a half pages of text and illustrated it with two pages of maps. He wrote in a relaxed and colloquial style, and the only specialist terms are names of different periods of Earth history:

> According to this idea, for example, millions of years ago the South American continental surface lay right next to the African continental surface, so that they formed a single large block that, beginning in the Cretaceous period, split into two parts that afterward, like ice floes in water, moved ever farther from one another. . . . Similarly, North America was located close to Europe. . . . In the same way it is assumed that Antarctica, Australia, and India lay adjoining South Africa, and with the latter and South America formed, until the beginning of the Jurassic a single large . . . continental area, which . . . split and disintegrated into its individual blocks that drifted away from one another in all directions. . . . The three maps of the Earth reproduced in Figures 1 and 2 show these developments during the Upper Carboniferous, Eocene and Lower Quaternary periods.[87]

The maps that Wegener refers to in this passage appear on pages 4 and 5 and are, after the Mercator map of the world, some of the most reproduced maps in the history of cartography and are likely known to every schoolchild. Beginning in 1922, they were copied and repeated in nearly every publication for, against, or describing the theory of continental displacements. The maps are not the whole theory, but they are the basic outlines of the theory. Wegener drew them using the new convention he adopted in 1921, with lines of latitude and longitude reflecting the *present* position of Africa. He also drew in, as stippled areas, shallow "epicontinental" seas, representing areas of continental crust inundated at one time or another by shallow water. These were not deep oceans; they were water-covered continental areas. There were no mountains, as they would be too small to see on this scale, but he did put in the world's great rivers and sketched the present-day outlines of the continents "only for identification."[88]

The maps have some wonderful features. First of all, they are on facing pages and have identical polar diameters. The maps on the left-hand page are elliptical maps using the Hammer projection, similar to the Mollweide but inflecting the parallels of latitude slightly to decrease distortion away from the center of the map. On the right-hand facing page are the "pictures of the globe" of the Lambert oblique projection, with both hemispheres (for each period) presented side by side. The Hammer projection allows one to see the entire ensemble of continents at a glance clustered near the center of the map with minimal distortion. The Lambert projection emphasizes the extent to which, in Wegener's theory, there has always been, since the first emergence of the land, a water hemisphere and a land hemisphere holding the single protocontinent.

The land-hemispheric part of the Lambert projections, with the North Pole obliquely facing the viewer, gives a clear visualization of that which Diener and others had found impossible to understand about the separation of the continents. Diener had claimed that Wegener's theory required a 35° longitudinal gap between Siberia and

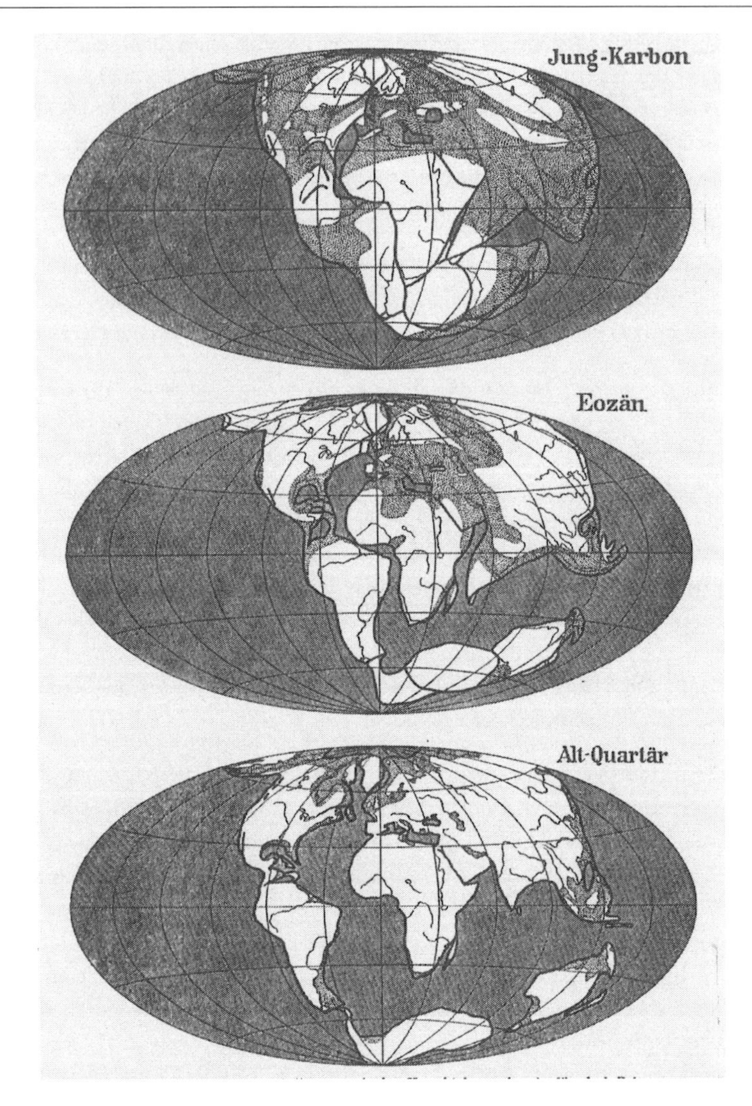

Wegener's maps of continental displacement, shown in the Hammer equal-area projection, from the third edition of Alfred Wegener, *Die Entstehung der Kontinente und Ozeane* (Braunschweig: Friedrich Vieweg & Sohn, 1922). The three periods of geological history shown are the late Carboniferous, the Eocene, and the older Quaternary. Following his practice, Wegener held Africa in its current position (latitude and longitude) and moved the other continents away from it. The map therefore shows relative rather than absolute continental displacements.

Alaska which would later have to be closed. Wegener's maps here make it easy to see that North and South America did not slide to the west uniformly but rotated away from Eurasia and Africa, first in the southern part, while Siberia and Alaska always remained connected by land or separated only by a shallow shelf sea, never by deep ocean.

The change in tone and attitude and the economy of expression are evident once again in the next chapter, in which he discusses the relationship of the displacement theory to the contraction theory (*Theorie*) and to the doctrines (*Lehre*) of land bridges and permanence. The change of emphasis here is subtle but telling: the shrinking of Earth was a general theory meant to explain everything—continents and oceans, con-

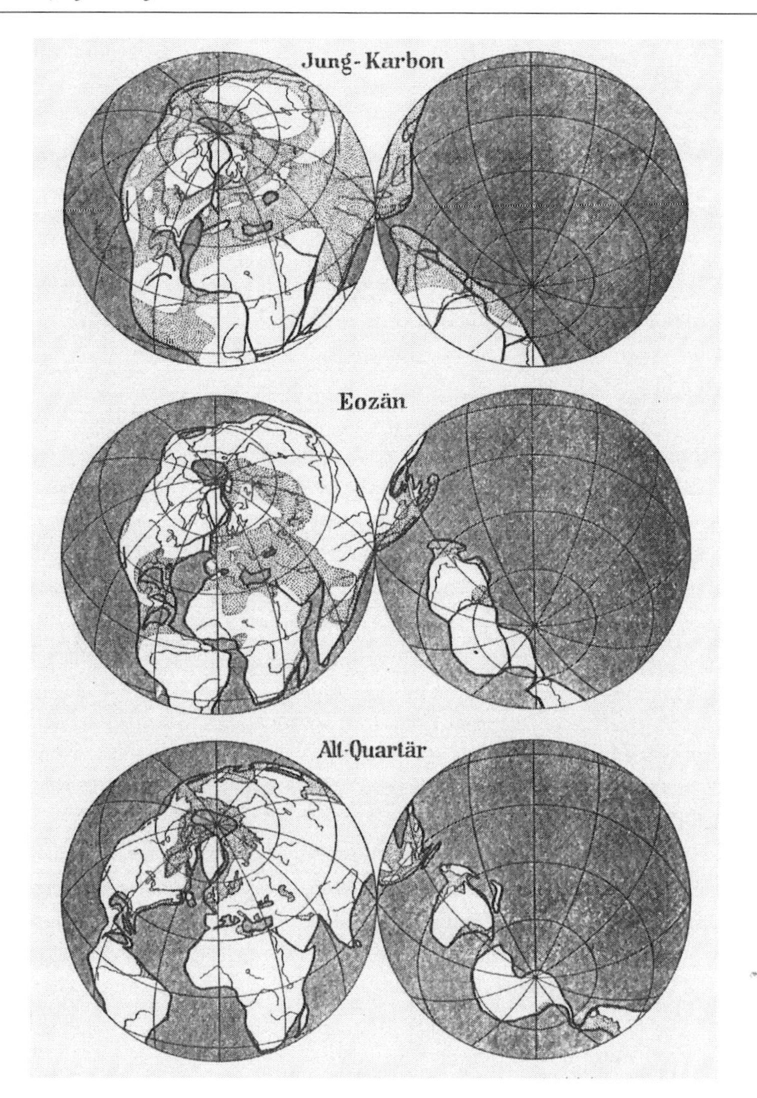

Wegener's maps of continental displacement, shown in the Lambert oblique equal-area projection, from the third edition of *Die Entstehung der Kontinente und Ozeane* (1922). The three periods of geological history shown are the late Carboniferous, the Eocene, and the older Quaternary. Following his practice, Wegener held Africa in its current position (latitude and longitude) and moved the other continents away from it. This map, like its partner above, shows relative rather than absolute continental displacements.

nection and lack of connection between landmasses, the origin of mountains. Therefore, Wegener said, "the historical service of this theory cannot be denied; and for a long time it allowed us to construct a serviceable synthesis of our geological knowledge. In consequence of the long period in which the contraction theory allowed the integration of a very large quantity of individual research results, the great simplicity of its fundamental conception and the multiplicity of its applications, it retains a strong hold even today."[89]

Contrasted with the contraction *theory*, the *doctrines* of great sunken land bridges and of the permanence of continents and oceans are much more limited in scope. "The relation of the displacement theory to both these doctrines is different from the contraction theory. To anticipate the results, the arguments that are led out into battle by both of these doctrines are correct. Each is based on only that portion of the facts necessary for a favorable judgment and receives its refutation immediately once the other portion of the facts is introduced. The displacement theory will fit the entire facts and therefore prepares the way for a reconciliation of these hostile doctrines in a manner which satisfies all reasonable demands of both parties."[90]

Wegener had now completely recast the theoretical situation once again. The contraction theory was a universal theory that explained all the known facts of geology and paleontology in terms of a single organizing principle: the contraction of Earth. The doctrine of great land bridges became a foundation of paleontology because its adherents assumed that the contraction theory was true, self-evidently so, and did not require further proof.[91] The contrary doctrine of permanence of continents and oceans was not a biological but a geophysical doctrine. The principle of the general permanence of the continental blocks was sound, wrote Wegener; the present continents, with but a few exceptions, have never in the history of Earth been part of the floor of the ocean but have been as they are in the present: continental platforms.[92] He continued, "While we thus must completely reject the contraction theory we need only to scale back the doctrine of land bridges, and the doctrine of permanence, to the conclusions that can be legitimately drawn from the argument advanced for them, in order to reconcile these apparently so opposed doctrines by means of the displacement theory. That theory says: land connections, yes, but not via former bridge-continents, rather by direct contact; permanence, not of individual continents and oceans as such, but of the oceanic areas and the continental areas taken as a whole."[93]

Wherever possible, Wegener defended his ideas with the ideas of his supporters and, preferentially, with ideas from his opponents. To point out the weaknesses of the idea that land bridges rise and sink "as needed," he cited the misgivings of Bailey Willis and Albrecht Penck, both strong supporters of the permanence of continents and oceans. When defending his choice of Airy's "roots of mountains" and the floating-continent version of isostasy, he used as an authority Schweydar's remarks in the Berlin symposium in February, not his own explanations.

These extracts from the first pages of Wegener's third edition convey the different tone of this version of his theory and the extent to which the role of arbiter has replaced the role of a partisan combatant. In place of brash certainty, startling novelty, and individual effort, we see generosity toward opponents, caution in assertion, gratitude for the efforts of the great geologists of the past, hope for reconciliation of views, awareness of difficulties, and finally the characterization of displacement as an idealistic attempt to bring back together the unity of geology lost in the fall of the contraction theory. This approach and voice continue in the next section of the book, entitled simply "Beweisführung" (Presentation of the evidence).

In this next section Wegener radically reformulated what counted in his theory as evidence (*Beweis*). He made a firm decision to strip out conjecture and inference to things that could not be seen, in favor of observational evidence. Everything outside the ensemble of visible evidence he moved to a different category: *Erläuterungen und Schlußfolgerungen*. There are a number of reasonable translations of these two words; many English translators like "explanations and conclusions." In this context, it would be more accurate to say "elucidations and inferences."

Into this new category of "elucidations and inferences" went all of Wegener's ideas about the viscosity of Earth and its paradoxical properties, temperature distributions in Earth's interior, Earth tides, transgressions and regressions of the seas caused by polar shifts, whatever might lie beneath the ocean floors, the character of the underside of the continental blocks, the way in which folds and rifts might take place in the displacement of continents, the form and structure of the continental shelves, the causes of island arcs and ocean trenches, the differences between the Atlantic and Pacific coasts, and finally all speculations concerning the mechanisms and forces bringing about displacement of the continents. Most or all of these had been offered in both the first and the second editions of the book as evidences; they were now remanded to the realm of interpretations.

This third edition of Wegener's *Origin of Continents and Oceans* was by far the most influential version of his theory ever published, as well as the only one widely translated and read outside Germany. His original articles in 1912 have only recently been translated into English, and the first and second editions of his book on continents and oceans have never been translated into English at all. The only other of his publications to find its way into English is a translation of *Die Entstehung der Mondkrater* by the geologist and historian of geology A. M. Celal Sengör.[94]

Wegener was generally well served by his first English translator, J. G. A. Skerl, and since Wegener reviewed the translation of the third edition of his book into English, he must bear the responsibility. Nevertheless, in a variety of ways the book gives a misleading impression of Wegener's tone and replaces the tentativeness of his approach with a certain vehemence, which works very much against Wegener's intention at the time. Throughout the book, Wegener used a variety of expressions to calibrate the depth of his confidence in his evidence: "these arguments imply," or "this evidence would be difficult to shake," or the "documented evidence for this is . . . ," or "This is to my knowledge the first certain evidence for the soundness . . . ," and so on. In each instance, Skerl translated the word in question as "proof": "this proves," "This is to my knowledge the first proof," the "paleontological and biological proofs," and so on. Similarly, he translated a phrase such as "confirmation of a long asserted principle" as "confirmation of the long postulated law." Talk of presentations, evidences, and principles is very different from talk of demonstrations, proofs, and laws. The latter suggest mathematical and logical certainty, whereas the former merely suggest the balance of evidence that would likely produce a favorable verdict.

This question of attitude toward evidence is also of some historical importance because in this edition Wegener was above all trying to convince *geologists*, an audience that he had nearly abandoned two years before. In 1920, geophysics explained everything, and every geological detail was but an idiosyncratic departure from some general geophysical principle. In 1922, the geophysical explanations are still there, but everywhere there are complications, exceptions, and reservations. For instance, in this edition, as in every previous version of the theory, Wegener drew a cartoon schematic cross section through a continental block in the form of a large cube, perfectly rectilinear, with sharp borders, immersed in a medium, and adjacent to a thin rectilinear section of ocean water. There are three substances (Sial, Sima, water), three densities, and three behaviors—no ambiguity. In this edition, however, this schematic is accompanied by the following gloss:

> It is essential to immediately exercise caution not to exaggerate this new conception of the ocean floors. As in the case of our comparison with tabular icebergs we must certainly also consider that the upper surface of the sea [here the Sima in which the Sial bergs float] between them can be again covered with new ice, and also further, that small fragments of the iceberg that have become detached from its upper surface or have broken off from its foot can also cover the surface. In a similar manner this will occur at many places on the ocean floors.[95]

This is a response to the charge by Penck that his theory could not be true because fragments of continental material are found scattered across large sections of the ocean floor. Wegener hoped that reservations like the above, which accounted for the overwhelming portion of added text in this edition, would make it more rather than less compelling to geologists. In the 1922 edition the world was again varied, local, and detailed: it was the world of geology, not the exclusive "bulk property" world of thermodynamics and gravity. To balance these many reservations, Wegener urged his audience to steer a middle course on the question of detail: "The ever-increasing stream of separate discoveries allows the picture of these connections to grow under our eyes, and today a very far-reaching agreement prevails already among various experts about the most important land bridges, although there are those also who cannot 'see the forest for the trees.'"[96]

Dividing the book into two separate sections, one concerned with firm observational data and the other with all other matters, allowed Wegener to go far beyond the "forest and the trees" to discuss things buried deep beneath the forest floor, as it were. Consider the following diagram, which appears as figure 30 of the new edition in a chapter titled "The Sialsphere." Here instead of the ice cube version of the floating continent, there is a "mash" of Sial and Sima, of the kind that Cloos had advocated in 1915. The diagram as drawn by Wegener (in the lower left of the accompanying figure) appears to be a merging of two different conceptions into a new view of the continental block. One of these is from a 1915 drawing by Dacqué of what he took to be a more plausible version of Wegener's continental blocks. The second, also by Dacqué, and also from his 1915 book, is a picture of the complex mixing of rock caused by lateral pressure in the formation of fold mountains. Wegener appears to have worked these visualizations together into a conception that united ideas about the continents with ideas about fold mountains.

"Such a structure," Wegener wrote concerning this picture, "of the continental blocks offers an explanation of many phenomena."[97] He then went on with a complex speculation to show how such a ragged trailing underlayer could explain sedimentary loading of shelf seas, predisposition to folding, extrusion of lava, fold structures incorporating different kinds of rock, island festoons, volcanic islands, and anomalous gravity data.[98] Removing these speculations to the latter part of the book (elucidations and inferences) would, he hoped, differentiate them from the "essentials of the theory" and allow him to continue to develop his ideas without damaging the presentation of secure evidence.

There were, however, two major areas of the theory in which Wegener was forced to give significant ground, based on research spurred by his own theory. The first of these was the geodetic evidence for shifts in latitude and longitude of the various continents. Wegener's claim for the unique advantage of the displacement theory, over all other theories of similar wide-ranging ambition, was that it could be tested by astronomical position finding, so this was a rather hard blow.[99]

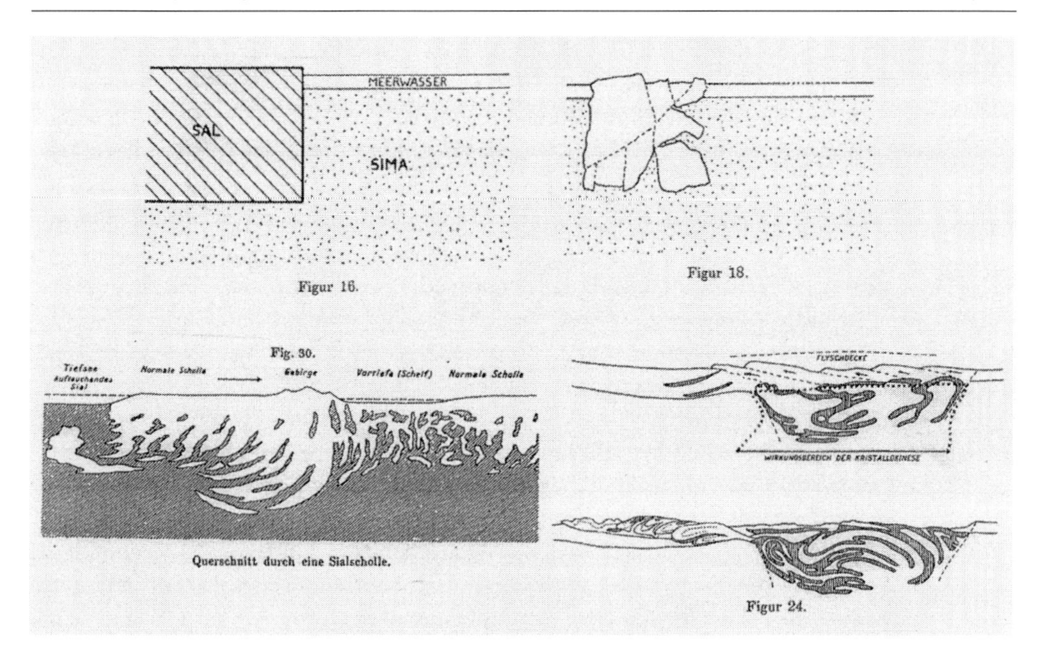

Top left: Wegener's 1912 cross section of the continents as redrawn by Edgar Dacqué in *Grundfragen und Methoden der Paläogeographie* (Jena: Gustav Fischer Verlag, 1915). *Top right*: Dacqué's suggestion of a more plausible version of a continental cross section, also from 1915. *Bottom left*: Wegener's cross section of the continents, from *Die Entstehung der Kontinente und Ozeane* (1922), which seems to combine some of Dacqué's ideas. *Bottom right*: genesis of fold mountains, under tangential pressure, again from Dacqué, *Grundfragen und Methoden der Paläogeographie*.

Since 1912 he had asserted that comparison of longitude observations made in East Greenland in 1823, 1870, and 1906/1907 showed the westward drift of the observing stations. In 1912 and 1915 he had also used telegraphic time signals exchanged between Cambridge, Massachusetts, and Greenwich, England, in 1866, 1872, and 1892; in 1920 he dropped the 1866 measurement but added the measurements made in 1914. These measurements appeared, he claimed, to show discernible shifts in the longitude of North America. Finally, in 1920, he had added observations made by the International Latitude Service suggesting a decrease in the latitude of these stations since 1870, lending support to his idea of a "flight from the poles."

In 1921–1922 he was forced to abandon *all* of these claims. The hardest was the claim for Greenland, since he had participated in the measurements and their publication had come from his expedition colleague and good friend Koch. Greenland was the only place, Wegener believed, still spreading fast enough in the present to provide really accurate measurements of displacement. In 1920 he had recalculated, using recent estimates of geological time, the spreading rates for displacement of all the major continental blocks, reducing most of these to centimeters per year; only in the far north—England, Iceland, Scotland, Labrador, Greenland—did he hold out for estimates of displacements of several *meters* per year.

The question of the accuracy of these measurements had dominated the discussion after his lecture in Frankfurt in May 1921, and subsequently Friedrich Burmeister (1890–1959), a geophysicist at the Geomagnetic Observatory in Munich and an expert on precise geophysical measurements, had written an article in *Petermanns* challenging the means by which Koch had reduced the estimate of error in his calculations. He had also raised the problem of making such measurements with single observers, far apart in time, under difficult conditions, without reporting the "personal equation" of each observer. The "personal equation" of an observer is the difference—often a matter of fractions of a second, but sometimes of several seconds—between the assessments by two different observers making the same observation with the same instrument of when an astronomical event (something reaching a maximum altitude, or passing a certain point) has actually happened. Burmeister pointed out that the personal equations alone of these observers could easily swamp the measurements.

There were other sources of error as well, including printing errors in the ephemerides that reported the expected, as opposed to the observed, time of a celestial event. Burmeister had ended his critique by saying that it should in no way be interpreted as a criticism of Wegener's hypothesis of displacements, merely an elucidation of the inaccuracy of the method used to measure them, and urging the promise of radio time signals to provide definitive evidence for the displacement of Greenland.[100]

Wegener accepted this critique, while suggesting that Burmeister's account of the inaccuracy of the observations was exaggerated. Hence, in this third edition he still presented the results for the Greenland measurements from 1823, 1870, and 1907, but he admitted that they were insufficiently accurate to claim that the displacements of Greenland had been demonstrated. Still, "if the exact quantitative proof [*Nachweis*] must be reserved for more accurate measurements, Koch would still have priority for the measurement of the change in the coordinates."[101]

For the case of the telegraphic time signals between Cambridge, Massachusetts, and Greenwich, England, the news was also bad. Galle, in his 1916 article on Wegener's hypothesis in *Deutsche Revue*, had said that the 1866 measurement Wegener had invoked was now considered to be unreliable; Wegener had therefore already discarded it in 1920. Further, in his 1912 paper Wegener had made a copying error for the 1892 measurement which made its difference from the 1872 measurement *ten times* the actual value. He had repeated this erroneous value in 1915 but corrected it in 1920. Wegener now acknowledged "the possibility that even after new measurements, the displacement is still too small to be detectable with any degree of certainty."[102]

In 1920, to shore up his measurement claims and to promote the idea of the pole-fleeing force suggested by Köppen, he had excitedly reported that the latitude stations in the Northern Hemisphere seem to be moving toward the equator. Now it appeared that all of these conjectural shifts were the result of similar errors made at all component observatories, the errors in question being insufficient correction for atmospheric refraction. If anything, the more recent measurements of latitude shifts in Europe and North America showed an *increase* of latitude.[103]

As if this were not enough, Wegener was forced into a major retreat (if not retraction) on the question of the displacing forces. First, it turned out that the pole-fleeing force was not even their idea: Köppen was not the first to think of it. In July 1920, Kohlschütter had told Köppen that the Hungarian geophysicist, Roland von Eötvös (1848–1919), had mentioned the possibility of such a force at the International Geodetic Conference in Hamburg in 1912.[104] Köppen had insisted to Wegener that he include this information in the third edition of his book, but Wegener, intensely loyal to Köppen, referred to Eötvös's work as a "brief, obscure reference."[105]

Wherever one wishes to assign priority, the idea of such a mechanism caught the imagination of a number of scientists, not least the young physicist Paul Epstein

(1883–1966), about to depart Germany for an appointment at Caltech. Epstein wrote a letter to the editor of *Die Naturwissenschaften* noting that not only were the readers of that journal well informed in the spring of 1921 about Wegener's theory, but in Berlin it had been hard to escape the public excitement over it. Nonscientists ("the general public"), Epstein wrote, seemed to be swept away by the idea, and so many geographers and geologists approved of it that one had to take it seriously. Yet, he said, the physicist still has a question: will the force he proposes to drive these continents work?

Epstein then carried out a rather simple and clever calculation involving the mass of a continental block, the distance between the floor of the ocean and the surface of the continental block, and the angular (rotational) velocity of Earth, to calculate the coefficient of viscosity of the Sima, based on an estimated velocity of the continents—using the extreme (as Wegener said) value of 33 meters (108 feet) per year. He obtained a value of the order of 10^{16}, roughly three times as great as the viscosity of steel at room temperature. He concluded, however, that based on the masses involved, "we can summarize the results in the statement that the centrifugal force of the rotation of the earth can produce a drift from the poles of the magnitude indicated by Wegener and, in fact, must produce it."[106]

This was not the only such support. Walter Lambert (1879–1968) of the U.S. Coast and Geodetic Survey published an article on the same topic in September 1921. He mentioned both Taylor and Wegener in a footnote, citing Taylor's paper of 1910, Wegener's 1912 paper, and Wegener's 1915 book.[107] The article was apparently based on an address he had given at a mathematical meeting in 1920, in Annapolis, Maryland. With all due respect for coincidence, Lambert had almost certainly seen Wegener's second edition with its postulation of a "pole-fleeing force"; it would have been impolitic for a U.S. Army lieutenant recently returned from service in France to give a positive discussion of a German theory: anti-Germanism was extremely powerful in the United States in the immediate postwar years.

Lambert came to the same conclusion as Epstein, that such a force was possible and indeed necessary: "the equatorward force is present, but whether it has had in geologic history an appreciable influence on the position and configuration of our continents, is a question for geologists to determine."[108] He also pointed out that "according to the classical theory a liquid, no matter how viscous, will give way before force, no matter how small, provided sufficient time be allowed for the force to act in. . . . But the viscosity of the liquid may be of a different nature from that postulated by the classical theory, so that the force acting might have to exceed a certain limiting amount before the liquid would give way before it, no matter how long the small force in question might act."[109]

That was the good news. The bad news was that Epstein also declared that in spite of the ability of such a force to move the continents toward the equator, it would be incapable of producing great fold mountain chains around the equator, since the force between the pole and the equator was equivalent to a "fall" of only 10–20 meters (33–66 feet), while raising up mountain chains constituted work against gravity on the scale of kilometers, for which the pole-fleeing force was not sufficient.[110] Even so, Epstein declared that his calculations should be in no way interpreted as opposition to Wegener's theory, which could certainly be true of Earth.

Schweydar had read Epstein and declared that Epstein's velocity for the continents was too great and his estimate for the coefficient of viscosity too small, such that the viscosity ought to be on the order of 10^{19} and the velocity of the continent something like 20 centimeters (8 inches) per year.[111] Further, he had reiterated his opposition to the notion that there was any westward component to the pole-fleeing force. Not only that, but he also argued that the masses of the continents were so inconsequential compared to the mass of Earth as a whole that they were unlikely to be responsible for the displacements of the pole of rotation, thus undercutting Wegener's entire theory of dynamic interplay; this was a severe blow. Lambert had come to the same conclusion, though Wegener did not mention this in his discussion of Lambert's work.[112]

Wegener therefore had to reformulate his position on the displacing forces. "It seems to me therefore highly probable that in fact, as a consequence of the pole-fleeing force, in the course of geological time the continents have undergone considerable displacement through the Sima. Contrarily, it now seems doubtful whether this force can also explain the equatorial fold mountains although perhaps Epstein's result is not yet the last word in this matter."[113] This was a considerable retreat from one of the main points of Wegener's theory, that the forces responsible for moving the continents were also the forces responsible for creating the great tertiary fold mountain belts. Moreover, having given up the notion that the *Westwanderung* of the Americas was a component of the pole-fleeing force, he was thrown back on highly speculative ideas concerning tides in the solid Earth, or perhaps flow in the Sima to move the continents westward, along with unnamed "cosmic forces." Finally, the dynamic interplay of pole displacement and continental motion was not only uncertain but likely improbable. Wegener put a brave face on it, but his *unified* pole-fleeing force was now effectively canceled out, and this aspect of his theory, like measurement, moved from *fact* to *inference*.

One last major subject remained for Wegener to rework for this edition: the paleoclimatic arguments. Most of the major novelty in the second edition came directly from work in this area with Köppen. All of it concerned, in one way or another, the displacement (or migration) of the pole. The pole-fleeing force had become the mechanism for the displacement of the continents, and such continental displacements the mechanism for the further migration of the pole. The linkage between them was the Ariadne's thread through the labyrinth of paleoclimatology: the explanation of the anomalies in the fossil record in the Northern Hemisphere, and even greater anomalies for the Southern Hemisphere in the Carboniferous. Working out the pole positions had absorbed most of the time it took Wegener to rewrite the book, and this remained unfinished as the manuscript deadline approached in April 1920.

Wegener and Köppen had continued to work together on this problem through the summer and fall of 1920, and they were still working on it when the Berlin symposium took place in February 1921. The work points they added to their plan, published with Köppen's second article in September 1921, show that they were still hard at work in the summer and fall of 1921, as does the content of the notes Wegener had continued to accumulate in his research notebook.

Early in the winter of 1922, however, they decided that the defense of the displacement theory and the role of the displacement theory as the "Ariadne's thread through the labyrinth of paleoclimatology" were two different undertakings. The former was a work of advocacy for a hypothesis, and the latter was a descriptive reconstruction of Earth's history based on the assumption that the theory was true. They resolved to continue their collaboration on the climates of Earth's past, but to move all but the bare essentials needed to defend the displacement theory out of the planned third edition and reserve the climate material for a separate, jointly authored book on past climates.[114] Wegener's notebook shows that throughout all of 1922 he continued to collect material on the climates of the past from a variety of sources, and he was still making extracts well after the third edition of his book had appeared.[115]

With the burden of a complete climatic history of Earth removed, Wegener was able to move forward rapidly, rewriting his chapter on paleoclimatology. Once again, compared with his coverage of the same material in 1920, his treatment is simpler and more colloquial, as well as both more cautious and more compelling. In 1920 he had attempted to provide evidence for all the pole positions from the Carboniferous forward. Now, he was concerned exclusively with two matters: the necessity of moving Earth's pole in order to explain the distribution of climate zones in the Tertiary, and the necessity of adding relative continental displacements to these migrations of the pole in the Carboniferous period in order to explain the ice and climate evidence without creating absurd contradictions.

"This is not," he began, "the place to wrap up the discussion of the complete problem of the climates of the past. It is nevertheless necessary for our presentation of the case [for continental displacement], to provide at least a rapid orientation. Only in this way will it become clear how the evidence of paleoclimatology verifies the displacement theory."[116] If, he continued, paleoclimatology is in a very undeveloped state, this is not from a lack of evidence. We have a great body of evidence from plant and animal fossils, and we have ways to turn this evidence into evidence of climate. We can relate temperature to vegetation; we know where trees will grow at all, we know where trees with annual rings will grow, we know that palm trees, of which we have many fossil remains, will only grow in a certain temperature regime as well. We know that corals are only found today in water, the temperature of which never falls below 20°C, and that reptiles cannot live in polar climates. If we take these species one by one, of course the indications would be very uncertain, because we see many plants and animals adapt to climates very different from the rest of their families. "But here, just as in the calculation of the path of a meteor from a great number of inexact estimates, the individual data can be quite uncertain, and contradictory, but when the ensemble is treated according to the theory of errors, a reliable result is obtained."[117]

To this organic evidence, Wegener wished to add "inorganic evidences of climate," by which he meant geology. Boulder-clay (till, ground moraine; German: *Blocklehme*) polished rock and striations in rock taken together indicate Inland Ice. Coal and peat, forming at different temperatures but wherever there is more rainfall than evaporation, and salt deposits, forming wherever there is more evaporation than rainfall, indicate wet and dry climates, respectively. Thick red sandstones with no fossils mean hot deserts; yellow colored sandstones, more temperate arid areas.[118]

Wegener is here leapfrogging over, and reserving for a later time, a huge number of controversies: where some particular plant or animal might have lived in the past, whether all supposedly glacial deposits are really glacial deposits, whether all coals are formed in the tropics in temperate regions, where salt deposits and massive sandstones come from, and so on. Even ignoring all this, he is saying, we still may infer the past from the present, as we know what sorts of climates the majority of certain groups of animals and plants live in today. Wherever coal forms, it has to be where there is more rain than evaporation; wherever salt forms, there has to be more evaporation than rain.

When, he continued, we take what we know about the environments of living animals and plants and the conditions under which glacial deposits, coal, salt, and sandstones form today and apply to the past this ensemble of evidence, they form a pattern suggesting shifts in latitude, as there are always different climates in every period, but they change location from one period to another. "It is not to be wondered that in the attempt to explain the systematic alteration of the Earth's climate in the past, recourse was had early, and increasingly, to changes in the positions of the Earth's pole." Whatever the imagined mechanism, "today the majority of geologists take the standpoint formulated in E. Kayser's *Lehrbuch*, that the assumption of a great Tertiary displacement of the poles is, in any case, 'difficult to avoid.' This can indeed be considered as established, in spite of the remarkable vehemence with which some of its opponents struggle against it."[119]

Once the movement of Earth's pole is established, he said, there is remarkably good agreement on the climate zones for the more recent periods, with a position for the North Pole at the beginning of the Tertiary in the neighborhood of the Aleutian Islands, and from there movement toward Greenland, where it is found in the Quaternary. However, "the situation is entirely different for the periods before the Cretaceous. Here not only do the views of the best-known authors differ widely, but every one of their reconstructions, consequential on their inattention to continental displacements, leads to hopeless contradictions, indeed contradictions of an order that they form a barrier to every conceivable placement of the pole."[120]

Finally, the point of his argument has arrived. "The riddle of the Carboniferous glaciation finds an extremely compelling solution in the displacement theory; once the different portions of the earth which today bear traces of ice action are pushed together concentrically around South Africa, the entire ice-covered area becomes no larger in extent than that of the Pleistocene glaciation [*diluviale Eisspuren*]."[121] Then, having redeployed the evidence from the second edition, with an increased reliance on the disposition of coal (divided into tropical and subtropical coal), the *Glossopteris* flora, trees with tree rings, salt, gypsum, and other desert deposits, he argues that this solves the problem of all the data for the Carboniferous.

All of this writing and thinking about the continents happened in the course of the summer and winter semesters of 1921–1922, against the background of full-time work at the observatory, research on climate, and university teaching. Now that he was pushing the work on climates of the past forward to a prospective volume on which he and Köppen would collaborate, he had reduced the number of major efforts from four to three, but there was still not enough time. The extensive commuting—among Großborstel, the university, the observatory, and the Geological Institute—was, with Kurt gone to Berlin, completely lost time. Living in a household with seven full-time residents (four adults and three children), with constant visits from colleagues, neighbors, and the other children of the elder Köppens, gave him no respite at home; he had to find a way out if he were ever to finish this book.

There was a way. Wegener and Kuhlbrodt had gone to sea briefly in the summer of 1921 to test the prototype of their new balloon theodolite, and they continued to work on it through the fall and winter. It was a good design, combining a standard balloon theodolite with a marine sextant. One need not go too much into the details, except to say that the sextant could be adjusted by sliding a large armature along an inscribed arc while looking through the telescope. This allowed very precise adjustments to happen rapidly and to be read off easily, in contrast with turning a thumbscrew on a standard instrument. Even better, a sextant had a split mirror that allowed one always to be looking both at the horizon and at a celestial object (in this case a balloon); histori-

Wegener and Kuhlbrodt's highly mobile theodolite for tracking pilot balloons on their sea voyage to Mexico and the United States in 1922. The instrument itself is on the left, and on the right is Wegener's photograph of Kuhlbrodt making an observation with it, using the novel control arm from a sextant allowing fine adjustments of the telescope with large adjustments on the arc. From Erich Kuhlbrodt and Alfred Wegener, "Pilotballonaufstiege auf einer Fahrt nach Mexico, März bis Juni 1922," *Archiv der Deutschen Seewarte* 30, no. 4 (1922).

cally its principal use has been the determination of latitude by measuring the maximum altitude of the Sun at noon.

Applied to the problem of following a pilot balloon, one could get a constant reading on the altitude of the balloon above the horizon without ever losing sight of the balloon, while simultaneously making fine adjustments of the telescope. With a second observer to record the angle at every moment and to keep track of the compass direction and speed of the ship, it was possible to determine within a few hundred meters the absolute altitude of a balloon and, from its motion, to track the direction and velocity of the upper atmospheric winds.

So at about the time that the intellectual work for the third edition was largely completed, January and February 1922, Wegener sought and obtained permission from Adm. Kohlschütter to make a voyage across the Atlantic with Kuhlbrodt and try out the instrument. There was no question of having a research vessel, and Germany was not allowed to have naval vessels at sea under the terms of the armistice ending the war, so Wegener and Kuhlbrodt hit upon the expedient of finding a ship headed for the Americas in the spring and early summer, on which they would be nominal members of the crew but devote themselves to their scientific work.

Germany's merchant fleet, though decimated by the demands of the Allies for reparations, still existed, and as a meteorologist at a naval observatory, Wegener hardly had to present credentials to make his case. The steamer *Sachsenwald* was scheduled to depart in mid-March, and it was no great work to sign on (nominally) as the purser, given his position and the backing from the Admiralty. Kuhlbrodt also took a nominal role as a member of the crew; they were, however, simply scientific passengers.[122]

The voyage would take about three months, with a scheduled return around 15 June. Wegener and Kuhlbrodt planned to send up two balloons a day while at sea, one in the morning and one in the afternoon, and one balloon a day while in harbor, the latter as a check on the velocity calculations concerning the ship, to provide correction for the balloon flights made while the ship was under way.

They also would take a long focal-length camera for photographing cloud forms and documenting the flights. Kuhlbrodt would take a series of water temperatures at different depths, and Wegener would take some preliminary measurements of gravity at sea.[123] This program still left many hours each day with no duties whatever: no teaching, no commuting, no family life, no visitors, no research trips to a library, no entertaining speakers at a colloquium, no conferences with PhD students, no correspondence, no filing of reports at the observatory. All of this time could be and would be devoted to the preparation of a consistent, clear, and balanced manuscript. Wegener was accustomed to writing at sea, was rarely seasick, and was used to cramped quarters and odd hours. What he needed was time alone.

17

The Paleoclimatologist

HAMBURG, 1922–1924

> Before beginning my lecture on the criteria by which ancient glaciations can be recognized I want to remind you of the curious fact that it would be impossible to give such a lecture without the existence of recent glaciers. If we did not live so near an Ice Age with glacier-ice on the poles and in the higher mountains, but in a warm period of Earth history with green trees right up to the polar regions, no one would be able to give a correct interpretation of the polishing and striations of rocks which we attribute to glacial action, for no one would imagine that ice can stream like a river and be a most important geological factor.
>
> Martin Schwarzbach (1963)

The *Sachsenwald* departed from Hamburg on 17 March 1922, with Wegener and Kuhlbrodt aboard, and made a rapid crossing to Havana in about five days. Sea conditions were rough, with stormy and rainy weather, which only improved when they entered the trade winds. After Havana, they hopped from port to port in Cuba and then headed for Veracruz in Mexico, where, arriving on 30 April, they were docked for eleven days. Wegener and Kuhlbrodt were welcomed there by the local scientific community, the Sociedad Cientifica Antonio Alzate, shown around the observatory, and given a demonstration of their observing methods, and both men were made honorary members of the society.[1] Leaving Mexico, they had nine more days at sea before arriving in New Orleans, on 20 May, and after a three-day layover there, they had a slow trip around Florida, arriving in Jacksonville (Fernandina) on the Atlantic Coast on 28 May, remaining there through 16 June.

There was plenty of writing time for Wegener, especially in these American ports. Though they had sent up pilot balloons each day while in port in Cuba and Mexico, the captain was afraid that somehow the Americans would find out that Wegener and Kuhlbrodt were not members of the ship's crew but scientific passengers, and in fact employees of the German Admiralty, and seize the ship; he therefore forbade them to send any balloons aloft or make any other measurements from the deck. On the return voyage, they had no better luck with weather than on the outgoing one. While conditions were less stormy than on the outbound voyage, there was a thick overcast, and almost all the balloons were immediately lost in the persistent low clouds.[2]

Wegener seems to have been invigorated by the voyage and to have enjoyed himself, even in the midst of the bad weather. Whenever he was in an especially good mood, he would write humorous doggerel of the kind favored by his father and his father-in-law, in the spirit of Christian Morgenstern's *Galgenlieder*. He produced a poem with illustrations for the children: "*Auf hoher See, Hipp Hipp Hurra, Wir fahren nach Amerika / Wir stampfen gegen Wind und See, sieben Meilen Fahrt, Herrjemineh!* [Across the sea, hip hip hurrah! We're traveling to America. We struggle against the wind and sea, just 7 miles, Oh my! Oh me!]."[3] That is probably enough to get the idea, and it explains why Else would generally run away whenever he recited poetry, either Morgenstern's or his own.

He arrived home on 18 June with the rest of the manuscript of the third edition, which he sent off to the press within days. While he had been away, there had been interesting news, exciting but unsettling. Gustav Hellmann (1854–1939) had decided to retire as head of the Prussian Meteorological Institute and as professor of meteorology at the University of Berlin. Wegener would certainly be one of the three candidates in the short list for the *Ruf* (call). The directorship of the institute and the professorship were tied together, and there was no question of having one without the other. The administrative work of running the institute, which Hellmann had headed since 1907, was even more burdensome than the administrative work Wegener faced at Hamburg. He was very worried he would be chosen; he hated the administrative work, and Else said that for him the appointment in Berlin would be "out of the frying pan and into the fire" (*vom Regen in die Traufe gekommen*).[4]

He wanted to be a professor, but not a professor with an institute. There was some talk by Lundager and Koch of creating a professorship for him in Denmark, but he was not sure he wished to leave Germany. If he were chosen for Berlin, he thought he must refuse. It would be an insult to Berlin and a considerable professional embarrassment for him to do so, and it might cause him to be passed over when a more desirable position became vacant. He decided that if he were chosen he would plead that his scientific and personal connections with Köppen and their current collaboration would make it impossible for him to accept—and hope that Berlin would believe him and that honor would be satisfied.[5]

The top-ranked candidate was August Schmauß (1877–1954), then in Munich and working at the Meteorological Institute, without a professorship. His interests and Wegener's were exactly parallel: they had studied atmospheric layering, thermodynamics, precipitation, and meteorological optics and acoustics. Quite independently of one another, they had both written papers on "the atmosphere as a colloid" in 1920, and Schmauß would later become famous for this work. The second-ranked candidate was Wegener's old colleague, Heinrich von Ficker (1881–1957), who had held the professorship in meteorology and geophysics at the University of Graz since 1911; here the overlap with Wegener's work was also great. These were both strong candidates, and how could Schmauß, in particular, refuse?

Wegener managed to calm down and convince himself that he was in the clear as far as Berlin was concerned, but, to everyone's surprise and Wegener's distress, Schmauß declined the offer. The University of Munich and the Meteorological Institute there did not want to lose him, and to keep him there, they created a new professorship in meteorology, which he took up in the late summer of 1922. Now everything depended on what Ficker decided, and no one knew what he was going to do.

An Austrian Summit Meeting

An opportunity to find out appeared to be in the offing. While Wegener was away at sea, an invitation had arrived from Exner and Ficker that Wegener should represent the German Naval Observatory at a scientific meeting to be held in Austria from 10 to 14 October 1922, to celebrate the twenty-fifth anniversary of the Austrian mountain meteorological station atop the Alpine peak Sonnblick, at 3,100 meters (10,171 feet) altitude. The invited guests (and their spouses) would stay in hotels in the resort town of Bad Gastein. The invitation specified that only those who felt capable of an eight-hour Alpine ascent to the peak should register for the meeting.[6] The Austrians wished to make a statement about the vitality of mountain meteorology and of Austrian science generally, even as they had gone from a great empire to a small country. It was an impressive list of invitees—Exner, Ficker, Schmauß, Defant, Schmidt, Geiger, Wegener, and others. Wegener accepted with alacrity and looked forward to the trip; Else, even more so.

The principal question facing Alfred and Else was how they should pay for their excursion. Since late 1921, Germany had entered an inflationary spiral that constantly devalued its currency; the proximate cause was that the Allies were demanding payment of war reparations in gold and would not accept German currency. The German government began to print money at a furious pace in order to buy foreign currencies that were acceptable in lieu of gold; this exacerbated the inflation.

Things were grim in Hamburg in summer 1922. The combination of Alfred's salary and Köppen's pension was "not enough for the simple essentials of life."[7] They let go what domestic help they had, they planted an extensive garden on open land at the balloon station next to the airport, and they worked hard to increase the output of their fruit trees. Else's mother tended a large flock of chickens for eggs and meat. They were getting up at four o'clock in the morning most days to pump water and bring the buckets into the house so they could warm up enough for them to wash the children when evening came. Just getting from one end of the day to the other took all their energy; Else said that their backs were constantly sore from hoeing and weeding the garden. Alfred reiterated his conviction that ten years in Hamburg would kill him.[8]

In spite of the money trouble, Alfred was determined to go to Austria for the meeting, and he told Else they should empty their bank account and use it to finance a month-long vacation in the Austrian Alps wrapped around the conference; if they just left the money in the bank, it would soon be worthless anyway. Else thought this reckless but was excited to go. Alfred's seagoing trip, she noted, had filled him with a wanderlust that made it difficult for him to stay at his desk. Administrative work had piled up while he was away, and it took him the rest of the summer to clear the backlog.

They were a bit short of outdoor gear for a hiking trip, and especially short of warm clothing. Alfred still had some of his expedition clothing, skis, and other mountain gear, but Else had to make the rest out of army surplus pants and coats (for him) and, for herself, to make winter clothing from wool she had bought to make a suit for Alfred. She had said that she didn't want to do it, but he said, "Why not? We can't afford to send it to the tailor anyway!"[9]

They left in late September, two weeks before the conference, a welcome relief after a month of hard work "putting by" the garden produce. They took the train to Munich, where they changed all of their remaining money into Austrian currency. For nearly two weeks they hiked in the Alps, and it was the most thrilling time of Else's life. She marveled at her husband's skill with an alcohol stove in freezing temperatures, his ability to make delicious meals out of almost nothing, and his idea to cook food in the morning and wrap it in wool so it would be warm at noon. Sometimes they had to sleep in makeshift shelters, because most of the Alpine huts were closed and locked at the end of September; sometimes they had to force open a window in a locked hut and get in that way.[10] They barely saw a soul for the last week of their hike and lost track of time, arriving in Bad Gastein a day early for the conference. They quickly found Exner and Ficker, who were staying at one of the great hotels, and who concurred that Alfred and Else should of course spend the night as their guests.[11]

On 9 October, comfortably settled in their hotel, they dined with the Exners and Fickers. Ficker turned to Alfred and Else over coffee and said, "I'm going to accept the offer from Berlin." He'd been mulling it over for a long time, and he wanted to return to Germany. Else said that at that moment she felt a heavy weight lift from her heart; they would not have to go to Berlin, and Alfred would not have to be embarrassed. The next morning, there was more. August Schmauß had arrived the previous evening, and after breakfast he took Wegener aside and told him that he would most likely be listed first as Ficker's successor (*Nachfolger*) in Graz: the professorship would probably be his if he wanted it. Things had already started to move, and Schmauß said that all things being equal, Wegener could probably expect a formal offer before the end of 1922.[12]

They were thrilled but hardly dared hope. The possibility of this appointment was a chain of a series of unpredictable events. There was Hellmann's decision to retire, Schmauß's decision to stay in Munich, and then Ficker, after much deliberation, deciding to go to Berlin. The chain was even longer than they knew. There was the matter of Victor Conrad (1876–1962), another strong candidate besides Wegener to replace Ficker. Conrad had been an *Extraordinarius* (associate professor) at the University of Czernowitz in Bukovina, on the eastern border of the Austro-Hungarian Empire. With the collapse of the empire, all subjects had until 1922 to relocate themselves freely within its old borders, after which mobility would be frozen. Conrad, who had lost his professorship and all his possessions, was in Vienna, working on seismology and making major strides in the understanding of the layering of the inside of Earth. He was a logical candidate for Graz as part of the larger "resettlement" of German and Austrian professors who had lost their posts. But Conrad was a Jew, and there was considerable agitation in Graz over the influx of Jewish students to the university. Admission at Graz was by competitive examination, and Jews from the eastern part of the empire were consistently beating out local students, especially for the medical school. The anti-Jewish feeling in Graz would have made it difficult to put Conrad on the list. This may also have affected Wegener's fate, though he was completely unaware of it. In the end, Conrad did not make the list. Wegener was to be listed first, Albert Defant (1884–1974) second, and Wilhelm Schmidt (1883–1936) third.[13]

However contingent the circumstances, if Wegener were actually able to become the professor of geophysics and meteorology at the Karl-Franzens-Universität in Graz, it would be the first academic or professional appointment of his career (since he first habilitated at Marburg) not made under emergency conditions. It could be his first real professorship (*Ordinarius*), and more importantly it would come without an institute, without administrative responsibilities, and with only opportunity for teaching and pursuing his own scholarly work. Of course, there would be negotiations over salary and pensions; nothing was certain in this precarious time, and he might well face the prospect of a salary lower than what he drew from the observatory. Still, it was what he and Else had always wanted and what all their family and friends had always wanted for them.

Climates of the Geological Past

When Wegener returned from Austria, there was the usual accumulation of administrative work, as well as a relapse into the severe conditions of life in Hamburg, as the German mark spiraled down toward worthlessness. Luckily, Koch and Lundager invited him to come to Copenhagen to give a series of lectures on his theory of displacements and the history of climate. The honorarium, payable in Danish kroner, would help a good deal, and Wegener's ability to give these lectures in Danish (he did) strengthened their bid to bring him to Copenhagen as a professor of geophysics. Wegener was also able (probably with the help of van Bemmelen) to get his work on lunar craters published in the Netherlands. While still closed out of international scientific cooperation with France, England, and the United States, there was no barrier to German scientific work appearing in Scandinavia or in the Netherlands, and Wegener was increasingly vigilant about making his science available for wider consideration.[14]

Wegener's major intellectual effort in late 1922, with the third edition in press, was his unfinished work with Köppen on climates of the past. He had already taken all but the barest summary of the evidence concerning past climates out of the book on continents and oceans. He had said in the introduction that he had done so because the amount of material at his disposal would have thrown the book out of balance.[15] Indeed, the third edition (1922) was only ten pages longer than its 1920 predecessor. While the evidence he published in 1922 concerning paleoclimates was a necessary part of the observational support for displacements, the history of Earth's climate, based on continental displacements and migration of the poles, was altogether *Beiwerk* (supplemental material) from the standpoint of the displacement theory. It was an interesting development of his ideas, but not a necessity in their demonstration or defense. As Wegener had said, the farther one came forward in time, the less one needed displacement theory to explain climate zones based on shifting *latitudes*, and the more one needed it to explain the pattern of geological structures and fossil remains based on shifting *longitudes*.

The proposed joint work on climates would be a different *kind* of book than *Die Entstehung der Kontinente und Ozeane*, in a different genre. It would not be a theory of origins (*Entstehung*), nor a review of basic questions (*Grundfragen*) or a summary of principles (*Grundzüge*), nor even a *Handbuch/Lehrbuch*, where one got the full range of mainstream opinion on a topic keyed to the relevant literature.

Wegener and Köppen were after a synthesis (*Synthese*) of available paleoclimate evidence, along the lines of what Sueß had done for geology in *Das Antlitz der Erde*. There, Sueß had assumed as a framework both Earth contraction and the sinking of great land bridges and then put the available evidence together in those terms. There were no "multiple working hypotheses" in *Das Antlitz der Erde*, and no one expected them: it was not a *Handbuch*. The book Wegener and Köppen were writing would begin from the premise that Wegener's theory of continental displacements was correct. This is a very important point: they would not attempt to argue for the reality of such displacements, any more than Sueß had argued *for* Earth contraction: Sueß had rather shown that if you assumed it, many disparate facts could be integrated and understood together. That would be the stance that Wegener and Köppen would take for climate history by assuming displacements.

They were entirely in agreement on the range of significant causes of climate change and their rank ordering. Wegener had, preparatory to this writing, put together an article for *Deutsche Revue* on the subject entitled "*Die Klimate der Vorzeit*" (The climates of the past), which was a preview of both the methods they would follow and the subjects they would cover. Köppen had listed a number of possible causes of climate change in his *Petermanns* article in 1921; Wegener now repeated and extended this material from a different standpoint. He divided these causes into "essential" and "likely." Of the essential causes, the first and most important was migration of the pole, and the second was continental displacements. These were both "progressive" through geologic time. The third most important cause, the distribution of land and water, was also essential but not directional in time. Rather, it was oscillatory. This question (of distribution of land and water) referred here to the alternating transgression and regression of shallow seas on continental surfaces, not an alternation of land and deep ocean. Wegener's fourth "essential" cause of climate change was "variation in the elements of Earth's motion." After that there came the three residual "likely" causes: variations in solar radiation, the absorption and emission of such radiation from Earth (what we would now call the "albedo"), and "cosmic radiation" (*Weltraumstrahlung*).[16]

The novel element (other than continental displacements) in this list of causes is "variation in the elements of Earth's motion." This refers not to the nineteenth-century tradition of astronomical causes of ice ages, as in the work of Joseph Adhémar (1797–1862) and James Croll (1821–1890), but specifically to the work of the Serbian mathematician, astronomer, and geophysicist Milutin Milankovich (1879–1958). Milankovich (by profession a civil engineer) had been captured early in the war by the Austro-Hungarian army and had spent four years in house arrest in Budapest, working by day in the reading room at the Hungarian Academy of Sciences on his passion for celestial mechanics. Milankovich had become very excited by the first accurate measurement of the solar constant in 1915, and he was using this measure of solar radiation to estimate the surface temperature on Mars. The red planet interested him especially because of its polar caps.[17] Using the techniques he developed in this work, as well as his love of calculation, he spent the years 1915–1918 working on an astronomical theory of the ice ages on Earth. This was the first version of a theory for which he would later become famous, a theory generally accepted today (with modifications) as the most likely explanation of the great advances of ice in the Paleozoic, the Permo-Carboniferous, and the Pleistocene.[18]

Milankovich had returned home after the war with his manuscript, and he published it in 1920 under the title *Théorie mathématique des phénomènes thermiques produits par la radiation solaire*.[19] By his own account, the work had received "a lukewarm response," but his Paris publisher had sent out copies to a variety of scientific institutions and "to some known scientists" to promote it.[20] One of these scientists was Wladimir Köppen.

Milankovich's theory concerns the variation in solar radiation reaching the surface of Earth: day and night, season to season, year to year. To explain long-term variation in solar radiation, he had settled, after a number of tries, on three controlling astronomical elements, all known to fluctuate on different timescales above 10,000 years. These were the precession of the equinoxes (whereby Earth's axis spins like a top completing one revolution every 26,000 years), then a much longer (96,000-year) cycle of the changing ellipticity of Earth's orbit, and finally a 41,000-year cycle in the tilt of Earth's axis. These three parameters each caused a different variation in solar radiation arriving on Earth, and thus they could either reinforce each other or cancel each other out. Milankovich had the astronomical data for these changes, which had been published by Ludwig Pilgrim (1849–1927), but he had to work out the thermal consequences of the interactions and develop the mathematics to do this successfully in a general way. The result of his calculations was a theory of Quaternary ice ages based on these fluctuating parameters over the past 130,000 years.

In September 1922, Milankovich received a letter from Köppen, who told him that he had read the book with interest and that he believed his theory to be correct and able to explain the main features of the climate changes in the Quaternary period. Köppen also informed Milankovich that he was now working with his son-in-law, Alfred Wegener, on a major project in paleoclimatology. Köppen had long been an advocate of astronomical "forcing" (as we now say) of meteorological phenomena, and he had worked long and hard on various ways to correlate weather with the eleven-year sunspot cycle known as the "Maunder Minimum."[21]

Köppen went much further than to simply praise Milankovich for his theory. He asked Milankovich if he would be willing to collaborate with them. Köppen wanted him to calculate the intensity of sunlight and its long-term changes in the Northern Hemisphere for the past 650,000 years. He would not have to take the effect of Earth's atmosphere into consideration, but Köppen wanted separate calculations for latitudes 55°, 60°, and 65° north. Köppen also told him that in his calculations he should concentrate on summer rather than winter sunlight, as he and Wegener were convinced (as are scientists today) that the key to ice ages is not colder winters but cooler summers that allow the persistence of ice and snow cover from year to year and accelerate the spread of ice by progressively reflecting more incident sunlight back into space.[22] Köppen asked him to produce a graph of the summer intensity of sunlight, as well as the amplitude and frequency of its variation, for sixty-five 10,000-year intervals: Köppen and Wegener would publish the resulting graph and attendant data in their planned book, and they wanted Milankovich to publish his mathematical apparatus and calculations separately, so that they might reference his publication.[23]

Milankovich agreed and went to work immediately to develop the attendant mathematics. By 13 November he had developed the equations to determine the fluctuations in solar radiation at the specified latitudes. He and Köppen exchanged ideas about the best way to graph this material so that it would clearly demonstrate the timescale, the amplitude, and the frequency. Milankovich's calculations showed that the oscillations in summer sunlight over long periods of time are not regular in their frequency or their amplitude, and he concluded that this irregularity was indeed caused by his three astronomical parameters, all of which were the result of the gravitational interaction of the Sun, the Moon, and the large planets. He then computed the values; the calculations took 100 days, and he finished them sometime in late February 1923.[24]

Enlisting Milankovich in this way emphasized the collaborative character of the planned work and the clear partition of authorship. Köppen and Wegener had decided that Wegener was to be responsible for assessing climate change from the Carboniferous and Permian through the Tertiary; Köppen would be in charge of the material on the Quaternary. Milankovich's results and theory would be restricted to Köppen's portion of the book.

Not only would the assignments differ, but also the approach. Wegener's aim, or the aim of Wegener's chapters, would be to establish the "Climate Belts" (*Klimagürtel*) for each of the major divisions of geological time, based on the geological and paleontological evidence, and then prepare a map of the results for each period. The map projection would be the oval "Hammer" view of Earth which Wegener had used in the third edition, and the conventions would be the same: the parallels of latitude and meridian of longitude would be those of the modern world, with Africa in its current position. The maps for successive geological periods would show the continuous dispersal of continents relative to Africa, and (very much to the point) the map would also show, as curved lines, the positions (relative to the present) of the equator, the lines of 30° and 60° latitude, and also the poles—the *Klimagürtel* for each period.

Köppen's aim, or the aim of his long chapter on the Quaternary, would be more detailed, in keeping with the greater volume and variety of evidence for this recent period. Rather than establishing "Climate Belts," he would produce a summary of the climate evidence and a variety of maps of the extent of the ice as it oscillated back and forth over the past half-million years. He would correlate this evidence with the theory of Milankovich and map the displacement of the poles throughout the Quaternary. Milankovich's work would take up perhaps one-fourth of the part of the book allotted to Köppen, and it would be restricted to the Northern Hemisphere. The chapter title, "Die Klimate der Quartärs" (plural), indicates the concentration on the frequent climate oscillations within the framework of the ice age.

Their focus on the Quaternary on the one hand and the importance of the Permo-Carboniferous glaciation for Wegener's displacement theory on the other, along with the possibility of even earlier ice ages in the Algonkian (Achaean), led Köppen and Wegener to believe they could safely assume that in all periods of Earth history "the same climate zones as today's have existed, namely, an equatorial rain zone, two arid zones, two rain-zones in temperate latitudes," in turn flanked by polar climates beyond them.[25] They also assumed that just as on the present Earth, the poles would lie 90° away from the equatorial rain zone and 60° away from the nearest arid zone. Their work would also reflect certain other conventions of Köppen's climatology concerning atmosphere and ocean circulation, as well as the tendency of dry zones to be interrupted on the eastern margin of continents by monsoon rains.[26]

If their climate belt scheme were true for the past as well as the present, then they should be able to map climate zones in each period of Earth history by having, for each zone, a uniquely distinguishing marker: something that unambiguously said "polar," or "arid," or "temperate rain," or "tropical." Most paleogeography and paleoclimate books had used fossil animals and fossil plants as such indicators, but, as Wegener pointed out in the paleoclimate chapter of his third edition, these were "general indicators" requiring large aggregations of data for a distinct signature. Further, the adaptive plasticity of both plants and animals made it difficult to rely on them alone. He suggested instead a greater reliance on geological indicators: desert deposits, coal, and glacial deposits, along with some particular plants (trees with rings, and the famous *Glossopteris* flora of the Southern Hemisphere). He had drawn a conjectural map using such geological makers and their distributions for the Carboniferous and Permian to show how these helped locate the equator and the poles.[27]

These classes of geological deposits (and a few fossils) would be for Köppen and Wegener what we now call "proxy" data that would "stand in for" a certain climate. This was a common practice in paleontology, where the occurrence of a certain fossil species was so reliably an indicator of a specific period of the past that it became an "index fossil." The same is true today in ecology, where some animal or plant becomes an "indicator species" for a certain ecological assemblage. There existed, in 1922, a huge, painstakingly assembled geological literature of where (on the current surface of Earth) geologists had found and mapped glacial deposits, arid formations, coal, massive limestones, and reef corals. Each of these layers in any given locale was assigned to a time in the past. Geologists (minus a few outliers) agreed that these geological markers unambiguously indicated not only certain kinds of climates but climates associated today with specific latitude zones between the pole and the equator. The key to a narrative history of the climate of the past would be to pull apart the geology books that tell us which rocks are in what place *now* and rewrite the story to say where they were *then*. "The climate history of any place on earth is, to a first approximation, the history of its position between the pole and the equator."[28]

The question was not whether such a reassembly of the data was possible but the security of the markers. This turns out not to have been a serious issue for most of the categories. There was in the 1920s, as there is now, agreement that salt deposits and gypsum are "evaporites" that can form only when there is an excess of evaporation over precipitation. They are evidence of arid conditions at the time in which they were deposited. There was (and is) agreement on so-called aeolian sandstones: that such formations were once windblown dunes, in which the bedding is so distinctive and characteristic that they may be mapped to show the direction of the prevailing winds at the time they were deposited. There was (and is) agreement that massive limestones from diatoms and reef limestones indicate tropical seas. There was no dispute that the rocks called "tillite," or "boulder clay," or *Blocklehme*—rough, poorly sorted material of every grain size from boulders down to individual plates of clay minerals—were remnants of glacier action.

It is a standard geological problem to identify such sediments and to distinguish them from their imitators. For instance, landslide deposits called "turbidites," which occur where the landslide takes place underwater—on the continental shelves—give convincing imitations of glacial deposits. Similar pitfalls await the casual reconnaissance of the other kinds of deposits as well. But by 1920 detailed geological mapping of the continental surfaces had already exceeded about 20 percent of the total area, and while a deposit here or there might be misidentified, this would not be likely across dozens or hundreds of locations. Moreover, the figure of 20 percent is misleadingly small: salt, gypsum, limestone, and coal are all economically important minerals for which prospectors (today called "economic geologists") competed avidly. Investigators were not limited to surface exposures either, as salt, limestone, coal, gypsum, and sandstone could be identified by drill cores in the many thousands of prospecting wells drilled in the first twenty years of the century. The data set was large enough that an occasional error would not falsify the entire picture.

The one climate marker that in 1920 still needed discussion and defense was coal. Almost everyone agreed that coal was the result of the burial of vegetation and imagined that the greatest coal beds, those laid down in the eponymous "Carboniferous," were the accumulation of huge quantities of vegetable matter in freshwater basins—coal bogs and swamps—where, through deep burial, compression, and heat, they were turned into various forms of coal.

The great paleobotanist Henry Potonié (1857–1913) had devoted a good part of his life and of his major work, *Lehrbuch der Paläobotanik*, to the theory of the development of coal. Potonié had demonstrated that coal is formed in a sequence humus → peat → lignite → bituminous coal, through burial, compression, and heat. He did this not just by arguing about it or citing geological literature, but by providing photographic tours of modern locales where peat and humus are being formed and buried. He showed that the kinds of plants that are found in coal are resident today in bogs, moors, reed beds, and swamps, and that the process of humus formation, burial, and transformation into combustible coal products can be observed in every stage today. While many coal beds are filled with the fossils of cycads and ferns, others are produced by wet temperate forests and even from marine plants such as kelp beds buried along the shore.[29]

Potonié had also insisted, contrary to prevailing textbook opinion, that coals could form in the tropics, because coals form where there are bogs, and bogs are demonstrated to exist in Sumatra, Ceylon, Central Africa, and South America; thus, there must be tropical coals.[30] The prejudice against this idea was that tropical heat must speed up the rotting of vegetation to the point where coal could not form. From Potonié's standpoint, as adopted by Köppen and Wegener, "the real contribution of such accumulation [of coal] to the question of climate lies not in the area of temperature, but of humidity. Before a basin can turn into a bog, it has to be full of fresh water, which only happens in the rainy zones of Earth, not in the arid regions. Neither can coals form in the dry zone of the horse latitudes, but only in the equatorial rain zone and in the two temperate zones."[31]

For Köppen and Wegener, coal formed both tropically and in temperate humid climates, even up to the postglacial "pluvial" zones at relatively high latitudes. Where such periglacial rain climates were not available to create bogs, they would form in preexisting topographic depressions. Thus, the Appalachian zone, which they would argue was the Carboniferous equator, was a tropical rain zone, where geologic folding created conditions for freshwater basins and coal formation. They argued that all one might infer from coal formation is abundant rainfall, and the *temperature* of such regions must be indirectly inferred from the fossil evidence, via systems of floral characters.[32]

The "big picture" of the history of climate was the latitude *and* rainfall regime for a region of Earth through time. As Franz Lotze (1903–1971) later pointed out in his classic *Steinsalze und Kalisalze* (Rock salt and potash) (1938), the worldwide deposits of salt and gypsum look randomly spread on all continents from the equator to very high latitudes. However, when we date the deposits by traditional methods of superposition and sequence—the stratigraphic correlation of known formations as age markers across wide areas—we can see that evaporite deposits for any period of Earth history form distinct zonal bands similar in width to that defining the limits of the evaporates today, that is, areas with a net excess of evaporation over precipitation. Moreover, these bands, when mapped on a globe, migrate through time from current polar latitudes (Carboniferous) to current equatorial latitudes (within 30° of the equator). Thus, either the whole planet was arid and dry in certain periods, allowing evaporites to form at the poles or very near them, or the latitudes have shifted.[33] The content of this judgment by Lotze is the more important because he was at the time an ardent opponent of Wegener's theory of displacements.

As with coal and evaporates, so it was also for sandstones and fossil sand dunes, as well as the history of deserts. Friedrich Solger's *Dünenbuch* (1910) was a survey of everything that was known about sand dunes and dune formation, including bedding structure, shape relative to the prevailing wind, and relationship of dunes to the coastline at the time of formation.[34] Solger asserted that, assuming that global atmospheric circulation remained constant in terms of the location of polar easterlies and the prevailing westerlies, fossil sand dunes could help to locate the latitude of the deposit when formed.[35]

Even more detailed support of the relationship between latitude and climate came from Johannes Walther (1860–1937), whose *Gesetz der Wüstenbildung* (The law of desert formation) (1912) was a major treatise on the character, mode of formation, and history of deserts throughout geologic history. The balance of evidence led him to conclude that relative climate change must take place through the displacement of Earth's pole—moving the equator and all flanking zones. There might be, he argued, absolute climate change—passage through different regions of space, variation in solar radiation—but the unambiguous signatures of past climates are left respectively by glacial climates, rainy climates, and desert climates, each of which has characteristic geomorphology, weathering, and sedimentation, including characteristic deposits.[36]

Of all their proposed markers, best established by 1920 was the association of "glacial till" (*Blocklehme*) with the presence of large-scale glacier action. Such deposits had been mapped all over Earth, everywhere associated with continental glaciation in both the recent and distant past. Wegener's 1921 opponent concerning the displacement hypothesis, Albrecht Penck, was the world leader in establishing this particular geological deposit as the signature of an ice age and in establishing a chronology of the European ice ages based on the appearance and disappearance of these deposits in the sedimentary record over the past half-million to one million years.[37]

All the works cited above constituted "proof of principle" for Köppen and Wegener in the preparation of their book; the actual labor on which they were embarked would be vastly more difficult and required great patience and industry. Beyond the general consensus on climate-based latitude zonation of glaciers, coal, and evaporate deposits, the argument of Köppen and Wegener's book required specific reference to the original field descriptions and maps—many thousands of locales for each of their markers. The few scattered plots on Wegener's maps for each period of "E" (Eis), "K" (coal), "G" (gypsum), "S" (salt), and "W" (*Wüstensandstein* [desert sandstone]) in the published work are an immense reduction of information contained in geological maps and field reports.

Beyond a few textbooks global in scope, the literature of geology more often expressed the history of a state or at most a subcontinent. The latter works typically discussed limited areas in great detail, and though they sometimes provided a cross-sectional diagram of the strata, they rarely had more than a local sketch map. Wegener therefore had to work from the descriptions in the texts treating the locale or the exposure and, from a latitude and longitude or the name of a town, find a way to plot the description, or the "center" of the distribution—which might actually be quite scattered and have no discernible center—on a globe or a map.

An example of the difficulty of extracting this information comes from one of Wegener's important sources, the two-volume treatise *Das Salz: Dessen Vorkommen und Verwertung in sämtlichen Staaten der Erde* (Salt: its occurrence and exploitation in all countries of the world) (1906–1909) by Josef Ottokar Freiherr von Buschmann (1854–1921).[38] This 1,200-page compendium of saltworks around the world, prepared by a finance minister, was considerably more interested in the cost per unit weight than in the geological age of the deposits or their particular mineralogy. Organized by national sovereignty, with deposits located only by the name of the nearest town, without a single map or illustration, and with an index that covered only the volume for Europe (and not the more useful volume for the rest of the world), it was extremely difficult to use—often indicating only in passing that a certain deposit in India was Miocene and one in New York State was Silurian.

Buschmann put tremendous effort into the survey, and if much of the information was of considerable antiquity, it was all that was available. It laid out schematically, in a way that one might discover over many long evenings of reading, what information existed on various deposits of sea salt, rock salt, gypsum, and other economic minerals of this group. Sometimes Wegener had to extract the information very indirectly—that a certain salt mine was near Mount Safed-Koh in Afghanistan, reached by going east out of town from Kabul,[39] or that in the western part of the Karroo in South Africa, in an area called Graff-Reinert, 322 kilometers (200 miles) from the coast, there were salt lakes at an elevation of 1,524 meters (5,000 feet). With the aid of an atlas and a geological handbook, one might be able to infer with some hazard a geological age for such deposits, but it was very boring and taxing work.

The burden of finding good data was immeasurably lightened by one particular series, the *Handbuch der Regionalen Geologie*. This ambitious series, designed and edited by the German stratigrapher and paleontologist Gustav Steinmann (1856–1929) and his younger colleague, the structural geologist Otto Wilckens (1876–1943), had begun appearing in the 1890s, and by 1921, twenty-one of the projected fifty-eight parts had appeared, roughly one per year.[40] The authors of each volume were recognized specialists who had been field geologists, typically for many years, in the regions they covered.

The handbooks, all edited and published in Germany, nevertheless were written in German, English, or French, depending on the author. The volume for West Africa was in French, the volume for North America in English, and the volumes for Syria, Arabia, and Mesopotamia in German. Moreover, field areas covered by the authors did not always correspond to their national or imperial domains: Kurt Leuchs (1881–1947), who wrote the volume on Central Asia, was an Austrian geologist; Robert Douvillé (1881–1914), who produced the volume on Spain, was a French paleontologist; Otto Nordenskjöld (1869–1928), who produced the volume on Antarctica, was a Swedish geologist and oceanographer.[41]

Some of the handbooks were so extensive that they had subeditors; the more intensively studied a country or region, the more likely it was to have very local experts. For instance, the volume entitled *The British Isles and the Channel Islands*, edited by J. W. Evans (1857–1930), had thirteen contributors. This particular volume, written in English and with exclusively English authors, had the strange destiny of having been published in Germany, by a German publisher, in the middle of the First World War (1917); it was not available for review in England until 1920–1921.[42]

If these volumes had a bias, it was the bias of explorers, geological travelers, and prospectors (economic geologists) in general: a strong focus on economically valuable and exploitable minerals; but all of the volumes were brought into line and balance by the general editor's insistence on a very strict format. Both the bias and format were very beneficial to Wegener: if there was coal, limestone, salt, or gypsum, the authors typically covered these deposits in detail.

Each volume began the same way, with an introductory essay titled "Regional Boundaries and General Character," accompanied by a sketch map of the area covered. The treatment was then divided into detailed studies of subregions. Each of these sections began with a "Morphological Overview," discussing the latitude and longitude, area covered, typical topography, presence or absence of mountains, maximum and minimum elevations, presence or absence of volcanic action, and so on. Then followed the "Stratigraphy and Petrology," presented from oldest to youngest using conventional geological time markers (codified in the late nineteenth century) giving the typical rock type and categorizing the fossil contents by species, where appropriate. Stratigraphic rocks came first, then eruptive and plutonic, again by time period. Finally, there was a section titled "Tectonics and Evolutionary History." The word "evolution" here did not mean biological evolution as in the origin of species, but the more general sense of "developmental evolution," that is, geological history.

Sometimes, in a way very useful for Wegener, the volumes would conclude with comparison with nearby regions outside the coverage of the book. For instance, Nordenskjöld's *Antarktis* (1913) contained an outline chart comparing West Antarctica with South America, from the Jurassic through the postglacial period, giving the names of the principal formations and whether the "facies" (type of rock) was transgressive or regressive—meaning whether the oceans were advancing or receding at that time—and where the history of these two regions was identical and where it was different.[43]

Because Wegener intended to draw a separate map for each period of geological history, the organization of these volumes of the series allowed him to dip into all the volumes and attend only to what they had to say about the Carboniferous, the Jurassic, or the Eocene. This was the very opposite end of the spectrum from Buschmann in terms of ease of use.

Wegener liked the design of the handbook so much that he adapted and copied it as the format for his work. Let us consider as an example chapter 3 of *Die Klimate der geologischen Vorzeit*, "Klimagürtel im Mesozoikum" (Climate bands of the Mesozoic).[44] Wegener treated the three standard time intervals of the Mesozoic in turn: the Triassic, the Jurassic, and the Cretaceous. For each of these sections, he considered the evidence in the same order: (1) traces of ice, (2) coal, (3) salt, gypsum, desert sandstone, (4) plant fossils, (5) animal fossils. Within each of these time divisions he considered the geographic regions in the same order: North America, Europe, Asia, and then South America, Africa, Oceania, and the polar regions.

The approach was unflinchingly empirical. Köppen and Wegener were subsequently criticized for having asserted that the climate zones of the past were like those of the present in every period of the past, including the presence of glacial ice, but this criticism is certainly not supported by the text. Examination of this chapter on the Mesozoic shows no attempt whatever to describe or map traces of glaciation. For the Triassic, Wegener said, "Unambiguous traces of ice in the Triassic are unknown."[45] For the Jurassic, Wegener wrote, "traces of ice are unknown."[46] For the Cretaceous, he wrote, "We must accept it as likely, that nothing like a polar climate in the modern sense is required to explain the data."[47] In Wegener's work on past climate, as in his work on atmospheric layering and continents and oceans, theory had to give way to data.

Response to Wegener's Work, 1922–1924

Beginning in 1918, Wegener collected every reference that pertained to his work directly, or to the question of isostasy, and referenced them in the interleaved sheets bound into a copy of his 1915 book. Of these, there is one brief notice concerning Wegener's work in French, from 1914, and three articles in English, on questions of isostasy, Earth contraction, and displacement of the pole, all of which appeared in 1914, and none of which mention Wegener. There are two references in Dutch, both of which mention Wegener and appeared during the war, and then one in Danish from 1917 (Andreas Lundager) and one in Swedish from 1919. Wegener had asked Hans Cloos in January 1920 whether he knew of anything else, so Wegener did not depend solely on his own search, and his notes appear to be a fairly complete list of literature discussing the displacement hypothesis. Thus, with the exception of three passing references to his work appearing in the scientific literature of neutral countries during the war, one of which was written by a close personal friend, there was no discussion whatever of the hypothesis of continental displacements outside the German-speaking world.[48]

The principal reason for this lack of discussion was the outbreak of the war itself and the severing of intellectual contact after 1914 between the English-, French-, Italian-, and Russian-speaking worlds on the one hand and the German-speaking world on the other. As we know, direct telegraphic communication between Germany and the United States was deliberately cut by the United States. Within the latter anti-German feeling reached near-hysterical levels during the war; in all of the Allied countries, possession of materials produced in Germany after 1914 became cause for suspicion and in some places was actually illegal. The combination of the inability to write back and forth with German colleagues and increasing hostility to things German rendered an open discussion of Wegener's hypothesis impossible anywhere outside Germany for the duration of the war, as Wegener had himself anticipated.

A second reason for this lack of consideration was structural. The German academic community was larger than the establishments of most of the Allied countries taken together, and vastly greater than those of the United States and Great Britain. American universities began granting doctoral degrees only in the 1870s; as late as 1930 there were only a handful of universities in the United States where one could obtain a PhD degree in geology. This meant that in the Allied countries, early in the war, most ordinary scientific enterprise gave way to scientific war work, leading to the postponement of ordinary scientific discourse. This was not so in Germany or the other Central Powers; as we have seen, Wegener was able to publish books on scientific subjects throughout the war, and only near the very end of the war did the pace of ordinary scientific publication slow down, and then for economic reasons (a shortage of paper) rather than ideological ones. The German home front characterized the war in terms of the defense of Germany's cultural superiority; much of this superiority was academic and scientific, and thus the vigorous prosecution of scientific research during the war coincided with Germany's war aims and was abundantly supported by the government in consequence.

This second reason may seem a redundant restatement of the first, but it is not and had the following consequence: since Wegener's hypothesis was widely discussed and argued about in Germany during the war, debate inside Germany was one cycle ahead of the debate in the rest of the world, and this lag time persisted throughout the 1920s. When Wegener read criticisms of his work, he did not respond to them in print except by revising the book into a new edition; when he saw critics of his work repeating statements he or others had already refuted, he referred them to the relevant literature (as he had done with Penck in Berlin in 1921) or just ignored them, waiting for his supporters to carry forward the corrections in other forums. This sometimes led to a situation, especially in the gap between the very forceful advocacy of the 1920 volume and the much more cautiously worded and better organized 1922 volume, where critics were busy refuting points that Wegener no longer claimed, or pointing out structural defects in the organization of the book which no longer existed.

There is a third reason: one can see in English and more so in American scientific writing even before the war an emergent hostility to theories of broad scope if they emanated from Germany. Joseph Barrell (1869–1919), a geologist at Yale University and, of his generation, the one with the best understanding of physics and of the need for geologists to understand physical arguments, wrote an article for *Science*, published in September 1914, immediately after the outbreak of the war, concerning theories of displacements of Earth's pole of rotation and their role in geological theory. He ended with the following observation:

> In closing this article it seems appropriate to indulge in a brief moralization. This paper does not contribute any new facts, but was written to show the untenableness of certain hypotheses, emanating in this instance from Germany, and in danger of spreading in America. . . . A more respectful reception has been given in this country to these hypotheses because they were voluminously presented in German and backed by the prestige of a German professorship. . . . But if the writer is not mistaken, in Germany, preeminently the land for science, voluminous presentation is a fashion, and around the large body of high grade work is a larger aureole of pseudoscience than is found in either England or America.[49]

Here Barrell shows the idiosyncratic hostility to speculative theory characteristic of the United States, but not of the British Empire, and therefore not of the Anglophone world generally. One recalls Darwin's famous remark in *The Descent of Man*: "False facts are highly injurious to the progress of science, for they often endure long; but false views, if supported by some evidence, do little harm, for everyone takes a salutary pleasure in proving their falseness; and when this is done one path toward error is closed and the road to truth is often at the same time opened."[50]

The early discussion of Wegener's work in England after the war shows the "salutary pleasure in proving . . . falseness" mentioned by Darwin, which was very much a part of British intellectual life. In June 1922 the Cambridge geographer Philip Lake (1865–1949) wrote a letter to *Nature*, on "Wegener's Displacement Theory," in which he noted, "Wegener's speculations have attracted so much attention that there must be many who would be glad to find some simple means of testing his fittings and coincidences for themselves."[51] He is no doubt referring to a four-page article by Wegener in the magazine *Discovery* for May 1922—Wegener's first appearance in English.[52] He then went on to suggest that Wegener's technique of making tracing-paper cutouts (which appears only in the 1920 edition) was cumbersome and difficult, and Lake urged his readers to use modeling wax or plasticine instead to make shapes of continents to move about the globe. Lake seems to have done this himself and to have followed Wegener's instructions that he test the fit of the continents. "This is not the place to discuss Wegener's views, but the use of triangular compasses seems to show a rather high degree of plasticity is necessary in the masses of 'Sial' in order to produce the coincidences on which he bases his calculation of the probability that his theory is correct."[53]

Six months later, in January 1923, Lake presented a paper before the Royal Geographical Society in which he set about to demolish Wegener's theory. The criticisms he offered were not new—Lake had access to the criticisms of Semper, Diener, and Soergel. Indeed, he could not have missed them because they are repeated in Wegener's 1920 and 1922 editions. Paying absolutely no attention to Wegener's discussion of these criticisms, he went on to repeat them: the Atlantic fit is no good; the bimodal distribution could occur in any case out of a "single equipotential layer" and proves nothing about Earth. The use of the *Glossopteris* flora to rearrange the continents of the Southern Hemisphere doesn't make the case, because *Glossopteris* flora is found in Siberia. These are all points to which Wegener had responded extensively in 1920 and 1922. Lake had read Wegener's third edition, but the more cautious tone of the book seems to have made no impression whatever on his hostility to the idea, which he describes throughout as made up of nonexistent *facts*.[54] He concluded, "From this brief account it will be clear that the geological features of the two sides of the Atlantic do not unite in the way that Wegener imagines, and if the continental masses ever were continuous they were not fitted as Wegener has fitted them."[55]

The discussion that followed is recorded along with the paper in the *Geographical Journal* and makes interesting reading, as some of Britain's most distinguished geoscientists were in attendance. The first speaker was stratigrapher and structural geologist George Lamplugh (1859–1926), who remarked,

> It may seem surprising that we should seriously discuss the theory which is so vulnerable in almost every statement as this of Wegener's. Yet Wegener's hypothesis is of real interest to geologists because it has struck an idea that has been floating in our minds for a long time. Mr. Lake has touched many basal points at which the theory will not hold water, and other flaws present themselves to the specialist in various particulars. But the underlying idea that the continents may not be fixed has in its favor certain facts which give every geologist a predilection toward it in spite of Wegener's failure to prove it. . . . We are discussing his [Wegener's] hypothesis seriously because we should like him to be right, and yet I am afraid we have to conclude, as Mr. Lake has done, that in essential points he is wrong. But the underlying idea may yet bear fruit.[56]

Lamplugh was followed by R. D. Oldham (1858–1936), a geologist and seismologist who had worked for many years in India. Oldham told the group that he was surprised that people found in Wegener's ideas a novelty, and he went on to describe in some detail the work of Osmond Fisher (which Wegener had discussed as early as 1912; Oldham had apparently not read Wegener). Oldham concluded by saying, "I should like to express a hope that in this discussion of Wegener's theory we will remember that the important question is not whether Wegener is right or wrong in his specific conclusions, but whether the continental masses have throughout all times maintained their present position relative to the poles."[57]

Oldham was followed in turn by Frank Debenham (1883–1965), a geologist on Scott's "Terra Nova" Antarctic Expedition, who had at this time just founded, at Cambridge, the Scott Polar Research Institute. Debenham congratulated Lake (his closest colleague in the room) for his "explosion" of Wegener's theory, adding that Lake's labor in demolishing Wegener had been much aided by Wegener himself: "he is not only a very bad advocate, but he changes his ground so peculiarly." After criticizing the Atlantic fit, the connection to South America, the Greenland measurements, and other points, he concluded, "However I believe we are all ready to be kind to the germ of the theory, in fact most people are rather anxious that something of the sort should be proved, and we have to thank Prof. Wegener a great deal in bringing it forward in offering himself as a target for bullets. Geographers and geologists generally, I think, have looked with a short vision up to the present day at a good many of the problems. We imagine we know all about the ordinary processes of the minor features of Earth. . . . But larger areas have been somewhat neglected."[58]

Harold Jeffreys (1891–1989), a young mathematician and geophysicist in the tradition of Sir George Darwin, brought up the question of mechanism and argued that Wegener's mechanisms were not only inadequate but also ridiculously so. Even he was willing to concede that there may have been forces, owing to changes in Earth's rotation through geologic time, "perfectly capable of splitting up continents as much as you want. Whether there has been a split, of the kind that this would give is a matter for the geologist to settle."[59]

The next speaker was Sir John Evans (1857–1930), a fellow of the Royal Society and president of the Geological Society. He observed,

> Dr. Lake has made a damaging attack upon the very elaborate theory Wegener has constructed. . . . I agree with what Dr. Lake has said about the absolute failure of Wegener to prove his case by demonstrating the continuation of the lines of mountains and foldings from one continent to another. . . . Nevertheless, I think that there is evidence of the drifting of continents the one from the other, and that the distances which now separate South America from Africa are very much greater than was formerly the case. . . . I think that, however open to criticism Dr. Wegener's views may be he has done a service in directing attention to these important problems which I believe it will repay our time and trouble to study.[60]

Charles S. Wright (1887–1975), another member of Scott's Antarctic Expedition, as well as a glaciologist and oceanographer, echoed what was increasingly the emergent theme of the discussion:

> There is only one thing I would like to say other than to congratulate Mr. Lake on his clear exposition. I do not think the method of criticism is quite the right one. With a hypothesis that touches so many sciences, I think the only effective attack is a criticism of the hypothesis treated as a whole, not on a few points of detail. Moreover, the hypothesis is intended as an explanation of facts. As Mr. Lake points out, these include some fictions, but one must still compare this hypothesis with any others which are in the field in regard to their capacity for explaining these same facts. This is, I believe, the only fair method of criticizing Wegener's proposals.[61]

The discussion was then concluded by the president, the Earl of Ronaldshay (Lawrence Dundas, 1876–1961), who had just returned from having served as governor of Bengal. He thanked Lake and noted humorously,

> The impression left on my mind by the discussion is that geologists, as a whole, regret profoundly that Professor Wegener's hypothesis cannot be proved to be correct. If that statement had not been made so emphatically and so frequently this afternoon I should have been inclined to judge, from the vigor and vim of the speeches with which they destroyed the theory, that their satisfaction had lain in a somewhat different direction. What the general feeling no doubt is, is this: that some theory of this kind is required to explain facts which have long been known to geologists, and while they feel bound to condemn this particular hypothesis as being one which is not capable of meeting this long felt want, they still hope that some other hypothesis of a kindred nature will be discovered which will satisfy their requirements.[62]

This meeting in January 1923 was the second symposium to be held on the subject of Wegener's work in England; the previous September the Geological Section of the British Association, meeting in Hull, was "the theater of a lively but inconclusive discussion on the Wegener hypothesis of the origin of continents."[63] W. B. Wright (1876–1939), who was at the meeting and very impressed (favorably so) by Wegener's argument, wrote up a summary of the discussion which appeared in *Nature* the following January.

Sir John Evans, who spoke in favor of drifting continents at the March 1923 meeting in Cambridge, had prepared an exposition of Wegener's theory, to be read at the meeting in Hull. Evans was away from England at the International Geological Congress in Belgium in August and September 1922, but his prepared remarks were a careful exposition of Wegener's theory. The discussion immediately followed the reading of Evans's paper, and Wright said that "the discussion brought forth a great diversity of opinion regarding the validity of the hypothesis, almost the only point on which there seemed to be any general agreement being an unwillingness to admit that the birth of the North Atlantic could have occurred at so late a date as the quaternary."[64] Wright also mentioned that P. G. H. Boswell (1886–1960) had announced at the meeting that "the forthcoming English edition of Dr. Wegener's book will afford an easy means of becoming acquainted with the leading features of the subject."[65]

The English translation of Wegener's third edition, announced by Boswell in September 1922, was almost certainly arranged for Wegener with Evans's help and that of P. G. H. Boswell himself. When the translation finally appeared in 1924, J. G. A. Skerl (b. 1898), the translator, thanked Boswell, C. P. Chatwin (1897–1975), and especially Evans, who, he said, "has smoothed away many difficulties, and considerably enhanced its [the translation's] value."[66] Evans reviewed the translation both before and after Wegener and wrote the introduction to the volume. In his introduction, appearing in 1924, Evans concluded his favorable summary of Wegener's work with the following statement: "I have elsewhere criticized some of the details of the author's conclusions. It would be out of place to repeat these criticisms here. My only care has been to ensure that in this translation he should be allowed to state his own case in his own way."[67]

Evans's comment that Wegener should be allowed to "state his own case in his own way" reflects Evans's conviction that Wegener had been pretty thoroughly misrepresented both at Hull and at Cambridge in 1922 and 1923, and that many of the objections of particular specialists concerning particular pieces of evidence might well look different and less decisive in the context of the entire theory and the evidence for it. One senses that Evans agreed with the position of Charles Wright, the Antarctic glaciologist, who had argued at Cambridge that the appropriate way to judge such a theory is not by nitpicking this or that minor flaw, but by looking at the ensemble of facts which the theory claimed to explain and comparing with other theories that claim to explain the same facts.

We have no record of how Wegener responded to these debates, or if he even saw them, though his relationship with Evans makes it quite likely that he did. Based on what we know about his published response in 1921 in Berlin to the criticisms of Koßmat and Penck, it is likely that in spite of the severity of the criticism over details in the British discussions, Wegener would have been extremely pleased by the outcome.

His main point was always to establish the idea of continental displacements as a working hypothesis. This certainly seems to have been achieved in Great Britain, and even his severest critics thanked him for bringing the idea forward. Moreover, Charles Wright had pointed out, as had R. D. Oldham, that the question was never whether *Wegener* was right or wrong, but *whether the continents move*, and that the appropriate mode of criticism was the comparison of Wegener's theory with all other theories of similar scope which claimed to explain the same facts. Many of the detailed criticisms of Lake and others would have been of small concern to Wegener since he had already responded to most of them in detail: he was accustomed now to the rhetorical tendency of geologists to cite those authorities who supported their position while ignoring those whose publications supported Wegener's view. Finally, a number of the speakers in the January 1923 symposium claimed British priority for the idea, and it is an unfailing marker of a theory in the ascendant when others claim they "had the idea first."

Since most (not all—Lake was a geographer) of the speakers in the 1923 forum on Wegener's theory at Cambridge were geologists, this is perhaps an appropriate place to say something about geology and geologists. First of all, geology is the most nationalistic of the sciences, because the mapping of each nation-state and imperial dominion is the responsibility of a government-sponsored geological survey, and thus a large proportion of a nation's geologists are at one or another time in the pay of their government. In Britain this meant getting approval from superiors before publishing anything at all, even if prepared on their own time, and this led to a profound conservatism and resistance to new ideas. Secondly, geologists, whether they worked for surveys or not, tend to be fiercely proprietary about their "field areas" and are quick to assert ownership of "their rocks" as their entrée into any larger theoretical debate in which these rocks play a role. In this they are similar to anthropologists, fiercely proprietary about "their people," meaning the subjects of their ethnological inquiries.

Geologists tend to "work small," as Frank Debenham had suggested at the Cambridge meeting, and are unaccustomed to dealing with hypotheses of large scope and scale. This is partly because of the tendency to frame geology as a science in which one's curiosity about local details has no lower bound, right down to the grain size and even the molecular structure of component minerals in the rocks in the specific strata in the specific locales the geologists claim as their "field areas."

Finally, geologists are faced with the paradox that while the published literature represents the official record of their science, they tend to be distrustful and even contemptuous of geological writers who "work by the book" rather than with "boots on the ground." Even Evans, vastly more sympathetic to Wegener both in general and in detail than most of the other British commentators in these years, noted that Philip Lake's distrust of Wegener's evidence for the tectonic character of northern South America was well merited, and that Evans was in a position to know, because he had collected the data himself in 1905.[68]

These issues concerning geology as a science came to the fore in another spirited and consequential defense of Wegener's ideas and even more of his approach, in August of 1922, when the International Geological Congress, a triennial event, met in Brussels, Belgium. The keynote address for the congress was to be given by the Swiss geologist Emile Argand (1879–1940), a student of the great Swiss proponent of the nappe theory, Maurice Lugeon (1870–1953). Argand was an astonishing polymath, with a legendary capacity for understanding complex three-dimensional geological structures and rendering them as multicolored tectonic maps. This capacity was recognized as early as 1913, when his cartography resulted in his being awarded the Spendiaroff Prize at the International Geological Congress meeting in Toronto.[69]

Argand had been working since that time to extend the mode of analysis used to understand the Alps to comprehend the geological evolution of Eurasia generally, and beginning in 1915 he had developed an approach that he called *tectonique embryonnaire* (embryotectonics), in which he sought to understand the history of a region by successively unfolding (mentally) all the deformations a region had experienced back to the original flat surface and mapping every intermediate interval of change, a technique known in English as "palinspastic" mapping. This led him to the organic analogy, in which one added the dimension of time to geologic structures as an essential part of the imaginative work of geological interpretation, and he understood fully that his ideas in this area would require a revolution in geological thought and practice.[70]

In 1916 Argand had obtained and read a copy of the first edition of Wegener's book. This was, strictly speaking, an illegal act. In Switzerland, as elsewhere in Europe and North America, anti-German feeling was strong during the war, and it was prohibited at that time to read or possess—in public or in private—materials printed in Germany. In Wegener's hypothesis of mobile continents Argand found a theoretical structure that, as Albert Carozzi has remarked, "led to a complete change in Argand's thinking. It provided him with a new doctrine—'mobilism'—that satisfied to a much greater extent the space and time requirements of his synthesis than the concept of the general contraction of Earth."[71] In spite of the prohibition on reading German materials, Argand gave a public address in November 1916 on Wegener's ideas, in Neuchatel, Switzerland.

On 10 August 1922, Argand gave the inaugural address to the International Geological Congress. His title was *La Tectonique de L'Asie* (Tectonics of Asia), and he spoke for *several hours*, providing a synthesis of global tectonics explicitly in light of Alfred Wegener's hypothesis of continental displacements. The address was too long, many of the delegates did not understand French, the acoustics of the room were terrible, and his tectonic map of Eurasia—the basis of his presentation—was barely visible.[72] Nevertheless, it was a stunning defense of Wegener—not vague thanks for producing an incorrect theory that might help to give birth to a correct theory, which seems to have been the view from England. Rather, he argued, "We have theories by the dozen, but their very number decreases the chance of agreement among them. Today, it seems that the dispute is focused on the theories that imply the fixism of continents and the hypothesis of large-scale drifting (*dérive*) of continents as visualized and powerfully presented by A. Wegener."[73]

Argand went on to develop an extremely interesting theoretical argument. He said that "fixism" and "mobilism" were not themselves theories but attitudes, and that "fixism" was a negative element common to several theories, essentially the absence of a position on how mobility of continental elements takes place. Fixism, Argand said, generally asserts the contraction theory, refuses to accept the implications of isostasy and the differentiation of ocean and continental crust, and embraces the inertial predisposition to see everything "encrusted" in place, changing in time vertically but not horizontally.[74]

On the other hand, he continued, continental drift addresses the question of mobility in a way that gives a much better account of the phenomena of geology while coming into a sound alliance with modern geophysics. He concluded his remarks on Wegener (but not his whole address) with the following judgment:

> The validity of the theory is nothing else but a capability of accounting for all the known facts at the time it is presented. In that respect, the theory of large-scale continental drift is of flourishing validity. In its incipient stages, it was aiming at the absolute; subsequently it gained a lot of strength and flexibility without sacrificing anything of its rational structure: on the contrary, it became enriched and increasingly in harmony with the vision that leads to the whole. This work of clarification and of improvement is very obvious throughout the sequence of works by A. Wegener. Strongly documented at the meeting points of geophysics, geology, biogeography, and

> paleoclimatology, it has not been refuted. One has to have searched at length for objections, and particularly to have found a few, in order to estimate properly the kind of immunity that distinguishes it, and that originates from a great flexibility combined with a great richness in operational possibilities. One thinks one has found a decisive objection and that after another one the whole theory will collapse. In fact, nothing of that sort happens; one has only forgotten one or several mechanisms. It is the protean resistance of a plastic universe. . . . Therefore, it is quite true to speak of the validity of this theory in the sense mentioned previously.[75]

Here Argand captures the inertial, conservative, local, vertical predisposition of geology, only partly overcome by the nappe theory in the Alps. Argand was among the first *geologists* to unite the spatial and temporal dimensions of the world, not in terms of a layer cake successively built up and torn down or folded in place, but in a picture of a restless and dynamic Earth constantly in motion both up and down and laterally on a scale of thousands of kilometers, an Earth that was not a series of still photographs but a moving picture, in which the still photos must be assembled into a movie and put in motion: Earth was not only kinematic but cinematic.

Argand's defense of Wegener was both powerful and poignant—powerful given the podium he used to defend Wegener and his ideas, and poignant because all scientists of the Central Powers had been excluded from this congress: there were no Germans, Austrians, or representatives of the other lands for which German was the principal scientific language. It was an act of considerable intellectual independence and moral courage for Argand to speak as he did in defense of a German idea in that venue at that time. It is quite likely that the inspiration for a French edition of Wegener's third edition was born at this meeting and may even have come from Argand: Swiss paleontologist Manfred Reichel's (1896–1984) translation of Wegener into French, like that of J. G. A. Skerl into English, would appear in 1924.[76]

Argand had developed a picture of the tectonic evolution of Asia, but he found himself with a pile of discrete facts and no motor to drive them into a dynamic unity. The idea of "mobility" was for Argand a motive, an inspiration to move forward. Nothing that he saw in Wegener made him reorganize the facts at his disposal: Wegener's work provided a dynamical motif and a context in which to develop his own insights.

It was left to Edgar Irmscher, Wegener's young colleague at Hamburg, to go forward to try to reorganize his field of study—the distribution and diffusion of vascular plants—in terms of Wegener's hypothesis. He had said (in the lecture that Wegener heard in 1919) that he intended to do this, and by 1922 he had completed this task to a first approximation in the publication of a book entitled *Pflanzenverbreitung und Entwicklung der Kontinente* (The diffusion of plants and the developmental history of the continents).[77] This was a rewriting of paleobotany from the perspective of the permanence theory to that of the displacement theory which he intended to be read as a complement to Wegener's work on continental displacements, as well as to Wegener and Köppen's forthcoming work in paleoclimatology.[78]

It was a very ambitious book, 250 pages in length, which could serve as an introduction to Wegener's theory, an introduction to paleobotany, and an introduction to the subject declared in the title: the diffusion of vascular plants. It concluded with a firm declaration that migration of the poles and large displacements of the continents were essential to the explanation of the current distribution of vascular plants, in combination with the plants' own "efforts" to spread their range, develop, and interbreed with other species. At the very end of the book, Irmscher included a note of thanks to Wegener for his support and his generous sharing of his data and his unpublished maps (referring to the maps for the forthcoming book on climates of the past). These maps show that Wegener, by mid-1922, had begun to think about how to represent paleopoles and paleoequators.[79]

It was very convenient for Wegener that Irmscher was able to finish and publish this book so quickly, allowing Wegener to cite it as a source and use Irmscher's mapping of the distribution of plants in his own book on climates of the geological past. The situation here was similar to that of Köppen and Milankovich, where they engaged in a collaboration, the younger author published separately and extensively based on his own research, but this work was prepared in collaboration with Köppen and Wegener, who could then cite work they had essentially "commissioned."

While these various symposia on his work, plans for translations of his books, and endorsement of his theory proceeded, Wegener remained steadily and methodically at work on *Klimate der geologischen Vorzeit* throughout the autumn of 1922, all of 1923, and into the spring of 1924. He and Köppen had settled on that title in order to distinguish their work from many other books entitled "Climates of the Past," "History of Earth," "Climates of Earth," "History of Climate," and so on. There had already been a book titled *Klimate der geologischen Vergangenheit* (*Climates of the Geological Past*), published by the Dutch paleontologist Eugène Dubois (1858–1940) in 1893, and republished in English under that title in 1895.[80] Dubois had a climate theory based on changes in solar radiation, as the Sun evolved into different phases of its history as a star. To avoid confusion, they chose another word for "the past": *die Vorzeit*, with its sense of past ages of Earth.

In working out the geologic history of climate, there were a few textbooks that Wegener depended on heavily to supplement and frame the information in the various volumes of the *Handbuch der Regionalen Geologie* and individual monographs from the geological literature. In 1922–1924 his most comfortable resource was *Erdgeschichte* (History of Earth) by Melchior Neumayr (1845–1890) and Viktor Uhlig (1857–1911), from the 1895 edition of that work.[81] Wegener referred to this book throughout as "Neumayer–Uhlig," reminiscent of the way classical scholars, then and now, refer to "Pauly–Wissowa," an encyclopedic work on Greek and Roman classical antiquity.

Wegener searched for a book with a detailed treatment of the regional geology of North America, a book that could play a similar role to that of Neumayr–Uhlig, and he decided on Thomas C. Chamberlin and Rollin D. Salisbury's three-volume *Geology* (1905–1907), a standard undergraduate textbook for many years after its publication.[82] It covered basic science, physical geology, paleontology, stratigraphy, tectonics, and large questions concerning the origin of Earth. If Chamberlin had been a European, he would probably have been called a "cosmic physicist" because although he had made a great reputation in unraveling the advance and retreat of the Quaternary ice sheets in the northern half of the United States, he was insistent that geology should extend itself back to the origin of Earth and the origin of the solar system. This was certainly congenial to Wegener, and it was an excellent choice for a standard North American reference. Almost as important for Wegener was Eliot Blackwelder's (1880–1969) *Handbuch* volume (1912) for the United States. Blackwelder had worked on both desert and glacial deposits, had traveled with Bailey Willis in China, and had written on Meteor Crater, Arizona; Wegener already knew of him.[83]

Wegener depended, of course, as he had since 1915, on the work of Dacqué, and both he and Köppen were also entirely confident in the paleobotanical treatise of Henry Potonié, brought up to date in a 1921 edition by Walther Gothan (1879–1954) and generally referred to as "Potonié–Gothan."

Wegener also depended heavily on the compendious treatise *Unsere Erde* (Our Earth), authored by the paleontologist and economic geologist Lukas Waagen (1877–1959) with the help of Jacob van Bebber (1847–1909), a meteorological colleague of Köppen at Hamburg, and also Damian Kreichgauer. It was an odd conjugation of scientists: a very young paleontologist, a very old meteorologist, and a monk who was even then turning away from the history of climate toward an attempt to translate the Mayan codices.[84] Yet there was a good deal of experience and novelty mixed together in this volume, which pursued the same theme and the same problems that interested Wegener and Köppen.

Certain figures—Chamberlin and Salisbury; Neumayer and Uhlig; Potonié and Gothan; Waagen, van Bebber, and Kreichgauer—Wegener employed as "authorities." They had written major texts of established reputation, and their citation structure was reliable. Often Wegener's reference to a primary monograph would be not to the monograph itself but to the page where it was cited in one of these "authorities."

He used the *Handbuch* volumes in the same way to cite literature, and even some of the these emerged as "authorities," especially Blackwelder for the United States, Paul Lemoine (1879–1940) on West Africa, and Max Blanckenhorn (1861–1947) for the Middle East and Egypt. Beyond these there were still others, such as Darashaw Wadia (1883–1969), an expert in many areas of geology, author of a major textbook on the stratigraphical history of India, and a great student of the structure of the Himalayas. Wadia had a long section in *Geology of India* (1919) on the *Glossopteris* and *Gangamopteris* florae and on economic geology, including the location of Gondwana coalfields and of gypsum.[85]

The tendency to cite authorities led to an uneven citation pattern in the finished book. Sometimes Wegener gave a direct page reference; sometimes he just referred to the name of his authority, expecting that the structure of these textbooks would carry the reader to the correct page (generally it does). At times this looseness of citation reflects the time pressure he felt while working on this book. The facsimile edition of his research notebook shows that he often would record the necessary geological information and the name of the author, but very rarely the full citation or the page number. This notebook, which he carried with him in his time-consuming commutes from one worksite to another, was a basis for rumination, reading, and thinking about the distributions of his geological markers, but if he decided to use them as citations, there was no time for him to go back and redo all the research to find the page numbers. Since, however, he was referring in almost every case not to a theory of Earth but to some list of geological strata, or the occurrence of coal, gypsum, desert sandstone, or some other marker, the exact page number for a reference was not as important as it would have been in a work that was stringing together arguments rather than plotting data points.

Wegener in 1922–1923 was overcoming his discomfort with geological reports now that he could "reduce" them as a physicist might reduce physical data, in order to plot them on his globe. With his tissue paper cutouts of the continents in their conjectural position for the period of time he was working on, he could "pinpoint" coal, gypsum, sandstone, coral, and traces of ice and mark them on his half-meter-diameter globe with flags of colored paper, as he had done in 1920. Wegener adapted his meteorological technique for mapping layer boundaries at *altitude* by temperature into a geological technique for mapping geological boundaries at *latitude* by using specific mineral markers. In a two-dimensional map, a latitude line appears as an "altitude" above or below the equator. Wegener established the distribution of coal, desert sandstone, gypsum, and traces of ice for the most securely documented and mapped locations. He discussed, for each period of geological time, each of these locations (each item that ended up as a plot point on a map) and referenced it to the literature. Each letter "K" (coal) indicated a geologically mapped coal deposit registered in the literature for that period of time; the same was true for all the other indicators.

Depending on how much coal, salt, gypsum, and desert sandstone he could reliably locate for a period, he might begin by establishing the equator directly, especially in periods for which he had a very large number of data points indicating tropical coal, as in the Carboniferous. He might also begin by establishing the spread of sandstone, salt, and gypsum in two widely separated regions—say, across North America and Europe on the one hand and middle South America and Central Africa on the other—and then using these data to establish latitudes 30° north and 30° south, interpolating the equator between them. From these 30° lines one could establish the 60° latitude line and the position of the poles. On the other hand, for heavily glaciated periods, the large number of markers indicating glacial ice might be the controlling factor, leading him to establish the 60° latitudes and pole positions *first*, and then 30° and equatorial rainfall zones later.

Working through Wegener's text and maps, one can see how important were Potonié's ideas about coal formation, because sometimes these coal deposits are so widely scattered that unless one accepted that coal could and would form in tropical, temperate, and even periglacial environments, it would be impossible to make sense of the distributions. Similarly, one may see that gypsum and sandstone were important indicators of *rainfall* and not temperature, as these are sometimes found in high-latitude deserts where precipitation is scarce.

Wegener's mapping was iterative and self-correcting, and sometimes the climate data also indicated the need to relocate continental positions and even to rotate continents. For instance, Wegener was able to use geological information on India for different periods to calculate that India had moved about 49° of latitude (5,400 kilometers [3,355 miles]) and rotated about 30° counterclockwise since the beginning of the Tertiary period.[86]

We have to think of these maps as the result of two huge data sets superimposed on one another. The first is the map of continental positions dictated by faunal and floral continuities and discontinuities, leading to the establishment of a time of separation of continental fragments and a rate of separation calculated backward from their current distance apart. This allowed the interpolation of intermediate positions for different periods of time since the separation. Superimposed on that fossil data are the geological data for latitude zones, which allowed calculation of the pole positions, this time following geological *conjunctions* more than floral or faunal *disjunctions*.

While Wegener assumed the "truth" of continental displacements in establishing the climates of the past, this book contains the most detailed reconstructions of continental displacement he ever produced. Prior to this effort he had mapped the Carboniferous, the Eocene (early Tertiary), and the Quaternary. In this book he calcu-

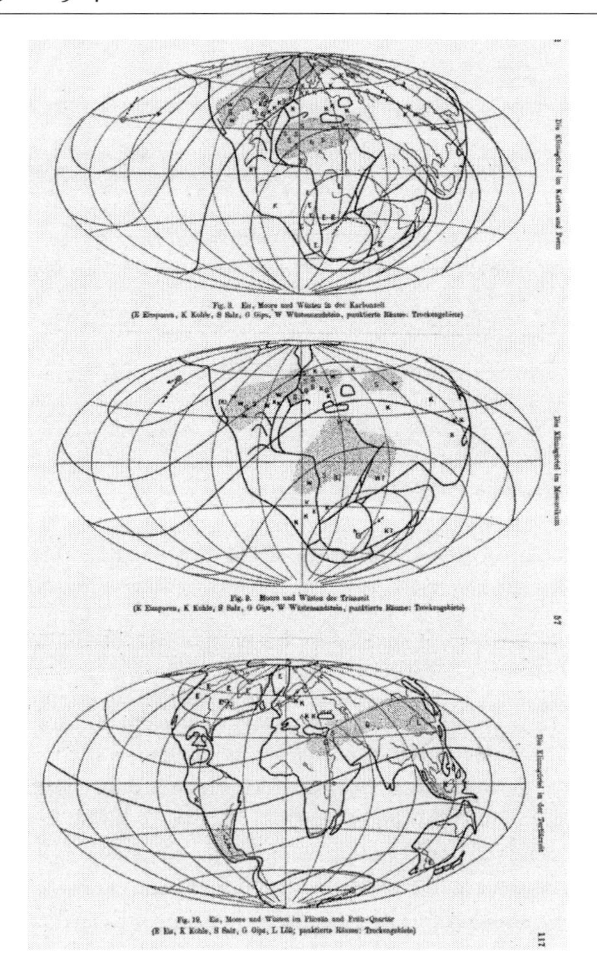

Wegener's climate maps of Earth (on a Hammer equal-area elliptical projection) for the Carboniferous, the Triassic, and the Pliocene and early Quaternary. E=traces of ice; K=coal; S=salt; G=gypsum; W=desert sandstone; *punktierte Räume* (dotted regions)=arid areas. The maps show the arrangement of the continents (with Africa arbitrarily in its current position), as well as the positions of the poles and the paleoequator, with arrows showing the direction and magnitude of the displacement of the pole during each period. Note the absence of traces of ice in the map for the Triassic. From Wladimir Köppen and Alfred Wegener, *Die Klimate der geologischen Vorzeit* (Berlin: Gebrüder Bornträger, 1924).

lated paleolatitudes and pole positions, as well as the arrangement and exact distance apart of the continents, for the Devonian, Carboniferous, Permian, Triassic, Jurassic, Cretaceous, Eocene, Miocene, and Pliocene and early Quaternary (the last two mapped as one period). He produced additional special maps using the same (Hammer) projection, for floral distributions in the Carboniferous and Permian, land and sea distributions in the Jurassic, and tropical corals in the Cretaceous.

The concluding chapter of this extraordinary research effort and imaginative tour de force consisted of only four pages, of which three are maps and charts of latitude and longitude data. Using the Lambert oblique projection, Wegener produced a map showing the path of the poles from the Devonian to the present, against a background showing the continents in bold outline (their current positions) and in shaded outline

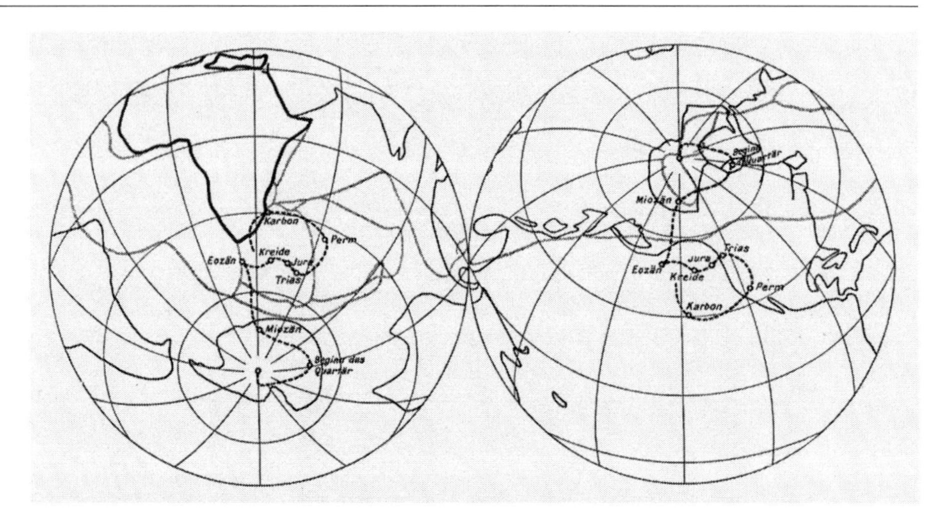

Wegener's map of the path of the poles (relative to Africa) from the Archaean to the Quaternary, on a Lambert oblique orthographic equal-area map. The current outlines are shown with solid lines, and their Carboniferous positions with shaded lines. The Southern Hemisphere is to the left, the Northern Hemisphere to the right. From Köppen and Wegener, *Die Klimate der geologischen Vorzeit.*

(their position in the Carboniferous). The polar displacements are shown relative to Africa fixed in its current position (*Polwege bezogen auf Afrika*).[87]

These pole-wandering paths are considerably shorter than those he had calculated and sketched in 1921 to be published along with Köppen's article for *Petermanns*. The distances between the Carboniferous and Permian pole positions are only about half of what they had been in 1921, and while he still mapped very large excursions of the pole in the Tertiary, similar large excursions of the pole for the Mesozoic (Triassic, Jurassic, and Cretaceous) had entirely disappeared, with the pole remaining at nearly the same latitude between the Triassic and the Eocene, and with a reduction of its shift in longitude from 75° down to about 45°.

In addition to this map, he produced a chart of the latitude and longitude (using the Greenwich Meridian and the current position of Africa) of the North and South Poles for every period from the Carboniferous to the Quaternary. He produced a graph of the latitude of Tokyo, Leipzig, Cairo, Punta Arenas (Chile), and Hobart (Tasmania) from the Precambrian (Algonkian) to the present. On the facing page he constructed a chart for the total change in latitude of twenty-seven locations on Earth since the Carboniferous. For variety and the interest of his readers he distributed these among North America, Europe, Asia, South America, Africa, Australia, and Antarctica, so one could find the change in location of (among others) New York, the Panama Canal, Madrid, Colombo (Sri Lanka), the Cape Colony (South Africa), Perth (Australia), and Mount Erebus, Antarctica's active volcano.[88]

It is of some interest to determine how accurate this method of work was, especially since so many of Wegener's opponents attacked his reconstructions as absurd. Paleomagnetic measurements made in the 1950s by Edward Irving, on a selection of rock samples from India, suggested that India had traveled 6,000 kilometers (3,728 miles) through 55° of latitude and 30° of counterclockwise rotation; Wegener's accuracy approached the limits of resolution of a novel geophysical technique thirty years after his publication. Irving did not publish these measurements at that time, but they cemented his conviction that latitude zonation was crucial to understanding the character and extent of continental mobility; Irving's research later became a crucial piece of evidence in the assemblage of scientific work which emerged as "plate tectonics" in the 1970s.[89]

In assessing the impact of Köppen and Wegener's book and its relationship to the theory of continental displacements, we should probably treat it, as we have already suggested, as two separate books, one on the "climates of the geological past" and one on the "climatology and causes of the North hemispheric Quaternary ice ages." The former book was by Alfred Wegener, and the latter by Wladimir Köppen and Milutin Milankovich. The two books under one cover differ in style, subject, method of approach, organization, citation conventions, depth of coverage, cartographic style, and mathematical content.

The former book is a descriptive and "morphological-empirical" reconstruction of continental positions, latitude zones, and movements of Earth's pole of rotation, based on geological and paleontological data. The latter book extends Köppen's scheme for the contemporary climate of Earth back one million years into the Quaternary and unites this climatic scheme with geological (and even Paleolithic archeological) data marking the advance and retreat of the ice sheets. These are, in turn, united and explained via Milankovich's mathematical theory of the astronomical forcing of ice ages by variation in the amount of solar radiation reaching the surface of Earth. One of these books starts on page 1 and extends through page 157; the other begins on page 158 and extends through page 256, plus the foldout at the end of the book graphing the advance and retreat of the ice sheets in rhythm with the fluctuations in solar radiation calculated by Milankovich.

Wegener and Köppen insisted in print, in person, and in correspondence that this was their joint work from the beginning to the end, and thus each accepted all the conclusions from the part written by the other. Yet, there was a serious theoretical contradiction between the first half of this book and the second, which had to be finessed.

In 1921, in Wegener's map for Köppen's article in *Petermanns*, the North Pole moved through a huge series of spiraling oscillations from the Bering Strait, vibrating between latitudes 70° and 80° north many times while shifting across 120° of longitude in a matter of one million years: more polar motion in the past one million years than in the previous 500 million.

These large polar oscillations were incompatible with the Milankovich theory, which required that Earth's pole of rotation should have changed its angle relative to the ecliptic (axial tilt, or obliquity of the ecliptic) by a very small amount: an oscillation of 2.4°, between 22.1° and 24.5°. This was required for the intersection (amplification or damping) of the various astronomical parameters controlling the amount of solar radiation reaching Earth's surface: any large excursion of the pole of rotation *relative to the fixed stars*—an *astronomical* pole shift—would have destroyed Milankovich's entire hypothesis.

The relationship in Wegener's theory between apparent motions of the pole and absolute motions of the pole—in conjunction with absolute movements of the continents and their motion relative to an Africa fixed in place—was one of the murkiest points of his theory and the most difficult to understand. Wegener believed in what we would now call "true polar wander," in which rearrangement of continental blocks could cause rapid and large excursions of Earth's pole of rotation, so that its pole of figure might coincide once again with its pole of inertia. Nowhere in his theory of the origin of continents and oceans (up to 1924) does Wegener make it at all clear whether his absolute polar motions are also *astronomical*, though his book often reads as if they are.

It would seem on the face of it that when Wegener says, as he did in 1922, that the data give "a North Pole at the beginning of the Tertiary in the neighborhood of the Aleutian Islands and from then a migration toward Greenland where it is to be found in the Quaternary," we are talking about a very large axial shift in a very short period of geological time, in which Earth's pole of inertia moved 10° or more and Earth's pole of figure raced to catch up with it.[90]

It is difficult to imagine this as other than an astronomical shift, since Wegener moved so firmly to contrast his theory of real pole shifts caused by separate continental motions with the theory of Kreichgauer, and earlier of John Evans, of the shift of the whole crust over the whole interior—a massive relative shift leaving the astronomical pole unchanged. On the other hand, Wegener had characterized the earlier history of Earth, of Pangäa, as a motion of the entire Sial sphere westward over the Sima sphere, and he had even speculated, as late as 1921, that the Atlantic split had been caused by the drag of the thicker Asian trailing side of the Pangäa continent, which allowed the Americas to break away and open the Atlantic. In 1924 these contradictions and confusions remained unresolved.

In order to bring his half of the book into correspondence with Köppen and Milankovich's half of the book, Wegener found it expedient and necessary to discard all the oscillations of the North Pole in the *Petermanns* map of 1921 and to push the extensive Tertiary excursions of Earth's pole of rotation back in time to the Pliocene and Early Quaternary. Moreover, because of the linkage in his theory between the motions of the pole and the motions of the continents, this meant that Wegener *had also to move the opening of the North Atlantic back in time to the Pliocene, and even the Miocene*. Indeed, his maps on pages 116 and 117 show the separation of Greenland from Canada and of Iceland from both Canada and Europe as already taking place in the later Tertiary. Additionally, these maps give a greater apparent distance between North America and Europe than that between South America and Africa in the Miocene. This is a very different picture of the timing and geometry of the continents than in any of his earlier versions of his theory, in which he insisted that the Atlantic had opened from south to north.[91]

Changing the theory in this direction did have advantages for Wegener, although neither here in *Klimate der geologischen Vorzeit* nor in any other venue in 1923 or 1924 did he announce it: he just changed it. *Nature* had reported in January 1923 that in the British meeting at Hull the previous September (1922) the only point of universal agreement was that Wegener had moved the major shift of continents in the Northern Hemisphere too far forward into the ice ages; everyone had found this improbable. Even scientists very much predisposed to accept Wegener's ideas, like Schweydar in 1921 at Berlin, had urged him to "slow down." He had done this for earlier periods by 1922, but not yet for the Quaternary. Now (in 1924) that Wegener assumed a stability of the pole of rotation throughout most of the past half-million years (the period of the great ice ages), it became much easier to keep the distance between Siberia and Alaska constant, or even diminishing, as North America rotated away from Greenland and Europe around a fixed rotational and inertial pole: if the changes were longitudinal and rotatory, there was no reason for the pole-fleeing force to act, and no reason for the continental shifts to cause a new pole position.

In retrospect, it is not clear how Wegener could have announced this major change in his theory. After all, the presentation of the "climates of the past" was to proceed by

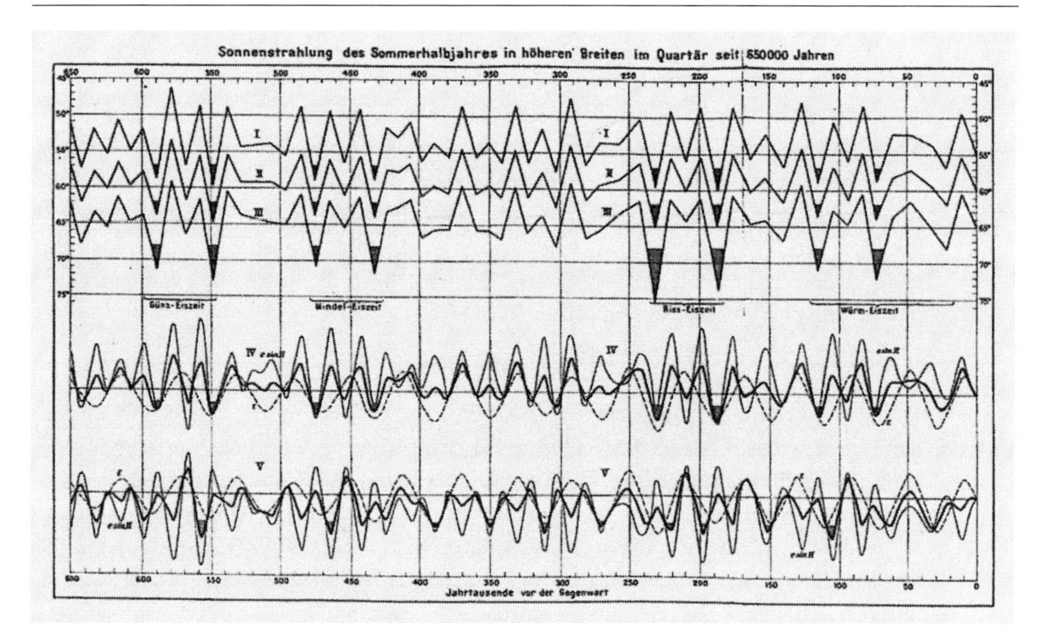

Köppen and Wegener's graphical rendering of Milankovich's calculations of the amount of sunlight reaching Earth in summer in high latitudes. Milankovich calculated the amounts for latitudes 55°, 60°, and 65° north at 10,000-year intervals from 650,000 BP to the present. These curves are shown as I, II, and III at the top of the map. Curves IV and V show (once again), in the dark line, the approximate value of solar radiation at 65° in the Northern and Southern Hemispheres, respectively. This line is superimposed on curves for the ecliptic angle (dashed line) and the eccentricity of the orbit times the longitude at perihelion (lighter solid line). The diagram aims to show that the last four glacial maxima coincide with the last four minima of summer sunshine, based on the reinforcing effect of the astronomical parameters, visible in the diagram. From Köppen and Wegener, *Die Klimate der geologischen Vorzeit*.

"assuming the correctness of Wegener's theory of continental displacements," and it is difficult to assume the correctness of a theory if that theory is undergoing major modifications at the time you are assuming its correctness. This is not impossible to do, but it is an unlikely premise for such an undertaking. Moreover, this relocation of the time of separation of the northern continents back into the Tertiary really did unseat the original areas of agreement between Wegener and Köppen in 1918. Köppen had found Wegener's theory of continents compelling not least because the clustering of Northern Hemisphere continents allowed a reconstruction of the southern borders (the terminal moraines) of the Quaternary ice sheets as a continuous line around a single northern paleocontinent. In Wegener's new construction of continental displacements, there had been considerable dispersal of the various Northern Hemisphere continental fragments before the ice age even began.

Wegener probably made these changes in the spring or summer of 1923. He would not by then have been finished with his mapping of Earth history, but Milankovich had completed his calculations in the late winter and early spring of 1923, and Wegener, Köppen, and Milankovich would have been working out the best way to graph the results at about this time. Milankovich's results were breathtaking when mapped against the geological record of advance and retreat of the ice sheet prepared by Albrecht Penck and Eduard Brückner (1862–1927)—the correspondences between ice maxima and "astronomically forced" solar minima were nearly perfect.

Wegener, ready as usual to surrender a theoretical idea in the face of convincing data, was willing to allow that during the Quaternary, for some reason or another, oscillations of the pole had been damped out over a period of a half-million years. In his larger scheme, this was no problem, because the earlier great excursions of the pole in the Tertiary were, he surmised, a response to the "flight from the poles" of continental fragments which had caused the great equatorial compression producing the Alps, Himalayas, and other mountains. His theory was flexible enough to encompass this change. As Argand had said, Wegener's theory reflected "the protean resistance of a plastic universe" to any general refutation.

The Call to Graz

In the winter of 1923, Heinrich von Ficker made good his promise of the previous autumn and resigned his professorship in Graz in order to accept the professorship in Berlin as Hellmann's succesor. The Graz faculty commission met on 9 March 1923 and recommended that Wegener be listed first as Ficker's successor. Their report speaks of Wegener's versatility, his success in many fields, his comprehensive training in physics and astronomy, his astonishing industry (*erstaunlicher Fleiß*), and his aptitude and productivity in the realm of meteorology.[92] It was at this meeting that it was decided not to list Victor Conrad, because of the growing anti-Semitic feeling in Graz and the near certainty that when the list was announced for the "call," if Conrad's name were on it, student groups would rise up in public opposition.

There were complications with the appointment, however, on Wegener's side. There was some difficulty over whether the call would go out as an appointment for an ordinary (full) professor or an extraordinary (associate) professor at a lower salary, with a contractual promise that within a few years the professorship would be converted via promotion to a full professorship. Wegener seems to have balked at this, though there's some confusion about the issue. Helmut Flügel, the leading authority on Wegener's tenure in Graz, says that Wegener actually wanted the call as extraordinary professor so that he could avoid administrative work. This seems improbable, in light of Wegener's letter in July 1923 to the Education Ministry in Vienna, in which he said he had serious reservations about accepting a call at that rank, "since I already occupy such a position at the University of Hamburg." Moreover, the salary offered was less than his salary as head of the meteorological division at the German Marine Observatory.[93]

The negotiations dragged on for the rest of 1923, while Wegener worked on the climate book with Köppen—when he could find the time away from his punishing schedule at the observatory and his teaching at the university. In the winter semester 1922/1923 he had given a course on the thermodynamics of the atmosphere, and in the summer semester 1923 a course on introduction to meteorology. These were offered in anticipation of the call to Graz and his need to have lectures in hand that were up to date. The difficulty of the negotiations with Graz is perhaps expressed in his turn away from introductory meteorology for the semester beginning in fall 1923 at Hamburg, with a return to a course in climatology. Wegener would not go to Graz except on his own terms, and things looked rather dark in the fall of 1923. If he were not going to Graz, he could at least support his climate research with his teaching at Hamburg.[94]

Wegener insisted that he be appointed as a full professor and that his salary be adjusted to reflect that he had already been (at Marburg, Dorpat, and Hamburg) an extraordinary professor for six years, and he insisted that for purposes of calculating his pension all his years of government service be counted back to his first year as an assistant at the Lindenberg Observatory eighteen years before. Else thought he was crazy to insist on this, and Felix Exner and Eduard Brückner in Vienna pleaded with him to back off on his demands, since his rank and salary would automatically rise within a year or two, saying that if he continued to be so "stiff-necked" about this he would ruin everything (*die ganze Sache aufs Spiel setze*).[95]

Negotiations between Wegener and the Austrian Education Ministry in Vienna had dragged on for nearly a year when in March 1924, to everyone's surprise, the ministry gave into all of Wegener's demands: a full professorship, a higher salary, and a pension benefit calculated from the time of Wegener's assistantship in Lindenberg in 1906.[96] The matter of the pension was of greater psychological import than even the salary for Wegener. He could remember that in 1906, when he had decided to go to Greenland with Mylius-Erichsen, his father had railed at him that instead of flying around in balloons and running off to the North Pole, he should be finding "pensioned employment," marrying, settling down, and starting his career. The other parts had come first: he had married, he had pursued his career relentlessly, and now he would finally be settled down, a full professor, pensioned at the same level as if he had taken his father's advice in 1906.

Alfred and Else were elated; the appointment would begin almost immediately, on 1 April 1924. Alfred would travel to Berlin to deliver his resignation from the Naval Observatory, and he wrote ahead to Ficker to tell him that he wanted to stop by and thank him in person. "This is the fulfillment of a wish I have had for as long as I can remember. If the education ministry knew how much I wanted the job in Graz, they wouldn't have given me even 12 years worth of pension benefit."[97]

Else accompanied him to Berlin and on the trip south to Graz. When they arrived in Graz, Else was stunned. Spring was so much farther ahead than it had been in Hamburg. The city was full of trees, the trees were full of birds, and there were parks everywhere. They climbed the clock tower and looked out over the city, and everywhere they looked they saw gardens. The Karl-Franzens University campus, unlike its brash, makeshift counterpart in Hamburg in the early 1920s, was architecturally unified, graceful, peaceful, and calm.[98]

They began immediately to look for a place to live, in the residential neighborhoods closest to the university. After years of train and streetcar travel, Alfred was determined to walk to work, as he had in Marburg. They hoped to buy a detached house but had no luck finding one in the first few days. Most of the owners were interested in a sale that contained a "swap" in which most of the cost of purchase would be made up by title to another house; there was considerable uncertainty about the stability of currency and a consequent conservatism among those who held real property.[99]

They needed a detached house with a lot of space rather than an apartment. During the protracted negotiations for the professorship, Alfred and Else had long discussions with her parents about what would happen if they should move. The Köppens had a large family, and the Hamburg house had long been a refuge for those having trouble, as well as a site for happy homecoming on holidays. But the unexpected death of one of Else's brothers and the decision of one of her sisters to move her family south to Bulgaria would leave the elder Köppens quite alone in Hamburg, if the Wegener family were to move away. They decided that if Alfred should win the professorship in Graz, they would all move together. Alfred and his father-in-law had a friendship and collaboration they were unwilling to sever, and having the grandparents around gave Alfred and Else a freedom of movement they could never achieve otherwise.

After a few days of looking, it was clear that nothing was available in April 1924, and Alfred moved into furnished rooms and began to write his lectures; to receive the salary of a professor beginning 1 April, he had to be in residence in Graz, though he thought he would not begin his full program of teaching until the fall of 1924. There was nothing to do but for Else to return to Hamburg without him and to set about finding a buyer for the house in the Violastraße. Summer was not too far away, and they would soon be back together. They were used to such separations and had spent as many years apart as together since their marriage. Graz looked like a way to bring that to an end, and Else returned to Hamburg full of hope for the future. Professor Wegener went to his rented rooms, spread his papers on his desk, lit a cigar, and began to write.

18

The Professor

GRAZ, 1924–1928

> [Wegener] reached his conclusions principally by instinctive, inner intuition and never, or very rarely, through a formal deduction from a formula, although he was certainly able to do that as well with great ease. I was often amazed at the reliability of his judgment, especially when he considered physical questions that lay far outside his particular area of expertise. He used to say, after a rather long pause lost in thought, "Yes, I think that's the case" [*Ja, ich glaube, die Sache ist so*], and most of the time it was so, even though it often took us many days afterward to find the evidence for it. He had an instinct for the truth that was rarely off the mark.
>
> HANS BENNDORF (1931)

After Else departed for Hamburg, Wegener turned his mind to his lectures and waited for the final proofs of *Klimate der geologischen Vorzeit* to arrive. He still had the notebook in which he had recorded material for the book, and after he and Köppen had sent off the manuscript, he cross-hatched in blue pencil page 152 of that notebook and wrote (also in blue) on the next page "*Notizen für die 4 Aufl d Entstehung d Kontinente u Ozeane.*" The blue pencil, in place of his normal graphite penciling, suggests that he was correcting proof at the time, which would put this in the late winter or early spring of 1924. While he (clearly) had planned a fourth edition immediately after the third—just as he had begun work on the third, even before the second was complete—Wegener would begin to collect data in this new division of the notebook only in the spring of 1927.[1]

In April 1924 he was poised once again between his career as an atmospheric physicist and his career as a theorist of continental displacements. Even the title of his professorship seemed to reflect this duality: "Meteorology and Geophysics." He was also, for the first time in almost a decade, in a pleasant and relaxed state of mind; except for a brief period of convalescent leave in 1915 and a few cold months in the winter of 1919, it had been nearly ten years since he had been able to work on his own science without administrative responsibilities and without supervising other men. Since 1914 he had been a combat infantry officer, an adjutant, an instructor in aerial navigation for zeppelin pilots, the head of a Field Weather Service outpost, the head of the Domestic Weather Service (*Hauptwetterwart für die Heimat*), and then head of the Field Weather Service for the Balkans.

After demobilization in the winter of 1919, he had enjoyed a few cold, intense, solitary months in Marburg, during which he had done fundamental work on lunar craters, the first original experimental/observational research since his traverse of Greenland in 1912–1913. This short sabbatical had been swallowed immediately, in April 1919, by five years of burdensome administrative and supervisory work at the German Marine Observatory. Almost all of the research and writing on continental displacements and the climates of the past he had accomplished in his "spare time" on evenings and weekends, on a schedule so grueling he swore that it was going to kill him.

Now, alone in Graz, he had time for reflection. His professorship was *in personam*, meaning that he was exempt from all the administrative tasks that would normally come with such a professorship. He need not take a turn as a dean or department chair, or take any role at all in university governance, including searching for new professors, overseeing promotions, approving and modifying degree programs, or serving on any of the astonishingly numerous committees that universities generate everywhere and in all periods of time.

Wegener had thought he would not begin his lectures until the fall, but shortly after he arrived at the Physical Institute, Hans Benndorf (1870–1953) approached him with a request that he begin lectures immediately in meteorology, as there was a pressing need, and Ficker had been gone almost a year. Benndorf was an experimental physicist and an expert on atmospheric electricity. He had been very anxious to bring Wegener to Graz, because the layering of the atmosphere, auroras, and the formation of precipitation in cold clouds—all of which Wegener had worked on—were all phenomena that involved electricity, interpreted not just as the accumulation of charge as in clouds and water droplets, or diurnal variations in the electrical state of the atmosphere, but also as ionization in the upper atmosphere. While Wegener's approach was thermodynamic and mechanical (his work on turbulence), as well as optical and acoustical, these were a wonderful complement to Benndorf's work on the electromagnetic states and behaviors of the atmosphere.

Wegener agreed to start teaching at once, and he began a course entitled "Introduction to Meteorology," which he had just given at Hamburg in the summer semester of 1923; thus, most of the lectures were "ready to go."[2] The lecture hall for meteorology was situated in the Physical Institute, and as Benndorf was the institute director (and the dean), they saw each other daily. Benndorf made a point to find Wegener after his lectures, and they typically talked for a half an hour. Wegener's introspective and potentially solitary character was obvious to everyone who knew him, and Benndorf was determined to draw him out. Their conversation was not small talk or hallway gossip; it was physics.[3]

Benndorf was ten years Wegener's senior and one of the *Exnerei*, the students of Franz Serafin Exner (1849–1926), and thus part of the dynastic tradition of the Exner family in Austrian intellectual life. Exner was a physicist and, until 1925, rector of the University of Vienna (he was also Felix Exner's uncle). In the 1920s and 1930s his students held most of the chairs of physics in Austrian universities. These included, in addition to Benndorf, Viktor Hess (1883–1964), Benndorf's closest colleague at Graz and the discoverer of cosmic rays (for which he would later win the Nobel Prize), and Marian Smoluchowski (1872–1917), famous for his work on Brownian motion. Erwin Schrödinger (1887–1961), just then achieving world fame as a pioneer in quantum physics, had begun his career as Exner's laboratory assistant.[4]

In replacing Ficker, Benndorf had been keen to find someone as broadly interested and confident in meteorology, and Wegener provided that and more. While Ficker was an excellent meteorologist, Wegener was truly both a meteorologist and a geophysicist and therefore a welcome complement to the physics faculty in Graz. In addition to Benndorf and Hess, this group included Michael Radakovic (1866–1934), a theoretical physicist known for his work on ballistics, and Karl Hillebrand (1861–1934), an observational astronomer.

Benndorf got Wegener to begin attending the *Institutstee* (institute tea), an afternoon tradition at Graz where the physics faculty and the graduate students (and even some undergraduates) would meet for tea (and tobacco) and present their work to one another, especially important for sharing problems "in real time," which advanced the work of everyone involved and gave the students a sense of how science progressed. It was also a social occasion, but work and relaxation mixed well at Graz.

Wegener was immediately popular with both the faculty and the students. He had always been solicitous of students and had a modest and informal manner. Benndorf was delighted to find that he also had a wonderful sense of humor and in fact was rather mischievous and fond of a good joke. He was popular with students not least because of his reputation and prowess as a polar explorer and as an Alpine climber, hiker, and excellent skier. Graz, close up against the mountains, attracted many students who came for these outdoor amenities, and they were excited to find a professor who was a robust outdoorsman and shared their enthusiasms as well as their professional interests.[5]

At these *Institutsteen* Benndorf had many opportunities to observe Wegener's manner of thinking, and he remarked that

> he reached his conclusions principally by instinctive, inner intuition and never, or very rarely, through a formal deduction from a formula, although he was certainly able to do that as well with great ease. I was often amazed at the reliability of his judgment, especially when he considered physical questions that lay far outside his particular area of expertise. He used to say, after a rather long pause lost in thought, "Yes, I think that's the case" [*Ja, ich glaube, die Sache ist so*], and most of the time it was so, even though it often took us many days afterward to find the evidence for it. He had an instinct for the truth that was rarely off the mark.[6]

By the time Wegener gave his inaugural lecture, on 10 May 1924, he was already integrated into the scientific community at Graz and a regular participant in the work of the Physical Institute. His interactions were mostly with other physicists and the geographers; things were so configured at Graz that meteorology, geography, and physics were in one part of the university and geology in quite another. Nevertheless, Wegener took the opportunity to try to draw in both of these audiences with a lecture entitled "The Theory of Continental Displacement, and Its Significance for the Exact and Systematic Geo-sciences."[7] This was presumably the same lecture that he had given under nearly the same title (in Danish) in Copenhagen the previous spring. It was a reprise of his premise that a complete theory of Earth required evidence from geology, paleontology, and biology on the one hand (systematic sciences) and from geophysics and geodesy on the other (exact sciences)—and that compartmentalization in science stood in the way of comprehensive understanding.[8]

While he was now a member of the Graz scientific community, he still had no permanent residence. Alfred and Else had agreed that during his initial term in Graz he would use his spare time to look for a place for them to live and would canvass the residential neighborhoods near the university. He looked without success, and he wrote to her that between all the talks and lectures he had to give, there was too little time. Since we know he was someone who could become so absorbed in his work that he would forget to eat to the point where he became ill, it is perhaps not surprising that he did not regularly jump up from his writing desk, put on his coat and hat, and begin chasing down real estate leads. He told Else that he would have to try again in the early fall.[9]

Summer 1924

In late June Alfred traveled home to Hamburg, carrying the suitcase and briefcase that had contained his worldly goods for the past two and half months. He and Else had decided to begin the summer vacation in Zechlinerhütte. Hilde was now almost ten; Käte, five; and Lotte, four. The younger girls barely knew their aunt Tony, nor had they any memory of Alfred's childhood home. It is a common and a sweet thing to want to take your children back to your own childhood, and that was Alfred's mission in this summer before their move to Austria.

On the train trip to Zechlinerhütte (in 1924 easily accessible by rail, in contrast to the 1890s) Alfred and Else discussed their future. She remembers that he was emphatic that he could no longer live in rented rooms and that he wanted her to come to Graz with the children and help find them a home. Whether or not they were able to sell the Hamburg house, they should pack up everything, have it shipped south, and hope for the best. After the move from Marburg to Hamburg, Wegener had never gone back, and he felt that way about Hamburg now; he wanted a fresh start.

They decided to take another sailing trip, of the kind they had taken Alfred's mother on in 1919, shortly before her death. They had no large sailboat now with a cabin, nor any money to rent one, and when they arrived at Zechlinerhütte, while the children played in the yard and orchard and ran about in the huge, open loft upstairs, Alfred and Else set about to patch and paint the open boat he and Kurt had rowed and sailed in as children.

There would be no hotel and restaurant stops on this trip; it was an expedition and a camping trip, with tent, blankets, and an alcohol stove that they had to unpack and set up each evening on the shore. It was crowded, with three adults (Aunt Tony came along), three children, and a good deal of gear in a very small boat. It was difficult to find a quiet place to camp, as many of the holiday boaters had brought along their gramophones and played music far into the night. That was a minor inconvenience; Alfred was in a good humor and had the children help him each night around a fire, composing doggerel about their adventures during the day, poems of the kind his father had composed for the entertainment of his own children under the same circumstances nearly forty years before. Alfred knew the waters well, where the pretty places were, as well as the quiet ones, and they found the trip as refreshing as always.[10]

They stayed on a while at *die Hütte* and returned to Hamburg in early September. In their absence it had still not been possible to sell the house, even though they had lowered the asking price significantly. While the hyperinflation was over and Germany had revalued the mark, it was difficult to sell and move because of the confusion over whether outstanding mortgages should be calculated in the old currency or the new. This problem was being resolved in the fall of 1924 just as they were trying to sell the house in the Violastraße, but no one knew the outcome as yet.

Returning from *die Hütte*, Wegener found a stack of letters from Koch in Copenhagen, about their work in Greenland. It had taken Koch nine years to complete the cartographic results from the Danmark Expedition of 1906–1908; Wegener had been waiting patiently for the results, hoping that they could support his claims for the drift of Greenland. We have already seen that these results came in for much opposition and dispute in the Berlin symposium on Wegener's work in 1921.[11]

None of the scientific results of their 1912–1913 traverse of the Greenland ice cap had appeared by 1924, and it was already eleven years since their return. Both Wegener and Koch had made preliminary and informal publication of some of the results, Koch on the glaciology and Wegener on the meteorology. Koch had, of course, written his popular account and published it before the war, and Else Wegener's translation had appeared in 1919. During the war, nothing could happen on the actual Greenland results; Wegener was a German military officer, and Koch the head of the Danish Air Force. It was impossible that they should regularly communicate, let alone collaborate.

Wegener very much wanted the scientific work to come out. It was meager in terms of their hopes and plans, but it now seemed more substantial than it had in 1913. Though he could work on it only sporadically during his time in Hamburg, he eventually completed the laborious calculations to reduce the barometric altitude measurements of the ice cap traverse and his meteorological observations for the overwintering.[12] He had sent these to Koch in the spring of 1924, imagining that Koch would have been, in the interim, working up the glaciology. Koch was able to proof Wegener's introductory remarks and to send news of their old expedition companions, including Achton Friis and Fritz Johansen (1882–1957), the expedition's zoologist, but that was all.

It soon became evident that Koch had nothing complete from the 1912–1913 expedition; he had only his rough notes and various initial drafts. He had nothing for the glaciology at all, none of the work on the structure and layering of the ice or of their measurements of fracture and flow, or work on the engrossing question of the "blue bands" within the ice layers and their meaning. He had done a draft of the topography of Dronning Louise Land, but not of the numerical data, either of the temperature measurements in the boreholes which had taken them all winter to perform or of the excavations of the snow and firn layers in the middle of the ice cap during the 1913 traverse. The one thing he had completed was the reduction of the measurements of the astronomical position finding of their locations going across the ice (the geodesy), which he had attacked first as the essential foundation for all the other calculations.

The first hint of this came in two letters from Koch in July and August 1924, thanking Wegener for the complete manuscript of his "meteorological observations during the traverse" and offering to find an independent publisher for it in Copenhagen, if Wegener could provide a typescript. This was odd; all of this work was supposed to come out, as did the results of every Danish expedition, in *Meddelelser om Grønland.* Then Koch thanked Wegener for an invitation to come to Graz and confer in the fall, but he demurred. Another letter followed, with chatty material about expedition plans of former comrades, including their plans to go to West Greenland, and a request for Wegener to find (for them) a copy of Erich von Drygalski's expedition reports from 1891–1893 for the west central Greenland coast.[13]

Wegener soon learned that Koch was unable (increasingly) to work concentratedly on anything that involved mental effort and especially calculation. He would sit at his desk and begin to work, then become dizzy, and, with increasing frequency, would actually faint. No psychological diagnosis is required here; this form of "syncope" (fainting) is a diagnostic marker, especially in a man of his age and with his physical history, for a condition called "aortic stenosis" and its associated pathology "congestive heart failure." Sitting posture, combined with the mental energy required for the calculations and the restriction of blood flow through the aorta, would reduce the blood flow to Koch's brain and cause him to faint. No doubt he had tried many times to complete the work, but each time he did, it brought on this condition, certainly a negative reinforcement.

First Year in Graz, 1924–1925

Alfred returned alone to Graz in early September 1924, with the understanding that Else and the children would come in early October and find a house as soon as possible. Then they would move in and "camp" until the furniture arrived. He did not wait in Hamburg to accompany them because he had been invited to deliver a major address at a scientific congress to be held in Innsbruck at the end of September.

Alfred was looking forward to the Innsbruck congress, and on the train south from Hamburg he was reminded of how much he loved mountainous terrain. The topography of Germany and Austria can best be understood from the standpoint of the United States if the United States were rotated 90° counterclockwise. Then the flat coastal plain of the eastern United States would be in the north and the Rocky Mountains in the south. In between would be the heavily cultivated agricultural area of the American Midwest.

As his train approached Munich, the terrain became hilly, but continuing south the landscape changed quite rapidly and remarkably, from rolling hills, to mountain foothills, and then to precipitous Alpine valleys settled as high as the topography would allow—large houses of white stucco with broad sloping roofs. Kurt Wegener once remarked that Graz was a backwater and therefore ideal for scientific work. One sensed how out of the way it was long before one arrived. On reaching Salzburg, a few hours east of Munich, the train was "cut": the main part of the "Austria Express" headed south, and a very modest segment, the "Steiermark Express," headed for Graz, with more and more Alpine stops along the way and then, once out of the scenic country, becoming a "flier" speeding down the last slope to the Styrian plain.

Graz is an old place with narrow streets and a distinctly southern feel, with the buildings bewilderingly clustered and the direction of the streets fairly random in their curves, following the early settlement and topography. For Wegener, after the gray of Hamburg, coming here through the Alps was a transformation—moving to a celebrated vacation spot from an industrial center on a flat coastal plain. It all put him back in the expansive mood of the summer sailing trip and away from the cares of Hamburg.

For the Innsbruck meeting, Benndorf (and Ficker) had planned for Wegener to give a plenary lecture on his displacement theory and his climate work as a way to position him, as well as Graz, in the complex politics of Austrian academic life and to mark out their contrast with Vienna, where Carl Diener had just replaced Franz Exner as *Rektor* of the university.

Although it is rarely mentioned today, as a condition of accepting his professorship in Graz, Wegener had become an Austrian citizen in mid-1924, along with Else and the girls.[14] Walking around Innsbruck with his colleagues, he was reminded again and again why he liked Austria and the "Austrian character" (*österreichisches Wesen*).

He and Köppen had both hoped that at the end of the war—with the collapse of both empires—the German-speaking countries could be integrated, and Köppen also hoped that this would be a prelude to something like a United States of Europe. When, in Innsbruck, Wegener learned that Albert Defant (who had been second to him in the call to Graz) was the primary candidate to replace him at the German Naval Observatory, his mind turned to this issue again.

He wrote to Köppen on 14 September from Innsbruck and remarked that the Viennese culture had produced a remarkable transformation in Hungary, Romania, and

Poland and could also work wonders for northern Germany, where the "embarrassing compulsion for official ceremonialism" and the general stiffness were obstructive and dispiriting. Northern Germany in general, and the observatory in particular, would benefit from "the example of a light-minded newcomer, and they will not get that without him [Defant]."[15]

This Innsbruck conference gives a sense of how Wegener had left behind a large, highly officious, somewhat impersonal Prussian universe and entered exactly the kind of relaxed, reduced, and intimate sphere that he wrote about to Köppen. If this informality was not the case, we should probably know nothing about Wegener's attendance at this conference, let alone what he said. As it turns out, there was, at Graz, a young man named Anton Rella (1888–1945), a newly appointed professor of mathematics. "Tonio," as he was known, had been Erwin Schrödinger's best friend at the *Gymnasium* in Vienna, and later on he had become close friends with Milutin Milankovich. Rella wrote Milankovich an excited letter in the summer of 1924, telling him about Wegener's impending lecture and urging him to travel with him to Innsbruck to hear it. The Graz faculty was very excited about this event, and Milankovich decided to attend. Rella had told him that "the highlight of this conference is going to be Wegener's new theory of continental drift."[16]

It is because of Milankovich that we know about Wegener at Innsbruck. The two had never met. Milankovich had come home from a business trip in the late spring of 1924 and found the page proofs of Köppen and Wegener's book on his desk, with a note from Köppen asking for him to review them. He had done so and thought little more about it until he got the letter from Rella; he decided immediately to go to Innsbruck.

When Milankovich and Rella arrived in Innsbruck in September 1924, they went looking to meet friends and then spent the evening moving from restaurant to restaurant, trying especially to find Wegener. They found him in a beer garden, drinking with Heinrich Ficker and brother Kurt. Milankovich remembers that even though Wegener was then forty-four, he was "slim and fit with a gentle expression in his gray eyes and a slightly melancholic smile around his thin lips; he looked considerably younger."[17] Rella and Milankovich sat down, and by his account the five of them sat talking until late into the night, arranging to meet for dinner the next night at a wine cellar just out of town.

Wegener's lecture took place the next day in the main lecture hall of the newly refurbished university (which had been occupied by the Italians during the First World War and rather badly damaged). Much to Milankovich's surprise and delight, when the lights went down and Wegener began to lecture, it was not just about the theory of continental displacements but also about the application of that theory to the climates of the past.

Wegener spoke using projection slides of his maps for each period of the past beginning with the Carboniferous, and he methodically explained his use of coal, salt, and ice to map the migration of the latitude zones. He continued through each of the eras of Earth history until he arrived at the Quaternary, at which point he began to talk about the importance of Milankovich's work, "which we included in a short chapter in our book." Wegener gave a detailed discussion of Milankovich's work and its correspondence with the geological sequencing of the ice ages worked out by Penck and Brückner, as well as the fineness of the correspondence between Milankovich's mathematics and the geological details. Milankovich was thrilled. It was the first time he had ever heard his work discussed in public. He said that Wegener's lecture made a great impact on the assembled audience and "prompted a lively discussion, in which I did not participate, about his theory."[18]

In the discussion following the lecture, Schweydar, who had spoken so favorably of Wegener's work in Berlin in 1921, now raised strong objections to the extensive displacement of the poles required by Wegener's theory, in light of work done in the previous few years in geophysics suggesting that these extensive motions were unlikely and perhaps impossible. Wegener had responded that he must defer to Schweydar and the other geophysicists, as "my mathematical knowledge is insufficient to disarm known experts." However, he added, "the geological evidence supports the claims I have made here."[19]

A few days later, Milankovich wrote to Wegener, saying that the lecture had stimulated his interest in a whole range of questions and asking him how he might obtain a copy of Wegener's work on the origin of continents and oceans. He also solicited a copy of Wegener's lecture so that he might publish it both in French and in Serbian. Strange as it may seem, given Milankovich's collaboration on the book on climates of the past, he had never read Wegener's theory, learning the details only during the Innsbruck lecture. Milankovich also asked Wegener whether it was possible that there had been a secular decrease in the solar constant (i.e., the amount of radiation emitted by the Sun) as part of the change in Earth's past climates.[20]

Wegener responded warmly on 6 October, begging a few days' grace in sending him a copy of the lecture, as his family was to arrive in two days. Wegener went on to say, "I would especially welcome it, if at some time you tackled the question of polar movement, even if a straightforward solution should not be reached. It would be very important if it were established at some time what, up till now, theoretical science has to say about this problem."[21]

On the question of the solar constant, Wegener thought that the presence of an ice age in the Archaean (Algonkian), along with the fact that "in the Mesozoic era, the climate of the then south polar region was milder than the present polar climate; all traces of inland ice are missing from this long period," argued against a long-term decrease in solar radiation. Moreover, Wegener said, "It often strikes me that both main periods, with great amounts of glaciation at the poles (Permocarboniferous and Pliocene-Quaternary) coincided with periods of intense mountain formation and rapid polar movement. I have a feeling that these things may have a common cause."[22]

Wegener needed geophysical and astronomical help. It was a curious reversal, when we consider that in 1912 and 1915, and even in 1920, the principal thrust of his theory of the origin of continents and oceans was to sort out candidate geological theories of the continents based on the adequacy of their geophysical foundation. Now, as the author of a thoroughly geological treatise on the climates of the geological past, he found his geophysical assumptions under stress. In 1921, Schweydar, Epstein, and Lambert had separately challenged the adequacy of the "pole-fleeing" force to raise mountains. Now, in mid-1924, Schweydar was questioning whether the poles could move at all in the way that Wegener needed them to move. Wegener had already been forced to reduce drastically the amount of pole shift in the Quaternary to accord with Milankovich's astronomical parameters. It was a serious quandary, as Milankovich's pole shifts were astronomical and absolute, and Wegener was now no longer sure what he meant by "pole-shift."

He now had several options to consider. How much had the poles actually shifted, and how much of the shift of latitude zones was a matter of continental displacement alone? The decision to measure displacements against an Africa fixed in place was a necessary heuristic device, but it seriously confused the issue of the absolute amount of continental motion, relative to a given pole position, and the absolute amount of displacement of the pole of rotation relative to the fixed stars. He still remained committed to the "verdict of the exact sciences" but was no longer certain what that verdict was.

Else arrived two days later with the children. She set about finding a house the very next day, 9 October 1924. Within a few days she had located a "nice house" in what she described as "the suburbs"—which did not mean much in a place as small as Graz, as it was a leisurely twenty-minute walk to the University Physical Institute and close to a primary school. Her efficiency in these matters (in comparison to Wegener's lack thereof) is sometimes startling. The house she chose was on a gently sloping and curving street, the Blumengaße; the house was No. 9. It was in a series of what we would now call "townhouses," sharing a continuous facade along the block, and built in 1910; No. 9 was on a corner lot and had windows on three sides. It had a high-pitched roof, a back garden, and a "daylight basement," so it was actually four stories tall.[23]

Finding enough money to buy it was a problem, as the Hamburg house had not yet sold, but Benndorf stepped in and informed Wegener that under such circumstances the Ministry for Instruction could provide an advance on his salary sufficient to purchase the house. Wegener traveled immediately to Vienna, and after two days of negotiations, he returned with enough money to buy the house. They moved in immediately, even before the furniture arrived, but by early November the house was, as Else said, "halfway livable" and ready to receive her parents.[24]

When Wladimir and Marie Köppen arrived in mid-November, the family, again united, seems to have made the transition to life in Graz rapidly and easily. Köppen had left all of his official and scientific correspondence, dating back well into the nineteenth century, to the Prussian State Library in Berlin, and they were happy to have it. He brought his extensive library to Graz, but he saw immediately that space in their new house would be at a premium; he therefore donated most of it, about 3,000 volumes in thirty-four large crates, to the Meteorological Library in the Physical Institute, which he had learned was rather limited in resources. In return, the university named him an honorary member of the faculty. After all the years he had lived in a distant suburb of Hamburg, he was delighted to be only twenty minutes or so away from the university and made the walk almost every day of the week. He also enjoyed the vegetation, which he said reminded him of the Crimea of his youth.[25]

The family quickly settled into a comfortable routine. Other than a shortage of money the first year because of the salary advance to buy the house, Else said, "Without any extraneous worries we could now set up our lives the way we liked. Alfred and I each had our own spheres of activity that left us completely fulfilled, and we were soon quite at home."[26]

One must remember that from Else's standpoint their lives had been one continuous emergency since his return from Greenland in 1913: Alfred's mother's stroke, the birth of their first child, Alfred twice wounded and once nearly dead of dysentery, with him being away from home for almost four full years. Then there was the death of Alfred's father, the death of one of her brothers, the arrival of two additional children, the death of Alfred's mother, and the very difficult years 1919–1924 in Hamburg, where, toward the end, they were nearly penniless and growing their own food in vacant lots while running a huge household always short of fuel, and with none of the domestic servants one might normally have associated with people of their social standing and professional rank.

For Else, Graz meant stability, adequate means, frequent and pleasurable vacations, a good home for the children, and a chance to see more of her husband, although he was never home as much as she wished. She remarked sardonically, concerning the roster of weekend hikes, handball games, cross-country ski tours, and Alpine vacations, "All of these undertakings were only the background music for my husband's real passion: the advancement of his scientific work."[27]

Wegener was available for some but not all of these undertakings. Else said that she often had to use all her powers of persuasion to pry him away from his writing desk, even on a Sunday morning. Typically on Sundays in good weather the Wegeners would get together with other families, for a stroll along a stream or to a park, often with a game of handball during the noontime picnic break. Team handball was a sport that Germany and Austria were mad about in the twenties and thirties, and its rules were just being codified. It was also playing a significant international role, as the reentry of Germany into international society and cooperation began with sporting contests, notably a team handball match between Germany and Belgium in 1925, an unspoken beginning to a difficult but necessary reconciliation. Wegener laughed at these games and hikes, calling them "*Tieftouren*," an untranslatable play on words which could mean either "hiking a stream" or "whirlwind tour," meaning that they were not proper mountain hikes and that they were too short to be interesting. Generally, he went along, if only "because he needed the fresh air and the exercise."[28]

Wegener's work rhythm in Graz was to pursue a project relentlessly, day and night until it was complete, or until he could make no further progress. When he finished, and for him finishing meant a manuscript of some sort to be sent to a journal, it was time to get away. Else said she could always count on getting a few days with him in the mountains every time he finished a big project. He felt, and she knew, that these trips were absolutely essential to clear his mind to get ready for new work. Moreover, in the Graz years, the family took regular summer vacations in the mountains, sometimes lasting several weeks, and she said that for him these vacations were *real* vacations, and that he never took scientific work along with him.

In the fall of 1924 Benndorf, who had been very impressed by Wegener's presentation in Innsbruck, asked him if he would consent to repeat it at greater length, in Graz, having time to dwell on both his theory of displacements and his work on climates of the past. Wegener agreed, delivering these lectures on 21 and 28 November 1924, followed by a full evening of discussion on 29 November. These were sponsored by the "Physics Section" of the Steiermark Scientific Association, and invitations were also sent to the members of the geology, zoology, botany, and geography sections, so it was in essence a "plenary session," for the entire scientific community.

Benndorf, already astonished by Wegener's intuition in problem solving at the *Institutstee*, was also impressed by Wegener's mode of presentation during these lectures. "He began without any introduction," Benndorf remembered, "with plain and simple words, almost sober and dry, and in the beginning, somewhat halting."[29] When he began, however, with systematic clarity to lay out the arguments first from geophysics and then from geology, paleontology, biology, and paleoclimatology, "he became more and more animated, his eyes flashing, and his hearers were carried away by the beauty, grandeur, and boldness of the intellectual structure he had created. Never have I been more aware of how inessential rhetoric is for the effect of a talk when its object has such a clear significance."[30]

Benndorf was even more impressed during the evening of discussion on the twenty-ninth. A number of objections were raised to Wegener's theory, and many of them, said Benndorf, "to my way of thinking focused on inessential points." Wegener answered every one of them without a trace of irritation, clearly, with careful deliberation. One felt, Benndorf said, the full weight of his mastery of the enormous range of scientific material he had brought together in the service of his theory.[31]

Though Wegener continued to follow developments with his hypothesis and to accept invitations to speak about his work on continents and oceans, especially on the climates of the past, the move to Graz coincided with a major shift in his scientific work. He had been completely consumed between 1919 and 1924 in developing and working out all the consequences of his theory of displacements. With the publication of *Klimate der geologischen Vorzeit,* he felt he could turn to other interests, long neglected.

He told Else when they moved to Graz that he would now have a chance to lay the foundations for the complete physics of the atmosphere he had wanted to produce as early as 1911. There was really nothing more for him to do with continental displacements in 1924; the book was being widely read and translated. *Klimate der geologischen Vorzeit* had not yet had time to make an impression. Irmscher had written to Wegener from Hamburg in early November 1924 to tell him that Gothan (the paleobotanist who had updated Potonié) had just published a book in which he had taken a position against Wegener, but without having seen the Köppen and Wegener climate book.[32] On the other side of the ledger, a Danish expedition to West Greenland in 1922 to update the geodetic measurements had determined a westward drift of Greenland corresponding to an annual westward displacement of more than 20 meters (66 feet) per year since 1873, for a total of 980 meters (3,215 feet), which corresponded very closely to Koch's East Greenland measurements.[33] Moreover, the International Geodetic and Geophysical Union Congress that had just met in Madrid in October 1924 had announced a plan to coordinate radio time signals in a network of circumglobal stations in order to unify geodetic and astronomical observations. This plan was publicized by some participants at the congress, but not by the congress itself, as a means to try to measure continental displacements using radio time signals.[34]

Owing to the combination of his teaching of meteorology at the University of Hamburg and his growing celebrity, the initial printing of his *Thermodynamik der Atmosphäre* (1911) had finally sold out, and his publisher wanted to issue a new printing. He consented, though it reminded him that that book was to have been only the first part of a general physics of the atmosphere which would also include mechanics, optics, acoustics, radiation, and atmospheric electricity. To this project he could now turn his attention.

It may come as something of a surprise that the author of a major theory of Earth's origin, physical constitution, and evolution, as well as an allied theory of global climate change throughout the whole span of Earth history, should suddenly cease to work on that problem and go off to do something quite different. Actually, for some scientists, this is not so unusual. Students were always arriving in Göttingen to study some problem with the mathematician David Hilbert (1862–1943), only to discover that he was no longer working in that area of mathematics and had moved along to something else. Wegener would continue to monitor the fate of his working hypothesis—his theory—but for the present he was no longer actively involved with its development.

He elected to teach, in the fall and winter of 1924/1925, an extensive elaboration of his *Optik der Atmosphäre*. Perhaps an encouragement to this work was that Felix

Exner had brought out a second edition of his 900-page treatise on meteorological optics in 1922. This book was known as "Pernter–Exner" because Exner had completed (in 1912) the manuscript left behind by Josef M. Pernter (1848–1908). It was a classic in the field and still in print, but Wegener thought that it needed work, especially in the area of mirages, his own specialty, and in the phenomenon of "halos," to which he had given a great deal of attention and on which he had published briefly during the war. In any case, Pernter–Exner was a gigantic reference work, and what was needed was a compact treatment that could go into a single-volume textbook on atmospheric physics.

Wegener had published an extensive mathematical treatment of mirages in 1918 and wanted to extend this kind of treatment to the phenomenon of halos. These "halos" are white or colored rings around the Sun or Moon, 22° away from the center of the object, and are connected with other associated but more rare phenomena: tangent arcs, sun pillars, and sun dogs (parhelia), all of which are complicated series of reflections and refractions of incident sunlight (or moonlight), dependent on the geometry of ice crystals in cirrus clouds in the very high atmosphere. The mathematics involved is not calculus but trigonometry, and the best way to present the results simply and clearly is not obvious. The basic theory was available in Pernter–Exner, but they were wrong about the geometry of ice crystals involved in some of these phenomena, and Wegener had worked it out correctly. Wegener deduced that a very rare arc, touching the 22° upper tangent arc above the Sun, was the result of a rare orientation of hexagonal, pencil-shaped ice crystals where the light refracted by a prism face was internally reflected by the prism base and then refracted again out of another face. These phenomena are known today as "Wegener arcs."[35]

The success of his work on halos was one of the first fruits of the *Institutstee* that Wegener regularly attended. He would produce the trigonometric formulas he had chosen to represent the phenomena and then show them to the assembled group. He got quite significant help from Radakovic, but the interest in and the grasp of the problems by the "assistants"—physicists on staff without professorial appointments—and even the graduate students, was a source of great surprise and pleasure to him, accustomed as he was to struggling alone with such problems. He made great progress with his theory of halos in the fall of 1924; the resulting work is still actively cited.[36]

His success in the work on optics, in which he felt supported and secure, encouraged him to pursue his more extensive plans for an atmospheric physics. The other problem that he had tackled during and immediately after the war, besides the high-altitude reflection and refraction of light by the ice crystals in cirrus clouds, was a matter of atmospheric acoustics: the problem of so-called outer zones of audibility, also known as "shadow zones." The phenomenon was well documented but poorly understood. Why was it that very loud noises—large explosions, meteor detonations, gunfire—could be heard a certain distance away from the source, then not heard for some tens of kilometers, and then heard again by others farther away? The concentric character of the zones was often complex, not a matter of a single incident zone, a shadow, and a reflected zone, but of alternate rings of silence and of transmitted sound.

Here again Wegener found support, although the problem was much more difficult than the question of the halos—even if that was difficult enough. For the halos, once one understood the geometry of the ice crystals (known from physical inspection) and understood the laws of reflection and refraction, the solution was foreordained if one had sufficient geometric imagination, persistence, and help. Pernter and Exner had gotten most of the answer correct, but not all of it. Wegener had finished the treatment of the topic, and the support of his group at Graz had been essential.

For the acoustical problem of the outer zone of audibility, the phenomenon was well documented, but the cause was much more difficult to discover. Something happened in the lower stratosphere, a phenomenon of the reflection and refraction of sound waves, but since pressure and density and temperature are all related, which of the three phenomena was causing this intermittent audibility? Was it the density of the atmosphere or its temperature? Was it a matter of the wind field and the shift from the turbulent troposphere to the laminar flow of the stratosphere? Here Wegener benefited again from his colleagues at the *Institutstee*. Wegener presented the problem several times, and the consensus answer was that, for the time being, no definitive solution was possible.

There was still some dispute about whether the reflection and refraction of sound, whatever the cause, happened at the boundary of the stratosphere (the tropopause) or within it. Wegener undertook a very clever study in which he showed that the distance from an explosion to the *äußere Hörbarkeitszone* (outer zone of audibility, as it was known) was different at different seasons of the year. Back in 1910 he had demonstrated that the stratosphere boundary was lower in winter and higher in summer. He demonstrated the same thing here using the same technique, suggesting that, indeed, whatever the cause (density or temperature or wind), the location of the deflecting surface was the tropopause.[37]

Thinking about temperature led him also to produce an interesting paper on the temperature of the atmosphere in the area beyond the stratosphere, what we would now call the mesosphere, and which Wegener took to be the boundary between the nitrogen atmosphere and the hydrogen atmosphere; it was also the level at which "noctilucent clouds" appeared, a phenomenon that had been studied intensively since Krakatoa. He extrapolated temperature curves from the decay of the temperature inversion within the stratosphere to argue that the temperature at 80 kilometers (50 miles) would be between −100°C (−148°F) and −110°C (−166°F), values that correspond to current estimates.[38]

Happy to be at home in his original career as an atmospheric physicist, a more expansive sense of having found a home persisted through the Christmas vacation of 1924, with concerts and parties with university colleagues. Alfred and Else found these events sufficiently jovial and informal to overcome Alfred's innate resistance to social life in almost any form beyond the sphere of family, close friends, and the exchange of work and ideas with colleagues.

Back at work in early 1925, Wegener received a request from Eduard Brückner (1862–1927), professor of geography in Vienna, to come and give "his lecture" on the climates of the past to the Geographical Institute in Vienna. Brückner had coauthored *Die Alpen im Eiszeitalter* with Penck in 1909, and before that he had been a colleague of Köppen at the old German Marine Observatory in the 1880s. He was extremely excited by what he saw to be the confirmation of his and Penck's dating of the ice ages through Milankovich's calculations, and he was willing to pay Wegener a handsome honorarium to come. These sorts of invitations were becoming frequent; some could be refused, but this one could not.[39]

The invitation is notable more for political reasons than for scientific; Wegener had made a splash at Innsbruck the previous September, and his lecture in Vienna in March apparently went off well, thus strengthening the geographers in Vienna against the geologists. The Innsbruck meeting had been a plenary congress of "naturalists," and the lecture in Vienna was before geographers; Wegener had yet to come into contact directly with Austria's leading geologists.

Finally, on 15 May 1925 the Geological Society (Vienna) held a symposium, of sorts, on Wegener's displacement theory. Otto Ampferer spoke on the subject of Wegener's displacement theory. The president of the society, Fritz Kerner (von Marilaun) (1866–1944), then spoke on the subject of Wegener and Köppen's book on the climates of the past. Wegener was quite conspicuously not invited to this session.

Wegener had depended since 1912 on Ampferer's (1906) idea of flow or "under-streaming" in a fluid (if extraordinarily viscous) subcrust. Where Ampferer had imagined such flowing currents acting locally as the mechanism for mountain formation, Wegener had taken this idea as hypothetical support for large-scale streaming of the Sima and as a possible mechanism for continental displacement. He had stressed this especially in the 1920 edition of *Die Entstehung der Kontinente und Ozeane*. In 1922 he had used such "streaming" again as a hypothesis subsidiary to *Polflucht* and *Westwanderung* to explain the displacement of India to the northeast (i.e., against the rotation of the globe), where it had collided with the Eurasian landmass.

The Vienna session of 15 May 1925 was attended, among others, by Felix Exner, Carl Diener (who had just become the *Rektor* of the University of Vienna), Lukas Waagen, Eduard Brückner, and Anton Handlirsch (1865–1935), all of whom spoke after the papers. Brückner, Exner, and Waagen gave support; Diener and Handlirsch (a famous insect paleontologist) spoke in opposition. Waagen added a manuscript to the reports of the session, which included his reflections on continental displacement as "an idea which shows up in minor publications by senior Austrian scientists" from the previous generation, and he added some cautiously supportive comments about Wegener's geophysics.[40]

Fritz von Kerner (1866–1944), the organizer, was a paleoclimatologist and son of the famous Austrian botanist Anton Kerner von Marilaun (1831–1898)—professorship in Austria was very much "going into the family business." Kerner had long been an advocate of temperature differences in different periods of Earth based on the distribution of land and water and had argued in 1895 that the Southern Hemisphere must have been warmer than the Northern Hemisphere in the Jurassic, because there was more land in the Southern Hemisphere at that time. This sort of argumentation about climate differences based on differences in temperature, given the distribution of land and water, Köppen and Wegener rejected in 1921 and again in 1924 as an unlikely cause of climate change, since it was a matter of a very few degrees of the mean annual temperature and expressed only at rather high latitudes.

Kerner no doubt had taken it very amiss that Wegener in 1922 had used him as an example of the kind of "impossible hypotheses" offered to explain away the extent of the ice cover in the Southern Hemisphere in the Carboniferous and Permian. Kerner had proposed that the widely distributed traces of ice were a question of local anomalies of the distribution of heat caused by cold currents in the ocean and "such like phenomena."[41]

Kerner did not seek to publish his paper from the symposium in May, but Ampferer's paper appeared in *Die Naturwissenschaften* in July 1925. In this paper, Ampferer made an attempt at a reconciliation of his earlier ideas and Wegener's continental displacements by postulating a global Earth convection, rising at the poles and sinking at the equator, which could drive the continents away from the poles and cause compressional mountain building at the equator. It is an interesting paper of the kind that makes people want to "find an idea ahead of its time," in this case a precursor of the mantle convection today believed by many to drive plate tectonics. Seen in context, it was an entirely speculative, hypothetical, qualitative extension of a very detailed mountain building theory he had offered and dropped eighteen years prior, unsupported by any new data or any quantitative argumentation.[42]

Ampferer's attempt to hitch his ideas to Wegener's was nevertheless significant. The well-worn expression "there is no such thing as bad ink" applies to Wegener's career at this time. Ampferer began his Vienna paper by remarking, as so many others had in 1923 and 1924, that Wegener's hypothesis was so well known that he need not review it to discuss it.

It is not nearly as important who (in the 1920s) supported Wegener and who opposed him as it is that Wegener's theory of continental displacements had thoroughly penetrated scientific discussions in geology, paleontology, climatology, geography, and geophysics in these few years. Some agreed, some claimed priority, some offered modifications; some opposed mildly, others violently. Even if Wegener was (as often as not) deemed to have "given the wrong answer," all but the most intransigent admitted that he was *asking the right question*. Wegener had succeeded in changing the question "Do continents move?" into the question "How, when, and where have they moved?" In the history of the sciences, many thousands of workers are busy at all times changing the answers; only a very few ever succeed, as Wegener did, in *changing the questions*.

Wegener's Atmospheric Physics, 1925–1926

At the end of May 1925, during the Whitsun (*Pfingsten*) vacation, Alfred and Else invited Felix Exner and his wife to stay with them for a few days in Graz. Theirs was a friendly but delicate relationship. They were working in the same field and had already reviewed each other's books in meteorology. Alfred wanted to talk to Felix Exner about his modifications to Exner's work in optics.

Exner had just finished the second edition of his book on dynamic meteorology (Wegener had reviewed the first edition in 1917) and wanted to be sure where Wegener stood in the debate between himself and Bjerknes, knowing that the Wegener and Bjerknes families were close. Wegener had made a point not to take sides in that dispute. He had made a mild criticism of Bjerknes's analogy between the wavelike disturbances of frontal boundaries and Helmholtz waves; on the other hand, he had studiously avoided taking sides between Exner and Bjerknes at the conference in Norway in 1919. Exner remained opposed to Bjerknes's idea of the origin of cyclonic storms as wavelike disturbances at frontal boundaries. Wegener's caution in not taking sides here had allowed Wegener and Exner to become friends. Recently, as a way to cement their personal goodwill, Exner had successfully nominated Wegener to be a member of the Academy of Sciences in Vienna, notably in the mathematical-physical section, not that for earth sciences.

Exner discovered, in the course of this friendly and productive visit, that the Wegeners had no plans to take a summer vacation that year because there was no extra money, given the purchase of the house. Exner immediately had a solution. There was a sanatorium on the summit of a mountain, the Stolzalpe (1,200 meters [3,937 feet]), in the Murau district of Styria. It was a tuberculosis sanatorium specializing in the treatment of children, following a treatment regimen in which it was believed that exposure to sunshine was an essential part of tuberculosis therapy.

Exner had been invited there a few years before to make measurements of sunshine, as the hospital wished to keep an accurate record of the intensity of their therapeutic sunlight; he and his family had stayed for free in a fire-damaged farmhouse on the sanatorium grounds. He suggested that Wegener and his family could stay there for free, if they could "rough it" with bedrolls and could cook on a Primus stove, bringing their own food. All that Wegener would have to do is take the sunlight measurements each day; the rest of their time would be their own, and the surroundings were beautiful. Of course, Else and the Wegener girls were experienced campers, boaters, and hikers, accustomed to sleeping in tents and on the ground. Alfred and Else accepted with alacrity; they would go at the end of the summer term.[43]

That summer Alfred offered an "Introduction to Oceanography" course. Its title in the catalog, *Einführung in die Meereskunde*, is an indication that it was already on the books, as the older word *Meereskunde* was being replaced everywhere by *Ozeanographie*.[44] Following the tradition of Graz, Alfred, as the newest member of the faculty, in spite of his exalted rank, had to teach the service courses needed in the curriculum in preference to his own work, at least in the summer semester. He undertook no new scientific investigations during this term, as he and Else had decided to repaint all the doors and window frames of their new home and to dismount the complexly filigreed shutters and paint them as well.

They spent most of August on the Stolzalpe, and while all the girls were vigorous hikers, both Alfred and Else were astonished by the tenacity of five-year-old Lotte, who could climb to the summit with her sisters, and even help cook lunch on the trail, and was also able to make the hike the full distance from the Stolzalpe to the summit of the Frauenalpe via the mountain road.

Alfred and Else stayed in Graz in early September only long enough to see the girls safely off to school in the care of their grandparents, and they immediately left again for an Alpine tour of their own, which was to culminate in yet another anniversary of yet another mountain observatory, the Zugspitze, on the southern border of Bavaria, to be held from 4 to 7 October 1925.

One mentions this otherwise unremarkable tour only because the weather turned wintry early, and they found one of the passes they had to cross deep in snow. Else said, "Despite the fact that the pass was deep in snow, Alfred, without hesitation, went forward and again and again unerringly to scrape the snow away from the trail markers while I, and a lanky Thuringer who had joined up with us, waited helplessly above."[45]

This reveals not so much what she knew about him but what she did not, in terms of what he had experienced in Greenland. Compared to the places he had been and what he had done, the things he did that gave her such pride and so surprised her—such as keeping food warm for a noon meal, or starting a stove on a cold morning, or finding the trail markers on a well-worn Alpine route in a foot or two of snow—were all child's play to someone who had walked 700 kilometers (435 miles) across Greenland at −30°C (−22°F).

When they arrived at the Zugspitze with Benndorf, Ficker, and Defant, they found good weather and yet another "Alpine Summit meeting." The meterologists had a sit-down meeting with Wegener and explained to him that in early July of that year the physicist Otto Lummer (1860–1925) had died suddenly. He was one of three general editors of the eleventh edition of *Müller-Pouillets Lehrbuch der Physik*, which was then undergoing a major revision. Lummer was not only one of the general editors but the editor for volume 5, part 1: *Physik der Erde*. That volume now had no lead editor. A few of the manuscripts had come in, but the volume was far behind schedule. The chapters were meant to be encyclopedia entries, many of them in excess of 100 pages.

Several of the authors were current or former members of the Graz faculty. Ficker was to do the main entry in meteorology, and Benndorf and Viktor Hess were to do the article on atmospheric electricity, a deceptively simple name for something that included Earth's electrical field, the ionization of the atmosphere, and cosmic radiation. Emil Wiechert (1861–1928), who had trained a generation of German geophysicists at the University of Göttingen, was to do geodesy and the figure of Earth; seismology was to be the task of one of his young students, Beno Gutenberg (1889–1960). The rest of the roster of authors was also impressive: Alfred Nippoldt (1874–1936) would write the article on terrestrial magnetism, and Gustav Angenheister (1878–1945), head of the geophysics division at Potsdam, would do the aurora and *Polarlicht*.

Benndorf and Ficker had discussed the matter with the other two general editors of the series, Arnold Eucken (1884–1950), a physical chemist at Göttingen, and Erich Waetzmann (1882–1938), and they agreed that Wegener would be an ideal choice to replace Lummer, not least because he had worked in almost every single field covered in the book. Benndorf and Ficker asked Wegener during the meeting whether he would be willing to take it on. The coverage could be expanded so that he could himself write the articles on atmospheric optics and atmospheric acoustics.[46]

This was an enormous undertaking, and whatever else Wegener decided to work on, it would likely be with him for years. On the one hand, it would slow down his plan to write a textbook on the physics of the atmosphere; on the other, it would put him in charge of a major reference book whose authors were some of the most celebrated scientists in Germany and Austria. Further, the editorial correspondence and a constant connection with the leading scientists in these fields could keep him up to date and solidify professional relationships somewhat endangered by the great distance of Graz from the rest of Germany. What Kurt had said about Graz being isolated and really good for some kinds of scientific work was true, but it made it very hard to get to libraries and to meetings.

Before the meeting had even ended, Wegener decided to take on the job, on the stipulation that he could widen the coverage a little bit to include the physics of glaciers and the physics of the oceans. These were topics that were of great interest to him and represented both the solid and liquid states of Earth's envelope of water; geophysics seemed incomplete to him unless Earth, its oceans, its ice cover, and its atmosphere could all be included. It had been thus in Rudzki's text in 1909 which had so influenced him, and it is difficult in any case to imagine that a polar scientist who had also written a theory of the origin of the oceans would want to edit a textbook on "physics of the Earth" which did not contain these topics. Moreover, he had just spent the period from April through July teaching an introduction to oceanography and was aware that this work needed to be brought up to date as well; the last general textbook of oceanography in Germany had been Krümmel's handbook from 1911. A formal letter soon came from Waetzmann, and Wegener accepted the terms, which included the expansion into glaciology and oceanography.

Benndorf later said that Wegener had not begun any fundamentally new work in Graz but had stuck with topics he had already worked on. He said that part of the fault was that "Wegener was not spared, in the 'handbook epidemic' that had broken out in Germany."[47] Benndorf here was perhaps expressing remorse at having gotten Wegener deeply into a monumental undertaking, in which Benndorf was already involved.

However, the *Handbuch* and *Lehrbuch* traditions, epidemic or not, were what kept Germany at the forefront of world science, especially in physics and chemistry, even while shut out of international scientific cooperation, and this was the sort of undertaking that made good use of Wegener's monumental appetite for work. Finally, at age forty-four (really almost forty-five), he was entering that phase of his career where, as he saw it, the moral responsibility of a scientist included the obligation to write summary works for the benefit of younger investigators—exactly the sort of work that had made his theory of displacements possible.

Back in Graz for the fall term by mid-October, Wegener began his lecture course entitled "Physics of the Atmosphere, Part 1," already planned before he had learned about the editorship. He had designed a full year of lectures that would cover the material for his planned textbook. Beginning in the autumn of 1925, he would lecture on mechanics, thermodynamics, and radiation; for the summer term in 1926 he planned to lecture on optics, acoustics, and atmospheric electricity. The textbook of atmospheric physics, of which these lectures were the foundation, was now postponed but not canceled, and the lectures would serve the subsidiary function of organizing his thinking about these topics.[48]

Immediately on returning from the meeting on the Zugspitze, he found a letter waiting for him from Berlin, announcing the death of Alfred Merz (1880–1925) from pneumonia, in Buenos Aries, Argentina. Merz was the lead scientist on the "German Atlantic Expedition," also known as the "Meteor" Expedition, following the oceanographic convention of naming the expedition after the ship. Merz had originally wanted an expedition in the Pacific, but funds were very limited, and he had to settle for the Atlantic. It was a combined expedition, planned to last two years, which would use sonar to establish depth profiles from the equatorial regions all the way to Antarctica. In addition, it would sample water for salinity to try to trace the flow of the currents in the Atlantic at various depths, and the scientific staff would also pursue maritime meteorology.

Wegener had been deeply interested in this expedition, which had departed Germany in April 1925. He had worked hard and successfully to get his old colleague Erich Kuhlbrodt a position on it as meteorologist.[49] Kuhlbrodt eventually sent up more than 800 balloons in the course of this expedition, tracking them with the theodolite that he and Wegener had designed and built. The expedition had been much on Wegener's mind in the summer of 1925 when he was teaching oceanography at Graz. He was extremely interested in what their profiles of the Atlantic floor would show. The expedition planned fourteen such transatlantic profiles, covering most of the coast of Africa and South America. The depth information, especially at the continental shelf margins, could be very important for his theory.

Because Merz was an Austrian and there were many obituaries to be written, Wegener got a request to write a brief obituary for him for *Meteorologische Zeitschrift*, and he complied, writing a two-page appreciation of Merz's life and work; they had met but did not know each other well.[50] This was, however, not the last request that would come to Wegener concerning Merz and the expedition. Merz had been the director of the Berlin Oceanographic Institute (founded by Penck), professor of oceanography at Berlin, and head of the Oceanographic Museum. Even though the expedition would proceed without him, he would have to be replaced; the institute and the university needed a professor of oceanography.

In November, Wegener received a letter from Albrecht Penck, asking him whether he would be interested in replacing Merz in Berlin. Penck praised Alfred as a great student of both the atmosphere and the solid Earth and wondered whether he could be enticed to add the ocean to his realm of expertise. Of course, said Penck, in addition to the professorship at Berlin he would have to run the institute, with all its administrative work, and be in charge of the outreach through the museum.[51]

He already knew that Wegener had been interested in the professorship at Graz precisely because of the lack of administrative responsibility. On the other hand, Penck stressed that in Germany there was no one within oceanography who was not "merely statistical," no one capable of moving oceanography in the geophysical direction, and this had led the search committee to look not only outside oceanography but also outside Germany for a successor.

Penck added at the end of the letter, "I've heard from your father-in-law how happy you are in Graz, and I can well understand this. I myself would also love to pursue my work in the midst of the mountains. But remember that old Hann moved from there back to Vienna. The atmosphere in Graz was, for him, too isolated. He longed for a different kind of fresh air, and then found what he was looking for again in a large city."[52]

Penck's letter was followed by another on the very next day, from Alfred Rühl (1882–1935), then an associate professor (*Extraordinarius*) at the University of Berlin in the Institute for Oceanography (*Meereskunde*) specializing in economic geography. They had been instructors together at Marburg, and Rühl wanted Wegener to know that the entire staff and faculty of the institute, as well as the Ministry for Education, were solidly behind the idea of bringing Wegener to Berlin as the director of the institute. The faculty was so excited by the prospect that they wanted to make the offer "*primo et unico*"—he would be the only candidate. Rühl knew, he said, of Wegener's "horror of all administrative duties" and knew that Penck had said these were an essential part of the job. Rühl wanted Wegener to know that these duties could be offloaded onto others, that Merz himself had done so, and that in the last two years before the beginning of his expedition he had not been impeded in his own scientific work by his title as director of the institute.[53] Rühl was very enthusiastic and pressed Wegener to come to Berlin.

Intellectually, the offer was extremely tempting, and he actually considered it for about ten days before turning it down. Penck had been very persuasive; Wegener was deeply interested in oceanography and had just finished teaching it. Penck had hit another nerve as well: Wegener had also begun to chafe at how far Graz was from large libraries, institutes, and scientific meetings in Germany, as well as at the expense of travel and the tiresomeness of wresting travel grants from the ministry in Vienna. Nevertheless, he wrote back to both Penck and Rühl on 18 November 1925 requesting that his name not be put in nomination to replace Merz.[54]

The Berlin job had many pitfalls and drawbacks, personal and professional, in spite of the intellectual stature it would have conferred and the advantages of Berlin. In the end, he used the same reasons to fend off the appointment which he had in readiness if he had been called to replace Gustav Hellmann at the Meteorological Institute in Berlin, namely, his close relationship and collaboration with Köppen. To this reason he was able to add that his parents-in-law had sold their house in Hamburg, that they were all living together in Graz, and that he could not in conscience now abandon the old pair to live alone so far away from their family. Köppen was now seventy-nine, he wrote, and too old to move again. However, he thanked Penck effusively and said, "I will not deny that your assumption that oceanography would entice me is entirely true, more true than perhaps you might imagine."[55]

With that behind him, he could turn to a very pressing issue at home: he had to teach the girls to ski. They were all three big enough now to ski cross-country, and they had proven themselves as hikers and climbers. In November, even before there was enough snow, he had them practicing the motions at home.[56] That Christmas, the whole family left for a ski vacation in the Ramsau, a high plateau between the Dachstein range and the Enns Valley. It offered unusually smooth terrain for this part of the world, had many cross-country trails, and included widely spaced villages. It was a favorite with many of the Graz faculty, who decamped Graz to stay in a chalet and gather together in the evening after a day of skiing. Benndorf had never learned to ski, and although Alfred was happy to teach him, he was reluctant. So, knowing this, Alfred gave him a short "book" he had just written: "Proven Methods for Old Men to Learn How to Ski." At fifty-four, Benndorf was only ten years Alfred's senior, but he enjoyed the joke and accepted the challenge; he and his family accompanied the Wegeners to Ramsau that Christmas and for years afterward.[57]

As winter turned to spring in 1926, Wegener found himself once again a full-time theoretical meteorologist. He was teaching meteorology and reviewing books in meteorology and geophysics—more reviews in 1926 alone than in the previous decade. The most difficult review was the one he had agreed to write of Felix Exner's second edition of his *Dynamische Meteorologie*; it was a better book than the first edition, but Exner had still not been able to bring himself around to accept Bjerknes's theory of frontal weather and of cyclones as wavelike disturbances moving along the "polar front" between the prevailing polar easterlies and the prevailing westerlies to their south. It was already clear to most meteorologists in Europe and elsewhere that Bjerknes had gone a long way toward solving "the cyclone problem." The best Exner could say about it was that the parts of the theory he disagreed with were the work of Bjerknes, and the parts he agreed with had been provided by Bjerknes's Austrian predecessors. There was actually a good deal to this criticism, as the animosity between Exner and Bjerknes had led the latter to be less than generous in citing Exner's and Margules's important work. Wegener made note of this dispute, praised the utility of the book and the superiority of the second edition over the first, made a few minor criticisms of Exner's treatment of precipitation, and was done.[58]

Wegener was willing to be gentle in the review with Exner, whom he had come to like, and to whom he felt indebted for his support in getting the job at Graz, the nomination to the Academy of Sciences, the summer vacation on the Stolzalpe, and other matters; Exner was also the editor of *Meteorologische Zeitschrift*. But Wegener had also to go on record, as a research meteorologist, and as a friend to Bjerknes, about his intellectual judgment of "where things stood" with the cyclone problem and with the theory of frontal weather.

Under such circumstances, Wegener turned to his favorite method of procedure: writing a review article about recent developments in the field. He set to work to write a brief (for him) history of dynamical meteorology from the beginning of the century. The centerpiece, or main line of discussion, was the career of Bjerknes, but set in the context of his time in Leipzig and of the contributions of many Germans and Austrians to the advancement of understanding of the cyclonic storms which had been the principal aim of meteorology for more than 100 years.[59]

He gave a very good and very fair summary of Bjerknes's career and thought, illustrated with nice line drawings of cyclonic waves, and left no doubt that this was the direction that forecast meteorology would take. Without criticizing Bjerknes directly, he managed to advance Exner's claims, by insisting that what Bjerknes was doing with his cyclone model was much more kinematic than dynamic; it concentrated on the phenomena of motion and their evolution but had not yet arrived at a clear and unambiguous treatment of the causes. Moreover, Wegener's treatment shows that he sided with Tor Bergeron and Sandström in arguing that the evolution of cyclones was not a matter of waves on a single continuous polar front, but that breaks occurred in this front. This, by the way, had been one of Exner's main points in his recent book on dynamical meteorology. Wegener ended his discussion by recommending Exner's book as a good place to go to understand, in great mathematical detail, the complexities of this problem.[60]

This was Wegener in teaching mode, encyclopedic mode, and *Handbuch* mode. Unsolved problems of meteorology were for him, at this point, less and less urgent challenges and more and more the perpetual status quo of every line of investigation—once one understood it deeply enough. Looking over the fields of science he had pursued, he could see everywhere that theoretical models *always* had to give way from their original beautiful simplicity to the empirical complexities the world offered. He was certainly willing to take this stance relative to his own work on continental displacements and on atmospheric acoustics (his current fascination in 1926). If he could accept this for himself, he could, in reviewing, comfortably suggest it to his colleagues as well.

This was an important development in his intellectual life, and it is not something every theorist experiences. Both Bjerknes and Exner were still "digging in" in defense of their own personal views of a large and complex problem—Bjerknes was, as always, relentlessly self-promotional, and Exner was chronically angry at being underappreciated. It was the same sort of dispute as that on ocean circulation which had poisoned relations between Brennecke at Hamburg and Merz at Berlin. Seen from inside, the issues were life and death to these men; from outside, it could often appear as the "pathology of small differences." Wegener saw this and took note that beyond a certain threshold of elaboration, any important theoretical idea would live or die on its capacity to generate empirical results that would, in turn, test its validity.

He was willing—it was certainly the role of a *Handbuch* (or *Lehrbuch*) editor to be willing—to offer a number of plausible solutions as wayposts for others finding their way forward. It is notable that he championed, in this review of the cyclone model in dynamic meteorology, the work of Lewis Fry Richardson (1881–1953), a British meteorologist and author of a complex mathematical treatise entitled *Weather Prediction by Numerical Process.*[61] Richardson had developed a series of fundamental equations which, with sufficient data and computing power, would allow one, if one knew the distribution of temperature and pressure across a broad area at a given time, to calculate a new temperature and pressure map at a later time. His initial results using real data had been disappointing, but Wegener said that both the optimists and pessimists were currently overstating the case, and the technique was promising even though difficulties were formidable. One notes that these equations by Richardson are the basic foundation of all computer forecast models in meteorology today.[62] Wegener's essential focus was always *die Arbeit* (the work) and how that was proceeding, much more than how his own ideas were faring.

Nearly twenty years earlier, when Wegener had given his *Antrittsvorlesung* on joining the faculty at Marburg, he had stressed that meteorology alternated between periods of gathering of data and synthesis into theory. Certainly his own career had shown this sort of alternation. During the war years, especially with regard to his work on tornadoes and waterspouts, it was a matter of data gathering; immediately after the war, of experimentation. The succeeding period of 1919–1924 was quite definitely a period of synthesis.

Now, in Graz, teaching and editing a major textbook/handbook, he was once again in gathering mode. He was working at his desk, not in the laboratory or in the field. The only empirical results he reported in his publications in 1926 were the quite ordinary sunlight intensity measurements he had made on the Stolzalpe in the summer of 1925 and a brief note on inversions based on a few photographs he had taken at the summit of the Zugspitze that autumn.[63]

Continental Displacements and Climates of the Past, 1926

The end of the summer semester in 1926 marked the end of Wegener's second full year in Graz. His whole life was in Austria now, both during the academic year and during his vacations—winter and summer. He was more deeply involved with family life than ever before, actually taking a hand in raising his children. The life they lived together in Graz had a measured and predictable routine unlike anything he had experienced since childhood. He had been out of Austria only once—a trip to Göttingen to give a lecture—if we do not count the outing on the Zugspitze. He had given a few lectures on his theory of displacements and climate, in Innsbruck, Vienna, and Graz. All his intellectual energy from the time of his arrival had been directed toward atmospheric physics and theoretical meteorology.

Some manuscript material for the Müller–Pouillet *Lehrbuch* was beginning to come in, mostly from Ficker on meteorology, but there was little to do there as Ficker was extremely competent and meticulous and needed no guidance in determining what were the important problems to discuss; this had all been laid out before Lummer's death. Wegener had spent some time finding authors to handle the chapter on glaciers and the chapter on oceans. He was extremely pleased that Hans Heß (1864–1940) accepted his invitation to write the chapter on the physics of glaciers. Wegener had admired Heß's work *Die Gletscher* (1904) so much that he had carried it with him to Greenland in 1912–1913. For the chapter on physics of the oceans, he turned to Hermann Thorade (1881–1945), an expert on ocean currents and a colleague at the German Naval Observatory in Hamburg. He hoped that he would be able to get Thorade to incorporate material from the Meteor Expedition, due back in 1927, before the book went to press.

He had given little thought during all this time to either climates of the past or displacements; at least if he gave such thought, it left little trace. From 1924 through early 1927 he made not a single entry in his notebook dedicated to the next edition of the book on continents and oceans. Since the autumn of 1924 he had published nothing on these topics, with the exception of a single (paid) article for a geography encyclopedia.[64] He had taught extensively and exclusively on meteorological topics for two full years, and all his publications (apart from two brief book reviews) for 1925 and 1926 dealt with the optics, acoustics, and thermodynamics of the atmosphere.

All this changed suddenly and unexpectedly in the summer of 1926. Wegener had entered into a correspondence with George Clarke Simpson (1878–1965), then the head of the British Meteorological Office; had Wegener succeeded Hellmann at Berlin, they would have been official counterparts. Simpson had defended Wegener's ideas at the 1923 meeting of the British Association for the Advancement of Science. Like Frank

Debenham and Charles Wright (both of whom had defended Wegener at the meeting of the Geographical Society of London in 1923), Simpson had also been on Scott's Antarctic Expedition. His career was an interesting parallel to Wegener's in this regard; Simpson had handled the temperature and wind observations, sent up balloons, and made electrical and magnetic observations at Cape Evans, only a few years after Wegener had filled that role on the Danmark Expedition in Greenland. When Wegener had been in the German Army Weather Service for the Balkans, Simpson had been his opposite number for the British Army in Mesopotamia.

Wegener had written to Simpson in July 1926. There was no summer vacation away from Graz that year for the Wegener family; Alfred and Else had scraped together enough money to make a major and much-needed addition to their house, including several upstairs rooms over a new breezeway entrance to their garden, and wanted to be there during the construction to supervise the work.[65] Wegener had wanted to know (from Simpson) the exact location of the astronomical observatories in Australia, at Adelaide and Sydney. These were part of the observation net for a planned major global experiment to fix the longitude of astronomical observatories by the exchange of radio time signals.[66]

The longitude program that prompted Wegener's letter to Simpson was the brainchild of a French general, Gustave Auguste Ferrié (1868–1932), who was himself a radio pioneer and now deeply involved in the establishment of an international network of radio stations at astronomical observatories, to coordinate their work and fix their longitudes with great exactness. Ferrié was the head of the International Commission on Longitudes by Radio, president of the French National Commission on Geodesy and Geophysics, and member of several other major organizations. The idea for his radio network had undergone initial planning at a meeting in Rome in 1922, had been ratified in Madrid in 1924 (Wegener had known of this work), and had been planned in detail after the International Astronomical Union meeting in Cambridge in July 1925.[67] The actual experiment to intensively measure longitudes at a global network of stations (thirteen in all) was to begin in October 1926 and last for two months.[68]

Wegener had thought that these observations might serve the subsidiary purpose of testing his theory of continental displacements, though the proceedings of the Cambridge meeting in 1925, like those of the Madrid meeting in 1924, do not mention continental displacements. He wanted to be able to compare the longitudes of these Australian stations in 1926 with longitudes measured by exchange of telegraphic time signals (before the First World War) between Australia, Hawaii, and Canada. Simpson was very forthcoming and sent Wegener a good deal of data. It was a conspicuous sign of thawing of relations between Britain and Austria (if not Germany) that they should be in correspondence at all.

Wegener had learned a year earlier of a full-page article in the *New York Times* entitled "Scientists to Test 'Drift' of Continents."[69] The article was by William Bowie (1872–1940) of the U.S. Coast and Geodetic Survey, who had, with John Hayford (1868–1925), done more than anyone else to establish the principle of isostasy, and on whose work Wegener had depended since 1912. In the *New York Times* article, Bowie represented Ferrié's international longitude network as an explicit attempt to measure not just continental drift but Wegener's theory of continental drift. (This was not actually the case at all, by the way.) The article was illustrated with Wegener's maps of the Carboniferous and the Quaternary, employing again the Hammer projections he had used in 1924. Wegener was identified in the article as "Professor Wegener of Austria," which was (strictly and recently) true, as well as much more politically acceptable in the United States in 1925 than "Professor Wegener of Germany."

In the article, Bowie gave a fair and clear summary of the outlines of Wegener's theory, stressing its dependence on the theory of isostasy (Bowie's "own" theory). Because, Bowie wrote, Wegener's theory predicts continual motions of the continents and because the radio measurements should be longitudinally accurate within about 10 feet (after two months of continuous measuring), it should be easy "after five, ten or some other number of years to make new determinations of longitude in exactly the same places. . . . If the new longitude for any place is found to differ more than twenty or twenty-five feet from the first determination, one would suspect that the change had been caused by earth movements rather than merely the unavoidable errors of observation."[70]

Wegener was "highly elated" to hear that such a test was to take place, but he was sorry to discover that almost all the stations would be either in midlatitudes or in the Southern Hemisphere. His theory supposed that most of the drift was currently happening farther north—Canada, Labrador, Greenland, Scotland. The tests as described in the *New York Times* by Bowie would not yield any results within a few years and would probably take "a century or more, before an appreciable change can be firmly established."[71] Of course, Wegener had no way of knowing that the International Commission on Longitudes by Radio had not planned this network to test his theory; that was Bowie's invention.

The "sudden development" in the summer of 1926 came from a different quarter: a letter from Willem A. J. M. van Waterschoot van der Gracht (1873–1943), a Dutch mining engineer and economic geologist who had worked in the Netherlands as well as in southern Europe, South and East Africa, Patagonia, Tierra del Fuego, and the Dutch East Indies. He was interested in everything of economic value: oil, coal, and mineral ores. He had made his reputation early, using German maps of gas wells drilled inside Germany to extrapolate to the Netherlands where and how deeply one should drill to find natural gas. This method resulted in a major discovery in 1906 and made him well known.[72]

In March 1915 he had accepted a commission from Royal Dutch Shell to look into the oil and coal resources of the Mid-Continent Region of the United States. He spent two years in the United States and then returned to the Netherlands, but in 1923 he came back to the United States, and by 1926 he was a vice president (in charge of exploration) for the Marland Oil Company, of Tulsa, Oklahoma. Marland at that time was one of the largest oil companies in the United States. Exploration geophysics was secretive work, and he published little, but he was extremely active and known for his desire to employ general theory to find new deposits.[73]

Van der Gracht told Wegener that he had organized a symposium to take place in New York City in November 1926 under the auspices of the American Association of Petroleum Geologists, on Wegener's theory of displacements. Van der Gracht described Wegener's as a "theory of the origin and movement of landmasses, both intercontinental and intra-continental."[74] By this he meant both the displacement of continents and the great compressive eras of mountain building, stressing the homology between Wegener's theory and that proposed by Argand to explain the Alps. Wegener could not attend the symposium, of course, not least because the United States and Germany had not yet resolved their differences over scientific cooperation. Nevertheless, Wegener could send a contribution updating his views as of the fall of 1926.

Wegener accepted the invitation to participate (remotely) and sent along two notes to be read at the meeting. Before we go into their content, it is interesting to see why and how the symposium was organized.

Van der Gracht was at this time working on a major study of North American geology, eventually published in the Netherlands in 1931 under the title "The Permo-Carboniferous Orogeny in the South Central United States."[75] He was aware of the controversy that had been going on in Europe for some years concerning Wegener's hypothesis. He was also aware that almost no public attention had been paid to Wegener in the United States. One might, as an instance of this, point to the difference between the reviews of Wegener's third edition offered by the British publication *Nature* and the American publication *Science* in 1925. The *Nature* review, extensive, respectful, and reflective, was the work of John Walter Gregory (1864–1929), a full professor at the University of Glasgow, a fellow of the Royal Society, and a distinguished field geologist who had worked on the East African rift valleys and in Australia.[76] On the other hand, the review in *Science* was by Frank A. Melton (1896–1985), a junior instructor at Columbia one year out of graduate school; his review was a single page in length and noted that only three North American authors had contributed to the debate: two Canadians and the U.S. mineralogist H. S. Washington (1867–1934), who had a PhD from Leipzig (1893).[77]

Since U.S. geologists, in their professional publications—the *Journal of Geology*, the *Bulletin of the Geological Society of America*, and the *American Journal of Science*—seemed determined to ignore both Wegener's theory and his entire approach to earth science, van der Gracht had organized the symposium to force the issue, inviting a panel of well-known geologists—opposed to, neutral on, and supportive of displacement—to speak. His purpose was pragmatic, and although he was sympathetic to Wegener, his interest in the theory was chiefly its economic utility:

> We petroleum geologists should feel a lively interest in this controversy. We realize that exploration for petroleum becomes ever more difficult, and that we are now looking for petroleum deposits of which there is very little, if any, indication at the surface. The days of mere hunting for structure are past, certainly for this country, although we feel convinced there exist many hidden pools which await our drill. Unless we leave their discovery to chance, we have to approach the problem of exploration ever more from a viewpoint of scientific research into the fundamentals of regional structural geology and resultant sedimentation. This is why the problem, which deals with the question of whether or not there has occurred and still occurs considerable drift in the outer shell of the earth, is of practical interest to us. If true, it must have affected sedimentation and deformation of the strata. But drift theory is very closely connected with the determination of the climate which various parts of the earth may have had in geological periods . . . [and] must have seriously influenced the character of the sediments and their value as source beds or reservoirs for petroleum. Such considerations should be taken seriously into account when planning explorations in more or less virgin territory. It is of particular interest for the study of foreign oil fields and of possibilities in remote, little-known regions of our globe.[78]

Wegener sent his contribution to the planned symposium to van der Gracht to translate. Wegener titled it "Two Notes Concerning My Theory of Continental Displacements," though van der Gracht translated this as "Continental *Drift*."[79] Wegener addressed two very specific matters. The first of these was the question of the so-called

Squantum Tillite. This apparently glacial deposit of pebbles, boulders, and clay—poorly sorted, and even with some apparent "varve" deposits, indicating periglacial lakes—occurred in beds 610 meters (2,000 feet) thick and was part of a larger deposit called the Roxbury Conglomerate, which had been mapped south from Massachusetts and west through the Appalachian Region. It seemed to correlate with other apparent glacial deposits of the same age in Kansas and Oklahoma. The problem for Wegener was that these deposits were dated from the Carboniferous and occurred almost exactly on the position of the equator which he had reconstructed for that period. He had raised this difficulty himself (in detail) in *Klimate der geologischen Vorzeit.*[80]

Wegener's own conviction was that these were pseudoglacial, and geological opinion today is that he was correct, that they represent marine landslide deposits, and in any case that they are hundreds of millions of years older than the Carboniferous period. However, in 1926 it was the most outstanding climatological anomaly of all the evidence he collected, and he took it to be a direct challenge to his theory. As he wrote, "if any of these conglomerates are truly glacial, they would be in flagrant contradiction to my conception. . . . Either one or the other of the previous conclusions must be wrong: either these conglomerates are only pseudo-glacial, or the geographic latitude of these areas was not 10 to 30°, but at least 60° in the Permo-Carboniferous."[81] He wanted this to be resolved and hoped that it could be done without reference to continental displacements. He said that it should be decided independently and impartially, and that this could only be done in America, disregarding all assumptions about other parts of the world.[82]

Wegener's second note concerned the question of measurement. He referenced his discovery of the plan by the International Astronomical Union, "of which I became aware through an article of W. Bowie in the New York Times."[83] He repeated his disappointment that no high-latitude stations were involved in this network, as well as his hope that stations could be established both in Greenland and in Madagascar, locations in which much greater shifts could be expected in a short period of time. "In any case," he wrote, "it is to be expected that if we concentrate on such regions as would be particularly susceptible to change under the drift theory, it would be possible much sooner to obtain positive results, and here, as everywhere else, the results of greater value are positive rather than negative."[84] What he meant here was that a positive result would be evidence of displacement, whereas a negative result would not show displacement but would also not rule it out; therefore, it was a choice, in the short term, between a confirmation and a longer wait. Then as now, "absence of evidence is not evidence of absence."

The symposium was held as planned on 15 November 1926 in New York. Most of the speakers opposed Wegener both in substance and in method. Notable among these were the Yale stratigrapher Charles Schuchert (1858–1942), William Bowie the geodesist, and Chester Longwell (1887–1975), another Yale geologist. This symposium, as well as the fate of Wegener's theory in North America in its aftermath, has a complicated history. It has been analyzed in greatest detail by historian Naomi Oreskes in her book *The Rejection of Continental Drift* (1999). The standard story in the United States, for some decades, was that North American geologists rejected continental drift in the 1920s (and thereafter) based on the sound and substantive geological objections to Wegener's ideas presented during this symposium. Oreskes argues that this is quite wrong.

Much more important than the issues raised in the symposium—the supposed lack of a suitable "mechanism" for drift, or a host of individual factual "mistakes" by Wegener—was a methodological stance, an American insistence on an interpretation of "multiple working hypotheses" that ruled out any compact presentation of any theory and its supporting evidence (and thus all displacement theory) as "unscientific" advocacy. The Americans blended this insistence (which they preached but rarely practiced) with an interpretation of "uniformitarianism" (a uniform pace and mode of geological change) which effectively ruled out any theory but the permanence of continents and oceans.

Oreskes shows that the North Americans repeated these heuristic and methodological principles—multiple hypotheses and uniformitarianism—as if they were a theory *of Earth* with empirical warrant, while treating their own picture of Earth (permanence of continents and oceans, with small land bridges appearing and disappearing) not as a theory but as a body of empirical fact. This rendered the community impervious to criticism and immobile in the face of changing methods and results in geology through the middle of the twentieth century. This sort of attitude is exactly the description that Emile Argand applied to proponents of continental permanence, which he called not so much a theory but an attitude toward theory, or rather a refusal to take a theoretical stance.

This sort of intransigence has also been emphasized by Robert P. Newman, who has pointed out that the "symposium," as discussed by most geologists in their publications, *never took place*. The only actual speakers were van der Gracht, Schuchert, Bowie, Longwell, Charles Berkey (1884–1955), and Andrew C. Lawson (1861–1952). The last two never submitted manuscripts for the published volume, which contained notes and papers by (nonattending) Stanford University geologist Bailey Willis (1857–1949); Chicago structural geologist Rollin T. Chamberlin (1881–1948), the son of T. C. Chamberlin; Edward W. Berry (1875–1945), a Baltimore paleobotanist; Joseph Singewald (1884–1963), a structural geologist from Johns Hopkins University; David White (1862–1935), an expert on coal and former chief geologist of the U.S. Geological Survey; and Frank Bursley Taylor (1860–1938), who had proposed drift independently of Wegener in 1910. Van der Gracht solicited manuscripts from Europe to complete the volume and published contributions by, in addition to Wegener, G. A. F. Molengraff (1860–1942), who had worked in South Africa and the East Indies; the British geologist J. W. Gregory; and the Irish geologist John Joly (1857–1933), an expert on radioactivity, who had his own theory of Earth, modeled on Wegener's approach, but with a different idea about the mechanism.[85]

Moreover, of the 240-page volume, published in 1928 by the American Association of Petroleum Geologists (in both the United States and England), pages 1–76 are an extensive introduction by van der Gracht to the problem of continental drift in general, including a detailed summary of Wegener's theory, but also summarizing the views of Frank B. Taylor; Reginald A. Daly (1871–1957), a Canadian geologist and author of *Our Mobile Earth* (1926); and John Joly, as expressed in his *Surface History of the Earth* (1925). These last two authors had theories of intracontinental mobility and the creation of mountain ranges by heating, asymmetric uplift of continents, and "continental sliding." At the end of the volume, between pages 197 and 226, van der Gracht published a summary of the major criticisms of Wegener presented in the papers, refuting them one after another.

Taken altogether, the published volume, in spite of the vigorous criticisms it contains of Wegener's hypothesis, had much more material defending and expounding Wegener than attacking him, given that van der Gracht was an outspoken proponent of continental drift. It was not really a North American response, since the convener and four of the authors were Europeans. Most importantly, as van der Gracht pointed out in his concluding remarks, "an outstanding feature of this symposium is that the majority of those contributors who attack Wegener's theory express themselves as not fundamentally opposed to the conception of such a thing as intra-and inter- continental drift, even on a considerable scale. This is a very important step forward."[86] It was indeed a step forward, and a large one; Wegener, had he attended, would have been as pleased at the outcome as he had been in Berlin in 1921. They were agreeing on displacement and disagreeing about the details.

Turning away from the misleadingly titled "symposium" to the more general and longer-lived North American debate over Wegener's theory, Naomi Oreskes has shown that this debate was quite different from the open consideration Wegener's theory clearly enjoyed in Germany as early as 1921; in Denmark, Holland, and England by 1922; and in Austria by 1924. Rather, it was a concerted and deliberate attempt to reject Wegener and his work as the kind of German "pseudo-science" of which Yale geologist Joseph Barrell, back in 1914, had warned his American colleagues to be wary.[87] The rejection of Wegener's theory in North America was the result of a crusade to forestall not just its acceptance but even its discussion. This crusade, led by Charles Schuchert and Bailey Willis, is amply documented by Naomi Oreskes through their correspondence—hence the apposite title of her book, *The Rejection of Continental Drift*.[88]

More recently, philosopher-historian Henry Frankel has completed a massive four-volume study, *The Continental Drift Controversy* (2012), covering from the beginnings until the 1980s. This is, by the way, an indispensable work for anyone interested in the relationship of Wegener's career to the emergence and acceptance of the theory of plate tectonics. Frankel devoted the first volume entirely to the debate about Wegener's hypothesis and its aftermath—a debate, we have often noted, in which Wegener took a very small part. Frankel's broad and deep analysis of the many hundreds of publications concerning Wegener's hypothesis demonstrates very convincingly (if one knows the European background) that the same arguments used against Wegener in the first decade of the theory's life, principally in Germany, were mobilized again and again by Wegener's opponents, who sought to discredit his expertise as well as dismiss his ideas. This is not to say that all the debates were like this; far from it—there was much good science in play, the issues were real, and the flaws in Wegener's arguments were apparent. What Frankel documents (as a side issue to his own deeper discussions) is the peculiar lag time from debate to debate.[89]

It appears that the consideration of this hypothesis, everywhere it went, followed the same trajectory on roughly the same timescale as the original German debates. In the middle 1920s, when Wegener's book was already in a third edition, many English and most North American opponents were still depending heavily on criticisms offered of the first edition of 1915, though we have seen that the hypothesis, in the form that Wegener was propounding it in 1920 and in 1922 (and after), was quite different from the hypothesis that brought about the initial criticism. Finally, the *Klimate der geologischen Vorzeit*, already out for two years before the New York "symposium," amounted to yet another major revision of Wegener's displacement theory, both in the timing and extent of continental motions and in the timing and amount of movement of the pole of rotation.

One of Frankel's most important claims is that, beginning in the middle of the 1920s, the debate over continental drift began to divide up into subcontroversies argued by subdisciplinary specialists. Frankel counts four such controversies as being of special significance: floral and faunal (biotic) disjunctions, the Permo-Carboniferous glaciation, the measurement controversy, and a controversy over the mechanism of drift.[90]

If this is so, and it appears to be so, this was not good news for Wegener's theory. Theories of great scope must, in consequence of their scope, lack precision. The more you explain in general, the less you explain in detail. This is true not only of theories of Earth but of the theory of gravity, the theory of electromagnetism, and the principle of conservation of energy. Wegener had been quite explicit in 1912 that he was proposing a unifying theory of Earth based less on the full ensemble of details of each component discipline than on solving evidentiary conflicts by an appeal to Earth's bulk (thermodynamic) properties. It was only later, as he refined his own theory, that he went more and more into the details of the various classes of evidence he had assembled. Still, compared to his geologist counterparts, there was a distinct lower bound to his curiosity about anomalies, and this shows clearly in Frankel's fine-grained analysis.

All this notwithstanding, van der Gracht did a great service to Wegener in organizing the symposium and ensuring its much-amplified publication. This is perhaps most notably so for a reason that van der Gracht never mentioned: he chose to illustrate his long introduction to "continental drift" not with the well-known maps from Wegener's 1922 edition or its 1924 English translation, but with almost every single map from *Die Klimate der geologischen Vorzeit*. He did not integrate these maps in any particular way with the discussion in his text; he simply reproduced and identified them. This was the only publication of these maps in an English-language book for some decades to come, providing the opportunity for Wegener's critics, had they wished to take advantage of it, to study Wegener's continental reconstructions, his pole positions, and his latitude-based climate zones from the standpoint of the most recent and complete version of his theory.

This was important for Wegener, as the core of the theory was in the maps, and *Die Klimate der geologischen Vorzeit* has never been translated into any language from the original German. This alone tells us a good deal about the status of the debate over Wegener's full theory in the later 1920s, considering that (by 1926) Wegener's book on continental displacements was already in English, French, Spanish, Russian, and Swedish. It also tells us that Wegener and Köppen may have played an unwitting role in the move toward subcontroversies, by minimizing the extent to which their joint work of 1924 contained yet another version of Wegener's displacement theory, differing in crucial respects from all previous versions, and yet presenting the work as a strictly paleoclimatic book based on a theory (displacements) already complete and independently published.

Well before the November 1926 symposium in New York, Köppen and Wegener's work was already thoroughly integrated into climatology in Europe. In the fall of 1926, Charles Brooks (1888–1957), the librarian of the British Meteorological Office and editor of the *Meteorological Magazine*, published a large book entitled *Climate through the Ages: A Study of the Climatic Factors and Their Variations*.[91] Brooks was a respected climatologist and the author of the previous study *The Evolution of Climate* (1922).[92] The latter was mostly a book about the Quaternary ice ages and the postglacial periods, and Brooks had planned a companion volume dealing with the mechanism of climate change and drawing on this book for illustrations, but "then came the work of Wegener on the theory of continental drift, and of Köppen and Wegener on the interpretation of the climatic record in terms of the travels of the continents across the parallels of latitude and this stimulated me to complete the investigation in order to examine Wegener's theory from the climate side."[93]

The first part of *Climate through the Ages* was, as Brooks said, "essentially a text-book of meteorology, in which, however, some of the constants of the ordinary meteorological text-book are treated as variables. Various theories of climatic change are discussed in successive chapters as they arise, but no attempt has been made to include all the theories which have been put forward from time to time."[94]

The second part of *Climate through the Ages*, nearly 100 pages in length, was titled "Geological Climates and Their Causes," and it was an extensive and detailed examination of Wegener's theory of continental displacements and Köppen and Wegener's work on climates of the past. Brooks said, in his preface, "I may, I hope, be excused for the length to which the discussion of continental drift has run. The theory is at present on its trial before the tribunal of the world's scientists, and the verdict appears to be wavering in the balance. Geology, naturally will be the final arbiter, but the voices of other sciences are not without some weight."[95]

Brooks's judgments are quite interesting. He accepted almost without question that stretching and compression of the crust, in conjunction with tidal forces, had led to the gradual drifting apart of the continents in an east-west direction. On the other hand, he found Wegener's latitude shifts to be much more problematic, arguing that "the only real evidence adduced in support of this view is climatological and, practically speaking, the climate of the Upper Carboniferous; the geological evidence is quite inadequate. . . . Wegener's explanation, though not probable, is possible."[96]

When, in chapter 13, Brooks undertook his detailed examination of the theory of continental drift, his framing shows the extent to which Wegener's thinking seemed unproblematic in its general outlines (if not proven in its details) from the standpoint of paleoclimatology as practiced in Great Britain. Brooks accepted the work of Theodor Arldt and of Edgar Dacqué (both deeply influential to Wegener) as showing the way paleoclimatology ought to be done from the standpoint of paleontology and biogeography.[97]

Brooks's approach also seems to demonstrate that Wegener's formulation of the problem of continents and oceans, his restriction of the alternatives to either displacement or permanence (no sunken land bridges), and his argument that the results of geology, paleontology, geophysics, geodesy, and paleoclimatology were all required to solve the problem of ancient climate had carried the day in how the question of past climate should be resolved. Wegener's work was not just a serious part of this discussion in Brooks's work; it was altogether the impetus for his book. Brooks had questions about mechanism, timing, and other issues concerning displacements, but the substance and tone of his treatment of Wegener show that of the forty-eight climate change theories Brooks listed, in eleven different categories, Wegener's theory was for him the most serious and important intellectual event in paleoclimatology in many years, and that no further progress was likely until it had been thoroughly evaluated and mastered by climatologists and put to a rigorous test.

As pleasant as this must have been for Wegener to read, it was also frustrating. What, for instance, could Brooks mean by "putting the theory to the test?" Did this mean an indefinite postponement of commitment on the geological model, or a certain span of time in which to compare Wegener's ideas with other ideas of similar scope? Brooks seemed to be waiting for the "community" to make up its mind, but there lay the problem: if every investigator is waiting for the "community" to make up its mind, this event will never happen, as everyone postpones commitment until everyone else is committed.

This was not just a geological problem: Wegener had once quoted Rudzki as saying that if climatologists found evidence of a shift of the pole, the geophysicist must surrender to their judgment. Harold Jeffreys had said in London in 1923 that the decision about Wegener's theory was a matter for geologists to decide. Brooks in 1926 had said of Wegener's theory, "Geology, naturally, will be the final arbiter, but the voices of other sciences are not without some weight."[98]

In Wegener's view, this was quite wrongheaded. In the Innsbruck symposium in 1924, when challenged by Schweydar, Wegener said that he stood by his geological reconstructions but that he had to defer on the question of the motion of the pole to the expertise of mathematical geophysicists more competent than he. In order to get out his views about the order of the sciences, and of who should be deferring to whom, he accepted an invitation from Eugenio Rignano (1870–1930), the editor of the multilingual popular science journal *Scientia*, published in Milan. Rignano was a well-known philosopher and author of the book *Psychology of Reasoning*, and he was interested in the phenomenon of selective attention. He published articles on all aspects of science in his journal, and he always had an eye toward vital (currently active) controversies, soliciting articles from major participants in such debates.

Wegener wrote an article based on his Graz lecture on the significance of the theory of displacements for the systematic and the exact sciences. Now, however, he came to have a somewhat different conclusion. Where before he expressed the need for all these sciences to collaborate, he now asserted that while collecting information from many fields was necessary, which he indeed had done, "nevertheless, I believe that the ultimate resolution of this question must come via geophysics, because only geophysics has sufficiently exact methods."[99] If geophysics were to decide that the displacement theory was false, then all the systematic sciences would have to go along with that judgment and would then have to find another explanation for their own facts.[100] He then went on to sketch how modern results in the measurement of gravity, in seismology, and in the study of rigidity and viscosity in solid mechanics all pointed unambiguously toward the impossibility of the sinking of continents, while they underlined the likelihood that floating continents might also displace laterally.

This can hardly be surprising; Wegener was, after all, a physicist, trained in physics and astronomy, and once again doing work of the most definitive kind in optics and acoustics at the time of this writing. He was trying to find a way out of the impasse created by the decision of the proponents of continental permanence to continue to assert their geological ideas in the face of their geophysical implausibility.

Looking Far North

The life of a theorist and the life of his theory do not necessarily run in tandem. While Wegener's theory was being debated in New York and treated in Great Britain by Brooks, Wegener the theorist was teaching (in the winter semester 1926/1927) "Introduction to Geophysics," which had less to do with his theory of displacements and past climates than with his editorial work on the *Lehrbuch der Physik*. Wegener had not yet begun to catch up on work bearing on his theory going back as far as 1924.[101]

As his editorial work pulled him away from continental displacements and climates of the past, a variety of independent events were also turning his mind back to the Arctic. The phenomenon of *Haupthalos* (parhelia), on which he had just spent so much time—studying the refraction of sunlight by ice crystals in the upper atmosphere—was something that was visible most frequently in high latitudes. He was, at this time, in correspondence with Carl Störmer (1874–1957), a Norwegian mathematician and investigator of the auroras who had found Wegener's work on the auroras essential reading and had solicited offprints of Wegener's work as part of a major study of the aurora which Störmer was pursuing in 1926.[102] Additionally, the Royal Danish Geographical Society had named Wegener an honorary member in November 1926. In December, Eduard Brückner wrote to Wegener to inform him that he had been named Austria's representative to the International Society for the Exploration of the Arctic by Airship (AEROARCTIC), whose president was Fridtjof Nansen.[103]

More powerful than all of these Arctic stimuli, however, was an appeal he received from Koch in the late fall of 1926. Koch's health was deteriorating rapidly, and he proposed that Wegener should take over the writing up of the glaciology from 1912–1913; Koch said he no longer had the strength or the powers of concentration to see it through.[104] Wegener wrote back with another suggestion: they would work on the glaciology together; Wegener would work up a draft from Koch's notes, and then they would meet and go over it extensively, to clear up disputed points. Koch agreed.[105]

Nothing could happen immediately with this project; Wegener was too deeply involved with his teaching and with the meteorological problems with which he had started to engage again. As we have seen, in his search for information about the "outer zone of audibility" he had come across data indicating a variation in altitude (with the seasons) of the reflecting layer in the stratosphere and had written a paper on this in 1925 for *Meteorologische Zeitschrift*.[106] He had also been working on the temperature profile of the atmosphere above 80 kilometers and its connection with his theory of elemental segregation of oxygen, nitrogen, and hydrogen at successive altitudes.

These two lines of investigation, the seasonal migration of the reflecting layer in the stratosphere and his continued concern with layering in the upper atmosphere above 80 kilometers, came together in late 1926 with his previous work on meteors, and in a most interesting way. He came across a paper in the *Proceedings of the Royal Society of London* showing that the visible path of "shooting stars" (*Sternschnuppen*) had different altitudes in summer and winter; in winter the frequency maximum for the disappearance of the trail was around 75 kilometers (47 miles), whereas in summer it was much higher, closer to 85 kilometers (53 miles). These phenomena of "law-like variability" had long fascinated him, and he wanted to see whether this variation in frequency maxima (yet another bimodal distribution) held for large meteors as well. He went to a new large-meteor catalog compiled by Nießl and Hoffmeister (in Vienna) in 1925 and divided up the data by season. He was able to determine quickly that such "remarkable regularity [in the endpoint of the visible path] was not limited to shooting stars, but also applies to large meteors," albeit with frequency maxima at different altitudes from the smaller meteors.[107]

There was now time for these investigations, the work on *Lehrbuch der Physik* and lectures notwithstanding. Wegener still had a voluminous correspondence, but he wrote no reports and filled out no forms; he had no long rail commutes each day, and instead of four work sites (as in Hamburg) he now had one—the university. This allowed him to expand his consideration of what otherwise might have seemed "side issues." His work on the length of meteor paths led to another investigation on the estimation of the speed of large meteors. The issue, too complex to go into here, had to do with the discrepancy between the estimated velocity of a meteor relative to Earth and its velocity relative to the Sun, which involved the matter of Earth's velocity around the Sun, the angle of incidence of the meteor path, and whether the meteor's orbit was elliptical, parabolic, or hyperbolic—the so-called Keplerian orbits. This work took Wegener back to the tools he had learned in graduate school estimating asteroid orbits, once again including a result with a bimodal distribution requiring an explanation.[108]

His editorial work on *Lehrbuch der Physik* also forged new intellectual alliances. Especially notable was Wegener's cordial and warming relationship with seismologist Beno Gutenberg (1889–1960). Both men had been in the Army Weather Service on the western front during the war. After the war, Gutenberg (a student of Wiechert) was one of those who lost his position at the University of Straßburg, and he had great difficulty finding another appointment—as did many other young academics, especially if they were Jews. He ended up taking over his father's factory in Darmstadt, borrowing seismographic records from the University of Frankfurt and doing theoretical seismology on evenings and weekends. Like Wegener, he was attracted to problems of discontinuities and layering in both the atmosphere and the solid Earth; they had much in common. Gutenberg had even worked on the problem of "the outer zone of audibility," and in the early 1920s he had followed up the work of Tams and Angenheister on the difference of seismic wave propagation beneath the continents and beneath the oceans; Gutenberg's work seemed to confirm that they were made of different materials.[109] Gutenberg and Wegener found themselves in the amusing position in 1926 that they were both editing major textbooks of geophysics for different publishers; Gutenberg was an author on seismology in Wegener's volume, and Wegener an author on atmospheric optics in Gutenberg's volume.

The winter term wound down quietly before Christmas in Graz in late 1926. Over Christmas the Wegener family left for their (now) annual ski vacation in Ramsau, and when they returned, the New Year's burden of correspondence was not as large as in some past years. Skiing with his children and with Else had made him think of all the time he had spent in Greenland and all the snow he had traversed. The cross-country trails in Ramsau were just right for the girls, but they didn't go anywhere; one just skied for a period of time and then went back to a hotel.

When he returned to Graz in January, he was moved to write a letter to Peter Freuchen, his meteorological assistant on the Danmark Expedition in 1906–1908. Freuchen wrote back warmly and extensively in February; he told Wegener of his life, that he had lived for many years in Greenland as a hunter and trapper and had married a Greenland woman, and that on an exploring expedition in Canada in 1923 he had been so badly frostbitten that one of his legs had to be amputated at midcalf. He was now making a living writing about his "exploits in the frozen North." Ironical and droll as always, he turned serious in talking about their time with Mylius-Erichsen and commented on how little attention had gone into understanding or preparing for the technical side of exploring in those days. He told Wegener that he had heard that new longitude measurements were to be made in Greenland from the southwestern city of Godthab (now Nuuk) and hoped that they supported his theory of *Kontinentaltrift*.[110]

Wegener continued his lectures on geophysics; the editorial work was dovetailing nicely with these, and he was learning a great deal about seismology. In addition, the work of Benndorf and Hess on the electrical phenomena, especially in the upper atmosphere, was helping him a good deal and encouraging him in the sort of work he had recently pursued on meteor trails and optical phenomena; it was good to feel himself part of a working group.

Things changed quite a bit toward the end of the semester at the Easter break. He sent off his chapter on optics to Gutenberg, and in April he received a note of thanks. Gutenberg thanked him not only for his work but also for the rapid turnaround time and the excellence of his figures; would that all authors were like him! He thought that Wegener's contribution probably would be one of the highlights of this section of the text, which was being published in segments and sold as separates. The cost of the whole book was very high, and it was an experiment by Gutenberg's publisher to let readers pick and choose the sections they needed and wanted. It had been a complaint of Wegener's that many valuable pieces of research were not read because they were buried in hugely expensive multitopical publications.

Turning from the editorial correspondence, Gutenberg said, "Now, about the displacement theory. . . ."[111] In 1926–1927 Gutenberg had become convinced through further analysis of seismic records of the ocean floors that Wegener's simple Sial/Sima model was not correct; instead, a sialic crust covered the whole Earth but thinned to about 5 kilometers (3 miles) or less under the Pacific, yet it was, by his estimate, much thicker—more than 30 kilometers (19 miles)—under the Atlantic. It occurred to Gutenberg that while Wegener was reading his work on seismology, he might not have been aware of the extent to which his most recent analyses seem to contradict some of the basic premises of Wegener's theory. He therefore made a point to direct Wegener's attention to the differences between the floors of the Atlantic and Pacific Oceans. Wegener had long argued that the Pacific floor was more isostatically compensated than the Atlantic, and this would explain some of the gravity measurements over the Atlantic, which appeared less dense than it should, if the floor of the ocean were really Sima. Gutenberg told Wegener that isostasy was probably not the cause of the difference between the Atlantic and Pacific floors, and that it was quite likely that most of Earth, except for the Pacific, was a shell of Sial. If the crust floated, and it probably did, Europe, North America, and the Atlantic floor composed a single sheet of thickness varying between 60 kilometers (37 miles) and 30 kilometers. Gutenberg's letter contained the data to be published in May in *Gerlands*.[112]

Wegener trusted Gutenberg's scientific acumen; he had read his works on seismology as they came out piecemeal, and he had actually reviewed them for *Geographische Zeitschrift*.[113] In this way he became aware of Gutenberg's sympathy for the idea of continental displacements and for motion of the pole. Yet this other, newer aspect of Gutenberg's work offered a fundamental challenge to Wegener's basic conception of the flotation of continents and the possibility of displacement: if Gutenberg was correct, then the continental surfaces and the ocean floor were the same Earth shell, floating on a denser medium below, and most of what he had written was now facing the "verdict of the exact sciences," which he had just declared, in *Scientia*, to be "final."

Wegener, deeply concerned by this development, began to fill his notebook with reading notes, rapidly surveying work on isostasy and the seismic study of the ocean floors going back to the time of Gutenberg's 1924 publication and even before. The theory of isostasy, he found, was in much better shape than he had left it in 1922. In 1924, a Finnish geodesist, Veikko Heiskanen (1895–1971), had published the results of an extensive gravity survey that demonstrated the extent and exactitude of isostatic compensation across northern Europe and Scandinavia and also seemed to resolve the character of that compensation strongly in favor of the "roots of mountains" hypothesis (Airy isostasy). This was the version that Wegener had depended on: that when mountains were thrust together, much in the way of pressure ridges in ice, most of the compressed material went down rather than up and then was lifted by flotation. Americans had preferred another hypothesis (Pratt isostasy), of a smoothly distributed increase in density down to a spherically symmetrical "surface of no strength."

Wegener copied out Heiskanen's judgment—"Today isostasy is no longer a hypothesis, but a confirmed theory"—and made another note: "s.95 [p. 95] from the discussion of the anomalies in the Caucasus, in the Alps, and in the USA show that the Airy [roots of mountains] view is at least as good or slightly better than the Pratt view [the uniform surface]. Gravity observations consistently show that the thickness of the crust of Earth measured from sea level in different parts of the earth corresponds to the Airy principle with a thickness between 30 and 80 km."[114]

Gutenberg, who, like Heiskanen, was using both gravity data and seismic data to gauge the thickness of the crust, seemed to be extending Heiskanen's estimate for the minimum thickness of continental crust (30 kilometers) from the continental platforms across the entire Atlantic basin. Wegener was looking at all the isostatic research he could locate and trying to determine the range of the most recent estimates of the thickness of the outer (Sial) crust. Along the way, he discovered much more.

Throughout April 1927 Wegener was busy finding out just how far behind the debate about his hypothesis he had fallen. His notes indicate that he first came into contact with Argand's publication supporting him (which had appeared in 1924) only in 1927.[115] He read for the first time Ampferer's paper from the 1925 Vienna symposium on his work, with its suggestion of global convection.[116] He read a major publication by the French geologist and proponent of the nappe theory Pierre Termier (1859–1930), through which he discovered the recent work of John Joly, who had developed a theory of continental displacement somewhat similar to his own and based on parts of it.[117]

At this time he also first read any of the work of R. A. Daly, the Canadian geologist who in 1926 had already published a book on the subject: *Our Mobile Earth*.[118] Wegener learned from a journal article of 1923 that Daly had proposed "the comparatively recent sliding of North America over the sunken crust of the old, Greater-Pacific basin," to which Wegener added the notation "!!." He also learned that there was an expanding literature hypothesizing the trapping of radioactive heat beneath the continents, melting the subcrust underneath continental surfaces, and allowing continents to "slide" laterally more easily.[119] One notes in passing how these late discoveries underscore Wegener's mental distance from the supposed "symposium" in New York in November 1926. Van der Gracht's summary of the displacement hypothesis in his address to the American Association of Petroleum Geologists in November 1926 contained extensive references to the works of both Daly and Joly, and Wegener was oblivious.[120]

In all, Wegener filled about thirty pages of his notebook in April of 1927, bringing himself up to date with most major publications on his theory, pro and con. The majority of these were geophysical publications with data and hypotheses concerning the possibility of lateral drift. Among them he found intriguing suggestions, such as that of the Prague geophysicist Adalbert Prey (1875–1949), who had published an article in 1926 in *Gerlands* hypothesizing a long-term increase in Earth's coefficient of viscosity which could have enabled, in earlier ages of Earth, much greater shifts of the continents than were now possible, because of Earth's lower viscosity.[121] Wegener also learned, in copying out his criticisms, that Carl Diener was still, after almost ten years, tirelessly devoting himself to refuting Wegener's hypothesis.[122]

Wegener began to draft a paper in response to Gutenberg's forthcoming work, finishing it at the very end of the winter semester in April 1927. He had read enough of the literature to be a good deal less alarmed than he had been initially, but his tone is sober and cautious.[123] While not criticizing the seismic work, he noted that the gravity profile of the Atlantic which Gutenberg employed (as a check on the seismology) had followed a course between the United States and Gibraltar which missed most of the deepest part of the Atlantic (below 5,000 meters [~16,000 feet]), which Wegener had claimed—based on correlation of dredge samples and echo profiles—to be the exposed Sima. Further, the ship track for the gravity measurements passed over a very wide portion of the Mid-Atlantic Ridge and across Azores, which, as the greatest fragments of Sial in the Atlantic, would have to distort the overall estimate of mean depth. This dependence on an anomalous transect had caused Gutenberg to give a much too large value for the mean thickness of Sial, and his work, far from a fundamental challenge, was such that "the displacement theory as presented to date requires only a minor correction."[124]

It may have been Wegener's wish that this was a "minor correction," but it was not. Even the map he provided in his article for *Gerlands* showed that more than half the Atlantic floor probably consisted of something lighter than Sima; he had been giving ground on this issue for some time, as more and more Sial was dredged from the Atlantic floor. In 1922 he had used the analogy that when an iceberg calves and drifts away, some surficial pack ice was left behind. Yet the amounts in play here were not fragments and not superficial—these were very large areas of the ocean floor, and the thicknesses postulated by Gutenberg would be very difficult to pare away.

Wegener might well have begun his work on a fourth edition immediately, but there was a problem: he had already promised Koch that he would complete all the documentation of their 1912–1913 expedition to Greenland. During Easter break between semesters in late April 1927, Wegener traveled to Copenhagen.[125] Wegener and Koch agreed on the nature of their collaboration. Wegener had already done the meteorology and would now take Koch's rough notes on glaciology and especially the calculations of the distribution of temperatures in the glacier ice boreholes and in the "firn layer," the porous recrystallized layer between the recent snow and *névé* (multiyear recrystallized snow) above and the glacial ice below.[126]

Wegener arrived in Copenhagen to find Koch very ill and depressed. As they went over the notes, Koch revealed to Wegener that he had run into insuperable difficulties in trying to reduce the temperature data, and that the equations he had developed and the methods that he had followed had failed utterly to produce a common standard of measurement that would allow him to finish the data and summarize it. Wegener said that this was "a severe blow to him [Koch]," so much so that he didn't even want to give Wegener the calculations he had made thus far for fear that Wegener would be carried down the same line of attack which he, after such a great expenditure of energy, had discovered to be worthless.[127]

Wegener began work on the glaciology almost immediately and could see that it was going to be a very large undertaking. His summer course was "General Meteorology," nothing that could distract him very much; this was fortunate. The physical descriptions of the glacier ice were not difficult to organize or supplement: qualitative descriptions of what the eastern margin of the ice had been like, a description of the character of the glaciers at their Winter Station, an account of the calving of the iceberg that had almost killed them in September 1912 as they were camped on their approach to the Inland Ice. The measurements of the movement of the glacier on which "Borg" (their prefabricated hut) had rested in 1912–1913 were interesting but not difficult.

The hard part was the *science*, and this is where Koch had done almost nothing. There was the question of the nature and geometry of the crevasses and the "blue-bands" that formed in the ice. There were other phenomena associated with melting, as well as miscellaneous observations of glacier ice out of which Koch and Wegener had both hoped to make some pattern. The data posing the greatest difficulty were the temperature measurements. It took weeks for Wegener to work out a scheme, but once he found the way, he was just left with the labor of calculation; this was unlike the geodetic work at which Koch was so skilled, and Wegener had always liked calculation, since the days of his dissertation on the Alfonsine Tables.[128]

So though the work might seem to have been drudgery, it was not for Wegener. He recalled much of the Greenland trip while working up these calculations. It was, as he had already remarked, one of the happiest periods of his life, and he could feel again what it was like to be out there doing science, and he also felt very powerfully the contrast between their serious scientific work in the winter of 1912–1913 and their trans-Greenland stunt in the summer. Even though they had taken extensive measurements of the altitude of the ice cap, it had mostly been a matter of living through it.

In early July Milutin Milankovich paid him a visit. Milankovich remembered meeting Wegener and Köppen at their garden gate (they had been reading in the back garden, but this gate was not built until the summer of 1926). They sat and talked for a while, and Wegener pressed Milankovich to see whether there was anything he could do to uncover the physical causes of the displacement of the pole. Milankovich remembers Wegener saying that as soon as he finished with Koch's report on the investigation of Greenland, he intended to begin immediately on the fourth edition of his book, but that he needed to understand the mechanical reason why the poles should shift. He felt now that it was not so much a shift of surface masses, but some instability deep inside Earth. To Milankovich he appeared supremely confident.[129]

While the continuing work on the glaciology was not intensive, it was certainly extensive. Under Wegener's hand the manuscript continued to grow. By the time the semester ended, in July, the final product of all the expedition results appeared destined to appear as two volumes, totaling probably 700 pages. Even so, at least he was (for the moment) the sole author; not so for his *Lehrbuch* volume. This had also grown apace, especially the contributions by Benndorf and Hess on various topics in atmospheric electricity, which (alone) looked to be at least 150 pages of printed text. Ficker had gone even further, and his section on meteorology was almost 175 pages long.[130] Wegener read these with extreme care: this was, after all, a textbook, and every formula and every diagram had to be correct. It now looked as if the completed book, in proof, might exceed 800 pages.

Wegener was very much looking forward to the summer vacation. He would not carry his work with him, and he and Else had decided that this summer, now that the girls were older and stronger, they would take their vacation in the Ramsau, where they skied in the winter, and would spend the summer mountaineering. Wegener was more in touch now with his old expedition colleagues in Copenhagen. He would have liked it if Koch could have come to the mountains, but he could not, and Freuchen had only one good leg. He invited Andreas Lundager, the botanist and his old friend from the Danmark Expedition, to come for a visit, which he did.[131]

Else remembers 1927 as a particularly wonderful summer in their rented house in the Ramsau. Her letters to her parents in Graz were so vivid and full of spirit that the old pair actually came to stay with them in the mountains. Köppen was delighted; in the move south in 1924 he had taken Marie on a vacation in the Alps, assuming that they would never be in the mountains again. Now Alfred and Else took both of them hiking, carrying folding canvas chairs so they would not have to sit on hard benches in the alpine huts, and so they could rest along the trail. She remembers that Alfred became insatiable in his desire to conquer every peak in the Dachstein, and that on rainy days he would pore over mountain guides, sometimes even leaving the family for a few days to climb in southern Germany.[132]

In early September 1927, Else returned to Graz with the girls, who had to begin school. Alfred went north to Riga, in Latvia, where he had been invited by the *Herder Gesellschaft* to give a series of lectures. Shortly after he had published his work on lunar craters (in 1921), he began receiving letters from an engineer in Riga who believed that a circular depression on the island of Ösel (now Sarimaa), filled with water and about 100 meters (328 feet) across, was likely an impact crater, and who wanted Wegener to come and investigate it. He repeated his invitation to Wegener in 1925 and again in 1927. Now that Wegener had travel money, as well as the possibility of an honorarium for the lectures, he could afford to go and investigate.

Wegener had other reasons to go north. In June he had received a letter from Johannes Letzmann (1885–1971), whom he had known during his short time in Dorpat in 1918. Letzmann, still a docent in meteorology at Dorpat, had read Wegener's book on tornadoes and waterspouts and had begun a series of field observations, some theoretical work, and experimental simulations of tornadoes; this had now spanned almost a decade. He wanted to come to Germany to advance his scientific career and get more training. He had written to Wegener and to Ficker both, and Ficker agreed with Wegener that they should do something for him if they could.[133]

In the middle of September Wegener went to Ösel to study the crater with his host, Rudolf Meyer. At the last moment, their party was joined by Ernst Kraus (1889–1970), a structural geologist from Riga. The outcome produced a scientific paper authored by Kraus, Meyer, and Wegener which was even longer than Wegener's Vieweg *book* on the origin of lunar craters. Meyer and Wegener agreed that it was an impact crater, using the various metrics developed by Ebert for the Moon; the result was that the depression was exactly the same shape as the Meteor Crater in Arizona, only one-tenth the size. Kraus, however, was convinced that it was the remains of a salt dome that had been washed away, and the ultimate publication included a reconnaissance, Kraus's view, and Wegener and Meyer's view. Wegener ruefully reflected that he seemed unable to generate support and consensus even where there was only a single geologist involved.[134]

The meeting with Letzmann after the lectures in Riga went well, and Wegener was impressed by the sophistication and depth of his work on tornadoes. They decided that Letzmann should come to Graz, and that they would spend the next year (1928) solving this problem—the origin and mechanics of tornadoes—together.[135]

While he was in Riga, Alfred received a letter from Else informing him that a large tornado had struck just outside Graz on 23 September, causing much destruction. She had gone out "into the field" with the girls to walk the path of the funnel cloud and had noted that the trees had fallen in a vortex pattern. When Wegener returned to Graz, he and Else spent some time walking this tornado path and talking to those who might have seen and experienced it; it was a reprise of their 1916 collaborative field trip to investigate the meteor impact in Hesse.[136] This cemented his resolve to bring Letzmann to Graz in the New Year; Letzmann had traveled extensively in the previous decade to study the path of every tornado he could reach, and now Wegener could see why.

At summer's end he had nearly completed the work on glaciology, especially the burdensome temperature calculations. He returned to this with renewed energy throughout October and early November 1927, and by late November he was able to send the promised manuscript to Koch in Copenhagen.

In tandem with his work on his lectures, the editorial work, and finishing off the manuscript to send to Koch, he had spent the fall filling his notebook with more and more information to incorporate in the next edition of his book on continents and oceans. Direct notations from primary sources in this notebook end with material published in December 1927; the remainder of the notebook consists of calculations and interleaving of correspondence.[137] One suspects that the volume of material he was reviewing had, by late 1927, far outgrown such a notebook, and Wegener's increasing use of the typewriter after 1925 makes it likely that he had many other notes for the fourth edition than we see in these handwritten references. Nevertheless, it is clear that he had fully embarked on a new edition of this book, having fulfilled the prediction he made to Milankovich that he could begin as soon as he had finished editing Koch's work.

Wegener heard nothing from Copenhagen after sending off the manuscript. Then, just before Christmas, he received letters, almost on the same day, from Koch and from Jens Lindhard (1870–1947), the physician from the Danmark Expedition. Koch was cheerful and rueful; he explained to Wegener that he was going to be too ill for the foreseeable future to work on the manuscript, but that he had gone to the Carlsberg Foundation for funds to recompense Wegener should he have to work up all the results on his own. He had asked for 10,000 kroner, but the fund could only provide 2,000; "would that be enough?"[138] Lindhard's letter was a good deal less cheerful. Koch was extremely ill and had been in and out of the hospital, and the prognosis was not good. Wegener could read between the lines: Koch was dying.[139] Lindhard wrote again on 12 January 1928 that Koch had been hospitalized and that the end was near. Koch died the next day.[140]

Wegener learned of Koch's death two weeks later, in a letter from his widow, Marie, who wrote to Wegener, "You have lost a beloved friend."[141] Lindhard wrote to Wegener on that same day (25 January 1928) to tell him that Koch had been buried and that his pallbearers were all colleagues from the Danmark Expedition.[142] Wegener was not there to carry the body, but he was still destined to carry Koch's weight. In early February Marie Koch sent to Wegener all of Koch's notes, notebooks, manuscripts, offprints, and documents. Wegener was to be Koch's literary executor; he would finish their joint work. He went through the documents quickly; they were rough, but he was up to the task. He would have to write up the material on the reconnaissance and supplement the informal mapping, but Koch's sketches were good enough for that. It would add a month or two to the labor.[143]

In early January 1928 Wegener had received a letter from his former student and colleague Johannes Georgi.[144] They had last corresponded in 1925 when Georgi had sent Wegener a manuscript he had prepared on the measurement of wind speed; in 1926–1927 Georgi had gone to Iceland to make upper air observations and had discovered some curious high-velocity winds in the upper atmosphere, what we now know as the "jet stream."[145] Subsequently, he had traveled to Iceland and Greenland on the *Meteor*, now returned from its two-year voyage to the South Atlantic. Wegener and Georgi had seen each other at a meeting on measuring wind speeds in Potsdam in July 1927.[146] Georgi, now back at the German Marine Observatory as head of the Instruments Section, told Wegener that he wanted to set up a base on the east coast of Greenland and overwinter to measure wind velocity and barometric pressure, and he also raised the question of a research station for overwintering on the Inland Ice. Both men were aware of the stirrings of plans to mount a second International Polar Year in 1932–1933, fifty years after the first such endeavor in 1882–1883. Whoever participated in or led such an expedition in Greenland in the next year or two would be positioned for a major role in the German planning of the Polar Year.

Wegener wrote back immediately, congratulating him on his "beautiful plans" and wishing him the best of luck. "You are entering," he said, "a fascinating area of research and your work in Iceland has already put you past the initial difficulties." As for the plan for the overwintering station on the Inland Ice, Wegener said, "That's an old plan of Freuchen's, Koch's and mine. If the war hadn't come we'd have accomplished it long ago. However, in the meanwhile Freuchen has lost a leg, Koch is now too old and in the hospital and I have my own problems [*und auch habe ich einen kleinen 'Knax'*] and am no longer a young man."[147]

Wegener told Georgi that he had planned to write an article on methods of work and a scientific program for such an expedition for the first issue of *Arktis*, the magazine of AEROARCTIC, to which organization Wegener was still the Austrian representative. He sketched out for Georgi the main theme: The expedition should use an airplane making repeated flights over the border of the Inland Ice to locate the best route onto the ice cap. Then a depot should be established at about an elevation of 2,000 meters (6,562 feet), and motorized sleds used to pull supplies and equipment to the middle of the ice. He noted, as a caution, that one would need a backup plan, that not everything could depend on the plane, and that one had to plan for a situation in which the plane crashed on its first flight.[148]

The ascent to the Inland Ice, Wegener wrote, should happen from the western coast, because it is so much easier, and he stressed that it was hardly possible to overstate the danger, cost, and risk of failure of an attempt to mount the ice from the east. Don't take my word for it, Wegener cautioned him, but read the accounts of previous expeditions. He was, of course, referring to his own terrible troubles in 1906–1908 and 1912–1913.[149]

He also briefly sketched the sort of work that should take place: climatology, aerology, glaciology, measurements of ice thickness, elevations, and measurements of gravity. He concluded by saying, "You can see how much your plan interests me, and it would naturally make me very happy if you would keep me up-to-date as your plans continue to develop. With greetings from our house to your house, faithfully yours, Alfred Wegener."[150]

With that, Wegener turned back to editing the very large pile of manuscripts for the *Lehrbuch* which had accumulated during his absence (the family had taken their traditional Christmas skiing vacation in the Ramsau). It was good to think of the next generation coming forward to take up the work in Greenland where he had left off, but it also made him wistful. Arctic research was, as he told Georgi, "a fascinating area," and as he reviewed his and Koch's glaciological work, he could see many more questions than answers. Moreover, he now had Hans Heß's manuscript on the physics of glaciers for the *Lehrbuch*; with a handful of exceptions, the literature Heß cited was all from the *nineteenth century*. What new information there was about temperature at depth in glaciers came from an article by Koch in 1913.[151] As for the glaciology of the great masses of Inland Ice in Greenland and Antarctica, Hess had almost nothing to say. Maybe there would be time later in the year for Wegener to actually write the article for *Arktis* on methods for an expedition like Georgi's, but for now there was the rest of the work from 1912–1913 and the new edition of the book on the origin of continents and oceans.

19

Theorist and Arctic Explorer

GRAZ AND GREENLAND, 1928–1929

The Newton of the displacement theory has not yet appeared. One need not fear that he will fail to arrive; the theory is still young and is today widely doubted, and, in the end, one can hardly blame the theorist if he hesitates to spend the time and trouble on the elucidation of a law when no consensus prevails about its validity. In any case, it is likely that the full solution of the problem of the forces [driving displacement] will be a long time coming.

ALFRED WEGENER (1928)

I have the conviction that the current preference for "economically valuable" research is as unwise as it is immoral. Unwise—for hundreds of examples show that any economically valuable discovery is based on numerous others that have no apparent economic benefits. But forego the latter, and the "economically valuable" research comes to a halt.—Immoral . . . because to the scientist the Holy of Holies must be discovery of the world around us and its laws, to which he dedicates his life. Utility is not his aim and it does not need to be, because he has already found another. If he switches to practical use, he abandons this sacred quest and has ceased to be a scientist in the true sense. Fortunately, the economic motivation that the scientists today assert to get the funds for their research are, in the majority of cases hypocrisy, something like Kepler, paying his way by casting horoscopes.

ALFRED WEGENER TO WILLI MEYER (1929)

Throughout the spring of 1928, Wegener could not stop thinking about the Arctic. It was not for lack of other work: he had published his description of the tornado that swept through the eastern part of Styria the previous September. This had turned his mind again to the theory of the mechanics of tornadoes and waterspouts, and he had written a highly original paper on their formation. He now believed that the genesis of tornadoes came from what he called a *Mutterwirbel* (mother-vortex) with a horizontal axis created by wind shear; he had discussed this at length with Letzmann, now in Graz to work with him; Letzmann agreed that this was a promising line of attack. This horizontally rotating vortex in the wind shear zone, at an altitude of 3–4 kilometers (1.9–2.5 miles), could then be pulled up to the vertical, where it connected with the powerful up- and downdrafts in a thunderstorm; this would increase the energy of the vortex, which would then travel along with the thunderhead. Once again, it was a phenomenon created at a discontinuity, with air masses of different velocities moving past each other, and creating waveforms that actually turned into vortices.[1]

In later February he had received the rest of Koch's notes and maps, and this cartographic and reconnaissance work reminded him again and again of what he had loved best about Greenland: the exploring. His mind kept turning to Georgi's recent plan; it was a good one, timely and likely to be productive. He also recalled the difference between his and Koch's difficult traverse in 1913 from east to west and de Quervain's relatively easy travels in that same year going from west to east, beginning from Disko Island near Jacobshavn. Georgi's instinct, to approach the Inland Ice from the east, in a section of coast that was not well known, was exactly the wrong approach. One should, Wegener now believed, begin any expedition from the *west*, and from a point within easy range of a major Danish settlement. Such an expedition could employ regular Danish commercial shipping and establish cooperation with the Danish government at trading stations and provincial outposts, to ease the logistics. Wegener also knew what Georgi did not yet know well enough: there is no need to look for danger in the Arctic; it comes to you on its own.

At just about the time that Marie Koch sent the parcel of her husband's diaries, notes, and maps, Wegener received a letter from Arthur Berson, who had been the head observer and aeronaut at Lindenberg twenty years prior. Berson was one of the leaders of AEROARCTIC and wanted Wegener to give an address at the second congress of that organization, to be held 18–23 June 1928 in Leningrad (St. Petersburg). Wegener was on the board of directors and a member of both its aerological-meteorological commission and its technical committee. Berson had written to Wegener to ask him for a title for his address and to promise that whatever he wrote should be published in *Arktis*, probably in the July 1928 issue. He noted, however, that what was important was that Wegener's voice be heard at the congress; he should put more energy into the lecture than the paper for *Arktis*.[2]

Wegener agreed and sent a title: "Working Conditions and Tasks for a Station on the Greenland Inland Ice." As German polar historian Cornelia Lüdecke has pointed out, "Wegener often went his own way," and this was one of those occasions.[3] Wegener not only sent the title to Berson but also told Erich Kuhlbrodt (now back from the Meteor Expedition) that he would give a lecture at Hamburg in September (at a major scientific meeting) on "Problems of Greenland's Inland Ice."[4] The aeronautical part of Wegener's "working conditions and plans" for Greenland had nothing to do with zeppelins (the whole thrust of AEROARCTIC). Rather, he urged that fixed-wing aircraft could carry out aerial reconnaissance of entry points to the interior of the ice and rapid surveys of the *Randzone* (the edge of the ice). Later in such an expedition, one or more such aircraft, with their wings removed, might serve as propeller-driven motor sleds to pull loads of material to the interior of Greenland.[5]

Wegener got the idea to employ airplanes as tractors on an ice cap from the experiences of Douglas Mawson (1882–1958), the British/Australian Antarctic explorer, who had taken a damaged Vickers airplane to Antarctica in 1912. Its wings had been damaged in an aerial flight demonstration the year before, but stripped of its wings and mounted with skis, it was to be Mawson's snow tractor. Apparently, the plane managed to pull one load in Antarctica before the engine seized up; it was a prototype aircraft from an earlier age of aviation, far inferior to the kind of airplanes being built in the 1920s. Moreover, Richard Byrd, an American navy officer who claimed in 1926 to have been the first man to fly over the North Pole, was in 1928 already in Antarctica with a trimotor aircraft similar to the one that he and Floyd Bennett had used in their flight to the North Pole; Bryd's expedition also had snow tractors.[6]

Wegener appears to have already been moving from championing airships to championing fix-winged aircraft when his confidence in the whole airborne approach to the Arctic was badly shaken by a series of events in May 1928. Following the successful trans-Arctic flight of the airship *Norge* in 1926, Umberto Nobile (1885–1978) attempted another trans-Arctic flight in 1928 in his airship *Italia*. The airship crashed and tore apart while returning from the pole; ten members of the crew, including Nobile, were thrown to the ice and badly injured, and some died. The remaining crew members were swept away with the buoyant remains of the airship and perished. The international, month-long rescue effort included Roald Amundsen himself, who died when the French seaplane carrying him to the rescue headquarters disappeared; his body was never found. A Swedish pilot eventually rescued Nobile, but on returning to pick up others on the ice, he crashed his plane. In the course of the ensuing month of this incredibly bungled rescue attempt, more men died.[7]

Shortly before these events, Wegener and Köppen had received a visit in Graz over the Easter break (the first weekend in April) from Wilhelm Meinardus (1867–1952), a geographer on the faculty of the University of Göttingen. They were all mourning their friend and colleague Emil Wiechert, who had died in March. Wegener was just then in the process of finding someone to finish Wiechert's chapter for the *Lehrbuch*; he was to have written the extensive chapter entitled "Mechanics and Thermodynamics of the Solid Earth." His final illness had prevented him from completing major sections, but Wegener had been able to convince Beno Gutenberg, who had already written subchapters on seismology, isostasy, and movements of Earth's axis, to finish the job. The timeline was short, but Gutenberg was grateful for the offer, not least because he was one of Wiechert's students and a possible successor to him at Göttingen.[8]

Meinardus, Köppen, and Wegener reminisced about *Altmeister* Wiechert, but that is not why Meinardus had come to Graz. He had been working for several years on "explosion seismology," in which one set off an array of explosive charges and recorded (with portable seismographs) the resulting reflection profiles from the subsurface. He had had a good field trial of this technique working with Hans Mothes (1902–1989) in 1926 on glaciers in the Austrian Alps, and Mothes had gone on to develop the technique extensively. Meinardus had come to ask Wegener whether he would be willing to lead a summer expedition to Greenland to test this technique on a larger scale, where the ice was perhaps hundreds of meters thick, rather than a few score. He was prepared to use his influence to obtain funding from the Emergency Committee for German Science (Notgemeinschaft der Deutschen Wissenschaft), the organization that provided the means for almost all of German science at this time.

Wegener seems to have accepted the offer nearly on the spot; not only that, but he replied that the summer expedition to test the seismological technique in 1929 should serve also as a reconnaissance expedition for a two-year expedition in 1930 and 1931 to solve the "full range of questions" concerning the Inland Ice of Greenland. Wegener was, by the common standard, too old for this kind of work. Yet he seems to have convinced Else that Greenland was by now so well known that it was fruitless to even attempt adventurous record-setting expeditions. Moreover, as leader of the expedition, the most strenuous physical tasks would fall to others and not to him. Finally, the technical side of polar travel had developed to the point where he would have a very different set of tools in order to overcome the obstacles and the distances.[9] Everything we know of every Arctic expedition to this date (1928) tells us that none of these three statements were very near the truth, and Wegener certainly knew that. It was, however, the story with which he overcame her natural objections, and he stuck to it.

Wegener was by nature restless and physically active. Each year in Graz he had chafed more at the isolation, and he traveled further afield to scientific meetings. Moreover, by the summer of 1927, he was pushing himself in a series of Alpine ascents in Austria and Germany, climbing peak after peak. Alfred's longest stay in any one place since leaving home in 1906 was the three years he spent in Hamburg between April 1919 and the April 1922 trip to Mexico. He had arrived in Graz in April 1924, and he had not been anywhere, with the exception of weekend trips and brief summer vacations, in *four years*. He was ready to move, and this was his best chance.

In April 1928, Wegener had accepted the role of contributing editor to *Arktis*, but following Meinardus's visit and the crash of the airship *Italia*, he withdrew his paper from the planned AEROARCTIC meeting in Leningrad and withdrew any plan to publish it in the journal *Arktis*. A later communication with Willi Meyer, author of a book on the Arctic crash of the zeppelin *Italia* (*Der Kampf um Nobile*, 1931), shows Wegener's concern that AEROARCTIC was more about economics and individual "milestone" competition than about science; he also withdrew at this time from a planned (polar) zeppelin flight he had nominally agreed to join, one that was to take place in 1929.[10]

By June 1928, at the latest, Wegener had committed completely to Meinardus's suggestion of a summer expedition in 1929, as well as to his own idea to plan and lead a full-year expedition to Greenland in 1930–1931, for which the 1929 trip would be a reconnaissance (*Vorexpedition*).

Already in June he had provided a sketch of his plans to Peter Freuchen, asking for advice, and by July he was already in correspondence with Georgi about how to meld the latter's plan to study the high-velocity current in the upper atmosphere (the jet stream) with the plan hatched by Wegener, Koch, and Freuchen before the war: three stations across Greenland at latitude 71° north, all three performing aerological work and tracking the weather.[11] Wegener was forced to approach Georgi immediately, as the latter had already submitted his own plan to the Notgemeinschaft for an overwintering station in 1929 north of Angmagssalik, with the idea of overwintering—himself—on the Inland Ice in 1930.[12] Wegener did not want these proposals to compete, and he offered Georgi the overwintering at the mid-ice station if he would agree to merge the plans. To this meteorological work Wegener wanted to attach a major glaciology program—Meinardus's seismology of the ice, but also glacier motion studies and analyses of the microcrystalline structure of névé, firn, and glacier ice.

Such a plan could not, Wegener's desire notwithstanding, entirely avoid commercial interests, nationalism, or personal ambition. Both the Americans and the British planned Greenland overwinterings in 1930–1931. The American expedition, modest in scope (two men) and sponsored by the University of Michigan, was inspired by William H. Hobbs (1864–1952) and planned meteorological and glaciological work with a specific aim: to test Hobbs's theory of the "glacial anticyclone." Hobbs believed that a giant ice cap like Greenland or Antarctica could create its own weather, and that the extremely cold air of the interior would create a permanent "blocking high pressure center" that would steer cyclonic storms away from the interior of Greenland, either to the north or to the south. To test this idea, the Americans would build a base in the Upernavik region at latitude 72° north on the west coast of Greenland.

The British planned the (privately funded) "British Arctic Air Route Expedition" for 1930–1931 to detect wind conditions in the interior of Greenland. With the Atlantic now "conquered" from the air, Europeans and Americans were preparing for commercial and (without saying so) military transatlantic flights. The Great Circle route across Greenland from North Europe seemed most economical of time and fuel; the question was the weather and wind conditions. The British expedition was planning some essential meteorological research, but it mostly planned to test the flight conditions. They would work from the southeastern coast of Greenland at a shore station near the settlement of Angmagssalik (now Tasillaq) at latitude 65° north, but they also planned an Inland Ice station for overwintering at an altitude of 2,600 meters (8,600 feet) about 160 kilometers (100 miles) from the coast.

By 1928 Germany had been scientifically isolated for almost fifteen years. The Meteor Expedition had been the first major scientific expedition to leave Germany since early 1914. A German expedition to Greenland would be the first land-based scientific effort outside Germany since Filchner's expedition had returned from Antarctica in 1912. The German scientific establishment in Berlin, including Albrecht Penck and the Antarctic explorer Erich von Drygalski, knew and respected Wegener. Drygalski knew, or suspected he knew, that Wegener was thinking of going to Greenland again because in July 1927 Wegener had written to him asking whether he had any leftover copies of the results of his two-year expedition to West Greenland in 1891–1893.[13] Actually, Wegener was just looking for any information to help flesh out his and Koch's glaciological work, but the result was the same: Drygalski wanted to get Wegener involved once again in Germany's polar plans.

Meinardus was the perfect envoy and intermediary to involve Wegener in Germany's polar future. Meinardus, like Wegener, was both a meteorologist and a geophysicist. He had written his doctoral dissertation in the 1890s under Köppen at Hamburg, work that is today credited with the discovery of what is now known as the Intertropical Convergence Zone. He had developed a great reputation as a climatologist, and he and Köppen were in 1928 planning joint work on a new edition of Köppen's climate handbook. When Drygalski had returned from the Antarctic in 1903, he had entrusted the working up of the expedition's data to Meinardus, and now, more than twenty years later, massive volumes of results continued to appear through the latter's industry and acumen. There was a confluence here of mutual understanding, experience, and respect. More to the point, Meinardus and Drygalski were interested in science, not adventure travel, and Wegener knew this. The thought of combining his own long-planned Greenland work with Meinardus's glacial seismology was almost too good to be true; a seismic profile of the Inland Ice could demonstrate the truth of isostasy with measurements. If the theory of "glacial loading" were true, Greenland should be shown by a seismic reflection profile to be a gigantic ice-filled bowl. The implications for displacement theory would also be substantial and positive.

A number of things now had to proceed concurrently. Wegener had to develop a plan both for the reconnaissance expedition and for the main expedition—one had to know what the main expedition was going to do before one could carry out a reconnaissance about where best to accomplish it. To flesh out this plan, he wanted to talk further with Freuchen, who, since the death of Koch, was the Arctic explorer in whom he had the greatest confidence. Until the loss of his leg in 1923, Freuchen had been exploring in the Arctic for eleven consecutive years. After talking to Freuchen, Wegener had to come to an agreement with Georgi about the way in which their expedition plans should be combined. He then had to fill out the scientific complement for the reconnaissance expedition with at least two other men and present a formal plan to the Notgemeinschaft for approval.

That, however, was not all. He still, as he said to Köppen, had to "liquidate" all his current writing projects. These included the work on the 1912–1913 expedition, on which he signed off in June. He was waiting only for Gutenberg to complete what Wiechert had left undone before the *Lehrbuch* could be sent to the press. The last and largest project he had to liquidate was the fourth edition of his book on the origin of continents and oceans.[14]

The Origin of Continents and Oceans, 1928

Wegener made it quite clear in the fourth edition of his book that the origin of continents and oceans belonged to his past and not to his future. This edition was something he had begun to work on in late 1927 before he decided to redirect his energy toward Greenland. Now, in the spring of 1928, it had the same status as his work on the 1912–1913 expedition to Greenland and his work on the geophysical *Lehrbuch*—it was something to finish as quickly as possible so that he could place his energy elsewhere. Before it was even complete, he had told Kurt that it was the last edition he would attempt; the component literatures were now too numerous and too specialized for him to master.[15] He repeated this judgment even more feelingly in a brief foreword, when he said, "At times . . . my spirits failed me during the revision of the book."[16]

The book was an uneasy hybrid, and he knew it. "Whereas," he wrote, "the earlier editions were for the most part simply a presentation of the theory and a collection of individual facts that spoke to its correctness, this new edition represents a transition to a synoptic review of the various new branches of research [inspired by the theory]."[17] The book was still about the origin of continents and oceans, but it might have been better titled *Introduction to the Displacement Theory*, or *Handbook of the Displacement Theory*, as it was much more an introduction to a field of study, or a synoptic collection of varying opinions on significant issues within that field, than a dedicated theoretical treatise. He wrote the book in sections, and the sections do not always agree with one another in their conclusions. The shell of advocacy from the earlier editions remains, but he interjected and modified so thoroughly, adding reservations and contrary opinions, that the overall tone in many passages is almost agnostic. Perhaps the best title would have been *Basic Questions Concerning the Displacement Theory*, as the finished work has many more "fundamental questions" than firm answers.

This is not to say that he had succumbed to doubt; he was as convinced as ever of the rightness of his central idea and equally convinced that the preponderance of the evidence in most of the fields of earth science still pointed toward his hypothesis rather than away from it—notwithstanding the abundance of cogent objections to his specific formulations. Sometime in early 1928 he had shown Benndorf a letter he had just received from Denmark indicating that the 1927 longitude measurements in Godthab, Greenland, seemed to confirm Jensen's 1922 estimate of a drift of Greenland to the west at 36 meters (118 feet) per year. Benndorf expressed astonishment that Wegener did not seem more pleased, to which Wegener laconically replied, "I've always known it was true; the question is now whether the others will finally believe it."[18]

Though fully convinced of the truth of his hypothesis and confident that these 1927 Greenland measurements had "put the theory on a different footing," now that there was "astronomical proof" of displacement, he was by 1928 tired of battling over numerous matters of detail marshaled by men who had become obsessed with opposing his theory, such as Georg Pfeffer and Carl Diener.[19] This sort of jousting had come to a head for him in late 1927 when he finally broke his long-standing rule of not engaging in disputes in learned journals with individual authors who opposed him.

He lost patience on the occasion of the publication in geophysical journals of two articles, as well as an ensuing book, by Hermann von Ihering (1850–1930), a German ornithologist and émigré to Brazil who had returned to Germany in 1920 after forty years abroad and had decided to spend his retirement years, beginning at age seventy-seven, in opposing Wegener. Wegener's rebuttal, published in *Zeitschrift für Geophysik*, pointed out seven major gratuitous (*willkürlich*) misquotations of Wegener's work by von Ihering, and he noted that the latter was still working from a copy of the second edition and had not seen either the third edition or the Köppen and Wegener climate book. Moreover, von Ihering had got it in his mind that Wegener and Irmscher (the botanist from Hamburg) were coauthors, though Irmscher's work was quite independent of Wegener's own from the standpoint of authorship.[20]

Wegener's article in rebuttal made an important subsidiary point. Though von Ihering was totally opposed to Wegener's notion of continental displacements, the data von Ihering had published in his recent book, *Die Geschichte des Atlantischen Ozeans* (1927), was in profound agreement with the basic contentions of Wegener and Köppen's book on the climates of the past.[21] This was something that Wegener had often seen before—that the empirical facts produced by a scientist's research and a scientist's theoretical commitments were entirely separate entities. Wegener was more than happy to infuriate von Ihering by including his researches in the fourth edition as evidence favoring displacement.[22] He would not, however, engage in disputation any longer, and his decision not to reply combatively (after writing about von Ihering) kept the fourth edition remarkably free of contentiousness, without avoiding disagreement.

The fourth edition of Wegener's book is nearly 100 pages longer than the third, making it even more difficult to summarize. Fortunately, the fundamental structure of the argument remains; most of the new material revises and extends evidence for the displacement theory, and the remainder tries to address remaining questions by giving a range of opinion. Thus, while preserving the notion that there are advocates of continental permanence, of land bridges, and even of the old and discredited version of the contraction theory, the book is less a contest between opposing viewpoints than it is a series of differences of opinions within what Henry Frankel has aptly called the "sub-controversies" emerging from Wegener's theory.[23]

One major shift, immediately evident, is Wegener's decision to move the geodetic argument—the actual measurement of displacement—closer to the front of the book, just after the historical introduction and the summary chapter entitled "Character of the Displacement Theory, and Its Relationship to Previously Dominant Ideas Concerning Alterations of Earth's Surface in Geologic Time."[24] In the first edition (1915) and also in the second edition (1920) the measurement data and claims for continental displacement had been given in the last chapter. In the third edition (1922) they were in the final chapter of "presentation of the evidence" (*Beweisführung*), just before the extensive and more speculative "elucidations."

Wegener wanted very much to claim that continental displacement had finally been measured, and therefore he began by saying, "We begin the presentation of the evidence with the proof [*Nachweis*] of contemporary continental displacement through repeated astronomical position-finding, on the grounds that recently (by this method) the first real proof of the contemporary displacement of Greenland—predicted by the displacement theory—has been furnished, and because this sort of quantitative corroboration will probably be considered, by the majority of scientists, to be the most reliable and exact evidence in favor of the theory."[25]

This initial paragraph somehow captures the flavor of the entire book. The long, convoluted sentence, the shift back and forth between the notion of proof and that of simple evidentiary corroboration, and the hopeful sense that a recent result would tip the balance in favor of measurement of drift all reflect the problems that Wegener was facing. Reviewing the geodetic argument, we can see that Wegener's problem was one that always seems to bedevil scientists looking for "long geophysical time-series." Whatever technique was being employed (whether in 1823 or in 1927) always seemed to seek its answers just at the instrument's limit of resolution. The reduction of systemic and personal errors always seemed to fall short of the desired quantity: the probable or possible error in every case exceeded the quantity being measured.

These geodetic measurements in Greenland involved the methods of lunar distances, lunar culminations, and stellar transits. They were recorded in successive decades by chronometer, telegraphic time signal, and finally radio signal, with constantly evolving zenith telescopes and transits. The stations from which they were made were imperfectly mapped and, in the case of the 1871 Germania Expedition, not known accurately within a kilometer. While the instruments employed in the course of a century of observations were increasingly sophisticated, only the 1927 measurements—made with a radio time signal and a zenith telescope with an impersonal micrometer that could measure the culmination independent of the observer's reaction to it—would even begin to meet a reasonable level of accuracy by the standards of the 1920s.

Wegener continued to press the question of measurement because he had always thought that it was the only true means to establish continental displacements as a fact rather than a hypothesis. Moreover, from the time of his graduate student years and his PhD dissertation, the painstaking "stitching together" of measurement protocols with different error margins, different measuring techniques, and different systems of reduction had been his stock-in-trade; in short, he believed *completely* in the ability of scientists to calibrate and make accurate observations with precision instruments.

In the next chapter, "Geophysical Arguments," a different series of problems appears. The original geophysical premise was that the continents were made of lighter material than the ocean floor in which they floated, and that they moved through the ocean floor like icebergs through water. The principle of isostasy governed the flotation, and the difference between short-term behavior of the subcrust (rigidity) and its long-term behavior (fluidity, viscosity) allowed the continents and oceans to react by fracture to short-term stresses and by displacement and flow to long-term stresses.

Now Wegener had to confront the complications and uncertainty in the analysis of the behavior of the crust and mantle (the Sial and Sima) brought forward by seismology, gravity, and magnetic surveys far more sophisticated than those available early in the century. While continuing to make the same argument about continental flotation which he had made in the first edition, he was forced to admit that the ocean floors perhaps had a partial layer of Sial—*whatever that was*—on them, since no one knew any longer precisely what rocks (not just densities but actual mineral content) dominated the composition of the "Sial" and the "Sima," nor was there any possibility of a consensus on the viscosity of Earth at any depth.[26] It is not that anything fundamentally disconfirmed his ideas, but that the forward march of geophysics, geochemistry, seismology, gravimetry, and solid mechanics had deprived his theory of the certainty attendant on its initial simplicity. On every topic there was now a range of opinions rather than a "yes" or a "no." Wegener was resolute in presenting these complexities and uncertainties, while maintaining that the general principles involved still pointed toward displacement and away from permanence.[27]

For instance, Wegener was able to argue that it was a matter of little consequence that the coefficient of plasticity or "stiffness" of Earth was that of steel, with a coefficient of 10^{11} in the mantle and 10^{12} in the core, since this stiffness, detected through seismology, is a measure of instantaneous response. On the other hand, the coefficient of viscosity—the parameter that would govern the ability of Earth to flow on long timescales, was estimated to be anywhere from 10^{13} to 10^{21}—a variation from scientist to scientist across *nine orders of magnitude*. Wegener was (as was his habit) simple and direct: "Under forces applied over geological time scales, the earth must behave as a fluid; for example, this is shown by the fact that its oblateness corresponds exactly to its period of rotation. But the critical point in time where elastic deformations merge into flow phenomena depends precisely on the viscosity coefficient."[28] In 1928 no one had empirical knowledge of what this was, either for Earth as a whole or for any particular layer of Earth.

There is a complex and highly interesting history attached to this debate, and one hopes that someday someone will write it. Suffice it to say that in the 1920s the debate over the states of matter (solid, liquid, gas) and their behavior under high temperature and confining pressure was increasingly framed in terms of the behavior of rocks at great depth. That is, fundamental questions in solid mechanics were being debated based on evidence and speculation concerning the interior of Earth, rather than on laboratory experimentation.

As Adrian Scheidegger has pointed out, of those branches of "continuous displacement theory" which were more or less well developed, it was the theory of "plasticity" that was most pertinent to discussions of displacements observed in Earth's crust. However, this theory of plasticity described the (observed) behavior of a metal during cold-working, and this is why there is always reference to the elasticity of steel in such discussions. Again, the results do not apply to inhomogeneous materials under high confining pressures many thousands of times greater than those available in a laboratory and acting on timescales millions of times longer than a human life.[29]

This shifting balance of certainty and uncertainty which Wegener encountered in the various subfields of earth science in 1927–1928 is even more evident in the next chapter, "Geological Arguments." Faced with geophysical uncertainty in his arguments—where certainty had prevailed before—Wegener was more and more forced to rely on geological, paleontological, and paleoclimatic evidence. Yet it was on the geological evidence that he had been most thoroughly battered by his critics. Most of the favorable geological evidence for displacement was from the Carboniferous, and there was not much of that—especially after his critics had finished picking away at the details; there remained the secure evidence of glacial deposits from a Carboniferous ice cap leaving remains on widely separated Southern Hemisphere continents, as well as the *Glossopteris* flora, some sediments, and some trend lines of mountain ranges.

In 1928, however, Wegener found himself in possession of a cornucopia of positive, detailed, globally distributed, and powerfully confirmatory *geological* evidence of exactly the sort of continental connections, and subsequent disruptions, that were the content and basis of his entire displacement theory. Foremost among these were the results of an extended reconnaissance in South America, funded by the Carnegie Institution of Washington and carried out in 1923 by Alexander du Toit (1878–1948).[30]

Du Toit was one of South Africa's leading geologists. He had been a member of the South African Geological Survey since 1909 and of its irrigation department since 1920 (the latter post gave him more flexibility in his fieldwork). He went to South America in 1923 specifically to test Wegener's displacement hypothesis from the standpoint of the field data in South America in comparison to the rocks he had worked on in South Africa for fifteen years. Because his funding sponsors were in the United States, he had to restate his objective in a way that did not mention Wegener directly, but the intention was always there.[31] Du Toit spent five months in the field between late June and late November 1923, in Argentina, Chile, Uruguay, and Brazil, everywhere accompanied by leading South American geologists and paleontologists. He covered a 45° arc of latitude in a 10° wide band of longitude along the eastern coast of South America.[32]

He presented his fieldwork in 1927 as *A Geological Comparison of South America with South Africa* and tried in the descriptive sections to strike as neutral a tone as possible (once again a requirement of his sponsors in the United States). Nevertheless, the book as a whole amounted to a full endorsement of Wegener's continental displacements as the "working hypothesis" with the most explanatory power. At the end of du Toit's synthesis of the stratigraphical and paleontological data he had collected and analyzed along the eastern coast of South America and in parts of the Andes, he wrote,

> That these two continents were intimately connected during several geological epochs will, I venture, be acknowledged after the perusal of the preceding pages, though the manner of such union would admittedly be speculative. In preparing this review an attempt was made first of all to write the historical account, irrespective of any hypothesis as to the manner of such union or of the ultimate mode of separation of the land-masses, though it became evident, as the data were assembled, that they pointed very definitely in the direction of the displacement hypothesis, and that they could most satisfactorily be interpreted in the light of that brilliant conception.[33]

He then reiterated his conviction of the superiority of Wegener's displacement hypothesis over all other explanations of the similarity between South America and Africa, in a chapter entitled "Bearing on the Displacement Hypothesis." Du Toit here wrote,

> Such points of resemblance have now become so numerous as collectively almost to exceed the bounds of coincidence, while they are, moreover, confined not to one limited region nor to one epoch, but implicate vast territories in the respective land-masses, and embrace times ranging from pre-Devonian almost to the Tertiary. Moreover, these so-called "coincidences" are of a stratigraphical, lithological, paleontological, tectonic, and climatic nature. . . . If . . . the two land-masses are pictured as having moved closer together . . . a great number of observations and deductions are now found to be brought into apparent harmony and these possible "coincidences" are disposed of in the simplest fashion.
>
> This is precisely what the displacement hypothesis effects, thereby providing a simple explanation of many otherwise puzzling observations. The fact that many eminent scientists have cast doubt upon its geophysical possibility should not be permitted to cloud the issue.[34]

Moving along to his conclusions, du Toit argued that it was of particular importance that his readers recognize the decisiveness of this *geological* evidence, "for those arguments based on zoo-distribution are incompetent to do so [i.e., decide], being as a rule equally, though more clumsily, explicable under the orthodox views involving lengthy land connections afterward submerged by the oceans."[35] Du Toit, however, concluded,

> Such an analysis [of the existing data], it can be affirmed, does not favor the notion of one or more relatively narrow connecting links or "land bridges" lasting down to the early Mesozoic, but on the contrary supports the presumption of some continuous land area embodying those sections of Gondwanaland that are now represented in these two continents. It furthermore strongly favors the admittedly revolutionary idea that geographically these two portions were appreciably closer in the past; indeed, the evidence is, I think of sufficient weight to warrant such a viewpoint being adopted as a working hypothesis at any rate . . . the displacement hypothesis, if not an actual explanation of the phenomena, would at least seem to contain more than a germ of the truth, despite its revolutionary and heterodox nature and apparent lack of agreement with geophysical considerations.[36]

Wegener recognized du Toit's book immediately for what it was: the most influential geological support he had ever received. He quoted from it at length in his "Geological Arguments," though he remarked, "If we wanted to cite every detail in the book which favors the theory we would have to translate it [the book] from start to finish."[37] Wegener then quoted du Toit's comment (given above) that all the data in South America pointed—from the beginning of his investigation—toward the displacement hypothesis; Wegener modestly left out du Toit's characterization of his work as "a brilliant conception."[38]

Wegener reproduced du Toit's map of the reassembly of South America with Africa, fully agreeing with him that a gap of 400–800 kilometers (250–500 miles) between coastlines was needed to accommodate present-day differences. "I am bound," wrote Wegener, "to agree completely with this point. Not only must there remain room between the two coastlines for the shelves that extend in front of them, but possibly even for the material composing the Mid-Atlantic Ridge [*Bodenschwelle*]." Wegener concluded this discussion thus: "I must admit that the reading of du Toit's book made an extraordinary impression on me, as I had scarcely dared to expect a geologically perfect match between these two continents."[39]

If this were the only geological support Wegener obtained between 1924 and 1928, it would still have been a stunning endorsement of his original conception and, moreover, would still have validated the very transatlantic correspondences scoffed at by British, German, and North American geologists. But it was *not* the only such support, for Wegener had by now read and digested Emile Argand's *Tectonique d'Asie*. Here again was geological support for his displacement hypothesis from an internationally well-known, if admittedly controversial, supporter. Wegener quoted Argand extensively, giving his principal results for the unified character of the great folds extending from the Alps through the Himalayas to the marginal ranges of East Asia.[40] He concluded by quoting Argand as follows: "The elegance with which the displacement theory explains these significant facts, which were not known when the theory was first advanced, is certainly a strong point in its favor. Strictly speaking, none of these facts really proves displacement theory or even the presence of sima [as a medium in which the mountains float] but they all fit in excellently with both ideas to an extent that makes them highly probable."[41]

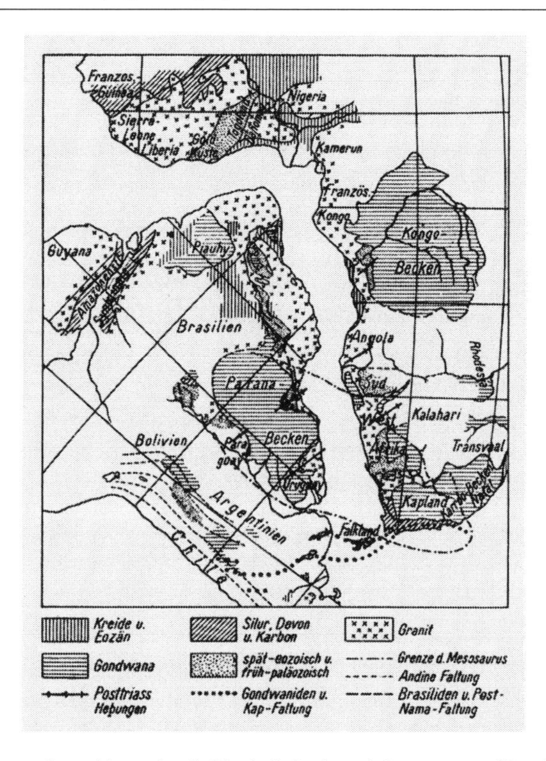

Wegener's redrawn map from Alexander du Toit's *A Geological Comparison of South America with South Africa* (Washington, DC: Carnegie Institution of Washington, 1927). Wegener was quite willing to leave the large "gap" between the continents in du Toit's resonstruction. From Alfred Wegener, *Die Entstehung der Kontinente und Ozeane*, 4th (completely revised) ed. (Braunschweig: Friedrich Vieweg & Sohn, 1929).

If there were no comparable supporters for Wegener's theory of the separation of the North Atlantic continents (North America from Europe) or for the separation of Madagascar and India, he could point nevertheless to an abundant literature on the East Indies, especially the Sunda Archipelago and the Celebes, produced and published by Dutch geologists, with increasing reliance on Wegener's displacement theory as the most likely explanation for their complex geological features.

From the time of Molengraaf's work on coral reefs in 1916, Wegener had always had supporters among those whose fieldwork covered the East Indies and Australasia. Van der Gracht, in the published version of his address to the American Association of Petroleum Geologists in New York in 1926, the "symposium" volume that had finally appeared in early 1928 (and that Wegener had seen), remarked that the explanation Wegener was able to give for the Sunda Archipelago's otherwise bewildering tectonic structure "is the reason why the Dutch geologists (Molengraaf, Brouwer, Wing Easton) who worked in the East Indies, are invariably favorably inclined to Wegener's hypothesis. I have also visited this area: the evidence is indeed striking."[42] To this roster of Dutch supporters Wegener was able to add G. L. Smit Sibinga (1895–1963), who had concluded that "a working hypothesis based on the fundamental ideas of Taylor and Wegener" was superior to any other mode of approach in explaining the complex geological facts of this region.[43]

One can imagine the pleasure and satisfaction that Wegener took in this transformation of his theory from an outlandish supposition to a real working hypothesis for geologists. In 1912 this had been the sum of his ambition, and whatever ground he had had to relinquish in 1920, 1922, and 1924 in terms of specific proposals and ideas, he could see, as he compiled (in 1928) the evidence in the various categories from the component sciences, that he was exactly where he had always wanted to be. He had everything except the absolute proof of ongoing displacement in the present, and until that absolute proof appeared, the most that his theory of continental displacements could achieve was the status of a working hypothesis. It was now a working hypothesis in Europe, in Africa, in South America, in Asia, and in Oceania, though still short of the mark in the United States and Canada.

There was no need any longer, under these circumstances, for a detailed rebuttal of his geological critics, or even to spend time discussing their concerns. He ended his consideration of the geological evidence with the following observation:

> If one surveys the results of this chapter, it is impossible to escape the impression that the displacement theory can today be regarded as well founded geologically even down to its detailed assertions. It is true that there are many opponents of the theory among geologists today, and objections have been raised on different topics by Soergel [35], Diener [108], Jaworski [109], W. Penck [111], A. Penck [110], Ampferer [68], Washington [113], Nölke [114] and many others. However, it can be said that generally speaking these objections—where they are not simply misunderstandings (as in the case of Diener, in particular)—concentrate on peripheral issues of minimal significance for the basic concepts of displacement theory.[44]

The next section of the book contains Wegener's two chapters on paleontological and biological evidence and on paleoclimatic evidence. The distinction here is somewhat artificial from the standpoint of subject matter, since the evidence for paleoclimate was also paleontological, especially where the *Glossopteris* flora was concerned. Why Wegener did this becomes somewhat clearer when we think of this book in its new format and with its new function as a means to address workers in different fields of earth science. The chapter on paleontology and biology is addressed to paleontologists and biologists, and the chapter on paleoclimate is addressed to climatologists; it is a matter of where the focal interest and background lay for each group.

These two subject matters (these two chapters) served different functions for the theory of continental displacements. Paleontology mostly served Wegener in showing that the fossil record demonstrates continuity between (now) widely separated continents in the past, continuity later disrupted as these separated continents underwent continued biological evolution and their flora and fauna continued to diverge. Paleontology was therefore mostly about disjunction. Paleoclimatology, on the other hand, did not support continental displacements as much as continental displacements made sense of otherwise irresolvable contradictions when using latitude zone reconstructions. Moreover, as we approach the present, and as the continental configurations of the recent past more and more approximate those of the present, continental displacement has much less to say about climate, as pole movements alone can then handle most of the variation. In the deep past, only displacing the continents from their present positions back to some earlier close conjunction made sense of the coal, desert, and ice distributions.

Wegener did not try to use the paleoclimate chapter to compress the complete contents of his work with Köppen on climates of the past (1924) into a form where it could be interpolated into the story of continents and oceans. While the chapter on paleoclimate surveyed the problem of climate zones and gave a substantial historical introduction, Wegener concentrated all his energy in showing how only a combination of continental displacement and displacements of the pole, superimposed on one another, makes sense of the differences between the Carboniferous and the Eocene. Thus, though he did include in the paleoclimate chapters his maps for the Carboniferous and Permian (using the Hammer elliptical projection) and a map using the same projection to show the distribution of ice traces in the Carboniferous—now on widely separated continents—he restricted his discussion to the evidence that shows the stark contrast between the Carboniferous Earth and that of the more recent geological past. Using the new radioactive dates from the middle 1920s, his maps contrast Earth of 320 million years ago with Earth 20 million years ago.[45]

The next two chapters, one on continental drift and polar wandering and one on the question of the displacing forces, are also to a certain extent artificially divided, as in the case of paleontology and paleoclimate. The first of these two chapters, entitled "Fundamentals [*Grundsätzliches*] of Continental Displacement and Polar Wandering," can be considered as a discussion of the "kinematics" of these topics, and the second chapter, "The Displacing Forces," as a consideration of the dynamics. That is to say that the first is about the reality of such motions without any question as to why or how they happen, and the next addresses the question of why and how these may happen (physically) without any reference to specific historical displacements.

The "Fundamentals" chapter has the dubious distinction of being the least clear thing Wegener ever wrote and published. As he was a master of scientific explication, one can only surmise that here he was either not sure or very conflicted in what he wanted to say. Wegener had asked Milankovich in 1927 to find him a deep physical explanation for what could move the pole of Earth, not in the simple oscillatory sense that governed a shift of a few degrees back and forth with a regular time interval reliably measurable over hundreds of thousands of years, but something that could actually displace the pole of rotation. Milankovich had been unable to do this, and Wegener could find nothing in his own knowledge of physics or in the work of his contemporary colleagues which could help him overcome his perplexity.

Wegener's editing (in summer 1928) of a manuscript by Gutenberg (part of the *Lehrbuch*), entitled "Movements of the Earth's Axis and Polar Wandering," brought things to a head for him and forced him finally to rethink everything he had said and believed about the motions of the pole. Gutenberg, in his contribution to the geophysics book, had approached the problem head-on. He gave a discussion of the precession of the pole and of its nutation: the first a gyroscopic 23,000-year cycle, and the second a perturbation played out over some hundreds of days. He then presented the question of pole wandering and the variation of climate.

While not criticizing or otherwise commenting on Köppen and Wegener, Gutenberg pointed out that their theory of the timing of the ice ages used a parameter involving very slight oscillations of the pole of a few degrees on a scale of a few tens of thousands of years. He also, however, included a diagram of the motion of the North Pole *relative to Europe* specified and mapped in Köppen and Wegener's work, which showed displacements of the pole of more than 20° of longitude in as little as 500,000 years. These two things within the covers of the same book stood in direct contradiction to one another.[46] Was it moving 3° or was it moving 20°, and could it be doing both at the same time? Gutenberg suggested that the 3° motion was real and that the 20° motion was an *apparent* pole wander, caused by the northerly motion of the whole Eurasian continent.[47]

Wegener began, therefore, to construct a new argument asserting that the expressions "continental displacement" and "polar wandering" were being used in a variety of "different and contradictory senses in the literature" that had appeared thus far, and that it was necessary to clear up these confusions, to specify the relationships, and to give a precise definition of both terms. One should say at the outset that a good deal of this confusion had to be owned by Wegener, as he had himself at different times used the notion of pole wandering and continental displacement in different senses. It is very likely that he was, in fact, talking chiefly about himself.[48]

Writing in the summer of 1928, Wegener argued that the assertions of displacement theory related entirely to relative displacement of the continents with respect to an arbitrarily chosen portion of the crust, namely, Africa. This was true for the book on the climates of the past, and it had been true for the 1922 edition of the book on the origin of continents and oceans, but not true of the 1920 edition, which clearly showed Africa moving north toward the equator from a more southerly position. Wegener wanted to emphasize now that the choice of a reference frame was entirely arbitrary, although in the past he had chosen Africa because it seemed to have moved absolutely the least of all the continents.

Wegener wanted to make a second argument: that his new definition of continental displacement as relative only has nothing to say about changes of the longitude of the pole, or of any changes of the substratum relative to the crust of Earth. He asserted that from the standpoint of his arbitrarily chosen frame of reference we might say that America has drifted westward, or that Africa has drifted eastward, or that both may have happened. This was a logical consequence of choosing an immovable Africa as the anchor of an arbitrary reference frame.[49]

But immediately one sees that this thoroughgoing relativism is in conflict with the sorts of geological ideas he originally hypothesized (now newly confirmed by du Toit) that the Andes in South America have been upthrust as a consequence of the resistance of the Sima through which the continent of South America moved as it shifted to the west and away from Africa. The disconnect here is startling, and it is not at all apparent how both things can be true, at least when offered as evidence for one and the same hypothesis. It is true that Wegener had once said that the Sima might have moved eastward underneath South America, and this would still create a situation in which the west coast of South America was crumpled. On the other hand, if Africa moved away to the east and South America remained stationary, the geological argument for the folding of South America's western margin disappeared, as it was no longer an "advancing edge." The physiographic and structural argument (upthrust mountains) appears to have been uncoupled from the argument for displacements.

There is more. Wegener then asserted that pole wandering, a "geological idea," was also only *relatively* determinable. He had said something like this before, meaning that the evidence for the supposed migrations of Earth's pole was inferred from fossil and geological (salts, coal, etc.) evidence of former climate. Now he was abandoning the whole idea of a detectable, directional, physical motion of the pole. Instead, Wegener asserted that an abstract system of parallels of latitude should be imagined to rotate relative to the full surface of the globe, or contrarily a rotation of the whole surface relative to the system of parallels; once again, everything was relative. Under these terms, any physical relationship of the crust of Earth to the interior, and any relationship between such a "geologically shifted" pole and the possible motion of the actual axis of spin of Earth, was entirely irrelevant for this (new) definition of "relative pole motion." There is no mistake here. Wegener says quite plainly, "Geophysics can give no opinion concerning the existence or possibility of such a movement."[50]

Once again, this is a sharp contrast with almost everything he had previously asserted about the movements of the pole. In the past versions of his theory there were real motions of the pole of rotation of Earth, and these actual physical motions were potentiated by the actual motion of continents, in a dynamic interplay that was supposed to have controlled the entire history of the crust, with rearrangements of the crust forcing changes in the position of the pole consequent upon a redistribution of surface masses, with the newly reconfigured oblateness of Earth (given its new pole position) creating still further opportunities for high-latitude continents to "flee from the poles" toward the equator and, in consequence, control the succession of oceanic transgressions and regressions onto the continental surfaces.

The contradiction to every previous version of his theory here is so stark that had one been given this chapter without Wegener's name on it, one would have to imagine that it had been written by quite another theorist. But Wegener is the author, and he is quite serious. He produced a map to indicate that all the pole positions he and Köppen had ever calculated were relative to an arbitrary African reference frame, and to demonstrate this, he produced a (parallel) Southern Hemisphere pole-wandering path for every period from the Cretaceous through the present, as it would appear if the reference frame were *South America* held in its current position, with Africa having drifted to the east. This path is entirely different from any pole-wandering path he had ever drawn. He was now saying, and we must be clear about this, that he doesn't know where the pole has moved, or where the continents have moved, except relatively speaking in their distance from an Africa arbitrarily held at its current latitude and longitude.[51] He says that it is impossible to know in what direction the pole has moved from one period to another because the continents are moving at the same time that the poles are moving. Therefore, he says, his maps of the continental displacements do not show their actual motions, and therefore the actual vector of the migrating pole cannot be known, only known relatively to an arbitrarily shifted continent, in the context of an arbitrarily fixed continental frame centered on Africa.

This is rather difficult to reconcile with his previous constructions. There is no hint in *Klimate der geologischen Vorzeit* that the pole positions are arbitrary or even relative; polar, temperate, desert, and equatorial geological markers uniquely determine them. In the maps, from one period to the next, we were always shown an arrow indicating the direction of motion of the pole. Moreover, in the time period between 1922 and 1924 Wegener altered the timing, position, and extent of the migrations of the pole in order to accommodate Milankovich's calculations of very slight polar motions in the Quaternary, without moving any of the continents differently. In 1924 the motions of the continents were relative to the fixed reference frame of Africa, but there was nothing to have us believe that the motions of the pole and the latitude zones were anything but displacements of the actual physical pole of rotation, via some motion of Earth's axis of rotation which was determinate and measureable.

Having proposed this new relative scheme, Wegener introduced two new concepts, this time of real and geophysical (not relative and geological) motions: *Krustenwanderung*, or "crustal wandering," and *Krustendrehung*, or "crustal rotation." The latter he imagined to be an overall rotation of the whole crust of Earth relative to the interior

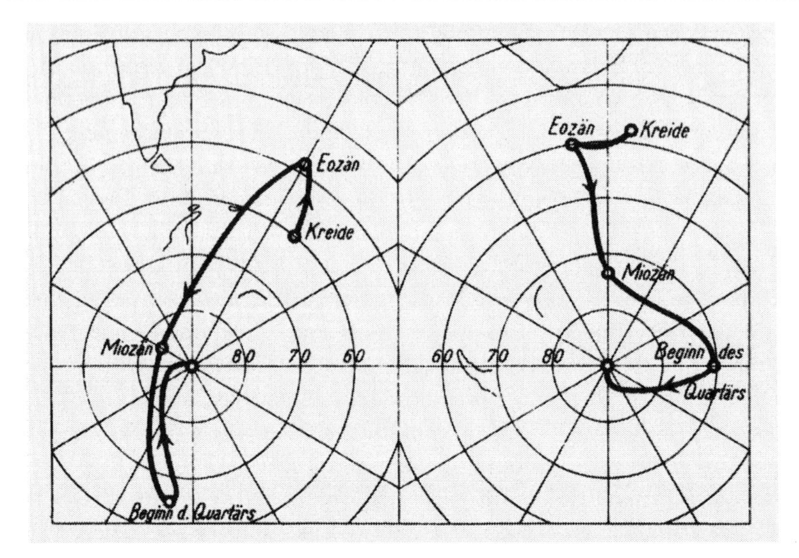

Wegener's map of the path of the South Pole since the Cretaceous. On the left is the shift of the pole seen from South America (in its current position); on the right, the shift of the pole seen from a stationary Africa. The point is that in this edition, all motions are relative to arbitrarily chosen reference frames. From Wegener, *Die Entstehung der Kontinente und Ozeane* (1928).

and axis of rotation, toward the west. This is obviously what he formerly called *Westwanderung.* The other of these two new concepts, the "crustal wandering," is a movement of the superficial crust equatorward, away from the axis of rotation. This is what he formally called *Polflucht,* and moreover these new names become a source of confusion because of the redefinition of *Wanderung* as a matter of *latitude* rather than *longitude.*[52]

He also said that in understanding what these motions mean, we have to separate the idea of an internal axial shift (*innere Achsenverlagerung*) from that of an *astronomical* axial shift relative to interstellar space.[53] These are very difficult concepts to discriminate, as it is hard to imagine an actual shift of Earth's axis which would not cause it also to point to a different position in interstellar space; Wegener obviously believed, for reasons not entirely clear, that this was a meaningful distinction.

From this point on, Wegener's argument is even less clear. He wishes to reintroduce his entire theory of transgressions and regressions of the oceans caused by the actual physical motion of the pole, but he wishes to have their driver, these "internal axial shifts," be held separate from any pointing of the axis of Earth very far away from the stars at which it currently points. So this internal axial shift does not appear to be what we would call today "inertial interchange true polar wander," in which Earth's axis, consequent on some internal mass shift, may point in any direction relative to the fixed stars. Instead, it is some unknown process whereby the crust does not slide relatively over the interior with respect to a fixed axis of rotation, but in which the axis of rotation shifts without somehow pointing anywhere else in space. Moreover, having made all these careful distinctions between superficial and geophysical polar wandering, he persisted in this same chapter in drawing maps giving vectors for the actual motion of the pole in the Devonian and the Carboniferous periods.[54]

With regard to this strange distinction of two kinds of pole movements, one can see that Wegener would be reluctant to move the astronomical pole—the ecliptic angle (the angular distance between where the axis points in space and the plane of Earth's orbit)—as any appreciable shift in the ecliptic angle would make it difficult to interpret the entire history of Earth's climate in terms of latitude zones defined by the kind of seasonality and distribution of precipitation which characterize Earth in the present. He was trapped between his own theory of displacements and motions of the pole of rotation on the one hand and the climate theory of Köppen on the other. These at one time had fit together, but since 1924 they could no longer easily be conjugated. The sticking point was the need to maintain the explanatory power of assuming latitude zones like those in the present—polar, temperate, and equatorial—at every point in the past, in conjunction with continental displacements, in order to make sense of the paleontological record. The depth of these contradictions had clearly not emerged in his mind until he attempted to rewrite the theory in 1928.

Had Wegener never embarked on a collaboration with his father-in-law and not embraced the idea that shifts of the pole were "givens" that needed to be "supplemented" by continental displacements in order to make sense of Gondwanaland (and the post-Carboniferous dispersion of this giant Southern Hemisphere continent), there is no reason why he could not, in 1928, have accomplished everything in his theory with continental displacements alone. One can see that in terms of another, but plausible, evolution of the theory of continents and oceans, the entire notion of large-scale motions of Earth's pole of rotation as an explanation for anything could have been discarded in favor of complex motions of the continents alone. The "theoretical cost" would have been high, but Wegener was already in the midst of yet another radical revision of his ideas; here the constraints on Wegener's theory were personal and biographical, much more than geological or geophysical.[55]

Toward the end of the chapter on displacements, Wegener ventured very gingerly into a way out of this difficult series of contradictions. The one way out appeared to be occasional but large excursions of Earth's pole of rotation relative to the fixed stars. Wegener was quite aware of the effect of such variations on Earth's climate system, but he then noted quite tentatively that from the Triassic through the early Tertiary there was no ice anywhere on Earth, although at least one of the poles was on land or near land most of the time, so that there would have been an opportunity to form ice sheets.[56] During this period, there was a marked advance of both animal and plant life toward the poles. "These variations in polar climate can be accounted for quite well by the assumption that the mean value of the 40,000 year fluctuation [of the angle of the axis, one of Milankovich's three parameters, that shifted between 21.5° and 24.5°] underwent significant changes during the course of Earth history, such that in periods when there was inland ice the oscillations were small and in the periods without ice—those periods in which organisms spread widely—the oscillations were large."[57] Wegener does not say how large, but his descriptions make it clear that these might be as high as 20° or more of tilt.

Wegener and Kepler

In the following chapter, "The Displacing Forces," Wegener began his discussion by pointing out that the "Newton of the displacement theory has not yet appeared."[58] Wegener characterized his arguments concerning the "ascertainment and establishment of relative continental displacements" as a purely empirical matter, arrived at by a consideration of an ensemble of evidence, without adopting any position on the causes. He then remarked that the formulation of the law of falling bodies and of planetary orbits had also been determined purely inductively, and only later did Newton arrive and show how to derive all of these from the single law of universal gravitation; this was the proper sequence: induction, *then* deduction.[59]

The law of falling bodies had been determined by Galileo, albeit purely kinematically, and the laws of planetary orbits, also kinematic, were the work of Johannes Kepler. Among Wegener's books left behind in Graz is a copy of Kepler's *Traum vom Mond* (Dream of the Moon), the Latin title of which is *Somnium*. Wegener's version was an annotated edition produced by Ludwig Günther in 1898.[60] In this manuscript, not published until after Kepler's death, the protagonist is transported to a point halfway between Earth and the Moon where he can see their relationship and compare their surface features. It also contains remarks on the formation of lunar craters by impact, astronomical position fixing, geomagnetism, the effects of solar and lunar tides on the surface of Earth, and other matters. Wegener's copy is heavily underlined and marked in the kind of blue pencil he used to correct proof.[61]

Kepler (1571–1630) had been a professor at Graz, and he had been a mathematician, physicist, and astronomer. He had produced a set of planetary tables, he had done fundamental work in optics and on the geometry of snowflakes, but above all he had solved the problem of the orbits of the planets and generated several empirical laws governing their motion. Kepler, in order to solve the problem of the motion of the planets around the Sun and to make his discovery that they moved in elliptical orbits with the Sun at one focus, used the observational data of his colleague and sometime employer, the Danish astronomer Tycho Brahe (1546–1601), after the latter's death.

Here was Wegener, 300 years later, in Graz, a mathematician, physicist, and astronomer, who had done work on optics and on ice crystals, who had written a book on the Moon and its craters, who had studied the lunar tides on Earth, and who had just spent a full year working up the data of his deceased Danish colleague, friend, and sometime employer, the geodetic astronomer and glaciologist Johan Koch. Among the many marked passages of Wegener's copy of Kepler's *Traum vom Mond* is one pointing out the extent to which Kepler's work was the absolutely necessary foundation for that of Newton. The passage is in Günther's commentary, following his discussion of Kepler's notion that the seas on the Moon must be attracted by the gravitational power of Earth just as those on Earth are attracted by the gravitational power of the Moon, and that the Sun, Moon, and Earth all work together: "I have treated the subject at some length to show how large a part Kepler played in the discovery of the gravitational law: he was Newton's teacher, and as Humboldt said, to call the laws 'Newton's' almost contains an injustice to the memory of this great man."[62]

Hence, Wegener was to play the part of Kepler and prepare the way for the Newton who would follow. To turn the motion of continents from a kinematic theory (showing the motions) to a dynamic theory (specifying the unifying forces) would be the work of another. The problem was not many problems but one: "It is clear from the outset that for this question of forces, the entire complex of continental displacements, crustal wandering, polar wandering, internal and astronomical axial displacements form a single problem."[63]

Wegener's consideration of these forces in the fourth edition was brief and mostly directed toward the clarification of the *Polflucht*, on which an interesting experiment had recently been done by a Dutch physicist, U. Ph. Lely; Wegener had repeated the experiment with Letzmann, who was then in Graz.[64] It was a simple lecture-demonstration using a cylindrical vessel placed on a turntable, with a flat cork with a nail driven into it used as a "continent." The vessel was then set in motion, and depending on whether the nail was up or down, the float either approached the upper margin of the fluid (the equator) or drifted toward the bottom of the vessel (the pole).[65] This was the effect of the distance in the "continent" between the center of mass and the center of buoyancy which Wegener had earlier investigated in detail.

In all, Wegener was willing to consider six different displacement forces, driving the five different kinds of movements (continental displacement, crustal wandering, polar wandering, internal axial displacement, and astronomical axial displacement). The pole-fleeing force that Wegener still believed could move the continents was (he was now convinced) not strong enough to create mountain ranges. Wegener briefly considered tidal friction (Earth tides) as a possible cause driving continents to the west, but he deemed the basis of this insufficiently established.[66]

He then considered a possibility offered by Schweydar which could drive the continents to the west and also depended on the attraction of the Sun and the Moon, which was that the axis of rotation with respect to the continents was different from the axis of rotation for Earth as a whole—such that the precession of the "continental axis" produced a very large force, much larger than the pole-fleeing force. He offered this only as a conjecture and said that he planned to develop it at a later time.[67]

In addition to these three forces, there was the possibility of the distortion of Earth's figure away from a rotational ellipsoid by streaming in the Sima, which would, under certain conditions, force a westward motion, though Wegener considered this unlikely. He also noted that several investigators were willing to invoke the idea of radioactive heating to produce sufficient fluidity in the Sima beneath the continents to allow convection currents to form; these might drive the floating continents.[68] Finally, as the sixth candidate force, Wegener considered that extensive excursions of the pole of rotation (however they were caused) would themselves become a force that, in conjunction with the pole-fleeing force, might "supply the energy required for folding."[69]

It is remarkable how easy and untroubled was this brief discussion compared to the tortured attempt in the previous chapter to reformulate what he meant by continental displacement and polar wandering. Wegener was perfectly content, in his Keplerian stance, to announce that "the problem of the forces which have produced and are producing continental displacement, except for the pole-fleeing force, already thoroughly investigated, is still in its infancy."[70] There was work enough for a Newton: "Continental displacement, splitting and lateral compression, earthquakes, volcanoes, the alternations of transgression and regression of the seas, and polar wandering are beyond doubt causally connected as part of a single grand ensemble. Their simultaneous intensification in certain periods of history shows this to be the case. Which of these are causes, and which of these effects, is something that only the future will reveal."[71]

The reader may be interested to know that "which of these are causes, and which effects" is a question still under discussion today. In the theory of "plate tectonics," the continents are generally believed to move laterally as they are "rafted" on lithospheric "plates" that float on a zone of weakness: an "asthenosphere." In the last quarter of the twentieth century there was general agreement that these plates moved in a convective cycle, in which molten material welled up at mid-ocean ridges, creating new "plate" segments that were eventually "subducted" through collision with other plates and drawn back down into Earth's interior.

While convection is considered today to be the principal driving force, it is not the only one, and all (save one) of the mechanisms proposed by Wegener in

1928—including heating from below, spreading and sliding, tidal forces, and even the pole-fleeing force—are still under active consideration as components of the forces driving the continents apart. Currently, no one considers the continents having a precessional axis of their own and different from that of Earth as a whole as a driving force, though since the later 1990s the notion of true polar wander ("inertial interchange true polar wander"), long rejected, and not a part of plate tectonic theory from the 1960s through the 1990s, has made a comeback as a companion to the convective plate motions in explaining the geological history of Earth.[72]

The motion of the plates in the contemporary theory also supposes the creation of new continents by the rifting and splitting of old ones in rift zones where Earth tears, suffers downfaulting, and eventually opens up enough that molten material can well up from below. These rifts can be detected by gravity anomalies, as well as by surface morphology. Twenty-first-century textbook and encyclopedia illustrations of this process look remarkably like Wegener's conception in 1928—a great deal more like Wegener's ideas than did such illustrations in the later part of the twentieth century.

Another hypothesis of continental motion proposed in the 1920s (which Wegener considered but to which he did not subscribe) is also still in play today. This was the idea of radioactively generated heat, trapped beneath the continents, forcing them upward and potentiating the subsequent "downhill" gravity sliding of tilted continental masses, as in the theories of R. A. Daly and John Joly. Robert Schwinner (1878–1953), Wegener's Graz colleague, had in 1920 fleshed out an idea (first proposed by Otto Ampferer) of convection currents in a "tectonosphere" about 100–120 kilometers (62–75 miles) thick, in which radioactively heated material rose and drove "plates of crust" laterally. When two such plates collided, one dived under another and descended into a *Verschluckungszone* ("swallowing zone"), where it melted and was reabsorbed into the tectonosphere, while creating alpine ranges at the surface. R. A. Daly borrowed this entire conception (without attribution) for his 1926 book *Our Mobile Earth*.[73]

The point of such comparisons is not to determine "who was right" or who most thoroughly prefigures present conceptions, but to show that Wegener's cautious agnosticism concerning actual driving forces—the "mechanism"—was in 1928 well considered. Something is moving the continents, there are a number of candidate hypotheses, and whatever force or forces explain these motions also probably explain ocean trenches, volcanism, the creation of mountain ranges, rift valleys, and the full ensemble of movements of the crust envisaged by Wegener between 1912 and 1928. What that "force" *is* (or those forces are) remains uncertain even today.

Expedition Planning, 1928–1929

Whatever else he felt about finishing this book—in proof by early September—Wegener also felt relief. All through the process of composition in the spring and early summer of 1928, he had been deeply involved in plans for Greenland. Nothing had diminished his fascination with the origin of continents and oceans, but dividing his energy between future and past investigations was a distraction and an obstruction. Already in the middle of June he had an expedition plan (several times revised) to send along to Peter Freuchen in Copenhagen. He had planned to travel there in mid-July to get Freuchen's advice and to begin to obtain the necessary permissions to mount his two scientific expeditions in Greenland—one in the summer of 1929, and a second and much larger expedition in 1930 and 1931.[74]

Wegener wrote to Johannes Georgi in July to tell him that he would come to Hamburg on his way back from Copenhagen and discuss the expedition with him.[75] Georgi was only eight years younger than Wegener, not quite his junior, but neither his contemporary. They had been close colleagues in Hamburg; their households had been side by side in Großborstel and their children playmates. We do not know what they discussed when Wegener arrived in Hamburg, though he had come principally to talk to the new director of the observatory to obtain Georgi's release so that he might be part of the expeditions. Their relationship was delicate; Georgi had wanted Wegener's support, but only as an endorsement for his own expedition to East Greenland. Wegener's large plan was a major shift for both men, though certainly more for Georgi than for Wegener; Georgi would no longer be a *Leiter* (leader), but only a *Teilnehmer* (participant).

Wegener now had a definite plan to fold Georgi's ideas together with Meinardus's seismology, along with his own earlier plan to set up West Greenland, Mid-Ice, and East Greenland observatories at the same latitude, in order to chart the transit of weather systems. We do not know what Georgi thought about all this, but he did agree to be part of it. It was a sensible course: Wegener had the intellectual and institutional power to make this happen, and he had extensive experience on the Inland Ice of Greenland; Georgi had none, did not speak Danish, and had never pulled a sledge, mounted an ice cap, or driven a team of dogs. He could surrender, at least for the time being, to an apprenticeship, however much he wanted an expedition of his own.

Wegener was also in Hamburg to get another permission: that of Georgi's wife. Georgi remembered many years later this kindness on Wegener's part: "I shall never forget his long and friendly conversation with my wife; how Wegener explained, when naturally she asked what risk was entailed for me, how he hoped to minimize the risk by careful planning keeping in mind his own experience there; how what one usually calls bad luck is very often only a result of errors or inadequacies in preparation."[76]

We have seen, in looking at the Danmark Expedition and Koch's 1912–1913 traverse, that expeditions like these begin long before they "begin" and end long after they "end." The total amount of labor expended on preparation before and publication after is at least as great and probably greater than one's efforts in the field. Preparations for a summer reconnaissance expedition in 1929 had to be under way by the summer of 1928 in order to secure funding and begin to choose participants. Before either of these could happen, there needed to be a plan to ask for funds and to recruit scientists to join.

While in Copenhagen in late July, Wegener had gone over the maps of West Greenland with Freuchen, looking for possible routes onto the Inland Ice. Freuchen also called in their old expedition comrade Aage Bertelsen, the doctor from the Danmark Expedition, who had been for some years the district physician in the Umanak District of West Greenland, and who had made many photographic trips along the coast. He showed Wegener some routes he thought possible in the Umanak Fjord. Freuchen remembered having found a usable valley in Ingnerit, one of the inner branches of the Umanak Fjord.[77]

From July 1928 on, Wegener was completely immersed in planning for Greenland, to the exclusion of all other scientific work except his remaining experiments with Letzmann on tornadoes. His courses for the winter semester of 1929–1930 were to be a colloquium on the theory of continental displacements (two hours a week) and an introductory course on aerology. Neither of these required special preparation, and he could throw his energy into the immense task of planning a multiyear expedition in the high Arctic.[78] As of 13 July 1928, he had one expedition member besides himself: Georgi. He had promises of funds but no commitments, and he had yet to obtain approval for his plans. In spite of this slim warrant, he was in fact now the leader of what was tentatively known as the "Inland Ice Expedition to Greenland."[79]

After his discussions with Freuchen and Bertelsen in Copenhagen and his meeting with Georgi in Hamburg, Wegener was ready to put together a final plan, and by 30 July he had it in shape for Freuchen to review. He asked Freuchen especially to look at the technical questions on methods of travel (*Reisemethode*), as he hoped to have a finished report submitted to the Notgemeinschaft (the funding agency) in Berlin by 20 August.[80]

The next day, 31 July, Wegener began to put together the rest of his expedition. He wrote to Fritz Loewe (1895–1974), a Berlin-trained meteorologist who had replaced Kurt Wegener as director of meteorological observations (kite and balloon flying) at Lindenberg in 1925. Kurt had great confidence in Loewe and had helped to select him; Wegener had also discovered that Loewe was an experienced and talented alpine climber. The latter skill would be an absolute prerequisite for participation in this expedition, as the only snow and ice experience (other than war service on the Russian front) young German scientists were likely to have would have been obtained in recreational climbing. Loewe was, as Wegener had been delighted to learn, an experienced radio operator; that had been his war service.

Wegener asked Loewe, whom he had yet to meet, whether he would be willing to join both the 1929 reconnaissance expedition and the 1930–1931 main expedition to Greenland. He sketched out for Loewe the scientific plan and some technical details of transport and travel. He then got directly to the point: would Loewe be willing to go climbing with him in the week of 5–11 August in the Ötztal Alps (Austria), where they would join Hans Mothes on the Hintereisferner glacier and learn from him the basics of ice thickness measurements using seismic/acoustic techniques?[81] Loewe was delighted to accept, and he agreed to meet Wegener in August.

Wegener had it in mind to have Hans Mothes (the seismologist) join the expedition to Greenland to be in charge of the seismic investigations, and during the Alpine tour with Loewe in early August (Else came along to climb with them) he made the request. Mothes thought it over but regretfully declined, so Wegener sent Loewe into the field again in late August and early September for another week of training with Mothes; Wegener had wrangled a stipend from the Notgemeinschaft for Loewe to work as Mothes's assistant. He added in his letter to Loewe as a postscript, "I hope you will not be angry with me for the way I order you about!"[82]

Wegener liked Loewe and trusted him enough already to ask his advice about someone to replace Mothes for the expedition, now that the latter had declined. Loewe made some suggestions, and Wegener finally decided on Ernst Sorge (1899–1946), a young geographer (a student of Penck) and a much-respected alpine climber who had been climbing since 1920 not just in Germany and Austria but also in Iceland. Like Sorge and Georgi, he had also seen war service. Wegener, on 8 September, invited Sorge to join the expeditions, and Sorge replied immediately in the affirmative.[83]

Now things could begin to move rather more rapidly, and as they did, the first hints of tension began to emerge between Wegener and Georgi on the matter of what should be in the final plan submitted to the Notgemeinschaft. Wegener and Georgi were in full agreement on the scientific plan, but Georgi had rather different ideas about logistics and transport: he wanted it to be a completely "modern" expedition, of the kind that was going forward in Antarctica with Byrd, with only motorized transport—airplanes and motor sleds or tractors.

Wegener's thinking was quite different. He had drawn up three different scenarios for the logistical structure of the expedition. All of these included an approach to the Inland Ice from the west coast of Greenland; the approach from the east was simply too difficult. It was too isolated, the crevasse fields were massive, and for some hundreds of miles the ice front was inaccessible, a cliff rather than a slope. The expedition would travel to Greenland by commercial Danish shipping and transfer the supplies to an ice-hardened coasting vessel in West Greenland; the expedition would also have a 9-meter (30-foot) boat with an 8-horsepower engine and a hull strong enough to push through light pack ice.

The logistical task was to move 70,000 kilograms (154,324 pounds) of equipment, food, housing, fuel, and accessory supplies from sea level to a winter station within the margin of the Inland Ice at an altitude somewhere between 500 and 1,000 meters (1,640–3,281 feet), perhaps 100 kilometers inland. From there, approximately 10,000 kilograms (22,046 pounds) would be transported 400 kilometers (249 miles) further, to a planned mid-ice station at an altitude of 3,000 meters (9,843 feet).

In his draft plan, Wegener considered three options. The first would be entirely traditional transport: Icelandic ponies and sleds of the kind he had used in 1912–1913, along with dogsleds of the kind used by most expeditions. The latter would be hired, with the dogs and drivers from among the local population.[84]

The second alternative was to supply the Inland Ice base using airplanes. Wegener clearly did not favor this alternative, as adding a pilot, a mechanic, and a replacement pilot would mean possibly three more members for the expedition, and with each additional person came thousands of kilograms of supplies. Wegener also doubted that a sufficiently flat landing strip could be found anywhere along the coast, and he was sure that, during the winter, flight would be impossible, as repeated experience had shown a tendency for the fuel lines to freeze at temperatures of −50°C and −60°C (−58°F to −76°F). Moreover, should something happen to the plane, repair in Greenland was impossible. Georgi was strongly in favor of employing aircraft, but as Wegener had pointed out to him the previous spring, the expedition had to be prepared for the plane to crash on its initial flight, and the inland base would still have to be supplied. This would mean horses, dogs, or both.[85]

The third alternative would be snow tractors or propeller-driven motor sleds, of the kind currently being used successfully for winter transport on land and sea ice in Finland and Sweden. The motor sleds were aerodynamic plywood skimobiles with push engines (the propeller and engine at the back), a carrying capacity of about 1,200 kilograms (2,646 pounds), and a maximum speed somewhere around 40 kilometers per hour (25 miles per hour). These would be of little or no use in the steep coastal zone, which would still require animal transport, but the flat and even surface of Greenland's interior made them technologically almost ideal. They required no landing and takeoff spot and could serve for regular communication in all but the depths of winter between the mid-ice station and the coast. The alternative—snow tractors—could pull uphill, but they were enormously heavy and slow and not terribly fuel efficient.

Here was the basis of the disagreement between Wegener and Georgi. Wegener could not see any way to plan an expedition that could avoid a catastrophic failure of supply without the use of pack animals. Georgi pleaded with him to submit a plan that did not mention the use of draft animals, either ponies or dogs. He wanted to use the reconnaissance expedition to find a place where they could land planes and drive motor sleds up onto the ice. He feared that the Notgemeinschaft, if offered the low-cost alternative of animal transport, would fail to fund either airplanes or motor sleds.

They discussed this in an exchange of correspondence in September, and Wegener finally emphatically refused to drop reference to animal transport. "I think that one of our most important technical tasks is closing the gap between the old methods of animal transport, and the new mechanical means of travel."[86] Georgi was apparently one of those men who cannot take no for an answer when they hear it the first time, or even the second. Wegener made it clear that his strong preference was for propeller sleds, but through the fall and into the winter of 1928, Georgi tried to press for the use of air transport for the central station, enlisting Walther Bruns (1889–1955), the general secretary of AEROARCTIC, to advance his case.[87]

Fortunately, Georgi and Wegener were at least agreed on the scientific program for both the reconnaissance expedition and the main expedition. This plan, as Wegener circulated it in the summer of 1928 and submitted it in the fall to the Notgemeinschaft, had six parts.[88] The first was the measurement of the thickness of the ice cap directly using the seismic technique of Mothes. These would be the first measurements of the full ice cap thickness ever taken. The second objective was an exact trigonometric leveling of the altitude of the ice surface above sea level in a profile from the coast to the center of the ice, to serve as an exact baseline for future measurements to determine whether the thickness of the ice was increasing or decreasing. The third objective, closely connected with the first two, was the use of a pendulum gravity-measuring device (a Sterneck pendulum, a tried-and-true instrument) along the same transect as the trigonometric measurements and the seismic measurements, to help determine the extent of isostatic compensation. There would also be direct glaciological measurements of ice temperatures in boreholes, as well as measurements of glacier motion.

The second set of objectives, developed at length, were those in meteorology, most of them directed to the fulfillment of Georgi's plans for an overwintering station in the middle of the ice and a station on the east coast of Greenland. The principal goal for these stations was to track the upper atmospheric winds and to examine whether or not there was such a thing as Hobbs's "glacial anticyclone," a permanent high-pressure vortex generated by the intense cold of the interior of Greenland, which would block the transit of cyclonic storms across the ice cap. Having three stations would allow meteorological observers on both coasts and in the middle at the same latitude to generate a series of simultaneous observations, which could demonstrate the existence or absence of such a blocking high-pressure center.

In addition to this, there were other more direct and local meteorological problems. No one knew how cold it got (and stayed) in the middle of the ice in the winter because no one had ever been there all winter. Neither was there any secure knowledge about the kind and quantity of precipitation. Atmospheric optics—measurement of radiation, mirages, and other matters of interest to Wegener—could also be handled at all three of the stations, although Wegener planned to make these measurements during the overwintering at the station on the west coast, to which he has assigned himself.

Wegener had assured Georgi that he would have a secure place as meteorologist at the central station, that he would be accompanied by a glaciologist/meteorologist, probably Sorge, and that they would be joined by a technical assistant (*Hilfsarbeiter*) who would also function as the radio operator. Wegener wanted Georgi to understand and to feel that his plan had not been hijacked and would not be diminished. As they worked through the plan together, this had to be evident to Georgi, as the overwhelming logistical cost of the expedition was directed to the creation, supply, and staffing of the station in the middle of the ice cap, where the observations would "belong" to Georgi, and where he would be an independent operator.

All of this moved from conjecture to fact in early November 1928, when the four members of the reconnaissance expedition, who would also form the leadership core of the 1930–1931 main expedition, met in Berlin with the Greenland Committee established by the director of the Notgemeinschaft, Friedrich Schmidt-Ott (1860–1956). The committee was well stacked in Wegener's favor. A good deal of political maneuvering, in which Wegener was not directly involved, had already been accomplished to keep the expedition from being postponed until the international polar year 1932–1933; there was agitation in some quarters that this should happen.[89]

Hergesell, Penck, Drygalski, Kohlschütter, Angenheister, Meinardus, and Defant were there to ensure the support of meteorology, geography, oceanography, geophysics, the Admiralty, and the German university community in general for Wegener's proposal. Meinardus had let it be known that he feared that the principal aim of the expedition, the glacial seismology, might later be excised from the Polar Year program, and everyone agreed that Wegener, who would be past the age of fifty in 1932, might not any longer be a candidate to lead an expedition that he clearly very much wanted to lead.[90]

The committee approved the scientific program and even extended support for the East Greenland station at Scoresby Sound, approving as well the transport scheme, with the motorboat, twenty-five Icelandic ponies, dogs and dogsleds, and two propeller sleds, which, the official report noted, the expedition leader had particularly specified as needed for the geophysical program on the Inland Ice once the mid-ice station had been established, especially because of the huge weight of food required (alternatively) to keep dogs on the ice over the winter.[91] Here was something to placate Georgi, who had wanted motorized transport, not ponies and dogs.

The *Vorexpedition* to Greenland, 1929

Everyone who knew Wegener, especially Else, Georgi, and Benndorf, remembered that the late fall of 1928 and the early spring of 1929 were for him times of furious and continuous activity: planning, building, buying, cajoling, traveling, meeting, negotiating, and above all writing letters, sometimes a dozen or more a day. The money from the Notgemeinschaft began to arrive in December 1928, and after the Christmas vacation in the Ramsau (he invited Georgi and his family, but we do not know whether they came), Wegener began to spend it. He bought as much modern lightweight camping equipment as he could get: no more reindeer-skin sleeping bags, but lightweight waterproof bags weighing under 3 kilograms (7 pounds), and lightweight waterproof tents with bamboo poles, flexible and light. He searched for the lightest instruments he could find, and he ordered ultralight Nansen sleds, modern versions of the lightweight man-hauled sledges used by Nansen in his first traverse of Greenland, for the reconnaissance expedition.

He had placed an order for a boat to be built already in the fall of 1928, at the same company in Copenhagen which had built the boat he and Koch had used in

1912–1913. It had a new design motor, built by the "Dan" company, a so-called hot bulb engine, intermediate between a gasoline and a diesel engine, which could burn almost anything: gasoline, heavy oil, kerosene, even blubber.[92] This boat, to be christened *Krabbe*, was a 9-meter cabin cruiser of the kind he had owned in Hamburg, and he specified that in addition to a reinforced deck, a large cargo hold, and bunks for four men, it should have a photographic darkroom, a necessity for an expedition that would be extensively documented with photographs. Wegener also placed an order for prefabricated huts from the same Danish company that had built the huts in 1912; in every case it was his instinct to go with what he knew, and there was also considerable goodwill to be generated in Denmark by purchasing these things there.

In mid-March 1929, shortly before the planned departure of the *Vorexpedition* from Copenhagen to Greenland, and with his bags (literally) already packed, he and Else traveled to Sweden and Finland to make the final decision between propeller-driven motor sleds (made in Finland) and continuous-tread tractors (made in Sweden). Wegener had hoped to test-drive the propeller sleds in Finland, but there was bad weather—rain and no snow—so he drove the sleds on sea ice rather than uneven ground. They seemed marvelous—although might prove finicky—but the continuous-tread tractors they examined in Sweden were so heavy that they made the propeller sleds the safer option, so he chose the latter. If a tread tractor were to tip into a crevasse, there would be no pulling it out. Else, who had been with him on this trip, traveled to Oslo to see the Bjerknes family, and Wegener headed to Copenhagen, in order to travel on to Greenland with his reconnaissance team: Georgi, Loewe, and Sorge.[93]

The *Vorexpedition* to Greenland was to last from late April until early October 1929. The principal object was to find a way onto the Inland Ice in west central Greenland, somewhere between Quervainshavn (N 69°) and Umanak Fjord (N 71°). In this region, unlike the dry coast farther south, the Inland Ice lies very close to the shore, and many glaciers flow directly to the sea from the Inland Ice. Farther north, in the Upernavik district (~N 72°), the Inland Ice also comes to the shore, but the annual timing of the "ice-out" (melting of the sea ice) was so late as to make it a doubtful landing point for an expedition.

In the expedition plan agreed upon with the Notgemeinschaft, the mid-ice station should be at latitude 71° north, in order to be in line with the east coast station at Scoresby Sound. The question was how best to approach the ice. The southerly candidate site, opposite Disko Island, was the starting point chosen by Alfred de Quervain in his successful 1912 traverse from west to east, thus the name Quervainshavn. From here, the planned expedition for 1930 would have to take a northeasterly route to reach latitude 71° north; the extra distance was not great, more or less the hypotenuse of a right triangle whose base was parallel to the coast and its altitude along the line of latitude 71° north.

The northerly route would be on the northern side of the Nugsuak (today Nugssuaq) Peninsula just to the north of Disko Island, in the Umanak District, where there were a number of glaciers from the Inland Ice emptying directly into the sea. This area was less well known, though Drygalski had investigated it, as he had the region just to the south, between 1891 and 1893.

The principal scientific aim of the reconnaissance, and the reason for this entire effort, was to test the seismic method of measuring ice thickness with explosion seismology. They would take 100 kilograms (220 pounds) of dynamite and their portable seismometer onto the ice (and firn) and determine whether the technique would work under such conditions. The subsidiary scientific aims of the expedition were to compare the extent of the glaciers mapped by Drygalski in 1891–1893 with their extent in 1929, and to make some boreholes in the ice and insert bamboo wands into them, so that they might return the following year and measure the growth or ablation of the ice at several locations and the depth of new snow. They would make meteorological observations and measure their altitude and position, and they would spend a week at Jacobshavn Glacier measuring its extent and motion.

In addition, there were two other sets of tests. The first was with equipment and food. Wegener remembered all too clearly the immense difficulty of men hauling sledges with loads in excess of 68 kilograms (150 pounds) during his reconnaissance with Koch in 1907 and again in 1912. In planning this expedition he did everything he could to conserve weight; in addition to the light sleeping bags and tents, he ordered sleds in two different models, one slightly sturdier, at 8 kilograms (18 pounds) per sled, and the other less robust, at 4.5 kilograms (10 pounds) per sled. The idea was that each man should pull no more than 45 kilograms (99 pounds), including the weight of his sled. Also undergoing tests were their footwear, their clothing, the lightweight instrumentation, and a variety of concentrated "sledging rations," including a kind of pemmican developed by Amundsen.

The second test was the men themselves. What were they like, and how would they respond to the hardship? Wegener had to learn what his physical capacities were at age forty-eight; certainly they would not be the same as when he was twenty-six (1906) or thirty-two (1912). As for Georgi, Loewe, and Sorge, they had to learn everything: how to sledge; how to drive dogs; how to sleep, work, and travel in bad weather when wet and cold; how to march twelve to fourteen hours a day and then to pitch camp, cook food, melt water, record their work in their diaries, take photographs, and make scientific observations. There were no alpine huts here, no day trips, and no backup; they would carry their lives with them on their sleds.

Wegener also had to observe them to see the extent to which they could give orders and take them, work cooperatively, follow instructions, and be willing to improvise. Finally, Wegener had to know how hardy they were psychologically. This pertained not just to the reconnaissance team but also to the core leadership cadre that would have to train the others.

Wegener and his companions left Copenhagen on 27 March and arrived in Holsteinborg, West Greenland, on 21 April. Georgi, Sorge, and Loewe were surprised to discover that Wegener was not only known and liked here but actually famous, and that he seemed to know everyone. The captain of their ship, the *Disko,* Wegener had known since 1912. When they were unloading their motorboat, the *Krabbe,* it hit the side of the ship and sprung a leak. They towed it to the shipyard in Holsteinborg, where it was fixed by Martin Hansen, the brother of one of Wegener's companions on the Danmark Expedition. The captain of the *Disko* thought they lacked enough mooring line and made them a gift of a 100-meter (328-foot) hawser that, as Wegener said, later proved essential; the district officer thought they ought to have a second anchor for their boat and made them a gift of one much stronger than the one they brought from Denmark.[94]

The warm welcome they experienced in Holsteinborg continued to follow them north. Georgi had immediately learned how to manage the engine, but they knew nothing of local waters and ice sailing, so the district officer arranged for a pilot to accompany them north to Egedesminde, and for a much lower rate than was customary if they would agree to carry the mail sack with them. On the twenty-eighth they ventured north to Godhavn, this time without a pilot, and on their arrival found that the governor of North Greenland was there to greet them, having received a telegram from the secretary for Greenland in Copenhagen. They found themselves guests in the governor's home, with an offer to leave their European clothing and belongings there during the expedition and to overwinter their boat until they were done with the reconnaissance expedition. The governor also arranged for a 60-kilometer (37-mile) dogsled trip with some experienced drivers so that Wegener's companions could feel what it was like to travel as a passenger on a sled before they had to pull one themselves.[95]

They sailed on their own to Jacobshavn, where they were met on 5 May by Tobias Gabrielsen (1878–1945), one of Wegener's companions on the Danmark Expedition, and one of the best dogsled drivers in Greenland, who had worked on many expeditions since that of 1906–1908. He was also an experienced machinist, quite familiar with the kind of engine on the *Krabbe*, and he joined the *Vorexpedition* as machinist and "ship's husband" to mind the boat while the expedition was reconnoitering in the interior; he would stay with them until October.[96]

Between 6 May and 17 June they explored the possibilities of a route to the interior from Quervainshavn. The details, many of them hair-raising and all of them good reading, make up the first part of Wegener's book on this expedition, *Mit Motorboot und Schlitten in Grönland* (1930).[97] Wegener had this book in mind from the beginning, and much of it is a fleshing out of his reports to the Notgemeinschaft with his own diary entries and some from Georgi, Loewe, and Sorge.

They were able, in the course of a sledge trip from 19 May to 11 June, to establish eight depots along a line northwest from their landing spot, marking these with black flags on flexible wands and building snow towers. They calculated that they had traveled 153 kilometers (95 miles) and achieved an altitude of about 2,100 meters (6,890 feet). They had strong winds almost all the time, often föhn, and they had rain and snow on eighteen of their twenty-three days. Although their equipment was good, they were often cold and wet, and all three of the younger men had their hands frostbitten, sometimes badly. Their equipment—sleeping bags, tents, and sleds—worked well, but the sledging rations, the "Amundsen pemmican," were indigestible, and they all suffered badly from fatigue and indigestion, especially on the way back to the coast.[98]

They concluded that it was a usable route, suitable for horses, with no mention of dogsleds or motor transport. There were 8 kilometers (5 miles) of rocky ground to traverse to get to the edge of the snow, and from there the path onto the ice itself was quite sudden and steep, almost a scarp, requiring detours that would add extra hours to the ascent. Moreover, once on the ice, in the first 100 kilometers or so, there was an extensive *Firnzone* of compact snow, and this was fissured with many small crevasses, into which they fell repeatedly up to their shoulders.[99]

Tobias Gabrielsen returned to pick them up with the *Krabbe* on 14 June, and they sailed to Jacobshavn, where they reloaded their supplies and left for Umanak. The ice conditions in the Umanak region were severe; the Rink Glacier, 60 kilometers (37 miles) to the north of Umanak, was one of the most productive in West Greenland. Moreover, the time of "ice-out" here was a full month later than in the area around Quervainshavn. Wegener and his expedition partners spent the next two weeks incessantly searching along the mainland shore, by boat and on foot, for any way up to the ice that some combination of men, horses, and dogs might traverse.[100]

They searched most extensively within the Umanak District, with its bewildering side channels, islands, peninsulas, and glacial outlets. Nothing was suitable: this glacier was too steep, that approach had too much slippery and bare rock, this other snow field had too many crevasses, the next glacier calved too many icebergs. They were becoming discouraged when, on a trip on the *Krabbe*, they motored into what looked like a "dead end," after having investigated (and found unsuitable) a location that Freuchen and Bertelsen had shown Wegener in Copenhagen.[101]

Straight ahead they saw a glacier, not very big, and not reaching all the way to the water, ending on a broad, curving gravel beach more than a mile in extent. They looked on their nautical chart, but the glacier and the arm of the fjord in which it lay had no name; they later learned that the local Greenlanders called it Kamarajuk (Black Bight). Wegener sent Georgi and Loewe ashore to investigate. They returned fourteen hours later with the "amazing report" that this small glacier emerged directly from the Inland Ice. This was just what they were looking for: a glacier connected directly to the Inland Ice which did not calve icebergs into the sea. The connection to the Inland Ice immediately at the top of the glacier meant that the "West Station" need not be 100 kilometers or more inland; it could be there at the coast. This would be an enormous saving of time and energy.[102]

The route they had found on Kamarajuk Glacier was not perfect, but it was an order of magnitude better than anything they had seen. The front of the glacier was only a few hundred yards from the beach. For the first 2 kilometers (1.2 miles) the slope was gentle, though from kilometer 3 to kilometer 6 it became a very steep ascent. At one point, on the left (north side), it was not really a glacier at all but an icefall, with a precipitous face and, in the late afternoon, catastrophically large avalanches of ice, with chunks the size of automobiles falling along a range of about 100 meters.[103] Here the right side of the glacier was more compact but deeply crevassed, creating a difficult and circuitous ascent. The best route was a series of meanders back and forth.

Wegener decided to test the glacier as an ascent route, and at the store in Umanak he bought some dogs, dog food, some ice axes, and a large tent. From there he motored the *Krabbe*, along with a local (for hire) schooner, to the village of Uvkusigssat at the mouth of Umanak Fjord, where he hired seven Greenlanders, their sleds, and their dogs, to help them ascend the glacier with their supplies and to travel with them on the Inland Ice to set up depots.[104] In this endeavor Wegener was aided by Johan Davidson, a Greenlander from Nugaitsiak and a respected dog driver who could help train Wegener's expedition colleagues in this art.

The weather was bad almost every day, with incessant rain and cold wind, but by 18 July they had landed themselves, the Greenlanders, the sleds, and the dogs on the beach at the foot of the Kamarajuk Glacier. The ascent of the glacier with their provisions and equipment turned out to be the most difficult work of the whole expedition. The Greenlanders wore kamiks, with smooth skin soles, and constantly fell on the ice. Wegener was the only one who spoke Danish, so his colleagues were unable to help or guide. The going was so steep that the Greenlanders veered off onto the moraine at the left side; here they could drive their dogs up the dirt slope as long as their load was no greater than 100 kilograms. Perhaps it would be possible to make a road here; the moraine was a huge gravel bar parallel to, but hundreds of meters higher than, the ice surface. Once across the moraine, there was a broad meadow sloping upward, with thin snowfields—a good intermediate depot for the main expedition—and it was possible

The *Krabbe* (*right*) and a hired schooner from Umanak (*left*) at anchor in Kamarajuk Fjord during the 1929 reconnaissance expedition. The photo is taken from the south side of the Kamarajuk Glacier. In the background is the (as-yet-unnamed) Alfred Wegener Peninsula. Photo courtesy of the Alfred Wegener Institute, Bremerhaven.

to walk on reasonably flat ground and then recross the moraine 2 kilometers further on, where the glacier was again smoothed out. They named this place Grünau (Green Meadow) and thought it would do well for the ponies.[105]

Over the next week, Wegener, Georgi, Sorge, Loewe, Davidson, and the seven Greenlanders hired from Uvkusigssat managed to cut a path along the least treacherous parts of the glacier and strew it with engine soot from the *Krabbe* in order to mark it and to help it melt. It was incredibly difficult and dangerous work; on the first day one of the dogs fell into a meltwater stream and was swept away to its death. All too often dogs fell into crevasses, and Sorge, Loewe, and the Greenlanders had to climb down in and carry them out. Once an entire span of twelve dogs disappeared into a crevasse, although all were rescued. It was clear that much help would be needed from the Greenlanders for the main expedition as well, as there were sections of the glacier and the moraine where men carrying backpacks seemed to be the safest way to move supplies.

As the summer wore on, Wegener began to suffer physically more than the younger men, though he still pulled his weight. While he did not complain about his heart "*Knax*," he suffered badly from what was then still known as "rheumatism"—arthritic and swollen joints, bursitis in his shoulders, and muscle pain and fatigue. He also had a toothache, about which nothing could be done for the time being. Still, once they were at the top of the glacier and had sent the Greenlanders back to Uvkusigssat on 30 July, they found that they were in a position to do some science in a beautiful place.

After camping for several days in wet and soggy snow, they moved their base camp to a small *nunatak* that they named Scheideck (The Divide), sitting on a ridge crest at 975 meters (3,199 feet) between the Kamarajuk and Kangerdluarsuk Glaciers. Wegener's preliminary report of this place to the Notgemeinschaft was, to be sure, a barebones tale of trouble encountered and distances covered, but the more ample account in his popular book *Mit Motorboot und Schlitten* is especially poignant with regard to Scheideck. What a relief it was, as Wegener wrote, "to be on dry land," to see plants growing, minuscule alpine flowers in the crevices of the rock. It was "a friendly welcoming place—how it makes our hearts swell! And what a view." The sky was a stratospheric blue overhead, with a softer azure to the west over the ocean and "on the other side, the inland ice!"[106] Wegener's exclamation points are the testament to the surpassing beauty of the setting. Scheideck was not a steep pinnacle but a long rectangular slab. The rock had a clean and unvarnished surface and was a delicate ocher-brown. It made them all a bit homesick, as the color is that of the stucco of a German, or Danish, or Austrian house, "homelike" and beckoning. It was especially beautiful under the midnight Sun, as there was a chiaroscuro effect and the rock had a perpetual alpenglow against the white intensity of the snow and the dark blue of the sky. It was a welcome rest stop for Wegener between the travail of portage up the glacier and the unknown dangers of the Inland Ice.

Over the next week, they were able to do some real science. Here it finally became a *scientific* reconnaissance, not merely the reconnaissance for the logistics and the transport. They were able to set off a dynamite charge in a borehole and measure the thickness of the ice with the portable seismometer. The work was maddening at first, as the prism had fallen out of the seismometer and had to be carefully built back in, and even though they had a "darkroom tent" designed to keep out all light from the photographic paper on the seismometer, the tent floor turned out to be of lighter material and reflected sunlight upward from the glacial ice on which they were working. Eventually, they solved this and other real problems; Sorge turned out to have a talent for using the seismometer, and he and Loewe were able to measure an ice thickness at Scheideck of about 300 meters (984 feet). These two men then moved their tent and apparatus north to the Kangerdluarsuk Glacier, fought a host of new difficulties, and set off a charge of dynamite showing that glacier to be 600 meters (1,969 feet) thick.[107]

On 21 August they moved their base of operations, using the dogsleds, to a depot 25 kilometers (16 miles) further inland. They were now in the *Firnzone*, where the recrystallized snow was very dense and interlayered with thin beds that had already converted to ice. "Here we now had to perform our most important measurements in order to determine whether our method of measuring the thickness of the ice also could be used in the Firn [i.e., not directly on glacial ice]."[108]

Sorge and Loewe were now a team, switching back and forth between handling the seismograph and exploding the dynamite. On 27 August, at 25 kilometers inland and an elevation of 1,570 meters (5,151 feet), they exploded their last 13.5 kilograms (30 pounds) of dynamite, and Wegener reported ecstatically, "This measurement was unbelievably successful."[109] They got an unambiguous measurement of an ice thickness of 1,200 meters (3,937 feet) and were able to see that the speed of propagation of the waves was the same in the firn as it was in the ice. This meant that as they got deeper into the interior of Greenland, and as the firn achieved greater and greater depth, they need not excavate down to the ice surface in order to get accurate measurements.

With these measurements complete, as well as some borehole measurements and measurements of ice ablation, they were ready to do something that Wegener had wanted more than anything else—a *Hundschlittenreise* (dogsled trip) into the interior. They made a few experimental trips from their current depot, and Wegener taught Loewe how to "drive." Davidson led the way so that Loewe's dogs could follow, and though they were heading into a biting wind, in blinding sunlight and through deep snowdrifts, they covered 26 kilometers (16 miles) in a little more than five hours, and for most of that time they were riding on their sleds.[110]

With these preliminaries accomplished, they set out to go as far into the interior of Greenland as they could, to mark the way with flags and snow cairns in order to be ready for the main expedition the following year. It was colder now, between −1°C (30°F) and −16°C (3°F), so they had to be careful not to get frostbitten. They had the advantage now that, in the wind, everything dried out, which meant they could sleep dry and warm. The days, however, were not warm, and the camping was difficult because of the blowing and drifting snow, a dry powder that they had to dig the sleds out of every morning. Even in the bright sunlight, the cold wind from the interior was so strong, and the dogs were moving so fast, that from time to time they had to jump off their sleds and run with the dogs in order to get warm.[111]

In four days they penetrated 154 kilometers (96 miles) into the interior. They turned around on 31 August and took only three days going out. The sky was clear, they were going downhill, they had the wind at their backs, and they did not have to stop and take measurements, as they could aim for the coastal mountains that they could see 100 kilometers away, knowing that they were in a direct line to Scheideck. It was good that this part of the trip was easy, because in the bright sunlight both Davidson and Wegener went snow-blind; Johann lost all his vision for two days, Wegener could barely see, and both men were in terrible pain.[112]

In the few days that remained to them after their return to Scheideck, they did as much science as they could (mapping, ice-boring, photographing), but time was running out. By 6 September they were packing up, and they spent the next two days carrying their instruments from Scheideck down to the shore; met there by Tobias and the *Krabbe*, they ferried the dogs back to Uvkusigssat and from there sailed to Umanak, arriving on 10 September.[113]

From this point on, Loewe and Wegener stayed with the boat, and Georgi and Sorge did most of the exploring. On 21 September they arrived in the vicinity of the Jacobshavn Eisstrom, probably the world's fastest-moving glacier. Georgi and Sorge had planned to investigate this area using the folding boats they had brought with them from Europe. It was late in the year, and the two men found themselves often traveling in darkness, sometimes finding ice where there had been water, and water where there had been ice. It was their chance for independent action, and they gloried in it, the danger notwithstanding. Georgi shot as much film as he could, and Sorge measured the speed of the Jacobshavn Glacier ice front from several points with the aid of a sextant. They returned on 30 September, after a difficult and dangerous (and exhausting) trip over new ice, where they had to abandon their boats, both of which were leaking so badly as to be useless for travel.[114]

The return sea voyage south to Godhavn was extremely treacherous, owing to a combination of new sea ice forming (just a skim, but making it hard to push ahead) and the peak of the season of iceberg calving, especially dangerous along the Jacobshavn front. Trying to stay inshore, to be out of the wind, and skirting the ice front, they hit a submerged chunk of blue ice which knocked the blades off their propeller.[115]

They were 80 kilometers (50 miles) across the way from Godhavn, in the dark, with no power, no radio, and no way to get help in a Beaufort 5 east wind (10 meters per second [20 miles per hour]). There were whitecaps and larger swells coming in, and they were making only 4 knots under their tiny sail. They approached the harbor at Godhavn in pitch dark, and Wegener said that if it hadn't been for Tobias's sense of direction, they would never have made it at all. Using their flashlight as a semaphore, they were able to contact the shore and were eventually towed in, after fourteen hours of rough cold transit (in ice) on a rough sea—one last scary adventure. On 8 October, with the *Krabbe* on shore until next year and well covered, they sailed for home on the *Gertrud Rask*, arriving in Copenhagen on 2 November. Wegener spent the voyage home writing up the book of their adventures, and his sense of satisfaction and pleasure increased with each page. They had *done* it. And next year, they would do it again.[116]

20

The Expedition Leader

GRAZ AND GREENLAND, 1929–1930

> The whole business is a big catastrophe and there is no use concealing the fact. It is now a matter of life and death. . . . I do not consider Sorge's plan of setting out on 20th October with man-hauled sledges feasible; they would not get through but be frozen to death—We shall do what we can and we need not yet give up hope of things going well. . . . Best wishes to all, and may we meet again happy and healthy.
>
> ALFRED WEGENER, 62 kilometers (39 miles) inland, to Karl Weiken at Scheideck, 28 September 1930

Wegener returned from Greenland in 1929 in an excellent mood, with a feeling that they had accomplished their aims. Else said that when she saw him on his arrival in Graz, he looked "completely rejuvenated. The hard physical work had done him nothing but good."[1] He had brought gifts from Tobias Gabrielsen, a pair of polar bear fur kamiks for Else, and a sealskin pillow decorated with colored leather, which gifts, Else said, made their presence known principally by "their terrible stench." She had already banished his dog whip and kamiks to the backyard, and these gifts soon followed.[2]

There was, however, news of trouble, and it was serious. Even though the Notgemeinschaft had authorized the expedition, the Reichstag had not appropriated the funds, and when Wegener stopped off at the headquarters of the Notgemeinschaft in Berlin on his way back to Graz, he was told that the expedition would have to be put off for at least a year. He wrote to Georgi on 6 November, saying that this news had taken him completely by surprise and that he had protested to the Notgemeinschaft, noting that they should have told him while he was in Greenland; if they had known that the main expedition was to be postponed, they would have conducted their research very differently in 1929.[3]

Hence, there would have to be a political struggle. Returning to Graz, he wrote to the Notgemeinschaft explaining all the reasons for not postponing the expedition; there were equipment and stores "on the ground" in Greenland; arrangements had been made with the Danish government for the next year, not the year after. They had left experiments on ice thickness "in progress." That these consisted entirely of bamboo wands in holes in the ground he did not elaborate; he needed all the ammunition he could find. Wegener quickly finished his remaining "preliminary report" and sent it off to Schmidt-Ott to show him what they had gone through to prepare the way for this expedition.[4]

How strange Wegener found it to be back in "civilization," with a course of lectures to teach in general meteorology, to be once more a professor, a father, a husband, a colleague, and to walk to work in clean dry clothes, past building after building the ocher color of Scheideck. There was interesting mail for him: a long letter from du Toit thanking him for a copy of the fourth edition, and noting how completely Wegener had reworked it to bring it up to date with "our current understanding." He also told Wegener he thought it likely that the drift of Australia to the east would soon be measured, and he urged him to come to the International Geological Congress in Pretoria, South Africa, in July 1929, so that "everyone will be able to meet the man who has revolutionized our understanding of the structure and history of the earth."[5] There was also a letter from Pierre Termier, who had recently called Wegener's theory a beautiful dream that vanishes like smoke when you try to capture it, inviting Wegener to Paris to be Austria's representative to the 100th anniversary of the Geological Society of France in June 1930.[6]

These were welcome invitations, but also souvenirs of a former world he no longer inhabited, in which he was an active theorist of continental displacements. He was still fascinated by the westward drift of Greenland and had often talked to Georgi, Loewe, and Sorge about it on the reconnaissance from which they had just returned. However, measurements of this drift would not be part of his *Hauptexpedition* in 1930–1931, even though he planned to bring a geodesist to do accurate latitude, longitude, and altitude measurements and they would have radios to do time signals. He had decided they could not possibly achieve the level of accuracy that the Danes had been working with in Gothaab since 1927, and he made this clear to Georgi.[7]

Wegener traveled to Berlin at the end of November to press his case. Schmidt-Ott told him that he would have to give a lecture before the ensemble of funding agencies and philanthropic individuals supporting the Notgemeinschaft, on the "economic goals of Greenland exploration." Of course, these were exactly the sorts of claims and pretensions he despised and found inimical to sound research. If one had to make such claims in order to fund research, one should admit to oneself the hypocrisy—of which he had written to Willi Meyer—of an activity much like that of "Kepler casting horoscopes." It was an agony for him to write the lecture and give it; he found duplicity physically painful. Else said this single lecture cost him more trouble and "headaches" than all the other expedition preparations combined.[8]

Wegener was able to press his political and "practical" case successfully. His lecture was reprinted in one of Berlin's major newspapers, the *Berliner Tageblatt*, as a lead story on the "economic goals of the exploration of Greenland."[9] He wrote to Georgi informing him of this and asking him to send some usable photos to illustrate the expedition report for the Notgemeinschaft.[10] Within a few days of this request, Wegener and the Notgemeinschaft were in serious negotiation, and Wegener estimated the cost of the expedition at 328,000 marks; the expedition was once again "on."[11]

By 9 December, Georgi had still not sent any photos, and Wegener had to renew the request twice.[12] This was another hint of trouble; they had agreed that they would jointly author a popular book on the reconnaissance expedition: this was *Mit Motorboot und Schlitten in Grönland*. Wegener would do most of the writing, but they were each to contribute something on their exploits and on their scientific work; authorship would be Alfred Wegener with "contributions" from the other three, with their names on the title page. Georgi continued to ignore the requests for photographs, nor did he comply with Wegener's requests that he send the complete draft of his summary of the expedition's meteorological work. Finally, Wegener had to write what was for him a rather stern letter, pointing out that the photographs did not belong to Georgi, but to the expedition, and that Georgi's contract with the expedition and the Notgemeinschaft made this quite clear.[13] He softened this with a Christmas greeting two days later, wishing him continued success and a 1930 as productive as their 1929, but Wegener was clearly troubled by Georgi's sense that the photographs were "his."[14]

Wegener claimed that throughout the 1929 reconnaissance expedition none of them had ever exchanged a harsh word, but now that the expedition was over, there *had* been such words. Loewe wrote to Wegener on 21 December 1929; his letter crossed in the mail with Wegener's demand that Georgi fulfill his obligations under the contract and release the photographs. Loewe told Wegener that he and Georgi had quarreled over Georgi's assertion that the Mid-Ice Station was "his" to establish and to manage. Loewe said that he would not submit to the notion that Georgi had any special status. He said he would either resign from the expedition altogether or join the separate expedition to the east coast of Greenland; alternatively, Wegener could assure him that the meteorological work would be so arranged that he would never be involved in Georgi's work.[15]

The evidence was there for Wegener to see: Georgi resented being Wegener's subordinate rather than coleader of an expedition that, in Georgi's mind, was the inspiration for Wegener's entire plan. Of course, Wegener had earlier had a plan of his own and had not seriously considered going to Greenland at all until Meinardus had contacted him during Easter week 1928. Behind Wegener's back, Georgi had now claimed to Loewe that he, Georgi, was "in charge" of the Mid-Ice Station; further, he was already truculently insubordinate in not sharing his photos with Wegener until the latter (essentially) waved the signed contract in his face. The situation concerning the Mid-Ice Station had indeed been ambiguous; Wegener had suggested to Georgi that he would be in charge of the meteorology there (i.e., it would be "his") but had never actually put him in charge of the scientific work of others, including Loewe.

Wegener did not want to lose Loewe, whom he liked and trusted, and so he invited him and his wife to come and stay with the Wegener family in Graz during the month of January. When they arrived, Wegener entered into deep discussions with Loewe over the arrangements and costs of food, clothing, apparatus, and so on, and they began to work out the actual listing and calculation of the cost of the things that they would need to go to Greenland. Loewe got the unspoken message: he also would play an important and leading role and not be subordinate to Georgi. Doubtless, Wegener filled Loewe in on the troubles with the photographs and also on the difficult history of the merging of Georgi's planned expedition with Wegener's own.[16]

Wegener, however, also needed Georgi. He had only a little more than two and a half months (by mid-January) to do everything. He had to find the rest of his scientific and technical staff and put them under contract; buy, assemble, and crate their food, clothing, and housing; pay for their scientific equipment; and decide the many thousands of minute details that go into such an expedition (e.g., how many boxes of nails and how many hammers would be needed, and how far apart the ringbolts on the packing crates should sit to fit onto the hooks on an Icelandic packsaddle). Of all the members of the expedition, Georgi was the logical second-in-command, even unofficially. He was at Hamburg, was a division chief, was an expert in instrumentation and photography, and had the resources of the German Naval Observatory behind him; Wegener had known him for almost twenty years. Even if Wegener had wanted Loewe for this role, Loewe was back at Lindenberg as an assistant—out in the country at the aerological station with neither time nor support for such an undertaking.

Wegener was in touch with Georgi almost every day and sometimes more than once a day from December 1929 until the day before their departure on 1 April 1930;

Wegener would often write a letter and then supplement it with a telegram that same day. Because the observatory in Hamburg had its own radiotelegraph office (Wegener had helped set it up), this was an easy means of communication. By comparison with this correspondence with Georgi, there are only a few letters from Wegener to Loewe, and almost nothing from or to Sorge. It seems, based on the scant evidence, that Georgi was correct in assuming that he was, at the least, Wegener's "right hand man."

The range of topics in Wegener's correspondence with Georgi covers every aspect of the expedition and also places Georgi as Wegener's closest coworker, at least during the logistic phase. Wegener constantly asked Georgi to make purchases, to help him select expedition members, to take over preparation for all the filming to be done on the main expedition, and to make sure everyone had a valid passport, and he worked through his organizational plans with Georgi on a daily basis.[17]

As time grew shorter, the expedition grew larger, and the amount of equipment and supplies to be gathered and shipped grew steadily from 65,000 kilograms (143,300 pounds), to 75,000 kilograms (165,347 pounds), to 90,000 kilograms (198,416 pounds), and finally to 120,000 kilograms (264,555 pounds), not counting the weight of the men and the twenty-five Icelandic ponies. For the latter Wegener contacted his old friend and comrade from 1912–1913, Vigfus Sigurdsson (1875–1950), who agreed to come along on the expedition until the end of October 1930 and would bring two assistants with him, one an experienced wrangler (Jon Jonsson) and the other a medical student (Gudmunder Gislason).[18] These three Icelanders, all of whom expected to depart for Iceland in October 1930, were not part of the eventual count of the expedition members who departed in April 1930 from Europe; the latter totaled fourteen for the West Station and Mid-Ice Station and three for East Greenland and Scoresby Sound.

Listing the members gives us a chance to see how Wegener had decided to partition the work. Wegener, of course, was the leader of the expedition, and his scientific role would be as a glaciologist. In this work he would be assisted by Fritz Loewe, keeping him close at hand at the West Station through the course of the winter. In order to satisfy Loewe's demand that he not have anything to do with Georgi, Wegener found another meteorologist, Rupert Holzapfel (1905–1960), a young Austrian, to handle the meteorological work at the West Station.

Georgi and Sorge would winter at the Mid-Ice Station 300–400 kilometers (186–249 miles) inland, with Georgi doing meteorology and Sorge doing studies of ice crystals, ice temperatures, and ice thickness measurements (the latter in the summer, with seismology). Wegener also hired a young geodesist, Karl Weiken (1895–1983), from Potsdam to do the gravity measurements and the trigonometric survey, and Kurt Wölcken (1904–1992) from the Geodetic Institute in Göttingen to measure (seismically) the ice thicknesses at the West Station and across the ice cap to the middle of the ice. These men would be assisted by Hugo Jülg (1902–1988), a middle school teacher, and George Lissey (1906–1964), an engineering student; all these—with the exception of Georgi and Sorge—would winter at the West Station.

The technical demands of the expedition were daunting enough that Wegener needed engineers, mechanics, and machinists; to commit to the propeller sleds meant committing to having expert mechanics to assemble them and make them work. The two leads here were Kurt Herdemerten (1900–1951), a mining engineer and an expert on dynamite and on building roads and shaft works, and Emil Friedrichs (1900–1982), an experienced mechanic from the observatory in Hamburg, who would take care of the engine on the *Krabbe* and any other mechanical work.

For the fall of 1930 only, Wegener induced Kurt Schif (1905–1990), an aeronautical engineer and friend of Kurt Wegener in Berlin, to take charge of the transport and assembly of the motor sleds; the sleds would be shipped in large crates from Finland, and the aircraft engines to power them would come from Germany. After Schif's departure in October 1930, the sleds would be in the hands of two other mechanics, Franz Kelbl (b. 1900) and Manfred Kraus (1904–1988), who would drive and repair them in the fall, spring, and summer and serve as radio operators for the expedition during the overwintering.

The life dates given here indicate an important aspect of Wegener's plan: he was training a new generation of German polar scientists. All of the scientific staff were PhD-holding professional scientists, were physically strong, and had to declare their willingness to participate in the backbreaking labor of transporting their equipment from the shore up to the proposed winter station at Scheideck, as well as in the massive effort to retransport 10,000 kilograms (22,046 pounds) of material to a series of depots on the ice and eventually to the Mid-Ice Station. Most of these men were half Wegener's age, and the rest fifteen to twenty years younger than he; this was quite deliberate on Wegener's part.[19]

Wegener did not compose a memoir concerning his preparations for this expedition; what we know about them comes from his correspondence and from the reports of others, especially Else and Georgi. Else worked by his side throughout the spring, often traveling with him to Vienna, Berlin, and Copenhagen; many of the arrangements required face-to-face contact and negotiation that could not proceed by correspondence—there was not enough time. In addition to securing everything needed by the expedition, Wegener also needed permission of the Education Ministry in Vienna to have a leave with pay and to obtain replacements to teach his courses while he was in Greenland.[20] In late January he and Else went to Vienna to arrange this, and they met with Exner to ask for his help. A week later, back in Graz, they learned that Exner had died of a heart attack at the end of the first week in February. "I could not believe it," said Else; "he was only four years older than Alfred."[21]

Exner's sudden death so soon after their meeting in Vienna must have shaken Wegener, and here he came closest to an expression of regret that he had taken on this expedition. On the first Sunday in March he took Else and the three girls cross-country skiing in the mountains near Graz. The girls were now fifteen (Hilde), twelve (Käte), and nine (Lotte).[22] On this occasion, remembered Else, "Alfred asked me quite seriously if we would all travel with him to Greenland, overwinter in Umanak, and in the spring make Greenland dogsled trips on the sea ice, which he described as the summit of bliss."[23] She was reluctant to agree, reminding him that their oldest daughter Hilde had a kidney infection for several months in the previous year and could not be so far away from medical care; it was too risky. Moreover, she and the girls would have to sit alone in Umanak in winter while Alfred was at the West Station. She added that she had no objection, however, to the family taking their summer vacation in 1931 in Greenland.[24]

By the end of March, things were moving along quite well, and Wegener could be encouraged even in the midst of his near exhaustion. The kinds of letters he was writing were now different, not so much new ordering as announcing delivery of material ready for shipment to Greenland, as in his letter to the Danish Ministry for Greenland in Copenhagen, stating that he had sent 22,000 kilograms (48,502 pounds) of compressed hay (for the ponies) from Pomerania (East Prussia) to Copenhagen to be put on board the *Disko*. This was the same ship he had used in 1929, and it would once again carry him to Greenland.[25]

Even in March 1930, and in spite of their constant collaboration, the problem with Georgi concerning authorship and photography rights had not yet been resolved. By the middle of February the book (*Mit Motorboot und Schlitten*) was ready to go to press, except that Georgi had still not sent the photographs, many times requested by Wegener. In fact, Wegener had to quell the suspicion of his publisher that he had another book in progress and was diverting the photographic material there.[26] Toward the end of the month (24 February), Wegener wrote again to the publisher regarding Georgi's desire to forward the photographs but withhold the copyright.[27] The publisher was clearly unwilling, but by 3 March Wegener seems to have resolved the problem by telling both Georgi and Sorge that they would get photo credit underneath each picture they had taken.[28]

That settled, now Georgi had a new demand: he wanted to reserve ownership rights in any photos and movie film that he might shoot on the main expedition; the Notgemeinschaft was absolutely unwilling to allow this. Georgi had refused to sign his expedition contract, hoping to force the hand of the authorities. Time was running out, and toward the middle of March, when everyone was departing either for Berlin or for Copenhagen to get ready for the final departure for Greenland, Wegener wrote to Georgi. Georgi remembered this letter many years later and quoted it thus: "We may sometimes have to give way despite our convictions, but when we come home triumphant, laden with new scientific knowledge, then such legal quibbles will be of as little importance as a scrap of paper!"[29] Wegener continued, "You will certainly not, because of a passing ill mood, make a decision which for many years, perhaps for your whole life, would cloud the memory of our expedition. I believe rather—at least I hope—that when you reach such a point you will bury the hatchet and grab a camera instead and re-affirm the principle which I too have used to smooth away many a difficulty during my expeditions: Whatever happens the cause must not suffer in any way."[30] On the night before their departure, Georgi signed his contract. Loewe, who had been waiting to see what Georgi "got" from Wegener, was satisfied, and he also signed. Now, they could all leave for Greenland.

Greenland, 1930–1931

They left for Greenland on 1 April 1930. Wegener wrote in his diary early in the morning: "Today at 10 AM the Disko will depart. . . . Some farewells, the group shot for the book and then, only then the thread is cut, then begins the expedition. I have the overwhelming feeling that I'm escaping from a swarm of bees. Uff!"[31] The "bees" were questions and requests: "Will the hay reach the boat in time? . . . In my contract we have to change this and that. . . . Shall we take the big barometer?" This diary entry, with twenty-five or thirty such humorous and exasperating questions, was his first record of the expedition under way.

After a stop in Reykjavík, Iceland, to pick up Vigfus, his assistants, and twenty-five ponies, they sailed directly to West Greenland, arriving in Holstensborg on 15 April, where everything was off-loaded onto the dock to await the *Gustav Holm*, an exploring ship with thick ice sheathing which would take them to Umanak. George Lissey, the engineering student from Hamburg whom Wegener had hired as a surveying assistant, spent his time on the dock counting the number of crates in their mountain of baggage and told Wegener that he had counted 2,500, which Wegener noted was "a figure that horrified everybody. But if we count each packet as weighing 45 kg that comes to 100,000 kg which is exactly what we're supposed to have."[32] The *Gustav Holm* was a smaller ship than the *Disko*, and it seemed impossible that they would get everything on board, but they did.

It took only two days (27–29 April) to get them to Godhavn, where the expedition really began. The news was not good about the ice further north, as it appeared that it was so thick that they would reach neither Umanak nor Kamarajuk in the ship. Wegener was pleased to see the *Krabbe*, launched the previous day and ready to come alongside. It was –16°C (3°F) and snowing. Wegener noted, "The Greenland summer is not here yet."[33] As soon as the *Krabbe* was ready, Loewe, Holzapfel, and Jülg, along with Friedrichs and Kraus (the machinist and the mechanic), departed for Quervainshavn. The first three were to make a dogsled journey to check the snow depth markers the reconnaissance expedition had left the previous year, and then drive their sleds north on the ice to Scheideck, while Friedrichs and Kraus would bring the boat north after dropping them off.[34]

On 4 May Wegener wrote, "Now the difficulties begin." The ice was too thick for them to force their way into Umanak Harbor. The governor of the colony at Umanak, Dan Møller, came out to the ship by dogsled to tell them that the ice extended quite far north and had not gone out at Kamarajuk or even Uvkusigssat. They sailed north and found this to be true, so on the next day they began to off-load everything onto the ice. A call went around to all the local Greenland settlements that the expedition was paying for dogsleds to carry the material to Uvkusigssat, 10 kilometers (6 miles) away over the ice, and soon, Wegener said, "Greenlanders had streamed over the ship."[35]

Sorge and Georgi, who knew the lay of the land, took off immediately with Greenlanders on dogsleds to check the ice conditions at Kamarajuk. Wegener was resigned: "Now we must make up by hard work what fortune has denied us."[36] He hoped at least that once there gear was on the beach at Uvkusigssat the pony sleds could haul it to Kamarajuk. But fortune had not yet finished with its denials, and Sorge and Georgi returned with bad news: the ice in the inner part of Kamarajuk Fjord was so rotten that you could put a stick through it. They immediately winched the huge crates holding the propeller sleds onto the ice, attached them to the dog teams, and made for Kamarajuk, 30 kilometers (18.5 miles) distant. They returned the next day, 7 May. They had just barely managed to get the propeller sleds on the beach, and on the return trip they had several times gone through the ice themselves; it was so thin that even the weight of the dogs could break it apart.[37]

Now they were on shore at Uvkusigssat and could do nothing but wait for the ice to go out. They established camp along the shore, near their mountains of goods, and were delighted with everything. The tents had wooden floors, real tables, and real bunks with mattresses of dry hay and down sleeping bags. They set up the lightproof seismograph tent for a dark room and strung up a radio antenna. They were snug, cooking, unpacking, and enjoying the warm temperatures, only –1°C (31°F). By the fourteenth Weiken was making gravity measurements, and Georgi had set up a workshop to repair all the meteorological instruments damaged in transit.[38]

Two weeks later, they were still stuck in Uvkusigssat. The ice had gone out around Umanak, and when the *Krabbe* arrived, they sailed it there to make arrangements with a Capt. Olson—whose schooner, the *Hvidfisken*, had wintered in Umanak—for their late-arriving gear to be shipped to Uvkusigssat. On 30 May Wegener finally could stand it no longer and made a trip to Kamarajuk, which took twelve hours, beginning with

Unloading the Gustav Holm at Uvkussigssat. The motor sleds have been taken off as well as much of the gear. In the foreground are some of the ponies taking a rest break between hauls across the six miles of ice to the shore. Photo courtesy of the Alfred Wegener Institute, Bremerhaven.

the dogsled, then pulling sledges themselves, going along the shore on a light wicker sledge, and finally inflating the collapsible boat to cross over the last section of open water. Everything was deep in snow, but they managed to push the motor sledges across a stream onto the snowfield at the foot of the glacier so that they would not be caught on open ground once the melt began.[39]

On 3 June, their twenty-fifth day of waiting, they learned that Loewe's sled at Quervainshavn had been too heavy and the snow too deep and he had turned back; Wegener sent the *Krabbe* south to pick him up. Wegener was now extremely nervous. By 13 June Wegener could once more stand it no longer, and when Capt. Olson arrived with the *Hvidfisken*, they agreed that two days later they would try to break through the ice by ramming it. This effort availed them nothing. Finally, on the thirty-eighth day in Uvkusigssat, they were roused at 1:00 a.m. by the Greenlander keeping watch, who shouted that the "ice is going out!"[40] They arrived to find only about 500 yards of ice left blocking the way into Kamarajuk, and they made several attempts to blow it apart with dynamite, all of which failed. Then suddenly, after twelve hours of work, the ice parted not far away from them, and the ice began to flow out of its own accord into the ocean. Wegener wrote, "How indifferent Nature is to our puny achievements."[41]

The Struggle to Ascend the Glacier

Not until 25 June were they able to ferry all of their 2,500 boxes to the beach at the foot of the glacier at Kamarajuk using *Krabbe* and *Hvidfisken* both. Now began the work that would take them the rest of the summer, getting their belongings to the top of the glacier, to Scheideck, and onto the Inland Ice. They would establish a series of

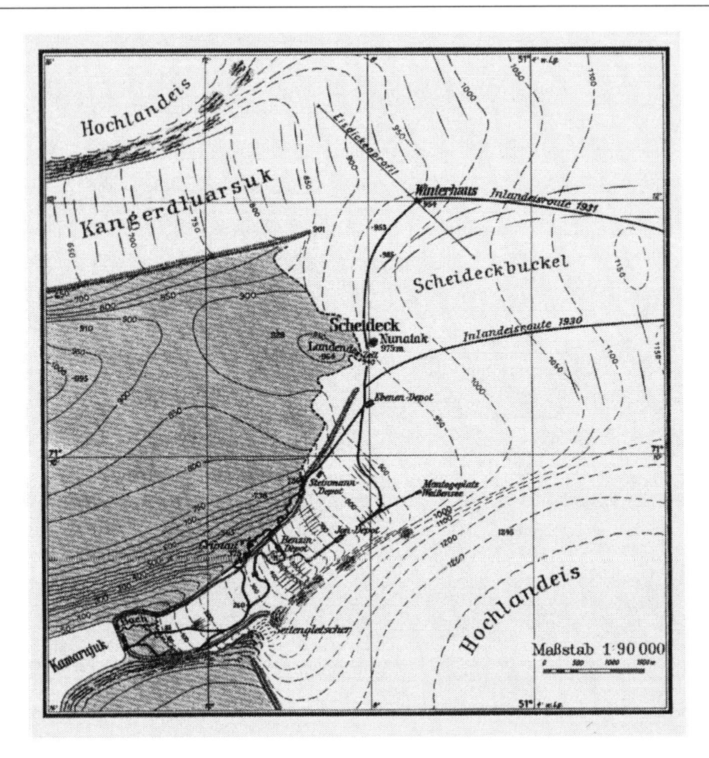

Map of the Kamarajuk Fjord and Glacier, showing the various routes and places discussed in the text. The ice-free land on the map is shaded. From Kurt Wegener, *Wissenschaftliche Ergebnisse der Deutschen Grönland-Expedition Alfred Wegener 1929 und 1930/31*, vol. 1, *Geschichte der Expedition* (Leipzig: F. A. Brockhaus, 1933).

depots on the glacier, using ponies with packsaddles, dogsleds, and themselves and the Greenlanders as porters. Herdemerten was going to try to blast a roadway in the middle section of the glacier, where it was serrated with crevasses and ice hummocks. George Lissey, ever enthusiastic, said, "We all became transport workers and slaves from sunset to sunrise—for we worked at night; during the day it was too hot for man and beast in the burning glare of the sun on the glacier. . . . To us Polar novices everything seemed very unlike our idea of an expedition; it was much more like erecting some building in the mountains. Scientific work was out of the question. Nothing but transport, transport, and still more transport."[42]

By 15 July, after three weeks of strenuous and very unpleasant work, enough of Georgi's gear had been carried to the top of the glacier for Wegener to be willing to send him out with ten Greenlanders and Loewe and Weiken. Loewe and four others would turn back at the 200-kilometer (125-mile) depot; Georgi, Weiken, and the rest would go on to the 400-kilometer (250-mile) point, where they would begin to establish the Mid-Ice Station.[43]

At this point Georgi made the request of Wegener that he be allowed to remain alone at the central station, now christened "Eismitte" (Mid-Ice). Georgi recalled that Wegener was reluctant to give him permission to stay alone at the 400-kilometer station. Georgi did not say why Wegener objected but suggested that the latter was short of men at the West Station, and that he (Georgi) had then made the argument that the meteorological instruments, once set up, would have to be tended.[44] There were

Ascending the Kamarajuk Glacier. The chaos of the moraine is to the left, and the narrow ice road may be seen just above the span of dogs. Note how small the load on the sled is: the steepness of the ascent meant innumerable small loads, taking much additional time. To the right of the man with the whip is visible a porter with a backpack—much of the 20,000 kilograms of material was carried up in this way. Photo courtesy of the Alfred Wegener Institute, Bremerhaven.

other reasons as well that Wegener might have been reluctant. He was not entirely convinced that the 400-kilometer point was the best location, and as late as 5 August he was thinking that the station should be moved back 50 kilometers (31 miles) toward the coast, to the 350-kilometer (217-mile) mark, to make transport easier; this would cut off 100 kilometers (62 miles) round-trip for each supply mission.[45] Moreover, for Georgi to leave for the Mid-Ice Station and not return to the west coast at all was essentially for him to "secede" from the expedition, relieving himself of all transport work, while the others would be forced to forgo their scientific work in order to support *him*. This is exactly the sort of outcome that Loewe had feared and objected to the previous December. Finally, Georgi seems to have proposed to Wegener that he (Georgi) should mount an expedition to East Greenland at the end of the following summer; it was psychologically indicative of Georgi's desire to get away completely from *this* expedition.[46]

Wegener also noted in his diary on 17 July that Georgi's sled trip had been held up because of "Georgi's endless packing (essential)."[47] When Georgi finally set off, at the end of the first day the Greenlanders said that the sleds were too heavy for the dogs in the deep snow, and they would not go further. Georgi, with the greatest reluctance, removed half a ton (450 kilograms) of supplies from the sleds.[48] When they had reached the halfway point a few days later, they had to reduce the loads again, leaving another 160 kilograms (350 pounds) at the 200-kilometer depot.[49] The snow was deep and soft, and the going was hard, quite different from the experiences reported by all the expeditions since 1912 about the middle of the ice cap. Everything at all times was more difficult than anyone had imagined; this seemed to be the fundamental law of this expedition.

After Georgi's sled group departed for the ice cap, for everyone else it was "more transport." By now the crevasse zone of the glacier had melted so much that it was no longer passable by ponies, and on 22 July Wegener began, with a large team of Greenlanders, to build a road over the moraine with a series of switchbacks on which the ponies could work as pack trains. By the twenty-sixth, it was all but finished, and Wegener was delighted. He enjoyed working with the Greenlanders, of whom he said, "It is splendid to see how the Greenlanders work. European workers would not have done so much; they would not have moved so quickly. So we have one [a road]; the result far surpasses my expectations."[50] Once across the moraine, they could circumvent the crevasses entirely, and it looked as if things were beginning to go well.

Expedition in Crisis—August 1930

Wegener's elation was short-lived. By 5 August he found that the morale of his expedition was beginning to fail, with much less work getting done each day, and he also found the first louse on his shirt—"nasty but not surprising, for living in such close contact with the Greenlanders makes infection almost inevitable."[51] Wegener was at this time overmastered by the sheer bulk of everything he had to do. He was the only one who could deal with the Greenlanders, and he had to pay them, drink coffee with them, negotiate with them, and direct their work. As a result, he was often at the beach, as well as having to take the *Krabbe* to get supplies at Umanak. Because he had to spend so much time on transport and building the road over the moraine, he had little time for large-scale organization, and while he left much to Sorge and Loewe, they complained (in their diaries) of a lack of direction and organization. Sorge said that Wegener always wants everyone to be "active" but doesn't take any time to assess the situation or actually consult. Loewe also complained that "he settles arrangements regarding work for the next day in private conversations of which others only learn by chance."[52]

Getting the motor sleds to the top of the glacier was a terrible job and took a week of effort by almost everyone; they had to be hand winched up a 70° slope. The men were doing no science and had not even (by the end of July) set up a meteorological station, though the instruments were there at Scheideck and Holzapfel was working hard at it.[53] There were no general meetings to organize the work, and Wegener began to see that things were indeed in danger of falling apart.

The ponies had begun to founder; they had already lost three. The loads were too heavy, the bolts on the boxes tore out, and everything had to be lashed with rope to their saddles. The men were also in trouble. Jülg was so overstressed that he was close to collapse, Lissey had sciatica, and Wölcken was so exhausted that he couldn't work and had to go to his tent. Vigfus had rheumatism and had lost two teeth while shoeing a pony that managed to kick him in the face. Jon Jonsson had terrible stomach pain and had begun to vomit blood. He would have to be evacuated to Umanak. Moreover, Wegener discovered (to his horror) on 9 August, when he returned to the beach after finally getting the motor sleds to the top of the glacier, that they had only twenty days of hay and five to ten days of fodder left, about which he exclaimed, "A catastrophe!"[54]

Wegener's very extensive diary entries for this period of time exhibit the same volatility as the weather and the work, as the fate of the expedition seemed constantly to hang in the balance because of the immense difficulties of transport. First, there was too much ice in the ocean to get through, and then too much snow melt on the glacier to make solid footing. The glacier road was finished, just in time to be abandoned for the road on the moraine.

Wegener was aware of what was happening. "So the crisis is still on or if you like, getting more intense. The expedition's prospects are clouded over. We are not going to deceive ourselves and go on working as if all were going well." Wegener wrote in his diary that same night (9 August), "I still feel rather desperate. The Mid Ice Station really depends on the motor sledges functioning, the transport journeys can only be carried out with difficulty, and now the pony stores are threatened. Since I found that louse the difficulties have grown in a very worrying way. Since I found that louse!"[55]

Wegener then, perhaps understandably, began to be away from the expedition and the glacier transport for longer and longer periods. He took more and more voyages in the *Krabbe*, to find hay, to make purchases in Umanak, to deliver Greenlanders back home who did not want to work any longer, and to try to hire their replacements. He spent days fussing with the *Krabbe*'s engine, which was misfiring. His diary is full of entries about his success and failure in acquiring hay for pony fodder. Back at Kamarajuk, things were still functioning, but the mood was black, and Wegener was more and more distant, as his notions of success and failure were now focused on how much hay he could collect for the remaining ponies.

After Wegener's departure, the men—injured, frustrated, overworked, and (in their own minds) leaderless—began to find fault with one another and to break down. Anyone who has discovered heretofore invisible faults in a friend on a weekend camping trip can readily extrapolate to the intensity of bad feeling generated in an arduous and dangerous polar expedition in utterly remote, foreign, and uncomfortable surroundings, with cold, wet, and lice joining anxiety, fear, backbreaking labor, perpetual discomfort, boring food, dirt, stench, snarling dogs, wind, rain, treacherous ice, equipment failure without hope of replacement, absence of adequate rest, absolute lack of privacy, homesickness and isolation, and the disconcerting effects of endless light, all in the presence of a group of peers not chosen by oneself, in a context in which failures of schedules and supply and exigencies of time, weather, and locale must pit the scientific responsibilities of expedition members—and the members themselves—against one another. The potential for bad feeling is enormous and has to be controlled by acts of self-discipline which further sap emotional and physical resources.

These tensions are exacerbated by the discovery of arrogance, cowardice, or just clumsiness and lack of stamina in a colleague. All of this devolves to the expedition leader, who is confidant to all injured parties, arbitrator of disputes, and a triage officer in the intellectual as well as physical realm. The realities are quite stark: polar travel, like Alpine mountaineering, is a mecca for those who would "test" themselves, and a gladiator's arena for those whose tests must involve the conquest of other men in the process. Under these circumstances, it is little wonder that Wegener sought the solace of his cabin in the *Krabbe* and spent his time at sea, going from village to village, looking for hay.

Returning from his various trips on 29 August (the hay was now acquired, and Loewe was taking Greenlanders home on the *Krabbe*), Wegener hiked up to Scheideck with Holzapfel to see how the work was going. He was enormously encouraged, as only "odds and ends" were left at the foot of the glacier and most of the contents of the intermediate depots had now been transferred close to the top; in spite of deterioration of the road on the moraine, they were able to repair it and keep the pony caravans coming. The ponies were now pastured at Grünau, a broad, flat grassy area just on the north side of the moraine. Things had improved without him. Wegener had said to Loewe in July, "I have the impression that the work would go better . . . if I were not here at all."[56] This had turned out to be the case, at least in August; without Wegener's ceaseless "encouragement," they could find their own speed and take charge of their own work. This is what Wegener had wanted; he had only to leave it alone for it to happen.

"When we arrived at Scheideck," Wegener wrote on 29 August, "we suddenly heard the hum of a motor. This was music to our ears! We stood rooted to the spot and listened reverently until the trial run was over. Twice more this music of the spheres resounded, and each time I was so struck by it that I stood still and listened till the engine was switched off. I had a feeling that a dream was coming true."[57] Schif had finally managed, with the help of his mechanics, to mount the aircraft engines on the propeller sleds and test the motors. This was already three weeks since the arrival of the propeller sleds at the top of the glacier, but Wegener could not contain his pleasure, in spite of the lost time.

> I am proud of the motor sledges, as their use in conjunction with dog sledges marks an important advance in polar exploration. We have certainly hit the mark with all our means of transport. The Icelandic ponies have also stood their first real test here with us. . . . Ponies on the glacier, dog sledges and motor sledges on the ice cap, this is the correct choice. Now we began at once a new epoch in polar exploration. Everything that we wish to try and can try must be tested thoroughly. What we are doing here points the way at once for future Antarctic exploration. How wonderful that it should fall to us to make this pioneering step; nay, in view of the many air disasters that have occurred in Polar Regions, I may say the redeeming step.[58]

That day, Wegener got his first ride on one of the propeller sleds in Greenland. He wrote in his diary that night, "Now the dream has really come true today I traveled on the inland ice comfortably sitting in a closed cabin and smoking my pipe."[59] He had missed the departure of the next sledge party for Eismitte, but at the 5-kilometer (3.1-mile) depot the propeller sleds caught up with Sorge, Wölcken, Jülg, and nine Greenlanders, twelve sleds in all. Wegener, who had been away and had not supervised the preparation or the loading of these sleds, or even the planning of this trip other than its approximate date, was not pleased. "Sorge's attempt to pack as much as he possibly can onto this trip goes too far, he has used up all the dogs." Now there were only two Greenlanders and five dogs left, not enough for the work still to be accomplished—one more thing for Wegener to worry about.[60]

The next few days, up until about the fifth or sixth of September, were a period of experimentation and fine-tuning of the engines, and Wegener had a chance to make another motor sled trip, as Schif, Kelbl, and Kraus worked with both propeller sleds to move as much fuel as possible to a depot 10 kilometers from Scheideck, past the worst of the crevasses.

Along with their joy that the sleds were working, and as they were learning to operate the engines, they made some unpleasant and troubling discoveries. It turned out that the propeller sledges could make almost no headway when facing a strong headwind, and when the wind blew, it generally blew from the center of the ice toward the

Polar Bear (Eisbar), one of the two propeller sleds, here operational on its first trip inland. The sleds worked well on flat and even ground but could not pull up slopes or against strong wind. This put the supply plan for the Mid-Ice Station far behind schedule. Photo courtesy of the Alfred Wegener Institute, Bremerhaven.

coast—meaning a headwind. Moreover, the sleds got stuck in deep snow and drifts all too easily, and in any combination of snow and wind they could not pull loads even up moderate slopes. They learned to zigzag, almost as if tacking, to get up the slopes at a walking pace. When they stopped, the skids immediately froze to the ice and had to be pried up with crowbars.

The sleds also used much more fuel than they had imagined, running only for about six hours per tank of fuel. This meant they would have to recalculate both the time schedule and the means of getting to Eismitte, as they would have to depot fuel at much closer intervals and then leapfrog their depots forward just to get to the 200-kilometer station.[61] It was fortunate for them that the route was so well marked, because visibility was often poor in the moving sleds. Wegener had established a protocol that a tall wand with a black flag be set up every half kilometer, and at every fifth kilometer a snow tower with a much larger black banner on it, stretched around four poles.

On 5 September they managed a trip 85 kilometers (53 miles) inland, and here there was good news. Once they were more than 48 kilometers (30 miles) inland, the ground became flat, making it easier for the sleds, and they also discovered that they had not overestimated the effective load, which really would be a half ton. Over the next seven days, the sleds covered more than 800 kilometers (500 miles). They established all the fuel depots for the trip to Eismitte, and at the 200-kilometer depot they took a ton of fuel oil and stove oil, the prefabricated hut (in sections) for the Eismitte station, and another 1,200 kilograms (2,646 pounds) of supplies, instruments, and food.[62]

Back at Scheideck, the winter snow had already begun. On 1 September they had a storm of sticky wet snow that made moving about, as Wegener put it, feel like a fly must feel walking on a sheet of flypaper.[63] More snow fell the next day, about 20 centimeters

(8 inches) in all.[64] Just as Wegener needed to focus more than ever ("it will be 1 October before we get everything to the Mid-Ice Station, and even then only by the skin of our teeth"), he had news that the *Disko* had arrived in Umanak and would leave in a few weeks: they would lose the two remaining Icelanders and Kurt Shif, leaving Kraus and Kelbl to handle the propeller sleds on their own.[65]

Then, to his consternation, Wegener discovered he would have guests. Jens Daugard-Jensen (1871–1938), director of administration for Greenland, whose help Wegener had depended on repeatedly for this expedition, had made the trip to see how the expedition was going, and he was to be accompanied by Peter Freuchen and the Danish archaeologist Helge Larsen.[66] As glad as he would be to see Freuchen, this would all be lost time. The scientific sled trip that Loewe and Weiken had planned to make a short distance inland to do some glaciology would now have to "be stripped of its purely scientific character, and turned into another piece of the transport."[67] He informed them of this on 4 September.

On 8 September he was still entranced by the propeller sleds; at one point they had taken him over the ice (with the wind at their backs, and going downhill on perfectly smooth ice and snow ground) at almost 70 kilometers per hour (43 miles per hour).[68] Yet it was increasingly apparent that the number of misfortunes these machines suffered and the delays endured in repairing them, along with their inability to perform in deep snow or strong wind, made it unfeasible for them to dependably supply the Mid-Ice Station with the necessary food and fuel before travel of any kind became absolutely impossible—sometime around the middle of October.

From Crisis to Catastrophe—September 1930

"What now?" Wegener had already written on 6 September. "The catastrophe has arrived."[69] Kraus might not be able to overwinter in the middle of the ice, and Georgi and Sorge would therefore have to do without radio contact, as neither of them knew Morse code, nor could they operate the radio efficiently. Wegener would have to abandon the attempt to deliver any inessentials, perhaps not even taking the prefabricated hut intended for Eismitte. Wegener decided that he would have to organize a fourth dogsled trip to carry food and fuel all the way to the middle of the ice.[70]

He was no longer sure that Georgi and Sorge would be able to overwinter, and he might have to pick them up and bring them back.

> The situation in which the expedition finds itself, in the face of the failure of these motor sleds is simply intolerable. We cannot recall the crew from the [Central] Station! That would be a catastrophe for the entire program. We must, no matter what the cost, put together a supply journey with sledges of such dimensions, that all the essentials for the overwintering can be carried by dogsleds alone. No matter what the cost. That is easily said! Where will we get the sleds? We need 15! Where will we get the men, and do all of this when time is so short? This is the biggest and most fatal crisis that the expedition will have to undergo.[71]

After the arrival and departure of his guests between 10 and 12 September (he was able to arrange a motor sled trip for Freuchen), Wegener turned all his attention to his new catastrophe, the failure to supply Eismitte in time to go into winter quarters when he had planned. He would now have to break two of the three rules he had many times repeated to Herdemerten, Weiken, and the others: (1) no more than three major sledge trips to the central part of the ice, and (2) no new travel after mid-September

Rueful Alfred Wegener facing the camera amid the catastrophic (his characterization) supply difficulties of September 1930, with winter closing in and insufficient food and fuel at the Mid-Ice Station. Photo courtesy of the Alfred Wegener Institute, Bremerhaven.

and absolutely no travel after 1 October.[72] He could observe the third rule he had made—that the propeller sledges were an experiment and were not to be depended on—only by breaking the other two.[73]

There was now very little time, and drivers and new dogs—as many as 130—needed to be found for this trip. Most of the young Greenlanders were on their way to Eismitte with Sorge's group, and they were not expected back until 20 September at the earliest. All the lightweight Nansen sleds were gone, and that meant using heavy Greenland sleds intended not for snow but for sea-ice travel; these could be exchanged with the lighter sleds being used by the third party on the way, if the parties crossed paths.[74]

On the seventeenth, the propeller sleds, fully loaded, left for Eismitte with Schif, Kelbl, Kraus, and Lissey. They were carrying only essential food and fuel, and Wegener intended for the fourth dogsled group to leave as soon as possible thereafter. At almost the same time, Weiken and Vigfus took six ponies with their sleds and transported most of the loads intended for the fourth dogsled journey to a depot 15 kilometers (9 miles) inland, so that the dogs would not have to pull heavy sleds over the sharp, late-summer ice of the crevasse zone and risk badly cutting their feet.[75]

On the eighteenth, with preparations well under way, Wegener told Loewe and Weiken that he had decided he would lead the fourth sledge journey. He saw that a number of decisions would have to be made at every point along the way and at Eismitte itself—decisions that were his responsibility. Moreover, he knew that it would be difficult to keep the Greenlanders going, as they did not like being on the Inland Ice, were suspicious of the size of the loads they would be pulling, and all thought that it was too late in the year for such a trip (they were right). Wegener's Danish, in any case, would be an important tool in convincing them to keep going during the inevitable negotiations en route. Wegener announced that he, Loewe, and thirteen Greenlanders, with fifteen sleds in all, would leave as soon as possible; Weiken would assume command of the expedition at the West Station.[76]

They set out on the morning of the twenty-first, and only a few miles out on the ice cap they encountered Wölcken, Jülg, and the Greenlanders from the third sledge journey. The returning party reported that they had met the propeller sleds at the 200-kilometer halfway depot, and that these sleds were preparing to go on to Eismitte. This was wonderful news! They also carried a letter from Georgi and Sorge in which they provided a list of the amount of food and fuel that they had, as well as their estimates of how many men could live for how many months with that amount of food and fuel—separate estimates for two men and three men. If three men were to overwinter, additional cases of food and fuel would be needed. They went on to say, "The fact that we don't have the propeller sleds puts us in a new and dangerous position. We both have come to an understanding that in case the propeller sleds cannot bring the Firnhaus [the prefabricated tent house of Wegener's design, weighing 500 kilograms (1,102 pounds)] we would still remain and overwinter if by 20 October we had received. . . . [what follows is a short list of items, petrol and ice bore some dynamite and cable several boxes marked with their name, seismometer tent, skis a tent, and a few other items] . . . If," the letter continued, "we don't have these items by 20 October here or have not heard anything from you otherwise, we will on that day depart with hand sleds." They hoped, they said in closing, each day to see the propeller sleds, "whereby all these problems will be solved."[77]

Wegener immediately dispatched sleds back to Kamarajuk to obtain the items on Georgi and Sorge's list which they had not brought with them. It was a reasonably short list, and the items, other than the extra petrol (which they already had planned to take with them), were not especially heavy. Wegener had the additional items in hand late on the twenty-second, and they started up again, heading inland. On this first day they covered almost 17 kilometers (10.5 miles). Of course, the sleds were not yet loaded and were easy for the dogs to pull.

On the next day, 23 September, they reached the 40-kilometer (25-mile) point and realized that they were too heavily laden; Wegener decided therefore to off-load everything not absolutely necessary. The next morning, as they awoke, they saw two tents and one motor sledge camped a mile or so ahead of them, and they learned that the sleds had met terrible weather and deep snow at the 200-kilometer depot. Though they might be only *one day* from Eismitte, they could not make the sleds go forward, and eventually they had to give up and turn around. All four men were now in one sled, having abandoned the other one at kilometer 50 (mile 31.5). They would try to make it back to Scheideck in the remaining propeller sled and then come back and tow both of the motorized sleds in with dogsleds.[78] At this point, one of the Greenlanders traveling with Wegener decided to quit then and there and left with the motor sled party for the coast.

Bad weather—headwinds and heavy snow—now plagued them for three days, and on the twenty-eighth all of the Greenlanders came to Wegener and Loewe's tent. They wanted to go home. They were cold, they did not have enough clothes, and they could see that the dogs could not pull even these loads in the soft snow. Loewe said that he could see their point, and the majority of them could not be dissuaded by money or argument. Eventually, Wegener talked four of the Greenlanders into going on, with an increase in pay.[79]

Before the eight Greenlanders left to return to the coast, Wegener sat down and wrote a letter to Weiken (the same letter quoted as the epigraph to this chapter). In it Wegener gave instructions for paying the Greenlanders who were coming out with the letter, and he said that he did not know whether Sorge and Georgi would be able to stay or would have to come back out with them. He said that he wanted the station manned through the winter, and that Georgi and Sorge had agreed to stay even without the hut. Then came the part about it being "a matter of life and death." Wegener was convinced that he had to go on and either meet Georgi and Sorge coming back or get there before they left, as he was sure they would die trying to haul sledges—they would certainly freeze to death.[80]

So they pushed on, and it took them eight days to cover a little less than 80 kilometers (50 miles). On 6 October, 151 kilometers (94 miles) from the coast, the remaining Greenlanders refused to go on. Wegener therefore wrote another letter to Weiken, with more instructions about paying the men coming out this time, and he told him that in the event that Sorge and Georgi did not want to stay on, he and Loewe would replace them and overwinter, even with a minimal scientific program reduced to some basic meteorology. He asked for a relief party to be sent to the 62-kilometer (38.5-mile) depot, leaving around 10 November and planning to stay until at least 1 December. Wegener told Weiken that he calculated that they had enough rations to keep going as long as they made 12 kilometers (7.5 miles) per day. Whoever came back from Eismitte, there would be three returning, as in the meantime Wegener had talked the youngest of his Greenlanders, Rasmus Villumsen (age twenty-two) from Uvkusigssat, into going the rest of the way with them. Rasmus would certainly come out with whichever two men made the return journey.[81]

By 10 October Rasmus Villumsen, Wegener, and Loewe had only covered 19 more kilometers (12 miles). They took a day of rest and discussed their situation. Loewe and Wegener agreed that they could not, at this rate, reach the station at Eismitte before Georgi and Sorge departed; Wegener was at his low point and suggested to Loewe that they might not make it to Eismitte at all, or even push forward to meet Sorge and Georgi successfully. They decided that if they had to turn back, they would do so at kilometer 230 (mile 143), based on their remaining rations.

On 14 October they took another one-day rest at the 200-kilometer depot. They now had too many dogs for the amount of dog food remaining, forcing them to kill some dogs, skin them, and store the meat to feed the remaining dogs. They also rearranged the depot and checked on the meteorological instruments. The latter seems to have put Wegener's mind at ease somewhat, and they talked about the question of whether Georgi and Sorge would leave Eismitte. Loewe clearly remembered that Wegener hoped they would not, while Loewe thought that they would. Wegener said he very much wanted them to stay on, because he wanted to go back to the West Station. He also told Loewe that, one way or another, they had to find their way to Sorge and Georgi, as the uncertainty about their fate, if Wegener and Loewe came back without them, would be "intolerably disturbing to the work of the station."[82] Loewe may not have known this, but Wegener was motivated here by his memory of the winter of 1907/1908, when the rescue party had returned without Mylius-Erichsen and Hoeg-Hagen. The Danmark Expedition had collapsed into recrimination and infighting that lasted until the following spring. Wegener never wanted to be part of anything like that again.

The three men now battled on through deep snow, taking ten days to reach milepost 208 on 24 October. Loewe said that by now their clothes and sleeping bags were getting wet, and he had developed very painful frostbite on the tips of his fingers, from the work of untangling the dog harness. The temperature had been falling steadily, and it was never higher than −40°C (−40°F). They had to have a talk about what to do. Had Georgi and Sorge left on the twentieth, as they said they would, they should have met each other by now, as the Eismitte crew would be traveling downhill with the wind at their backs and would have had to cover only 67 kilometers (42 miles) in six days, or only 7 miles per day. Loewe was sure they had decided to stay and overwinter. Rasmus at this point had lost all hope and wanted to turn around, but Wegener would not hear of it so close to the end of the trip, and eventually Rasmus relented.[83]

In the later afternoon of the twenty-fifth, as they were still camped at the 208-mile point, Wegener suggested to Loewe that they go for a walk. It was between three and four in the afternoon, already dusk at this time of year, but the wind had dropped. They could see night advancing toward them, a clearly demarcated line on the ice to the east. Wegener then began to talk to Loewe in a way Loewe had never heard before ("Wegener spoke openly as he did but rarely"), about the nature and destiny of mankind. As they walked back and forth, Wegener told him of his belief that there was purpose behind human evolution, that human liberation was a direct result of the growth of knowledge, and that the expansion of human knowledge was the ideal that inspired all his actions. "As the darkness meanwhile fell," Loewe remembered, "and sledges, dogs, and camp lay as dark shadows under the glittering firmament, the gleaming arch of the northern lights, a symbol of such faith as his, led our gaze to its colored bands and along its mazy forms into infinity."[84]

Wegener had seen how depressed Loewe was, and he took him for this walk not least to inspire him to go on. One cannot help remembering Wegener's Christmas stargazing walk alone in the Greenland darkness in 1906, a quarter of a century earlier. He had then said that someday, if he led such an expedition, he would do what he could to inspire the others, and that he would survive, that these stars were "his stars." Now, back in Greenland, only a week before his fiftieth birthday, he had remained true to this faith: he had survived, and if he no longer felt he possessed these stars, he felt himself part of a scheme of evolutionary development in which all of this merged into one.

So they pushed on, with Wegener doing most of the outside work with Rasmus, as Loewe by the twenty-seventh had lost all feeling in his toes. He says that Wegener massaged his feet for "hours on end," but to no avail, as the feeling in his toes never returned. He also remembered that "Wegener's energy was marvelous—he was always the first up—and so was the skill with which he avoided getting frostbitten although he did a lot of work with his hands bare."[85]

From 26 to 30 October the mean temperature was −50°C (−58°F). The dog pemmican had frozen so hard that it had to be broken with an ax. Their breath immediately froze into ice crystals and fell to the ground. There were only a few hours of daylight, and they always pitched their tent in the dark. The tent could not be kept warm even with the stove burning. They ran out of dog food on 28 October, at 377 kilometers

(234 miles) inland, and ran out of fuel for their stove the next day. On the morning of 30 October, at a temperature of −52°C (−62°F), they arrived at Eismitte. They had been traveling at high altitude on the ice cap—2,000 to nearly 3,000 meters (6,500–9,800 feet)—at temperatures far below zero, in deep snow, against strong winds, for forty days.

Rasmus arrived first, and Wegener and Loewe a few minutes later. The arriving party discovered that Sorge and Loewe had excavated a huge multiroom dwelling in the firn, using saws to cut rectilinear walls. They had cut and repositioned squares of the dense recrystallized snow as if they were wooden blocks. They had used packing cases to make tables, desks, and flooring; they had a workbench, a balloon room, and storerooms. They had built a tower above ground for weather observation and for launching balloons and had established a complete station for the meteorological instruments. They had erected their tent inside the main cavern to keep meltwater off them as they worked.

It was −5°C (23°F) in the underground shelter—47°C (85°F) warmer than where the three travelers had just been; it felt nearly tropical to them. Sorge and Georgi saw at once how badly Loewe was frostbitten on his toes, fingers, and face, and they massaged his feet and hands and put him in a dry and warm sleeping bag. Sorge remembers most of all Wegener's miraculous appearance: "he looked as fresh, as happy, and fit as if he had just been for a walk. . . . Wegener kept exclaiming, 'You *are* comfortable here! You *are* comfortable here!' over and over again."[86]

Wegener was flooded with relief and even with a sense of joy. From the moment he entered their ice cave, he knew that Georgi and Sorge intended to stay, that the expedition would be a success, that they would survive the winter, that they would overwinter in the middle of the ice cap, and that all the effort and suffering and pain had been worth it. Wanting to be sure of what his emotions already told him, he asked them directly: if you think it's too risky, if you don't want to stay on, I'll stay here with Loewe. No, they responded; they would stay. It is exactly what he wanted most to hear. He wanted to go back to the coast to be at the West Station; he wanted to be able to tell the scientists there that Georgi and Sorge were well, were dug in, and were determined to succeed.

Both Georgi and Sorge later remembered that he questioned them closely for many hours, as he ate and drank coffee. They rested in between bouts of eating and coffee drinking. Wegener wrote for hours in his diary. He asked to see their meteorological records and all the scientific observations they had made up until then regarding the strata in the ice, and he made detailed notes. Sorge was surprised to hear Wegener speak of a second springtime crossing of the ice, not to Scoresby Sound and the Eastern Station; that had always been in the plan. This was a new crossing that Sorge had not heard of, south to Angmagssalik. This was the plan that Georgi had presented to Wegener before he left on 15 July. Wegener asked them for a list of supplies that such a trip would require, and he stuck that list into his diary, promising to send those supplies in the spring.[87] This was Wegener's way of saying "thank you" to Georgi for what he had done, in spite of all the failures of transport, to make the expedition a success by staying at Eismitte.

They calculated the remaining food and fuel at the station. Wegener suggested ("he thought it best")—he always "ordered" by asking—that Loewe should stay; if anything should happen on the way back, such as if the dogs should die, or if they had to ski or walk out, then Loewe would surely die. In fact, there was enough food and fuel that they could keep Loewe and also spare two cases of food and a canister of stove fuel—weighing 300 kilograms (185 pounds); with Wegener and Rasmus each driving a sled with only half that load plus the dog food, and dividing the remaining seventeen dogs between them, they should be able to move very fast, going downhill with the wind at their backs.[88]

On the night of 31 October, with Rasmus dozing by the stove and Loewe sleeping fitfully, bothered by his (intensely painful) frostbitten feet, now that they had warmed up, Wegener spoke for quite a while with Georgi and Sorge. Georgi remembered this conversation so well that thirty years later he could quote it verbatim. "I know you are downhearted," he said, "and because you don't have your equipment, your scientific work will be sketchy compared to what you wanted. But the fact that you have spent the winter here in the middle of Greenland, even without any particular results in research, doing only the simplest and most routine measuring, is something which is worth all that has gone into this expedition."[89] Georgi added to his recollection, "Who knows if it was not this encouraging word that helped us psychologically to get through that winter?"[90]

The next morning, 1 November 1930, was Wegener's fiftieth birthday. They celebrated with him. Wegener and Rasmus put on their sledging clothes, went upstairs, and roused the dogs and fed them. The dogs were sluggish but responded to encouragement and food that had been thawed and wetted inside Eismitte. Georgi brought out the movie camera, Sorge grabbed a still camera, and then Sorge photographed Georgi filming Wegener and Rasmus. Then Sorge took a portrait of Wegener and Rasmus. Wegener looked determined; he had already been outside for an hour or more, and his mustache was frozen with ice; Rasmus looked anxious to get going. They harnessed the dogs and waved farewell, and Georgi filmed them disappearing in the half-light; after only a few seconds, they were over a hummock, down a slope, and gone.

Wegener and Rasmus were now in a race for their lives. The temperatures were still brutally low, around −50°C. Wegener's plan, as he had described it to Sorge and Georgi, was to race with two sleds as far as the halfway point and then consolidate the remaining healthy dogs onto one sled that Rasmus would drive, while Wegener skied. Things, however, went bad quickly, and dogs began to die, unable to pull, and had to be abandoned earlier than they had imagined and hoped. Already at the 284-kilometer (177-mile) point they tossed away a case of dog pemmican. At the 254-kilometer (158-mile) point, just short of halfway, Wegener abandoned his sledge, and from that point he was on skis. This was probably 8 or 9 November.[91]

Perhaps four days later, at kilometer 189 (mile 118), they camped for the night. In the evening, with the stove going, with his heavy clothes off, and brushed free of snow, perhaps while waiting for dinner or making notes in his diary, Wegener suffered a massive heart attack and fell over dead, his eyes wide open. Just like that, his life came to an end. Under the immense stress of skiing at high altitude for days on end, his injured heart simply gave out.

Rasmus, who at twenty-two had already seen many men die, prepared Wegener for his grave. He dug a hole in the snow, and into that hole he put down a sleeping bag and a reindeer skin, and on top of that he put Wegener's body, sewn into two sleeping bag covers. On top of that he put Wegener's fur clothing and another reindeer skin and then buried him in the snow. He then stuck his skis into the snow to mark the spot, along with a broken ski pole. Rasmus then took Wegener's pipe and tobacco, his diary, and his fur gloves and pushed on toward the coast and safety.

Alfred Wegener and Rasmus Villumsen preparing to depart the Mid-Ice Station and race back to the coast. This photo was taken was on 1 November 1930, Wegener's fiftieth birthday. It was −50°C (−58°F). Photo courtesy of the Alfred Wegener Institute, Bremerhaven.

Rasmus made it as far as kilometer 168 (mile 105), and he camped there for several days, leaving behind a hatchet that Wegener and he had taken from Eismitte for chopping dog food.[92] From there he vanished. He never made it back to the coast, and all trace of him, as well as of Wegener's diary containing all his entries from 10 September onward, disappeared into the ice with him. His body has never been recovered.

No one knew they were dead until 7 May, when the first propeller sled arrived at Eismitte, driven by Kraus, with Johann Villumsen, Rasmus's brother, as a passenger. They barely made it, as they had a fuel leak, but Kraus accelerated and, approaching Eismitte, saw two figures waving to them in the snow. Only two! Kraus wrote, "I turn [the sled] in a circle and stop the sledge on the hard snow of its own track. Gears in neutral; I leap out and throw my arms around Sorge. In one breath we ask for Wegener. The silent answer tells us both all."[93]

That night the dogsled party under the command of Weiken also arrived. Weiken saw the tents pitched beside the tower of Eismitte but felt "a sinister stillness, not a soul to be seen." He rushed up to the tent shouting "What's the matter!?" Loewe limped out and said, "Wegener and Rasmus left for the West on the first of November, so they are dead." Kraus set up the radio and was able to contact Godhavn by noon on 8 May, with a message to be relayed to Germany, announcing that Wegener was dead.[94]

The propeller sleds left immediately for the coast on 9 May, taking Loewe with them. They arrived at the West Station thirty-four hours later, having covered the same

Wegener's grave, May 1931. "Wir haben Wegener gefunden. Tot im Eis." (We have found Wegener. Dead in the ice.) Photo courtesy of the Alfred Wegener Institute, Bremerhaven.

distance (only sixteen hours of running time) that Wegener had covered in thirty-nine days. Weiken, Sorge, and five Greenlanders with dogsleds left Eismitte that same day, and a few days later, on 12 May, they saw at kilometer 189 Wegener's skis crossed in the snow, with a broken ski pole between them. They dug down less than a meter below the snow surface of November 1930 and there found Wegener's body. "Wegener's eyes were open, and the expression on his face was calm and peaceful, almost smiling. His face was rather pale, but looked younger than before. There were small frost bites on the nose and hands, such as are usual in journeys like this."[95]

They examined him and deduced that Rasmus must have taken his pipe, tobacco, and diary; noting that there was no snow on his beard and mustache and that his clothes were dry, they determined that he must have died in his tent. The Greenlanders with Weiken and Sorge sewed up his body in the bags, as before, and reburied him in the ice. They built a large mound of blocks of firn and covered it with a Nansen sledge, making a cross of his broken ski pole.[96]

While the search for his body was still under way, the *New York Times* ran a front-page headline on Sunday, 10 May 1931: "Wegener Given up as Lost in Greenland's Ice Fields; Three Aides Found Safe." This headline and the breathless dispatch that accompanied it contained many details of Wegener's struggle to resupply Eismitte. It included pictures of the pony sleds, photos of sledges at Eismitte, and a picture of Wegener, in his furs, standing next to one of the propeller sleds (a photo taken the previous August).[97] The following Thursday, 14 May 1931, another headline followed: "Wegener Gave Up His Life to Save Greenland Aides; Left so Food Would Last."[98]

Weiken, Sorge, and the Greenlanders, having reburied Wegener in the ice, arrived at the West Station on 16 May, and two days later they sent a telegram to Schmidt-Ott at the Notgemeinschaft, detailing the finding of Wegener's body and describing their unsuccessful attempts to follow Rasmus's trail.[99] Schmidt-Ott sent the telegram, along with his condolences, to Else in Graz on 19 May.[100]

Two days later, the story exploded onto the front page of newspapers around the world; there was not a major newspaper in Europe or North America that did not carry the report. It was once again front-page news in the *New York Times* on Thursday, 21 May 1931: "Wegener's Body Found in Greenland Waste; Died Peacefully, Buried in Furs by Native." It was the full story of the finding of Wegener's body, and the "native" was soon identified in the story as Rasmus, who was described as Wegener's "faithful companion" and praised for the care with which he had buried Wegener "deep in the snow." Interestingly enough, Wegener was identified in the story not only as a great Arctic explorer (he was being compared everywhere to the other greats who had lost their lives in the Arctic, including Scott and Amundsen) but also as "a noted geologist."[101]

The following day, the *New York Times* wound down its coverage with yet another front-page headline story: "Wegener's Widow Asks Burial in Greenland Hills for Him." The story notes that there had been a plan in Germany to send a battleship to bring Wegener's body home and to bury him with military honors, but that Else "has decided that the body of her husband should not be brought back to Germany but buried in the mountains in Greenland at a spot overlooking the vast stretches of the Inland Ice where he met his death. Germany's president, von Hindenburg, and its Chancellor, Brüning, and other high officials have sent letters of condolence to his widow." The *Times* also proclaimed, "The German press and official quarters deeply mourn the death of Dr. Wegener who is praised as a genius with an incomparable imagination."[102]

In Berlin, Kurt Wegener was packing his Arctic clothing and equipment; the expedition plan specified that if anything "happened" to Alfred, Kurt would take over as the leader of the expedition. Further, when the expedition was done, he was to take charge of editing the scientific reports, compiling the history of the expedition, and, as importantly, editing the film of the expedition. With the latter work some years off, Kurt was, in the meantime, hurrying to Copenhagen to catch the first ship to West Greenland.

Alfred Wegener devoted his entire life to the study of surfaces of discontinuity, searching for the invisible causes that bring about sharp changes in the behavior of matter and energy. There is no discontinuity more abrupt and complete than that between life and death. For Wegener there was no slow decline in power, no retirement, no slide into obscurity; he died at the peak of his power, ambition, and scientific productivity. He died in the midst of *die Arbeit*—the work—in which he so deeply believed, and for which he was willing to give his life.

In the very week that Wegener's body was discovered, on the other side of Greenland, Augustine Courtauld (1905–1959) was rescued from the Inland Ice station where he had wintered alone for five months as the solo meteorologist for the British Arctic Air Route Expedition. News of his rescue and its aftermath often appeared on the same page in the *New York Times* as news of Wegener's death. When Courtauld was asked whether it had been "worth it," he responded, "The objects of the Expedition, like most objects of most Expeditions were the means of living the life [we] liked to live, rather than ends in themselves."[103] This is not exactly how Wegener felt, but it is close. Arctic exploration and polar science are, as he warned Georgi, seductive and intoxicating pursuits, and it is not uncommon for polar explorers to continue to return to the north (or the south) until they meet their deaths in the midst of this work, or, like Freuchen, become so badly injured that they cannot continue. One is never too old. Amundsen was fifty-six when he disappeared in 1928 in the course of the rescue of the crew of the zeppelin *Italia*.

Wegener's 1930–1931 expedition was the last pulse of the great age of heroic exploration. With his death on the ice cap, he joined a long sequence of Arctic explorers who met their end in the midst of their polar work—he was immediately added to the list that includes Sir John Franklin, Amundsen, and Scott. He is today, in histories of polar travel, also classed with Mawson, Shackleton, Nansen, and Rasmussen. Wegener was, along with Rasmus Villumsen (1908–1930), one of the last great casualties of that age.

Wegener's expedition was Keplerian, a watershed event bridging the old (sailing ships, ponies, dogs, exploration of uncharted territory) and the new (motorboats, motorized sleds, radio communication). For the last two years of his life, Wegener increasingly saw his own work in this light and was not unhappy about it. Transitions, of course, sometimes just happen, but they go better when they are planned. As a young man he had thought of himself as a revolutionist; as a mature man he viewed life, human history, science, and the cosmos itself from the standpoint of the theory of evolution, in which every step is necessary, and every step is meaningful, no matter how slow and insignificant it may appear at the time.

Wegener did not live a hero's life and would have scorned the idea, but it was not within his power to keep himself from becoming a hero after his death. Today, his reputation persists as Germany's greatest polar explorer, and Germany's Institute for Polar and Marine Research is named for him. So is the peninsula in Greenland just to the north of Kamarajuk Glacier, now the Alfred Wegener Halvø, 71.1° north, 51.8° west.

There is a "Wegener" crater on the Moon, at 45° north, 114° west. It is slightly irregular, about 90 kilometers (56 miles) in diameter, and (perhaps fittingly) on the *dark* side of the Moon. There is also a Wegener crater on Mars, 64° south, 355°, filled with dunes and changing color each year, from summer black to winter frost and back again. Wegener would have been embarrassed by this honor publicly but enjoyed it privately, as he would have enjoyed having his name given to an asteroid: "29227 Wegener," an object with a regular period, a known eccentricity, and an inclination known to five decimal places. He would have, had he lived to see it, quite likely amused himself by checking the accuracy of the orbital calculations and trying to figure out a means to get there.

Epilogue

> The biography of a man of science can never end with his death. If his contributions to science are important enough to warrant a detailed account of his life, then his effect upon science does not end when he dies. The facts he discovers are rapidly assimilated and become part of the mainstream of science. When, as was the case with Faraday, the factual discoveries were the results of a radically new vision of physical reality the process of assimilation is not so rapid.
>
> L. PEARCE WILLIAMS, *Michael Faraday* (1964)

Kurt Wegener sailed to Greenland in June 1931, and in July he took over the expedition. This proved to be a matter of some controversy, as Loewe had assumed tacit leadership from the time of the discovery of Wegener's body in May. The expedition members, used to Alfred Wegener's and then Loewe's rather loose and free-form leadership, chafed under Kurt Wegener's strong direction, and there were a number of disputes, both personal and professional.[1]

During the summer of 1931, the seismic exploration and gravity surveys proved that Greenland was an ice-filled bowl and that "glacial isostasy" and the larger principle of isostasy were *true*—answering the question that Nansen had posed fifty years before. The bulk of meteorological and glaciological results of the expedition and its geodetic measurements of altitudes of the ice cap were unremarkable, though they went into the mainstream of polar meteorology and glaciology. There was one exception. The three stations at latitude 71° north were able to track the west-to-east travel of cyclonic storm systems throughout the autumn and winter of 1930 and the spring and summer of 1931. The steady passage of these systems was strong evidence against the existence of any persistent high-pressure area, or "glacial anti-cyclone," in the middle of the ice cap. Thus, two of the major aims of the expedition, both with theoretical significance, came to pass as part of Wegener's legacy.

While at Eismitte in winter and spring 1930/1931, Georgi discovered that he had 40 percent more food and fuel than he had told Wegener. There was, after all, sufficient means of existence for all five men (including Wegener and Rasmus) to winter over. As Cornelia Lüdecke has pointed out in a long examination of the factors leading to Wegener's death, it is hard to imagine how one could be wrong by a factor of two under such circumstances.[2]

This (along with several other issues) led to a long and bitter dispute—*die Schuldfrage* (the question of guilt)—driven less by public accusations than by Georgi's increasing fear that he had somehow caused Wegener's death. His published defenses and self-exculpations (against no specific charges) finally led to a legal dispute with Kurt, and eventually a bitter dispute with Else, over the way Georgi was "presented" in the film of the expedition and in the accounts of the expedition (both official and nonofficial) published in 1932–1934. These exchanges became so nasty and shrill that in 1937 a court ordered all parties to desist.

From the very beginning Else had tried to prevent this from happening, and in October 1931 she had written to the Notgemeinschaft begging them to make sure that "the silly question of guilt" would not become part of the legacy of the expedition. In consequence, Schmidt-Ott traveled to Copenhagen to meet the returning expedition members in November 1931, securing a promise from them not to discuss details of expedition events and especially of Wegener's death.[3] Georgi, always jealous of his reputation, and now besieged by feelings of guilt and shame, continued to publish defenses, which only made his position worse.

Else edited the story of the expedition, with chapters written by each of the scientific and technical participants (including Georgi): *Alfred Wegeners Letzte Grönlandfahrt*. By the early 1940s it had gone through fourteen editions and had been translated into Danish (in 1933) by Andrea Lundager (with a foreword by Rasmussen) and into English (in 1939) as *Greenland Journey*.[4]

Michael Spender, writing in the *Geographical Journal* in 1934, reviewed Kurt Wegener's official account of the expedition, the volume edited by Else, and Georgi's separate account of the expedition: *Im Eis Vergraben*. Spender remarked, "Dr. Georgi's book . . . is not so happy in its achievement [as the other two volumes]. On the disinterested reader the undercurrent of polemic will produce quite the opposite effect to that intended by the author. . . . If Dr. Georgi still thinks that anyone believes that the burden of the tragedy rests upon him, then he would have done better to let the account published by Kurt Wegener speak for him than to publish this book 2 years later."[5]

In 1932 Sorge and Loewe went back to Umanak (along with their wives) as "scientific advisers" on an Arnold Fanck action film, *S.O.S. Eisberg*, about the search for the lost papers of a famous scientist missing on an ice cap. The film starred, among other contemporary action heroes, Leni Riefenstahl. It is available on DVD in both German and English versions and begins with two cigar-smoking gents in an explorer's club strolling past portraits of the "greats" of polar exploration, pausing before the portrait of Wegener, last in the sequence, whereupon one of the men says, "and then Wegener, the greatest of them all."[6]

After Wegener's Death

Kurt took over Alfred's professorship in Graz in 1931; he lived "in rooms" in Graz's main hotel, rather than joining the Wegener household on the Blumengaße (by then renamed "Alfred Wegener-Gaße"). He edited the expedition's final reports, appearing over the course of a decade beginning in 1933. In 1935 he published *Vorlesungen über Physik der Atmosphäre* (Lectures on atmospheric physics) under his and his brother's names, and he also oversaw another (5th) edition of *Die Entstehung der Kontinente und Ozeane*, with unchanged text but a larger bibliography.[7] After the *Anschluss* (the assimilation of Austria into Nazi Germany), he resigned his professorship and went to Argentina, not returning to take it up again until 1952.

As for Wegener's daughters, Hilde became a nurse and died in 1936 of typhus, while nursing victims of an epidemic. Lotte, in 1938, married the famous mountaineer Heinrich Harrer (1912–2006), who later became the tutor of the Dalai Lama and wrote *Seven Years in Tibet*. The marriage ended in divorce. Käte, in 1939, married Siegfried Uiberreither, Nazi governor of Styria and a notorious SS officer who fled into hiding after the Second World War to avoid extradition to Yugoslavia for war crimes.

Marie Köppen, Else's mother, died in 1939, and Wladimir Köppen died the following year. He was still correcting proofs of his climate handbook during his final illness, at which time he sent a telegram to his publisher: "Please hurry. Am dying." In 1955 Else wrote his biography with help from Erich Kuhlbrodt, under the pen name Else Wegener-Köppen: *Wladimir Köppen: Ein Gelehrtenleben für die Meteorologie*.[8]

In 1960 Else wrote a book about her husband, a narrative of his life with extracts from his expedition journals and letters: *Alfred Wegener: Tagebücher, Briefe, Erinnerungen*.[9] Most of what we know of Wegener's personal life comes from this one source. The majority of Wegener's papers in Graz were lost or destroyed in the chaos after the war. Eventually, Else moved to the Tyrol; in later years she and Kurt lived together as brother and sister. Else died in 1992 at the age of 100.

Wegener's Scientific Legacy

Wegener is unquestionably the progenitor of the theory of the origin of continents and oceans by means of continental drift. No other scientist in the late nineteenth and early twentieth centuries considered the matter in anywhere near the detail or with anything like the persistence that Wegener showed between 1912 and 1928. No other "mobilist" thinker went as far as Wegener to assemble evidence from different scientific fields in the service of his conception, and no other thinker has had his name more closely associated with the idea. No other proponent of continental drift had more written about his ideas, pro and con, or had as many symposia, articles, and books devoted to his ideas. From 1915 until the later 1930s the phrases "Wegener's theory" and "continental drift" were virtually synonymous.

The parallel with Charles Darwin is suggestive. Like Wegener, Darwin wrote a book of origins: the origin of species by natural selection. No other thinker in the latter part of the nineteenth century assembled as much evidence from different scientific fields in support of the theory of evolution, or explored the matter in greater detail, or showed greater persistence in developing and refining his ideas. No other thinker had his name more closely associated with the idea of the evolution of species. To this day the phrases "Darwin's theory" and "evolution" are virtually synonymous.

Here, however, the parallelism ends. While the term "Darwinian evolution" has managed to gather under its banner everything from Mendelian inheritance to the double helix, and while almost every evolutionary biologist in the world today would comfortably describe him- or herself as a "Darwinian," this was not the fate of Wegener's theory of continental drift. No earth scientist has ever, I believe, described her- or himself as "Wegenerian," and almost none of the developers of plate tectonics saw themselves as working directly in the tradition and style of research associated with Wegener's name.

Thus, while there is no question that the idea of continental drift belongs to Wegener every bit as much as the idea of natural selection belongs to Darwin, the investigators (working mostly after 1950) who developed the theory of plate tectonics have not chosen to see their work as a continuation of his. This is a crucial point. Discovery, however we are to define it (and it is a very complex issue), can rarely be assigned to a specific place and time and usually consists of a chain of coordinated events. Helge Kragh and Robert Smith have pointed out that "the discovery of a phenomenon (or object, or relationship) is not identical to the incorporation of the phenomenon into the body of scientific knowledge. Discovery accounts cannot be only intellectual accounts of how an idea entered a scientist's mind, they must also include a social history of how the discovery claim became accepted by the scientific community."[10]

The stipulation that a discovery account must also include the social history of the acceptance of the discovery claim by the scientific community is our key to why Wegener may be accorded priority in the theory of continental drift, but not in the theory of plate tectonics, though the latter incorporates many elements of the former. I have here written the story of Wegener's life, including both an intellectual account of how the idea entered his mind and a social history of how that claim was treated by the scientific community during his lifetime. This social history is *not* equivalent to a story of acceptance, and the historical record is absolutely clear and unequivocal that Wegener's version of continental drift was never "accepted by the scientific community." It was never more than what he hoped it would be: a working hypothesis that accounted for most of the facts of geology, paleontology, and geophysics better than any other theory of its time, and a way of organizing work in the earth sciences which could stimulate productive research.

The question of how the scientific community came to accept the theory that the continents move has been definitively answered by Henry Frankel in his four-volume history *The Continental Drift Controversy* (2012).[11] Frankel very appositely subtitled his first volume *Wegener and the Early Debate*. It begins with Wegener and traces the controversy about moving continents up to the outbreak of the Second World War. Wegener is mentioned so many times in Frankel's first volume that the entries take up a full column of the index and are divided into multiple subheadings. By contrast, in Frankel's second volume, *Paleomagnetism and the Confirmation of Drift*, which picks up the story after the Second World War, consideration of Wegener has shrunk to sixteen entries for "Wegener, Alfred" in the space of 500 pages.[12] This seems to me a very clear delineation of the outcome: already by 1946 Wegener's own theoretical construct of drift was no longer the focus of direct attention for either proponents or opponents of moving continents.

Within the realm of atmospheric physics Wegener had two lasting achievements notable enough to bear his name. The first is the theory of the formation of precipitation in cold clouds. This theory, once attributed to Tor Bergeron and W. Findeisen alone, is now correctly denominated the Wegener–Bergeron–Findeisen process, based on Bergeron's testimony that he discovered the idea in Wegener's *Thermodynamik der Atmosphäre* in 1918–1919. The second is a rare parhelion (a refracted arc adjacent to the orb of the Sun) known as a Wegener arc.

Additionally, Wegener was among the first to discern the correct explanation for the formation of tornadoes, their relationship to a squall line of thunderstorms, and the penetration of the top of a tornado into a "supercell" within a cumulonimbus cloud. His catalog of tornadoes and waterspouts in Europe, *Wind- und Wasserhosen in Europa* (1917), is still actively cited by tornado researchers.[13]

His theory of atmospheric layering, from his own standpoint one of his proudest scientific achievements, and also the inspiration for the idea of a layered solid Earth that could permit continental displacement, has not fared so well. While he was correct in determining that the stratosphere ended somewhere around 80 kilometers (50 miles) above Earth's surface, where a new layer began (a layer now known as the mesosphere), his notion that the atmospheric layers were defined by chemical abundances successively of oxygen, nitrogen, and hydrogen has turned out to be false, as the atmosphere is well mixed up to the mesosphere. Moreover, his conjectural light element, geocoronium, does not exist, and its supposed spectral line is part of the oxygen spectrum in the very high atmosphere.

In the study of meteors, Wegener was among the first to give the correct interpretation of the color of meteor trails as an indication of the altitude of the meteor. He was also right about the craters of the Moon being impact phenomena, and "right for the right reason"—that being his morphological-empirical method of comparison of the structure of impact craters with the structure of volcanoes.

Wegener was right about many things, but being right and being important are very different things. Even in the areas of his greatest effort and interest, his work (with the exception of that on continental displacements) fell out of the citation cascade very soon after his death. Wegener had several things working against him and his legacy. Not least of these was the precipitous decline of the role of German as the international scientific language of the earth sciences in the 1920s and after. It is not generally remembered now that instruction in the German language ceased during World War I in the United States, and the language never regained its scientific prominence in this area. Between 1918 and 1950, the proportion of German language earth science publications cited in the United States fell from about 50 percent to about 5 percent. You cannot be influential if you cannot be read, and other than the third edition of his book on continents and oceans, almost nothing that he wrote was translated into any other major language until the 1970s.

In addition to the decline of German as a scientific language, Wegener was denied another important source of continuity for his ideas: none of his doctoral students ever became a university professor. Most of his students, given his preference for "cosmic physics" and the scarcity of university professorships in meteorology, were destined for careers as secondary school teachers (*Gymnasium*). He never founded, nor did he attempt to found, a tradition of research with a distinctive methodology, or distinctive content, in any of the many disciplines he pursued.

Finally, and so obviously that one almost forgets to mention it, all of his scientific work was prematurely deprived of its strongest advocate with his death in 1930 on the Greenland ice cap. Had he not died as he did, his family history, his heavy smoking, and his heart defect all point toward a death in his later sixties or early seventies, which would have taken him into the post–Second World War period: he would have been seventy years old in 1950, and by then interest in his theories was already reviving, and he might have played a role. Wegener always showed great flexibility in modifying his views based on new research, and he seemed to take an almost proprietary interest in any work that he could see had been inspired by his own, even if it seemed to modify his previous views in some crucial aspect. The work on continental displacements in the later 1930s and early 1940s would certainly have found favor in his eyes, and he would likely have continued to modify his own views in the direction the community was going. His conception of science was always social and evolutionary, and he believed above all in the value of "the work" (*die Arbeit*), far beyond anything that might attach to his own credit or name.

Notes

Preface

1. Else Wegener, *Alfred Wegener: Tagebücher, Briefe, Erinnerungen* (Wiesbaden: F. A. Brockhaus, 1960); Ulrich Wutzke, *Durch die weiße Wüste: Leben und Leistungen des Grönlandforschers und Entdeckers der Kontinentaldrift Alfred Wegener*, Petermann ed. (Gotha: Justus Perthes Verlag, 1997).

2. Alfred Wegener, *Thermodynamik der Atmosphäre* (Leipzig: J. A. Barth, 1911).

3. I owe this observation to Thomas Hankins, my dissertation supervisor and the biographer of Jean d'Alembert and William Rowan Hamilton.

4. Claude Lévi-Strauss, *La Pensée Sauvage* (Paris: Plon, 1962), 340ff.

5. Mott T. Greene, "Writing Scientific Biography," *Journal of the History of Biology* 40 (2007): 727–759.

Chapter 1. The Boy: Berlin and Brandenburg, 1880–1899

1. Demolished by the East German government in 1950.

2. Destroyed in World War II by Allied bombing.

3. Destroyed in 1944. Mario Krammer, *Berlin im Wandel der Jahrhunderte: Eine Kulturgeschichte der deutschen Hauptstadt* (Berlin: Rembrandt-Verlag, 1956), 80. The redbrick buildings remain.

4. "Alfred's Jugend," typescript by Kurt Wegener (ca. 1950), Heimatmuseum Neuruppin, 1.

5. Else Wegener, *Alfred Wegener: Tagebücher, Briefe, Erinnerungen* (Wiesbaden: F. A. Brockhaus, 1960), 11.

6. Destroyed in World War II by Allied bombing.

7. "Alfred's Jugend," 1.

8. Else Wegener's genealogical record of the Wegener family is deposited in the Heimatmuseum, Neuruppin, along with the family photographic album and some mementos. This material supplements the collection at the Wegener Gendenkstätte, Zechlinerhütte.

9. E. Wegener, *Alfred Wegener: Tagebücher, Briefe, Erinnerungen*, 10.

10. Richard Wegener, *Begriff und Beweis der Existenz Gottes bei Spinoza* (Berlin: Druck der Nauk'schen Buchdruckerei, 1873).

11. "Alfred's Jugend," 1.

12. Richard Wegener, *Poetische Fruchtgarten* (Göthen: Paul Schettler's Erben, 1895).

13. Richard Wegener, *Aufsätze zur Litteratur* (Berlin: Wallroth, 1882).

14. Richard Wegener, *Die Bühneneinrichtung des Shakepeareschen Theaters nach den zeitgenössischen Dramen* (Halle: Max Niemeyer, 1907).

15. R. Wegener, *Poetische Fruchtgarten*, 42–43.

16. Ibid., 43–45.

17. "Alfred's Jugend," 1.

18. Kurt remembers Käte dying at the age of eight, which would have put the event in 1887; Else Wegener, in her 1960 memoir, is quite definite that Kurt was six and Alfred four when this happened. This is more consistent with Kurt's comment that "they scarcely remembered her."

19. E. Wegener, *Alfred Wegener: Tagebücher, Briefe, Erinnerungen*, 10.

20. Philip A. Ashworth, "Berlin," in *Encyclopedia Britannica* (New York: Encyclopedia Britannica, 1910), III, 789.

21. Martin Schwarzbach, *Alfred Wegener: The Father of Continental Drift* (Madison, WI: Science Tech. Inc., 1986), 241.

22. "Alfred's Jugend," 3.

23. It is today a memorial museum to Alfred Wegener, and the Wegeners senior, Tony Wegener, and Kurt Wegener are all buried in the family plot in the village churchyard nearby. There is also a gravestone memorializing Alfred Wegener himself.

24. E. Wegener, *Alfred Wegener: Tagebücher, Briefe, Erinnerungen*, 11.

25. Theodor Fontane, *Graffschaft Rüppin, Wanderungen durch die Mark Brandenburg* (Stuttgart: Cotta, 1909–1912).

26. "Alfred's Jugend," 1.

27. E. Wegener, *Alfred Wegener: Tagebücher, Briefe, Erinnerungen*, 11.

28. "Alfred's Jugend," 3.

29. Ibid., 4.

30. E. Wegener, *Alfred Wegener: Tagebücher, Briefe, Erinnerungen*, 11.

31. George Coore, "Education: National Systems," in *Encyclopedia Britannica* (New York: Encyclopedia Brittanica, 1910), VIII, 967.

32. E. Wegener, *Alfred Wegener: Tagebücher, Briefe, Erinnerungen*, 11.

33. Ibid.

34. "Alfred's Jugend," 2.

35. E. Wegener, *Alfred Wegener: Tagebücher, Briefe, Erinnerungen*, 12.

36. "Alfred's Jugend," 2.

37. Ibid., 4.

38. E. Wegener, *Alfred Wegener: Tagebücher, Briefe, Erinnerungen*, 12.

39. "Alfred's Jugend," 3. The phrase "*nulla dies sine linea*" was a Latin proverb, perhaps based originally on a Greek one, and it supposedly takes its origin from the habits of the famous painter Apelles. Here is the account of Pliny the Elder, Hist. Nat., Bk. 35.84: "It was the regular custom of Apelles to never pass a day that was so busy that he didn't keep his artistic hand in by drawing a line—something which passed from him into a proverb." So the "*linea*" in question was originally not a line of verse or writing of any sort, but a line of drawing. I thank my colleague David Lupher for this reference and discussion.

40. "Alfred's Jugend," 1.

41. E. Wegener, *Alfred Wegener: Tagebücher, Briefe, Erinnerungen*, 12.

42. Friedrich Diesterweg, *Populäre Himmelskunde und mathematische Geographie*, 12th and 13th eds. (Berlin: E. Goldschmidt, 1890); M(ax). W(ilhelm). Meyer, *Das Weltgebäude: Eine gemeinsverstandliche Himmelskunde* (Leipzig: Bibliographisches Institut, 1898).

43. "Alfred's Jugend," 3.

Chapter 2. The Student: Berlin-Heidelberg-Innsbruck-Berlin, 1899–1901

1. Christa Jungnickel and Russell McCormmach, *Intellectual Mastery of Nature: Theoretical Physics from Ohm to Einstein*, vol. 2, *The Now Mighty Theoretical Physics, 1870–1925* (Chicago: University of Chicago Press, 1986), 26–28.

2. Wegener's academic records are in the "Wegenerarchiv" in the library and archive of the Alfred-Wegener-Institut für Polar- und Meeresforschung in Bremerhaven.

3. Kurt-R. Biermann, "Weierstrass, Karl Theodor Wilhelm," in *Dictionary of Scientific Biography*, ed. Charles Coulston Gillispie (New York: Charles Scribner's Sons, 1980), 222; and H. Boerner, "Schwarz, Hermann Amandus," in *Dictionary of Scientific Biography*, passim.

4. Walther Lietzmann, *Aus Meinen Lebenserinnerungen* (Göttingen: Vandenhoek & Ruprecht, 1960), 22–23.

5. Ibid., 23.

6. Ibid., 22.

7. Else Wegener, *Alfred Wegener: Tagebücher, Briefe, Erinnerungen* (Wiesbaden: F. A. Brockhaus, 1960), 12; J(ohannes) Georgi, "Memories of Alfred Wegener," in *Continental Drift*, ed. Stanley Keith Runcorn, *International Geophysics Series* (New York: Academic Press, 1962), 312.

8. Jungnickel and McCormmach, *Intellectual Mastery of Nature*, 2:213.

9. Jürgen Hamel and Klaus-Harro Tiemann, "Der Vertretung der Astronomie an der Berlin Universität in den Jahren 1810 bis 1914," *Vorträge und Schriften (Archenhold-Sternwarte Berlin-Treptow)* 69 (1988): 25.

10. Kurt Lambeck, *The Earth's Variable Rotation: Geophysical Causes and Consequences* (Cambridge: Cambridge University Press, 1980), 85–86; Adolf Marcuse, *Die Erdmessungs-expedition nach den Hawaiischen Inseln* (Berlin: W. Portmetter, 1892), 18; Adolf Marcuse, *Ergebnisse er polhöhenbestimmungen in Berlin ausgeführt in den Jahren 1889, 1890, 1891 am Universal-Transit der Königlichen Sternwarte*, Centralbureau der Internationalen Erdmessung, Neue Folge der Veröffentlichungen, no. 6 (Berlin: G. Reimer, 1902), 29.

11. Lietzmann, *Aus Meinen Lebenserinnerungen*, 22.

12. Hamel and Tiemann, "Der Vertretung der Astronomie," 7, 25.

13. Adolf Marcuse, *Die Hawaiischen Inseln von Dr. Adolf Marcuse: Mit vier karten und vierzig Abbildungen nach photographischen Original-Aufnahmen* (Berlin: R. Friedlander & Sohn, 1894), 186.

14. Lietzmann, *Aus Meinen Lebenserinnerungen*, 24.

15. F. Fraunberger, "Quincke, Gerog Hermann," in *Dictionary of Scientific Biography*, 241.

16. Ibid., 242.

17. Wilhelm Valentiner, *Atlas des Sonnensystems: 25 Abbildungen in Lichtdruck* (Lahr: M. Schauenberg, 1884); Wilhelm Valentiner, *Handwörterbuch der Astronomie*, vol. 4 (Leipzig: Barth, 1897–1902).

18. The author can vouch for this characterization of such organizations from personal experience as a member and officer of the ΑΔΦ fraternity at Columbia College in New York City between 1963 and 1967.

19. "Strafs-Verfugung" (Summons) in the Wegener Nachlass, Alfred-Wegener-Institut für Polar- und Meeresforschung Bremerhaven. I am indebted to Ulrich Wutzke's transcription in Ulrich Wutzke, *Der Forscher von der Friedrichsgracht: Leben und Leistung Alfred Wegeners*, 1st ed. (Leipzig: Brockhaus, 1988), 15, 17.

20. E. Wegener, *Alfred Wegener: Tagebücher, Briefe, Erinnerungen*, 12.

21. Lietzmann, *Aus Meinen Lebenserinnerungen*, 23.

22. Julius Bauschinger, *Die Bahnbestimmung der Himmelskörper* (Leipzig: W. Engelmann, 1906); Julius Bauschinger, ed., *J. H. Lambert's Abhandlungen zur Bahnbestimmung der Cometen: Insignores orbitae comitarum proprietates (1761); Observations sur l'Orbite apparente des Cometes (1771); Auszüge aus den "Beiträgen zum Gebrauche der Mathematik,"* Ostwalds Klassiker der exakten Wissenschaften, no. 133 (Leipzig: W. Engelmann, 1902).

23. Julius Bauschinger and Jean Peters, *Logarithmisch-trigonometrische Tafeln mit acht Dezimalstellen ehthaltend die Lograrithmen aller Zahlen von 1 bis 200000 und die Logarithmen der trigonometrishce Funktionen für jede Sexagesimalsekunde des Quadranten, mit Unterstützung der Kgl. Preussischen Akademie der Wissenschaften in Berlin und der Kais. Akademie der Wissenschaften in Wien*, 2 vols. (Leipzig: W. Engelmann, 1910–1911).

24. Ibid., 257.

25. Gisela Kutzbach, *The Thermal Theory of Cyclones: A History of Meteorological Thought in the Nineteenth Century*, Historical Monograph Series (Boston: American Meteorological Society, 1979), 225.

26. A. Kh. Khrgian, *Meteorology: A Historical Survey (Ocherki razvitiya meteorologii)*, trans. Ross Hardin, 2nd ed., rev. Kh. P. Pogosyan (Jerusalem: Israel Program for Scientific Translations, 1970), 272.

27. Registration information from Wegener's student records in the library of the Alfred Wegener Institute, Bremerhaven.

28. Josef Blaas, *Geologischer Führer durch die Tiroler und Vorarlberger Alpen*, 7 vols. (Innsbruck: Wagner'schen Universitäts-Buchhandlung, 1902).

29. Personal communication from Peter Misch (1983), Professor Emeritus of Structural Geology at the University of Washington, recalling his student days with Hans Stille, before the advent of geochemistry as a regular university subject gave many geologists the means and confidence to work metamorphic rocks into the historical narrative of the evolution of a mountain chain.

30. The following account is expanded from the scanty information in E. Wegener, *Alfred Wegener: Tagebücher, Briefe, Erinnerungen*, 13.

31. Alan Blackshaw, *Mountaineering*, 2nd rev. ed. (Baltimore: Penguin, 1970), 464–465.

32. E. Wegener, *Alfred Wegener: Tagebücher, Briefe, Erinnerungen*, 13.

33. Franz Hubmann, *Dream of Empire: The World of Germany in Original Photographs, 1840–1914* (London: Routledge & Kegan Paul, 1973), 186–187.

34. Colmar Freiherr von der Goltz, *Das Volk in Waffen, ein Buch über Heerwesen und Kriegsführung unserer Zeit*, Funfte umgearbeitete und verbesserte Aufl. ed. (Berlin: R. V. Decker, 1899), 276.

35. Ibid., 183–187.

36. Ibid., 56.
37. Ibid., 164ff.
38. Ibid., 162–164.
39. Alfred Kelly, *The Descent of Darwin: The Popularization of Darwinism in Germany, 1860–1914* (Chapel Hill: University of North Carolina Press, 1981), 72–73.

Chapter 3. The Astronomer: Berlin, 1901–1904

1. Herbert Schnädelbach, *Philosophy in Germany, 1831–1933*, trans. Eric Matthews (Cambridge: Cambridge University Press, 1984), 139.
2. Rüdiger Safranski, *Martin Heidegger: Between Good and Evil*, trans. Ewald Osers (Cambridge, MA: Harvard University Press, 1998), 49.
3. Robert Musil, *The Man without Qualities*, trans. Eithne Wilkens and Ernst Kaiser, 1st English ed., 3 vols., vol. 1 (London: Secker & Warburg, 1954), 59–60.
4. H. P. Rickman, "Dilthey, Wilhelm," in *The Encyclopedia of Philosophy*, ed. Paul Edwards (New York: Macmillan, 1967), vol. 2, 403–406.
5. William Kluback and Martin Weinbaum, *Dilthey's Philosophy of Existence: Introduction to Weltanschauungslehre* (New York: Bookman Associates, 1957), 21–27.
6. Schnädelbach, *Philosophy in Germany, 1831–1933*, 147.
7. Rickman, "Dilthey, Wilhelm," 405. For a longer discussion of these ideas see H. P. Rickman, ed., *Dilthey: Selected Writings* (Cambridge: Cambridge University Press, 1976).
8. Friedrich Paulsen, *Einleitung in die Philosophie*, 3rd revised and enlarged ed. (Stuttgart: Cotta, 1895).
9. L. E. Loemker, "Pauslen, Friedrich," in *Encyclopedia of Philosophy*, vol. 6, 60.
10. Friedrich Paulsen, *Introduction to Philosophy*, trans. Frank Thilly, 2nd American from 3rd German ed. (New York: Henry Holt, 1895), vii.
11. Ibid., v–vi.
12. Ibid., 25.
13. Ibid., 27.
14. Ibid., 32.
15. Ibid., xiii.
16. Ibid., 19.
17. Loemker, "Pauslen, Friedrich," vol. 6, 60–61.
18. Paulsen, *Introduction to Philosophy*, 182.
19. Ibid., 227.
20. Alfred Kelly, *The Descent of Darwin: The Popularization of Darwinism in Germany, 1860–1914* (Chapel Hill: University of North Carolina Press, 1981), 83–84.
21. Paulsen, *Einleitung in die Philosophie*, 423: "*Erkenntnis, auch Erfahrungserkenntnis, ist nicht ein passiv aus der außeren Wirklichkeit ausgenommener Inhalt, sondern ein Erzeugnis spontaner Tatigkeit der Seele*" (Knowledge, even empirical knowledge, is not the passively received contents of the external world, but a spontaneous creation of the spirit). *Seele* (soul) also can mean "mind," or "intellect," or even "heart." A translator in these situations is always left with the choice of keeping this rich German synonymy or making it sound more normal and less "lofty" to English speakers, whose habit it is to keep the mind, the soul, the spirit, and the heart anatomically and functionally distinct. A less lofty translation of the same phrase, just as exact and more prosaically Anglophone, would be "knowledge is a spontaneous activity of the mind."
22. Wilhelm Förster, *Sammlung populärer astronomischer Schriften* (Berlin: F. Dümmlers, 1878); Wilhelm Förster, *Sammlung von Vorträgen und Abhandlungen*, vol. 3, *Wissenschaftliche Erkenntnis und sittliche Freiheit* (Berlin: F. Dümmlers, 1890).
23. Jürgen Hamel and Klaus-Harro Tiemann, "Der Vertretung der Astronomie an der Berlin Universität in den Jahren 1810 bis 1914," *Vorträge und Schriften (Archenhold-Sternwarte Berlin-Treptow)* 69 (1988): 11.
24. Wegener Journal entry, 8 Mar. 1907, Wegener Nachlass, DMH 1968 594/5.
25. Antonia Schlette, "Chamberlain, Houston Stewart," in *Encyclopedia of Philosophy*, vol. 2, 72.

26. Houston Stewart Chamberlain, *Foundations of the Nineteenth Century*, ed. and trans. John Lees, vol. 2 (London: John Lane, 1912), 272–273.

27. Ibid., 268–272.

28. Ibid., 272–273.

29. Ibid., 297.

30. Kelly, *Descent of Darwin*, 82–83.

31. Ibid., 86.

32. Wilhelm Bölsche, *Die Naturwissenschaftlichen Grundlagen der Poesie: Prolegomena einer realistischen Ästhetik*, trans. Johannes J. Braakenburg (1887; repr., Tübingen: Max Niemeyer Verlag, 1976), chap. 6 passim.

33. Ibid., 13.

34. Ibid., 117.

35. Ibid., 160.

36. Georg Uschmann, "Haeckel, Ernst," in *Dictionary of Scientific Biography*, ed. Charles Coulston Gillispie (New York: Charles Scribner's Sons, 1980), vol. 6, 6–10.

37. Ernst Haeckel, *The Riddle of the Universe*, trans. Joseph McCabe (New York: Harper & Brothers, 1900).

38. Schnädelbach, *Philosophy in Germany, 1831–1933*, 97.

39. Rollo Handy, "Haeckel, Ernst," in *Encyclopedia of Philosophy*, vol. 3, 400.

40. Wilfried Schröder, "Wegener's Work Included Studies of Noctilucent Clouds, Auroras," *EOS (Transactions of the American Geophysical Union)* 80 (1999): 357, 361; Wilfried Schröder, "Wilhelm Foerster und die Entwicklung der solar-terrestrischen Physik," *Die Sterne* 59, no. Heft 6 (1983): 348–352.

41. Christa Jungnickel and Russell McCormmach, *Intellectual Mastery of Nature: Theoretical Physics from Ohm to Einstein*, vol. 2, *The Now Mighty Theoretical Physics, 1870–1925* (Chicago: University of Chicago Press, 1986), 255.

42. Max Planck, *Vorlesungen über Thermodynamik* (Leipzig: Verlag von Veit & Comp., 1897).

43. Enrico Fermi, *Thermodynamics* (New York: Prentice-Hall, 1937).

44. Planck, *Vorlesungen über Thermodynamik*, v.

45. The "third law" of thermodynamics, the Nernst heat theorem, was not produced by Nernst until 1906 and not included in Planck's lectures until the third edition of 1910. Even there the treatment is given in what amounts to an appendix to the final topic of chap. 6: "Applications to Special States of Equilibrium."

46. H. C. Van Ness, *Understanding Thermodynamics* (New York: McGraw-Hill, 1969), chap. 1 passim.

47. C. B. P. Finn, *Thermal Physics* (London: Routledge & Kegan Paul, 1986), 2.

48. Planck, *Vorlesungen über Thermodynamik*, 148.

49. Ibid., 63.

50. Svante August Arrhenius, *Lehrbuch der kosmischen Physik* (Leipzig: S. Hirzel Verlag, 1903).

51. Hamel and Tiemann, "Der Vertretung der Astronomie," 20.

52. Wilhelm von Bezold, *Gesammelte Abhandlungen aus den Gebieten der Meteorologie und Erdmagnetismus* (Braunschweig: Friedrich Vieweg & Sohn, 1906).

53. H(ans) Benndorf, "Alfred Wegener," *Gerland's Beiträge zur Geophysik* 31 (1931): 338.

54. Joh(annes) Georgi, "Alfred Wegener (†) zum 80. Geburstag," *Polarforschung Beiheft* 2 (1960): 1–102.

55. Erich von Drygalski, *Grönland-Expedition der Gesellschaft für Erdkunde zu Berlin, 1891–1893* (Berlin: W. H. Kühl, 1897); Erich von Drygalski, *Zum Kontinent des eisigen Südens: Deutsche Südpolarexpedition Fahrten und Forschung des "Gauss," 1901–1903* (Berlin: Georg Reimer, 1904).

56. A typescript of the Edda is in the Wegener collection (uncatalogued in 1994) at the Heimatmuseum, Neuruppin.

57. Alfred Wegener, "Sieben Tage im Boot: Bericht über eine Reise von Zechlinerhütte nach Plau und zurück," holograph manuscript. Original unknown, photocopy in collection of Alfred-Wegener-Schule, Berlin-Dahlem (1904). I am indebted to Ulrich Wutzke for alerting me to this manuscript's existence.

58. Hamel and Tiemann, "Der Vertretung der Astronomie," 28–29. There were thirty-four doctoral degrees in astronomy granted at Berlin between 1885 and 1914. Of these, two were philosophical and one (Wegener's) historical.

59. Simon Newcomb, *The Reminiscences of an Astronomer* (New York: Houghton Mifflin, 1903). Newcomb gives an engaging account of this search. See also R. C. Archibald, "Simon Newcomb," *Biographical Memoirs, National Academy of Sciences* 17 (1924): 19–69.

60. Alfred Wegener, *Die Alfonsinischen Tafeln für den Gebrauch eines modernen Rechners: Inaugural-Dissertation zur Erlangung der Doktorwürde genehmigt von der philosophischen Facultät der Friedrich-Wilhelms-Universität zu Berlin* (Berlin: E. Eberling, 1905), 6.

61. Ibid. The calculation appears on p. 28 of Wegener's book.

62. Ibid., 3.

63. Ibid., 7.

64. Ibid., 129–185.

65. Ibid. The manuscript is dated February 1905.

66. Ibid., 79, 81.

Chapter 4. The Aerologist: Lindenberg, 1905–1906

1. Much of what Wegener did in this year has to be inferred from other sources. Chief among these is a lavishly produced volume, appearing on the occasion of Aßmann's retirement in 1915: Richard Aßmann, *Das Königlich Preußische Aeronautische Observatorium Lindenberg* (Braunschweig: Friedrich Vieweg & Sohn, 1915).

2. Robert Marc Friedman, *Appropriating the Weather: Vilhelm Bjerknes and the Construction of a Modern Meteorology* (Ithaca, NY: Cornell University Press, 1989), 48. The remark is from a letter of Vilhelm Bjerknes to Hugo Hergesell in 1926 recollecting a conversation with Kohlrausch in 1900, and it implies both the lack of respectability and rigor in meteorological work viewed from the standpoint of physics at the turn of the century and the onerous character of the paperwork at state meteorological services, likely to interfere fatally with one's research.

3. Gisela Kutzbach, *The Thermal Theory of Cyclones: A History of Meteorological Thought in the Nineteenth Century*, Historical Monograph Series (Boston: American Meteorological Society, 1979), 231.

4. Friedman, *Appropriating the Weather*, 77. Vilhelm Bjerknes found Aßmann in despair over these problems during a 1911 visit to Lindenberg and wrote about the matter to his wife in Bergen.

5. Richard Aßmann, *Das Königlich Preußische Aeronautische Observatorium Lindenberg* (Braunschweig: Friedrich Vieweg & Sohn, 1915), 73.

6. Ibid., 69–85.

7. Ibid., 73.

8. Ibid., 169.

9. Ray Monk, *Ludwig Wittgenstein: The Duty of Genius* (New York: Free Press, 1990), 28.

10. Aßmann, *Das Königlich Preußische Aeronautische Observatorium Lindenberg*, 176.

11. Ibid., 186.

12. Ibid., 183.

13. Karl Schneider-Carius, *Weather Science and Weather Research: History of Their Problems and Findings from Documents during Three Thousand Years* (New Delhi: Indian National Scientific Documentation Center [for NOAA and NSF], 1975), 320–322.

14. A. Kh. Khrgian, *Meteorology: A Historical Survey (Ocherki razvitiya meteorologii)*, trans. Ross Hardin, 2nd ed., rev. Kh. P. Pogosyan (Jerusalem: Israel Program for Scientific Translations, 1970), 269.

15. Ernest Rutherford and Frederick Soddy, "Radioactive Change," *Philosophical Magazine* 5, no. 6 (1903).

16. Ernest Rutherford, *Radioactive Transformations* (New York: Charles Scribner's Sons, 1906), 187–188.

17. George Ohring, "A Most Surprising Discovery," *Bulletin of the American Meteorological Society* 45 (1964): 13.

18. Ibid., 13; W. E. Knowles Middleton, *Invention of the Meteorological Instruments* (Baltimore: Johns Hopkins University Press, 1969), 301–305; Kutzbach, *Thermal Theory of Cyclones*, 181ff.

19. Ohring, "Most Surprising Discovery."

20. Léon Teisserenc de Bort, "Étude de l'atmosphere dans la verticale par cerfs-volants et ballons-sondes," *Journal de Physique* 9 (1900).

21. Léon Teisserenc de Bort, "Variations de la température de l'air libre dans la zone comprise entre 8 km et 13 km d'altitude," *Comptes Rendus à l'Académie des Sciences à Paris* 134 (1902): 987; Ohring, "Most Surprising Discovery," passim; Kutzbach, *Thermal Theory of Cyclones*, 183.

22. Schneider-Carius, *Weather Science and Weather Research*; Ohring, "Most Surprising Discovery," 353–356; Richard Aßmann, "Über die Existenz eines wärmeren Luftstromes in der Höhe von 10 bis 15 km," *Sitzungsberichte der Preußischen Akademie der Wissenschaften zu Berlin* (1902).

23. Aßmann, *Das Königlich Preußische Aeronautische Observatorium Lindenberg*, 87.

24. Ibid., 265.

25. Kurt Wegener and Alfred Wegener, "Die Temperatur im oberen Luftschichten im April 1905," *Das Wetter* 22 (1905).

26. Alfred Wegener, "Studien über Luftwogen," *Beiträge zur Physik der freien Atmosphäre* 2 (1906): 55–72.

27. Ibid.

28. Alfred Wegener, "Blitzschlag in einen Drachenaufstieg am Königlichen Aeronautischen Observatorium Lindenberg," *Das Wetter* 22, no. 7 (1905).

29. Cornelia Lüdecke, *Die deutsche Polarforschung seit der Jahrhundertwende und der Einfluß Erich von Drygalskis*, vol. 158, *Berichte zur Polarforschung* (Bremerhaven: Alfred-Wegener-Institute für Polar- und Meeresforschung, 1995), 156.

30. Alfred Wegener, "Über die Flugbahn des am 4. Januar 1906 in Lindenberg aufgestiegenen Registrierballons," *Beiträge zur Physik der freien Atmosphäre* 2 (1906).

31. Ibid., 33. See the notation in the table between eighty-one and eighty-five minutes of flight: "Fallschirm war unsichtbar" (parachute was invisible).

32. Ibid.

33. Ibid., 30.

34. Ibid., 31n1. The de Quervain theodolite was a popular and widely used instrument in succeeding years, in the form manufactured by Bosch.

35. Aßmann, *Das Königlich Preußische Aeronautische Observatorium Lindenberg*, 272.

36. Wilhelm von Bezold, *Gesammelte Abhandlungen aus den Gebieten der Meteorologie und Erdmagnetismus* (Braunschweig: Friedrich Vieweg & Sohn, 1906), 1.

37. Wegener, "Studien über Luftwogen," 55.

38. Aßmann, *Das Königlich Preußische Aeronautische Observatorium Lindenberg*, 272.

39. Alfred Wegener, "Bericht über Versuche zur astronomischen Ortsbestimmungen im bemannten Freiballon," *Ergebnisse der Arbeiten am Königich Preußische Aeronautische Observatorium, Lindenberg* (1906), contains an account of the flight, also available in abbreviated form in Ulrich Wutzke, *Der Forscher von der Friedrichsgracht: Leben und Leistung Alfred Wegeners*, 1st ed. (Leipzig: Brockhaus, 1988), 23–24.

40. Alfred Wegener, "Astronomische Ortsbestimmungen in Luftballon," *Illustrierte Aeronautische Mitteilungen* 6 (1906), contains a full account of the flight, on which the following summary is based; it is also more easily available in Else Wegener, *Alfred Wegener: Tagebücher, Briefe, Erinnerungen* (Wiesbaden: F. A. Brockhaus, 1960).

41. Wegener, "Astronomische Ortsbestimmungen in Luftballon." 116.

42. E. Wegener, *Alfred Wegener: Tagebücher, Briefe, Erinnerungen*, 14–15.

43. Knowles Middleton, *Invention of the Meteorological Instruments*, 305.

44. G(eorg). Amdrup, "Report on the Danmark Expedition to the North-East Coast of Greenland 1906–1908," *Meddelelser om Grønland* 41, no. 1 (1913).

45. Ibid., 55.

46. Wegener to Paulsen, 1 Nov. 1905, DP Copenhagen DEA, 26/136.

47. Paulsen to Mylius-Erichsen, 3 Nov. 1905, DP Copenhagen DEA, 26/107.

48. Amdrup, "Report on the Danmark Expedition," 41.

49. Ibid., 41–43.

50. Wegener to Mylius-Erichsen, 14 Nov. 1905, DP Copenhagen DEA, 26/136.

51. Wegener to Mylius-Erichsen, 26 Nov. 1905, DP Copenhagen DEA, 26/132. The letter is actually dated 26 December, but internal evidence and the succeeding correspondence with Paulsen indicate that the letter had to have been sent in November.

52. Wegener to Mylius-Erichsen, 18 Dec. 1905, DP Copenhagen DEA, 26/134.

53. Wegener, "Studien über Luftwogen," 55.

54. Ibid., 57.

55. Ibid., 71.

56. Wegener to Paulsen, 23 Mar. 1906, DP Copenhagen DEA, 26/132.

57. Ibid.

Chapter 5. The Polar Meteorologist: Greenland, 1906

1. Erich von Drygalski, *The Southern Ice-Continent: The German South Polar Expedition aboard the Gauss 1901–1903*, trans. M. M. Raraty (Bluntisham: Bluntisham Books and Erskine Press, 1989), 17–27.

2. Cornelia Lüdecke, *Die deutsche Polarforschung seit der Jahrhundertwende und der Einfluß Erich von Drygalskis*, vol. 158, *Berichte zur Polarforschung* (Bremerhaven: Alfred-Wegener-Institute für Polar- und Meeresforschung, 1995), 12–14, gives a list of the expeditions. Seventeen of these expeditions went north; the remainder, south.

3. Alfred Lansing, *Endurance* (New York: McGraw-Hill, 1959), 17; and Roland Huntford, *Shackleton* (London: Carroll & Graf, 1998), 402.

4. Wegener to Wladimir Köppen, 28 Mar. 1906, DMH 1968-595/1. This is Document 001-1906 in Ulrich Wutzke, "Alfred Wegener: Kommentiertes Verzeichnis der schriftlichen Dokumente seines Lebens und Wirkens," *Berichte zur Polarforschung* 288 (1998): 1–144.

5. Alfred Wegener, "Drachen- und Fesselballonaufstiege ausgeführt auf der Danmark-Expedition 1906–1908," *Meddelelser om Grønland* 42, no. 1 (1909): 7.

6. Georg Lüdeling, "Die Luftelektrischen Messungen ausgeführt von A. Wegener auf der Danmark-Expedition 1906–1908," *Meddelelser om Grønland* 42, no. 2 (1911): 80.

7. Wegener to Aßmann, 29 Mar. 1906, Lindenberg Observatory Archive no. 48, W 003-1906.

8. The account of the argument over the trip is from Else Wegener, *Alfred Wegener: Tagebücher, Briefe, Erinnerungen* (Wiesbaden: F. A. Brockhaus, 1960), 16. The appeal for a leave of absence was made formally on 30 March 1906, Wegener to Aßmann, W 004-1906, Geheimes Staatsarchiv Preußischer Kulturbesitz Berlin, abt. Merseberg.

9. Alfred Wegener, "Studien über Luftwogen," *Beiträge zur Physik der freien Atmosphäre* 2 (1906): 55n1. "The above work has my brother as author. He was not able to complete it, having been offered the opportunity, on astonishingly short notice, to take part in a Danish expedition to Northeast Greenland, on which he would be able to carry out kite and balloon ascents. It has fallen to me to complete it and publish it . . . Kurt Wegener."

10. Richard Assmann, *Das Königlich Preußische Aeronautische Observatorium Lindenberg* (Braunschweig: Friedrich Vieweg & Sohn, 1915), 89; Kurt Wegener to Aßmann, 4 Apr. 1906, Lindenberg Observatory Archive no. 51, W 005-1906.

11. The following account of the flight is compiled from Ulrich Wutzke, *Der Forscher von der Friedrichsgracht: Leben und Leistung Alfred Wegeners*, 1st ed. (Leipzig: Brockhaus, 1988), 25–27; and Alfred Wegener, "Astronomische Ortsbestimmungen des Nachts bei der Ballonfahrt vom 5. bis 7. April 1906," *Illustrierte Aeronautische Mitteilungen* 10, no. 6 (1906): 205–207.

12. Wegener to Aßmann (telegram), 7 Apr. 1906, Lindenberg Observatory Archive no. 5, W 006-1906.

13. Wutzke, *Der Forscher von der Friedrichsgracht*, 26. Wegener did not realize the size of the record until somewhat later. Writing to Mylius-Erichsen later that week, he estimated that they had broken the record by eleven hours, but no matter: the achievement stood.

14. Wegener to Mylius-Erichsen, 10 Apr. 1906, DP Copenhagen DEA, 26/131.

15. Wegener, "Drachen- und Fesselballonaufstiege ausgeführt auf der Danmark-Expedition 1906–1908," 7.

16. Wegener to Köppen, 19 May 1906, DMH 1969-595/5, W 009-1906.

17. Patent als Leutnant der Reserve, 21 May 1906, HN W 010-1906.

18. Wegener to Bidlingsmaier, 29 Mar. 1906, Alfred Wegener Institute Bst Potsdam (Deutsche Sudpolar Expedition, Acta betr. Magnetica Abt. 1), W 002-1906; W(alter). Brückmann, "Magnetische Beobachtungen der Danmarks-Expedition," *Meddelelser om Grønland* 42, no. 8 (1914): 595–597.

19. Wegener to Bidlingsmaier, 6 June 1906, AWI Bst Potsdam (Deutsche Sudpolar Expedition Acta betr. Magnetica Abt. 1), W 013-1906, 014-1906.

20. The expedition eventually produced an archive of over 9,000 photographs. The glass negatives are stored at the Dansk Polarcenter in Copenhagen (Arktisk Institute, Danmarks Expedition Archive). The expedition's scientific results, published in Meddelelser om Grønland, contain hundreds of these photographs. Hundreds more are available in Achton Friis, *Im Grönlandeis mit Mylius-Erichsen: Die Danmark-Expedition 1906–1908*, trans. Friedrich Stichert, unaltered 2nd (1913) German ed. of 1909 Danish original ed. (Leipzig: Otto Spamer Verlag, 1910).

21. Else Wegener-Köppen, *Wladimir Köppen: Ein Gelehrtenleben für die Meteorologie*, ed. H. W. Frickhinger, Grosse Naturforscher (Stuttgart: Wissenschaftliche Verlagsgesellschaft, 1955), 96.

22. Wegener to Wladimir Köppen, 21 June 1906, DMH 1968-595/2, W 015-1906.

23. Ibid.

24. Ibid. For a fuller account of the Köppen household in these years see Wegener-Köppen, *Wladimir Köppen*, passim.

25. Peter Freuchen, *Vagrant Viking: My Life and Adventures*, trans. Johan Hambro (London: Victor Gollancz, 1954), 74.

26. Wegener kept irregular journals, written in pencil in a clear hand, throughout the expedition—in addition to his scientific notebooks. These journals are in the Handschriftensammlung of the Deutsches Museum in Munich and are catalogued as follows: 24 June–11 Nov. 1906: 1968 594/5; 6 Dec. 1906–26 Mar. 1907: 1968 594/4; 29 Mar.–29 May 1907 (contains passages from some of J. P. Koch's journal as well): 1968 594/11; 12 June–22 Nov. 1907: 1968 594/7; 23 Nov. 1907–6 May 1908 (all after 1 Jan. 1908 in Danish): 1968 594/1; 7 May–6 Aug. 1908: 1968 594/2. Long extracts from these are published in E. Wegener, *Alfred Wegener: Tagebücher, Briefe, Erinnerungen*, with ellipses. I shall henceforth cite them only as "Wegener's Tagebuch" with the date of the entry.

27. G(eorg). Amdrup, "Report on the Danmark Expedition to the North-East Coast of Greenland 1906–1908," *Meddelelser om Grønland* 41, no. 1 (1913): 47.

28. Wegener's Tagebuch, 24 June 1906.

29. Ibid. See also Freuchen, *Vagrant Viking*, 73–74; Friis, *Im Grönlandeis mit Mylius-Erichsen*, 8–11; and Amdrup, "Report on the Danmark Expedition," 50ff.

30. Friis, *Im Grönlandeis mit Mylius-Erichsen*, 8–12; Freuchen, *Vagrant Viking*, 74.

31. Drygalski, *Southern Ice-Continent*, x.

32. Roald Amundsen, *The South Pole: An Account of the Norwegian Antarctic Expedition in the "Fram," 1910–1912*, trans. A. G. Chater (London: J. Murray, 1912), acknowledgments. Amundsen used his acknowledgments to punish as well as to thank, naming the firms whose products had failed him, as well as those that had worked satisfactorily.

33. Friis, *Im Grönlandeis mit Mylius-Erichsen*, 12–14.

34. Wegener's Tagebuch, 8 July 1906.

35. Ibid., 11 July 1906.

36. Freuchen, *Vagrant Viking*, 75.

37. Amdrup, "Report on the Danmark Expedition," 50ff.

38. Wegener's Tagebuch, 18 July 1906.

39. Ibid.

40. Amdrup, "Report on the Danmark Expedition," 50ff.

41. Wegener's Tagebuch, 25 July 1906.

42. Ibid., 29 July 1906.

43. Ibid., 31 July 1906.

44. Ibid.

45. Moritz Lindeman and Otto Finsch, *Die Zweite Deutsche Nordpolarfahrt in den Jahren 1869 und 1870 unter Führung des Kapitän Koldewey* (Leipzig: F. A. Brockhaus, 1875); Fridtjof Nansen, *The First Crossing of Greenland*, 2 vols. (New York: Longmans, Green, 1890).

46. Wegener's Tagebuch, 1 Aug. 1906.
47. Ibid., 4 Aug. 1906.
48. Ibid., 6 Aug. 1906.
49. The anecdote about Christiansen and the hat and the characterization of Wegener are from Friis, *Im Grönlandeis mit Mylius-Erichsen.*
50. Wegener's Tagebuch, 7 Aug. 1906.
51. Ibid.
52. The following account is corroborated from several sources, including Wegener's journal; Freuchen, *Vagrant Viking*, 77; and Friis, *Im Grönlandeis mit Mylius-Erichsen*, 51–54.
53. Friis, *Im Grönlandeis mit Mylius-Erichsen*, 51–54.
54. Freuchen, *Vagrant Viking*, 77.
55. Wegener's Tagebuch, 14 Aug. 1906.
56. Amdrup, "Report on the Danmark Expedition," 72–73.
57. Wegener's Tagebuch, 14 Aug. 1906.
58. Ibid.
59. Amdrup, "Report on the Danmark Expedition," 72.
60. Wegener's Tagebuch, 25 Aug. 1906.
61. Ibid., 1 Sept. 1906.
62. Ibid.
63. Ibid., 7 Sept. 1906.
64. Ibid., 5 Sept. 1906.
65. Ibid., 7 Sept. 1906.
66. Ibid., 14 Sept. 1906.
67. Ibid., 13 Sept. 1906.
68. Amdrup, "Report on the Danmark Expedition," 73.
69. Freuchen, *Vagrant Viking*, 79.
70. Amdrup, "Report on the Danmark Expedition," 72–76.
71. Friis, *Im Grönlandeis mit Mylius-Erichsen*, 255. Thostrup even invented a machine to make the ropes and dog harnesses for the sleds, with a foot treadle to drive it.
72. Wegener's Tagebuch, 22 Sept. 1906.
73. Ibid.
74. Ibid.; Wegener, "Drachen- und Fesselballonaufstiege ausgeführt auf der Danmark-Expedition 1906–1908," 23ff.
75. Wegener's Tagebuch, 26 Sept. 1906.
76. Amdrup, "Report on the Danmark Expedition."
77. Wegener's Tagebuch, 12 Oct. 1906, notes, "I have arranged with Koch that we will go with Mylius-Erichsen to Sabine island, Koch will anchor the longitude net with observation of the culminations of the pole star, and I will make absolute magnetic determinations on the German expedition's observation pillar."
78. Wegener's Tagebuch, 23 Sept. 1906.
79. Wegener, "Drachen- und Fesselballonaufstiege ausgeführt auf der Danmark-Expedition 1906–1908," 19.
80. Wegener's Tagebuch, 5 Oct. 1906.
81. Ibid., 15 and 18 Oct. 1906.
82. Ibid., 22 Oct. 1906.
83. Ibid., 26 Oct. 1906.
84. Ibid.
85. Ibid., 30 Oct. 1906.
86. Ibid.
87. Ibid., 2 Nov. 1906.
88. Amdrup, "Report on the Danmark Expedition," 91.
89. Wegener's Tagebuch, 6 Nov. 1906.
90. Ibid.
91. Ibid., 6 Dec. 1906.

92. Amdrup, "Report on the Danmark Expedition." "Sledge Journey to Pendulum Islands," pp. 91–95, is a verbatim transcript of Koch's diary. This was Koch's entry for 15 Nov. 1906.

93. Friis, *Im Grönlandeis mit Mylius-Erichsen*, 230.

94. Wegener's Tagebuch, 6 Dec. 1906.

95. Amdrup, "Report on the Danmark Expedition," 92.

96. Ibid., 93.

97. Wegener's Tagebuch, 6 Dec. 1906.

98. Alfred Wegener, *Magnetische Observationen 1906*, DP Copenhagen DEA, 153, 24 sides, unpaginated, Wützke 021-1906, 1906. Wegener's diary for this entry consists entirely of telegraphic notation about food until the arrival at Germania Haven on 21/22 November.

99. Amdrup, "Report on the Danmark Expedition," 93n. The two dogs made their way back to the ship on their own in January 1907—a trip of 120 kilometers (75 miles).

100. Wegener's Tagebuch, 6 Dec. 1906.

101. Ibid.

102. Friis, *Im Grönlandeis mit Mylius-Erichsen*, 234. See also Amdrup, "Report on the Danmark Expedition," 94.

103. Alfred Wegener, *Magnetische Beobachungsergebnisse*, unpaginated notebook, Danmark Ekspedition no. 153, Dansk Polarcenter, W 021-1906, 1906.

104. Ibid.

105. Wegener's Tagebuch, 6 Dec. 1906.

106. Friis, *Im Grönlandeis mit Mylius-Erichsen*, 236.

107. Wegener's Tagebuch, 6 Dec. 1906.

108. Ibid., 15 Dec. 1906.

109. Ibid.

110. Hermann Günzel, *Alfred Wegener und sein Meteorologisches Tagebuch der Grönland-Expedition 1906–1908* (Marburg: Universitätsbibliothek, 1991), 69.

111. Adolf Marcuse, *Handbuch der geographischen Ortsbestimmung fur Geographen und Forschungsresienden: Mit 54 in de Text eingedruckten Abbildungen und 2 Sternkarten* (Braunschweig: Friedrich Vieweg & Sohn, 1905). Quoted in Günzel, *Alfred Wegener und sein Meteorologisches tagebuch*, 71.

112. Günzel, *Alfred Wegener und sein Meteorologisches Tagebuch*, 71. The works referred to are standard references: Josef Pernter, *Meteorologische Optik* (Wien: Braumuller, 1902); Marcuse, *Handbuch der geographischen Ortsbestimmung fur Geographen und Forschungsresienden*; and Karl Koss, *Zeit und Orts-Bestimmungen* (Wien: K. K. Hof & Statsdruckerei, 1901).

113. Wegener's Tagebuch, 21 Dec. 1906.

114. Friis, *Im Grönlandeis mit Mylius-Erichsen*, 249.

115. Ibid., 248.

116. Wegener's Tagebuch, 25 Dec. 1906.

117. Ibid.

118. Ibid.

Chapter 6. The Arctic Explorer (1): Greenland, 1907–1908

1. Wegener's Tagebuch, 1 Jan. 1907, DMH 1968 594/4, W 016-1906.

2. Ibid.

3. Ibid., 13 Jan. 1907.

4. Ibid., 14 Jan. 1907.

5. Alfred Wegener, "Drachen- und Fesselballonaufstiege ausgeführt auf der Danmark-Expedition 1906–1908," *Meddelelser om Grønland* 42, no. 1 (1909): 18.

6. Wegener's Tagebuch, 8 Feb. 1907.

7. Ibid., 1 Feb. 1907.

8. Ibid., 17 Feb. 1907.

9. Erich von Drygalski, *The Southern Ice-Continent: The German South Polar Expedition aboard the Gauss 1901–1903*, trans. M. M. Raraty (Bluntisham: Bluntisham Books and Erskine Press, 1989); quoted by Raraty (pp. iv–v) from Erich von Drygalski, *Grönland-Expedition der Gesellschaft für Erdkunde zu Berlin 1891–1893* (Berlin: W. H. Kühl, 1897), 60–61.

10. Wegener's Tagebuch, 17 Feb. 1907.

11. Hermann Günzel, *Alfred Wegener und sein Meteorologisches Tagebuch der Grönland-Expedition 1906–1908* (Marburg: Universitätsbibliothek, 1991), 76. This book, in several hands, contains information about outstanding weather events in the course of the expedition. It was discovered in 1988 in the papers of Walther Brand, a Marburg scientist who helped Wegener write up some of his expedition results, by the Marburg bookdealer John Wilcockson, who understood its historical importance and saw that it found its way into the university library, where an excellent facsimile edition was prepared by Hermann Günzel. My thanks to Herr Günzel for allowing me to examine the original diary in Marburg in the summer of 1999.

12. G(eorg). Amdrup, "Report on the Danmark Expedition to the North-East Coast of Greenland 1906–1908," *Meddelelser om Grønland* 41, no. 1 (1913): 107–110.

13. See chap. 2.

14. Wegener's Tagebuch, 9 Mar. 1907.

15. This measurement was off by almost a full degree.

16. Actually N 81°41′.

17. Amdrup, "Report on the Danmark Expedition," 111–121.

18. Wegener's Tagebuch, 5 Apr. 1907.

19. Ibid., 2 Apr. 1907.

20. Amdrup, "Report on the Danmark Expedition," 115.

21. Wegener's Tagebuch, 16 Apr. 1907.

22. Ibid., 18 Apr. 1907.

23. Amdrup, "Report on the Danmark Expedition," 143.

24. Wegener's Tagebuch, 30 Apr. 1907.

25. Ibid., 4 May 1907.

26. Ibid., 5 May 1907.

27. Ibid.

28. Ibid., 29 May 1907.

29. Achton Friis, *Im Grönlandeis mit Mylius-Erichsen: Die Danmark-Expedition 1906–1908*, trans. Friedrich Stichert, unaltered 2nd (1913) German ed. of 1909 Danish original ed. (Leipzig: Otto Spamer Verlag, 1910), 353.

30. Wegener's Tagebuch, 27 May 1907.

31. Ibid., 30 May 1907.

32. Ibid., 27 May 1907.

33. Magnetische Apr. 1907, Single Sheet, DP Copenhagen DEA, 157.

34. Friis, *Im Grönlandeis mit Mylius-Erichsen*, 422.

35. Ibid., 423–424.

36. Wegener, "Drachen- und Fesselballonaufstiege ausgeführt auf der Danmark-Expedition 1906–1908," 6.

37. Ibid., 9.

38. Günzel, *Alfred Wegener und sein Meteorologisches Tagebuch*, 77.

39. Ibid., 79.

40. Wegener's Tagebuch, 31 Jan. 1908. Deutsches Museum, Munich, Handschriftensammlung, 1968 594/1.

41. Wegener, "Drachen- und Fesselballonaufstiege ausgeführt auf der Danmark-Expedition 1906–1908," 6.

42. Amdrup, "Report on the Danmark Expedition," 163.

43. Wegener's Tagebuch, 8 Sept. 1907.

44. Ibid.

45. Ibid., 13 Sept. 1907.

46. Wegener, "Drachen- und Fesselballonaufstiege ausgeführt auf der Danmark-Expedition 1906–1908," 19.

47. Günzel, *Alfred Wegener und sein Meteorologisches tagebuch*, 8 Nov. 1907.

48. Peter Freuchen, *Vagrant Viking: My Life and Adventures*, trans. Johan Hambro (London: Victor Gollancz, 1954), 84–85.

49. Peter Freuchen, *Arctic Adventure: My Life in the Frozen North* (New York: Farrar & Rinehart, 1935), 16.

50. Wegener to Freuchen at Pustervig, 12 Nov. 1907, DP Copenhagen DEA, 230, W 004-1907.

51. Wegener to Freuchen at Pustervig, 27 Dec. 1907, DP Copenhagen DEA, 230, W 009-1907. R. P. W. Lewis, ed., *Meteorological Glossary*, 6th ed. (London: HMSO, 1991), s.v. "aurora": "Numerous reports . . . of aurora reaching almost to the ground in great displays are generally discredited."

52. Wegener to Freuchen at Pustervig, 27 Dec. 1907, DP Copenhagen DEA, 230, W 010-1907.

53. Wegener to Friis, n.d., DP Copenhagen DEA, 154, W 004-1908.

54. Wegener's Tagebuch, 31 Jan. 1908.

55. Ibid.

56. Ibid., 1 Feb. 1908.

57. Ibid., 2 Feb. 1908.

58. Ibid., 13 Apr. 1908.

59. Amdrup, "Report on the Danmark Expedition," 184–185.

60. Ibid., 184.

61. Wegener's Tagebuch, 14 Apr. 1908.

62. Wegener, "Drachen- und Fesselballonaufstiege ausgeführt auf der Danmark-Expedition 1906–1908," 19.

63. Wegener's Tagebuch, 24 Jan. 1908.

64. Ibid., 9 May 1908.

65. Ibid., 10 May 1908.

66. Ibid.

67. Ibid., 13 May 1908.

68. Ibid., 21 May 1908.

69. Ibid., 26 May 1908.

70. Ibid.

71. Ibid., 1 June 1908.

72. Ibid.

73. Ibid., 4 June 1908. Harald Hagerup (1877–1947) was, like Freuchen, supposed to help with the scientific measurements. Like Freuchen, he was an adventurer who had been in Greenland before and had signed on as a stoker.

74. Wegener's Tagebuch, 10 June 1908.

75. Freuchen, *Vagrant Viking*, 88.

76. Wegener, "Drachen- und Fesselballonaufstiege ausgeführt auf der Danmark-Expedition 1906–1908," 6.

77. Wegener's Tagebuch, 3 July 1908.

78. Ibid., 10 July 1908.

79. Wegener to the Danmark Ekspedition Komitee, København, 17 July 1908, Summary of Scientific Work on the Expedition, DP Copenhagen DEA, 185, W 005-1908.

80. Amdrup, "Report on the Danmark Expedition," 240.

81. Freuchen, *Vagrant Viking*, 88.

82. Ibid.

Chapter 7. The Atmospheric Physicist (1): Berlin and Marburg, 1908–1910

1. A. Kh. Khrgian, *Meteorology: A Historical Survey (Ocherki razvitiya meteorologii)*, trans. Ross Hardin, 2nd ed., rev. Kh. P. Pogosyan (Jerusalem: Israel Program for Scientific Translations, 1970), 290.

2. Else Wegener-Köppen, *Wladimir Köppen: Ein Gelehrtenleben für die Meteorologie*, ed. H. W. Frickhinger, Grosse Naturforscher (Stuttgart: Wissenschaftliche Verlagsgesellschaft, 1955), 93.

3. Khrgian, *Meteorology*, 290.

4. See chap. 4.

5. Khrgian, *Meteorology*, 295.

6. Gisela Kutzbach, *The Thermal Theory of Cyclones: A History of Meteorological Thought in the Nineteenth Century*, Historical Monograph Series (Boston: American Meteorological Society, 1979), 233.

7. Petra Seibert, "Hann's Thermodynamic Foehn Theory and Its Presentation in Meteorological Textbooks in the Course of Time," ICHM, www.meteohistory.org/2004polling_preprints/seibert.

8. Else Wegener, *Alfred Wegener: Tagebücher, Briefe, Erinnerungen* (Wiesbaden: F. A. Brockhaus, 1960), 65.

9. Alfred Wegener, "Vorläufiger Bericht über die Drachen- und Fesselballonaufstiege der Danmark-Expedition nach Nordostgrönland," *Meteorologische Zeitschrift* 26 (1909): 23.

10. Ibid.

11. Ibid. This picture appears in chap. 6.

12. Eventually published in Alfred Wegener, "Über die Ableitung von Mittelwerten aus Drachenaufstiegen ungleicher Höhe," *Beiträge zur Physik der freien Atmosphäre* 3 (1910). For Aßmann's interest in the problem see Wegener to Danmark Ekspedition Komittee, 25 Jan. 1909, DP Copenhagen DEA, 311, W 001-1909.

13. Hugo Hergesell, *Observationen in bemannten, unbemannten Ballons und Drachen sowie auf Berg- und Wolkenstationen im Jahre 1907*, Veröffentlichen der Internationalen Kommission für wissenschaftliche Luftshiffahrt (Straßburg: DuMont Schauburg, 1909).

14. Wegener to Johannes Warming, 12 Oct. 1908, DP Copenhagen DEA, 311, W 007-1908.

15. E. Wegener, *Alfred Wegener: Tagebücher, Briefe, Erinnerungen*, 66.

16. Quoted in Wegener-Köppen, *Wladimir Köppen*, 102–103.

17. Gustav Angenheister, *A History of the Samoan Observatory from 1902 to 1921* (Wellington: Geophysics Division, New Zealand Dept. of Scientific and Industrial Research, 1978).

18. Richard Wegener, *Die Bühneneinrichtung des Shakepeareschen Theaters nach den zeitgenössischen Dramen* (Halle: Max Niemeyer, 1907).

19. Wegener to Johannes Warming, 12 Oct. 1908, DP Copenhagen DEA, 311, W 007-1908.

20. Ibid.

21. Wegener to Komitee der Danmark-Expedition, 22 Oct.1908, DP Copenhagen DEA, 311, W 008-1908.

22. E. Wegener, *Alfred Wegener: Tagebücher, Briefe, Erinnerungen*, 136; see also Wegener to Köppen, 6 Nov. 1910, DMH 1968 595/12, W 027-1910.

23. Invar is, for this reason, also used for balance wheels in clocks and for the structural components in laser systems: it expands or contracts about one part per million per degree Celsius—whereas brass expands and contracts at nineteen times that value.

24. Details concerning these meteorographs may be found in W. E. Knowles Middleton, *Invention of the Meteorological Instruments* (Baltimore: Johns Hopkins University Press, 1969), 291ff.

25. Alfred Wegener, "Drachen- und Fesselballonaufstiege ausgeführt auf der Danmark-Expedition 1906–1908," *Meddelelser om Grønland* 42, no. 1 (1909): 18–20.

26. Ibid., 21.

27. Ibid., 23.

28. Alfred Wegener, "Zur Schichtung der Atmosphäre," *Beiträge zur Physik der freien Atmosphäre* 3 (1910): 33.

29. Wegener, "Drachen- und Fesselballonaufstiege ausgeführt auf der Danmark-Expedition 1906–1908," 21.

30. Wegener to Komitee der Danmark Ekspedition, 27 Nov. 1908, DP Copenhagen DEA, 311.5, W 012-1908.

31. Wegener to Komitee der Danmark Ekspedition, 31 Oct. 1908, DP Copenhagen DEA, 311.3, W 010-1908.

32. Wegener, "Zur Schichtung der Atmosphäre," 31–32.

33. Ibid., 31.

34. Ibid.

35. Ibid.

36. Wegener, "Über die Ableitung von Mittelwerten aus Drachenaufstiegen ungleicher Höhe."

37. Wegener to Danmark Komitee, 25 Jan. 1909, DP Copenhagen DEA, 311.6, W 009-1909.

38. Förster to Richarz, 8 Feb. 1909, HSM [307d Nr. 269].

39. Aßmann to Richarz, 9 Feb. 1909, HSM [307d Nr. 269], W 003-1909.

40. Aßmann to Richarz, 9 Feb. 1909, HSM [307d Nr. 269], W 004-1909.

41. Wegener to Aßmann, 12 Feb. 1909, Observatorium Lindenberg [67], W 008-1909.

42. Wegener, "Über die Ableitung von Mittelwerten aus Drachenaufstiegen ungleicher Höhe"; and Wegener to Aßmann, 12 Feb. 1909, Observatorium Lindenberg [67], W 008-1909.

43. Wegener to Philosophische Fakultät, Marburg, HSM [307d Nr. 269], W 006-1909.

44. This paper would be published, only slightly modified, as Alfred Wegener, "Probleme der Aerologie," *Das Wetter* 11 (1909).

45. Ibid., 241.

46. Ibid., 251.

47. Ibid., 253ff.

48. Ibid., 255.

49. See, e.g., Khrgian, *Meteorology*; Karl Schneider-Carius, *Weather Science and Weather Research: History of Their Problems and Findings from Documents during Three Thousand Years* (New Delhi: Indian National Scientific Documentation Center [for NOAA and NSF], 1975); Robert Marc Friedman, *Appropriating the Weather: Vilhelm Bjerknes and the Construction of a Modern Meteorology* (Ithaca, NY: Cornell University Press, 1989); Kutzbach, *Thermal Theory of Cyclones*.

50. Lewis F. Richardson's book *Weather Prediction by Numerical Process* (London: Cambridge University Press, 1922) contains the equations that today guide most computer-based model forecasts.

51. Wegener to Aßmann, 25 Mar. 1909, Observatorium Lindenberg [34], W 012-1909.

52. Wegener to Danmark Ekspedition Komitee, 29 Mar. 1909, DP Copenhagen DEA, 311.8, W 013-1909.

53. Richarz to Wegener, May 1909, HSM [307d Nr. 269].

54. Wegener to Danmark Ekspedition Komitee, 17 Apr. 1909, DP Copenhagen DEA, 311.9, W 016-1909.

55. Wegener to Danmark Ekspedition Komitee, 25 May 1909, DP Copenhagen DEA, 311.10, W 019-1909.

56. Alfred Wegener, "Über eine eigentümliche Gesetzmässigkeit der oberen Inversion," *Beiträge zur Physik der freien Atmosphäre* 3 (1910): 206.

57. Ibid., 206.

58. Ibid., 209.

59. Ibid.

60. Wegener, "Über eine eigentümliche Gesetzmässigkeit der oberen Inversion," 214.

61. Aßmann to Wegener, 20 July 1909, DMH 1968 595/7 N 1/7, W 024-1909.

62. Ibid.

63. Ibid.

64. Ibid.

65. Wegener to Richard and Anna Wegener, DMH 1968 595/7 N 1/7, W 025-1909.

66. Wegener to Aßmann, 27 July 1909, Observatorium Lindenberg [96], W 027-1909.

67. Ibid.

68. Wegener to Aßmann, 18 Aug. 1909, Observatorium Lindenberg [97], W 030-1909.

69. Wegener to Danmark Ekspedition Komitee, 22 Oct. 1909, DP Copenhagen DEA, 311.15, W 032-1909.

70. Ulrich Wutzke, "Alfred Wegener als Hochschullehrer," *Zeitschrift für geologische Wissenschaften* 25, nos. 5/6 (1997): 557.

71. Cornelia Lüdecke, *Die deutsche Polarforschung seit der Jahrhundertwende und der Einfluß Erich von Drygalskis*, vol. 158, *Berichte zur Polarforschung* (Bremerhaven: Alfred-Wegener-Institute für Polar- und Meeresforschung, 1995), 13ff.

72. Khrgian, *Meteorology*, 279–280.

73. Wegener to Danmark Ekspedition Komitee, 1 Nov. 1909, DP Copenhagen DEA, 311.15, W 033-1909.

74. Wegener to Philosophiches Facultät, 2 Nov. 1909, HSM [307d Nr. 269], W 034-1909.

75. Wegener to Kayser, 19 Nov. 1909, HSM [307d Nr. 269], W 036-1909.

76. Wegener to Köppen, 13 Nov. 1909, DMH 1968 595/8 N 1/8, W 035-1909.

77. Wegener to Danmark Ekspedition Komitee, 7 Dec. 1909, DP Copenhagen DEA, 311.17, W 037-1909.

78. Wegener to his parents, 14 Dec. 1909, DMH 1968 595/6 N 1/6, W 038-1909.

79. Ibid.

80. Ibid.

81. Richard Assmann, *Das Königlich Preußische Aeronautische Observatorium Lindenberg* (Braunschweig: Friedrich Vieweg & Sohn, 1915), 273.

82. Wegener to Richarz, 18 Feb. 1910, HSM [307d Nr. 269], W 005-1910.

83. Alfred Wegener, "Über eine neue fundamentale Schichtgrenze der Erdatmosphäre," *Beiträge zur Physik der freien Atmosphäre* 3 (1910).

84. Ibid., 225.

85. Ibid. Italics in original.

86. Ibid.

87. Ibid., 227–228.

88. Ibid., 230.

89. Ibid., 229–231.

90. Ibid., 231.

91. Alfred Wegener, "Mylius Erichsens 'Danmark'-Expedition nach Nordost Grönland 1906–1908," *Mathematisch-Naturwissenschaftliche Blätter* 6, no. 8 (1909).

92. Wegener to Danmark Ekspedition Komitee, 3 Mar. 1910, DP Copenhagen DEA, 311.20, W 007-1910.

93. Wegener to Freuchen, 11 Mar. 1910, DP Copenhagen, Peter Freuchen no. 1, W 008-1910.

Chapter 8. The Atmospheric Physicist (2): Marburg, 1910

1. Karl Schneider-Carius, *Weather Science and Weather Research: History of Their Problems and Findings from Documents during Three Thousand Years* (New Delhi: Indian National Scientific Documentation Center [for NOAA and NSF], 1975), 554.

2. Wilhelm von Bezold, *Gesammelte Abhandlungen aus den Gebieten der Meteorologie und Erdmagnetismus* (Braunschweig: Friedrich Vieweg & Sohn, 1906), 91–127.

3. Alfred Wegener, "Probleme der Aerologie," *Das Wetter* 11 (1909): 241–255.

4. Gisela Kutzbach, *The Thermal Theory of Cyclones: A History of Meteorological Thought in the Nineteenth Century*, Historical Monograph Series (Boston: American Meteorological Society, 1979). This excellent monograph is an indispensable source for the history of atmospheric physics, and the interested reader is referred here for a detailed treatment of this problem.

5. Ibid., introduction and synopsis, sec. 1.1.

6. Wegener to Köppen, 1910. I have not been able to locate the original of this letter. This and many essential letters to Köppen and to Else Wegener apparently exist only in quoted form in her 1960 memoir of her husband, published on the thirtieth anniversary of his death. I have cross-checked all extant letters to Köppen and to Else where originals exist of documents she quotes. She often edited heavily but never fabricated, and there is every reason to accept these documents as real. See Else Wegener, *Alfred Wegener: Tagebücher, Briefe, Erinnerungen* (Wiesbaden: F. A. Brockhaus, 1960), 75.

7. E. Wegener, *Alfred Wegener: Tagebücher, Briefe, Erinnerungen*, 75.

8. J(ohannes) Georgi, "Memories of Alfred Wegener," in *Continental Drift*, ed. Stanley Keith Runcorn (New York: Academic Press, 1962), 312–313.

9. Ibid.

10. Ibid., 313.

11. Ibid.

12. Alfred Wegener, *Thermodynamik der Atmosphäre* (Leipzig: J. A. Barth, 1911), 331, iii.

13. Kutzbach, *Thermal Theory of Cyclones*, 244.

14. Wegener, *Thermodynamik der Atmosphäre*, iii.

15. These developments are covered in considerable detail in chap. 6 of Kutzbach, *Thermal Theory of Cyclones*.

16. Susan Schlee, *The Edge of an Unfamiliar World: A History of Oceanography* (New York: E. P. Dutton, 1973), 171–173.

17. Kutzbach, *Thermal Theory of Cyclones*, 191.

18. Ibid., 192.

19. Alfred Wegener, "Über die Eisphase des Wasserdampfes in der Atmosphäre," *Meteorologische Zeitschrift* 27 (1910): 451–459; Alfred Wegener, "Die Größe der Wolkenelemente," *Meteorologische Zeitschrift* 27 (1910): 354–361.

20. Wegener, "Über die Eisphase des Wasserdampfes in der Atmosphäre," 451.

21. A good discussion of contemporary views of the state of the science appears in Cleveland Abbe, "Meteorology," in *Encyclopaedia Britannica* (New York: Encyclopaedia Britannica, 1911).

22. Vincent J. Schaefer and John A. Day, *A Field Guide to the Atmosphere* (Boston: Houghton Mifflin, 1981), 64–65.

23. Wegener, "Über die Eisphase des Wasserdampfes in der Atmosphäre," 455.

24. For Bezold's views see Bezold, *Gesammelte Abhandlungen aus den Gebieten der Meteorologie und Erdmagnetismus*, 184.

25. For a simple explanation see Schaefer and Day, *Field Guide to the Atmosphere*, 76–77.

26. Wegener, "Über die Eisphase des Wasserdampfes in der Atmosphäre."

27. Wegener, "Die Größe der Wolkenelemente," 360.

28. Ibid., 356–357.

29. Schneider-Carius, *Weather Science and Weather Research*, 321–322.

30. Robert Marc Friedman, *Appropriating the Weather: Vilhelm Bjerknes and the Construction of a Modern Meteorology* (Ithaca, NY: Cornell University Press, 1989), 77.

31. Wegener to Kayser, 27 Apr. 1910, HSM [307d Nr. 269], W 012-1910.

32. Wegener to Danmark Ekspedition Komitee, 27 Apr. 1910, DP Copenhagen DEA, 311.24, W 016-1910.

33. For instance, Alfred Wegener, "Mit Mylius-Erichsen in Grönland," *Umschau* 11 (1908): 1011–1016.

34. Karl Stuchtey and Alfred Wegener, "Die Albedo der Wolken und der Erde," *Nachtrichten der Kgl. Gesellschaft zu Göttingen, mathematisches-naturwissenschaftliches Klasse*, no. 3 (1911): 209–235.

35. Wegener to Danmark Ekspedition Komitee, 3 Sept. 1910, DP Copenhagen DEA, 311.25, W 018-1910.

36. Wegener to Danmark Ekspedition Komitee, 24 Sept. 1910, DP Copenhagen DEA, 311.26, W 019-1910.

37. Wegener to Köppen, 26 Sept. 1910, DMH 1968 595/11, 1/11, W 020-1910.

38. E. Wegener, *Alfred Wegener: Tagebücher, Briefe, Erinnerungen*, 67.

39. Ibid.

40. Ibid.

41. Ibid.

42. Wegener to Danmark Ekspedition Komitee, 12 Oct. 1910, DP Copenhagen DEA, 311.27, W 021-1910.

43. Wegener to Danmark Ekspedition Komitee, 5 Nov. 1910, DP Copenhagen DEA, 311.29, W 026-1910.

44. Ibid.

45. Wegener to Köppen, 6 Nov. 1910, DMH 1968 595/12 N 1/12, W 027-1910.

46. Ibid.

47. There is a brief appreciative biography by Else Wegener; Else Wegener-Köppen, *Wladimir Köppen: Ein Gelehrtenleben für die Meteorologie*, ed. H. W. Frickhinger, Grosse Naturforscher (Stuttgart: Wissenschaftliche Verlagsgesellschaft, 1955).

48. Wegener to Köppen, 14 Nov. 1910, DMH 1968 595/13 N 1/13, W 029-1910.

49. Ibid.

50. Ibid.

51. Wegener to Köppen, 17 Nov. 1910, DMH 1968 595/14 N 1/14, W 031-1910.

52. Wegener to Köppen, 28 Nov. 1910, DMH 1968 595/16 N 1/16, W 034-1910.

53. Wegener to Köppen, 1 Dec. 1910, DMH 1968 595/17 N 1/17, W 035-1910.

54. Alfred Wegener, "Untersuchungen über die Nature der obersten Atmosphärenschichten," *Physiikalosche Zeitschrift* 12 (1911): 220.

55. Wegener to Köppen, 1 Dec. 1910, DMH 1968 595/17 N 1/17, W 035-1910.

56. Wegener, "Untersuchungen über die Nature der obersten Atmosphärenschichten."

57. Wegener to Köppen, 1 Dec. 1910, DMH 1968 595/17 N 1/17, W 035-1910.

58. Wegener to Köppen, 21 Dec. 1910, DMH 1968 595/18 N 1/18, W 036-1910.

59. E. Wegener, *Alfred Wegener: Tagebücher, Briefe, Erinnerungen*, 69.

60. Edward N. Lorenz, "The Evolution of Dynamic Meteorology," in *Historical Essays on Meteorology, 1919–1995*, ed. James Rodger Fleming (Boston: American Meteorological Society, 1996), 3.

61. Wegener, *Thermodynamik der Atmosphäre*, v.

62. Svante August Arrhenius, *Lehrbuch der kosmischen Physik* (Leipzig: S. Hirzel Verlag, 1903).

63. Wegener, *Thermodynamik der Atmosphäre*, 6–7.

64. Ibid., 7–23.

65. Ibid., chaps. 4 and 5.

66. Ibid., sec. 4, chaps. 9–15 passim.

67. See, e.g., the drawings on page 207 of cumulus and stratus clouds and then the photographic plate of the same, facing page 208, from which it was drawn. Ibid., 207–208.

68. Ibid., 324.

Chapter 9. At a Crossroads: Marburg, 1911

1. Richard Andree and Albert Scobel, *Andrees Allgemeiner Handatlas in 139 Haupt- und 161 Nebenkarten; nebst vollständigem alphabetischem Namensverzeichnis*, 5th revised and expanded; Jubilee ed. (Bielefeld: Velhagen & Klassing, 1907).

2. Else Wegener, *Alfred Wegener: Tagebücher, Briefe, Erinnerungen* (Wiesbaden: F. A. Brockhaus, 1960), 75. Like the rest of the correspondence between Wegener and his wife, the extracts in this volume are the only source of these letters, which were not deposited in the collection at Munich in 1968. For other versions of this anecdote see Alfred Wegener, *Die Entstehung der Kontinente und Ozeane*, 3rd (completely revised) ed. (Braunschweig: Friedrich Vieweg & Sohn, 1922); or Alfred Wegener, *Die Entstehung der Kontinente und Ozeane*, 4th (completely revised) ed. (Braunschweig: Friedrich Vieweg & Sohn, 1929).

3. Wegener to Köppen, 6 Jan. 1911, DMH 1968 596/1 N 1/20, W 003-1911.

4. Ibid.

5. E. Wegener, *Alfred Wegener: Tagebücher, Briefe, Erinnerungen*, 69. This is another one of those letters that only exist in the context of Else Wegener's memoir.

6. Karl Stuchtey and Alfred Wegener, "Die Albedo der Wolken und der Erde," *Nachtrichten der Kgl. Gesellschaft zu Göttingen, mathematisches-naturwissenschaftliches Klasse*, no. 3 (1911): 209–235.

7. "Editorial," *Nature* 86 (27 Apr. 1911): 275–276.

8. Wegener to Köppen, 3 Apr. 1911, DMH 1968 596/3 N 1/22, W 018-1911.

9. Alfred Wegener, *Neuere Forschungen auf dem Gebiete der Atmosphäreschen Physik*, ed. Emil Abderhalden, Fortschritte der Naturwissenschaftlichen Forchung (Berlin: Urban & Schwarzenberg, 1911).

10. J. P. Koch and A. Wegener, "Die glaciologischen Beobachtungen der Danmark-Expedition," *Meddelelser om Grønland* 46 (1912): 1–79.

11. See Ejnar Mikkelsen, *Farlig Tomandsfaerd (A Dangerous Two-Man Journey)* (Copenhagen: Gyldendal, 1962).

12. Fridtjof Nansen, *Auf Schneeschuhen durch Grönland*, translation of *Paaski over Grønland*, original Norwegian ed., 2 vols. (Hamburg: J. F. Richter, 1897); Fridtjof Nansen, *The First Crossing of Greenland*, 2 vols. (New York: Longmans, Green, 1890).

13. Nansen, *First Crossing of Greenland*, 1:79.

14. Ibid., 1:69.

15. Ibid., 1:468–469.

16. Ibid., 2:496.

17. Ibid., 2:444ff.

18. Ibid., 2:454.

19. Ibid., 2:456.

20. Ibid., 2:445.

21. Oswald Heer, *Die Urwelt der Schweiz* (Zurich: Verlag von Friedrich Schulthess, 1865).

22. Nansen, *First Crossing of Greenland*, 2:455.

23. Ibid., 2:458.

24. Ibid., 2:460.

25. Wegener to Köppen, 3 Apr. 1911, DMH 1968 596/3 N 1/22, W 018-1911.

26. Robert Marc Friedman, *Appropriating the Weather: Vilhelm Bjerknes and the Construction of a Modern Meteorology* (Ithaca, NY: Cornell University Press, 1989), chap. 3 passim.

27. Ibid.

28. Wegener to Köppen, 2 May 1911, DMH 1968 596/4 N 1/23, W 019-1911.

29. Ibid.

30. Ibid.

31. E. Wegener, *Alfred Wegener: Tagebücher, Briefe, Erinnerungen*, 71.

32. Ibid.

33. Ibid.

34. Ibid.

35. Ibid., 72.

36. Ibid.

37. Ibid. Else Wegener's 1960 memoir is our only source of this information; the letter does not survive in any of the best-known collections of Wegener's papers, including that at the Zechlinerhütte Museum, where one might have expected to find it. Else Wegener did not deposit her personal correspondence with her husband at the Deutsches Museum in Munich when the rest of the letters and diaries were deposited in 1968. Their provenance and whereabouts are unknown, although they may still be held by the family. None of the biographers seem to have seen these letters, and they do not appear in Ulrich Wutzke's checklist of letters by, to, and about Wegener.

38. Wegener to Köppen, 30 June 1911, DMH 1968 596/5 N 1/24, W 022-1911.

39. Ibid.

40. Ibid.

41. Ibid.

42. Wegener to Köppen, 6/7 July 1911, DMH 596/7 N 1/7, W 024-1911.

43. Ibid.

44. Ibid.

45. Wegener to Köppen, n.d., Charlottenburg, July 1911, DMH 596/6 N 1/25, W 023-1911.

46. Ibid.

47. Wegener refers to this visit in Wegener to Köppen, 15 Oct. 1911, DMH 1968 596/9 N 1/28, W 026-1911.

48. Alfred Wegener, "Die Windverhältnisse in der Stratosphäre," *Meteorologische Zeitschrift* 28 (1911): 271–273; and Wegener to Köppen, 3 Nov. 1911, DMH 1968 596/12 N 1/31, W 030-1911.

49. Wegener to Köppen, 3 Nov. 1911, DMH 1968 596/12 N 1/31, W 030-1911.

50. Wegener to Köppen, 21 Nov. 1911, DMH 1968 596/14 N 1/3, W 032-1911. See Otto Krümmel, *Handbuch der Ozeanographie*, 2nd rev. ed., vols. 1 and 2 (Stuttgart: Verlag von J. Engelhorn, 1907, 1911).

51. Wegener to Köppen, 21 Nov. 1911, DMH 1968 596/14 N 1/3, W 032-1911.

52. Ibid.

53. Ibid.

54. Ibid.

55. Ibid.

56. Hans Reck, "Die Geologie Islands in ihrer Bedeutung für Fragen der allgemeinen Geologie," *Geologische Rundschau* 2, nos. 5–6 (1911): 302–314.

57. Erich Krenkel, "Die Entwicklung der Kriedformationen auf dem afrikanischen Kontinente," *Geologische Rundschau* 11, nos. 5–6 (1911): 330–366.

58. Ibid.

59. Alfred Wegener, *The Origin of Continents and Oceans* (New York: E. P. Dutton, 1924). The first appearance of this historical remark (here p. 3 of the English translation) comes in the preface to Wegener, *Die Entstehung der Kontinente und Ozeane* (1922, 3rd German ed.).

60. Aart Brouwer, "Was veranlaßste Alfred Wegener zum Studium der Kontinentverschiebung?," *Geologische Rundschau* 72, no. 2 (1983): 739–741.

61. Konrad Keilhack, "Über postglaciale Meeresablangerungen in Island," *Zeitschrift der deutschen geologischen Gesellschaft* 36 (1884): 145–160. See also Eduard Sueß, *The Face of the Earth*, trans. Hertha Sollas, 4 vols. (Oxford: Clarendon, 1904–1909), 2:482.

62. See Mott Greene, *Geology in the Nineteenth Century* (Ithaca, NY: Cornell University Press, 1982).

63. Wegener to Köppen, 6 Dec. 1911, DMH 1968 596/17 N 1/36, W 037-1911.

64. E. Wegener, *Alfred Wegener: Tagebücher, Briefe, Erinnerungen*, 75.

65. Wegener to Köppen, 6 Dec. 1911, DMH 1968 596/17 N 1/36, W 037-1911.

66. Ibid.

67. Ibid.

68. Walter Moore, *Schrödinger: Life and Thought* (Cambridge: Cambridge University Press, 1989).

69. Wegener to Köppen, 6 Dec. 1911, DMH 1968 596/17 N 1/36, W 037-1911.

70. Ibid.

Chapter 10. The Theorist of Continental Drift (1): Marburg, December 1911–February 1912

1. Wegener to Köppen, 6 Dec. 1911, DMH 1968 596/16 N 1/35, W 037-1911.

2. Thomas S. Kuhn, *The Structure of Scientific Revolutions* (Chicago: University of Chicago Press, 1962); Thomas S. Kuhn, *The Copernican Revolution: Planetary Astronomy in the Development of Western Thought* (Cambridge, MA: Harvard University Press, 1957), 297.

3. Alfred Wegener, *Die Alfonsinischen Tafeln für den Gebrauch eines modernen Rechners: Inaugural-Dissertation zur Erlangung der Doktorwürde genehmigt von der philosophischen Facultät der Friedrich-Wilhelms-Universität zu Berlin* (Berlin: E. Eberling, 1905).

4. Wegener to Köppen, 6 Dec. 1911, DMH 1968 596/16 N 1/35, W 037-1911.

5. Wegener to Köppen, 17 and 29 Jan. 1912, DMH 1968 597/3 and 6 N 1/38 and 1/141, W 002-1912 and 003-1912.

6. Wegener to Köppen, 24 Feb. 1912, DMH 1968 597/7/5 N 1/40, W 009-1912.

7. Eduard Sueß, *The Face of the Earth*, trans. Hertha Sollas, 4 vols. (Oxford: Clarendon, 1904–1909).

8. Otto Krümmel, *Handbuch der Ozeanographie*, 2nd rev. ed., vols. 1 and 2 (Stuttgart: Verlag von J. Engelhorn, 1907, 1911).

9. Theodor Arldt, *Handbuch der Palaeogeographie*, 2 vols. (Leipzig: Gebrüder Borntraeger, 1919–1922), 1647.

10. M[aurycy] P. Rudzki, *Physik der Erde* (Leipzig: Chr. Herm. Tachnitz, 1911).

11. Emanuel Kayser, *Lehrbuch der allgemeinen Geologie*, 3rd ed., 2 vols. (Stuttgart: Ferdinand Enke, 1909).

12. Eduard Sueß, *Die Entstehung der Alpen* (Vienna: Wilhelm Braumuller, 1875), 144.

13. Mott Greene, *Geology in the Nineteenth Century* (Ithaca, NY: Cornell University Press, 1982). For an extended discussion of Sueß and his impact see especially chaps. 6–8 passim.

14. Krümmel, *Handbuch der Ozeanographie*, 1:24–43.

15. Ibid., 1:39. Hettner (1859–1941) was a German geographer who forcefully advocated the integrated causal study of large regions, the origins of their topography, and their lifeworlds.

16. Sueß, *Face of the Earth*, 4:544. See also Wilfried Schröder, "Emil Wiechert und seine Bedeutung für die Entwicklung der Geophysik zur exacten Wissenschaft," *Archive for History of Exact Sciences* 27, no. 4 (1982): 369–389.

17. Arldt, *Handbuch der Palaeogeographie*.

18. Ibid.; see the rear endpaper plates.

19. Ibid., plates 19–23.

20. Kayser, *Lehrbuch der allgemeinen Geologie*, passim.

21. Wegener to Warming, 7 Jan. 1912, DP Copenhagen DEA, 311, W 001-1912.

22. Ibid.

23. Else Wegener, *Alfred Wegener: Tagebücher, Briefe, Erinnerungen* (Wiesbaden: F. A. Brockhaus, 1960), 79; contains a digest of a letter from Wegener to his parents on this matter, no longer extant.

24. Alfred Wegener, "Die Entstehung der Kontinente," *Petermanns Mitteilungen* 58 (1912): 185–195, 253–256, 305–309, 185.

25. Wegener to Köppen, 24 Feb. 1912, DMH 1968 597/5 N 1/40, W 009-1912.

26. Wegener, "Die Entstehung der Kontinente," 185.

27. Ibid.

28. Ibid. Wegener here cites the applicable pagination and near-identical wording of his critique from Rudzki, *Physik der Erde*, 210–212.

29. Wegener, "Die Entstehung der Kontinente," 185. For a further discussion of Taylor's views see Greene, *Geology in the Nineteenth Century*, 280–284; and for the original arguments see Frank Bursley Taylor, "Bearing of the Tertiary Mountain Belts on the Origin of Earth's Plan," *Geological Society of America, Bulletin* 21 (1910): 179–226.

30. Taylor, "Bearing of the Tertiary Mountain Belts," 206, 216, 219–223.

31. Wegener, "Die Entstehung der Kontinente," 186.

32. Ibid. The quotation is from Albert Heim, *Untersuchungen über den Mechanismus der Gebirgsbildung; im Anschluss an die geologische Monographie der Todi-Windgällen Gruppe*, 2 vols. (Basel: Benno Schwabe, 1878), 2:237.

33. Wegener, "Die Entstehung der Kontinente," 186.

34. Ibid.

35. Wegener cites John Joly, *Radioactivity and Geology* (London: Constable, 1909), but the information is likely from Rudzki, *Physik der Erde*, 122–125.

36. Wegener, "Die Entstehung der Kontinente," 186.

37. The reference is to John F. Hayford, *The Figure of the Earth and Isostasy from Measurments in the United States* (Washington, DC: U.S. Government Printing Office, 1909).

38. Wegener, "Die Entstehung der Kontinente," 187.

39. Ibid.

40. Ibid., 187–188.

41. For the history of this concept see Greene, *Geology in the Nineteenth Century*, 238–250, 267–269.

42. See, e.g., Naomi Oreskes, *The Rejection of Continental Drift: Theory and Method in American Earth Science* (New York: Oxford University Press, 1999); Marco Segala, *La Favola della Terra Mobile: La controversia sulla teoria della deriva dei continenti* (Bologna: Il Mulino, 1990); Anthony Hallam, *An Outline of Phanerozoic Biogeography* (New York: Oxford University Press, 1994); Martin Schwarzbach, *Alfred Wegener und die Drift der Kontinente*, ed. Heinz Degen, vol. 42, Grosse Naturforscher (Stuttgart: Wissenschaftliche Verlagsgesellschaft mbH, 1980); Henry Frankel, "Alfred Wegener and the Specialists," *Centaurus* 20 (1976): 305–324; H[omer] E. Le Grand, *Drifting Continents and Shifting Theories* (New York: Cambridge University Press, 1988).

43. See his "Cosmic Physics" introduction to his geology textbook, defining Earth as a planet among other planets. Kayser, *Lehrbuch der allgemeinen Geologie*, introduction, 1:1.

44. Wegener, "Die Entstehung der Kontinente," 189.

45. Andrija Mohorovicic, "Potres od 8 X 1909," *Godishje Isvjesce Zagrebakog Meteoroloskog Opservatorija za godinu 1909 {avail as Jahrbuch der Meteorologische Observatorium, Agram[Zagreb] für 1909}* (1910); Wegener, "Die Entstehung der Kontinente," 190.

46. Wegener, "Die Entstehung der Kontinente," 190.

47. Ibid., 191.

48. Ibid.

49. Ibid.

50. Kayser, *Lehrbuch der allgemeinen Geologie*, 1.

51. R[eginald] A. Daly, *Strength and Structure of the Earth* (New York: Prentice Hall, 1940).

52. Harold Jeffreys, *The Earth: Its Origin, History, and Physical Constitution* (Cambridge: Cambridge University Press, 1929), 181.

53. Jeffreys and Daly (see previous notes) had very different notions about how viscosity, plasticity, and liquidity ought to be defined.

54. Wegener, "Die Entstehung der Kontinente," 192.

55. Adrian E. Scheidegger, *Principles of Geodynamics*, 2nd ed. (New York: Academic Press, 1963), 148.

56. Rudzki, *Physik der Erde*, 504.

57. Wegener, "Die Entstehung der Kontinente," 193.

58. Ibid.

59. Sueß, *Face of the Earth*, 1:604.

60. Wegener, "Die Entstehung der Kontinente," 194.

61. Ibid.

62. Ibid.

Chapter 11. The Theorist of Continental Drift (2): Marburg, February–April 1912

1. Alfred Wegener, "Die Entstehung der Kontinente," *Petermanns Mitteilungen* 58 (1912): 253–306.

2. Ibid., 253.

3. Eduard Sueß, *The Face of the Earth*, trans. Hertha Sollas, 4 vols. (Oxford: Clarendon, 1904–1909), 4:270–275, 282–286.

4. Wegener, "Die Entstehung der Kontinente."

5. Ibid., 254.

6. Ibid.

7. Ibid.

8. Sueß, *Face of the Earth*, 2:75ff., 3:386–387.

9. Wegener, "Die Entstehung der Kontinente," 254.

10. Frank Bursley Taylor, "Bearing of the Tertiary Mountain Belts on the Origin of Earth's Plan," *Geological Society of America, Bulletin* 21 (1910): 216–217, plate 4.

11. Wegener, "Die Entstehung der Kontinente," 254.

12. Ibid., 255.

13. Ibid.

14. Ibid., 255n.

15. Ibid., 255.

16. Ibid.

17. Wegener assumed that his audience would understand the absurdity of this contention. We may note that, for instance, glacial deposits within 30° of the equator, transferred to the current Northern Hemisphere, would require glacial phenomena for a Northern Hemisphere ice age as far south as Los Angeles, California. The actual limit of West Coast glaciation is instead at the northern border of Washington State, 48° away from the equator.

18. Wegener, "Die Entstehung der Kontinente," 256.

19. Ibid., 256.

20. Sueß, *Face of the Earth*, 4:73–82.

21. Wegener, "Die Entstehung der Kontinente," 305.

22. Ibid., 306.

23. Fridtjof Nansen, *Auf Schneeschuhen durch Grönland*, translation of *Paaski over Grønland*, original Norwegian ed., 2 vols. (Hamburg: J. F. Richter, 1897); Fridtjof Nansen, *The First Crossing of Greenland*, 2 vols. (New York: Longmans, Green, 1890), 444–497. For Nathorst, see 456ff.

24. There is an extensive discussion of this material in chap. 9.

25. Wegener, "Die Entstehung der Kontinente," 307.

26. Ibid.

27. Giovanni Schiaparelli, *De la rotation de la terre sous l'influence des actions géologiques* (St. Petersburg, 1889), 31–32.

28. Ibid., 32.

29. Wegener, "Die Entstehung der Kontinente," 307.

30. These figures, as well as the entire argument in this section, are taken verbatim, or very nearly, from M[aurycy] P. Rudzki, *Physik der Erde* (Leipzig: Chr. Herm. Tachnitz, 1911), 208.

31. Wegener, "Die Entstehung der Kontinente," 307.

32. Ibid., 308.

33. C. A. Schott, *The Telegraphic Longitude Net of the United States and Its Connection with That of Europe 1866–1896*, Annual Report of the Director, U.S. Coast and Geodetic Survey, for 1897 (Washington, DC: GPO, 1898), appendix 2.

34. Wegener, "Die Entstehung der Kontinente," 307.

35. Ibid.

36. Felix Maria Exner, "Referat über Wegener: Thermodynamik der Atmosphäre," *Meteorologische Zeitschrift* 28, no. 12 (1911): 389–390.

37. See Deborah R. Coen, *Vienna in the Age of Uncertainty* (Chicago: University of Chicago Press, 2007), 282; Berta Karlik and Erich Schmid, *Franz Serafin Exner und sein Kreis: Eig Beitrag zur Geschichte der Physik in Österreich* (Wien: Österreischen Akademie der Wissenschaften, 1982); A. Kh. Khrgian, *Meteorology: A Historical Survey (Ocherki razvitiya meteorologii)*, trans. Ross Hardin, 2nd ed., rev. Kh. P. Pogosyan (Jerusalem: Israel Program for Scientific Translations, 1970), 231; and Paul A. Hanle, "Indeterminacy before Heisenberg: The Case of Franz Exner and Erwin Schrödinger," *Historical Studies in the Physical Sciences* 10 (1979): 235.

38. Else Wegener, *Alfred Wegener: Tagebücher, Briefe, Erinnerungen* (Wiesbaden: F. A. Brockhaus, 1960), 79. This letter does not survive in the original but was quoted extensively and approvingly by Else Wegener.

39. We know of this only through an exchange of correspondence between the philosophical faculty at Marburg and the minister of education, attempting to create a post with sufficient remuneration to make it attractive for Wegener to stay at Marburg. Geheimes Staatsarchiv Preußischer Kulturbesitz Berlin (Merseburg). See Wutzke 1998 004-1913.

40. The Leipzig "search" was a formality, and Bjerknes was the only real candidate. See Robert Marc Friedman, *Appropriating the Weather: Vilhelm Bjerknes and the Construction of a Modern Meteorology* (Ithaca, NY: Cornell University Press, 1989), 84ff.

41. Wegener to Köppen, 29 Jan. 1912, DMH 1968 597/6 N 1/41, W 003-1912. Wegener expressed embarrassment that Köppen quoted Voeikov's comment to him in a previous letter.

42. E. Wegener, *Alfred Wegener: Tagebücher, Briefe, Erinnerungen*, 79. This letter no longer survives.

43. Wegener to Köppen, 17 Jan. 1912, DMH 1968 597-3 N 1/38, W 002-1912; italics in the original.

44. Ibid.

45. Exner, "Referat über Wegener," 389.

46. Coen, *Vienna in the Age of Uncertainty*, 289.

47. Friedman, *Appropriating the Weather*, 199.

48. Exner, "Referat über Wegener," 389–390.

49. Wegener to Köppen, 17 Jan. 1912, DMH 1968 597/3 N 1/38, W 002-1912.

50. Wegener to Köppen, 29 Jan. 1912, DMH 1968 597/6 N 1/41, W 003-1912.

51. Wegener to Ludwig Darmstädter, 15 Feb. 1912, Prussian State Library Berlin, Sammlung Darmstädter, W 004-1912; Wegener to Köppen, 16 Feb. 1912, DMH 1968 597/6 N 1/37, W 005-1912.

52. Wegener to Elstner, 12 Apr. 1912, HSM [307d Nr. 269], W 015-1912.

53. Ibid.

54. Ibid.; Resolution of the Faculty [Elstner], 19 May 1912, HSM [307d Nr. 269], W 015-1912.

55. Wegener to Köppen, 16 Feb. 1912, DMH 1968 597/64 N 1/39.

56. E. Wegener, *Alfred Wegener: Tagebücher, Briefe, Erinnerungen*, 80. Else recalled the eclipse being on the sixteenth; it was actually on the seventeenth.

57. Ibid.

58. Wegener to Köppen, 16 Feb. 1912, DMH 1968 597/6 N 1/37, W 005-1912.

59. Wegener to Elstner, 15 June 1912, HSM [307d Nr. 269], W 026-1912.

60. E. Wegener, *Alfred Wegener: Tagebücher, Briefe, Erinnerungen*, 80.

61. Ibid. See also Ulrich Wutzke, *Durch die weiße Wüste: Leben und Leistungen des Grönlandforschers und Entdeckers der Kontinentaldrift Alfred Wegener*, Petermann ed. (Gotha: Justus Perthes Verlag, 1997), 89.

62. E. Wegener, *Alfred Wegener: Tagebücher, Briefe, Erinnerungen*, 80–81.

63. (Kaptajn) J. P. Koch, *Gennem den Hvide Ørken: Den danske Forskningsrejse tvaers over Nordgrønland 1912/13* (Kjøbenhavn: Gyldendalske Boghandel Nordisk Forlag, 1913), 110.

Chapter 12. The Arctic Explorer (2): Greenland, 1912–1913

1. Wegener began to keep a daybook starting on 7 June, and he would keep this diary and an observational journal until the arrival on the west coast of Greenland in mid-July 1913. These journals he kept in pencil in the small-format notebooks he had used in 1906–1908, comprising nearly 600 pages. Three of the volumes are his expedition diary; the fourth is the "Journal of Observations" and is partly in Danish, as indeed are parts of the other journals.

All are in the Wegener Nachlass, Deutsches Museum, Munich. Tagebuch, 7 June–15 Sept. 1912: 594/8; Tagebuch, 17 Sept. 1912–18 Apr. 1913: 594/9; Tagebuch, 19 Apr.–17 July 1913, 594/10; Beobachtungs-Journal, 13 Apr.–17 July 1913. In the back of this volume is an essay on the Jacobshavn Eisstrom: 594/13.

2. Wegener to his parents, 4 July 1912, quoted in Else Wegener, *Alfred Wegener: Tagebücher, Briefe, Erinnerungen* (Wiesbaden: F. A. Brockhaus, 1960), 81. The original letter is lost.

3. Wegener recorded this trip in his journal: 19–29 June 1912. Koch worked up his notes into an article for *Petermanns Mitteilungen*. A detailed account drawn from these sources can be found in Ulrich Wutzke, *Durch die weiße Wüste: Leben und Leistungen des Grönlandforschers und Entdeckers der Kontinentaldrift Alfred Wegener*, Petermann ed. (Gotha: Justus Perthes Verlag, 1997), 93–96.

4. Wegener to his parents, 1 July 1912. This letter does not survive and is quoted from E. Wegener, *Alfred Wegener: Tagebücher, Briefe, Erinnerungen*, 81.

5. Wegener to his parents, 4 July 1912, quoted from ibid.

6. Ibid.

7. Wegener to Elstner, 4 July 1912, HSM [307d Nr. 269], W 033-1912.

8. Wegener to his parents, 21 July 1912, in E. Wegener, *Alfred Wegener: Tagebücher, Briefe, Erinnerungen*, 81. Koch's article appeared as Johan Peter Koch, "Die dänische Expedition nach Königen-Luise-Land und quer über das nordgröndlandische Inlandeis 1912/1913. I. Die Reise durch Island 1912," *Petermanns Geographische Mitteilungen* 58, no. 2 (1912): 185–189.

9. Wegener to his parents, 21 July 1912, in E. Wegener, *Alfred Wegener: Tagebücher, Briefe, Erinnerungen*, 81.

10. (Kaptajn) J. P. Koch, *Gennem den Hvide Ørken: Den danske Forskningsrejse tvaers over Nordgrønland 1912/13* (Kjøbenhavn: Gyldendalske Boghandel Nordisk Forlag, 1913), 7.

11. Ejnar Mikkelsen, *Farlig Tomandsfaerd (A Dangerous 2-Man Journey)* (Copenhagen: Gyldendal, 1962), chap. 13.

12. Koch, *Gennem den Hvide Ørken*, 9.

13. Wegener's Tagebuch, 27 July 1912.

14. Johan Peter Koch, *Durch die weiße Wüste: Die dänische Forschungsreise quer durch Nordgrönland 1912–1913*, trans. Else Wegener (Berlin: Springer, 1919), 10.

15. Ibid.

16. The following account is assembled from Wegener's Tagebuch, 21 July 1912–8 Sept. 1912; and Koch, *Durch die weiße Wüste*, 59–67. The entries in the book are collations of Wegener's and Koch's diaries for those dates.

17. Koch, *Durch die weiße Wüste*, 21, 12 Aug. 1912.

18. These accounts collate mishaps from 31 July to 8 September, recorded in Wegener's Tagebuch.

19. Wegener's Tagebuch, 5 Aug. 1912.

20. Ibid., 8 Aug. 1912.

21. Koch, *Durch die weiße Wüste*, 54.

22. Wegener's Tagebuch, 8 Sept. 1912.

23. Ibid., 12 Sept. 1912.

24. Koch, *Durch die weiße Wüste*, 76–78; Wegener's Tagebuch, 15 Sept. 1912.

25. Koch, *Durch die weiße Wüste*, 79.

26. Ibid.

27. Wegener's Tagebuch, 17 Sept. 1912.

28. Ibid., 18 Sept. 1912.

29. Koch, *Durch die weiße Wüste*, 81; the expedition report states that it was a torn muscle. J. P. Koch and A. Wegener, "Wissenschaftliche Ergebnisse der dänischen Expedition nach Dronning Louises-Land und quer über das Indlandeis von Nordgrönland 1912–13 unter Leitung von Hauptmann J.P. Koch. Abteilung 1–2," *Meddelelser om Grønland* 74 (1930): 24.

30. Wegener's Tagebuch, 18 Sept. 1912.

31. Ibid., 22 Sept. 1912.

32. Koch, *Durch die weiße Wüste*, 91; Koch, *Gennem den Hvide Ørken*, 103.

33. Koch, *Durch die weiße Wüste*, 88.

34. Wegener's Tagebuch, 30 Sept. 1912.

35. Ibid.

36. Ibid.

37. Ibid., 1 Oct. 1912.

38. Ibid., 7 Oct. 1912.

39. Ibid.

40. Koch, *Gennem den Hvide Ørken*, 126–128.

41. Koch, *Durch die weiße Wüste*, 111.

42. Wegener's Tagebuch, 28 Oct. 1912.

43. Koch, *Durch die weiße Wüste*, 115; Wegener's Tagebuch, 27 Oct. 1912.

44. Koch, *Gennem den Hvide Ørken*, 137

45. Wegener's Tagebuch, 29 Oct. 1912.

46. Koch, *Gennem den Hvide Ørken*, 139.

47. Wegener's Tagebuch, 5 Nov. 1912.

48. Ibid.

49. Ibid.

50. Ibid., 6 Nov. 1912.

51. Ibid., 11 Nov. 1912.

52. There is an account of this work in Koch, *Durch die weiße Wüste*, 122–123; a much more detailed account in Koch and Wegener, "Wissenschaftliche Ergebnisse der dänischen Expedition," 48–54; and ancillary material in Wegener's Tagebuch, 30 Nov. and 1 Dec. 1912.

53. Koch, *Durch die weiße Wüste*, 123. This instrument is today mounted on the wall in the display case in the conference room of the Royal Danish Geographical Society in Copenhagen.

54. Koch and Wegener, "Wissenschaftliche Ergebnisse der dänischen Expedition," 200ff.; E. Wegener, *Alfred Wegener: Tagebücher, Briefe, Erinnerungen*, 107.

55. Koch and Wegener, "Wissenschaftliche Ergebnisse der dänischen Expedition," 6ff.

56. Wegener's Tagebuch, 1 Dec. 1912; Hans Heß, *Die Gletscher* (Braunschweig: Friedrich Vieweg & Sohn, 1904).

57. Koch and Wegener, "Wissenschaftliche Ergebnisse der dänischen Expedition," 6–7.

58. Ibid.

59. E. Wegener, *Alfred Wegener: Tagebücher, Briefe, Erinnerungen*, 108.

60. Koch, *Durch die weiße Wüste*, 148.

61. Wegener's Tagebuch, 3 May 1913.

62. Koch, *Durch die weiße Wüste*, 190.

63. Ibid., 191.

64. Wegener's Tagebuch, 13 May 1913.

65. Ibid., 15, 16, and 19 May 1913.

66. Koch, *Durch die weiße Wüste*, 208–209.

67. Ibid., 210.

68. Ibid.

69. Wegener's Tagebuch, 12 June 1913.

70. Ibid., 25 June 1913.

71. Ibid., 26 June 1913.

72. Ibid.

73. Ibid., 29 June 1913.

74. Ibid., 4 July 1913.
75. Ibid.
76. Ibid.
77. Ibid., 5 July 1913.
78. Koch, *Durch die weiße Wüste*, 233.
79. Ibid., 235; Wegener's Tagebuch, 9 July 1913.
80. Wegener's Tagebuch, 9 July 1913.
81. Koch, *Durch die weiße Wüste*, 236.
82. Ibid., 238–239.
83. Ibid.
84. Ibid., 241.
85. Ibid., 242.
86. Ibid.
87. Ibid., 245.
88. Wegener's Tagebuch, 19 July 1913.
89. Koch, "Die dänische Expedition nach Königen-Luise-Land," 247.
90. Wegener's Tagebuch, 17 July 1913.
91. E. Wegener, *Alfred Wegener: Tagebücher, Briefe, Erinnerungen*, 131.
92. Børge Fristrup, *The Greenland Ice Cap* (Seattle: University of Washington Press, 1966), 250.
93. E. Wegener, *Alfred Wegener: Tagebücher, Briefe, Erinnerungen*, 133.
94. Der Königliche Kurator der Universität Philosophische Facultät, 27 Sept. 1912, HSM [307d Nr. 269].
95. Wegener to Marie Köppen, Aug. 1913. Like most of the personal correspondence reported by Else Wegener, this letter does not survive and is quoted from E. Wegener, *Alfred Wegener: Tagebücher, Briefe, Erinnerungen*, 133.

Chapter 13. The Soldier: Marburg and "The Field," 1913–1915

1. Copies of the citation for "Ritter des Dannebrogs-Ordens" (18 Oct. 1913) are in the Wegenerarchiv at the AWI, Bremerhaven, as well as in the Heimatmuseum, Neuruppin.
2. Else Wegener, *Alfred Wegener: Tagebücher, Briefe, Erinnerungen* (Wiesbaden: F. A. Brockhaus, 1960), 134.
3. This letter survives only in an excerpt from ibid., 134. Wegener to Köppen, Oct. 1913.
4. E. Wegener, *Alfred Wegener: Tagebücher, Briefe, Erinnerungen*, 135.
5. Ibid.
6. Ibid.
7. Wegener to Köppen, 4 Jan. 1914, DMH 1968 598/2 N 1/42, W 001-1914.
8. E. Wegener, *Alfred Wegener: Tagebücher, Briefe, Erinnerungen*, 132.
9. Alfred Wegener, "Staubwirbel auf Island," *Meteorologische Zeitschrift* 31 (1914).
10. Alfred Wegener, "Über turbulente Bewegungen in der Atmosphäre," *Meteorologische Zeitschrift* 29 (1912): 49–59.
11. Ibid., 54.
12. Ibid., 57.
13. Ibid., 53n60.
14. Karl Schneider-Carius, *Weather Science and Weather Research: History of Their Problems and Findings from Documents during Three Thousand Years* (New Delhi: Indian National Scientific Documentation Center [for NOAA and NSF], 1975), 360–363.
15. Wegener, "Über turbulente Bewegungen in der Atmosphäre," 52.
16. Ibid., 55.
17. Wegener to Köppen, 16 Jan. 1914, DMH 1968 598/2 N 1/43, W 003-1914.
18. Ibid.
19. Wegener to Köppen, 4 Jan. 1914, DMH 1968 598/2 N 1/42, W 001-1914.
20. E. Wegener, *Alfred Wegener: Tagebücher, Briefe, Erinnerungen*, 77.
21. For a summary of this material see H. P. H., "Carl Theodor Albrecht," *Monthly Notices of the Royal Astronomical Society* 86, no. 4 (1916): 282–284.

22. Ibid., 283.

23. Alfred Wegener, *Die Entstehung der Kontinente und Ozeane* (Braunschweig: Friedrich Vieweg & Sohn, 1915), iii.

24. Wegener to Komitee der Danmark-Expedition, 3 Mar. 1914, DP Copenhagen DEA, 311, W 009-1914.

25. Alfred Wegener, "Durch Grönlands Eiswüste," *Himmel und Erde* 26 (1914): 453–462, 498–511.

26. E. Wegener, *Alfred Wegener: Tagebücher, Briefe, Erinnerungen*, 135.

27. Proclamation by the Kurator of Marburg University, 17 Feb. 1914, HSM [307d Nr. 269], W 005-1914.

28. Wegener to Komitee der Danmark-Expedition, 3 Mar. 1914, DP Copenhagen DEA, 311, W 009-1914.

29. E. Wegener, *Alfred Wegener: Tagebücher, Briefe, Erinnerungen*, 136–137.

30. Wegener to Dekan der philosophischen Facultät, Marburg, HSM [307d Nr. 269], W 013-1914.

31. Wegener to Köppen, 31 May 1914, DMH 1968 598/3 N 1/44, W 015-1914.

32. (Kaptajn) J. P. Koch, *Gennem den Hvide Ørken: Den danske Forskningsrejse tvaers over Nordgrønland 1912/13* (Kjøbenhavn: Gyldendalske Boghandel Nordisk Forlag, 1913).

33. E. Wegener, *Alfred Wegener: Tagebücher, Briefe, Erinnerungen*, 136.

34. Köppen to Wegener, 23 June 1914, DMH 1968 598/8 N 1/95, W 018-1914.

35. Wegener to Köppen, 23 July 1914, DMH 1968 598/5 N 1/46, W 020-1914.

36. Wegener to Julius Bauschinger, DMH 1968 603/8 N 1/46, W 019-1914.

37. Wegener to Köppen, 31 July 1914, DMH 1968 598/9 N 1/46, W 021-1914.

38. E. Wegener, *Alfred Wegener: Tagebücher, Briefe, Erinnerungen*, 139.

39. (General) James E. Edmonds, *Military Operations: France and Belgium, 1914* (London: MacMillan, 1922), 36; and E. Wegener, *Alfred Wegener: Tagebücher, Briefe, Erinnerungen*, 139–140.

40. E. Wegener, *Alfred Wegener: Tagebücher, Briefe, Erinnerungen*, 140. This account is based on a letter or series of letters from Alfred to Else Wegener. They are not included in the Munich archive or in Ulrich Wutzke's checklist of Wegener documents. Else Wegener seems to have held her correspondence with Alfred after the deposit of letters in Munich in 1968; their current location is unknown.

41. Edmonds, *Military Operations*, 457.

42. Ibid.

43. E. Wegener, *Alfred Wegener: Tagebücher, Briefe, Erinnerungen*, 141–144.

44. Ibid., 140.

45. Ibid.

46. Medical certificate Charlottenburg, 18 Dec. 1914, Wegenerarchiv, AWI Bremerhaven.

47. Alfred Wegener, "Neuere Forschungen auf dem Gebiet der Meteorologie und Geophysik," *Annalen der Hydrographie und Maritimen Meteorologie* 43 (1915): 168.

48. Ibid., 163.

49. Ibid., 165.

50. Ibid.

51. Ibid., 167.

52. Ibid., 168.

53. Ibid.

54. Ibid.

55. Hans Cloos, *Gespräch mit der Erde: Geologische Welt-und Lebesfahrt* (München: R. Piper, 1947), 98.

56. Ibid., 305–317.

57. Ibid., 329.

58. Ibid.; see also Alfred Wegener, *Die Entstehung der Kontinente und Ozeane*, ed. Eilhard Wiedemann, 2nd (completely revised) ed. (Braunschweig: Friedrich Vieweg & Sohn, 1920), vi.

59. Alfred Wegener, *Die Entstehung der Kontinente und Ozeane* (Braunschweig: Friedrich Vieweg & Sohn, 1915), iv.

60. Cloos, *Gespräch mit der Erde*, 329.

61. Wegener, *Die Entstehung der Kontinente und Ozeane* (1915), 1.

62. Ibid., 2.

63. Ibid., 6; italics added.

64. Joseph Loukaschewitsch, *Sur le méchanisme de l'écorce terrrestre et l'origine des continents* (St. Petersburg: Imprimérie Russo-Francaise, 1911).

65. Wegener, *Die Entstehung der Kontinente und Ozeane* (1915), 13.

66. Bailey Willis, *Research in China: In Three Volumes and an Atlas*, 4 vols. (Washington, DC: Carnegie Institution, 1907).

67. Wegener, *Die Entstehung der Kontinente und Ozeane* (1915), 39–41.

68. Ibid., 42.

69. Ibid.

70. Helmut W. Flügel, "Wegener-Ampferer-Schwinner: Ein Beitrag zur Geschichte der Geologie in Österreich," *Mitteilung der Österreichischen Geologischen Gesellschaft* 73 (1980).

71. Otto Ampferer, "Über das Bewegungsbild von Faltengebirgen," *Jahrbuch der k.k geol. Reichsanstalt* 56 (1906).

72. Mott Greene, *Geology in the Nineteenth Century* (Ithaca, NY: Cornell University Press, 1982), 270–272, contains a discussion of these rather complicated issues.

73. Wegener, *Die Entstehung der Kontinente und Ozeane* (1915), 44, 46.

74. Ibid., 54.

75. Ibid., 57–59.

76. Ibid., 57.

77. Ibid., 85ff.

78. Ibid., 59.

79. Ibid., 67.

80. Paul Lemoine, *Afrique Occidentale*, Heft 14 (vol. 7, pt. 6a), Handbuch der Regionalen Geologie (Heidelberg: Carl Winter—Universitätsverlag, 1913); see also Patrick Marshall, *New Zealand and Adjacent Islands*, vol. 7, pt. 1, Handbuch der Regionalen Geologie (Heidelberg: Carl Winter—Universitätsverlag, 1912); and Paul Marshall, *Oceania*, Heft 9 (vol. 7, pt. 2), Handbuch der Regionalen Geologie (Heidelberg: Carl Winter—Universitätsverlag, 1911).

81. Wegener, *Die Entstehung der Kontinente und Ozeane* (1915), 77.

82. Ibid.

83. E. Wegener, *Alfred Wegener: Tagebücher, Briefe, Erinnerungen*, 143. This letter does not survive.

84. Ibid.

85. Edgar Dacqué, *Grundlagen und Methoden der Paläogeographie* (Jena: Gustav Fischer Verlag, 1915).

86. Ibid., 4.

87. Edgar Dacqué, *Der Descendenzgedanke und seiner Geschichte vom Altertum bis zur Neuzeit* (Munich: Reinhardt, 1903).

88. Dacqué, *Grundlagen und Methoden der Paläogeographie*, 93.

89. Ibid., 93–100, 119–124.

90. Ibid., 178–185.

91. Damian Kreichgauer, *Die Äquatorfrage in der Geologie* (Steyl: Verlag Missiondruckerei, 1902).

92. Dacqué, *Grundlagen und Methoden der Paläogeographie*, 101ff.

93. Wegener, *Die Entstehung der Kontinente und Ozeane* (1915), iii.

94. Ibid.

95. Wegener to Köppen, 11 Jan. 1915, DMH 1968 599/1 N 1/47, W 002-1915.

96. Wegener to Köppen, 21 Jan. 1915, DMH 1968 599/3 N 1/49, W 004-1915.

97. Alfred Wegener, "Zur Frage der atmosphärischen Mondgezeiten," *Meteorologische Zeitschrift* 32 (1915).

98. Wegener to Köppen, 22 Feb. 1915, DMH 1968 599/4 N 1/50, W 006-1915.

99. Wegener to Köppen, 24 Feb. 1915, DMH 1968 599/5 N 1/50, W 007-1915.

100. Alfred Wegener, "Über den Farbenwechsel der Meteore," *Das Wetter*, Sonderheft (Aßmann Festschrift, 1915).

101. Berlin, Minesterium der geistlichen und Unterrichts-Angellegenheiten zu Universitäts-Kurator, Marburg, 30 Oct. 1914, Geheimes Staatsarchiv, Berlin, Abt. Merseberg (Rep 76-Va, Tit. 4, No. 13, Bd 4, 231), W 024-1914.

102. 20 Mar. 1915, Universitäts-Kurator, Marburg, HM Neuruppin Bst. Zechlinerhütte; and E. Wegener, *Alfred Wegener: Tagebücher, Briefe, Erinnerungen*, 148.

103. Wegener to Köppen, 16 Apr. 1915, DMH 1968 599/7 N 1/53.

104. Robert Marc Friedman, *Appropriating the Weather: Vilhelm Bjerknes and the Construction of a Modern Meteorology* (Ithaca, NY: Cornell University Press, 1989), 103–105.

105. Ibid., 133; and Wegener to Köppen, 16 Apr. 1915, DMH 1968 599/7 N 1/53.

106. Friedman, *Appropriating the Weather*, 56.

107. Alfred Wegener, "Astronomische Ortsbestimmungen in Luftballon," *Illustrierte Aeronautische Mitteilungen* 6 (1906).

108. Wegener to Dean of the Faculty, Marburg, 12 May 1915, HSM [307d Nr. 269], W 013-1915.

109. 21 June 1915, Charlottenburg, Militärärtzliches Attest, Dr. R. Friedlander, Staff and Battalion Medical Officer, Replacement Battalion of the Queen Elizabeth Grenadier Guards Regiment #3, AWI, Bremerhaven, Wegenerarchiv.

110. E. Wegener, *Alfred Wegener: Tagebücher, Briefe, Erinnerungen*, 188.

111. 21 June 1915, Charlottenburg, Militärärtzliches Attest.

112. 19 July 1915, Attestation, Dr. Hildebrand, Marburg, AWI Bremerhaven, Wegenerarchiv.

113. Wegener to Köppen, 13 July 1915, DMH 1968 559/8 N 1/54, W 016-1915.

114. Ibid.

115. Wegener to Köppen, 30 July 1915, DMH 1968 599/9 N 1/55, W 018-1915.

116. Wegener to Köppen, 7 Aug. 1915, DMH 1968 599/10 N 1/56, W 019-1915.

117. Wegener to Köppen, Aug. 1915, DMH 1968 599/11 N 1/5, W 020-1915.

118. Wegener to Köppen, 13 July 1915, DMH 1968 559/8 N 1/54, W 016-1915.

119. Wegener to Köppen, 30 July 1915, DMH 1968 599/9 N 1/55, W 018-1915.

120. Wegener to Köppen, 7 Aug. 1915, DMH 1968 599/10 N 1/56, W 019-1915.

121. E. Wegener, *Alfred Wegener: Tagebücher, Briefe, Erinnerungen*, 149.

Chapter 14. The Meteorologist: "In the Field," 1916–1918

1. Wegener to Dekan der philosophischen Facultät, 3 Oct. 1915, HSM [307d Nr. 269], W 021-1915.

2. Wegener to Richarz, 6 Nov. 1915, Phillips-Universität, Marburg, Fachbereich Physik (Prof. W. Walcher), W 022-1915.

3. Wegener to Köppen, DMH 1968 600/1 N 1/59, W 001-1916.

4. Else Wegener, *Alfred Wegener: Tagebücher, Briefe, Erinnerungen* (Wiesbaden: F. A. Brockhaus, 1960), 152.

5. Ibid., 151.

6. A single airport currently serves Mulhouse, France; Basel, Switzerland; and Freiburg, Germany.

7. The book was finished in 1917 but not published until after the war. See Kurt Wegener, *Vom Fliegen* (Charlottenburg: Druckerei der Idflieg, 1918).

8. Alfred Wegener, *Wind- und Wasserhosen in Europa*, Sammlung Wissenschaft Bd. 60 (Braunschweig: Friedrich Vieweg & Sohn, 1917), viii.

9. Wegener to Köppen, 2 Feb. 1916, DMH 1968 600/1 N 1/58, W 001-1916.

10. Wegener to Köppen, 11 Mar. 1916, DMH 1968 600/2 N 1/59, W 002-1916.

11. Wegener to Köppen, 30 Mar. 1916, DMH 1968 600/3 N 1/60, W 003-1916.

12. Wegener, *Wind- und Wasserhosen in Europa*, 58ff.

13. Wegener to Köppen, 6 Apr. 1916, DMH 1968 600/4 N 1/61, W 006-1916.

14. M(aurycy) Rudzki, "Referat über A. Wegener Die Entstehung der Kontinente und Ozeane (Vieweg, 1915)," *Die Naturwissenschaften* 4, no. 2 (1916).

15. Alexander Tornquist, *Geologie. 1. Allgemeine Geologie* (Leipzing: Engelmann, 1916), 30, 511.

16. Franz Koßmat, *Paläogeographie: Geologische Geschichte der Meere und Festländer*, 2nd ed. (Berlin: J. G. Göschen, 1916), 72, 140.

17. Alfred Wegener, *Die Entstehung der Kontinente und Ozeane* (Braunschweig: Friedrich Vieweg & Sohn, 1915), iii.

18. Wegener to Köppen, 6 Apr. 1916, DMH 1968 600/4 N 1/61, W 006-1916.

19. E. Wegener, *Alfred Wegener: Tagebücher, Briefe, Erinnerungen*, 149–150.

20. Aßmann to Ministerium der geistlichen und Unterrichts-Angelegenheiten, 22 Dec. 1915, Geheimes Staatsarchiv Berlin, Abt Merseberg Rep 76-Va, Sect 12, Tit. 4, n 2; 17:129–130, W 023-1915.

21. Alfred Wegener, *Das detonierende Meteor vom 3 April 1916 3 1/2 Uhr nachmittags in Kurhessen*, vol. 14, *Schriften der Gesellschaft zur Beförderung der gesamten Naturwissenschaften zu Marburg* (1917).

22. Ibid., 1.

23. Ibid., 2–4.

24. Ibid., 48.

25. E. Wegener, *Alfred Wegener: Tagebücher, Briefe, Erinnerungen*, 150.

26. Wegener, *Das detonierende Meteor vom 3 April 1916*, 14:66.

27. Ibid., 14:67.

28. Alfred Wegener, "Akustik der Atmosphäre," in *Physik der Erde* [Vol. 5, Part 1 of Müller-Pouillet's *Leherbuch der Physik*, 11th ed.], ed. Alfred Wegener, *Müller-Pouillet's Lehrbuch der Physik* (Braunschweig: Friedrich Vieweg & Sohn, 1928), 184.

29. Wegener, *Das detonierende Meteor vom 3 April 1916*, 14:32–33; see also Julius Bauschinger, *Die Bahnbestimmung der Himmelskörper* (Leipzig: Wilhelm Englemann, 1906), for the techniques of calculation.

30. Wegener, *Das detonierende Meteor vom 3 April 1916*, 14:8.

31. E. Wegener, *Alfred Wegener: Tagebücher, Briefe, Erinnerungen*, 151; and Wegener to Köppen, 7 Aug. 1917, DMH 1968 601/4 N 1/75, W 021-1917.

32. E. Wegener, *Alfred Wegener: Tagebücher, Briefe, Erinnerungen*, 150.

33. Wegener, *Wind- und Wasserhosen in Europa*, 6–7.

34. Wegener to Köppen, 5 July 1916, DMH 1968 600/5 N 1/62, W 009-1916.

35. Wegener to Köppen, 12 July 1916, DMH 1968 600/8 N 1/65, W 010-1916.

36. Ibid.

37. Wegener to Köppen, 19 July 1916, DMH 1968 600/9 N 1/66, W 011-1916.

38. Wegener to Köppen, 5 July 1916, DMH 1968 600/10 N 1/67, W 013-1916.

39. Wegener, *Wind- und Wasserhosen in Europa*, 180.

40. Ibid., 181.

41. Ibid.

42. Ibid., 224–225; boldface in original.

43. Ibid., 293.

44. Ibid., 294–299.

45. Nikolai Dotzek, "Tornadoes in Germany," *Atmospheric Research* 56, nos. 1–4 (2001).

46. Alfred Wegener, "Äußere Hörbarkeits-Zone und Wasserstoff-Sphäre," *Meteorologische Zeitschrift* 33 (1916).

47. Wegener to Köppen, 3 Aug. 1916, DMH 1968 600/10 N 1/67, W 013-1916.

48. Ibid.

49. Wegener to Köppen, 14 Aug. 1916, DMH 1968 600/11 N 1/68, W 015-1916.

50. Wegener to Köppen, 21 Aug. 1916, DMH 1968 600/12 N 1/69, W 017-1916.

51. 20 Sept. 1916, Minister der geistlichen und Unterrichts-Angelegenheiten, Patent als Professor, Heimatmuseum Neuruppin, Bestand Zechlinerhütte, W9, W 020-1916.

52. E. Wegener, *Alfred Wegener: Tagebücher, Briefe, Erinnerungen*, 152.

53. Ibid.

54. Ulrich Wutzke, "Alfred Wegener als Hochschullehrer," *Zeitschrift für geologische Wissenschaften* 25, nos. 5/6 (1997): 558.

55. E. Wegener, *Alfred Wegener: Tagebücher, Briefe, Erinnerungen*, 62.

56. E. Neuhaus, *Die Wolken in Form, Färbung und Lage als lokale Wetterprognose* (Zürich: Orell Füßli, 1914).
57. Ernst Mylius, *Wetterkunde für den Wassersport* (Berlin: Wedekind, 1914).
58. Wegener to Köppen, 20 Jan. 1917, DMH 1968 601/2 N 1/73, W 002-1917.
59. Ibid.
60. F. Richarz to Wegener, 12 Mar. 1917, DMH 1968 601/2 N 1/73, W 004-1917.
61. Wegener, *Das detonierende Meteor vom 3 April 1916*, 14:n.p., insert tag opposite title page.
62. Andreas Lundager, "Kontinenters og Oceaners Opstaaen," *Geografisk Tidskrift* 24 (1917/1918): 68.
63. E. Dacqué to Wegener, 2 May 1917, DMH 1968 603/35, W 010-1917.
64. Derek J. de Solla Price, "Networks of Scientific Papers," *Science* 149, no. 3683 (1965): 513.
65. Wegener to Köppen, 15 May 1917, DMH 1968 601/3 N 1/74, W 014-1917.
66. Ibid.
67. Ibid.
68. Alfred Wegener, "Referat über F.M.Exner: Dynamische Meteorologie, Mit 68 Fig. in Text 8°, X u. 308 S. Leipzing und Berlin 1917 (Teubner)," *Annalen der Hydrographie und Maritimen Meteorologie* 45 (1917).
69. Ibid.
70. Wegener to Köppen, 15 May 1917, DMH 1968 601/3 N 1/74, W 014-1917.
71. Karl Andrée, "Alfred Wegeners Hypothese von Horizontalverschiebung der Kontinentalschollen und das Permanenzproblem in Licht der Pälaogeographie und dynamische Geologie," *Petermanns Geographische Mitteilungen* 63 (1917).
72. E. Mylius to Wegener, 23 May 1917, DMH 1968 601/18 N 1/187, W 015-1917.
73. K. Wegener to A. Wegener, 31 May 1917, DMH 1968 601/6 N 1/111, W 016-1917.
74. Edgar Dacqué, *Grundlagen und Methoden der Paläogeographie* (Jena: Gustav Fischer Verlag, 1915), 181.
75. Ibid., 183–184.
76. Ibid., 171.
77. Andrée, "Alfred Wegeners Hypothese," 50.
78. Ibid., 80.
79. Ibid., 50–53.
80. Ibid., 51.
81. Ibid., 77, 80.
82. Ibid., 53.
83. Ibid., 77.
84. Carl Diener, "Die Großformen der Erdoberfläche," *Mitteililungen der kaiserlich-königlichen geographischen Gesellschaft, Wien* 58 (1915).
85. Max Semper, "Was ist ein Arbeitshypothese?," *Centralblatt für Mineralogie, Geologie, und Paläontologie* (1917).
86. Ibid., 163.
87. Wolfgang Soergel, *Das Problem der Permanenz der Ozeane und Kontinente* (Stuttgart: Schweizerbart, 1917).
88. Wolfgang Soergel, "Die atlantische 'Spalte': Kritische Bemerkungen zu A. Wegeners Theorie von der Kontinentalveschiebung," *Zeitschift der deutschen geologischen Gesellschaft. Monatsberichte* 68 (1916).
89. Edgar Dacqué, "Referat über W. Soergel Das Problem der Permanenz der Ozeane und Kontinente. (Habilitationsvortrag). Stuttgart (Schweizerbart,)1917," *Geographische Zeitschrift* 24 (1918).
90. Andrée, "Alfred Wegeners Hypothese," 81.
91. E. Wegener, *Alfred Wegener: Tagebücher, Briefe, Erinnerungen*, 152.
92. Ibid., 153.
93. Wegener to Köppen, 7 Aug. 1917, DMH 1968 601/4 N 1/75, W 021-1917.
94. E. Wegener, *Alfred Wegener: Tagebücher, Briefe, Erinnerungen*, 153.
95. Wegener to Köppen, 11 and 17 Sept. 1917, DMH 1968 601/5 N 1/76, W 021-1917; and 1968 601/6 N 1/77, W 026-1917.

96. E. Wegener, *Alfred Wegener: Tagebücher, Briefe, Erinnerungen*, 153.

97. Wegener to Köppen, 11 and 19 Sept. 1917, DMH 1968 601/5 N 1/76, W 021-1917; and 1968 601/7 N 1/78, W 027-1917.

98. Wegener to Köppen, 13 Sept. 1917, DMH 1968 602/12 N 1/91, W 024-1917.

99. E. Wegener, *Alfred Wegener: Tagebücher, Briefe, Erinnerungen*, 154.

100. Wegener to Köppen, 1 Jan. 1918, DMH 1968 602/1 N 1/78, W 001-1918.

101. Wegener to Köppen, 19 Jan. 1918, DMH 1968 602/4 N 1/783, W 004-1918.

102. Wegener transcribed Pfeffer's letter; it appears as a part of the interleaved criticisms of his 1915 book, which he prepared sometime in 1919, and has been published online as Jutta Voss-Diestelkamp and Reinhard A. Krause, "Transkription der handschriftlichen Bemerkungen in Alfred Wegener: Die Entstehung der Kontinente und Ozeane. 1 Auflage 1915," Alfred Wegener Institute, www.awi.de/fileadmin/user_upload/Discover/History_of_Polar_Research/Famous_Scientists/TranskriptionNotizen_EntstehungderKontinenteundOzeane.pdf.

103. Wegener to Köppen, 25 Dec. 1917, DMH 1968 601/8 N 1/79, W 034-1917.

104. Ibid.

105. Ibid.

106. Wegener to Köppen, 9 Jan. 1918, DMH 1968 602/2 N 1/81, W 002-1918.

107. Wegener to Köppen, 21 Mar., 17 Apr., 10 May 1918, DMH 1968 602/7 N 1/86, W 011-1918; 1968 602/8 N 1/87, W 014-1918; 1968 602/9 N 1/88, W 017-1918.

108. Wegener to Köppen, 16 and 21 Mar., 17 Apr. 1918, DMH 1968 602/6 N 1/85, W 010-1918; 1968 602/7 N 1/86, W 011-1918; 1968 602/8 N 1/87, W 014-1918.

109. Wegener to Köppen, 17 Apr. 1918, DMH 1968 602/8 N 1/87, W 014-1918.

110. E. Wegener, *Alfred Wegener: Tagebücher, Briefe, Erinnerungen*, 156.

111. Ibid., 155.

112. 17 May 1918, Philosophischen Facultät der Universität, HSM [307d Nr. 269], W 018-1918.

113. 18 July 1918, Marburg, Rektor der Königlichen Universität, Heimatmuseum Neuruppin, Bestand Zechlinerhütte, W7.

114. E. Wegener, *Alfred Wegener: Tagebücher, Briefe, Erinnerungen*, 155.

115. Wutzke, "Alfred Wegener als Hochschullehrer," 559–561.

116. Kurt Wegener to Alfred Wegener, 26 Oct. 1918, DMH 1968 602/31 N 1/114, W 037-1918.

117. Köppen to Wegener, 6 Sept. 1918, DMH 1968 602/23 N 1/7107, W 029-1918.

118. Wegener to Carl Becker, 22 Sept. 1918, Gesamt Staatsarchiv Berlin I, HA, Rep. 92, C. H. Becker Nr. 4961, W 032-1918.

119. Wegener to Köppen, 1 Nov. 1918, DMH 1968 602/13 N 1/92, W 038-1918.

120. Kurt Wegener to Alfred Wegener, 26 Oct. 1918, DMH 1968 602/31 N 1/114, W 037-1918.

121. Wutzke, "Alfred Wegener als Hochschullehrer," 561.

Chapter 15. The Geophysicist: Hamburg, 1919–1920

1. Ulrich Wutzke, "Alfred Wegener als Hochschullehrer," *Zeitschrift für geologische Wissenschaften* 25, nos. 5/6 (1997): 559.

2. Universitätskurator, Marburg to Wegener, 25 Jan. 1919, HSM [307d Nr. 269], W 001-1919.

3. Alfred Wegener, *Die Entstehung der Kontinente und Ozeane* (Braunschweig: Friedrich Vieweg & Sohn, 1915), 52.

4. Else Wegener, *Alfred Wegener: Tagebücher, Briefe, Erinnerungen* (Wiesbaden: F. A. Brockhaus, 1960), 157. This letter does not survive.

5. Ibid., 157–158.

6. Else Wegener-Köppen, *Wladimir Köppen: Ein Gelehrtenleben für die Meteorologie*, ed. H. W. Frickhinger, Grosse Naturforscher (Stuttgart: Wissenschaftliche Verlagsgesellschaft, 1955), 87.

7. Ibid.

8. E. Wegener, *Alfred Wegener: Tagebücher, Briefe, Erinnerungen*, 158.

9. Ibid.

10. Ibid.

11. Letter from Wladimir Köppen to Else Wegener, cited in ibid. This letter does not survive.

12. Ibid.

13. Alfred Wegener, "Über die planmäßige Auffindung des Meteoriten von Treysa," *Astronomische Nachrichten* 207 (1918); Alfred Wegener, "Einige Hauptzüge aus der Natur der Tromben," *Meteorologische Zeitschrift* 35 (1918); Alfred Wegener, "Elementare Theorie der atmosphärischen Spiegelungen," *Annalen der Physik* 57 (1918).

14. Alfred Wegener, "Versuche zur Aufsturtz-Theorie der Mondkrater," *Nova Acta, Abhandlung der Kaiserlichen Leopoldinisch- Carolinischen Deutschen Akademie der Naturforscher* 106 (1920).

15. Alfred Wegener, "1. Über Luftwiderstand bei Meteoren. 2. Versuche zur Aufsturtz-Theorie der Mondkrater," *Sitzungsberichte der Gesellschaft der gesamten Naturwissenschaften zu Marburg* (1919).

16. Mott T. Greene, "Alfred Wegener and the Origin of Lunar Craters," *Earth Sciences History* 17, no. 2 (1998): 118–120.

17. Alfred Wegener, "Die Entstehung der Kontinente," *Petermanns Mitteilungen* 58 (1912): 185.

18. Alfred Wegener, *Die Entstehung der Mondkrater* (Braunschweig: Friedrich Vieweg & Sohn, 1921), 26.

19. Ibid., 7.

20. Ibid.

21. Ibid., 8.

22. Ibid.

23. Ibid., 11–14.

24. Ibid.

25. Ibid., 23.

26. G. K. Gilbert, "The Moon's Face: A Study of the Origin of Its Features," *Bulletin of the Philosophical Society of Washington* 12 (1893).

27. Hermann Ebert, "Über die Ringgebirge des Mondes," *Sitzungsberichte der Physikalische-Medizinische Soc. Erlangen* (1890).

28. Wegener, *Die Entstehung der Mondkrater*, 26.

29. Ibid., 31, 128.

30. Greene, "Alfred Wegener and the Origin of Lunar Craters."

31. Wegener, *Die Entstehung der Mondkrater*, 37.

32. Greene, "Alfred Wegener and the Origin of Lunar Craters," 132; Wegener, *Die Entstehung der Mondkrater*, 40.

33. Wegener, "1. Über Luftwiderstand bei Meteoren. 2. Versuche zur Aufsturtz-Theorie der Mondkrater."

34. E. Wegener, *Alfred Wegener: Tagebücher, Briefe, Erinnerungen*, 153.

35. Ibid., 161, 166.

36. Paul Pummerer, "Zur Erweiterung des tägliches Wetterberichtes der Deutschen Seewarte," *Annalen der Hydrographie und Maritimen Meteorologie* 50, no. 1 (1922): 21.

37. Ibid.

38. E. Wegener, *Alfred Wegener: Tagebücher, Briefe, Erinnerungen*, 166.

39. Ibid.; and Kurt Wegener, *Vom Fliegen* (Charlottenburg: Druckerei der Idflieg, 1918).

40. Wegener to Dean of the University, Hamburg, 18 July 1919, Staatsarchiv, Universität Hamburg, Mathematisches-Wissenschafliches Klasse Nr. 238. I am indebted to Walter Lenz of Hamburg University for this otherwise unknown letter, which contains news of the refusal of the call from Karlsruhe.

41. E. Wegener, *Alfred Wegener: Tagebücher, Briefe, Erinnerungen*, 167.

42. Notiz über ein Segeltour vom 27 Aug.–6 Sept. 1919, Heimatmuseum Neuruppin, Bestand Zechlinerhütte; and Ulrich Wutzke, *Durch die weiße Wüste: Leben und Leistungen des Grönlandforschers und Entdeckers der Kontinentaldrift Alfred Wegener*, Petermann ed. (Gotha: Justus Perthes Verlag, 1997), 142.

43. Dean of the University to Alfred Wegener, 27 Sept. 1919, Staatsarchiv, Universität Hamburg, Mathematisches-Wissenschafliches Klasse Nr. 238.

44. E. Wegener, *Alfred Wegener: Tagebücher, Briefe, Erinnerungen*, 165.

45. Alfred Wegener, *Die Entstehung der Kontinente und Ozeane*, ed. Eilhard Wiedemann, 3rd (completely revised) ed. (Braunschweig: Friedrich Vieweg & Sohn, 1922), 55.

46. Edgar Irmscher, *Pflanzenverbreitung und Entwicklung der Kontinente: Studien zur genetischen Pflanzengeographie* (Hamburg: Institut für allgemeine Botanik, 1922), 217.

47. In 2005 the Alfred Wegener Institute in Bremerhaven published in facsimile and in transcription two long-lost research notebooks belonging to Alfred Wegener. These were presented to the institute by Wegener's grandson Dr. Günther Schönharting, of Eichhofen, Bavaria. These consist of a 4″×8″ quadrille-ruled hardbound notebook with the words "Continental-Displacements" written on the cover, containing research notes, and a rebound copy of the first edition of Wegener's book *Die Entstehung der Kontinente und Ozeane* (1915), with interleaved blank sheets containing detailed references to comments and criticisms concerning Wegener's book. There is no doubt of the originality of these documents, as they are clearly in Wegener's distinctive hand. See Jutta Voss-Diestelkamp and Reinhard A. Krause, "Transkription der handschriftlichen Bemerkungen in Alfred Wegener: Die Entstehung der Kontinente und Ozeane. 1 Auflage 1915," Alfred Wegener Institute, www.awi.de/fileadmin/user_upload/Discover/History_of_Polar_Research/Famous_Scientists/TranskriptionNotizen_EntstehungderKontinenteundOzeane.pdf; and Reinhard Krause and Jörn Theide, eds., *Kontinental-Verschiebungen: Originalnotizen und Literaturauszüge*, vol. 516, *Berichte zur Polar- und Meeresforschung* (Bremerhaven: Alfred-Wegener-Institut für Polar- und Meeresforschung, 2005).

48. Wladimir Köppen, "Über Isostasie und die Entstehung der Kontinente," *Verhandlungen des Naturwissenschaften Vereins in Hamburg* 26 (1918).

49. Ibid., 43.

50. Ibid.

51. W[ladimir] Köppen, "Über Isostasie und die Entstehung der Kontinente," *Geographische Zeitschrift* 25 (1918).

52. Wutzke, "Alfred Wegener als Hochschullehrer," 561.

53. E. Wegener, *Alfred Wegener: Tagebücher, Briefe, Erinnerungen*, 161.

54. Ibid., 167.

55. J(ohannes) Georgi, "Memories of Alfred Wegener," in *Continental Drift*, ed. Stanley Keith Runcorn, *International Geophysics Series* (New York: Academic Press, 1962), 310.

56. Alfred Wegener, *Die Entstehung der Kontinente und Ozeane*, 2nd (completely revised) ed., Die Wissenschaft (Braunschweig: Friedrich Vieweg & Sohn, 1920), vi.

57. Ibid.

58. E. Wegener, *Alfred Wegener: Tagebücher, Briefe, Erinnerungen*, 162.

59. Thomas C. Chamberlin and Rollin D. Salisbury, *Geology*, 2nd ed., 3 vols. (New York: Holt, 1905–1907).

60. Oswald Heer, *Die Urwelt der Schweiz* (Zurich: Verlag von Friedrich Schulthess, 1865).

61. Wladimir Köppen, *Versuch einer Klassifikation der Klimate: Vorzugsweise nach iheren Beziehungen zur Pflanzenwelt* (Leipzig: B. G. Teubner, 1901).

62. John P. Snyder, *Flattening the Earth: Two Thousand Years of Map Projections* (Chicago: University of Chicago Press, 1993), 76.

63. Alfred Wegener, "Ueber die Entwicklung der kosmischen Vorstellungen in Der Philosophie," *Mathematisch-Naturwissenschaftliche Blätter, Berlin* 3 (1906): 82.

64. Köppen, "Über Isostasie und die Entstehung der Kontinente," 47.

65. Edgar Dacqué, *Grundlagen und Methoden der Paläogeographie* (Jena: Gustav Fischer Verlag, 1915), 302–375.

66. Ibid., 371.

67. Ibid., 375.

68. Wegener, *Die Entstehung der Kontinente und Ozeane* (1920), vi.

69. Wegener to Cloos, 13 Jan. 1920, Universität Freiburg, Geologen Archiv, Geol. Vereinigung 11872, W 001-1920; Wegener to Cloos, 26 June 1920, and 13 July 1920, AWI Bremerhaven Wegenerarchiv, W 002, 003-1920.

70. Andreas Galle, "Entfernen sich Europa und Nordamerika voneinander?," *Deutsche Revue* 41, no. 1 (1916).

71. Heinrich Schmitthenner, "Dacqué, E. Geographie der Vorwelt," *Geographische Zeitschrift* 25 (1919): 279.

72. Wegener, "Elementare Theorie der atmosphärischen Spiegelungen."
73. Wegener, *Die Entstehung der Kontinente und Ozeane* (1920), 10.
74. Ibid.
75. Ibid.
76. Ibid.
77. Ibid., 11.
78. Ibid.
79. Ibid., 13.
80. Ibid., 15.
81. Ibid.
82. Ibid., 16–21.
83. Ibid., 12
84. Ibid., vi.
85. Richard Hartshorne, "The Nature of Geography: A Survey of Current Thought," *Annals of the Association of American Geographers* 29, nos. 3–4 (1939): 286.
86. Georg Gerland, "Vorwart der Herausgebers," *Gerlands Beiträge zur Geophysik* 1, no. 1 (1887): xxxv.
87. Bailey Willis, "A Theory of Continental Structure Applied to North America," *Geological Society of America, Bulletin* 18 (1907): 392.
88. Wegener, *Die Entstehung der Kontinente und Ozeane* (1920), chap. 3 passim.
89. Ibid., 38–39.
90. Ibid., 31.
91. Ibid., 33.
92. Ibid., 34.
93. Ibid., 35.
94. Ibid.
95. Ibid., 54.
96. Ibid., 58.
97. Ibid.
98. Ibid.
99. Ibid.
100. Ibid., 92.
101. Ibid., 62.
102. Ibid.
103. Theodor Arldt, *Handbuch der Paläogeographie*, 2 vols. (Leipzig: Gebrüder Borntraeger, 1917–1922).
104. Wegener, *Die Entstehung der Kontinente und Ozeane* (1920), 64; original in Arldt, *Handbuch der Paläogeographie*, 1:278–281.
105. Wegener, *Die Entstehung der Kontinente und Ozeane* (1920), 67.
106. Voss-Diestelkamp and Krause, "Transkription der handschriftlichen Bemerkungen in Alfred Wegener," 165.
107. Wegener, *Die Entstehung der Kontinente und Ozeane* (1920), 92.
108. Ibid., 101.
109. Ibid., 93.
110. Ibid., 94–95.
111. E. Wegener, *Alfred Wegener: Tagebücher, Briefe, Erinnerungen*, 162.
112. Wegener, *Die Entstehung der Kontinente und Ozeane* (1920), 100; and Dacqué, *Grundlagen und Methoden der Paläogeographie*, 432.
113. Emanuel Kayser, *Lehrbuch der Geologie*, 5th rev. ed., 2 vols. (Stuttgart: Ferdinannd Enke, 1918), ii.
114. Wegener, *Die Entstehung der Kontinente und Ozeane* (1920), 106–107.
115. Ibid., 118.
116. Ibid., 119.
117. Ibid.

118. Ibid., 120.

119. Ibid.

120. W[ladimir] Köppen, "Ursachen und Wirkungen der Kontinentalverschiebungen und Polwanderungen," *Petermanns Geographische Mitteilungen* 67 (1921): 145.

121. Wegener, *Die Entstehung der Kontinente und Ozeane* (1920), 121.

122. Ibid., 122–124.

123. Ibid., 119.

124. Ibid., 125.

125. Ibid., 126.

126. Alfred Wegener, "Referat über J.P. Koch: Nordgrönlands Trift nach Westen," *Astronomische Nachrichten* 208 (1919).

127. Wegener, *Die Entstehung der Kontinente und Ozeane* (1920), 126–127.

128. Ibid., 130.

129. Köppen, "Ursachen und Wirkungen der Kontinentalverschiebungen und Polwanderungen."

130. Wutzke, "Alfred Wegener als Hochschullehrer," 561.

Chapter 16. From Geophysicist to Climatologist: Hamburg, 1920–1922

1. Else Wegener-Köppen, *Wladimir Köppen: Ein Gelehrtenleben für die Meteorologie*, ed. H. W. Frickhinger, Grosse Naturforscher (Stuttgart: Wissenschaftliche Verlagsgesellschaft, 1955), 14, 24–25.

2. W[ladimir] Köppen, "Die Wärmezonen der Erde, nach der Dauer der heissen, gemässigten und kalten Zeit, und nach der Wirkung der Wärme auf die organische Welt betrachtet," *Meteorologische Zeitschrift* 1 (1884).

3. Wladimir Köppen, *Versuch einer Klassifikation der Klimate: Vorzugsweise nach iheren Beziehungen zur Pflanzenwelt* (Leipzig: B. G. Teubner, 1901).

4. Wladimir Köppen, *Klimakunde I. Allgemeine Klimalehre*, [1st ed. 1899, 2nd ed. 1906, repub. 1911] with a new afterword ed. (Leipzig: Göschen, 1918).

5. C. E. P. Brooks, *Climate through the Ages: A Study of the Climatic Factors and Their Variations* (New York: R. V. Coleman, 1926). This contemporary volume is a standard source for the range of climate theories, with interesting commentary.

6. Else Wegener, *Alfred Wegener: Tagebücher, Briefe, Erinnerungen* (Wiesbaden: F. A. Brockhaus, 1960), 168.

7. Robert Marc Friedman, *Appropriating the Weather: Vilhelm Bjerknes and the Construction of a Modern Meteorology* (Ithaca, NY: Cornell University Press, 1989), 196.

8. Alfred Wegener, *Die Entstehung der Kontinente und Ozeane*, 2nd (completely revised) ed., Die Wissenschaft (Braunschweig: Friedrich Vieweg & Sohn, 1920), 111 and 115.

9. W[ladimir] Köppen, "Polwanderung, Verschiebungen der Kontinente und Klimageschichte," *Petermanns Geographische Mitteilungen* 67 (1921). The article did not appear until January–March 1921 but was written in the summer of 1920.

10. Wegener-Köppen, *Wladimir Köppen*, 182–185.

11. Köppen, "Polwanderung, Verschiebungen der Kontinente und Klimageschichte," 2. The quotation from Rudzki is from M[aurycy] P. Rudzki, *Physik der Erde* (Leipzig: Chr. Herm. Tachnitz, 1911), 209.

12. E. Wegener, *Alfred Wegener: Tagebücher, Briefe, Erinnerungen*, 167.

13. Ibid., 166.

14. Friedman, *Appropriating the Weather*, 199.

15. Ibid. For a more detailed treatment of the contraction of science in Austria see Deborah R. Coen, *Vienna in the Age of Uncertainty* (Chicago: University of Chicago Press, 2007).

16. Tor Bergeron, "Some Autobiographical Notes in Connection with the Ice Nucleus Theory of Precipitation Release," *Bulletin of the American Meteorological Society* 59, no. 4 (1972): 390.

17. E. Wegener, *Alfred Wegener: Tagebücher, Briefe, Erinnerungen*, 165.

18. Eric Mills, *The Fluid Envelope of Our Planet: How the Study of Ocean Currents Became a Science* (Toronto: University of Toronto Press, 2009). Chapter 5 contains an excellent discussion of this dispute in its contemporary context.

19. Karl Frisch, "Die Inversionflächen in der freien Atmosphäre," *Annalen der Hydrographie und Maritimen Meteorologie* 50 (1922).

20. Else remembered that it was Ångstrom, but he had been dead forty-five years by this time.

21. E. Wegener, *Alfred Wegener: Tagebücher, Briefe, Erinnerungen*, 165.

22. Alfred Wegener, "Frostübersättigung und Cirren," *Meteorologische Zeitschrift* 37 (1920); and Alfred Wegener, "Turbulenz und Kolloidstruktur der Atmosphäre," *Meteorologische Zeitschrift* 37 (1920).

23. Ulrich Wutzke, "Alfred Wegener als Hochschullehrer," *Zeitschrift für geologische Wissenschaften* 25, nos. 5/6 (1997): 561.

24. Wegener to Cloos, 13 July 1920, and 1 October 1920, AWI Bremerhaven Wegenerarchiv, W 003, 006-1920.

25. Köppen, "Polwanderung, Verschiebungen der Kontinente und Klimageschichte," 6–8.

26. Ibid., 6.

27. Damian Kreichgauer, *Die Äquatorfrage in der Geologie* (Steyl: Verlag Missiondruckerei, 1902), figure on title page.

28. David Greenhood, *Mapping* (Chicago: University of Chicago Press, 1964), 166.

29. Köppen, "Polwanderung, Verschiebungen der Kontinente und Klimageschichte," 2–3.

30. Alfred Wegener, "Versuche zur Aufsturtz-Theorie der Mondkrater," *Nova Acta, Abhandlung der Kaiserlichen Leopoldinisch- Carolinischen Deutschen Akademie der Naturforscher* 106 (1920); and Alfred Wegener, "Die Aufsturzhypothese der Mondkrater," *Sirius* 53 (1920).

31. Alfred Wegener, *Die Entstehung der Mondkrater* (Braunschweig: Friedrich Vieweg & Sohn, 1921), 13.

32. Ibid., 19.

33. Ibid., 8.

34. Ibid., 21ff.

35. Ibid., 18–19.

36. Alfred Wegener, "Die Entstehung der Mondkrater," *Naturwissenschaften* 9 (1921); and Alfred Wegener, "Das Antlitz des Mondes," *Umschau* 25 (1921).

37. Franz Koßmat, *Paläogeographie: Geologische Geschichte der Meere und Festländer*, 2nd ed. (Berlin: J. G. Göschen, 1916), 72.

38. Anonymous, "Deutsche Meteorologische Gesellschaft (Berliner Zweigverein.) Nachrichten," *Die Naturwissenschaften* 9, no. 18 (1921).

39. Alfred Wegener, "Die Theorie der Kontinentalverschiebungen," *Zeitschrift der Gesellschaft für Erdkunde zu Berlin*, nos. 1, 2 (1921), 90.

40. Ibid., 92.

41. Ibid., 96.

42. Ibid., 98.

43. Ibid., 101–103.

44. Ibid., 103.

45. Ibid.

46. Franz Koßmat, "Erörterungen zu A. Wegeners Theorie der Kontinentalverschibungen," *Zeitschift der Gesellschaft für Erdkunde zu Berlin*, nos. 1, 2 (1921), 104.

47. Ibid., 105.

48. Ibid., 109.

49. Ibid., 106.

50. Ibid., 110.

51. Albrecht Penck, "Wegeners Hypothese der kontinentalen Verschiebungen," *Zeitschrift der Gesellschaft für Erdkunde zu Berlin*, nos. 1, 2 (1921), 111.

52. Ibid., 112.

53. Ibid., 113, 117.

54. Ibid.

55. Ibid., 115–117.

56. Ibid., 119.

57. Ibid., 120.

58. Wilhelm Schweydar, "Bemerkungen zu Wegeners Hypothese der Verschiebung der Kontinente," *Zeitschrift der Gesellschaft für Erdkunde zu Berlin*, nos. 1, 2 (1921), 121.

59. Ibid., 123.

60. Ibid.

61. Ibid., 125.

62. Alfred Wegener, "Schlußwort," *Zeitschrift der Gesellschaft für Erdkunde zu Berlin*, nos. 1, 2 (1921), 125.

63. Ibid., 127.

64. Ibid.

65. Ibid.

66. Ibid., 128.

67. Ibid., 130.

68. Otto Baschin, "Gesellschaft für Erdkunde zu Berlin, Nachrichten," *Die Naturwissenschaften* 9, no. 18 (1921): 219–220.

69. Bruno Schulz, "Die Alfred Wegnerische Theorie der Entstehung der Kontinente und Ozeane," *Die Naturwissenschaften* 15 (1921): 250.

70. Reinhard Krause and Jörn Theide, eds., *Kontinental-Verschiebungen: Originalnotizen und Literaturauszüge*, vol. 516, *Berichte zur Polar- und Meeresforschung* (Bremerhaven: Alfred-Wegener-Institut für Polar- und Meeresforschung, 2005).

71. J(ohannes) Georgi, "Memories of Alfred Wegener," in *Continental Drift*, ed. Stanley Keith Runcorn, *International Geophysics Series* (New York: Academic Press, 1962), 315.

72. Alfred Wegener, "Die aerologischen Flugzeugaufstiege der Deutschen Seewarte im Jahre 1921," *Annalen der Hydrographie und Maritimen Meteorologie* 50 (1922): 113–120.

73. Wutzke, "Alfred Wegener als Hochschullehrer," 561.

74. Georgi, "Memories of Alfred Wegener," 315.

75. Otto Krümmel, *Handbuch der Ozeanographie*, 2nd rev. ed., vols. 1 and 2 (Stuttgart: Verlag von J. Engelhorn, 1907, 1911). Vol. 1, p. 13 cited in W[ladimir] Köppen, "Ursachen und Wirkungen der Kontinentalverschiebungen und Polwanderungen," *Petermanns Geographische Mitteilungen* 67 (1921): 147–148.

76. Köppen, "Ursachen und Wirkungen der Kontinentalverschiebungen und Polwanderungen," 194.

77. It is consequential because it determines the length and intensity of the seasons of the year. If Earth's axis were exactly perpendicular to the plane of the orbit around the Sun, there would be no seasons at all, because all places on Earth would receive exactly the same amount of sunlight all year round; the present obliquity (inclination) is approximately 23.5°.

78. Alfred Wegener, "Die Theorie der Kontinentalverschiebungen," *Verhandlungen der 20. Deutschen Geographentag* 20 (1921); W[ladimir] Köppen, "Über die Kräfte, welche die Kontinentalverschiebungen und Polwanderungen Bewirken," *Geologische Rundschau* 11 (1921).

79. Ulrich Wutzke, *Durch die weiße Wüste: Leben und Leistungen des Grönlandforschers und Entdeckers der Kontinentaldrift Alfred Wegener*, Petermann ed. (Gotha: Justus Perthes Verlag, 1997), 147.

80. W[ladimir] Köppen, *Die Klimate der Erde* (Berlin: Walter de Gruyter, 1923), iii–vi.

81. Krause and Theide, *Kontinental-Verschiebungen*, 1–311.

82. Alfred Wegener, *Die Entstehung der Kontinente und Ozeane*, 3rd (completely revised) ed., Die Wissenschaft (Braunschweig: Friedrich Vieweg & Sohn, 1922), v.

83. Mathematisch- Naturwissenschaftliche Facultät der Hamburgischen Universität to Wegener, 6 Aug. 1921, HM Neuruppin, Bestand Zechlinerhütte, W 13, W 003-1921.

84. Wutzke, "Alfred Wegener als Hochschullehrer," 561–562.

85. Wegener, *Die Entstehung der Kontinente und Ozeane* (1922), v.

86. Ibid., 1.

87. Ibid., 1–2.

88. Ibid., 6, caption to fig. 1.

89. Ibid., 8.

90. Ibid., 12.

91. Ibid., 13–14.

92. Ibid., 14.
93. Ibid., 18–19.
94. Alfred Wegener, "The Origin of Lunar Craters," *Moon* 14 (1975).
95. Wegener, *Die Entstehung der Kontinente und Ozeane* (1922), 11–12.
96. Ibid., 12.
97. Ibid., 105.
98. Ibid., 106.
99. "Tested," not "proved," as Skerl translated it.
100. Friedrich Burmeister, "Die Verschiebung Grönlands nach der astronomischen Längebestimmungen," *Petermanns Geographische Mitteilungen* 67 (1921): 226–227.
101. Wegener, *Die Entstehung der Kontinente und Ozeane* (1922), 80–81.
102. Ibid., 82.
103. Ibid.
104. Köppen, "Über die Kräfte, welche die Kontinentalverschiebungen und Polwanderungen Bewirken," 149.
105. Wegener, *Die Entstehung der Kontinente und Ozeane* (1922), 132.
106. P. S. Epstein, "Über die Polflucht der Kontinente," *Die Naturwissenschaften* 9, no. 25 (1921): 502.
107. Walter D. Lambert, "Some Mechanical Curiosities Connected with the Earth's Field of Force," *American Journal of Science* 2 (1921): 137n.
108. Ibid., 138.
109. Ibid.
110. Epstein, "Über die Polflucht der Kontinente," 502.
111. Schweydar, "Bemerkungen zu Wegeners Hypothese der Verschiebung der Kontinente," 125.
112. Lambert, "Some Mechanical Curiosities," 157–158.
113. Wegener, *Die Entstehung der Kontinente und Ozeane* (1922), 136.
114. Ibid., v, 75.
115. Krause and Theide, *Kontinental-Verschiebungen*, 4.
116. Wegener, *Die Entstehung der Kontinente und Ozeane* (1922), 62.
117. Ibid.
118. Ibid.
119. Ibid., 64–65.
120. Ibid., 66.
121. Ibid., 67.
122. Hamburg Seemansamt, Anmusterung als Proviantmeister auf der "Sachsenwald," AWI Bremerhaven Wegenerarchiv, W 001-1922.
123. Wutzke, *Durch die weiße Wüste*, 149.

Chapter 17. The Paleoclimatologist: Hamburg, 1922–1924

1. Antonio Alzate (1737–1799) was a priest, cartographer, and meteorologist. Sociedad Cientifica Antonio Alzate, certificate of corresponding membership for A. Wegener, Wegenerarchiv, Alfred Wegener Institute, Bremerhaven.
2. This summary is from Ulrich Wutzke, *Durch die weiße Wüste: Leben und Leistungen des Grönlandforschers und Entdeckers der Kontinentaldrift Alfred Wegener*, Petermann ed. (Gotha: Justus Perthes Verlag, 1997), 148ff., from the fuller account in Erich Kuhlbrodt and Alfred Wegener, "Pilotballonaufstiege auf einer Fahrt nach Mexico, März bis Juni 1922," *Archiv der Deutschen Seewarte* 30, no. 4 (1922).
3. Else Wegener, *Alfred Wegener: Tagebücher, Briefe, Erinnerungen* (Wiesbaden: F. A. Brockhaus, 1960), 167.
4. Ibid., 168.
5. Ibid., 168–169.
6. Anonymous, "Meteorological Meeting on the Sonnblick," *Bulletin of the American Meteorological Society* 3, no. 5 (1922): 78.
7. E. Wegener, *Alfred Wegener: Tagebücher, Briefe, Erinnerungen*, 168.

8. Ibid.

9. Ibid., 169.

10. Ibid., 170.

11. Ibid., 171.

12. Ibid.

13. Bundesministerium fur Unterricht to Wegener, 20 Mar. 1924, Osterreichisches Staatsarchiv Wien 5 Graz Philos. 6738-I 1924, W 004-1924; see also Helmut W. Flügel, *Alfred Wegeners Vertraulicher Bericht über die Grönland-Expedition 1929* (Graz, Austria: Akademische Druck und Verlagsanstalt, 1980).

14. Alfred Wegener, "Het onstaan van de Kraters op de Maan," *Wetenschapplijke Bladen* 2 (1922). The Danish lectures were published twice; see also Alfred Wegener, *Tre Foredrag Holdte i Danmarks Naturvideskabelige Samfund 1922. I. Kontinenternes Forskydning. II. Jordskorpens Natur. III. Fortidens Klimater* (Copenhagen: Danmarks Naturvidenskabelige Samfund, 1923).

15. Alfred Wegener, *Die Entstehung der Kontinente und Ozeane*, 3rd (completely revised) ed., Die Wissenschaft (Braunschweig: Friedrich Vieweg & Sohn, 1922), v.

16. Alfred Wegener, "Die Klimate der Vorzeit," *Deutsche Revue* 47, no. 4 (1922): 40.

17. Milutin Milankovitch, *Milutin Milankovic 1879–1958*, ed. Vasko Milankovic (Katlenburg-Lindau: European Geophysical Society, 1995), 40.

18. John Imbrie and Katherine Imbrie, *Ice Ages: Solving the Mystery* (Short Hills, NJ: Enslow, 1979), 102–103.

19. Milutin Milankovitch, *Théorie mathématique des phénomènes thermiques produits par the raditation solaire* (Paris: Gauthier-Villars, 1920).

20. Milankovitch, *Milutin Milankovic 1879–1958*, 61.

21. Ibid. The holograph letter from Köppen appears as "Annex 1" at the end of this volume.

22. Ibid., 66.

23. Ibid.

24. Ibid., 67.

25. Wladimir Köppen and Alfred Wegener, *Die Klimate der geologischen Vorzeit* (Berlin: Gebrüder Bornträger, 1924), 2.

26. Ibid., 3.

27. Wegener, *Die Entstehung der Kontinente und Ozeane*, 67ff.

28. Köppen and Wegener, *Die Klimate der geologischen Vorzeit*, 3.

29. Henry Potonié and Walther Gothan, *Lehrbuch der Paläobotanik*, 2nd rev. ed. (Berlin: Gebrüder Borntraeger, 1921), 41ff.; see also Henry Potonié, *Die Entstehung der Steinkohle und verwandter Bildungen einschliesslish des Petroleums*, 4th ed. (Berlin: Gebrüder Borntraeger, 1907).

30. Köppen and Wegener, *Die Klimate der geologischen Vorzeit*, 8.

31. Ibid., 8–9.

32. For an interesting account see James F. Schopf, "Coal, Climate and Global Tectonics," in *Implications of Continental Drift to the Earth Sciences*, ed. D. H. Tarling and S. K. Runcorn (London: Academic Press, 1973).

33. Franz Lotze, *Steinsalze und Kalisalze Geologie*, vol. 3.1, *Die Wichtigsten Lagerstätten der "Nicht-Erze"* (Berlin: Gebrüder Borntraeger, 1938).

34. F. Solger, "Geologie der Dünen," in *Dünenbuch*, ed. F. Solger et al. (Stuttgart: Ferdinand Enke, 1910), 172.

35. Ibid., 31ff.

36. Johannes Walther, *Das Gesetz der Wüstenbildung in Gegenwart und Vorzeit* (Leipzig: Quelle & Meyer, 1912), 296–299; see also Johannes Walther, *Geschichte der Erde und des Lebens* (Leipzig: Verlag von Veit & Comp., 1908).

37. Albrecht Penck and Eduard Brückner, *Die Alpen im Eiszeitalter*, 3 vols. (Leipzig: Chr. Herm. Tauchnitz, 1909); see also Albrecht Penck, "Attempt at a Classification of Climate on a Physiographic Basis [Versuch einer Klimaklassification auf physiographischer Grundlage, 1906]," in *Climatic Geomorphology*, ed. Edward Derbyshire (New York: Harper & Row, 1973).

38. Josef Ottokar Freiherrn von Buschmann, *Das Salz: Dessen Vorkommen und Verwertung in sämtlichen Staaten der Erde*, 2 vols. (Leipzig: Wilhelm Engelmann, 1906–1909).

39. Ibid., 1:38.

40. George W. ["G. W. L."] Lamplugh, "Review of *Handbuch der Regionalen Geologie: The British Isles*. Heft 20 iii, Band I, 1917 ed. J.W. Gregory," *Nature* 105 (1921): 356.

41. Kurt Leuchs, *Zentralasien*, Heft 19 (vol. 5, pt. 7), *Handbuch der Regionalen Geologie* (Heidelberg: Carl Winter—Universitätsverlag, 1916); Robert Douvillé, *Espagne*, Heft 7 (vol. 3, pt. 3), *Handbuch der Regionalen Geologie* (Heidelberg: Carl Winter—Universitätsverlag, 1911); Otto Nordenskjöld, *Die nordatlantischen Polarinseln*, Heft 24 (vol. 4, pt. 2b), *Handbuch der Regionalen Geologie* (Heidelberg: Carl Winter—Universitätsverlag, 1921); Otto Nordensklöld, *Antarctis*, vol. 8, pt. 6, *Handbuch der Regionalen Geologie* (Heidelberg: Carl Winters Universitätsbuchhandlung, 1913).

42. Lamplugh, "Review of *Handbuch der Regionalen Geologie*," 356.

43. Nordensklöld, *Antarctis*, 14.

44. Köppen and Wegener, *Die Klimate der geologischen Vorzeit*, 55–93.

45. Ibid., 55.

46. Ibid., 65.

47. Ibid., 80.

48. Jutta Voss-Diestelkamp and Reinhard A. Krause, "Transkription der handschriftlichen Bemerkungen in Alfred Wegener: Die Entstehung der Kontinente und Ozeane. 1 Auflage 1915," Alfred Wegener Institute, www.awi.de/fileadmin/user_upload/Discover/History_of_Polar_Research/Famous_Scientists/TranskriptionNotizen_EntstehungderKontinenteundOzeane.pdf.

49. Joseph Barrell, "The Status of Hypotheses of Polar Wanderings," *Science* 40, no. 1027 (1914): 320.

50. Charles Darwin, *The Descent of Man and Selection in Relation to Sex* (London: John Murray, 1871), 385.

51. Philip Lake, "Wegener's Displacement Theory," *Nature* 110, no. 2750 (1922): 77.

52. Alfred Wegener, "The Origin of Continents and Oceans," *Discovery* 3, no. 5 (1922).

53. Lake, "Wegener's Displacement Theory," 77.

54. Philip Lake, "Wegener's Hypothesis of Continental Drift," *Geographical Journal* 61 (1923): 185–186.

55. Ibid., 187.

56. Ibid., 188.

57. Ibid., 188–189.

58. Ibid., 190–191.

59. Ibid.

60. Ibid., 192–193.

61. Ibid., 193.

62. Ibid.

63. W[illiam] B[ourke] Wright, "The Wegener Hypothesis," *Nature* 111, no. 2775 (1922): 30.

64. Ibid.

65. Ibid., 31.

66. Alfred Wegener, *The Origin of Continents and Oceans*, trans. John George Anthony Skerl (London: Methuen, 1924), xii. It was published simultaneously in New York by E. P. Dutton.

67. Ibid.

68. Lake, "Wegener's Hypothesis of Continental Drift," 192.

69. Emile Argand, *Tectonics of Asia*, trans. Albert V. Carozzi (New York: Hafner, 1976). Carozzi's translation is keyed (with the original pagination inserted in the English text in parentheses) to the French original, which may be found in Emile Argand, "La Tectonique de l'Asie," *Proceedings of the XIIIth International Geological Congress* 1, pt. 5 (1924), xvi–xvii.

70. Argand, *Tectonics of Asia*, xvi–xvii.

71. Ibid.

72. Ibid., xviii.

73. Ibid., 124; and Argand, "La Tectonique de l'Asie," 289.

74. Argand, *Tectonics of Asia*, 125; and Argand, "La Tectonique de l'Asie," 290.

75. Argand, *Tectonics of Asia*, 127–128; and Argand, "La Tectonique de l'Asie," 292.

76. Alfred Wegener, *La Genèse des Continents et des Océans*, trans. Manfred Reichel, Collection de monographies scientifiques étrangeres, 6 (Paris: Librarie Scientifique Albert Blanchard, 1924).

77. Edgar Irmscher, *Pflanzenverbreitung und Entwicklung der Kontinente: Studien zur genetischen Pflanzengeographie* (Hamburg: Institut für allgemeine Botanik, 1922).

78. Ibid., 24.

79. Ibid., 228–229.

80. Eugen Dubois, *Die Klimate der geologischen Vergangenheit und ihre Beziehung zur Entwicklungsgeschichte der Sonne* (Nijmegen: H. C. A. Thieme, 1893); see also Eugène Dubois, *The Climates of the Geological Past and Their Relation to the Evolution of the Sun* (London: S. Sonnenschein, 1895).

81. Melchior Neumayr and Viktor Uhlig, *Erdgeschichte*, 2nd ed., 2 vols. (Leipzig: Verlag des Bibliographischen Instituts, 1895).

82. Thomas C. Chamberlin and Rollin D. Salisbury, *Geology*, 2nd ed., 3 vols. (New York: Holt, 1905–1907).

83. Eliot Blackwelder, *United States of North America*, vol. 8, pt. 2, *Handbuch der Regionalen Geologie* (Leipzig: G. E. Stechert, 1912).

84. Lukas Waagen, J. van Bebber, and P. Kreichgauer, *Unsere Erde: Der Werdegang des Erdballs und seiner Lebewelt, seine Beschaffenheit und seine Hüllen* (München: Allgemeine Verlagsgesellschaft, 1909).

85. D. N. Wadia, *Geology of India*, 4th ed. [1st ed. 1919] (New Delhi: Tata McGraw Hill, 1975).

86. Edward Irving, "The Role of Latitude in Mobilism Debates," *Proceedings of the National Academy of Sciences* 102, no. 6 (2005): 1824.

87. Köppen and Wegener, *Die Klimate der geologischen Vorzeit*, 155.

88. Ibid., 156–157.

89. Irving, "Role of Latitude in Mobilism Debates," 1824.

90. Wegener, *Die Entstehung der Kontinente und Ozeane*, 65.

91. Köppen and Wegener, *Die Klimate der geologischen Vorzeit*, 116–117.

92. Flügel, *Alfred Wegeners Vertraulicher Bericht über die Grönland-Expedition 1929*; contains a complete account of the deliberations.

93. Wegener to Bundesministerium fur Unterricht, Wien, 27 July 1923, Osterreichisches Staatsarchiv Wien, 5-C [Graz 13047], W 002-1923.

94. Ulrich Wutzke, "Alfred Wegener als Hochschullehrer," *Zeitschrift für geologische Wissenschaften* 25, nos. 5/6 (1997): 561–562.

95. E. Wegener, *Alfred Wegener: Tagebücher, Briefe, Erinnerungen*, 172.

96. Bundesministerium to Wegener, 20 Mar. 1924, Osterreichisches Staatsarchiv Wien, 5 [Graz Philos. 6738-1 19245], W 004-1924.

97. Wegener to Ficker, 23 Mar. 1924, Staatsbibliothek Preußischer Kulturbesitz Berlin, Sammlung Darmstädter, [1924.25 15] W 005-1924.

98. E. Wegener, *Alfred Wegener: Tagebücher, Briefe, Erinnerungen*, 173.

99. Ibid.

Chapter 18. The Professor: Graz, 1924–1928

1. Reinhard Krause and Jörn Theide, eds., *Kontinental-Verschiebungen: Originalnotizen und Literaturauszüge*, vol. 516, *Berichte zur Polar- und Meeresforschung* (Bremerhaven: Alfred-Wegener-Institut für Polar- und Meeresforschung, 2005), 165.

2. Ulrich Wutzke, "Alfred Wegener als Hochschullehrer," *Zeitschrift für geologische Wissenschaften* 25, nos. 5/6 (1997): 562.

3. H(ans) Benndorf, "Alfred Wegener," *Gerlands Beiträge zur Geophysik* 31 (1931): 355.

4. A fascinating account of these intellectual filiations is found in Berta Karlik and Erich Schmid, *Franz Serafin Exner und sein Kreis: Ein Beitrag zur Geschichte der Physik in Österreich* (Wien: Österreischen Akademie der Wissenschaften, 1982). I thank Professor Sigfried Bauer (emeritus in Wegener's chair at Graz) for the gift of this helpful volume.

5. Benndorf, "Alfred Wegener," 356–357.

6. Ibid., 356.

7. Wutzke, "Alfred Wegener als Hochschullehrer," 562.

8. Alfred Wegener, "KontinentforskydningsTheorien og dens Betydning for de systematiske og de eksakte Naturvidesnkaber," *Naturens Verden* 7 (1923).

9. Else Wegener, *Alfred Wegener: Tagebücher, Briefe, Erinnerungen* (Wiesbaden: F. A. Brockhaus, 1960), 173.

10. Ibid.

11. J(ohann) P(eter) Koch, "Survey of Northeast Greenland," *Meddelelser om Grønland* 46 (1917); Alfred Wegener, "Referat über J.P. Koch: Nordgrönlands Trift nach Westen," *Astronomische Nachrichten* 208 (1919); Albrecht Penck, "Wegeners Hypothese der kontinentalen Verschiebungen," *Zeitschrift der Gesellschaft für Erdkunde zu Berlin*, nos. 1, 2 (1921).

12. These calculations and results took up forty pages for the altitude measurements and 173 pages for the meteorology; J. P. Koch and A. Wegener, "Wissenschaftliche Ergebnisse der dänischen Expedition nach Dronning Louises-Land und quer über das Inlandeis von Nordgrönland 1912–13 unter Leitung von Hauptmann J.P. Koch. Abteilung 1–2," *Meddelelser om Grønland* 74 (1930).

13. Koch to Wegener, 29 July 1924, DMH 1968 604/14 N 1/125; and Koch to Wegener, 4 Aug. 1924, DMH 1968 604/15 N 1/126, W 008, 009-1924.

14. E. Wegener, *Alfred Wegener: Tagebücher, Briefe, Erinnerungen*, 175.

15. Ibid.; Wegener to Köppen, 14 Sept. 1924. Else gives the year as 1919, but this is certainly wrong as this was an event from 1924 when Defant was a professor in Innsbruck, and Wegener, in September 1919, was already established in Hamburg and not writing to Köppen. The referenced letter does not survive, like many of the personal letters quoted in her memoir.

16. Milutin Milankovitch, *Milutin Milankovic 1879–1958*, ed. Vasko Milankovic (Katlenburg-Lindau: European Geophysical Society, 1995), 76.

17. Ibid., 78.

18. Ibid., 81.

19. Ibid., 84–85.

20. Milankovich to Wegener, 27 Sept. 1924, DMH 1984 603/45/1 N 1/159, W 013-1924.

21. Wegener to Milankovich, 6 Oct. 1924. The letter and its translation are published on pages 163–165 with facsimile of the original in Milankovitch, *Milutin Milankovic 1879–1958*.

22. Ibid.

23. E. Wegener, *Alfred Wegener: Tagebücher, Briefe, Erinnerungen*, 174.

24. Ibid.

25. Else Wegener-Köppen, *Wladimir Köppen: Ein Gelehrtenleben für die Meteorologie*, ed. Dr. H. W. Frickhinger, Grosse Naturforscher (Stuttgart: Wissenschaftliche Verlagsgesellschaft, 1955), 142–144.

26. E. Wegener, *Alfred Wegener: Tagebücher, Briefe, Erinnerungen*, 174.

27. Ibid.

28. Ibid.

29. Benndorf, "Alfred Wegener," 356.

30. Ibid.

31. Ibid.

32. Irmscher to Wegener, 7 Nov. 1924, DMH 1968 603/49/1, W 015-1924.

33. P. F. Jensen, "Ekspeditionen til Vestgrønland Sommeren 1922," *Meddelelser om Grønland* 63 (1922): 283.

34. William Bowie, "The Results of the Meeting of the Section of Geodesy of the International Geodetic and Geophysical Union of Interest to Astronomers," *Popular Astronomy* 33 (1925); Alfred Wegener, "Die geophysikalischen Grundlagen der Theorie der Kontinentalverschiebungen," *Scientia* 41 (1927): 116.

35. Alfred Wegener, "Theorie der Haupthalos," *Archiv der deutschen Seewarte* 43, no. 2 (1925).

36. Robert Greenler, *Rainbows, Halos, and Glories* (Cambridge: Cambridge University Press, 1980), 29.

37. Alfred Wegener, "Die äußere Hörbarkeitszone und ihre periodische Verlagerung im Jahreslauf," *Meteorologische Zeitschrift* 42 (1925).

38. Alfred Wegener, "Die Temperatur der obersten Atmosphären-Schichten," *Meteorologische Zeitschrift* 42 (1925).

39. Bruckner to Wegener, 28 Feb. 1925, DMH 1968 603/54/2 N 1/206, W 002-1925.

40. Bruckner to Wegener, 26 May 1925, DMH 1968 603/543/3 N 1/206, W 005-1925.

41. Alfred Wegener, *Die Entstehung der Kontinente und Ozeane*, 3rd (completely revised) ed., Die Wissenschaft (Braunschweig: Friedrich Vieweg & Sohn, 1922), 67.

42. Otto Ampferer, "Über Kontinentalverschiebungen," *Die Naturwissenschaften* 31, Juli (1925).

43. E. Wegener, *Alfred Wegener: Tagebücher, Briefe, Erinnerungen*, 177.

44. Wutzke, "Alfred Wegener als Hochschullehrer," 563.

45. E. Wegener, *Alfred Wegener: Tagebücher, Briefe, Erinnerungen*, 177.

46. Alfred Wegener, ed., *Müller-Pouillets Lehrbuch der Physik.V, 1.Physik der Erde*, 11th ed. (Braunschweig: Friedrich Vieweg & Sohn, 1928).

47. Benndorf, "Alfred Wegener," 358.

48. Wutzke, "Alfred Wegener als Hochschullehrer," 563.

49. Kuhlbrodt to Wegener, 30 Sept. 1924, DMH 1968 603/25 N 1/181, W 014-1924.

50. Alfred Wegener, "Alfred Merz," *Meteorologische Zeitschrift* 42 (1925).

51. Penck to Wegener, 7 Nov. 1925, AWI Bremerhaven Wegenerarchiv, W 012-1925.

52. Ibid.

53. Rühl to Wegener, 8 Nov. 1925, AWI Bremerhaven Wegenerarchiv, W 013-1925.

54. Wegener to Rühl, 18 Nov. 1925, AWI Bremerhaven Wegenerarchiv, W 014-1925; Wegener to Penck, 18 Nov. 1925, AWI Bremerhaven Wegenerarchiv, W 015-1925.

55. Wegener to Penck, 18 Nov. 1925, AWI Bremerhaven Wegenerarchiv, W 015-1925.

56. E. Wegener, *Alfred Wegener: Tagebücher, Briefe, Erinnerungen*, 176.

57. Benndorf, "Alfred Wegener," 357.

58. Alfred Wegener, "Referat über Felix M. Exner: Dynamische Meteorologie," *Die Naturwissenschaften* 14 (1926).

59. Alfred Wegener, "Ergebnisse der dynamischen Meteorologie," *Ergebnisse der exakten Naturwissenschaften* 5 (1926).

60. Ibid., 124.

61. Lewis F. Richardson, *Weather Prediction by Numerical Process* (London: Cambridge University Press, 1922).

62. Wegener, "Ergebnisse der dynamischen Meteorologie," 104–105.

63. Alfred Wegener, "Messungen der Sonnenstrahlungen am Sanitorium Stolzaple," *Meteorologische Zeitschrift* 43 (1926); Alfred Wegener, "Photographien von Luftspiegelungen an der Alpenkette," *Meteorologische Zeitschrift* 43 (1926).

64. Alfred Wegener, "Paläogeographische Darstellung der Theorie der Kontinentalverschiebungen," in *Enzyklopädie der Erdkunde*, ed. Oskar Kende (Leipzig: Deuticke, 1926).

65. Helmut W. Flügel, *Alfred Wegeners Vertraulicher Bericht über die Grönland-Expedition 1929* (Graz, Austria: Akademische Druck und Verlagsanstalt, 1980). The plans for this addition are included as plates in the back of Flügel's book and dated 1926.

66. G. C. Simpson to Wegener, 19 Aug. 1926, DMH 1968 603/43/2 N 1/150, W 005-1926.

67. Naomi Oreskes, *The Rejection of Continental Drift: Theory and Method in American Earth Science* (New York: Oxord University Press, 1999), 233.

68. Anonymous, "International Astronomical Union, Cambridge, July 14–22, 1925," *Observatory: A Monthly Review of Astronomy* 48, no. 615 (1925): 254.

69. William Bowie, "Scientists to Test 'Drift' of Continents," *New York Times*, 6 Sept. 1925, 5.

70. Ibid.

71. Alfred Wegener, "Two Notes Concerning My Theory of Continental Drift," in *Theory of Continental Drift: A Symposium on the Origin and Movement of Land Masses Both Inter-Continental and Intra-Continental, as Proposed by Alfred Wegener*, ed. W. A. J. M. Van Waterschoot van der Gracht (Tulsa: American Association of Petroleum Geologists, 1928), 102.

72. A. A. Thiadens, "Waterschoot van der Gracht, Willem Anton Joseph Maria van (1873–1943)," in *Bigrafisch Woordenboek van Nederland* (Bronvermelding: Huygens ING, 2012), www.historici.nl/Onderzoek/Projecten/BWN/lemmata/bwn1/waterschoot.

73. Ibid.

74. W. A. J. M. Van Waterschoot van der Gracht, ed., *Theory of Continental Drift: A Symposium on the Origin and Movement of Land Masses Both Inter-Continental and Intra-Continental, as Proposed by Alfred Wegener* (Tulsa: American Association of Petroleum Geologists, 1928), iii.

75. Ibid.; Thiadens, "Waterschoot van der Gracht, Willem Anton Joseph Maria van (1873–1943)."

76. J[ohn] W[alter] Gregory, "Continental Drift: Review of *The Origin of Continents and Oceans* by Alfred Wegener," *Nature* 115, no. 2886 (1925).

77. Frank A. Melton, "*The Origin of Continents and Oceans* by Alfred Wegener," *Science* 62, no. 1592 (1925).

78. Van Waterschoot van der Gracht, *Theory of Continental Drift*, 5–6.

79. Wegener, "Two Notes Concerning My Theory of Continental Drift," 97–103.

80. Wladimir Köppen and Alfred Wegener, *Die Klimate der geologischen Vorzeit* (Berlin: Gebrüder Bornträger, 1924), 33–34.

81. Wegener, "Two Notes Concerning My Theory of Continental Drift," 98.

82. Ibid.

83. Ibid., 102.

84. Ibid., 103.

85. Robert P. Newman, "American Instransigence: The Rejection of Continental Drift in the Great Debates of the 1920s," *Earth Sciences History* 14 (1995).

86. Van Waterschoot van der Gracht, *Theory of Continental Drift*, 197.

87. Joseph Barrell, "The Status of Hypotheses of Polar Wanderings," *Science* 40, no. 1027 (1914): 340.

88. Oreskes, *Rejection of Continental Drift*.

89. Henry R. Frankel, *The Continental Drift Controversy*, 4 vols. (New York: Cambridge University Press, 2012), 1:chaps. 3 and 4 passim.

90. Ibid., 1:81.

91. C. E. P. Brooks, *Climate through the Ages: A Study of the Climatic Factors and Their Variations* (New York: R. V. Coleman, 1926).

92. C. E. P. Brooks, *The Evolution of Climate* (London: Benn Brothers, 1922).

93. Brooks, *Climate through the Ages*, 5.

94. Ibid., 7.

95. Ibid., 11–12.

96. Ibid., 12.

97. Ibid., chap. 12, esp. 234ff.

98. Ibid., 11.

99. Wegener, "Die geophysikalischen Grundlagen der Theorie der Kontinentalverschiebungen," 103.

100. Ibid.

101. Krause and Theide, *Kontinental-Verschiebungen*, 167ff.

102. Carl Störmer to Wegener, 27 Oct. and 15 Nov. 1926, DMH 1968 603/44/1 and 44/2 N 1/145, W 008, 009-1926.

103. Bruckner to Wegener, 10 Dec. 1926, DMH 1968 603/53/5, W 015-1926; E. Wegener, *Alfred Wegener: Tagebücher, Briefe, Erinnerungen*, 183.

104. E. Wegener, *Alfred Wegener: Tagebücher, Briefe, Erinnerungen*, 182.

105. Ibid.

106. Wegener, "Die äußere Hörbarkeitszone und ihre periodische Verlagerung im Jahreslauf."

107. Alfred Wegener, "Anfangs-und Endhöhen großer Meteore," *Meteorologische Zeitschrift* 44 (1927): 282.

108. Alfred Wegener, "Die Geschwindigkeit großer Meteore," *Naturwissenschaften* 15 (1927): passim.

109. Leon Knopoff, "Beno Gutenberg. June 4, 1889–January 25, 1960," *Biographical Memoirs, National Academy of Sciences* 79 (1999).

110. Freuchen to Wegener, 12 Feb. 1927, DP Copenhagen [Peter Freuchen, No. 1] W 003-1927.

111. Gutenberg to Wegener, 7 Apr. 1927, DMH 1968 604/4, W 004-1927.

112. Ibid.

113. Alfred Wegener, "Referat über B. Gutenberg: Lehrbuch der Physik," *Geographische Zeitschrift* 33 (1927).

114. Krause and Theide, *Kontinental-Verschiebungen*, 324 (Wegener's p. 157).

115. Ibid., 327 (Wegener's p. 159).

116. Ibid., 332 (Wegener's p. 161).

117. Ibid., 338 (Wegener's p. 164).

118. Reginald Aldworth Daly, *Our Mobile Earth* (New York: Charles Scribner's Sons, 1926).

119. Krause and Theide, *Kontinental-Verschiebungen*, 335–336 (Wegener's pp. 162–163).

120. Van Waterschoot van der Gracht, *Theory of Continental Drift*, 38, 52.

121. Krause and Theide, *Kontinental-Verschiebungen*, 362 (Wegener's p. 176).

122. Ibid., 354 (Wegener's p. 172).

123. Alfred Wegener, "Der Boden des Atlantischen Ozeans," *Gerlands Beiträge zur Geophysik* 17 (1927). *Gerlands* noted that the paper was received on 20 April, which meant that it had been finished no later than the weekend at the end of the term, 17 and 18 April, and mailed on Monday the nineteenth.

124. Ibid., 311.

125. E. Wegener, *Alfred Wegener: Tagebücher, Briefe, Erinnerungen*, 182.

126. Koch and Wegener, "Wissenschaftliche Ergebnisse der dänischen Expedition," 7.

127. Ibid.

128. Ibid., 3–4.

129. Milankovitch, *Milutin Milankovic 1879–1958*, 86. Milankovich remembers this as having taken place in 1925 and provides some incidental detail to support this, but it could not have been because in 1925 Wegener was not working on Koch's material, nor had he plans for a fourth edition. Milankovich is very good about content and often very poor about dates; one feels he wrote his memoirs without consulting written records, but using only recollection.

130. Wegener, *Müller-Pouillets Lehrbuch der Physik.V, 1.Physik der Erde*, vii–ix, xiv–xvi.

131. Lundager to Wegener, 10 Oct. 1927, DMH 1968 604/21 N 1/160, W 014-1927.

132. E. Wegener, *Alfred Wegener: Tagebücher, Briefe, Erinnerungen*, 176.

133. Letzmann to Wegener, 7 June 1927, and Ficker to Wegener, 20 June 1927, DMH 1968 603/41 N 1/168 and 1968 603/56, W 008, 010-1917.

134. E. Kraus, R. Meyer, and Alfred Wegener, "Untersuchungen über den Krater von Sall auf Ösel," *Gerlands Beiträge zur Geophysik* 20 (1928); E. Kraus and R. Meyer, "Nachtrag zu: Untersuchungen über den Krater von Sall auf Ösel," *Gerlands Beiträge zur Geophysik* 20 (1928). See also Mott T. Greene, "Alfred Wegener and the Origin of Lunar Craters," *Earth Sciences History* 17, no. 2 (1998): 134–135.

135. Richard E. Peterson, "Johannes Letzmann: A Pioneer in the Study of Tornadoes," *Weather and Forecasting* 7 (1992): 177–178.

136. E. Wegener, *Alfred Wegener: Tagebücher, Briefe, Erinnerungen*, 183.

137. Krause and Theide, *Kontinental-Verschiebungen*, passim.

138. Koch to Wegener, 18 Dec. 1927, DMH 1968 604/19 N 1/130, W 018-1927.

139. Lindhard to Wegener, 18 Dec. 1927, DMH 1968 604/23, N 1/139, W 017-1927.

140. Lindhard to Wegener, 12 Jan. 1928, DMH 1968 604/24 N 1/140, W 001-1928.

141. Marie Koch to Wegener, 25 Jan. 1928, DMH 1968 604/29, N 1/133, W 003-1928.

142. Lindhard to Wegener, 25 Jan. 1928, DMH 1968 604/25 N 1/141, W 004-1928.

143. Marie Koch to Wegener, 25 Jan. 1928, DMH 1968-604/28 N 1/132, W 006-1928.

144. Wegener to Georgi, 2 Jan. 1928, AWI Bremerhaven, Nachlaß Georgi.

145. J(ohannes) Georgi, "Memories of Alfred Wegener," in *Continental Drift*, ed. Stanley Keith Runcorn, *International Geophysics Series* (New York: Academic Press, 1962), 317.

146. Wegener to Bundesministerium fur Unterricht, Wien, 2 July 1927, Osterreichisches Staatsarchiv, Wien 26 G 3 27796-I/2 1927, W 011-1927.

147. Wegener to Georgi, 2 Jan. 1928, AWI Bremerhaven, Nachlaß Georgi, W 002-1928.

148. Ibid.

149. Ibid

150. Ibid.

151. H. Heß, "Physik der Gletscher," in Wegener, *Müller-Pouillets Lehrbuch der Physik.V, 1.Physik der Erde*, 355–397.

Chapter 19. Theorist and Arctic Explorer: Graz and Greenland, 1928–1929

1. Alfred Wegener, "Beiträge zur Mechanik der Tromben und Tornados," *Meteorologische Zeitschrift* 45 (1928): 201.

2. Berson to Wegener, 23 Feb. 1928, DMH 1968 604/2 N 1/191, W 007-1928.

3. Cornelia Lüdecke, *Die deutsche Polarforschung seit der Jahrhundertwende und der Einfluß Erich von Drygalskis*, vol. 158, *Berichte zur Polarforschung* (Bremerhaven: Alfred-Wegener-Institute für Polar- und Meeresforschung, 1995), 41.

4. Wegener to Kuhlbrodt, 28 Mar. 1928, DMH 1968 604/3 N 1/180, W 010-1928.

5. Kurt Wegener, *Wissenschaftliche Ergebnisse der Deutschen Grönland-Expedition Alfred Wegener 1929 und 1930/31*, vol. 1, *Geschichte der Expedition* (Leipzig: F. A. Brockhaus, 1933), 6–8. See also Wegener to Georgi, 15 Jan. 1928, AWI Bremerhaven, Nachlaß Georgi, W 002-1928.

6. Bryd's claim to be the first to fly over the North Pole had been a great blow to AEROARCTIC, as their airship *Norge* was already planning to cross the Arctic from Svalbard to Alaska. It did so in May 1928, flying over the North Pole, but after Byrd had accomplished the feat.

7. A good contemporary account of this debacle and its aftermath is Willy Meyer, *Der Kampf um Nobile: Versuch einer objektiven Darstellung und Wertung der Leistung des italienischen Luftschiffers* (Berlin: Gebrüder Radetzki, 1931).

8. Alfred Wegener, ed., *Müller-Pouillets Lehrbuch der Physik.V, 1.Physik der Erde*, 11th ed. (Braunschweig: Friedrich Vieweg & Sohn, 1928), v–vi.

9. Else Wegener, *Alfred Wegener: Tagebücher, Briefe, Erinnerungen* (Wiesbaden: F. A. Brockhaus, 1960), 184.

10. Cornelia Lüdecke, "Vor 100 Jahren: Grönlandexpedition der Gesellschaft für Erdkunde zu Berlin (1891,1892–1893) unter der Leitung Erich von Drygalskis," *Polarforschung* 60, no. 3 (1990): 41ff.

11. Wegener to Freuchen, 17 June 1928 and 14 July 1928, DP Copenhagen, Peter Freuchen No. 1, W 010 and 014-1928; and Wegener to Georgi, 10 July 1928, AWI Bremerhaven, Nachlaß Georgi, W 012-1928.

12. E. Wegener, *Alfred Wegener: Tagebücher, Briefe, Erinnerungen*, 188.

13. Wegener to Drygalski, 25 July 1927, DMH 1968 603/11, W 012-1927; Erich von Drygalski, *Grönland-Expedition der Gesellschaft für Erdkunde zu Berlin 1891–1893* (Berlin: W. H. Kühl, 1897).

14. E. Wegener, *Alfred Wegener: Tagebücher, Briefe, Erinnerungen*, 189.

15. Alfred Wegener, *Die Entstehung der Kontinente und Ozeane*, 4th (completely revised) ed. (Braunschweig: Friedrich Vieweg & Sohn, 1929). Preface: "Alfred Wegener," by Kurt Wegener, vi.

16. Ibid., foreword (by Alfred Wegener), x.

17. Ibid., 2.

18. H(ans) Benndorf, "Alfred Wegener," *Gerlands Beiträge zur Geophysik* 31 (1931): 359.

19. Wegener, *Die Entstehung der Kontinente und Ozeane*, 29.

20. Alfred Wegener, "Bemerkungen zu H. v Ihering's Kritik der Theorien der Kontinentalveschiebungen," *Zeitschrift für Geophysik* 4 (1928): 47.

21. Ibid.

22. Wegener, *Die Entstehung der Kontinente und Ozeane*, 100.

23. Henry R. Frankel, *The Continental Drift Controversy*, 4 vols. (New York: Cambridge University Press, 2012), 1:81ff.

24. Wegener, *Die Entstehung der Kontinente und Ozeane*, 4.

25. Ibid., 22.

26. Ibid., 53, 55.

27. Ibid., chap. 4, passim.

28. Ibid., 55.

29. Adrian E. Scheidegger, *Principles of Geodynamics*, 2nd ed. (New York: Academic Press, 1963), 106ff.

30. Alex. L. (Alexander Logie) du Toit, *A Geological Comparison of South America with South Africa* (Washington, DC: Carnegie Institution of Washington, 1927).

31. Naomi Oreskes, *The Rejection of Continental Drift: Theory and Method in American Earth Science* (New York: Oxord University Press, 1999), 157–167.

32. Du Toit, *Geological Comparison of South America with South Africa*, introduction, passim.

33. Ibid., 97.

34. Ibid., 109–110.

35. Ibid., 118.

36. Ibid.

37. Wegener, *Die Entstehung der Kontinente und Ozeane*, 68.

38. Ibid.

39. Ibid., 74.

40. Ibid., 86–90.

41. Ibid., 89.

42. W. A. J. M Van Waterschoot van der Gracht, ed., *Theory of Continental Drift: A Symposium on the Origin and Movement of Land Masses Both Inter-Continental and Intra-Continental, as Proposed by Alfred Wegener* (Tulsa: American Association of Petroleum Geologists, 1928), 57.

43. Wegener, *Die Entstehung der Kontinente und Ozeane*, 94.

44. Ibid., 98.

45. Ibid., 24–25.

46. Wladimir Köppen and Alfred Wegener, *Die Klimate der geologischen Vorzeit* (Berlin: Gebrüder Bornträger, 1924), 227; and Beno Gutenberg, "Bewegungen der Erdaschse und Polwanderung," in Wegener, *Müller-Pouillets Lehrbuch der Physik*, 717–719.

47. Gutenberg, "Bewegungen der Erdaschse und Polwanderung," 719.

48. Wegener, *Die Entstehung der Kontinente und Ozeane*, 152.

49. Ibid.

50. Ibid.

51. Ibid., 153–154.

52. Ibid., 157.

53. Ibid., 161.

54. Ibid., 166–167.

55. Ibid.

56. Ibid., 170.

57. Ibid., 171.

58. Ibid., 172.

59. Ibid.

60. Ludwig Günther, *Keplers Traum vom Mond* (Leipzig: B. G. Teubner, 1898).

61. I am much indebted to Walter Hoflechner for providing me this book to examine during my stay in Graz in 1989.

62. Günther, *Keplers Traum vom Mond*, 52.

63. Wegener, *Die Entstehung der Kontinente und Ozeane*, 173.

64. Reinhard A. Krause, "Die Polfluchtkraft: Die LELY-Versuch—Vergessene Begriffe der Geologiegeschichte—," *Polarforschung* 76, no. 3 (2006).

65. Wegener, *Die Entstehung der Kontinente und Ozeane*, 178.

66. Ibid., 181.

67. Ibid., 182.

68. Ibid., 184.

69. Ibid.

70. Ibid., 185.

71. Ibid.

72. Joseph L. Kirschvink, Robert L. Ripperdan, and David E. Evans, "Evidence for a Large-Scale Reorganization of Early Cambrian Continental Masses by Inertial Interchange True Polar Wander," *Science* 277 (1997).

73. Reginald Aldworth Daly, *Our Mobile Earth* (New York: Charles Scribner's Sons, 1926), 269; Helmut W. Flügel, "Wegener-Ampferer-Schwinner: Ein Beitrag zur Geschichte der Geologie in Österreich," *Mitteilung der Österreichischen Geologischen Gesellschaft* 73 (1980).

74. Wegener to Freuchen, 17 June 1928, Dansk Polarcenter, Copenhagen, Peter Freuchen No. 1, W 012-1928.

75. Wegener to Georgi, 10 July 1928, AWI Bremerhaven, Nachlaß Georgi, W 013-1928.

76. J(ohannes) Georgi, "Memories of Alfred Wegener," in *Continental Drift*, ed. Stanley Keith Runcorn, *International Geophysics Series* (New York: Academic Press, 1962), 318.

77. Helmut W. Flügel, *Alfred Wegeners Vertraulicher Bericht über die Grönland-Expedition 1929* (Graz, Austria: Akademische Druck und Verlagsanstalt, 1980), 53.

78. Ulrich Wutzke, "Alfred Wegener als Hochschullehrer," *Zeitschrift für geologische Wissenschaften* 25, nos. 5/6 (1997): 563.

79. Alfred Wegener, "Denkschrift über Inlandeis-Expedition nach Grönland," *Deutsche Forschung (Aus der Arbeit der Notgemeinschaft der Deutschen Wissenschaft)*, no. 2 (1928).

80. Wegener to Freuchen, 30 July 1928, DP Copenhagen, Peter Freuchen No. 1, W 015-1928.

81. Wegener to Loewe, 31 July 1928, Heimatmuseum, Neuruppin, [W 121] W 016-1928.

82. Wegener to Loewe, 26 Aug. 1928. This letter is in a private collection, that of P. Schulze, in Berlin. W 017-1928.

83. Wegener to Ernst Sorge, 8 Sept. 1928, AWI Bremerhaven, Nachlaß Loewe, W 020-1928.

84. K. Wegener, *Wissenschaftliche Ergebnisse der Deutschen Grönland-Expedition Alfred Wegener 1929 und 1930/31*, 1:6–8; and E. Wegener, *Alfred Wegener: Tagebücher, Briefe, Erinnerungen*, 185–188. These were preliminary drafts, not the published version of Wegener, "Denkschrift über Inlandeis-Expedition nach Grönland." Kurt and Else put them in their publications to show that Alfred had considered a variety of modern forms of transport.

85. Ibid.

86. Wegener to Georgi, 21 Sept. 1928, AWI Bremerhaven, Nachlaß Georgi, W 022-1928.

87. W. Bruns to Wegener, 20 Dec. 1928, AWI Bremerhaven, Nachlaß Georgi, W 043-1928.

88. Wegener, "Denkschrift über Inlandeis-Expedition nach Grönland." I have here given a summary of the main points of the document.

89. Lüdecke, *Die deutsche Polarforschung seit der Jahrhundertwende und der Einfluß Erich von Drygalskis*, 158.

90. Flügel, *Alfred Wegeners Vertraulicher Bericht über die Grönland-Expedition 1929*, 9–10.

91. K. Wegener, *Wissenschaftliche Ergebnisse der Deutschen Grönland-Expedition Alfred Wegener 1929 und 1930/31*, 1:9–10.

92. Wegener, "Denkschrift über Inlandeis-Expedition nach Grönland," 201.

93. E. Wegener, *Alfred Wegener: Tagebücher, Briefe, Erinnerungen*, 190.

94. Flügel, *Alfred Wegeners Vertraulicher Bericht über die Grönland-Expedition 1929*, 31. Flügel here publishes his transcriptions of Wegener's letters to the Notgemeinschaft, which were not intended for publication at the time. Wegener's superscript on each letter specifies this condition, which was part of his publishing contract with Belhagen and Klasing for the book *Mit Motorboot und Schlitten in Grönland*. All of the references past p. 10 to this short book by Helmut Flügel (*Vertraulicher Bericht*) are to Wegener's verbatim account in the letters to the Notgemeinschaft, not to a narrative by Flügel himself.

95. Ibid., 35.

96. Ibid.

97. Alfred Wegener, *Mit Motorboot und Schlitten in Grönland* (Bielefeld: Belhagen & Klasing, 1930).

98. Flügel, *Alfred Wegeners Vertraulicher Bericht über die Grönland-Expedition 1929*, 38–48.

99. Ibid.

100. Ibid., 49ff.

101. Ibid., 53–54.

102. Ibid.

103. This describes the glacier as I climbed it in 1989, but the general structure is confirmed here by aerial photos (1949) and still photos from the various books on the expedition.

104. Flügel, *Alfred Wegeners Vertraulicher Bericht über die Grönland-Expedition 1929*, 58.
105. Ibid., 59ff.
106. Wegener, *Mit Motorboot und Schlitten in Grönland*, 104ff.
107. Flügel, *Alfred Wegeners Vertraulicher Bericht über die Grönland-Expedition 1929*, 64.
108. Ibid., 65.
109. Ibid.
110. Ibid., 66.
111. Ibid., 69.
112. Benndorf, "Alfred Wegener," 361–362.
113. Flügel, *Alfred Wegeners Vertraulicher Bericht über die Grönland-Expedition 1929*, 72.
114. Ibid., 74.
115. Ibid.
116. Ibid., 75.

Chapter 20. The Expedition Leader: Graz and Greenland, 1929–1930

1. Else Wegener, *Alfred Wegener: Tagebücher, Briefe, Erinnerungen* (Wiesbaden: F. A. Brockhaus, 1960), 201.
2. Ibid.
3. Wegener to Georgi, 6 Nov. 1929, AWI Bremerhaven, Nachlaß Georgi, W 039-1929.
4. Wegener to Schmidt-Ott, 11 Nov. 1929, OStA Wien [26 G 3 35505-1/2 1929], W 043-1929; includes third part of the report on the Greenland reconnaissance.
5. Alexander du Toit to Wegener, 23 Apr. 1929, DMH 1968 603/34, W 025-1929.
6. Pierre Termier to Wegener, 30 May 1929, DMH 1968 603/38 N 1/201, W 027-1929.
7. J(ohannes) Georgi, "Memories of Alfred Wegener," in *Continental Drift*, ed. Stanley Keith Runcorn, *International Geophysics Series* (New York: Academic Press, 1962), 319.
8. E. Wegener, *Alfred Wegener: Tagebücher, Briefe, Erinnerungen*, 202.
9. Wegener to Georgi, 1 Dec. 1929, AWI Bremerhaven, Nachlaß Georgi, W 045-1929.
10. Wegener to Georgi, 1 Dec. 1929, AWI Bremerhaven, Nachlaß Georgi, W 046-1929.
11. Wegener to Notgemeinschaft Berlin, 3 Dec. 1929, W 047-1929.
12. Wegener to Georgi, 9 Dec. 1929, AWI Bremerhaven, Nachlaß Georgi, W 049, 050-1929.
13. Wegener to Georgi, 23 Dec. 1929, AWI Bremerhaven, Nachlaß Georgi, W 052-1929.
14. Wegener to Georgi, 25 Dec. 1929, AWI Bremerhaven, Nachlaß Georgi, W 053-1929.
15. Cornelia Lüdecke, "Lifting the Veil: The Circumstances That Caused Alfred Wegener's Death on the Greenland Icecap, 1930," *Polar Record* 36, no. 197 (2000): 150. See also Loewe to Wegener, 21 Dec. 1929, AWI Bremerhaven, Nachlaß Loewe, Box 2, 18/II/2.
16. E. Wegener, *Alfred Wegener: Tagebücher, Briefe, Erinnerungen*, 203.
17. There are so many letters in this sequence that it would be useless to cite them all; in any case, they are available online and by pdf download, indexed by date and topic, often with a brief extract, helpfully annotated by Ulrich Wutzke in his checklist of archival documents pertaining to Wegener's life. Just listing these letters takes up pages 104–122 of the 144 pages of Wegener's entire Nachlaß. Other than the letters that Else saved herself and the collection of not quite 100 letters he sent to Köppen between 1906 and 1919, these letters to Georgi make up the bulk of his surviving correspondence. Since the checklist is available electronically and all the letters are in the same place—the Alfred Wegener Institute Georgi Nachlaß—I will cite them henceforth using only the date and Wutzke's numbering, the same system I have cited in the beginning of this book, as in "Wegener to Georgi, 9 Dec. 1929, W 049-1929."
18. Wegener to Vigfus Sigurdsson, late Jan. 1930, Bundesarchiv Koblenz [R73/255], W 019-1930.
19. See Ulrich Wutzke, *Durch die weiße Wüste: Leben und Leistungen des Grönlandforschers und Entdeckers der Kontinentaldrift Alfred Wegener*, Petermann ed. (Gotha: Justus Perthes Verlag, 1997), passim; see also Else Wegener, ed., *Greenland Journey: The Story of Wegener's German Expedition to Greenland in 1930–31 as Told by Members of the Expedition and the Leader's Diary*, trans. Winifred M. Deans (London: Blackie & Sons, 1939), 9–10. For the convenience of readers I will cite the English translation of her account of the expedition rather than the German original.

20. Helmut W. Flügel, *Alfred Wegeners Vertraulicher Bericht über die Grönland-Expedition 1929* (Graz, Austria: Akademische Druck und Verlagsanstalt, 1980), 22–23.
21. E. Wegener, *Alfred Wegener: Tagebücher, Briefe, Erinnerungen*, 204.
22. Ibid.
23. Ibid.
24. Ibid.
25. Wegener to Grønlands Styrelse, 20 Mar. 1930, Dansk Rigsarkivet Copenhagen, Grl. St J Nr. 349/1930.8, W 095-1930.
26. Wegener to Verlag Velhagen, 18 Feb. 1930, Bundesarchiv Koblenz R 73/244, W 048-1930.
27. Wegener to Verlag Velhagen, 24 Feb. 1930, AWI Bremerhaven, Nachlaß Georgi, W 053-1930.
28. Wegener to Verlag Velhagen, 3 Mar. 1930, Bundesarchiv Koblenz R 73/244, W 078-1930.
29. Georgi, "Memories of Alfred Wegener," 320.
30. Ibid.
31. *Tagebücher*, Apr. 1930–Sept. 1930, DMH, NL 001/014, 1 Apr., hereafter Wegener's Tagebuch.
32. Wegener's Tagebuch, 20 Apr. 1930.
33. Ibid., 29 Apr. 1930.
34. Ibid., 2 May 1930.
35. Ibid., 4 May 1930.
36. Ibid.
37. Ibid., 7 May 1930.
38. Ibid., 14 May 1930.
39. Ibid., 31 May 1930.
40. E. Wegener, *Greenland Journey*, 26.
41. Ibid., 27.
42. Ibid., 30.
43. Ibid., 79ff.
44. Ibid., 82.
45. Wegener's Tagebuch, 5 Aug. 1930.
46. Wegener to Georgi, 31 July 1930, AWI Bremerhaven, Nachlaß Georgi, W 116-1930.
47. Wegener's Tagebuch, 17 July 1930.
48. E. Wegener, *Greenland Journey*, 83.
49. Ibid., 87.
50. Wegener's Tagebuch, 26 July 1930.
51. Ibid., 5 Aug. 1930.
52. Cited and quoted in Lüdecke, "Lifting the Veil," 150–151.
53. Wegener to Georgi, 31 July 1930, AWI Bremerhaven, Nachlaß Georgi, W 116-1930.
54. Wegener's Tagebuch, 9 Aug. 1930.
55. Ibid.
56. Lüdecke, "Lifting the Veil," 151, quoting Loewe's Tagebuch, 5 July 1930, AWI Bremerhaven, Nachlaß Loewe.
57. Wegener's Tagebuch, 29 Aug. 1930.
58. Ibid., 29–30 Aug. 1930.
59. Ibid., 30 Aug. 1930.
60. Ibid.
61. E. Wegener, *Greenland Journey*, 68–69.
62. Ibid.
63. Wegener's Tagebuch, 1 Sept. 1930.
64. Ibid., 2 Sept. 1930.
65. Ibid., 1 Sept. 1930.
66. Børge Fristrup, *The Greenland Ice Cap* (Seattle: University of Washington Press, 1966), 97.
67. Wegener's Tagebuch, 1 Sept. 1930.
68. Ibid., 8 Sept. 1930.

69. Ibid.

70. Ibid., 6 Sept. 1930.

71. Ibid.

72. Lüdecke, "Lifting the Veil," 142; and Kurt Wegener, *Wissenschaftliche Ergebnisse der Deutschen Grönland-Expedition Alfred Wegener 1929 und 1930/31*, vol. 1, *Geschichte der Expedition* (Leipzig: F. A. Brockhaus, 1933), 39.

73. Lüdecke, "Lifting the Veil," 142.

74. E. Wegener, *Greenland Journey*, 100.

75. Ibid., 103.

76. Wegener's Tagebuch, 18 Sept. 1930.

77. Georgi and Sorge to Wegener, quoted in K. Wegener, *Wissenschaftliche Ergebnisse der Deutschen Grönland-Expedition Alfred Wegener 1929 und 1930/31*, 1:38–39.

78. E. Wegener, *Greenland Journey*, 105.

79. Ibid., 106.

80. Wegener to Weiken, 28 Sept. 1930, transcribed in ibid., 113.

81. Wegener to Weiken, 6 Oct. 1930, transcribed in ibid., 117–118.

82. Loewe's recollections quoted in ibid., 172–173.

83. Ibid., 175.

84. Ibid., 174.

85. Ibid., 176.

86. Ernst Sorge, "Winter at *Eismitte*," in *Greenland Journey: The Story of Wegener's German Expedition to Greenland in 1930–31 as Told by Members of the Expedition and the Leader's Diary*, ed. Else Wegener (London: Blackie & Sons, 1939), 182.

87. Ibid.

88. Ibid.

89. Georgi, "Memories of Alfred Wegener," 322.

90. Ibid.

91. My account is reconstructed from our only source: Ernst Sorge and Karl Weiken, "The Finding of Alfred Wegener's Body," in E. Wegener, *Greenland Journey*, 198–204.

92. Ibid., 200–201.

93. E. Wegener, *Greenland Journey*, 168.

94. Ibid., 169.

95. Sorge and Weiken, "The Finding of Alfred Wegener's Body," 200.

96. Ibid.

97. *New York Times*, 10 May 1931, 1.

98. *New York Times*, 14 May 1931, 1.

99. Flügel, *Alfred Wegeners Vertraulicher Bericht über die Grönland-Expedition 1929*, 25.

100. Ibid.

101. *New York Times*, 21 May 1931, 1.

102. *New York Times*, 22 May 1931, 1.

103. F. Spencer Chapman, *Watkins' Last Expedition* (London: Chatto & Windus, 1934), introduction by Augustine Courtauld, p. 13.

Epilogue

1. Cornelia Lüdecke, "Lifting the Veil: The Circumstances That Caused Alfred Wegener's Death on the Greenland Icecap, 1930," *Polar Record* 36, no. 197 (2000): 144.

2. Ibid., 143.

3. Ibid., 144.

4. Else Wegener, ed., *Alfred Wegeners Letzte Grönlangfahrt: Die Ergebnisse der Deutschen Grönlandexpedition 1930/31 Geschildert von seinen Reisegefährten und nach Tagebüchern des Forschers* (Leipzig: F. A. Brockhaus, 1932); Else Wegener, ed., *Alfred Wegeners Sidste Grønlandsfærd: Beretningen om den berømte Forskers Ekspedition 1930–31 skildert af hans Ekspeditionkammerater og efter Forskerens efterlade Dagbøger*, Danish edition of German original, 7th ed. (København: H. Hirschsprungs Forlag, 1933); Else Wegener, ed., *Greenland Journey: The Story of Wegener's German*

Expedition to Greenland in 1930–31 as Told by Members of the Expedition and the Leader's Diary, trans. Winifred M. Deans (London: Blackie & Sons, 1939).

5. Michael Spender, "Alfred Wegener's Greenland Expeditions 1929 and 1930–31," *Geographical Journal* 84 (1934): 521.

6. Arnold Fanck, *S.O.S. Eisberg* (KINO International, 1933).

7. Alfred Wegener and Kurt Wegener, *Vorlesungen über Physik der Atmosphäre* (Leipzig: Johan Ambrosius Barth, 1935).

8. Else Wegener-Köppen, *Wladimir Köppen: Ein Gelehrtenleben für die Meteorologie,* ed. H. W. Frickhinger, Grosse Naturforscher (Stuttgart: Wissenschaftliche Verlagsgesellschaft, 1955).

9. Else Wegener, *Alfred Wegener: Tagebücher, Briefe, Erinnerungen* (Wiesbaden: F. A. Brockhaus, 1960).

10. Helge Kragh and Robert W. Smith, "Who Discovered the Expanding Universe?," *History of Science* 41 (2003): 144.

11. Henry R. Frankel, *The Continental Drift Controversy,* 4 vols. (New York: Cambridge University Press, 2012), 1:603–604.

12. Ibid., 2:525.

13. Alfred Wegener, *Wind- und Wasserhosen in Europa,* Sammlung Wissenschaft Bd. 60 (Braunschweig: Friedrich Vieweg & Sohn, 1917).

Bibliographical Essay

For reasons of space, a complete bibliography is not included herewith but is instead available online at www.press.jhu.edu. I urge my readers to follow the citations in the notes to each of the chapters and to use this brief essay only for a general orientation to Wegener's life and science.

There are four archives with significant collections of Wegener materials. The most important of these is the Wegener Nachlaß in the Deutsches Museum (Handschriftensammlung) in Munich, holding the bulk of Wegener's surviving letters and his expedition journals. These documents are identified in notes with the designation "DMH," followed by the accession numbers and document locators. There is also an important Wegener archive at the Alfred Wegener Institute's library in Bremerhaven. All documents from this collection are identified as "AWI Bremerhaven." There is a Wegener Nachlaß at the Heimatmuseum, Neuruppin (Germany), and its satellite, the Wegener-Gedenkstätte in Zechlinerhütte (the former Wegener family home); all documents from these collections are denominated "HM, Neuruppin." These contain documents pertaining to Wegener's early life, some early holograph manuscripts by Wegener, and the family photo album. HSM is Hessisches Staatsarchiv Marburg.

The Dansk Polarcenter in Copenhagen, now home to the Arktisk Institut, contains Wegener's correspondence concerning the Danmark Expedition of 1906–1908 and the Koch–Wegener *Durchquerungsexpedition* of 1912–1913. The Arktisk Institut's collection of the papers of Dan Møller (a Danish colonial official in Umanak) includes much correspondence concerning Wegener's last expeditions to Greenland in 1929 and in 1930–1931. In the notes these are identified as "DP Copenhagen," with the document number following.

Ulrich Wutzke has prepared an annotated checklist of Wegener's known surviving correspondence (letters to and from) and documents pertaining to his institutional careers—appointments, promotions, and so on: *Alfred Wegener: Kommentiertes Verzeichnis der schriftlichen Dokumente seines Lebens und Wirkens*, Berichte zur Polarforschung, vol. 288 (Alfred-Wegener-Institut für Polar-und-Meeresforschung, 1998). This document is available as a downloadable pdf at doi:hdl:10013/epic.10291.d007. References to all letters and documents concerning Wegener in the main body of my text contain the identifying number and year assigned by Wutzke. Wutzke's detailed analysis of Wegener's academic career is "Alfred Wegener als Hochschullehrer," *Zeitschrift für geologische Wissenschaften* 25, nos. 5/6 (1997).

Wegener's colleague in Graz, Hans Benndorf, prepared the standard list of Wegener's published works: "Alfred Wegener," *Gerlands Beiträge zur Geophysik* 31 (1931): 337–377. It is reprinted in the 1980 reissue of Wegener's 1st and 4th editions: Alfred Wegener, *Die Entstehung der Kontinente und Ozeane*, vols. 1 and 4, Auflage Herausgegeben und mit einer Einleitung und einem Nachwort von Andreas Vogel (Braunschweig: Friedrich Vieweg & Sohn, 1980). Benndorf's obituary essay, accompanying the list of publications, contains much of interest about Wegener's years in Graz. The work of Helmut Flügel also contains much material about Wegener's Graz years: see his *Geologie und Paläontologie an der Universität Graz 1761.1976* (Graz: Akademische Druck-u. Verlagsanstalt, 1977); "Wegener-Ampferer-Schwinner: Ein Beitrag zur Geschichte der Geologie in Österreich," *Mitteilung der Österreichen Geologischen*

Gesellschaft 73 (1980): 237–254; and *Alfred Wegeners Vertraulicher Bericht über die Grönland-Expedition 1929* (Graz: Akademische Druck-u. Verlagsanstalt, 1980).

The main source of biographical information concerning Wegener's life and career is the memoir written by his wife: Else Wegener, *Alfred Wegener, Tagebücher, Briefe, Erinnerungen* (Wiesbaden: F. A. Brockhaus, 1960). She also wrote a biography of her father, Wladimir Köppen, which contains much additional detail on Wegener's life: Else Wegener-Köppen, *Wladimir Köppen: Ein Gelehrtenleben für die Meteorologie* (Stuttgart: Wissenschaftliche Verlagsgesellschaft, 1955). The following two works by Johannes Georgi were written in parallel and perhaps in competition with Else Wegener, and though informative, they should be used with caution: "Alfred Wegener zum 80. Geburtstag," *Polarforschung* 2 (1960): S1-102; and "Memories of Alfred Wegener," in *Continental Drift*, edited by Stanley Keith Runcorn, 309–324 (New York: Academic Press, 1962). The popular biographies by Martin Schwarzbach—*Alfred Wegener, the Father of Continental Drift* (Madison, WI: Science Tech. Inc., 1986)—and Hans Günther Körber—*Alfred Wegener* (Potsdam: BSB B. G. Teubner, 1980)—both derive from Else Wegener's account, though supplemented by additional research. The same may be said of Christine Reinke-Kunze's *Alfred Wegener, Polarforscher und Entdecker der Kontinentaldrift* (Berlin: Birkhaüser, 1994). My views on writing the life stories of scientists are in Mott T. Greene, "Writing Scientific Biography," *Journal of the History of Biology* 40 (2007): 727–759.

The Wegener biography by Ulrich Wutzke, *Durch die Weiße Wüste: Leben und Leistung des Grönlandforschers und Entdeckers der Kontinentaldrift Alfred Wegener* (Gotha: Justes Perthes, 1997), while following Else Wegener closely, gives an expanded account of Wegener's travels and expeditions, though not too much detail on his science.

Cornelia Lüdecke has put Wegener in the context of the history of German polar science in her *Die deutsche Polarforschung seit der Jahrhundertewende und die Einfluß Erich von Drygalskis*, Berichte zur Polarforschung, vol. 158 (Alfred-Wegener-Institut für Polar-und-Meeresforschung, 1995). Her account of Wegener's last expedition, "Lifting the Veil: The Circumstances That Caused Alfred Wegener's Death on the Greenland Icecap, 1930," *Polar Record* 36 (2000): 139–154, is very valuable and essential reading.

For a history of the continental drift controversy during and especially after Wegener's life, the principal source is Henry Frankel, *The Continental Drift Controversy*, 4 vols. (New York: Cambridge University Press, 2012). For opposition to Wegener in the United States and Great Britain, a very interesting perspective is available in Naomi Oreskes, *The Rejection of Continental Drift: Theory and Method in American Earth Science* (New York: Oxford University Press, 1999). This is complemented by Robert P. Newman's "American Intransigence: The Rejection of Continental Drift in the Great Debates of the 1920s," *Earth Sciences History* 14 (1995): 62–83. A technical account of the relative adequacy of Wegener's ideas compared to other geodynamic hypotheses gets an interesting treatment in Adrian E. Scheidegger, *Principles of Geodynamics*, 2nd ed. (New York: Academic Press, 1963). Still useful is Homer E. LeGrand's *Drifting Continents and Shifting Theories* (New York: Cambridge University Press, 1988).

As a general orientation to the history of the earth sciences pertinent to Wegener's career, I recommend my own book, Mott T. Greene, *Geology in the Nineteenth Century: Changing Views of a Changing World* (Ithaca, NY: Cornell University Press, 1982). The best source for geophysics is Stephen G. Brush, *A History of Modern Plan-*

etary Physics, 3 vols. (New York: Cambridge University Press, 1996). For oceanography, see Susan B. Schlee, *The Edge of an Unfamiliar World: A History of Oceanography* (New York: E. P. Dutton, 1973). For meteorology, readers should consult A. Kh. Khrgian, *Meteorology: A Historical Survey*, 2nd ed. (Jerusalem: Israel Program for Scientific Translations, 1970); Karl Schneider-Carius, *Weather Science and Weather Research* (New Delhi: Indian Nat. Sci. Doc. Ctr., 1975); Gisela Kutzbach, *The Thermal Theory of Cyclones: A History of Meteorological Thought in the Nineteenth Century* (Boston: American Meteorological Society, 1979); and Robert Marc Friedman, *Appropriating the Weather: Vilhelm Bjerknes and the Construction of Modern Meteorology* (Ithaca, NY: Cornell University Press, 1989). For the history of meteorological instrumentation, please see W. E. Knowles Middleton, *Meteorological Instruments* (Toronto: University of Toronto Press, 1941). For the history of solid mechanics, the main source is James F. Bell, *The Experimental Foundations of Solid Mechanics* (New York: Springer, 1984). For the history of geography and for geomorphology, by far the best is Robert Beckinsale and Richard Chorley, *The History of the Study of Landforms; or, The Development of Geomorphology*, vol. 3 (New York: Routledge, 1991). For meteoritics, my essay "Alfred Wegener and the Origin of Lunar Craters," *Earth Sciences History* 17 (1998): 111–138, gives some detail, and so does Kathleen Mark, *Meteorite Craters* (Tucson: University of Arizona Press, 1987). Those interested in map projections and their history should consult John P. Snyder, *Flattening the Earth: Two Thousand Years of Map Projections* (Chicago: University of Chicago Press, 1993); and David Greenhood, *Mapping* (Chicago: University of Chicago Press, 1964).

There is little yet written on the history of climatology, though Pamela Robinson, "Paleoclimatology and Continental Drift," in *Implications of Continental Drift to the Earth Sciences*, edited by D. H. Tarling and S. K. Runcorn, 451–476 (London: Academic Press, 1973), is interesting, as is Judith Totman Parrish, "A Brief Discussion of the History, Strengths, and Limitations of Conceptual Climate Models for Pre-Quaternary Time," in *Paleoclimates and Their Modeling with Special Reference to the Mesozoic Era*, edited by J. R. L. Allen (London: Chapman & Hall, 1993).

The *Dictionary of Scientific Biography* remains invaluable, and toward the end of my research I found it wise not to turn up my nose at *Wikipedia*, especially for supplementary biographical data on figures not included in printed works.

Index